Sarcopenia

Sarcopenia

Molecular, Cellular, and Nutritional Aspects—Applications to Humans

Edited by
Dominique Meynial-Denis

CRC Press
Taylor & Francis Group
Boca Raton London New York

CRC Press is an imprint of the
Taylor & Francis Group, an **informa** business

CRC Press
Taylor & Francis Group
6000 Broken Sound Parkway NW, Suite 300
Boca Raton, FL 33487-2742

First issued in paperback 2021

© 2020 by Taylor & Francis Group, LLC
CRC Press is an imprint of Taylor & Francis Group, an Informa business

No claim to original U.S. Government works

ISBN 13: 978-1-03-208428-2 (pbk)
ISBN 13: 978-1-4987-6513-8 (hbk)

Dedication

This book is dedicated to Professor Antoine Puigserver who, by encouraging me to sit a competitive recruitment examination in 1986, set me off on my career as a nutritionist. It is also in memory of Dr. Maurice Arnal, in whose laboratory in Clermont-Ferrand. I had my first research post. Gifted with imagination and vision and exemplars of scientific rigor, both were renowned scientists in the field of nutrition, to which they made outstanding contributions. In spirit, they were humanists and, at all times, retained a sense of humility. In 1992, Dr. Arnal created the Human Nutrition Research Center in Clermont-Ferrand, which is now one of the most active centers in France. Professor Puigserver was actively involved in student teaching and was a director of the PhD program in Nutrition at the University of Aix-Marseille.

It was a privilege and a pleasure for me to have known or worked with two men of such exceptional qualities and throughout my career as a scientist I have attempted to apply their high standards to all areas of my work.

Contents

SECTION I Basics of Sarcopenia: Definition and Challenges of Sarcopenia Research

SECTION II New Data on Sarcopenia

SECTION III Molecular and Cellular Aspects of Sarcopenia

SECTION IV   Alterations in Muscle Protein Turnover in the Aging Process

SECTION V   Recent Advances Limiting Sarcopenia and Supporting Healthy Aging

SECTION VI   Applications

*Part 1: Muscle Impairments or Diseases due to the Frailty Induced
    by Sarcopenia*

Part 2: Complications due to Sarcopenia in Acute or Chronic Diseases

Preface

This book brings together advanced expertise on the most significant aspects of sarcopenia and its health implications. Its aim is to present comprehensive, up-to-date coverage of research in this field. It is didactically orientated and addressed to nutritionists and clinicians with a research interest in aging and, in particular, in sarcopenia, age-related muscle wasting and strength loss, to graduate students in nutrition, medical students, and postdoctoral researchers in nutrition, biology, and medicine, and to other care professionals looking for a comprehensive update on sarcopenia and its effects on the appearance of age-related diseases. Readers will be interested in being informed about the most advanced areas of experimental research into sarcopenia and in the definition of the molecular targets that can be used to limit it or combat it.

Such a book can only come into being through the efforts of many people. I would like to acknowledge the contribution of those who enthusiastically accepted to participate in this project and their understanding and cooperation during the preparation of the book. I must also thank Naji N. Abumrad, who from the outset understood the importance of this book in raising awareness of sarcopenia and its risks worldwide and agreed to write the introduction. I am deeply grateful to Maurice Arnal without whose initial efforts and foresight, the project would not have gotten off the ground. His help and support throughout my career up to his death in 2000 was invaluable. I am greatly indebted to Antoine Puigserver who believed in me and gave me the confidence to sit the competitive examination that allowed me to take up research in Clermont-Ferrand. My thanks go to Anthony A. Vandervoort, a firm supporter who was instrumental in turning my thoughts on sarcopenia into a book. I am also grateful to Alfonso Cruz-Jentoft, a coordinator of the European Working Group on Sarcopenia in Older People (EWGSOP), for providing recent data on sarcopenia, which is a heavy burden for patients and healthcare systems in Europe. I would like to thank Etienne Lefai for sharing his personal data on hibernating brown bear. I would also like to thank André Mazur, Director of the Human Nutrition Unit to which I belong, for allowing me to embark on this project. I am grateful to Léa Bara, Catherine Chabant, and Baptiste Cartayrade for secretarial assistance in the preparation of this book. Finally, I would like to thank my husband, Christian, and my daughters, Audrey-Marie and Marie-Anaïs, for their support and patience during the hours spent on preparing this book and not with them.

Dominique Meynial-Denis
Clermont-Ferrand, France

Editor

Dominique Meynial-Denis studied Biochemistry and Molecular Biology at the University Paul Sabatier of Toulouse, France, and obtained her PhD degree on intermolecular interactions between drug and plasma proteins followed by magnetic resonance spectroscopy (MRS) at the same university in 1985. Since 1986, she has worked as a scientist at the National Institute of Agricultural Research (INRA) in Clermont-Ferrand in a Department focusing on Human Nutrition. Consequently, she became a nutritionist and specialized her research on sarcopenia and aging in 1994. She applied MRS to metabolic pathways of amino acids in muscle during aging. Dr. Meynial-Denis received a second PhD in 1998 on amino acid fluxes throughout skeletal muscle during aging. More recently, she has mainly been interested in the effect of glutamine supplementation in advanced age. She is a member of the French Society of Enteral and Parenteral Nutrition (SFNEP), the European Society of Clinical Nutrition and Metabolism (ESPEN), and the International Association of Gerontology and Geriatrics (IAGG). She is a regular referee to different international nutrition journals.

Contributors

Naji N. Abumrad
Department of Surgery
Vanderbilt University Medical Center
Nashville, Tennessee

Stephen E. Alway
Division of Exercise Physiology
West Virginia University School of Medicine
and
Center for Cardiovascular and Respiratory
    Sciences, and Mitochondria, Metabolism,
    and Bioenergetics
West Virginia University School of Medicine
Morgantown, West Virginia

Coralie Arc-Chagnaud
INRA, UMR 866 Dynamique Musculaire
    et Métabolisme
Université de Montpellier
Montpellier, France
and
Freshage Research Group
Department of Physiology
University of Valencia, CIBERFES, INCLIVA
Valencia, Spain

Thiago Gonzalez Barbosa-Silva
Department of Propedeutics
Federal University of Pelotas School of
    Medicine
Pelotas, Brazil

Heike A. Bischoff-Ferrari
Department of Geriatrics and Aging
    Research
University of Zürich and University Hospital
    of Zürich
Zürich, Switzerland

Thomas Brioche
INRA, UMR 866 Dynamique Musculaire
    et Métabolisme
Université de Montpellier
Montpellier, France

Katie Brown
School of Health Studies of Memphis
Memphis, Tennessee

Thomas W. Buford
Department of Medicine School of Medicine
University of Alabama at Birmingham
Birmingham, Alabama

Riccardo Calvani
Fondazione Policlinico Universitario
    "Agostino Gemelli" IRCSS
and
Institute of Internal Medicine and Geriatrics
Università Cattolica del Sacro Cuore
Rome, Italy

Heather N. Carter
Muscle Health Research Centre
York University
Toronto, Canada

Matteo Cesari
Geriatric Unit
Fondazione IRCCS Ca' Granda Ospedale
    Maggiore Policlinico
and
Department of Clinical Sciences and
    Community Health
University of Milan
Milano, Italy

Nashwa Cheema
Muscle Health Research Centre
York University
Toronto, Ontario, Canada

Angèle Chopard
INRA, UMR866 Dynamique Musculaire
    et Métabolisme
Université de Montpellier
Montpellier, France

Bess Dawson-Hughes
Bone Metabolism Laboratory
Jean Mayer USDA Human Nutrition Research
    Center on Aging
Tufts University
Boston, Massachusetts

Christopher S. Fry
Department of Nutrition and Metabolism
University of Texas Medical Branch
Galveston, Texas

Giacomo Garibotto
Division of Nephrology, Dialysis
    and Transplantation
University of Genoa

and

IRCCS Ospedale Policlinico San Martino
Genoa, Italy

Ted G. Graber
Division of Rehabilitation Science
Department of Nutrition and Metabolism
University of Texas Medical Branch
Galveston, Texas

David A. Hood
Muscle Health Research Centre
School of Kinesiology and Health Science
York University
Toronto, Canada

Ander Izeta
Tissue Engineering Group
Instituto Biodonostia

and

Department of Biomedical Engineering
    and Science
School of Engineering
Tecnun-University of Navarra
San Sebastian, Spain

Andrew S. Layne
Department of Aging and Geriatric Research
University of Florida
Gainesville, Florida

Emanuele Marzetti
Fondazione Policlinico Universitario "Agostino
    Gemelli" IRCSS
Rome, Italy

Nicole Mazara
Department of Human Health and Nutritional
    Sciences
College of Biological Sciences
University of Guelph
Guelph, Canada

Robert Memelink
Department of Nutrition and Dietetics
Faculty of sports and Nutrition
Amsterdam University of Applied Sciences
Amsterdam, The Netherlands

Dominique Meynial-Denis
Unité de Nutrition Humaine, UMR 1019,
    Department AlimH INRA, CRNH
    Auvergne
Clermont-Ferrand, France

Neia Naldaiz-Gastesi
Tissue Engineering Group
Instituto Biodonostia

and

Neuromuscular Diseases Group
Instituto Biodonostia
San Sebastian, Spain

and

CIBERNED
Instituto de Salud Carlos III
Madrid, Spain

Allan F. Pagano
EA3072, Mitochondries, Stress Oxydant
    et Protection Musculaire
Faculté des Sciences du Sport
Université de Strasbourg
Strasbourg, France

Aaron Persinger
School of Health Studies of Memphis
Memphis, Tennessee

Daniela Picciotto
Division of Nephrology, Dialysis
    and Transplantation
University of Genoa
and
IRCCS Ospedale Policlinico San Martino
Genoa, Italy

Geoffrey A. Power
Department of Human Health and Nutritional
    Sciences
College of Biological Sciences
University of Guelph
Guelph, Ontario, Canada

Carla M.M. Prado
Division of Human Nutrition
Department of Agricultural, Food and
    Nutritional Sciences
University of Alberta
Edmonton, Alberta, Canada

Melissa Puppa
School of Health Studies of Memphis
Memphis, Tennessee

Guillaume Py
INRA, UMR 866 Dynamique Musculaire
    et Métabolisme
Université de Montpellier
Montpellier, France

Blake B. Rasmussen
Department of Nutrition & Metabolism
Sealy Center on Aging
University of Texas Medical Branch
Galveston, Texas

Lisa M. Roberts
Department of Medicine
University of Alabama at Birmingham
Birmingham, Alabama

Kunihiro Sakuma
Institute for Liberal Arts
School of Environment and Society
Tokyo Institute of Technology
Tokyo, Japan

Maturin Tabue-Teguo
CHU de Guadeloupe
Université des Antilles
Pointe-à-Pitre, France
and
INSERM 1219, Bordeaux Population Health
    Research Center
University of Bordeaux
Bordeaux, France

Paolo Tessari
Metabolism Division
Department of Medicine
University of Padova
Padua, Italy

Michael Tieland
Department of Nutrition and Dietetics
Faculty of sports and Nutrition
Amsterdam University of Applied Sciences
Amsterdam, The Netherlands

Inez Trouwborst
Department of Nutrition and Dietetics
Faculty of sports and Nutrition
Amsterdam University of Applied Sciences
Amsterdam, The Netherlands

Bruno Vellas
Gérontopôle, Centre Hospitalier Universitaire
    de Toulouse
INSERM UMR1027, Université de Toulouse III
    Paul Sabatier
Toulouse, France

Amely Verreijen
Department of Nutrition and Dietetics
Faculty of sports and Nutrition
Amsterdam University of Applied Sciences
Amsterdam, The Netherlands

Daniela Verzola
Division of Nephrology, Dialysis and
    Transplantation
University of Genoa
and
IRCCS Ospedale Policlinico San Martino
Genoa, Italy

Manlio Vinciguerra
Division of Medicine
Institute for Liver and Digestive Health
University College London
London, United Kingdom

Hidetaka Wakabayashi
Department of Rehabilitation Medicine
Yokohama City University Medical Center
Yokohama, Japan

Akihiko Yamaguchi
Department of Physical Therapy
Health Sciences University of Hokkaido
Hokkaido, Japan

Peter J.M. Weijs
Department of Nutrition and Dietetics
Faculty of sports and Nutrition
Amsterdam University of Applied Sciences

and

Department of Nutrition and Dietetics
Amsterdam University Medical Centers
Amsterdam, The Netherlands

Introduction

Sarcopenia is defined as an age-related decline in muscle mass and muscle function. The original report on loss of muscle mass in the extremities with old age was published by Critchley (1931). Rosenberg in 1989 first coined the word "Sarcopenia" derived from Greek *sarx, sark-* "flesh" + *penia* "poverty" (Rosenberg, 1989). However, Rosenberg attributed the first sarcopenia description to Nathan Shock who reported in the 1970s a cross-sectional study with observations derived from at least two decades of age-related decline in function "affecting ambulation, mobility, energy intake, overall nutrient intake and status, independence and breathing" (Rosenberg, 1997). Rosenberg went on to question whether sarcopenia is "a process of normative aging" or a disease. His question was finally answered in September 2016, when sarcopenia was recognized as a disease state with its own *International Classification of Disease, Tenth Revision, Clinical Modification (ICD-10-CM)* code M62.84 (www.prweb.com-prweb13376057) (Anker et al., 2016). This designation is in many respects similar to the much earlier recognition of osteoporosis as a disease state (Roman et al., 2013; Argiles and Muscaritoli, 2016). The designation of osteoporosis led to extensive research in the field accelerating the development of earlier diagnostic techniques (Chapter 2) and therapeutic approaches. While we hope the same will happen with developments in the area of sarcopenia, there is ample evidence that interest in this field has been expanding. The following figure is an updated format (Cao and Morley, 2016) showing the number of publications on sarcopenia in PubMed from 1993 through 2018. It is important to note that the definition of sarcopenia has now been expanded to describe muscle loss that is not purely age related, but that is associated with a sedentary life style and inactivity (Chapter 13) commonly seen in the elderly and even younger subjects leading to related alterations in muscle fiber contractility (Chapter 16). The muscle losses in mass and function have also been observed in high proportion in various chronic diseases such as sarcopenic dysphagia (Chapter 17), pulmonary and cardiovascular diseases, cancer, chronic kidney disease (Chapter 18), and Parkinson's disease (Chapter 19).

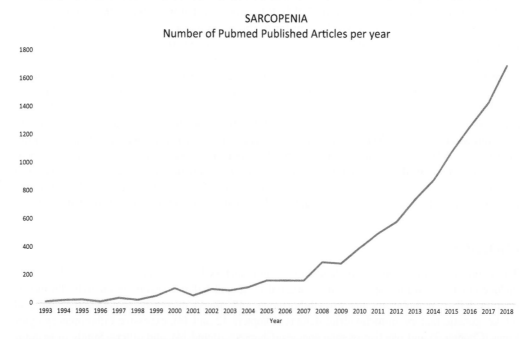

SARCOPENIA
Number of Pubmed Published Articles per year

Aging associates with significant alterations in the body that complicates the definition of sarcopenia (Chapter 3). Age-related changes in body composition include an increase in body fat, both intramuscular and extramuscular, associated with modest losses in muscle mass (Chapter 4), and this phenomenon gets much worse in what has been described as sarcopenic obesity (Chapter 20) or in very advanced age-related conditions such as frailty (Chapter 3). Recently Bischoff-Ferrari examined the various definitions used for sarcopenia and established that the definition of muscle mass index previously used by Baumgartner et al. (1998) (muscle mass (kg)/height (m)^2 less than two standard deviations below the mean of a young reference population) was a robust predictor of falls in men and women combined (Chapter 1). They also argued that this definition would optimally identify the people at risk who would benefit from early intervention(s) before they become frail. This, however, requires further testing in large populations at risk to identify its validity.

PREVALENCE

The prevalence of sarcopenia in the world is high and goes hand in hand with the aging population. In Europe, the overall estimated prevalence rates of sarcopenia in the elderly are expected to rise from estimates of 11.1%–20.2% in 2016 to 12.9%–22.3% in 2045. These estimates translate into a near 64% increase in numbers of individuals with sarcopenia from nearly 20 million to over 32.3 million in 2045 (Ethgen et al., 2017). In the United States, almost 12%–14% of the US population (approaching 50 million) is above the age of 65, and this number is expected to more than double by 2050 (McArdle and Jackson, 2011). It is estimated that sarcopenia will affect a third of people over the age of 60 years and nearly 50% of those over the age of 80 years (Baumgartner et al., 1998). The Foundation for the National Institutes of Health (FNIH) Sarcopenia Project recently determined that 5% of adults equal or greater than 60 years of age had weak muscle strength and 13% had intermediate muscle strength (Studenski et al., 2014). Weak muscle strength is clinically relevant because it is associated with mobility impairment. Loss of muscle strength with aging results in frailty (Chapter 3) leading to an elevated risk of suffering a fall, difficulty recovering from illness, prolongation of hospitalization, and long-term disability requiring assistance in daily living. Further, the reduction of muscle mass and physical strength leads to diminished quality of life, loss of independence, and mortality (McLean et al., 2014; Studenski et al., 2014). It is expected that worldwide, the number of persons afflicted with sarcopenia will be similar to those observed in the United States, and it is estimated that around 800 million persons are expected to be frail in 2025.

The health cost attributable to sarcopenia worldwide is likely to be staggering. The most cited direct estimate of the economic burden of sarcopenia in the United States for the year 2000 was $18.5 billion (around 1.5% of total healthcare expenditure) (Janssen et al., 2004). However, it should be acknowledged that direct individual assessment of healthcare costs has been very difficult to obtain. A recent review of the published literature by Bruyère et al. identified the various difficulties encountered in obtaining such estimates (Cruz-Jentoft et al., 2019). These include and are not limited to lack of alignment with guidelines or recommendations for the definition of sarcopenia; use of heterogeneous population with data derived from hospitalized patients. This invariably involves the cost of major surgical procedures; failures to account for long-term consequences of frailty such as fractures, fall, or loss of autonomy; and differences in the various statistical methods used, and so on (Bruyère et al., 2019).

ETIOLOGY

The etiology of sarcopenia is unclear with several factors having been implicated. These may include change(s) in muscle innervation, altered hormonal balance involving anabolic hormones (insulin, IGF-1, anabolic steroids, growth hormone, and others), decreased protein and amino acid intake, genetic factors, mitochondrial defects (Chapters 12 and 14), excessive inflammatory cytokines (Chapter 7) and reactive oxygen species (Chapters 10 and 14), and others. Singly or in combination, these factors lead to the development of "anabolic resistance" (Rennie, 2009), defined as the inability of muscle to regulate maintenance of protein homeostasis (Chapter 11) in response to feeding and exercise (Volpi et al., 2001; Prod'homme et al., 2005).

Several of the anabolic signals in stimulating muscle protein synthesis involve multiple pathways and in particular the mammalian/mechanistic target of rapamycin (mTOR) nutrient signaling proteins, mTORC1 and mTORC2 (Chapters 5 and 6) (Wang and Proud, 2006; Haran et al., 2012). These pathways are involved in improving the efficiency of ribosomal biogenesis and ultimately translation (Wang and Proud, 2006; Vary et al., 2007). Both human (Rasmussen et al., 2006) and rodent (Wang and Proud, 2006; Vary et al., 2007) studies have shown a suppressed ability of insulin and branched-chain amino acids (BCAAs) to initiate protein translation, which appear to be mediated by mTOR signaling (Bodine et al., 2001; Cuthbertson et al., 2005; Wang and Proud, 2006). Further, Rasmussen et al. (2006) demonstrated additional decreased ability of vasodilation in muscles of older individuals in response to insulin potentially leading to decreased nutrient delivery to tissues. This information in conjunction with a better understanding of the molecular changes (Chapters 4 through 10) involved in the muscle maintenance, hypertrophy, and wasting have expanded our potential for the development of newer therapeutic approaches for treating sarcopenia

PREVENTION AND TREATMENT OF SARCOPENIA

At present, resistance exercise and adequate protein intake are the hallmark of prevention and treatment of sarcopenia (Chapter 15). Progressive resistance training is safe and effective and needs to be disseminated to the rapidly aging population. Several other alternative approaches have been proposed. For example, high-dose testosterone has been proposed, as this mode of therapy results in increased muscle mass, with some evidence of a lesser increase in muscle strength; however, potential serious side effects of this treatment mitigate its dissemination to the general population in need. Other drugs that have been proposed include selective androgen receptor molecules, ghrelin agonists, myostatin antibodies, activin IIR antagonists, angiotensin-converting enzyme inhibitors, beta antagonists, and fast skeletal muscle troponin activators. Other approaches such as nutritional supplements might be beneficial. Notably, branched-chain amino acids, amino acid metabolites, such as the leucine metabolite β-hydroxy-β-methyl butyrate (HMB), have been proposed. Other therapeutic approaches being tested, at least in rodents and other animal species, include the use of satellite cells, considered as skeletal muscle stem cells (Chapters 8 through 10), which can be used as a potential cell source for skeletal muscle repair in vivo (Brack and Rando, 2012). Skeletal muscle multipotent progenitor cells that have a higher doubling potential than satellite cells have been considered; however, these cells do not have the same efficiency as satellite cells in differentiating into skeletal muscle fibers (Wilschut et al., 2008).

Doctor Dominique Meynial-Denis deserves tremendous of credit for assembling a distinguished group of investigators who have been pioneers in establishing the definition, etiology, and impact of sarcopenia on motility and quality of life as well its relationships to many other diseases. This book is a must-read for every student, basic science investigator, epidemiologist, and other healthcare professionals who is interested in learning about the impact of age-related muscle wasting and loss of muscle function on various disease entities. In this regard, it is very hard to overstate the impact of Dr. Meynial-Denis' and her team's work. I can foresee a future where investigators will be able to deliver extrinsic factors that rejuvenate muscle tissue by "introducing pro-youthful factors or abrogating pro-aging factors" (Laviano, 2014).

Naji N. Abumrad

REFERENCES

Anker SD, Morley JE, von Haehling S. 2016. Welcome to the ICD-10 code for sarcopenia. *J Cachexia Sarcopenia Muscle.* 7(5):512–514.

Argiles J, Muscaritoli M. 2016. The three faces of sarcopenia. *J Am Med Dir Assoc.* 17:471–472.

Baumgartner RN, Koehler KM, Gallagher D, et al. 1998. Epidemiology of sarcopenia among the elderly in New Mexico. *Am J Epidemiol.* 147:755–763.

Bodine SC, Stitt TN, Gonzalez M, et al. 2001. Akt/mTOR pathway is a crucial regulator of skeletal muscle hypertrophy and can prevent muscle atrophy in vivo. *Nat Cell Biol.* 3(11):1014–1019.

Brack AS, Rando TA. 2012. Cell stem. *Cell.* 10:504–514.

Bruyère O, Beaudart C, Ethgen O, et al. 2019. The health economics burden of sarcopenia: A systematic review. *Maturitas.* 119:61–69.

Cao L, Morley JE. 2016. Sarcopenia is recognized as an independent condition by an international classification of disease, tenth revision, clinical modification (ICD-10-CM) code. *J Am Med Dir Assoc.* 17(8):675–677.

Critchley M. 1931. The neurology of old age. *Lancet* 1:1221–1230.

Cruz-Jentoft AJ, Bahat G, Bauer J, et al. 2019. Sarcopenia: Revised European consensus on definition and diagnosis. *Age Ageing.* 48(1):16–31.

Cuthbertson D, Smith K, Babraj J, et al. 2005. Anabolic signaling deficits underlie amino acid resistance of wasting, aging muscle. *FASEB J.* 19(3):422–424.

Ethgen O, Beaudart, C, Buckinx F, Bruyère O, Reginster JY. 2017. The future prevalence of sarcopenia in Europe: A claim for public health action. *Calcif Tissue Int.* 100(3):229–234.

Haran PH, Rivas DA, Fielding RA. 2012. Role and potential mechanisms of anabolic resistance in sarcopenia. *J Cachexia Sarcopenia Muscle.* 3(3):157–162.

http://www.census.gov/population/www/projections/summarytathe bles.html.

Janssen I, Shepard DS, Katzmarzyk PT, Roubenoff, R. 2004. The healthcare costs of sarcopenia in the United States. *J Am Geriatr Soc.* 52:80–85.

Laviano A. 2014. Young blood. *N Engl J Med.* 371:573–575. doi:10.1056/NEJMcibr1407158.

McArdle A, Jackson MJ. 2011. Sarcopenia – Age-related muscle wasting and weakness, p. 318. Gordon S. Lynch (ed.). Springer Science.

McLean RR, Shardell MD, Alley DE, et al. 2014. Criteria for clinically relevant weakness and low lean mass and their longitudinal association with incident mobility impairment and mortality: The Foundation for the National Institutes of Health (FNIH) Sarcopenia Project. *J Gerontol A Biol Sci Med Sci.* 69(5):576–583.

Prod'homme M, Balage M, Debras E, et al. 2005. Differential effects of insulin and dietary amino acids on muscle protein synthesis in adult and old rats. *J Physiol.* 563(Pt 1):235–248.

Rasmussen BB, Fujita S, Wolfe RR, et al. 2006. Insulin resistance of muscle protein metabolism in aging. *FASEB J.* 20(6):768–769.

Rennie MJ. 2009. Anabolic resistance: The effects of aging, sexual dimorphism, and immobilization on human muscle protein turnover. *Appl Physiol Nutr Metab.* 34(3):377–381.

Roman D, Mahoney K, Mohamadi A. 2013. Sarcopenia: What's in a name? *J Am Med Dir Assoc.* 14:80–82.

Rosenberg IH. 1989. Summary comments: Epidemiological and methodological problems in determining nutritional status of older persons. *Am J Clin Nutr.* 50:1231–1233.

Rosenberg IH. 1997. Sarcopenia: Origins and clinical relevance. *J Nutr.* 127:990S–991S.

Studenski SA, Peters KW, Alley DE, et al. 2014. The FNIH Sarcopenia Project: Rationale, study description, conference recommendations, and final estimates. *J Gerontol A Biol Sci Med Sci.* 69(5):547–558.

Vary TC, Anthony JC, Jefferson LS, et al. 2007. Rapamycin blunts nutrient stimulation of eIF4G, but not PKCepsilon phosphorylation, in skeletal muscle. *Am J Physiol Endocrinol Metab.* 293(1):E188–E196.

Volpi E, Sheffield-Moore M, Rasmussen BB, Wolfe RR. 2001. Basal muscle amino acid kinetics and protein synthesis in healthy young and older men. *JAMA.* 286(10):1206–1212.

Wang X, Proud CG. 2006. The mTOR pathway in the control of protein synthesis. *Physiology (Bethesda).* 21:362–369.

Wilschut KJ, Jaksani S, Van Den Dolder J, Haagsman HP, Roelen BA. 2008. Isolation and characterization of porcine adult muscle-derived progenitor cells. *J Cell Biochem.* 105:1228–1239.

Section I

Basics of Sarcopenia

Definition and Challenges of Sarcopenia Research

1 Definitions of Sarcopenia

Heike A. Bischoff-Ferrari and Bess Dawson-Hughes

CONTENTS

1.1 INTRODUCTION: WHAT IS SARCOPENIA? WHY IS IT IMPORTANT, AND WHERE ARE THE CHALLENGES?

The Western World population is aging rapidly, and the number of seniors aged 70 and older is predicted to increase from 25% to 40% by 2030 (Eberstadt & Groth, 2007; EC, 2006; Europe; Eurostat, 2006; Lee, 2007), as is the number of seniors with mobility disability, physical frailty, and resulting consequences, such as falls, fractures, and loss of autonomy (Book, 2013; Commission, 2007b; Visser & Schaap, 2011). This causes enormous challenges to the individual, health economy, and societies as a whole, further magnified by unmet therapeutic needs for seniors and neglect of functional endpoints in today's medical care systems (CDC, 2013; Commission, 2007a; Evans, 2012; Janssen, Shepard, Katzmarzyk, & Roubenoff, 2004; Motion, 2013; Sayer, 2010; WHO, 2004).

A condition that is considered central to the development of physical frailty and its consequences is sarcopenia (Abellan van Kan et al., 2009; Baumgartner et al., 1998; Cruz-Jentoft et al., 2010; Delmonico et al., 2007; Delmonico et al., 2009; Fielding et al., 2011; Fried et al., 2001; Janssen et al., 2004; Morley et al., 2011; Morley, Baumgartner, Roubenoff, Mayer, & Nair, 2001; Motion, 2013; Sayer, 2010; Studenski, 2009; Vellas et al., 2013; Visser, 2009; Visser & Schaap, 2011), the loss of muscle mass, and strength (Rosenberg, 1997). However, to date, the development of effective treatments for sarcopenia is delayed by regulatory and consensual obstacles on how to define and measure sarcopenia as a medical condition in the senior population (Motion, 2013; Visser, 2009).

This chapter will summarize available definitions of sarcopenia and opportunities to move forward on an international consensus on an operational definition of sarcopenia.

1.2   CHANGES OF MUSCLE HEALTH WITH AGE

Muscle is composed of two basic fiber types, and a differential loss of these fibers have been described with age (Lexell & Downham, 1992). With aging, there is preferential loss of Type II fibers, related to a decline in the Type II fiber stem cell or satellite cell population (Verdijk et al., 2007). Type II fibers have fast contraction time, high force production, and low resistance to fatigue (<5 min) and are needed for fast reactions in the prevention of a fall. Type I fibers decline less with age; have a slow contraction time, low force production, and high resistance to fatigue (hours); and are needed for endurance tasks (Lexell & Downham, 1992). It has been suggested that vitamin D deficiency and the decline of the specific receptor for vitamin D (VDR) with age may be key drivers of the preferential loss of Type II muscle fibers with age (Bischoff-Ferrari, 2012; Bischoff-Ferrari et al., 2004; Ceglia et al., 2013).

Moreover, age-related changes in muscle health include a decline in muscle mitochondrial protein synthesis (Nair, 2005) and a decline in motor neurons, reducing the signal for muscle contraction (Brown, 1972). Finally, an age-related anabolic resistance at the muscle cell level (Breen & Phillips, 2011; Guillet et al., 2004; Murton, 2015; Rennie, 2009) has been suggested as a significant contributor to sarcopenia and frailty, which is enhanced by a decrease in protein intake with age (Kobayashi et al., 2013; Murton, 2015; Volpi & Rasmussen, 2000). The overall loss of muscle mass from age 20 to 80 has been found to be about 40%, preferentially at the lower extremities (Koopman & van Loon, 2009).

Conceptually, the age-related loss in muscle mass is exceeded by the loss of muscle power and strength (Barry & Carson, 2004). Goodpaster et al. (2006) assessed rates of loss of leg lean mass by dual energy X-ray absorptiometry and of knee extensor strength in 1,880 men and women aged 70–79 years in the Health ABC study. In the men and the women, muscle mass declined by an average of 1% per year, whereas strength declined by approximately 3% per year (Goodpaster et al., 2006). Decreased muscle mass and decreased strength impair physical function, which is thought to cause an increased risk of falls and to precede physical disability and frailty (Lang et al., 2010).

1.3   CONSEQUENCES OF SARCOPENIA

By all international working groups, falls in senior adults have been accepted as a key and severe complication of sarcopenia (Cruz-Jentoft et al., 2010; Fielding et al., 2011; Lang et al., 2010; Stevens, Corso, Finkelstein, & Miller, 2006). Notably, one in three adults aged 65 years and one in two aged 80 years and older fall each year (Tinetti & Williams, 1997; Tromp et al., 2001). Falls cause moderate to severe injuries in 20%–30% of cases, which result in functional impairment and increase the risk of nursing home admission and mortality (Sterling, O'Connor, & Bonadies, 2001; Tinetti & Williams, 1997). Other complications of sarcopenia are a reduced quality of life, hip fractures, frailty, loss of autonomy, and mortality (Abellan van Kan et al., 2009; Baumgartner et al., 1998; Cruz-Jentoft et al., 2010; Dawson-Hughes & Bischoff-Ferrari, 2016; Delmonico et al., 2007, 2009; Fielding et al., 2011; Fried et al., 2001; Janssen et al., 2004; Morley et al., 2011; Morley et al., 2001; Motion, 2013; Rizzoli et al., 2013; Sayer, 2010; S. Studenski, 2009; Vellas et al., 2013; Visser, 2009; Visser & Schaap, 2011). Figure 1.1 shows the potential consequences of sarcopenia and their trajectory toward loss of autonomy.

FIGURE 1.1   Consequences of sarcopenia.

1.4 THE HEALTH ECONOMIC BURDEN OF SARCOPENIA

Depending on the definition of sarcopenia, about 5%–13% of seniors aged 60–70 years and 11%–50% of seniors aged 80 years and older are affected (Bischoff-Ferrari et al., 2015b; Cruz-Jentoft et al., 2014; Perez-Zepeda, Gutierrez-Robledo, & Arango-Lopera, 2013; Rizzoli et al., 2013; von Haehling et al. 2010). Based on a representative sample of U.S. adults age 60 years and older, the estimated direct healthcare cost attributable to sarcopenia in 2000 was $18.5 billion, which represented about 1.5% of total healthcare expenditures for that year (Janssen et al., 2004). According to Janssen et al. (2004), the assessment of 10% reduction in sarcopenia prevalence would result in savings of $1.1 billion (dollars adjusted to 2000 rate) per year in U.S. healthcare costs.

1.5 CURRENT DEFINITIONS OF SARCOPENIA

Seven operational definitions of sarcopenia (Baumgartner et al., 1998; Cruz-Jentoft et al., 2010; Cruz-Jentoft et al., 2019; Delmonico et al., 2007, 2009; Fielding et al., 2011; Morley et al., 2011; Muscaritoli et al., 2010) have been published, and two main concepts have been proposed, one that is based on low muscle mass alone (Baumgartner et al., 1998; Delmonico et al., 2007, 2009) and the other that requires both low muscle mass and decreased performance in a functional test within a composite definition of sarcopenia (Cruz-Jentoft et al., 2010, 2019; Fielding et al., 2011; Morley et al., 2011; Muscaritoli et al., 2010). Furthermore, two related definitions have been published to date by the Foundation for the National Institutes of Health Biomarkers Consortium Sarcopenia Project (Studenski et al., 2014). Table 1.1 provides an overview of the instrument library of the published definitions of sarcopenia.

1.5.1 THE FIRST THREE OPERATIONAL DEFINITIONS FOCUSED ON LOW MUSCLE MASS BY DUAL-ENERGY X-RAY ABSORPTIOMETRY (DEXA) ALONE

The first definition was proposed by Baumgartner et al. (1998) and is based on the appendicular skeletal lean mass (ALM). According to this definition, individuals whose ALM divided by height squared $\left(\frac{\text{ALM (kg)}}{\text{Height}^2 \ (\text{m}^2)} \right)$ is two or more standard deviations below sex-specific means of the Rosetta study (Gallagher et al., 1997) reference data set are defined as sarcopenic. The 1996 Rosetta study included 284 participants (148 women and 136 men, white and African American, and average age 47.6 years) (Gallagher et al., 1997). Cutoffs for the definition based on data from Rosetta study reference data set are ≤ 7.26 kg/m^2 for men and ≤ 5.45 kg/m^2 for women.

TABLE 1.1
Instrument Library of Sarcopenia Definitions

Definition	ALM	TBLM	Muscle Quality	Fat Mass	Grip Strength	Gait Speed
Baumgartner	X					
Delmonico I	X					
Delmonico II	X			X		
EWGSOP1*	X				X	X
EWGSOP2 (Revision to 1)	X		X		X	X
Fielding	X					X
Morley	X					X
Muscaritoli		X				X
Studenski 1	X					
Studenski 2	X				X	

Note: *Cruz-Jentoft is the coordinator of the European Working Group on Sarcopenia in Older People (EWGSOP).

The second definition was proposed by Delmonico et al. in 2007 and is calculated in the same way, but the reference population and threshold for low appendicular lean mass is different. According to this definition, individuals whose ALM divided by height squared ($\frac{\text{ALM (kg)}}{\text{Height}^2 \text{ (m}^2)}$) is below the 20th percentile of sex-specific distribution of the reference population in the Health ABC study (Delmonico et al., 2007) are defined sarcopenic. The 2000 baseline cohort of the Health ABC study included 1,761 participants (992 men and 769 women and 25–44 years old) (Pichard et al., 2000). Cutoffs for the definition based on data from the Health ABC study reference data set are ≤ 7.25 kg/m^2 for men and ≤ 5.67 kg/m^2 for women.

The third definition was proposed by Delmonico et al. in 2009 and uses the same reference population (Health ABC) as in *Delmonico 1* but extends to gender-specific residuals that also include fat mass (Delmonico et al., 2007). Linear models regressing ALM by height and total body fat mass were fit separately to the data from men and women. For men, the resulting model was ALM $(\text{kg}) = -22.59 + 24.21 \times \text{height} (\text{m}) + 0.21 \times \text{total fat mass} (\text{kg})$. For women, the resulting model was ALM $(\text{kg}) = -13.21 + 14.76 \times \text{height} (\text{m}) + 0.23 \times \text{total fat mass} (\text{kg})$. These models, when used on the population under study, give predicted values for ALM for each individual. Then for each individual, a residual is calculated as a difference between observed and predicted ALM. A positive residual indicates that the person is relatively more muscular than predicted by the model, and a negative residual suggests that this person is less muscular than predicted. People whose residuals fall below the 20th percentile of the sex-specific distribution are defined as "sarcopenic."

1.5.2 ADDITIONAL FOUR OPERATIONAL DEFINITIONS FOCUSED ON BOTH LOW MUSCLE MASS BY DEXA AND DECREASED PERFORMANCE IN A FUNCTIONAL TEST

The fourth definition was proposed by Cruz-Jentoft et al. (2010) for the European Working Group on Sarcopenia in Older People (EWGSOP1). In addition to low ALM, the definition also requires decreased gait speed performance and/or decreased grip strength. Sarcopenia is defined by two conditions: (1) $\frac{\text{ALM}}{\text{height}^2} \leq 7.26$ kg/m^2 for men and ≤ 5.54 kg/m^2 for women and (2) gate speed < 0.8 m/s and/or low grip strength < 30 kg for men and < 20 kg for women. As the authors do not provide a definite recommendation on the cutoff for low appendicular mass, we used the cutoffs proposed by Baumgartner based on the Rosetta study.

The European Working Group on Sarcopenia in Older People revised their recommendations in early 2018 (EWGSOP2) (Cruz-Jentoft et al., 2019). In the updated definition, the focus is on low muscle strength as a key characteristic of sarcopenia, plus it uses detection of low muscle quantity and quality to confirm the sarcopenia diagnosis. Further, the revised definition identifies poor physical performance as indicative of severe sarcopenia and provides clear cutoff points for measurements of variables that identify and characterize sarcopenia (Cruz-Jentoft et al., 2019).

The fifth definition was proposed by Fielding et al. (2011) for the International Working Group on Sarcopenia. In addition to low ALM using Health ABC as the reference population, the definition also requires decreased gait speed performance. Sarcopenia is defined by two conditions: (1) $\frac{\text{ALM}}{\text{height}^2} \leq 7.23$ kg/m^2 for men and ≤ 5.67 kg/m^2 for women and (2) gate speed < 1 m/s.

The sixth definition was proposed by Morley et al. (2011) for the Society for Sarcopenia, Cachexia, and Wasting Disorders "Sarcopenia with limited mobility." In addition to low ALM, the definition requires low gait speed or a distance less than 400 m on the 6-min walk. Sarcopenia is defined by two conditions: (1) $\frac{\text{ALM}}{\text{height}^2}$ of 2 standard deviations (SDs) of more below the mean of healthy individuals between 20 and 30 years of age of the same ethnic group, using NHANES IV (Kelly, Wilson, & Heymsfield, 2009) as the reference population. The 2008 NHANES IV cohort includes 2,402 participants (1,035 men and 1,367 women and 20–30 years old) (Kelly et al., 2009). Sarcopenia is defined by two conditions: (1) appendicular muscle mass ≤ 6.81 kg/m^2 for men and ≤ 5.18 kg/m^2 for women and (2) gate speed < 1.0 m/s.

The seventh definition was proposed by Muscaritoli et al. (2010) for the Special Interest Groups "cachexia-anorexia in chronic wasting diseases" and "nutrition in geriatrics." In addition to low lean mass, the definition also requires low gait speed. To define low muscle mass, like other authors, a −2SD cutoff is proposed, however related to total body skeletal lean mass (TLM) as opposed to appendicular lean mass. For gait speed, they propose < 0.8 m/s as a cutoff for "low gait speed." For the low lean mass cutoff Muscaritoli et al. refer to a paper by Janssen et al. (2002). In order to standardize TLM measurements, Janssen et al. propose using a skeletal mass index (SMI): $SMI = \frac{SM}{Body\ Mass} \times 100$. Sarcopenia is defined by two conditions: (1) SMI ≤ 37% for men and SMI ≤ 28% for women and (2) gate speed < 0.8 m/s.

1.5.3 Two Related Definitions

The two related definitions were proposed by Studenski et al. (2014) for the Foundation for the National Institutes of Health Biomarkers Consortium Sarcopenia Project. "*Studenski 1*" refers to low lean mass and "*Studenski 2*" refers to low lean mass contributing to weakness. To define low lean mass, the authors use ALM adjusted for body mass index (ALM_{BMI}), and weakness is defined based on the grip strength. As a reference data set for the ALM_{BMI} thresholds, the authors use a pooled data set of 26,625 community-dwelling older adults age 65 years and older (11,427 men and 15,198 women) from nine studies (the Study of Osteoporotic Fractures [SOF], the Osteoporotic Fractures in Men [MrOS] Study, the Framingham study, the UConn clinical trials, the Boston Puerto Rican Health Study, the Rancho Bernardo Study, the InChianti, Health ABC, and the AGES-Reykjavik study). The "*Studenski 1*" definition of low lean mass alone is defined by $ALM_{BMI} < 0.789$ men and $ALM_{BMI} < 0.512$ for women. The "*Studenski 2*" definition of low lean mass contributing to weakness requires two conditions: (1) grip strength <26 kg for men and <16 kg for women and (2) $ALM_{BMI} < 0.789$ men and $ALM_{BMI} < 0.512$ for women.

1.6 FINDING AN OPERATIONAL DEFINITION FOR SARCOPENIA

As pointed out in the introduction of this chapter, to date the development of effective treatments for sarcopenia is delayed by regulatory and consensual obstacles on how to define and measure sarcopenia as a medical condition in the senior population (Motion, 2013; Visser, 2009).

Thus, ongoing research efforts aim to identify "the operational definition" suitable to define sarcopenia in clinical care and as an accepted endpoint in clinical trials for novel treatment strategies to prevent and treat sarcopenia. A first attempt to assess the comparative effectiveness of the available definitions of sarcopenia in the prediction of the rate of falling has been published recently (Bischoff-Ferrari et al., 2015a). Based on a cohort of 445 community dwelling seniors aged 65 years and older followed for 3 years, the authors compared the prevalence of sarcopenia and comparative performance in fall prediction of the seven operational definitions of sarcopenia (Baumgartner et al., 1998; Cruz-Jentoft et al., 2010; Delmonico et al., 2007, 2009; Fielding et al., 2011; Morley et al., 2011; Muscaritoli et al., 2010) and two related definitions (Studenski et al., 2014) outlined in Section 1.5 of this chapter. Table 1.1 shows the instrument library (measurement components) of the current sarcopenia definitions (Bischoff-Ferrari et al., 2015a).

Across the seven available definitions of sarcopenia plus the two related definitions, the prevalence of sarcopenia varied between 2.5% and 27.2% among women and 3.1% and 20.4% among men. The overall pattern suggested that the composite definitions were more conservative in defining sarcopenia compared with definitions that used ALM alone. Only one composite definition, the one by Muscaritoli et al. provided a higher prevalence of sarcopenia based on the total lean mass plus low gait speed identifying 20.4% of men and 26.2% of women as sarcopenic. Among the three definitions that used low lean mass alone based on ALM, the prevalence of sarcopenia in men and women combined varied between 11% (Baumgartner) and 21.4% (Delmonico 2). Among the other four definitions (excluding Muscaritoli) that used low lean mass based on ALM in combination with

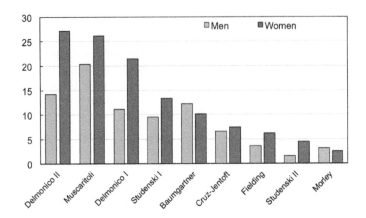

FIGURE 1.2   Prevalence of sarcopenia in community-dwelling seniors by definition and in percent.

decreased performance in a functional test, the prevalence of sarcopenia in men and women combined varied between 2.7% and 7.1% (see Figure 1.2; Bischoff-Ferrari et al., 2015a).

Falling is a widely accepted serious and clinically and economically relevant consequence of sarcopenia (Cruz-Jentoft et al., 2010; Fielding et al., 2011). The recent study on seven operational definitions of sarcopenia (Baumgartner et al., 1998; Cruz-Jentoft et al., 2010; Delmonico et al., 2007, 2009; Fielding et al., 2011; Morley et al., 2011; Muscaritoli et al., 2010) and two related definitions (Studenski et al., 2014) outlined in Section 1.5 and also compared the prospective rate of falls in sarcopenic versus nonsarcopenic individuals (Bischoff-Ferrari et al., 2015a). The rate of falls was best predicted by the Baumgartner definition based on the low ALM alone (Risk Ratio = 1.54; 95% confidence interval [CI]: 1.09–2.18) and the EWGSOP1 composite definition of low ALM plus low gait speed or decreased grip strength (Risk Ratio = 1.82; 95% CI: 1.24–2.69). For all other sarcopenia definitions but Delmonico 2 and Muscaritoli, sarcopenic individuals had a nonsignificantly higher rate of falls compared with the nonsarcopenic individuals (see Figure 1.3). For the two related definitions of Studenski 1 and 2, seniors with low lean mass had a neutral and those with low lean mass contributing to weakness had a nonsignificantly lower rate of falls.

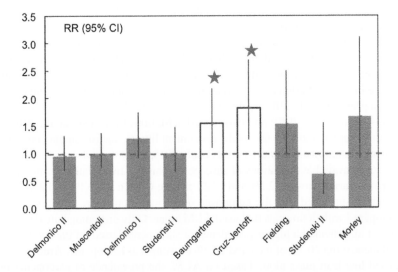

FIGURE 1.3   Rate of falls in sarcopenic versus nonsarcopenic community-dwelling seniors and by definition of sarcopenia.

In summary, based on the cohort of 445 community dwelling seniors aged 65 years and older followed for 3 years, the definitions of Baumgartner referring to low appendicular lean mass alone and EWGSOP1 requiring both low appendicular lean mass and decreased performance in gait speed and/or grip strength best predicted the rate of falls among sarcopenic versus nonsarcopenic community-dwelling seniors (Bischoff-Ferrari et al., 2015a). With the same cutoff for low appendicular lean mass, the additional requirement of decreased function in the EWGSOP1 definition increased the prediction of the rate of falls among sarcopenic individuals from an odds ratio of 1.54 (Baumgartner) to 1.82 (EWGSOP1) and reduced the respective prevalence of sarcopenia from 11% (Baumgartner) to 7.1% (EWGSOP1).

1.7 CONCLUSION AND PERSPECTIVES

After being coined by Rosenberg as "paucity of flesh" (Rosenberg, 1997), the definition of sarcopenia was initially related to the loss of muscle mass alone. Then clinical observations in aging cohorts emerged suggesting that muscle mass loss alone may not predict future strength decline (Visser, Deeg, Lips, Harris, & Bouter, 2000; Visser & Schaap, 2011). These important observations supported the rationale for use of composite definitions of sarcopenia, which required both low muscle mass and the presence of decreased strength/gait performance to define sarcopenia (Abellan van Kan et al., 2009; Visser, 2009). However, with the composite sarcopenia definition that purportedly captures decreased muscle mass and quality, several aspects may need to be considered. *First*, as decreased function is an integral part of these definitions, a better prediction of functional decline with the composite definition of sarcopenia, compared with the low lean mass alone, may be considered flawed. *Second*, by making it necessary that both criteria of low lean mass and decreased functional performance are met, most likely a progressed state of sarcopenia is identified, and the opportunity of early diagnosis of sarcopenia may be missed. *Third*, the requirement of both low lean mass and decreased performance in a functional test identifies very few community-dwelling seniors as sarcopenic (Scott et al., 2014) as demonstrated in the recent comparative performance investigation (Bischoff-Ferrari et al., 2015a) outlined in Section 1.6 of this chapter.

Alternatively, defining sarcopenia by low lean mass alone may be worth re-visiting for the following arguments. *First*, with low muscle mass alone, as proposed earlier (Cruz-Jentoft et al., 2010; Murphy et al., 2014), the opportunity of capturing an early disease stage is available. Also, the introduction of multiple cutoffs to define the extend of muscle mass loss would provide the prospect of defining a spectrum of disease, such as pre-sarcopenia and sarcopenia (Cruz-Jentoft et al., 2010; Murphy et al., 2014), similar to the definition of osteopenia and osteoporosis based on DEXA bone mineral density (Kanis, 1994). *Second*, by defining sarcopenia based on DEXA muscle mass alone, inter-rater variability and the influence of the patient's motivation would be eliminated from the diagnosis of sarcopenia. It is conceivable that regulatory agencies prefer an operator-independent and exact technology-based definition of sarcopenia. *Third*, based on the findings outlined in Section 1.6 of this chapter, a significant 54% increased rate of falls among sarcopenic individuals based on the Baumgartner definition of low muscle mass alone is clinically relevant, especially as falls have multifactorial causes. *Fourth*, the composite definition likely reflects a progressed disease stage reflected by the low prevalence of sarcopenia described by the EWGSOP1 (7.1%) and Fielding (5%) definitions compared with a useful prevalence based on the Baumgartner sarcopenia definition of low muscle mass alone (11%; Figure 1.2).

In contrast, however, the measurement of ALM by DEXA has shown inconsistent correlations with functional decline (Cawthon et al., 2015; Schaap, Koster, & Visser, 2013), which supports the search for new methods, such as the D-3 creatine dilution (Cawthon et al., 2018; Clark et al., 2018), to better reflect the lean muscle mass. Thus, until such methods are broadly available, a prudent approach for the definition of sarcopenia may need to include decreased performance next to a decline in DXA-based ALM among older adults.

In summary, getting to an internationally accepted operational definition of sarcopenia is a very timely effort and pivotal basis for the development of treatments against sarcopenia. Within this context, selecting a definition of sarcopenia requires balancing the potential benefit of including functional measures along with measurements of ALM against the conceptual difficulties and impact on the captured disease stage related to their inclusion. The functional measurements are harder to standardize than DXA-based ALM measurements, and they add considerable variability (see Figure 1.2). In contrast, DXA-based ALM measurements are estimates of ALM, are also prone to measurement variability, and have shown inconsistent correlations with functional decline (Cawthon et al., 2015; Schaap et al., 2013). However, by four of the five composite definitions including both DXA-based ALM decline and a decline in functional performance, the prevalence of sarcopenia was quite low, ranging from 2.7% to 7.1% (Bischoff-Ferrari et al., 2015a). A very low prevalence limits opportunity for early identification and application of prevention strategies. It also poses challenges in developing effective global strategies to reduce sarcopenia. For example, it may discourage drug development initiatives in low-prevalence environments because the interventions would have limited application there. From another point of view, this is not to downplay the importance of functional assessment among older adults at risk of sarcopenia. Functional performance measures are a core element in the development of treatment plans for sarcopenic seniors to prevent the progression of sarcopenia and its consequences.

Relevant to the future research efforts, a spectrum of operational definitions for sarcopenia is missing, that is, a health state for being at risk for sarcopenia (i.e., "pre-sarcopenia") and a health state for disease progression (i.e., "severe sarcopenia") for those at high risk to develop physical frailty. A spectrum of operational definitions for sarcopenia would be valuable for developing strategies that can be implemented early among those at risk for sarcopenia and prevent further disease progression among those with disease activity.

REFERENCES

Abellan van Kan, G., Andre, E., Bischoff Ferrari, H. A., Boirie, Y., Onder, G., Pahor, M.,... Vellas, B. (2009). Carla Task Force on Sarcopenia: Propositions for clinical trials. *J Nutr Health Aging, 13*(8), 700–707.

Barry, B. K., & Carson, R. G. (2004). The consequences of resistance training for movement control in older adults. *J Gerontol A Biol Sci Med Sci, 59*(7), 730–754.

Baumgartner, R. N., Koehler, K. M., Gallagher, D., Romero, L., Heymsfield, S. B., Ross, R. R.,... Lindeman, R. D. (1998). Epidemiology of sarcopenia among the elderly in New Mexico. *Am J Epidemiol, 147*(8), 755–763.

Bischoff-Ferrari, H. A. (2012). Relevance of vitamin D in muscle health. *Rev Endocr Metab Disord, 13*(1), 71–77. doi:10.1007/s11154-011-9200-6.

Bischoff-Ferrari, H. A., Borchers, M., Gudat, F., Durmuller, U., Stahelin, H. B., & Dick, W. (2004). Vitamin D receptor expression in human muscle tissue decreases with age. *J Bone Miner Res, 19*(2), 265–269. doi:10.1359/jbmr.2004.19.2.265.

Bischoff-Ferrari, H. A., Orav, J. E., Kanis, J. A., Rizzoli, R., Schlogl, M., Staehelin, H. B.,... Dawson-Hughes, B. (2015a). Comparative performance of current definitions of sarcopenia against the prospective incidence of falls among community-dwelling seniors age 65 and older. *Osteoporos Int, 26*(12), 2793–2802. doi:10.1007/s00198-015-3194-y.

Bischoff-Ferrari, H. A., Orav, J. E., Kanis, J. A., Rizzoli, R., Schlogl, M., Staehelin, H. B., ... Dawson-Hughes, B. (2015b). Comparative performance of current definitions of sarcopenia against the prospective incidence of falls among community-dwelling seniors age 65 and older. *Osteoporos Int*, doi:10.1007/s00198-015-3194-y.

Book, S. (2013). Alliance for Aging Research. http://www.agingresearch.org.

Breen, L., & Phillips, S. M. (2011). Skeletal muscle protein metabolism in the elderly: Interventions to counteract the "anabolic resistance" of ageing. *Nutr Metab (Lond), 8*, 68. doi:10.1186/1743-7075-8-68.

Brown, W. F. (1972). A method for estimating the number of motor units in thenar muscles and the changes in motor unit count with ageing. *J Neurol Neurosurg Psychiatry, 35*(6), 845–852.

Cawthon, P. M., Blackwell, T. L., Cauley, J., Kado, D. M., Barrett-Connor, E., Lee, C. G.,... Orwoll, E. S. (2015). Evaluation of the usefulness of consensus definitions of sarcopenia in older men: Results from the observational osteoporotic fractures in men cohort study. *J Am Geriatr Soc, 63*(11), 2247–2259. doi:10.1111/jgs.13788.

Cawthon, P. M., Orwoll, E. S., Peters, K. E., Ensrud, K. E., Cauley, J. A., Kado, D. M.,... Osteoporotic Fractures in Men Study Research Group. (2018). Strong relation between muscle mass determined by D3-creatine dilution, physical performance and incidence of falls and mobility limitations in a prospective cohort of older men. *J Gerontol A Biol Sci Med Sci. 74*(6), 844–852. doi:10.1093/gerona/gly129.

CDC. (2013). The state of aging and health in America 2013. http://www.who.int/kobe_centre/ageing/ahp_vol5_glossary.pdf.

Ceglia, L., Niramitmahapanya, S., da Silva Morais, M., Rivas, D. A., Harris, S. S., Bischoff-Ferrari, H.,... Dawson-Hughes, B. (2013). A randomized study on the effect of vitamin D(3) supplementation on skeletal muscle morphology and vitamin D receptor concentration in older women. *J Clin Endocrinol Metab, 98*(12), E1927–E1935. doi:10.1210/jc.2013-2820.

Clark, R. V., Walker, A. C., Miller, R. R., O'Connor-Semmes, R. L., Ravussin, E., & Cefalu, W. T. (2018). Creatine (methyl-d3) dilution in urine for estimation of total body skeletal muscle mass: Accuracy and variability vs. MRI and DXA. *J Appl Physiol (1985), 124*(1), 1–9. doi:10.1152/japplphysiol.00455.2016.

Commission, E. (2007a). Healthy ageing: A keystone for a sustainable Europe. http://ec.europa.eu/health/archive/ph_information/indicators/docs/healthy_ageing_en.pdf.

Commission, E. (2007b). Healthy Ageing: A keystone of a sustainable Europe. http://ec.europa.eu/health/ph_information/indicators/docs/healthy_ageing_en.pdf.

Cruz-Jentoft, A. J., Baeyens, J. P., Bauer, J. M., Boirie, Y., Cederholm, T., Landi, F.,... Zamboni, M. (2010). Sarcopenia: European consensus on definition and diagnosis: Report of the European working group on sarcopenia in older people. *Age Ageing, 39*(4), 412–423. doi:afq034 [pii] 10.1093/ageing/afq034 [doi].

Cruz-Jentoft, A. J., Bahat, G., Bauer, J., Boirie, Y., Bruyere, O., Cederholm, T.,... the Extended Group for EWGSOP2. (2019). Sarcopenia: Revised European consensus on definition and diagnosis. *Age Ageing, 48*(1), 16–31. doi:10.1093/ageing/afy169.

Cruz-Jentoft, A. J., Landi, F., Schneider, S. M., Zuniga, C., Arai, H., Boirie, Y.,... Cederholm, T. (2014). Prevalence of and interventions for sarcopenia in ageing adults: A systematic review. Report of the International Sarcopenia Initiative (EWGSOP and IWGS). *Age Ageing, 43*(6), 748–759. doi:10.1093/ageing/afu115.

Dawson-Hughes, B., & Bischoff-Ferrari, H. (2016). Considerations concerning the definition of sarcopenia. *Osteoporos Int, 27*(11), 3139–3144. doi:10.1007/s00198-016-3674-8.

Delmonico, M. J., Harris, T. B., Lee, J. S., Visser, M., Nevitt, M., Kritchevsky, S. B.,... Newman, A. B. (2007). Alternative definitions of sarcopenia, lower extremity performance, and functional impairment with aging in older men and women. *J Am Geriatr Soc, 55*(5), 769–774. doi:JGS1140 [pii] 10.1111/j.1532-5415.2007.01140.x [doi].

Delmonico, M. J., Harris, T. B., Visser, M., Park, S. W., Conroy, M. B., Velasquez-Mieyer, P.,... Goodpaster, B. H. (2009). Longitudinal study of muscle strength, quality, and adipose tissue infiltration. *Am J Clin Nutr, 90*(6), 1579–1585. doi:ajcn.2009.28047 [pii] 10.3945/ajcn.2009.28047 [doi].

Eberstadt, N., & Groth, H. (2007). *Europe's Coming Demographic Challenge: Unlocking the Value of Health.* Washington, DC: American Enterprise Institute for Public Policy Research.

EC. (European Commission, Directorate-General for Economic and Financial Affairs). European economy: Special Report n 1/2006.

Europe, C. O. Recent demographic developments in Europe 2005. (2006). *European base de déveine de population Committee of the Council of Europe.* Strasbourg: Council of Europe Publishing.

Eurostat. (2009). First demographic estimates for 2005: Statistics in focus. Accessed October 10, 2019. http://epp.eurostat.ec.europa.eu/cache/ITY_OFFPUB/KS-NK-06-001/EN/KS-NK-06-001-EN.PDF.

Evans, W. J. (2012). Endpoints and indicators for the older population. http://www.ema.europa.eu/docs/en_GB/document_library/Presentation/2012/04/WC500125114.pdf.

Fielding, R. A., Vellas, B., Evans, W. J., Bhasin, S., Morley, J. E., Newman, A. B.,... Zamboni, M. (2011). Sarcopenia: An undiagnosed condition in older adults. Current consensus definition: Prevalence, etiology, and consequences. International working group on sarcopenia. *J Am Med Dir Assoc, 12*(4), 249–256. doi:S1525-8610(11)00019-3 [pii] 10.1016/j.jamda.2011.01.003 [doi].

Fried, L. P., Tangen, C. M., Walston, J., Newman, A. B., Hirsch, C., Gottdiener, J.,... McBurnie, M. A. (2001). Frailty in older adults: Evidence for a phenotype. *J Gerontol A Biol Sci Med Sci, 56*(3), M146–M156.

Gallagher, D., Visser, M., De Meersman, R. E., Sepulveda, D., Baumgartner, R. N., Pierson, R. N.,...
Heymsfield, S. B. (1997). Appendicular skeletal muscle mass: Effects of age, gender, and ethnicity.
J Appl Physiol (1985), *83*(1), 229–239.

Goodpaster, B. H., Park, S. W., Harris, T. B., Kritchevsky, S. B., Nevitt, M., Schwartz, A. V.,... Newman, A. B.
(2006). The loss of skeletal muscle strength, mass, and quality in older adults: The health, aging and
body composition study. *J Gerontol A Biol Sci Med Sci*, *61*(10), 1059–1064. doi:61/10/1059 [pii].

Guillet, C., Prod'homme, M., Balage, M., Gachon, P., Giraudet, C., Morin, L.,... Boirie, Y. (2004). Impaired
anabolic response of muscle protein synthesis is associated with S6K1 dysregulation in elderly humans.
FASEB J, *18*(13), 1586–1587. doi:10.1096/fj.03-1341fje.

Janssen, I., Heymsfield, S. B., & Ross, R. (2002). Low relative skeletal muscle mass (sarcopenia) in older persons is associated with functional impairment and physical disability. *J Am Geriatr Soc*, *50*(5), 889–896.

Janssen, I., Shepard, D. S., Katzmarzyk, P. T., & Roubenoff, R. (2004). The healthcare costs of sarcopenia in
the United States. *J Am Geriatr Soc*, *52*(1), 80–85. doi:52014 [pii].

Kanis, J. A. (1994). Assessment of fracture risk and its application to screening for postmenopausal osteoporosis: Synopsis of a WHO report. WHO Study Group. *Osteoporos Int*, *4*(6), 368–381.

Kelly, T. L., Wilson, K. E., & Heymsfield, S. B. (2009). Dual energy X-Ray absorptiometry body composition
reference values from NHANES. *PLoS One*, *4*(9), e7038. doi:10.1371/journal.pone.0007038.

Kobayashi, S., Asakura, K., Suga, H., & Sasaki, S., & Three-generation Study of Women on, Diets and Health
Study Group. (2013). High protein intake is associated with low prevalence of frailty among old Japanese
women: A multicenter cross-sectional study. *Nutr J*, *12*, 164. doi:10.1186/1475-2891-12-164.

Koopman, R., & van Loon, L. J. (2009). Aging, exercise, and muscle protein metabolism. *J Appl Physiol
(1985)*, *106*(6), 2040–2048. doi:10.1152/japplphysiol.91551.2008.

Lang, T., Streeper, T., Cawthon, P., Baldwin, K., Taaffe, D. R., & Harris, T. B. (2010). Sarcopenia: Etiology,
clinical consequences, intervention, and assessment. *Osteoporos Int*, *21*(4), 543–559. doi:10.1007/
s00198-009-1059-y.

Lee, R. D. (2007). *Global Population Aging and Its Economic Consequences*. Washington, DC: AEI Press,
2007.

Lexell, J., & Downham, D. (1992). What is the effect of ageing on type 2 muscle fibres? *J Neurol Sci*, *107*(2),
250–251.

Morley, J. E., Abbatecola, A. M., Argiles, J. M., Baracos, V., Bauer, J., Bhasin, S.,... Anker, S. D. (2011).
Sarcopenia with limited mobility: An international consensus. *J Am Med Dir Assoc*, *12*(6), 403–409.
doi:S1525-8610(11)00142-3 [pii] 10.1016/j.jamda.2011.04.014 [doi].

Morley, J. E., Baumgartner, R. N., Roubenoff, R., Mayer, J., & Nair, K. S. (2001). Sarcopenia. *J Lab Clin Med*,
137(4), 231–243. doi:S0022-2143(01)80110-4 [pii] 10.1067/mlc.2001.113504 [doi].

Motion, A. I. (2013). Aging in Motion: The facts about sarcopenia. Accessed October 10, 2019. http://www.
aginginmotion.org/wp-content/uploads/2011/04/sarcopenia_fact_sheet.pdf.

Murphy, R. A., Ip, E. H., Zhang, Q., Boudreau, R. M., Cawthon, P. M., Newman, A. B.,... Health, Aging, and
Body Composition Study. (2014). Transition to sarcopenia and determinants of transitions in older adults:
A population-based study. *J Gerontol A Biol Sci Med Sci*, *69*(6), 751–758. doi:10.1093/gerona/glt131.

Murton, A. J. (2015). Muscle protein turnover in the elderly and its potential contribution to the development
of sarcopenia. *Proc Nutr Soc*, *74*(4), 387–396. doi:10.1017/S0029665115000130.

Muscaritoli, M., Anker, S. D., Argiles, J., Aversa, Z., Bauer, J. M., Biolo, G.,... Sieber, C. C. (2010). Consensus
definition of sarcopenia, cachexia and pre-cachexia: Joint document elaborated by Special Interest
Groups (SIG) "cachexia-anorexia in chronic wasting diseases" and "nutrition in geriatrics." *Clin Nutr*,
29(2), 154–159. doi:10.1016/j.clnu.2009.12.004.

Nair, K. S. (2005). Aging muscle. *Am J Clin Nutr*, *81*(5), 953–963.

Perez-Zepeda, M. U., Gutierrez-Robledo, L. M., & Arango-Lopera, V. E. (2013). Sarcopenia prevalence.
Osteoporos Int, *24*(3), 797. doi:10.1007/s00198-012-2091-x.

Pichard, C., Kyle, U. G., Bracco, D., Slosman, D. O., Morabia, A., & Schutz, Y. (2000). Reference values of
fat-free and fat masses by bioelectrical impedance analysis in 3393 healthy subjects. *Nutrition*, *16*(4),
245–254.

Rennie, M. J. (2009). Anabolic resistance: The effects of aging, sexual dimorphism, and immobilization on
human muscle protein turnover. *Appl Physiol Nutr Metab*, *34*(3), 377–381. doi:10.1139/H09-012.

Rizzoli, R., Reginster, J. Y., Arnal, J. F., Bautmans, I., Beaudart, C., Bischoff-Ferrari, H.,... Bruyere, O.
(2013). Quality of life in sarcopenia and frailty. *Calcif Tissue Int*, *93*(2), 101–120. doi:10.1007/
s00223-013-9758-y [doi].

Rosenberg, I. H. (1997). Sarcopenia: Origins and clinical relevance. *J Nutr*, *127*(5 Suppl), 990S–991S.

Sayer, A. A. (2010). Sarcopenia. *BMJ*, *341*, c4097.

Schaap, L. A., Koster, A., & Visser, M. (2013). Adiposity, muscle mass, and muscle strength in relation to functional decline in older persons. *Epidemiol Rev, 35*, 51–65. doi:10.1093/epirev/mxs006.

Scott, D., Hayes, A., Sanders, K. M., Aitken, D., Ebeling, P. R., & Jones, G. (2014). Operational definitions of sarcopenia and their associations with 5-year changes in falls risk in community-dwelling middle-aged and older adults. *Osteoporos Int, 25*(1), 187–193. doi:10.1007/s00198-013-2431-5.

Sterling, D. A., O'Connor, J. A., & Bonadies, J. (2001). Geriatric falls: Injury severity is high and disproportionate to mechanism. *J Trauma, 50*(1), 116–119.

Stevens, J. A., Corso, P. S., Finkelstein, E. A., & Miller, T. R. (2006). The costs of fatal and nonfatal falls among older adults. *Inj Prev, 12*(5), 290–295.

Studenski, S. (2009). What are the outcomes of treatment among patients with sarcopenia? *J Nutr Health Aging, 13*(8), 733–736.

Studenski, S. A., Peters, K. W., Alley, D. E., Cawthon, P. M., McLean, R. R., Harris, T. B.,... Vassileva, M. T. (2014). The FNIH sarcopenia project: Rationale, study description, conference recommendations, and final estimates. *J Gerontol A Biol Sci Med Sci, 69*(5), 547–558. doi:10.1093/gerona/glu010.

Tinetti, M. E., & Williams, C. S. (1997). Falls, injuries due to falls, and the risk of admission to a nursing home. *N Engl J Med, 337*(18), 1279–1284. doi:10.1056/NEJM199710303371806 [doi].

Tromp, A. M., Pluijm, S. M., Smit, J. H., Deeg, D. J., Bouter, L. M., & Lips, P. (2001). Fall-risk screening test: A prospective study on predictors for falls in community-dwelling elderly. *J Clin Epidemiol, 54*(8), 837–844.

Vellas, B., Pahor, M., Manini, T., Rooks, D., Guralnik, J. M., Morley, J.,... Fielding, R. (2013). Designing pharmaceutical trials for sarcopenia in frail older adults: EU/US task force recommendations. *J Nutr Health Aging, 17*(7), 612–618. doi:10.1007/s12603-013-0362-7 [doi].

Verdijk, L. B., Koopman, R., Schaart, G., Meijer, K., Savelberg, H. H., & van Loon, L. J. (2007). Satellite cell content is specifically reduced in type II skeletal muscle fibers in the elderly. *Am J Physiol Endocrinol Metab, 292*(1), E151–E157. doi:10.1152/ajpendo.00278.2006.

Visser, M. (2009). Towards a definition of sarcopenia—results from epidemiologic studies. *J Nutr Health Aging, 13*(8), 713–716.

Visser, M., & Schaap, L. A. (2011). Consequences of sarcopenia. *Clin Geriatr Med, 27*(3), 387–399.

Visser, M., Deeg, D. J., Lips, P., Harris, T. B., & Bouter, L. M. (2000). Skeletal muscle mass and muscle strength in relation to lower-extremity performance in older men and women. *J Am Geriatr Soc, 48*(4), 381–386.

Volpi, E., & Rasmussen, B. B. (2000). Nutrition and muscle protein metabolism in the elderly. *Diabetes Nutr Metab, 13*(2), 99–107.

von Haehling, S., Morley, J. E., & Anker, S. D. (2010). An overview of sarcopenia: Facts and numbers on prevalence and clinical impact. *J Cachexia Sarcopenia Muscle, 1*(2), 129–133. doi:10.1007/s13539-010-0014-2.

WHO. (2004). A Glossary of Terms for Community Health Care and Services. http://www.who.int/kobe_centre/ageing/ahp_vol5_glossary.pdf.

Section II

New Data on Sarcopenia

2 Models of Accelerated Sarcopenia

Andrew S. Layne, Lisa M. Roberts, and Thomas W. Buford

CONTENTS

2.1 INTRODUCTION

Sarcopenia, the age-related loss of muscle mass and quality, is a major healthcare concern for older adults. The condition is associated with the development of functional disability (Janssen et al. 2004; Visser et al. 2005) and may lead to the loss of independence for afflicted individuals. Because of the costs associated with caring for an individual with compromised function, sarcopenia has been linked to elevate healthcare costs (Janssen et al. 2004). Moreover, the absolute costs associated with sarcopenia are likely to rise sharply in the coming decades considering that the total number of persons over 65 years of age is expected to double over the next 25 years (Federal Interagency 2009). Hence, additional knowledge of mechanisms underlying sarcopenia development is necessary to advance prevention and treatment efforts that will improve the quality of life for millions of older adults.

To date, numerous age-related mechanisms have been described that result in the loss of muscle mass. Briefly, these mechanisms include increases in oxidative stress and pro-inflammatory cytokine production and decreases in the production of anabolic hormones such as testosterone. More specifically, myocyte apoptotic signaling (Marzetti et al. 2008), altered protein synthesis and/or turnover (Combaret et al. 2009), and impaired satellite cell (SC) function (Hepple 2006) play a role in age-related muscle loss. Further complexity is added to this discussion by the number of upstream factors that may affect each of these specific mechanisms. For example, proteolytic signaling may be stimulated by a number of factors including catabolic hormones (Ma et al. 2003), pro-inflammatory cytokines (Li and Reid 2000; Tsujinaka et al. 1996), or denervation (Sacheck et al. 2007). Moreover, ubiquitin–proteasome system (UPS) upstream factors can modulate multiple regulatory pathways. For example, tumor necrosis factor alpha (TNFα) can either stimulate proteolysis through the UPS or induce apoptosis via the death-receptor pathway.

As a result of these age-related changes, individuals tend to lose muscle mass at a rate of 1%–2% per year after the age of 50 years (Lauretani et al. 2003; Hiona and Leeuwenburgh 2008; Marcell 2003). This decline is primarily due to the progressive atrophy and loss of type II muscle fibers and motor neurons (Larsson, Sjödin, and Karlsson 1978; Tomlinson, Irving, and Rebeiz 1973). These changes occur in conjunction with increased infiltration of noncontractile material such as adipose and connective tissues (Brooks and Faulkner 1994; Goldspink et al. 1994; Goodpaster et al. 2008; McNeil et al. 2005; Petersen et al. 2003). Together, these changes contribute to declines in functional capacity of the muscle that contributes to functional disability (Evans 1997; Janssen, Heymsfield, and Ross 2002; Mühlberg and Sieber 2004; Rolland et al. 2008; Visser et al. 2005).

While impaired locomotion is certainly the hallmark concern of sarcopenia, muscle atrophy may impair other physiological functions such as glucose regulation, hormone production, and cellular communication. Moreover, muscle tissue provides the body's only major "reserve" of readily available amino acids. Thus, inadequate muscle mass prior to the onset of a disease condition may be dangerous in patients who need a large protein reservoir to recover. As a result, patients with sarcopenia prior to disease diagnosis may face impaired recovery from surgery (Rutan and Herndon 1990; Watters et al. 1993) or increased risk of mortality (Prado et al. 2009). Still, while contributions of sarcopenia to functional impairments are well documented, data regarding the importance of skeletal muscle in the recovery from life-threatening situations, such as severe burns or traumatic surgeries, are few. Like a standardized definition, further study of the clinical impact of sarcopenia in these stressful situations could improve physician awareness of the problem.

In an effort to increase understanding of the relationship between behavioral and disease states with aging, we review here extensive evidence demonstrating that unhealthy lifestyle and/or the presence of chronic diseases are capable of accelerating muscle loss in middle-aged and older adults. Because elevated risk of co-morbid diseases and behavioral factors such as sedentary lifestyle are established phenomena in older individuals, we suggest that these factors should be increasingly recognized and independently accounted for in the study of sarcopenia. Within this chapter, we will discuss a number of diseases and behavioral conditions that can theoretically accelerate the progression of sarcopenia. The secondary purpose of such discussion is to highlight specific areas where further research is needed to enhance understanding of the interactions between age and factors that may accelerate sarcopenia.

2.1.1 Challenges in Defining Sarcopenia

Previously, we discussed the need for an established clinical definition of sarcopenia (Buford et al. 2010a; Buford 2017). At the time, we discussed the need for a standardized definition of sarcopenia, which was necessary to establish a medical diagnosis for the condition. Without a definition, the ability of physicians to recognize and appropriately treat sarcopenia would remain poor. Previously, this same challenge faced clinicians treating patients with cachexia (Evans et al. 2008b). Since our original publication, a number of definitions of sarcopenia have been proposed (Cruz-Jentoft et al. 2010; Muscaritoli et al. 2010; Fielding et al. 2011; Studenski et al. 2014; Morley et al. 2011; Chen et al. 2014). Figure 2.1 and Table 2.1 show summaries of Medline publications relating sarcopenia to various conditions. In contrast to the original proposed definition of sarcopenia that included only age-related muscle atrophy (Roubenoff and Hughes 2000), more recent definitions also include measures of muscle strength (i.e., grip strength) and/or physical function (i.e., gait speed). For example, the recently proposed Foundation for the National Institutes of Health Biomarkers Consortium Sarcopenia Project (FNIH) criteria defines sarcopenia as low

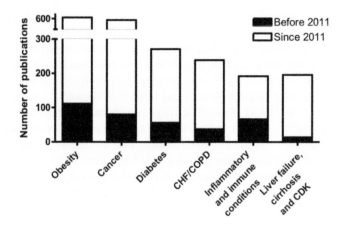

FIGURE 2.1 Comparison of number of Medline publications related to sarcopenia and each condition (e.g., "sarcopenia" and "obesity") before and after publication of the original conceptual review on accelerated sarcopenia, as of December 2017. (From Buford, T.W. et al., *Ageing Res. Rev.*, 9, 369–383, 2010a.)

TABLE 2.1
Number of Medline Publications for Sarcopenia and Each Condition Since 2011

Condition	Before 2011	Since 2011	Additional References
Obesity	111	499	Buch et al. (2016); Kalinkovich and Livshits (2016); Kob et al. (2015)
Cancer	80	507	Christensen et al. (2014); Mendes et al. (2015)
Diabetes	56	215	Cleasby, Jamieson, and Atherton (2016); Sinclair and Rodriguez-Mañas (2016); Umegaki (2016)
Inflammatory conditions	30	66	Bilski et al. (2013); Geremia et al. (2014)
CHF	24	96	Collamati et al. (2016); Ebner et al. (2014)
Liver failure and cirrhosis	4	102	Sinclair et al. (2016); Dasarathy (2016)
CKD	10	80	Kooman et al. (2014)
COPD	13	69	Langen et al. (2013)

Abbreviations: RA, rheumatoid arthritis; HIV, acquired human immunodeficiency virus; IBD, inflammatory bowel diseases; CHF, chronic heart failure; CKD, chronic kidney disease; COPD, chronic obstructive pulmonary disease.

muscle strength (hand grip strength <26 kg for men and <16 kg for women) and low lean mass (appendicular lean mass adjusted for body mass index [ALM_{BMI}] <0.789 for men and <0.512 for women) associated with low gait speed (<0.8 m/s) (Studenski et al. 2014).

However, the utility of these definitions has been questioned, as comparative studies showed a wide range in prevalence of sarcopenia depending on the definition used (Dupuy et al. 2015; Dam et al. 2014; Bijlsma et al. 2013). Others have shown that current definitions of sarcopenia are limited in their ability to predict functional disability (Dupuy et al. 2015) or falls (Dawson-Hughes and Bischoff-Ferrari 2016; Bischoff-Ferrari et al. 2015). Therefore, it is possible that consideration of other contributory factors such as co-morbid conditions may improve identification of persons at greatest risk of functional decline due to sarcopenia. However, this approach is somewhat controversial, as the underlying mechanisms leading to age-related muscle loss may be very different compared with disease-mediated wasting (Hepple 2012). Some authors have expressed concern that use of the term "sarcopenia" to describe aging- and disease-related muscle may hinder efforts to characterize molecular mechanisms underlying the sarcopenia phenotype and inhibit the development of treatments (Dardevet et al. 2012). In an effort to maintain sarcopenia as an explicitly age-related condition while also describing muscle loss that is attributable to both age and disease, a number of terms have been coined including myopenia (Fearon, Evans, and Anker 2011), skeletal muscle function deficit (Correa-de-Araujo and Hadley 2014), and "secondary sarcopenia"—sarcopenia that results from lack of physical activity (PA), disease, or malnutrition (Wakabayashi and Sakuma 2014). However, we have argued that aging is inherently intertwined with increased risk of developing numerous chronic diseases (Buford et al. 2012a). Thus, separating the unique contributions of aging and disease to muscle atrophy is likely impractical—if not impossible—in clinical research and patient care. As a result, we believe that ignoring the contribution of chronic diseases will limit, rather than enhancement, our understanding of the biological causes of sarcopenia (Dardevet et al. 2012).

2.2 BEHAVIOR-MEDIATED PATHWAYS

Chronological aging and genetics, significant contributors to the onset, and progression of sarcopenia are beyond human control and are therefore considered nonmodifiable risk factors for sarcopenia. However, humans often contribute to sarcopenia progression and/or disease development through lifestyle choices. A wealth of data exist that demonstrate beneficial effects of exercise or nutritional interventions to functioning of older individuals (Goodpaster et al. 2008; Kalapotharakos et al. 2007; Manini and Pahor 2009; Sugawara et al. 2002). Yet many older adults make poor dietary choices and engage in inadequate amounts of PA. In the United States, approximately 80% of individuals over the age of 65 years do not engage in regular PA (Fahey, Insel, and Roth 2018). Furthermore, 88% do not perform muscle strengthening exercises, known to provide many beneficial effects to skeletal muscle. Meanwhile, the US Department of Agriculture's Healthy Eating Index indicates that approximately 80% of older individuals could benefit from improvements in their diet including increased protein and fiber intake and decreased consumption of saturated fats (Federal Interagency 2009). Although some degree of sarcopenia is unavoidable in all older adults, poor dietary and PA choices might compound the problem (Buford et al. 2010b).

2.2.1 PHYSICAL INACTIVITY

Exercise training is one of the simplest, most feasible, and inexpensive strategies available to combat the onset of sarcopenia and reduce the rate of functional decline. Although protein degradation is often accelerated in sedentary older adults, the neuromuscular system of these individuals retains a tremendous ability to adapt in response to heavy loading (reviewed in Narici et al. 2004). Experimentally based studies support a dose-dependent increase in lean mass and physical function as a result of acute resistance exercise (He and Baker 2004; Hillsdon et al. 2005; Leveille et al. 1999; Manini and Pahor 2009). Resistance exercise provides a host of physiological benefits to muscle including, but not limited to, reduced inflammation (Greiwe et al. 2001), increased mitochondrial function (Melov et al. 2007), improved myogenic signaling (Kosek et al. 2006), and SC activity (Mackey et al. 2007). Even adults who are sedentary into their 80s demonstrate a sharp increase in muscle mass and strength following short-term resistance training (Fiatarone et al. 1994). Incredibly, these results occur with nominal time commitment as data indicate maintenance of muscle strength with one day of strength building per week (Trappe, Williamson, and Godard 2002). Recent studies have also indicated that aerobically focused training programs are capable of preserving muscle mass in older adults (Chomentowski et al. 2009; Goodpaster et al. 2008). Aerobic exercise provides physiological benefits that conserve muscle mass through improved muscle blood flow (Currie, Thomas, and Goodman 2009), improved mitochondrial function (Barbieri et al. 2015), decreased oxidative stress (Bouzid et al. 2015; Bloomer et al. 2005), and decreased Glucocorticoid (GC) sensitivity (Duclos et al. 2001). However, aerobically focused training is likely most effective for preserving muscle mass and strength when combined with resistance exercise (Marcell, Hawkins, and Wiswell 2014; Landi et al. 2014). Nevertheless, these studies indicate that short-term exercise training can either increase muscle mass or delay its decline in older adults.

The extent to which chronic training contributes to the prevention of sarcopenia is not clearly delineated. To date, no conclusive data are available to determine whether chronic training attenuates the progression of sarcopenia or simply provides an upward shift in the amount and quality of muscle with a similar rate of decline over time. The best attempts to understand the effects of chronic training on muscle changes have studied master athletes. These athletes are the most physically active older adults and provide a unique opportunity to study the potential benefits of regular exercise on the prevention of sarcopenia (Hawkins, Wiswell, and Marcell 2003). Master athletes appear to have greater muscle strength and power than their sedentary peers (Pearson et al. 2002). However, despite regular and typically life-long training, these athletes experienced age-related declines similar to sedentary controls for both peak power (1.3% vs. 1.2%) and force (0.6% vs. 0.5%) (Pearson et al. 2002). These data suggest that training may not be sufficient to alter the trajectory of skeletal muscle loss in older adults. Unfortunately, no longitudinal data exist to directly compare age-related changes in muscle mass or myofiber number between sedentary individuals and master athletes. This information is critical to determine whether chronic training can alter the rate of muscle loss in older adults or simply adjust the starting point.

In addition to questions concerning the effectiveness of exercise in combating sarcopenia, a pertinent question in the context of the present review is if sedentary lifestyle accelerates the loss of skeletal muscle mass in older adults. For over a quarter century, experts have differed in opinions regarding the contribution of a sedentary lifestyle age-related muscle loss (reviewed in Faulkner et al. 2007). These differences in opinion are not surprising given inherent difficulties understanding the heterogeneity of PA patterns across the lifespan as well as challenges in measuring free-living activity. These challenges can be addressed through the development and standardization of assessment techniques for PA and the energy expended during such activity. Longitudinal studies have utilized doubly labeled water (Manini et al. 2009), accelerometry (Shephard et al. 2013), and self-reported activity (Mijnarends et al. 2016) to compare activity and energy expenditure among older adults. These studies indicate that older persons with higher daily activity had the greatest amount of lean mass. However, the rate of muscle loss in more active individuals is similar to those with lower levels of PA, suggesting that physical inactivity may not directly accelerate the trajectory of sarcopenia. However, a sedentary lifestyle certainly adds to the risk of numerous pathological conditions (Arsenault et al. 2010; Venables and Jeukendrup 2009; White et al. 2011), many of which do appear to accelerate sarcopenia progression. As such, identifying strategies to reduce sedentary time (Aggio et al. 2016; Gianoudis, Bailey, and Daly 2015) and improve exercise adherence will potentially reduce the incidence and impact of sarcopenia.

2.2.2 UNDER-NUTRITION AND OBESITY

A proper diet is critically important to slow sarcopenia progression, as well as to maintain overall healthy aging (Houston et al. 2008; Paddon-Jones and Rasmussen 2009). Traditional discussions of dietary importance in sarcopenia focused on inadequate intake and the frail phenotype. Insufficient caloric intake, also known as the anorexia of aging (Morley et al. 2005), is common among older adults secondary to the loss of appetite, gastrointestinal changes, altered taste and smell, social changes, and economic limitations (Wysokiński et al. 2015; Bales and Ritchie 2002). This may lead to the development of sarcopenia through a lack of protein intake and subsequent decreases in muscle protein synthesis (Friedman, Campbell, and Caradoc-Davies 1985; Waters et al. 2003). Furthermore, anorexia of aging results in a reduced overall caloric intake, which may lead to insufficient intake of macro and micronutrients (e.g., polyunsaturated fats and vitamin D) that are linked to sarcopenia and frailty (Bruyere et al. 2016; Smith et al. 2015; Di Girolamo et al. 2014). Moreover, older individuals suffering from protein energy malnutrition (PEM) are unlikely to gain muscle mass and strength while engaging in resistance training (Rolland et al. 2008). PEM may also severely compromise quality of life and the ability to thrive, especially when combined with co-morbid diseases (Vetta et al. 1999).

As a result, insufficient protein intake appears to be a critical factor for sarcopenia development in older adults. Recent data indicate that lean mass in older adults is significantly and positively associated with dietary protein intake (Anton et al. 2018; Houston et al. 2008). Over the course of 3 years, the individuals who consumed more than the recommended dietary allowance (RDA) for protein of 0.8 g/kg/day experienced the smallest losses of lean mass (Houston et al. 2008). In contrast, persons who experienced the most significant muscle atrophy consumed protein in quantities either at or below the RDA. Based on this report and others suggesting that the current RDA for protein intake may not be adequate to maintain optimal skeletal muscle health in older adults (Evans et al. 2008a; Look AHEAD Research Group et al. 2007; Morais, Chevalier, and Gougeon 2006; Sood, Baker, and Coleman 2008), numerous consensus reports now suggest that protein intake for healthy older adults should be 1.0–1.2 g/kg/day (Volpi et al. 2013; Bauer et al. 2013; Deutz et al. 2014). Moreover, older adults with acute or chronic health problems may need 1.2–1.5 g/kg/day (Bauer et al. 2013). The timing and type of protein intake by older adults may be critical to maintenance of muscle mass. Reports indicate that aging does not reduce muscle protein synthesis in response to a high-protein (25–30 g) meal (Symons et al. 2009; Paddon-Jones and Rasmussen 2009; Volpi et al. 2001). However, protein synthesis declines in the elderly when meals contain less protein or are ingested in conjunction with carbohydrates (reviewed in Paddon-Jones and Rasmussen 2009). Subsequently, Paddon-Jones and Rasmussen (2009) suggested that sufficient protein with each meal should be encouraged more so than simply an overall increase in daily protein intake. However, to date little is known about the optimal protein feeding patterns. Studies have shown that older adults need at least 25–30 g of protein and 2.5–2.8 g of leucine at each meal to optimally stimulate muscle protein synthesis (Paddon-Jones and Rasmussen 2009), while others have shown that a single high-protein meal during the day is sufficient to induce muscle protein synthesis and preserve lean mass (Bouillanne et al. 2013). However, in a cross-sectional study of protein feeding patterns, Bollwein et al. (2013) noted that a skewed protein intake (majority of daily intake at midday meal) was significantly associated with an increased incidence of frailty, while total protein intake was not associated with frailty. The relationship between sarcopenia and frailty will be covered in depth in Chapter 3.

In addition to protein timing, the type of protein may be an important consideration for maximizing postprandial protein synthesis in older adults. Indeed, essential amino acids—that is, leucine, isoleucine, and valine—may be the most critical for stimulation of protein synthesis in older adults (Dillon et al. 2009; Henderson, Irving, and Nair 2009). Most notably, a number of studies suggest that leucine may be the most critical amino acid for stimulation of muscle protein synthesis (Anthony et al. 2002; Combaret et al. 2005; Holeček et al. 2001; Katsanos et al. 2006). Circulating levels of leucine and other essential amino acids can be increased with supplementation with whey protein (Forbes et al. 2013), and a number of studies indicate that whey protein is effective for increasing muscle mass in older adults (Bauer et al. 2015), particularly when combined with weight training (Naclerio and Larumbe-Zabala 2016). Moreover, dietary supplementation of omega-3 polyunsaturated fats (Di Girolamo et al. 2014), vitamin D (Wagatsuma and Sakuma 2014), and probiotics (Steves et al. 2016) may further enhance the anabolic response to protein feeding, perhaps by reducing inflammation (Zanetti, Harris, and Dawson-Hughes 2014). Despite these data, further investigation is needed to determine the optimal protein intake measures—both during meals and through supplementation—to combat sarcopenia. Regardless of the above, these measures will likely need to be combined with exercise to be effective (Verhoeven et al. 2009).

In addition to PEM, caloric insufficiency (CI) in older adults can accelerate the progression of sarcopenia. This CI often stems from the anorexia of aging, a common phenomenon in older individuals (Visvanathan and Chapman 2009). Many factors contribute to decreased food intake in the elderly, including loss of appetite, gastrointestinal changes, altered taste and smell, social changes, and economic limitations (reviewed in Bales and Ritchie 2002). CI often results in unintentional weight loss that includes the loss of lean mass. Significant increases in circulating pro-inflammatory cytokines and cortisol due to CI are among the mechanistic triggers thought to contribute to this loss of lean mass (Bales and Ritchie 2002; Douyon and Schteingart 2002; Bouchard, Dionne, and Brochu 2009).

Like CI, excess caloric intake that results in obesity may also accelerate sarcopenia (Bouchard, Dionne, and Brochu 2009; Jarosz and Bellar 2009). This concept may seem counter-intuitive as obese older adults clearly have higher amounts of muscle than their nonobese counterparts (Villareal et al. 2004). However, muscle quality in obese individuals is poor due to increases in intramuscular adipose tissue (Villareal et al. 2004), which may contribute to muscle weakness, frailty, and disability (Blaum et al. 2005). As a result, a new concept was created highlighting sarcopenic obesity that was initially defined as appendicular skeletal muscle mass $\times$ height^{-2} of two standard deviations below a young reference group (Baumgartner et al. 2004).

Sarcopenic obesity is associated with increases in both cortisol and pro-inflammatory cytokines, factors that promote muscle catabolism, abdominal fat accumulation, and the development of insulin resistance (Epel 2009). Fat accumulation in the muscle (myosteatosis) has been linked with decreased muscle function and muscle quality (Manini et al. 2007; Visser et al. 1998; Goodpaster et al. 2001). In preclinical models of myosteatosis, increased fatty infiltration of the muscle was associated with skeletal muscle insulin resistance, thus interfering with anabolic signaling in the muscle (Rivas et al. 2016). Furthermore, myosteatosis appears to facilitate type II to type I fiber type conversion (Mastrocola et al. 2015), which may lead to a reduction in skeletal muscle force and power production. Finally, intermuscular fat is associated with increased inflammation in surrounding skeletal muscle (Khan et al. 2015), which could excessively stimulate proteolysis and perhaps myonuclear apoptosis (Kob et al. 2015).

Obesity may also contribute to accelerated muscle loss through high levels of oxidative stress (Furukawa et al. 2004). High-calorie meals, particularly those consisting of quickly absorbable foods and drinks, have been found to produce abnormally large elevations in blood glucose and triglycerides (O'Keefe and Bell 2007), which are linked to the increased generation of reactive oxygen species (ROS). Thus, repeated consumption of high-calorie, energy-dense meals may not only promote weight gain but also increase oxidative damage to tissues (O'Keefe and Bell 2007). Although data in support of this hypothesis are scarce, oxidative stress derived from caloric excess may contribute to the obese sarcopenic etiology. Over the course of the aging process, an accumulation of DNA alterations and mutations from ROS accumulation can lead to impairments in protein synthesis and ATP generation and thus compromise cell viability (Hiona and Leeuwenburgh 2008). Furthermore, ROS accumulations may contribute to telomere shortening (von Zglinicki et al. 1995), dysregulation of Ca^{++} release from the sarcoplasmic reticulum, apoptosis (Adhihetty et al. 2007), NF-κB activation (Gloire, Legrand-Poels, and Piette 2006), and UPS upregulation (Li et al. 2003). Despite these suggestive data, further study of the impact of oxidative stress on the development and acceleration of sarcopenia due to obesity is necessary to establish a causal relationship.

Collectively, studies to date suggest that both under-nutrition and obesity can contribute to the sarcopenic process. Thus, a critical window of optimal caloric intake exists for maintaining the health of skeletal muscle. Therefore, older adults should consume calories in balance with caloric expenditure to maintain weight and to ensure adequate protein intake. Moreover, the current RDA for protein of 0.8 g/kg/day may not be adequate for the elderly. Older individuals may also derive enhanced benefit from consuming dietary supplements such as leucine, vitamin D, and Omega 3s, but future randomized controlled trials are needed to confirm this hypothesis.

2.3 DISEASE-MEDIATED PATHWAYS

Although chronic disease conditions are common in elderly patients, gerontologists and geriatricians have yet to fully understand the role that specific diseases may have on the development and trajectory of sarcopenia. Traditionally, scientists have studied sarcopenia primarily in healthy older adults to isolate the effects of aging per se on the skeletal muscle. However, these studies exclude a significant portion of older adults who are afflicted with a co-morbid disease. Here, we discuss, both mechanistically and morphologically, how several of these diseases might contribute to accelerate the progression of sarcopenia. Notably, these disease-mediated models are congruent with the established definition of cachexia (Evans et al. 2008b). Indeed, cachexia represents a primary accelerating model of sarcopenia. In contrast, cachexia can occur in individuals of all ages, which distinguishes it from sarcopenia. Much like sarcopenia research has been focused around aging, research on cachexia has been associated primarily with effects of disease. We argue that future sarcopenia research should investigate the interaction between disease and aging processes.

2.3.1 Cancer

Cancer is perhaps the most well-known pathological condition that induces muscle atrophy (Tisdale 2003). Cancer incidence and survivorship are highest in adults over the age of 60 years (Jemal et al. 2007; Wingo et al. 1998). Moreover, the likelihood of developing breast, colon, leukemia, lung, melanoma and non-Hodgkin lymphoma, prostate, and uterine cancers is higher with the increasing age (Brenner 2002). Many, but not all cancers, are associated with some form of muscle atrophy (Tisdale 2003), and this atrophy is associated with an elevated risk of death (Dewys et al. 1980; Mantovani et al. 2001). Therefore, older persons with sarcopenia are likely to experience a severe drop in body mass prior to and following cancer diagnosis (Lundholm et al. 1976). In children, cancer often leads to long-term muscle deficits that manifest in adulthood (Ness et al. 2013, 2007; Oeffinger et al. 2006; Warner 2008).

A similar pattern of muscle loss in older adults accelerates the progression of sarcopenia compared with age-matched cancer-free counterparts (Xiao et al. 2016; Freedman et al. 2004; Awad et al. 2012). Cancer-related muscle atrophy is typically associated with the development of cachexia, a serious condition during which patients rapidly lose large amounts of body mass with proportional decreases in muscle and adipose tissue (Tisdale 2009). As such, the onset of cancer in late adulthood has important implications for the development and progression of sarcopenia. Here we briefly describe the most common mechanisms underlying cancer-related muscle atrophy and call for future research into the progression of sarcopenia in older cancer survivors.

The presence of a tumor initiates a cascade of events leading to muscle wasting often labeled as the cancer cachexia syndrome. In response to the tumor, the acute phase response is responsible for the production of mediators such as serum amyloid A, C-reactive proteins, fibrinogen, and other proteins that are rapidly synthesized in the liver for the purpose of limiting tumor injury. The liver uses endogenous amino acids to produce these proteins, and skeletal muscle provides the largest and most accessible depot in the body. However, the process is quite inefficient. For example, 2.6 g of muscle protein are required to synthesize 1 g of fibrinogen (Preston et al. 1998; Reeds, Fjeld, and Jahoor 1994). Subsequently, muscle atrophy develops very rapidly during cachexia.

Pro-inflammatory cytokines are among the most potent catabolic triggers regulating cancer cachexia (Deans et al. 2006; Mantovani et al. 2000). While cytokines are released systemically, their local actions on specific tissues are unique. Cancer cachexia seems to specifically target the skeletal muscle since the visceral protein compartment is preserved even when weight loss reaches 30% (Fearon 1992; Preston et al. 1987). For example, TNFα is highly selective in targeting myosin heavy chains for breakdown in both cell culture and tumor models of wasting (Acharyya et al. 2004). As a result, reduction of pro-inflammatory cytokine expression in patients with cancer is a priority. However, randomized controlled trials of anti-TNF therapies etanercept (Jatoi et al. 2007) and infliximab (Wiedenmann et al. 2008; Jatoi et al. 2010) failed to prevent skeletal muscle wasting in patients with cancer-associated cachexia. Thus, blockade of a single pro-inflammatory cytokine appears to be insufficient for preventing cytokine-induced muscle wasting in patients with cancer, possibly due to functional redundancy of many cytokines. Accordingly, studies have shown that reducing inflammation with nonsteroidal anti-inflammatory drugs such as ibuprofen and celecoxib may prevent weight loss and improve lean muscle mass, particularly when combined with other treatments (Couch et al. 2015). Furthermore, a broad-spectrum immunomodulatory drug OHR118 targeting the actions of both IL6 and TNFα has shown promise for improving appetite and maintaining weight in patients with advanced-stage cancer (Chasen, Hirschman, and Bhargava 2011).

In addition to direct effects on muscle, cytokines are also capable of crossing the blood–brain barrier and altering the function of hunger regulatory systems (Gutierrez, Banks, and Kastin 1993). Subsequently, patients with cancer with cachexia develop anorexia at an incidence rate of 15%–40%. This anorexia is a major contributor to muscle wasting and is thought to originate from elevated levels and sensitivity to cytokines (Rubin 2003). Data indicate that food intake is suppressed when TNFα and IL1 are administered either centrally or peripherally (Ramos et al. 2004). These cytokines appear to act on the ventromedial nucleus, a brain structure critical for the regulation of body mass, to stimulate early development of satiety. This early satiety may result from cytokine-mediated alterations in production of satiety-related hormones including neuropeptide-Y (appetite stimulant) and pro-opiomelanocortin (appetite suppressant) and hypothalamic resistance to peripherally produced appetite regulators such as leptin and ghrelin (Patra and Arora 2012; Mendes et al. 2015). In addition to cytokine-mediated anorexia, many patients with cancer are at increased risk for under-nutrition due to treatment side effects (e.g., nausea, dysphagia, dysgeusia, malabsorption), depressed mood, and presence of solid tumor—particularly in cancers of the head, neck, and gastrointestinal tract (Nicolini et al. 2013).

In addition to changes in caloric intake, elevations in resting energy expenditure may contribute to cancer-related muscle atrophy. The presence of hypermetabolism may be cancer-type specific with high prevalence found in lung and pancreatic cancer (Fredrix et al. 1991a, 1991b). In contrast, no substantial changes in metabolism are typically observed in gastric or colorectal cancer (Fredrix et al. 1991a, 1991b). Cancer-induced hypermetabolism may be linked to elevated thermogenesis through mitochondrial uncoupling, in which the proton gradient across the inner membrane is dissipated without ATP synthesis. In animal models of cachexia, gene expression of uncoupling protein (UCP)-2 and UCP-3 in skeletal muscle is reportedly upregulated (Bing et al. 2000). Increases in UCPs may stem from tumor-related increases in TNFα, as intravenous injections of the cytokine appear to increase both UCP-2 and UCP-3 in the skeletal muscle (Sanchís et al. 1998). Thus, hypermetabolism may be involved in cachexia through high energy demand for mitochondrial respiration that leads to an inefficient synthesis of ATP.

Over the past few decades, the early detection of cancer improved surgical techniques; the development of more effective antineoplastics and improvements in radiochemotherapy protocols have substantially increased chances of tumor eradication. Such advances have increased patient survival, and clinicians are now faced with increased focus on management of postcare quality of life. Older cancer survivors are faced with major challenges in overcoming the loss in muscle caused by the disease and the battles against sarcopenia, both of which have no known cures. In addition, a growing body of evidence suggests that the presence of sarcopenia may impair recovery from surgical procedures (Boer et al. 2016) and results in dose-limiting chemotherapy toxicity (Antoun et al. 2010; Barret et al. 2014). As such, sarcopenia is widely recognized as a critical prognostic indicator of cancer survival (Shachar et al. 2016). Additional work is needed on the effects of cancer on the aging process that includes understanding how mechanisms of cachexia differ in aged persons, determining recovery patterns in aged muscle, and developing interventions aimed at preserving muscle tissue following cancer diagnosis.

2.3.2 CHRONIC INFLAMMATORY AND IMMUNE CONDITIONS

Several immune-related disorders—including inflammatory bowel diseases (IBDs) (Schneider et al. 2008; Bryant et al. 2015; Subramaniam et al. 2015; Zhang et al. 2016), rheumatoid and spondyloarthritis (Baker et al. 2015; Dao, Do, and Sakamoto 2011; Delgado-Frias et al. 2015; Doğan et al. 2015; Giles et al. 2008; Aguiar et al. 2014), multiple sclerosis (MS) (Formica et al. 1997; Kent-Braun et al. 1997; Wens et al. 2014), idiopathic inflammatory myopathies (Hanaoka et al. 2015), and human immunodeficiency virus (HIV) (Scherzer et al. 2011; Szulc et al. 2010; Grinspoon, Mulligan, and Department of Health and Human Services Working Group on the Prevention and Treatment of Wasting and Weight Loss 2003)—are associated with reduced muscle mass and function. Many of these conditions can onset at any age. However, like cancer, these conditions may lead to premature loss of muscle mass, which can accelerate functional decline later in life. For example, several reports indicate that sarcopenia—defined as appendicular skeletal muscle mass >2 standard deviations below the mean of young, healthy adults—is present in up to ~60% of patients with ulcerative colitis and Crohn's disease (Holt et al. 2016; Zhang et al. 2015; Schneider et al. 2008). Importantly, these patients were relatively young (mean ages 32–44 years), highlighting the critical need for interventions to prevent muscle wasting among persons with these conditions.

Although a detailed discussion of the pathogenesis of the inflammatory response to immune disorders is beyond the scope of this chapter, many excellent reviews have detailed the role of innate and adaptive immunity initiating an inflammatory cascade leading to chronic inflammation (Madaro and Bouche 2014; Hemmer, Kerschensteiner, and Korn 2015; Smolen, Aletaha, and McInnes 2016). Many of these conditions are thought to be triggered by environmental stressors, genetic factors, or gut bacteria. For example, many autoimmune diseases are thought to result from a dysfunctional innate immunity response to gut microbiota, which triggers an aggressive adaptive immune response (Knights, Lassen, and Xavier 2013; Huang and Chen 2016). This response can injure organs and tissues such as the liver, joints, blood vessels, and skeletal muscle and results in increased systemic inflammation. As a result, a number of circulating cytokines are chronically upregulated in persons with autoimmune disorders, including IL6, IL1β, and TNFα (Strober et al. 2010). Circulating levels of these cytokines are associated with muscle loss and weakness (Toth et al. 2005; Zhou et al. 2016). In some instances, autoimmune disorders can lead to acute myositis (Shimoyama et al. 2009; Voigt et al. 1999; Christopoulos et al. 2003; Szabo et al. 2009), which is typically resolved following anti-inflammatory treatment or bowel resection.

In addition to the direct effects of disease activity and inflammation on skeletal muscle, many inflammatory and autoimmune conditions are associated with physical inactivity (Learmonth and Motl 2016; Vancampfort et al. 2016; Tew, Jones, and Mikocka-Walus 2016), which may increase the risk of sarcopenia for persons with these conditions. For example, individuals with MS are 2.5 times more likely than healthy, age-matched controls to report insufficient PA (active energy expenditure <7 kcal/kg/week) (Motl et al. 2015). This lack of PA may be caused by several factors including, but not limited to, fatigue, pain, depressed mood, drug side effects, lack of self-efficacy, disease activity, and related symptoms, and lack of guidelines from exercise and healthcare professionals (van Zanten et al. 2015; Rehm and Konkle-Parker 2016; Tew, Jones, and Mikocka-Walus 2016). Studies of PA in persons with inflammatory and autoimmune conditions have shown that PA is associated with decreased disease activity, increased self-efficacy, improved mood, and better quality of life (Sandberg et al. 2014; Metsios, Stavropoulos-Kalinoglou, and Kitas 2015; Learmonth and Motl 2016; Verhoeven et al. 2016; Packer, Hoffman-Goetz, and Ward 2010; Nathan et al. 2013).

Importantly, PA is also associated with favorable changes in body composition including increased lean mass and reduced body fat. Many autoimmune and inflammatory disorders including IBD, rheumatoid arthritis (RA), and HIV are associated with increased abdominal obesity (Giles et al. 2008, 2010; Freitas et al. 2012; Andrade et al. 2015), which likely contributes to chronic inflammation and muscle dysfunction and increases risk for developing co-morbid conditions such as diabetes and cardiovascular disease (Dirajlal-Fargo et al. 2016; d'Ettorre et al. 2014). Taken together, PA is an important intervention for preserving muscle mass and function and for preventing development of other conditions, which may exacerbate sarcopenia progression in persons with inflammatory and autoimmune conditions.

Persons with inflammatory and autoimmune conditions may also be at increased risk of sarcopenia due to malnutrition. Individuals with IBD are particularly susceptible to malnutrition, with an incidence rate of up to ~70% of patients with IBD (Mijač et al. 2010). A number of factors contribute to malnutrition including reduced caloric intake including treatment side effects, reduced gastrointestinal tract motility, and depressed mood. Furthermore, chronic inflammation can increase energy expenditure and induce anorexia–cachexia by altering expression and response to appetite and satiety-regulating hormones (see Section 5.1). Persons with IBD may also experience malabsorption of nutrients (secondary to surgical resection of the small intestine, inflammation, and reduced intestinal nutrient absorption sites), and loss of micronutrients due to blood loss and/or chronic diarrhea (Massironi et al. 2013). Protein-energy malnutrition has also been described (Donnellan, Yann, and Lal 2013). As a result, persons with autoimmune and inflammatory conditions are likely to benefit from nutritional therapy and supplementation. However, more trials are needed to determine optimal nutritional interventions to counteract muscle wasting in these patients.

Future research is also needed to determine the long-term effects of biologics (e.g., anti-TNFα therapy for inflammatory conditions) and highly active antiretroviral therapy (HAART) for HIV on skeletal muscle mass in persons with immune and inflammatory conditions. Although anti-TNFα therapy is promising for improving body composition and fat free mass, most trials to date are relatively small and short term (<12 weeks) (Peluso and Palmery 2016). Additionally, more trials are needed to determine the effects of combined treatment with nutrition and exercise interventions aimed at decreasing inflammation, which will likely be more effective than single interventions alone (Benatti and Pedersen 2015; Perandini et al. 2012).

2.3.3 HYPOXIA-RELATED DISEASES

Although cardiovascular and respiratory diseases, such as chronic heart failure (HF), peripheral arterial disease (PAD), and chronic obstructive pulmonary disease (COPD), may begin to develop during middle age, the highest burden of symptoms is suffered by elderly patients (Ito and Barnes 2009). Importantly, older patients with any of these conditions typically experience muscle wasting in magnitude of 10%–40% greater than healthy age-matched controls (Anker et al. 1999; Clyne et al. 1985; Gosselink, Troosters, and Decramer 1996; Hambrecht et al. 2005; Marquis et al. 2002; McDermott et al. 2007; Regensteiner et al. 1993). Furthermore, hypoxic disease-related atrophy is associated with reduced strength, culminating in impaired muscle function (Schulze et al. 2004). Thus, interventions for these patients are needed to slow or reverse the expansion of the sarcopenic population.

One of the most likely triggers of muscle atrophy common to patients with HF, PAD, and COPD is chronic and/or intermittent hypoxia. This degree of hypoxia exposure is sufficient to cause muscle fiber atrophy, partially via suppression of mRNA translation and subsequent protein synthesis (Bigard et al. 1991; Hoppeler et al. 1990; Wust et al. 2009). In fact, hypoxia causes hypophosphorylation of mammalian target of rapamycin (mTOR) and its downstream effectors eukaryotic translation initiation factor 4E binding protein 1 (4EBP1), the 70-kDa ribosomal protein S6 kinase (p70S6K), ribosomal protein S6 (RPS6), and eukaryotic translation initiation factor 4G (eIF4G) (Arsham, Howell, and Simon 2003). Research is mixed regarding age-related changes in protein synthesis with studies showing both a decrease (Balagopal et al. 1997; Welle et al. 1993, 1994; Welle, Thornton, and Statt 1995; Yarasheski, Zachwieja, and Bier 1993) and no change (Volpi et al. 1999; Volpi et al. 2000; Yarasheski et al. 1999). In either case, hypoxia in combination with advanced age may significantly suppress protein synthesis and could explain the accelerated atrophy in patients with HF, PAD, and COPD.

Another potential causal mechanism of hypoxia-related atrophy receiving significant attention is the chronic overproduction of pro-inflammatory cytokines. An array of evidence indicates that levels of these cytokines are elevated, and related to muscle atrophy in patients with HF, PAD, and COPD (Anker et al. 1999; Di Francia et al. 1994; Signorelli et al. 2003, 2007; Van Helvoort et al. 2006). Similarly, hypoxia-related conditions are associated with increased expression of angiotensin II (AngII) (Shrikrishna et al. 2012). Ang II may further contribute to skeletal muscle atrophy by suppressing the Akt/mTOR signaling pathway and by increasing muscle expression of caspase 3 and components of the UPS (Song et al. 2005).

Hypoxia may also mediate a number of physiological mechanisms that ultimately lead to nutritional deficiencies and subsequent muscle wasting in patients with CHF, PAD, and COPD (Saitoh et al. 2016; Rahman et al. 2016; Brostow et al. 2016; Ferreira et al. 2012). For example, hypoxia can reduce appetite through multiple mechanisms, including decreased acetylated ghrelin, HIF1α-dependent increases in circulating leptin, and direct effects on activity of brain structures associated with appetite (Palmer and Clegg 2014; Bailey et al. 2015; Yan et al. 2011). Furthermore, this reduction in nutrient intake may be further exacerbated by lack of blood flow to intestinal tract and intestinal edema, which can lead to malabsorption of nutrients including protein and fat (Celik et al. 2010; Sandek et al. 2007; King, Smith, and Lye 1996; King et al. 1996). Nutrient malnutrition may be further compounded by increased anabolic resistance in response to protein feeding in patients with CHF (Toth et al. 2010) as well as hypermetabolism due to increased inspiratory muscle workload (Saitoh, dos Santos, and von Haehling 2016). Finally, medications commonly prescribed to treat hypertension in these persons—particularly Angiotensin-converting enzyme (ACEi)—have been shown to reduced appetite and nutrient intake in preclinical models (Mul et al. 2013; Santos et al. 2008). However, studies on ACEi use in humans are conflicting, as ACEi use is associated with weight maintenance (Schellenbaum et al. 2005; Anker et al. 2003) and better functional outcomes in response to exercise training (Buford et al. 2012b).

While PAD-mediated sarcopenia shares the mechanistic bases of chronic HF and COPD, patients with PAD also experience impaired muscle mitochondrial function. Interestingly, patients with PAD show accelerated mitochondrial dysfunction through downregulation of electron transport chain complexes I, III, and IV in skeletal muscle compared with age-matched controls (Pipinos et al. 2006). Moreover, markers of oxidative damage (Pipinos et al. 2006) and frequency of the mitochondrial DNA 4977 bp deletion (Bhat et al. 1999) are reportedly increased in the skeletal muscle of patients with PAD. Despite this evidence, mitochondrial dysfunction may simply be a consequence of the pathology, and thus additional research is needed to establish a causal link.

2.3.4 KIDNEY DISEASE AND LIVER FAILURE

Older persons are especially susceptible to renal failure due to the high prevalence of arteriosclerosis, hypertension, and diabetes mellitus (DM). Patients with chronic kidney disease (CKD) typically present with reduced exercise tolerance, due to the concurrence of muscle atrophy, anemia, cardiac dysfunction, inadequate nutrition, inactivity, and psychological factors such as negative mood states (Adams and Vaziri 2006). Importantly, low exercise capacity is also a powerful independent predictor of mortality in patients with end-stage renal disease (Sietsema et al. 2004). Similar to sarcopenia, the loss of muscle mass in patients with CKD appears to be primarily due to type II fiber atrophy. In middle-aged predialytic patients, type IIa and IIx fiber cross-sectional area was 25%–30% smaller than in healthy controls (Sakkas et al. 2003). This CKD-related atrophy appears to be largely influenced by altered protein turnover rates as the synthetic rate of muscle contractile and mitochondrial proteins was reduced in middle-aged patients with CKD compared with healthy controls (Adey et al. 2000). Uremia is common among persons with CKD and likely contributes to the observed reduction in protein synthesis rate. For example, in a preclinical model, acute uremia significantly attenuated mTOR, S6K1, and 4e-BP1 phosphorylation in response to leucine feeding, additionally, uremia increased skeletal muscle mRNA transcripts associated with inflammation and autophagy while reducing IGF-1 transcription (McIntire et al. 2014).

Furthermore, rates of muscle protein degradation are significantly accelerated in the presence of CKD, mainly due to enhanced activation of the UPS (Du et al. 2005). Systemic levels of pro-inflammatory cytokines are elevated in uremic subjects, negatively affecting insulin and IGF1 signaling and promoting UPS activation (Zanetti, Barazzoni, and Guarnieri 2008). Similarly, metabolic acidosis, which is highly prevalent among patients with CKD, can induce UPS upregulation and increased branch amino acid oxidation in skeletal muscle (Holeček et al. 2001; Bailey et al. 1996). These alterations in protein metabolism in CKD may be further exacerbated in patients undergoing dialysis. For example, protein degradation is increased while protein synthesis is suppressed for at least 2 h following a dialysis session (Ikizler et al. 2002). Additionally, approximately 10 g of amino acids is lost during hemodialysis, which may further contribute to protein-calorie wasting and long-term muscle loss (Raj et al. 2005).

Similar to chronic renal insufficiency, liver failure is also associated with significant loss of muscle mass (Tessari 2003). Protein-calorie malnutrition is thought to be a major determinant of cirrhosis-related muscle wasting (Muller 2007). However, altered protein metabolism also contributes to muscle atrophy in patients with liver failure (Tessari 2003), as significant muscle mass loss and increased myofibril degradation occur despite normal food intake (Lin et al. 2005). In particular, metabolism of branched-chain amino acids (BCAA) may be significantly altered in cirrhosis and liver failure secondary to hyperammonemia (Romero-Gómez et al. 2004; Souba, Smith, and Wilmore 1985) and liver glycogen depletion. Skeletal muscle uptake of circulating BCAA is increased in persons with cirrhosis and is used for ammonia detoxification (Holecek 2015). Furthermore, cirrhosis and liver failure mimic starvation as protein and fat stores are inappropriately used for gluconeogenesis (Sinclair et al. 2016; Dasarathy 2012). These mechanisms likely contribute significantly to skeletal muscle wasting. In addition, inflammatory markers are increased in persons with cirrhosis and liver failure, which results in muscle proteolysis (Tessari 2003), decreased production of anabolic hormones (Moller and Becker 1992), increased anabolic resistance, and increased metabolic rate.

As described above, the bulk of research concerning regulation of muscle mass in patients with CKD and liver failure has focused on alterations in protein metabolism. However, many other important questions remain unexplored, including the roles of myonuclear apoptosis, myogenic signaling, and muscle regeneration in muscle atrophy of individuals afflicted with these diseases. In addition, further study of muscle bioenergetics is needed in each of these disease conditions. Evidence suggests that mitochondrial DNA (mtDNA) is depleted in patients with both CKD and cirrhosis (Lim, Cheng, and Wei 2000; Pesce et al. 2002), suggesting that impairments in muscular bioenergetic efficiency may contribute to the pathogenesis of uremic and/or hepatic myopathies.

2.3.5 DM Type 2

Over 422 million persons worldwide, or 8.5% of the adult population, are currently afflicted with type 2 DM (Roglic 2016). Of those with the condition, approximately 25%–30% are over 60 years of age (Danaei et al. 2009). Along with the well-characterized effects of insulin resistance on muscle (DeFronzo and Tripathy 2009), sarcopenia and type 2 DM share a common set of lifestyle factors, including physical inactivity and a poor diet that contribute to accelerate muscle atrophy. Moreover, a "vicious cycle" can ensue in which each condition accelerates the progression of the other condition. Given the well-known effects of these factors on skeletal muscle health, a large body of evidence now exists describing the link between metabolic syndrome, DM and sarcopenia, and the topic has been expertly reviewed elsewhere (Cleasby et al. 2016; Umegaki 2016; Sinclair and Rodriguez-Mañas 2016). A brief overview of the mechanisms linking DM to sarcopenia is given later.

DM may accelerate the development of age-associated changes in body composition through a number of mechanisms. For example, insulin resistance decreases the activity of anabolic hormones that activate the phosphotidyl-inositol-3-kinase (PI3K) pathway, thus reducing muscle protein synthesis (Guttridge 2004; Morley 2008) and perhaps muscle mass (Shishikura et al. 2014). Moreover, changes in skeletal muscle metabolism may also lead to the greater extramyocellular lipid content in diabetic individuals and thereby influence muscle function and quality (Sakkas et al. 2006), possibly by increasing myonuclear apoptosis (Peterson et al. 2008; Kob et al. 2015). Various neuropathies associated with diabetes and the resultant decrease in motor end plates can cause weakness, ataxia, and poor coordination, thereby playing an important role in the pathogenesis of physical function decline and sarcopenia (Morley 2008). Finally, diabetic patients often have accelerated progression of atherosclerosis (Morley 2000), which can decrease peripheral blood flow, resulting in poor muscle perfusion (Morley et al. 2005). Collectively, the physiological changes associated with DM, in combination with changes from associated lifestyle behaviors, may lead to decreased muscle quality and accelerated sarcopenia progression.

Not only does diabetes lead to muscle atrophy, but muscle atrophy also contributes to diabetes progression through a reduction in insulin sensitivity (Kolterman et al. 1980). At the cellular level, accumulating evidence suggests that impairments in mitochondrial functioning may underlie the development and progression of both sarcopenia and type 2 diabetes (Plaza 2002). Suboptimal mitochondrial function appears to impair β-cell functioning (Kaufman, Li, and Soleimanpour 2015; Ma, Zhao, and Turk 2012; Janikiewicz et al. 2015; Porte and Kahn 1991) as well as skeletal muscle health by increasing myofiber susceptibility to apoptosis (Adhihetty et al. 2007; Koves et al. 2005). Evidence also exists to indicate that mitochondrial dysfunction may affect glucose transport throughout the body and precede the development of DM (Lee et al. 1998). Given the important role mitochondrial function appears to have in skeletal muscle quality and glucose/insulin regulation, the mitochondria may represent a key target for future interventions for both sarcopenia and DM. At present, however, additional research is needed to clearly demonstrate the role that mitochondrial dysfunction plays in the etiology of both conditions.

2.3.6 Neurodegenerative Conditions

Neurodegenerative conditions—including brain injury (stroke) and neurological diseases (Alzheimer's, Parkinson's)—may occur at virtually any age. However, the risk of developing these conditions is greatly increased with age. For example, stroke risk doubles for every decade of life beyond 40 years (Hirtz et al. 2007), while prevalence of Alzheimer's increases from 300 per 100,000 for individuals 60–69 years to 10,800 per 100,000 for those >80 years (Victor and Ropper 2002). Persons with these conditions are at risk for lean tissue loss (Kim et al. 2015; Springer et al. 2014; Aadal, Mortensen, and Nielsen 2015; English et al. 2010; Ryan et al. 2002, 2011; Jørgensen and Jacobsen 2001; Buffa et al. 2014; Auyeung et al. 2008; Burns et al. 2010). For example, one-quarter of stroke victims lost >3 kg of body mass 4 months poststroke, with an average loss of 8.3 kg after 1 year (Jonsson et al. 2008). Importantly, mortality rate for those with >3 kg weight loss poststroke was 14% compared with 4% for those who maintained weight. These neurological conditions are also associated with impaired physical function and incident disability (Auyeung et al. 2008; Kuo et al. 2007; Garber and Friedman 2003; Schilling et al. 2009; Kelly-Hayes et al. 2003).

There are several mechanisms common to stroke, Parkinson's disease, and Alzheimer's disease that may contribute to accelerated development of sarcopenia, including impaired motor function and PA levels, inflammation, and malnutrition. Neurological disorders may contribute to loss of muscle mass through central and peripheral nervous system degradation. Alzheimer's and Parkinson's diseases result in dysfunction and degeneration of numerous brain structures involved in motor planning and control, including the primary motor cortex (Suva et al. 1999; Vidoni et al. 2012; Pennisi et al. 2002; Horoupian and Wasserstein 1999; Lindenbach and Bishop 2013), striatum

(Del Campo et al. 2016), and substantia nigra (Schneider et al. 2006; Burns et al. 2005; Dauer and Przedborski 2003; Galvan and Wichmann 2008). Similarly, stroke often damages motor control portions of the brain leading to disability and paralysis. In fact, 25% of stroke survivors report impaired ability to perform activities of daily living, and 30% are unable to walk without assistance (Kelly-Hayes et al. 2003). Alterations in these critical brain structures may affect intracortical and corticospinal connectivity (Russ et al. 2012; Clark and Taylor 2011) and reduce central activation of the neuromuscular system through impaired cortical excitability and corticospinal output (Oliviero et al. 2006; Plow et al. 2013; Moreno Catala, Woitalla, and Arampatzis 2013). Accordingly, atrophy of brain white matter volume is significantly associated with lean mass in persons with Alzheimer's disease (Burns et al. 2010), suggesting that impaired central connectivity may directly influence muscle wasting.

It is also likely that physical function deficits resulting from impaired skeletal muscle activation leads to a reduction in PA (van Nimwegen et al. 2011; Fertl, Doppelbauer, and Auff 1993; Dontje et al. 2013). Interestingly, skeletal muscle in persons with stroke shows a phenotypic shift with a reduction and selective atrophy of type 1 fibers (Slager, Hsu, and Jordan 1985). While this phenotypic shift is typically reversed in sarcopenia (Verdijk et al. 2014), a shift toward type II fibers is consistent with muscle adaptations to prolonged inactivity (Borina et al. 2010). It should also be noted that sparing of skeletal muscle mass has been reported in the early and middle stages of Parkinson's disease. Although many factors may account for this finding, including long-term Levadopa therapy, it was suggested that increased activity due to resting tremors exerts a protective effect on muscle mass (Barichella et al. 2016). This finding further supports the hypothesis that reduced PA increases risk for the development of sarcopenia in persons with neurodegenerative conditions.

Neurodegenerative conditions are also associated with increased central and peripheral expression of many pro-inflammatory cytokines that may contribute to sarcopenia progression. For example, persons with Alzheimer's and Parkinson's diseases have elevated circulating levels of IL-1β, IL-6, and TNFα relative to healthy counterparts (Alvarez et al. 1996; Licastro et al. 2000; Singh and Guthikonda 1997; Dobbs et al. 1999; Blum-Degena et al. 1995; Reale et al. 2009). While chronic inflammation is a risk factor for several neurodegenerative conditions, mounting evidence suggests that central inflammation and acute brain injury may exacerbate peripheral cytokine production. In a preclinical model, injection of IL-1β into rat brain induced systemic expression of TNFα (Campbell et al. 2007), which in turn activates microglia in the brain, resulting in the exacerbation of ongoing neural damage and central cytokine production (Pott Godoy et al. 2008; Offner et al. 2006). Importantly, pro-inflammatory cytokines—particularly TNFα—are potent stimulators of muscle proteolysis via activation of the UPS (Llovera et al. 1997). Thus, central inflammation resulting from neurodegenerative conditions may be directly linked to muscle wasting. In support of this hypothesis, central administration of IL-1β increased muscle gene expression of MAFbx, MuRF1, and FOXO1 (Braun et al. 2011). However, more data are needed to determine whether chronic inflammation seen in neurodegenerative conditions is causally related to skeletal muscle loss.

Malnutrition is common among older adults with neurodegenerative conditions and is potential factor contributing to high rates of sarcopenia among this population. Although there are many potential causes of malnutrition in this population including disease-related olfactory impairments (Naudin and Atanasova 2014), changes in appetite secondary to chronic central inflammation, and impaired short-term memory, dysphagia is perhaps the most common cause of malnutrition among persons with neurodegenerative conditions. While dysphagia is relatively common among older adults, prevalence estimates for dysphagia in older adults with neurological conditions are as high as 80% (Meng, Wang, and Lien 2000; Kalf et al. 2012; Daniels, Brailey, and Foundas 1999). Swallowing is a tightly coordinated process involving more than 20 head and neck muscles and numerous central and peripheral subsystems (Dodds 1989). Lesions and degeneration in associated brain structures are common in neurodegenerative conditions and lead to swallowing difficulty (Suntrup et al. 2013; Humbert et al. 2010). As a result, many older adults with neurodegenerative diseases have reduced caloric intake and tend to avoid certain foods—particularly solid foods—resulting in muscle loss (Sura et al. 2012; Lorefält, Granérus, and Unosson 2006). Recent research indicates that muscles involved in swallowing may also atrophy in older adults. Function of the tongue is critical for proper swallowing, as the tongue is involved in preparing food for swallowing and for transferring the food bolus to the pharynx (Dodds 1989). In older adults, sarcopenia is associated with decreased tongue pressure and swallowing dysfunction (Maeda and Akagi 2015; Machida et al. 2016). Therefore, it is possible that there is a vicious cycle whereby dysphagia-induced malnutrition leads to atrophy and dysfunction of swallowing muscles, leading to further exacerbation of swallowing problems (Tamura et al. 2012).

2.4 CONCLUSIONS AND FUTURE DIRECTIONS

Sarcopenia is a growing societal healthcare problem due to the rapid expansion of the elderly population and the limited number of therapeutic approaches to the problem. Although the biologic and epidemiologic intricacies of the condition cannot be fully covered in a single review, we have provided here a wealth of information detailing the immense complexity of the problem. Sarcopenia is more than simply muscle atrophy attributable to the effects of chronological aging. Rather, this condition is accelerated by unhealthy lifestyle behaviors and comorbid conditions including cancer, chronic inflammation, neurodegeneration, PAD, hypoxia-related diseases, organ failure, and DM. The complexity of sarcopenia is further complicated that each of these factors affects muscle atrophy in a slightly different manner. Due to this complexity, combating the growth of the sarcopenic population will take a substantial unified effort from experts in gerontological and skeletal muscle research, clinical prevention and rehabilitation, and drug discovery and development.

Many previous authors have provided data and interpretation critical to understanding the basic mechanisms responsible for muscle atrophy at old age. Figure 2.2 shows a proposed model of mechanisms associated with accelerated sarcopenia. These mechanisms range from systemic changes such as reduced production of anabolic hormones to cellular mechanisms governing myofiber size and viability. These data, together with the discovery of yet unknown pathways, are critical to the creation of new therapeutic approaches. However, we argue that these approaches should be developed within the context of the paradigm proposed here.

This paradigm focuses on teasing out factors that accelerate the progression of sarcopenia. Current and past literatures on sarcopenia have primarily classified factors such as sedentary lifestyle or chronic disease under a large umbrella known as "age-related factors." However, these factors are not common to all elderly and/or sarcopenic individuals. Within clinical study cohorts, a large degree of heterogeneity exists that complicates the work of clinical and translational scientists studying sarcopenia. Furthermore, variability in lifestyle and disease conditions induces different atrophy-related mechanisms and degrees of atrophy. Therefore, global therapeutic approaches are unlikely to prove equally effective in all models of sarcopenia. For example, exercise may serve as an adequate therapeutic approach for inactivity-related sarcopenia. However, other interventions may be required to combat sarcopenia in individuals with comorbid conditions. Subsequently, well-controlled investigations are needed to compare skeletal muscle atrophy in "healthy" older adults to atrophy in various patient populations. Such investigations will enable scientists to identify, develop, and appropriately administer therapeutic approaches specific to each model of sarcopenia. Due to the critical need for improved study of the sarcopenic condition and the heterogeneity of sarcopenia development, this paradigm has significant potential to improve the translation of research findings into more effective preventive and/or treatment strategies. In conclusion, we believe that factors that theoretically accelerate the loss in muscle mass should be routinely considered in the design of future investigations on this subject.

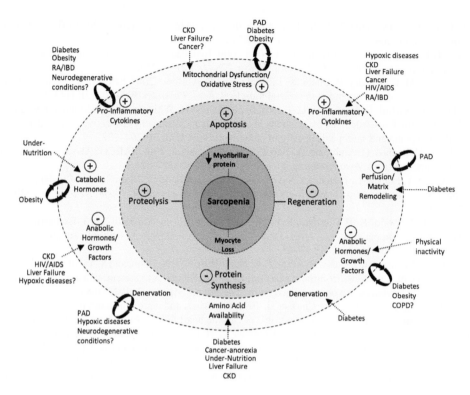

FIGURE 2.2 Mechanisms associated with accelerated sarcopenia due to disease conditions and behaviors. Although aging contributes to the presence of these atrophy-related mechanisms, these behaviors and diseases enhance their activity. The figure is interpreted directionally from outer to inner rings (light to dark gray) ultimately leading to myocyte and myofibrillar protein loss. Plus signs (+) indicate upregulation and negative signs (–) equate to downregulation of a specific pathway. The figure represents a step-down approach to how certain disease conditions and behaviors can modulate four of the major mechanistic pathways involved in sarcopenia: apoptosis, proteolysis, regeneration and protein synthesis. This figure is not intended to be exhaustive or detailed (i.e., a signaling pathway) and thus it may be prone to misrepresentation because of the complexity of interactions involved with multidimensional conditions. However, the model is drawn to illustrate the interactions that serve to feed the major pathways that accelerate sarcopenia. These interactions include but are not limited to changes in: pro-inflammatory cytokines, anabolic hormones, denervation, mitochondrial function and tissue blood perfusion. Some pathways have feed-forward properties where disease etiology exacerbates the condition (e.g., denervation worsens muscle perfusion in PAD). PAD: Peripheral Arterial Disease, CKD: Chronic Kidney Disease. RA: Rheumatoid Arthritis. IBD: Inflammatory Bowel Disease. (Modified from Buford, T.W. et al., *Ageing Res. Rev.*, 9, 369–383, 2010a.)

ACKNOWLEDGMENTS

We would like to thank S.D. Anton, A.R. Judge, E. Marzetti, S.E. Wohlgemuth, C.S. Carter, C. Leeuwenburgh, M. Pahor, and T.M. Manini for their contributions to the initial conceptual review on this topic (Buford et al. 2010a).

REFERENCES

Aadal, L., J. Mortensen, and J. F. Nielsen. 2015. Weight reduction after severe brain injury: A challenge during the rehabilitation course. *J. Neurosci. Nurs.* 47 (2): 85–90.

Acharyya, S., K. J. Ladner, L. L. Nelsen, J. Damrauer, P. J. Reiser, S. Swoap, and D. C. Guttridge. 2004. Cancer cachexia is regulated by selective targeting of skeletal muscle gene products. *J. Clin. Invest.* 114 (3): 370–378.

Adams, G. R. and N. D. Vaziri. 2006. Skeletal muscle dysfunction in chronic renal failure: Effects of exercise. *Am. J. Physiol. Renal Physiol.* 290 (4): F753–F761.

Adey, D., R. Kumar, J. T. McCarthy, and K. S. Nair. 2000. Reduced synthesis of muscle proteins in chronic renal failure. *Am. J. Physiol. Endocrinol. Metab.* 278 (2): E219–E225.

Adhihetty, P. J., M. F. O'Leary, B. Chabi, K. L. Wicks, and D. A. Hood. 2007. Effect of denervation on mitochondrially mediated apoptosis in skeletal muscle. *J. Appl. Physiol. (1985)* 102 (3): 1143–1151.

Aggio, D. A., C. Sartini, O. Papacosta, L. T. Lennon, S. Ash, P. H. Whincup, S. G. Wannamethee, and B. J. Jefferis. 2016. Cross-sectional associations of objectively measured physical activity and sedentary time with sarcopenia and sarcopenic obesity in older men. *Prev. Med.* 91: 264–272.

Aguiar, R., T. Meirinhos, C. Ambrósio, A. Barcelos, and J. Sequeira. 2014. SARCOSPA-sarcopenia in spondyloarthritis patients. *Acta Reumatológica Portuguesa* 39 (4): 322–326.

Alvarez, X. A., A. Franco, L. Fernández-Novoa, and R. Cacabelos. 1996. Blood levels of histamine, IL-1β, and TNF-α in patients with mild to moderate Alzheimer disease. *Mol. Chem. Neuropathol.* 29 (2–3): 237–252.

Andrade, M. I., R. Maio, K. F. Dourado, P. F. Macedo, and A. C. Barreto Neto. 2015. Excessive weight—Muscle depletion paradox and cardiovascular risk factors in outpatients with inflammatory bowel disease. *Arq. Gastroenterol.* 52 (1): 37–45.

Anker, S. D., A. Negassa, A. J. Coats, R. Afzal, P. A. Poole-Wilson, J. N. Cohn, and S. Yusuf. 2003. Prognostic importance of weight loss in chronic heart failure and the effect of treatment with angiotensin-converting-enzyme inhibitors: An observational study. *The Lancet* 361 (9363): 1077–1083.

Anker, S. D., P. P. Ponikowski, A. L. Clark et al. 1999. Cytokines and neurohormones relating to body composition alterations in the wasting syndrome of chronic heart failure. *Eur. Heart J.* 20 (9): 683–693.

Anthony, J. C., A. K. Reiter, T. G. Anthony, S. J. Crozier, C. H. Lang, D. A. MacLean, S. R. Kimball, and L. S. Jefferson. 2002. Orally administered leucine enhances protein synthesis in skeletal muscle of diabetic rats in the absence of increases in 4E-BP1 or S6K1 phosphorylation. *Diabetes* 51 (4): 928–936.

Anton, S. D., A. Hida, R. Mankowski et al. 2018. Nutrition and exercise in sarcopenia. *Curr. Protein. Pept. Sci.* 19 (7): 649–667.

Antoun, S., V. E. Baracos, L. Birdsell, B. Escudier, and M. B. Sawyer. 2010. Low body mass index and sarcopenia associated with dose-limiting toxicity of sorafenib in patients with renal cell carcinoma. *Ann. Oncol.* 21 (8): 1594–1598.

Arsenault, B., J. Rana, I. Lemieux, J. Despres, J. Kastelein, S. Boekholdt, N. Wareham, and K. Khaw. 2010. Physical inactivity, abdominal obesity and risk of coronary heart disease in apparently healthy men and women. *Int. J. Obes.* 34 (2): 340–347.

Arsham, A. M., J. J. Howell, and M. C. Simon. 2003. A novel hypoxia-inducible factor-independent hypoxic response regulating mammalian target of rapamycin and its targets. *J. Biol. Chem.* 278 (32): 29655–29660.

Auyeung, T. W., T. Kwok, J. Lee, P. C. Leung, J. Leung, and J. Woo. 2008. Functional decline in cognitive impairment—The relationship between physical and cognitive function. *Neuroepidemiology* 31 (3): 167–173.

Awad, S., B. H. Tan, H. Cui, A. Bhalla, K. C. Fearon, S. L. Parsons, J. A. Catton, and D. N. Lobo. 2012. Marked changes in body composition following neoadjuvant chemotherapy for oesophagogastric cancer. *Clin. Nutr.* 31 (1): 74–77.

Bailey, D. P., L. R. Smith, B. C. Chrismas, L. Taylor, D. J. Stensel, K. Deighton, J. A. Douglas, and C. J. Kerr. 2015. Appetite and gut hormone responses to moderate-intensity continuous exercise versus high-intensity interval exercise, in normoxic and hypoxic conditions. *Appetite* 89: 237–245.

Bailey, J. L., X. Wang, B. K. England, S. R. Price, X. Ding, and W. E. Mitch. 1996. The acidosis of chronic renal failure activates muscle proteolysis in rats by augmenting transcription of genes encoding proteins of the ATP-dependent ubiquitin-proteasome pathway. *J. Clin. Invest.* 97 (6): 1447–1453.

Baker, J. F., J. Long, S. Ibrahim, M. B. Leonard, and P. Katz. 2015. Are men at greater risk of lean mass deficits in rheumatoid arthritis? *Arthrit. Care Res.* 67 (1): 112–119.

Balagopal, P., O. E. Rooyackers, D. B. Adey, P. A. Ades, and K. S. Nair. 1997. Effects of aging on in vivo synthesis of skeletal muscle myosin heavy-chain and sarcoplasmic protein in humans. *Am. J. Physiol.* 273 (4 Pt 1): E790–E800.

Bales, C. W. and C. S. Ritchie. 2002. Sarcopenia, weight loss, and nutritional frailty in the elderly. *Annu. Rev. Nutr.* 22 (1): 309–323.

Barbieri, E., D. Agostini, E. Polidori, L. Potenza, M. Guescini, F. Lucertini, G. Annibalini, L. Stocchi, M. De Santi, and V. Stocchi. 2015. The pleiotropic effect of physical exercise on mitochondrial dynamics in aging skeletal muscle. *Oxid. Med. Cell. Longev* 2015: 917085.

Barichella, M., G. Pinelli, L. Iorio, E. Cassani, A. Valentino, C. Pusani, V. Ferri, C. Bolliri, M. Pasqua, and G. Pezzoli. 2016. Sarcopenia and dynapenia in patients with parkinsonism. *J. Am. Med. Dir. Assoc.* 17: 640–646.

Barret, M., S. Antoun, C. Dalban, D. Malka, T. Mansourbakht, A. Zaanan, E. Latko, and J. Taieb. 2014. Sarcopenia is linked to treatment toxicity in patients with metastatic colorectal cancer. *Nutr. Cancer* 66 (4): 583–589.

Bauer, J. M., S. Verlaan, I. Bautmans et al. 2015. Effects of a vitamin D and leucine-enriched whey protein nutritional supplement on measures of sarcopenia in older adults, the PROVIDE study: A randomized, double-blind, placebo-controlled trial. *J. Am. Med. Dir. Assoc.* 16 (9): 740–747.

Bauer, J., G. Biolo, T. Cederholm, M. Cesari, A. J. Cruz-Jentoft, J. E. Morley, S. Phillips, C. Sieber, P. Stehle, and D. Teta. 2013. Evidence-based recommendations for optimal dietary protein intake in older people: A position paper from the PROT-AGE study group. *J. Am. Med. Dir. Assoc.* 14 (8): 542–559.

Baumgartner, R. N., S. J. Wayne, D. L. Waters, I. Janssen, D. Gallagher, and J. E. Morley. 2004. Sarcopenic obesity predicts instrumental activities of daily living disability in the elderly. *Obes. Res.* 12 (12): 1995–2004.

Benatti, F. B. and B. K. Pedersen. 2015. Exercise as an anti-inflammatory therapy for rheumatic diseases [mdash] myokine regulation. *Nat. Rev. Rheumatol.* 11 (2): 86–97.

Bhat, H. K., W. R. Hiatt, C. L. Hoppel, and E. P. Brass. 1999. Skeletal muscle mitochondrial DNA injury in patients with unilateral peripheral arterial disease. *Circulation* 99 (6): 807–812.

Bigard, A. X., A. Brunet, C. Guezennec, and H. Monod. 1991. Effects of chronic hypoxia and endurance training on muscle capillarity in rats. *Pflügers Archiv* 419 (3–4): 225–229.

Bijlsma, A., C. Meskers, C. Ling, M. Narici, S. Kurrle, I. Cameron, R. Westendorp, and A. Maier. 2013. Defining sarcopenia: The impact of different diagnostic criteria on the prevalence of sarcopenia in a large middle aged cohort. *Age* 35 (3): 871–881.

Bilski, J., A. I. Mazur-Bialy, M. Wierdak, and T. Brzozowski. 2013. The impact of physical activity and nutrition on inflammatory bowel disease: The potential role of cross talk between adipose tissue and skeletal muscle. *J. Physiol. Pharmacol.* 64 (2):143–155.

Bing, C., M. Brown, P. King, P. Collins, M. J. Tisdale, and G. Williams. 2000. Increased gene expression of brown fat uncoupling protein (UCP)1 and skeletal muscle UCP2 and UCP3 in MAC16-induced cancer cachexia. *Cancer Res.* 60 (9): 2405–2410.

Bischoff-Ferrari, H. A., J. E. Orav, J. A. Kanis, R. Rizzoli, M. Schlogl, H. B. Staehelin, W. C. Willett, and B. Dawson-Hughes. 2015. Comparative performance of current definitions of sarcopenia against the prospective incidence of falls among community-dwelling seniors age 65 and older. *Osteoporos. Int.* 26 (12): 2793–2802.

Blaum, C. S., Q. L. Xue, E. Michelon, R. D. Semba, and L. P. Fried. 2005. The association between obesity and the frailty syndrome in older women: The women's health and aging studies. *J. Am. Geriatr. Soc.* 53 (6): 927–934.

Bloomer, R. J., A. H. Goldfarb, L. Wideman, M. J. McKenzie, and L. A. Consitt. 2005. Effects of acute aerobic and anaerobic exercise on blood markers of oxidative stress. *J. Strength Cond. Res.* 19 (2): 276–285.

Blum-Degena, D., T. Müller, W. Kuhn, M. Gerlach, H. Przuntek, and P. Riederer. 1995. Interleukin-1β and interleukin-6 are elevated in the cerebrospinal fluid of Alzheimer's and de novo Parkinson's disease patients. *Neurosci. Lett.* 202 (1): 17–20.

Boer, B. C., F. de Graaff, M. Brusse-Keizer, D. E. Bouman, C. H. Slump, M. Slee-Valentijn, and J. M. Klaase. 2016. Skeletal muscle mass and quality as risk factors for postoperative outcome after open colon resection for cancer. *Int. J. Colorectal Dis.* 31: 1117–1124.

Bollwein, J., R. Diekmann, M. J. Kaiser, J. M. Bauer, W. Uter, C. C. Sieber, and D. Volkert. 2013. Distribution but not amount of protein intake is associated with frailty: A cross-sectional investigation in the region of nürnberg. *Nutr. J.* 12 (1): 1.

Borina, E., M. Pellegrino, G. D'Antona, and R. Bottinelli. 2010. Myosin and actin content of human skeletal muscle fibers following 35 days bed rest. *Scand. J. Med. Sci. Sports* 20 (1): 65–73.

Bouchard, D. R., I. J. Dionne, and M. Brochu. 2009. Sarcopenic/obesity and physical capacity in older men and women: Data from the Nutrition as a Determinant of Successful Aging (NuAge)—the Quebec longitudinal study. *Obesity* 17 (11): 2082–2088.

Bouillanne, O., E. Curis, B. Hamon-Vilcot, I. Nicolis, P. Chrétien, N. Schauer, J. Vincent, L. Cynober, and C. Aussel. 2013. Impact of protein pulse feeding on lean mass in malnourished and at-risk hospitalized elderly patients: A randomized controlled trial. *Clin. Nutr.* 32 (2): 186–192.

Bouzid, M. A., E. Filaire, A. McCall, and C. Fabre. 2015. Radical oxygen species, exercise and aging: An update. *Sports Med.* 45 (9): 1245–1261.

Braun, T. P., X. Zhu, M. Szumowski et al. 2011. Central nervous system inflammation induces muscle atrophy via activation of the hypothalamic-pituitary-adrenal axis. *J. Exp. Med.* 208 (12): 2449–2463.

Brenner, H. 2002. Long-term survival rates of cancer patients achieved by the end of the 20th century: A period analysis. *Lancet* 360 (9340):1131–1135.

Brooks, S. V. and J. A. Faulkner. 1994. Skeletal muscle weakness in old age: Underlying mechanisms. *Med. Sci. Sports Exerc.* 26 (4): 432–439.

Brostow, D. P., A. T. Hirsch, M. A. Pereira, R. L. Bliss, and M. S. Kurzer. 2016. Nutritional status and body composition in patients with peripheral arterial disease: A cross-sectional examination of disease severity and quality of life. *Ecol. Food Nutr.* 55 (1): 87–109.

Bruyere, O., E. Cavalier, F. Buckinx, and J. Y. Reginster. 2016. Relevance of vitamin D in the pathogenesis and therapy of frailty. *Curr. Opin. Clin. Nutr. Metab. Care.* 20: 26–29.

Bryant, R., S. Ooi, C. Schultz, C. Goess, R. Grafton, J. Hughes, A. Lim, F. Bartholomeusz, and J. Andrews. 2015. Low muscle mass and sarcopenia: Common and predictive of osteopenia in inflammatory bowel disease. *Aliment. Pharmacol. Ther.* 41 (9): 895–906.

Buch, A., E. Carmeli, L. K. Boker, et al. 2016. Muscle function and fat content in relation to sarcopenia, obesity and frailty of old age—An overview. *Exp. Gerontol.* 76: 25–32.

Buffa, R., E. Mereu, P. Putzu, R. Mereu, and E. Marini. 2014. Lower lean mass and higher percent fat mass in patients with Alzheimer's disease. *Exp. Gerontol.* 58: 30–33.

Buford, T. W. 2017. Sarcopenia: Relocating the forest among the trees. *Toxicol. Pathol.* 45 (7): 957–960.

Buford, T. W., E. Marzetti, T.M. Manini. 2012a. Commentary on "Muscle atrophy is not always sarcopenia": On muscle atrophy, aging, and disease. *J. Appl. Physiol.* 113(4): 680–684.

Buford, T. W., M. B. Cooke, T. M. Manini, C. Leeuwenburgh, and D. S. Willoughby. 2010b. Effects of age and sedentary lifestyle on skeletal muscle NF-kappaB signaling in men. *J. Gerontol. A Biol. Sci. Med. Sci.* 65 (5): 532–537.

Buford, T. W., S. D. Anton, A. R. Judge, E. Marzetti, S. E. Wohlgemuth, C. S. Carter, C. Leeuwenburgh, M. Pahor, and T. M. Manini. 2010a. Models of accelerated sarcopenia: Critical pieces for solving the puzzle of age-related muscle atrophy. *Ageing Res. Rev.* 9 (4): 369–383.

Buford, T. W., T. M. Manini, F. Hsu, M. Cesari, S. D. Anton, S. Nayfield, R. S. Stafford, T. S. Church, M. Pahor, and C. S. Carter. 2012b. Angiotensin-converting enzyme inhibitor use by older adults is associated with greater functional responses to exercise. *J. Am. Geriatr. Soc.* 60 (7): 1244–1252.

Burns, J. M., D. K. Johnson, A. Watts, R. H. Swerdlow, and W. M. Brooks. 2010. Reduced lean mass in early alzheimer disease and its association with brain atrophy. *Arch. Neurol.* 67 (4): 428–433.

Burns, J. M., J. E. Galvin, C. M. Roe, J. C. Morris, and D. W. McKeel. 2005. The pathology of the substantia nigra in Alzheimer disease with extrapyramidal signs. *Neurology* 64 (8): 1397–1403.

Campbell, S. J., R. M. Deacon, Y. Jiang, C. Ferrari, F. J. Pitossi, and D. C. Anthony. 2007. Overexpression of IL-1β by adenoviral-mediated gene transfer in the rat brain causes a prolonged hepatic chemokine response, axonal injury and the suppression of spontaneous behaviour. *Neurobiol. Dis.* 27 (2): 151–163.

Celik, T., A. Iyisoy, U. C. Yuksel, and B. Jata. 2010. The small intestine: A critical linkage in pathophysiology of cardiac cachexia. *Int. J. Cardiol.* 143 (2): 200–201.

Chasen, M., S. Z. Hirschman, and R. Bhargava. 2011. Phase II study of the novel peptide-nucleic acid OHR118 in the management of cancer-related anorexia/cachexia. *J. Am. Med. Dir. Assoc.* 12 (1): 62–67.

Chen, L., L. Liu, J. Woo, P. Assantachai, T. Auyeung, K. S. Bahyah, M. Chou, L. Chen, P. Hsu, and O. Krairit. 2014. Sarcopenia in Asia: Consensus report of the Asian working group for sarcopenia. *J. Am. Med. Dir. Assoc.* 15 (2): 95–101.

Chomentowski, P., J. J. Dube, F. Amati, M. Stefanovic-Racic, S. Zhu, F. G. Toledo, and B. H. Goodpaster. 2009. Moderate exercise attenuates the loss of skeletal muscle mass that occurs with intentional caloric restriction-induced weight loss in older, overweight to obese adults. *J. Gerontol. A Biol. Sci. Med. Sci.* 64 (5): 575–580.

Christensen, J. F., L. W. Jones, J. L. Andersen, G. Daugaard, M. Rorth, and P. Hojman. 2014. Muscle dysfunction in cancer patients. *Ann. Oncol.* 25 (5): 947–958.

Christopoulos, C., S. Savva, S. Pylarinou, A. Diakakis, E. Papavassiliou, and P. Economopoulos. 2003. Localised gastrocnemius myositis in Crohn's disease. *Clin. Rheumatol.* 22 (2): 143–145.

Clark, B. C. and J. L. Taylor. 2011. Age-related changes in motor cortical properties and voluntary activation of skeletal muscle. *Curr. Aging Sci.* 4 (3): 192–199.

Cleasby, M. E., P. M. Jamieson, and P. J. Atherton. 2016. Insulin resistance and sarcopenia: Mechanistic links between common co-morbidities. *J. Endocrinol.* 229 (2): R67–R81.

Clyne, C. A., H. Mears, R. O. Weller, and T. F. O'Donnell. 1985. Calf muscle adaptation to peripheral vascular disease. *Cardiovasc. Res.* 19 (8): 507–512.

Collamati, A., E. Marzetti, R. Calvani, et al. 2016. Sarcopenia in heart failure: Mechanisms and therapeutic strategies. *J. Geriatr. Cardiol.* 13 (7): 615–624.

Combaret, L., D. Dardevet, D. Bechet, D. Taillandier, L. Mosoni, and D. Attaix. 2009. Skeletal muscle proteolysis in aging. *Curr. Opin. Clin. Nutr. Metab. Care* 12 (1): 37–41.

Combaret, L., D. Dardevet, I. Rieu, M. Pouch, D. Béchet, D. Taillandier, J. Grizard, and D. Attaix. 2005. A leucine-supplemented diet restores the defective postprandial inhibition of proteasome-dependent proteolysis in aged rat skeletal muscle. *J. Physiol. (Lond.)* 569 (2): 489–499.

Correa-de-Araujo, R. and E. Hadley. 2014. Skeletal muscle function deficit: A new terminology to embrace the evolving concepts of sarcopenia and age-related muscle dysfunction. *J. Gerontol. A Biol. Sci. Med. Sci.* 69 (5): 591–594.

Couch, M. E., K. Dittus, M. J. Toth, M. S. Willis, D. C. Guttridge, J. R. George, E. Y. Chang, C. G. Gourin, and H. Der-Torossian. 2015. Cancer cachexia update in head and neck cancer: Pathophysiology and treatment. *Head Neck* 37 (7): 1057–1072.

Cruz-Jentoft, A. J., J. P. Baeyens, J. M. Bauer et al. 2010. Sarcopenia: European consensus on definition and diagnosis: Report of the European working group on sarcopenia in older people. *Age Ageing* 39 (4): 412–423.

Currie, K. D., S. G. Thomas, and J. M. Goodman. 2009. Effects of short-term endurance exercise training on vascular function in young males. *Eur. J. Appl. Physiol.* 107 (2): 211–218.

d'Ettorre, G., G. Ceccarelli, N. Giustini, C. M. Mastroianni, G. Silvestri, and V. Vullo. 2014. Taming HIV-related inflammation with physical activity: A matter of timing. *AIDS Res. Hum. Retroviruses* 30 (10): 936–944.

Dam, T. T., K. W. Peters, M. Fragala et al. 2014. An evidence-based comparison of operational criteria for the presence of sarcopenia. *J. Gerontol. A Biol. Sci. Med. Sci.* 69 (5): 584–590.

Danaei, G., A. B. Friedman, S. Oza, C. J. Murray, and M. Ezzati. 2009. Diabetes prevalence and diagnosis in US states: Analysis of health surveys. *Popul. Health Metr.* 7 (1): 1.

Daniels, S. K., K. Brailey, and A. L. Foundas. 1999. Lingual discoordination and dysphagia following acute stroke: Analyses of lesion localization. *Dysphagia* 14 (2): 85–92.

Dao, H. H., Q. T. Do, and J. Sakamoto. 2011. Abnormal body composition phenotypes in Vietnamese women with early rheumatoid arthritis. *Rheumatology (Oxford)* 50 (7): 1250–1258.

Dardevet, D., I. Savary-Auzeloux, D. Remond et al. 2012. Commentaries on viewpoint: Muscle atrophy is not always sarcopenia. *J. Appl. Physiol.(1985)* 113 (4): 680–684.

Dasarathy, S. 2012. Consilience in sarcopenia of cirrhosis. *J. Cachexia Sarcopenia Muscle* 3 (4): 225–237.

Dasarathy, S. 2016. Cause and management of muscle wasting in chronic liver disease. *Curr. Opin. Gastroenterol.* 32 (3): 159–165.

Dauer, W. and S. Przedborski. 2003. Parkinson's disease: Mechanisms and models. *Neuron* 39 (6): 889–909.

Dawson-Hughes, B. and H. Bischoff-Ferrari. 2016. Considerations concerning the definition of sarcopenia. *Osteoporosis Int.* 27 (11): 1–6.

Deans, D., S. Wigmore, H. Gilmour, S. Paterson-Brown, J. Ross, and K. Fearon. 2006. Elevated tumour interleukin-1β is associated with systemic inflammation: A marker of reduced survival in gastro-oesophageal cancer. *Br. J. Cancer* 95 (11): 1568–1575.

DeFronzo, R. A. and D. Tripathy. 2009. Skeletal muscle insulin resistance is the primary defect in type 2 diabetes. *Diabetes Care* 32 Suppl 2: S157–S163.

Del Campo, N., P. Payoux, A. Djilali et al. 2016. Relationship of regional brain beta-amyloid to gait speed. *Neurology* 86 (1): 36–43.

Delgado-Frias, E., M. A. Gonzalez-Gay, J. R. Muniz-Montes, M. A. Gomez Rodriguez-Bethencourt, A. Gonzalez-Diaz, F. Diaz-Gonzalez, and I. Ferraz-Amaro. 2015. Relationship of abdominal adiposity and body composition with endothelial dysfunction in patients with rheumatoid arthritis. *Clin. Exp. Rheumatol.* 33 (4): 516–523.

Deutz, N. E., J. M. Bauer, R. Barazzoni, G. Biolo, Y. Boirie, A. Bosy-Westphal, T. Cederholm, A. Cruz-Jentoft, Z. Krznariç, and K. S. Nair. 2014. Protein intake and exercise for optimal muscle function with aging: Recommendations from the ESPEN expert group. *Clin. Nutr.* 33 (6): 929–936.

Dewys, W. D., C. Begg, P. T. Lavin, P. R. Band, J. M. Bennett, J. R. Bertino, M. H. Cohen, H. O. Douglass Jr., P. F. Engstrom, E. Z. Ezdinli, and J. Horton. 1980. Prognostic effect of weight loss prior to chemotherapy in cancer patients. *Am. J. Med.* 69 (4): 491–497.

Di Francia, M., D. Barbier, J. L. Mege, and J. Orehek. 1994. Tumor necrosis factor-alpha levels and weight loss in chronic obstructive pulmonary disease. *Am. J. Respir. Crit. Care Med.* 150 (5 Pt 1): 1453–1455.

Di Girolamo, F. G., R. Situlin, S. Mazzucco, R. Valentini, G. Toigo, and G. Biolo. 2014. Omega-3 fatty acids and protein metabolism: Enhancement of anabolic interventions for sarcopenia. *Curr. Opin. Clin. Nutr. Metab. Care* 17 (2): 145–150.

Dillon, E. L., M. Sheffield-Moore, D. Paddon-Jones, C. Gilkison, A. P. Sanford, S. L. Casperson, J. Jiang, D. L. Chinkes, and R. J. Urban. 2009. Amino acid supplementation increases lean body mass, basal muscle protein synthesis, and insulin-like growth factor-I expression in older women. *J. Clin. Endocrinol. Metab.* 94 (5): 1630–1637.

Dirajlal-Fargo, S., A. R. Webel, C. T. Longenecker, B. Kinley, D. Labbato, A. Sattar, and G. A. McComsey. 2016. The effect of physical activity on cardiometabolic health and inflammation in treated HIV infection. *Antivir. Ther.* 21 (3): 237–245.

Dobbs, R., A. Charlett, A. Purkiss, S. Dobbs, C. Weller, and D. Peterson. 1999. Association of circulating TNF-α and IL-6 with ageing and parkinsonism. *Acta Neurol. Scand.* 100 (1): 34–41.

Dodds, W. J. 1989. The physiology of swallowing. *Dysphagia* 3 (4): 171–178.

Doğan, S. C., S. Hizmetli, E. Hayta, E. Kaptanoğlu, T. Erselcan, and E. Güler. 2015. Sarcopenia in women with rheumatoid arthritis. *Health* 13: 14.

Donnellan, C. F., L. H. Yann, and S. Lal. 2013. Nutritional management of crohn's disease. *Therap. Adv. Gastroenterol.* 6 (3): 231–242.

Dontje, M., M. de Greef, A. Speelman, M. van Nimwegen, W. Krijnen, R. Stolk, Y. Kamsma, B. Bloem, M. Munneke, and C. van der Schans. 2013. Quantifying daily physical activity and determinants in sedentary patients with Parkinson's disease. *Parkinsonism Relat. Disord.* 19 (10): 878–882.

Douyon, L. and D. E. Schteingart. 2002. Effect of obesity and starvation on thyroid hormone, growth hormone, and cortisol secretion. *Endocrinol. Metab. Clin. North Am.* 31 (1): 173–189.

Du, J., Z. Hu, and W. E. Mitch. 2005. Molecular mechanisms activating muscle protein degradation in chronic kidney disease and other catabolic conditions. *Eur. J. Clin. Invest.* 35 (3): 157–163.

Duclos, M., J. B. Corcuff, F. Pehourcq, and A. Tabarin. 2001. Decreased pituitary sensitivity to glucocorticoids in endurance-trained men. *Eur. J. Endocrinol.* 144 (4): 363–368.

Dupuy, C., V. Lauwers-Cances, S. Guyonnet, C. Gentil, G. Abellan Van Kan, O. Beauchet, A. Schott, B. Vellas, and Y. Rolland. 2015. Searching for a relevant definition of sarcopenia: Results from the cross-sectional EPIDOS study. *J. Cachexia, Sarcopeni. Muscle* 6 (2): 144–154.

Ebner, N., S. Elsner, J. Springer, and S. von Haehling. 2014. Molecular mechanisms and treatment targets of muscle wasting and cachexia in heart failure: An overview. *Curr. Opin. Support. Palliat. Care.* 8 (1): 15–24.

English, C., H. McLennan, K. Thoirs, A. Coates, and J. Bernhardt. 2010. Reviews: Loss of skeletal muscle mass after stroke: A systematic review. *Int. J. Stroke* 5 (5): 395–402.

Epel, E. S. 2009. Psychological and metabolic stress: A recipe for accelerated cellular aging? *Hormones (Athens)* 8 (1): 7–22.

Evans, M. D., R. Singh, V. Mistry, K. Sandhu, P. B. Farmer, and M. S. Cooke. 2008a. Analysis of urinary 8-oxo-7, 8-dihydro-purine-2′-deoxyribonucleosides by LC-MS/MS and improved ELISA. *Free Radic. Res.* 42 (10): 831–840.

Evans, W. 1997. Functional and metabolic consequences of sarcopenia. *J. Nutr.* 127 (5 Suppl): 998S–1003S.

Evans, W. J., J. E. Morley, J. Argilés, C. Bales, V. Baracos, D. Guttridge, A. Jatoi, K. Kalantar-Zadeh, H. Lochs, and G. Mantovani. 2008b. Cachexia: A new definition. *Clin. Nutr.* 27 (6): 793–799.

Fahey, T. D., Insel, P., and Roth, T. 2018. *Fit & Well.* 13th ed. New York: McGraw-Hill.

Faulkner, J. A., L. M. Larkin, D. R. Claflin, and S. V. Brooks. 2007. Age-related changes in the structure and function of skeletal muscles. *Clin. Exp. Pharmacol. Physiol.* 34 (11): 1091–1096.

Fearon, K. C. 1992. The mechanisms and treatment of weight loss in cancer. *Proc. Nutr. Soc.* 51 (02): 251–265.

Fearon, K., W. J. Evans, and S. D. Anker. 2011. Myopenia—A new universal term for muscle wasting. *J. Cachexi. Sarcopeni. Muscle* 2 (1): 1–3.

Federal Interagency Forum on Aging-Related Statistics. accessed October 14, 2009, http://www.agingstats.gov/agingstatsdotnet/Main_Site/Data/2008_Documents/tables/Tables.aspx.

Ferreira, I. M., D. Brooks, J. White, and R. Goldstein. 2012. Nutritional supplementation for stable chronic obstructive pulmonary disease. *Cochrane Library* Vol. 12, page: CD000998.

Fertl, E., A. Doppelbauer, and E. Auff. 1993. Physical activity and sports in patients suffering from Parkinson's disease in comparison with healthy seniors. *J. Neural Transm. Park. Dis. Dement. Sect.* 5 (2): 157–161.

Fiatarone, M. A., E. F. O'Neill, N. D. Ryan, K. M. Clements, G. R. Solares, M. E. Nelson, S. B. Roberts, J. J. Kehayias, L. A. Lipsitz, and W. J. Evans. 1994. Exercise training and nutritional supplementation for physical frailty in very elderly people. *N. Engl. J. Med.* 330 (25): 1769–1775.

Fielding, R. A., B. Vellas, W. J. Evans, S. Bhasin, J. E. Morley, A. B. Newman, G. A. van Kan, S. Andrieu, J. Bauer, and D. Breuille. 2011. Sarcopenia: An undiagnosed condition in older adults. current consensus definition: Prevalence, etiology, and consequences international working group on sarcopenia. *J. Am. Med. Dir. Assoc.* 12 (4): 249–256.

Forbes, S. C., L. McCargar, P. Jelen, and G. J. Bell. 2013. Dose response of whey protein isolate in addition to a typical mixed meal on blood amino acids and hormonal concentrations. *Int. J. Sport Nutr. Exerc. Metab.* 24: 188–195.

Formica, C., F. Cosman, J. Nieves, J. Herbert, and R. Lindsay. 1997. Reduced bone mass and fat-free mass in women with multiple sclerosis: Effects of ambulatory status and glucocorticoid use. *Calcif. Tissue Int.* 61 (2): 129–133.

Fredrix, E. W., P. B. Soeters, M. J. Rouflart, M. F. von Meyenfeldt, and W. H. Saris. 1991b. Resting energy expenditure in patients with newly detected gastric and colorectal cancers. *Am. J. Clin. Nutr.* 53 (5): 1318–1322.

Fredrix, E., E. Wouters, P. Soeters, A. Van Der Aalst, A. Kester, M. Von Meyenfeldt, and W. Saris. 1991a. Resting energy expenditure in patients with non-small cell lung cancer. *Cancer* 68 (7): 1616–1621.

Freedman, R., N. Aziz, D. Albanes, T. Hartman, D. Danforth, S. Hill, N. Sebring, J. Reynolds, and J. Yanovski. 2004. Weight and body composition changes during and after adjuvant chemotherapy in women with breast cancer. *J. Clin. Endocrinol. Metab.* 89 (5): 2248–2253.

Freitas, P., D. Carvalho, A. Santos, M. Matos, A. Madureira, R. Marques, E. Martinez, A. Sarmento, and J. Medina. 2012. Prevalence of obesity and its relationship to clinical lipodystrophy in HIV-infected adults on anti-retroviral therapy. *J. Endocrinol. Invest.* 35 (11): 964–970.

Friedman, P. J., A. J. Campbell, and T. H. Caradoc-Davies. 1985. Prospective trial of a new diagnostic criterion for severe wasting malnutrition in the elderly. *Age Ageing* 14 (3): 149–154.

Furukawa, S., T. Fujita, M. Shimabukuro, M. Iwaki, Y. Yamada, Y. Nakajima, O. Nakayama, M. Makishima, M. Matsuda, and I. Shimomura. 2004. Increased oxidative stress in obesity and its impact on metabolic syndrome. *J. Clin. Invest.* 114 (12): 1752–1761.

Galvan, A. and T. Wichmann. 2008. Pathophysiology of parkinsonism. *Clin. Neurophysiol.* 119 (7): 1459–1474.

Garber, C. E. and J. H. Friedman. 2003. Effects of fatigue on physical activity and function in patients with Parkinson's disease. *Neurology* 60 (7): 1119–1124.

Geremia, A., P. Biancheri, P. Allan, G. R. Corazza, and A. Di Sabatino. 2014. Innate and adaptive immunity in inflammatory bowel disease. *Autoimmun. Rev.* 13 (1): 3–10.

Gianoudis, J., C. Bailey, and R. Daly. 2015. Associations between sedentary behaviour and body composition, muscle function and sarcopenia in community-dwelling older adults. *Osteoporosis Int.* 26 (2): 571–579.

Giles, J. T., M. Allison, R. S. Blumenthal, W. Post, A. C. Gelber, M. Petri, R. Tracy, M. Szklo, and J. M. Bathon. 2010. Abdominal adiposity in rheumatoid arthritis: Association with cardiometabolic risk factors and disease characteristics. *Arthritis Rheum.* 62 (11): 3173–3182.

Giles, J. T., S. M. Ling, L. Ferrucci, S. J. Bartlett, R. E. Andersen, M. Towns, D. Muller, K. R. Fontaine, and J. M. Bathon. 2008. Abnormal body composition phenotypes in older rheumatoid arthritis patients: Association with disease characteristics and pharmacotherapies. *Arthritis Rheum.* 59 (6): 807–815.

Gloire, G., S. Legrand-Poels, and J. Piette. 2006. NF-κB activation by reactive oxygen species: Fifteen years later. *Biochem. Pharmacol.* 72 (11): 1493–1505.

Goldspink, G., K. Fernandes, P. E. Williams, and D. J. Wells. 1994. Age-related changes in collagen gene expression in the muscles of mdx dystrophic and normal mice. *Neuromuscular Disord.* 4 (3): 183–191.

Goodpaster, B. H., C. L. Carlson, M. Visser, D. E. Kelley, A. Scherzinger, T. B. Harris, E. Stamm, and A. B. Newman. 2001. Attenuation of skeletal muscle and strength in the elderly: The health ABC study. *J. Appl. Physiol. (1985)* 90 (6): 2157–2165.

Goodpaster, B. H., P. Chomentowski, B. K. Ward, A. Rossi, N. W. Glynn, M. J. Delmonico, S. B. Kritchevsky, M. Pahor, and A. B. Newman. 2008. Effects of physical activity on strength and skeletal muscle fat infiltration in older adults: A randomized controlled trial. *J. Appl. Physiol. (1985)* 105 (5): 1498–1503.

Gosselink, R., T. Troosters, and M. Decramer. 1996. Peripheral muscle weakness contributes to exercise limitation in COPD. *Am. J. Respir. Crit. Care Med.* 153 (3): 976–980.

Greiwe, J. S., B. Cheng, D. C. Rubin, K. E. Yarasheski, and C. F. Semenkovich. 2001. Resistance exercise decreases skeletal muscle tumor necrosis factor alpha in frail elderly humans. *FASEB J.* 15 (2): 475–482.

Grinspoon, S., K. Mulligan, and Department of Health and Human Services Working Group on the Prevention and Treatment of Wasting and Weight Loss. 2003. Weight loss and wasting in patients infected with human immunodeficiency virus. *Clin. Infect. Dis.* 36 (Suppl 2): S69–S78.

Gutierrez, E. G., W. A. Banks, and A. J. Kastin. 1993. Murine tumor necrosis factor alpha is transported from blood to brain in the mouse. *J. Neuroimmunol.* 47 (2): 169–176.

Guttridge, D. C. 2004. Signaling pathways weigh in on decisions to make or break skeletal muscle. *Curr. Opin. Clin. Nutr. Metab. Care* 7 (4): 443–450.

Hambrecht, R., P. C. Schulze, S. Gielen, A. Linke, S. Mobius-Winkler, S. Erbs, J. Kratzsch, A. Schubert, V. Adams, and G. Schuler. 2005. Effects of exercise training on insulin-like growth factor-I expression in the skeletal muscle of non-cachectic patients with chronic heart failure. *Eur. J. Cardiovasc. Prev. Rehabil.* 12 (4): 401–406.

Hanaoka, B. Y., L. C. Cleary, D. E. Long, A. Srinivas, K. A. Jenkins, H. M. Bush, C. P. Starnes, M. Rutledge, J. Duan, and Q. Fan. 2015. Physical impairment in patients with idiopathic inflammatory myopathies is associated with the American College of Rheumatology Functional Status measure. *Clin. Rheumatol.* 34 (11): 1929–1937.

Hawkins, S. A., R. A. Wiswell, and T. J. Marcell. 2003. Exercise and the master athlete—A model of successful aging? *J. Gerontol. A Biol. Sci. Med. Sci.* 58 (11): M1009–M1011.

He, X. Z. and D. W. Baker. 2004. Body mass index, physical activity, and the risk of decline in overall health and physical functioning in late middle age. *Am. J. Public Health* 94 (9): 1567–1573.

Hemmer, B., M. Kerschensteiner, and T. Korn. 2015. Role of the innate and adaptive immune responses in the course of multiple sclerosis. *Lancet Neurol.* 14 (4): 406–419.

Henderson, G. C., B. A. Irving, and K. S. Nair. 2009. Potential application of essential amino acid supplementation to treat sarcopenia in elderly people. *J. Clin. Endocrinol. Metabol.* 94 (5): 1524–1526.

Hepple, R. T. 2006. Dividing to keep muscle together: The role of satellite cells in aging skeletal muscle. *Sci. Aging Knowledge Environ.* 2006 (3): pe3.

Hepple, R.T. 2012. Muscle atrophy is not always sarcopenia. *J. Appl. Physiol. (1985)* 113 (4): 677–679.

Hillsdon, M. M., E. J. Brunner, J. M. Guralnik, and M. G. Marmot. 2005. Prospective study of physical activity and physical function in early old age. *Am. J. Prev. Med.* 28 (3): 245–250.

Hiona, A. and C. Leeuwenburgh. 2008. The role of mitochondrial DNA mutations in aging and sarcopenia: Implications for the mitochondrial vicious cycle theory of aging. *Exp. Gerontol.* 43 (1): 24–33.

Hirtz, D., D. J. Thurman, K. Gwinn-Hardy, M. Mohamed, A. R. Chaudhuri, and R. Zalutsky. 2007. How common are the "common" neurologic disorders? *Neurology* 68 (5): 326–337.

Holecek, M. 2015. Ammonia and amino acid profiles in liver cirrhosis: Effects of variables leading to hepatic encephalopathy. *Nutrition* 31 (1): 14–20.

Holeček, M., L. Šprongl, I. Tilšer, and M. Tichý. 2001. Leucine and protein metabolism in rats with chronic renal insufficiency. *Exp. Toxicol. Pathol.* 53 (1): 71–76.

Holt, D. Q., B. J. Strauss, K. K. Lau, and G. T. Moore. 2016. Body composition analysis using abdominal scans from routine clinical care in patients with crohn's disease. *Scand. J. Gastroenterol.* 51 (7): 842–847.

Hoppeler, H., E. Kleinert, C. Schlegel, H. Claassen, H. Howald, S. Kayar, and P. Cerretelli. 1990. II. morphological adaptations of human skeletal muscle to chronic hypoxia. *Int. J. Sports Med.* 11 (S 1): S3–S9.

Horoupian, D. S. and P. H. Wasserstein. 1999. Alzheimer's disease pathology in motor cortex in dementia with lewy bodies clinically mimicking corticobasal degeneration. *Acta Neuropathol.* 98 (3): 317–322.

Houston, D. K., B. J. Nicklas, J. Ding, et al. 2008. Dietary protein intake is associated with lean mass change in older, community-dwelling adults: The health, aging, and body composition (health ABC) study. *Am. J. Clin. Nutr.* 87 (1): 150–155.

Huang, Y. and Z. Chen. 2016. Inflammatory bowel disease related innate immunity and adaptive immunity. *Am. J. Transl. Res.* 8 (6): 2490.

Humbert, I. A., D. G. McLaren, K. Kosmatka, M. Fitzgerald, S. Johnson, E. Porcaro, S. Kays, E. O. Umoh, and J. Robbins. 2010. Early deficits in cortical control of swallowing in Alzheimer's disease. *J. Alzheimers Dis.* 19 (4): 1185–1197.

Ikizler, T. A., L. B. Pupim, J. R. Brouillette, D. K. Levenhagen, K. Farmer, R. M. Hakim, and P. J. Flakoll. 2002. Hemodialysis stimulates muscle and whole body protein loss and alters substrate oxidation. *Am. J. Physiol. Endocrinol. Metab.* 282 (1): E107–E116.

Ito, K. and P. J. Barnes. 2009. COPD as a disease of accelerated lung aging. *CHEST J.* 135 (1): 173–180.

Janikiewicz, J., K. Hanzelka, K. Kozinski, K. Kolczynska, and A. Dobrzyn. 2015. Islet β-cell failure in type 2 diabetes—Within the network of toxic lipids. *Biochem. Biophys. Res. Commun.* 460 (3): 491–496.

Janssen, I., D. S. Shepard, P. T. Katzmarzyk, and R. Roubenoff. 2004. The healthcare costs of sarcopenia in the united states. *J. Am. Geriatr. Soc.* 52 (1): 80–85.

Janssen, I., S. B. Heymsfield, and R. Ross. 2002. Low relative skeletal muscle mass (sarcopenia) in older persons is associated with functional impairment and physical disability. *J. Am. Geriatr. Soc.* 50 (5): 889–896.

Jarosz, P. A. and A. Bellar. 2009. Sarcopenic obesity: An emerging cause of frailty in older adults. *Geriatr. Nurs.* 30 (1): 64–70.

Jatoi, A., H. L. Ritter, A. Dueck, P. L. Nguyen, D. A. Nikcevich, R. F. Luyun, B. I. Mattar, and C. L. Loprinzi. 2010. A placebo-controlled, double-blind trial of infliximab for cancer-associated weight loss in elderly and/or poor performance non-small cell lung cancer patients (N01C9). *Lung Cancer* 68 (2): 234–239.

Jatoi, A., S. R. Dakhil, P. L. Nguyen, J. A. Sloan, J. W. Kugler, K. M. Rowland, G. S. Soori, D. B. Wender, T. R. Fitch, and P. J. Novotny. 2007. A placebo-controlled double blind trial of etanercept for the cancer anorexia/weight loss syndrome. *Cancer* 110 (6): 1396–1403.

Jemal, A., R. Siegel, E. Ward, T. Murray, J. Xu, and M. J. Thun. 2007. Cancer statistics, 2007. *CA Cancer J. Clin.* 57 (1): 43–66.

Jonsson, A. C., I. Lindgren, B. Norrving, and A. Lindgren. 2008. Weight loss after stroke: A population-based study from the Lund Stroke Register. *Stroke* 39 (3): 918–923.

Jørgensen, L. and B. Jacobsen. 2001. Changes in muscle mass, fat mass, and bone mineral content in the legs after stroke: A 1 year prospective study. *Bone* 28 (6): 655–659.

Kalapotharakos, V., I. Smilios, A. Parlavatzas, and S. P. Tokmakidis. 2007. The effect of moderate resistance strength training and detraining on muscle strength and power in older men. *J. Geriatr. Phys. Ther.* 30 (3): 109–113.

Kalf, J., B. De Swart, B. Bloem, and M. Munneke. 2012. Prevalence of oropharyngeal dysphagia in Parkinson's disease: A meta-analysis. *Parkinsonism Relat. Disord.* 18 (4): 311–315.

Kalinkovich, A., and G. Livshits. 2017. Sarcopenic obesity or obese sarcopenia: A cross talk between age-associated adipose tissue and skeletal muscle inflammation as a main mechanism of the pathogenesis. *Ageing Res. Rev.* 35: 200–221.

Katsanos, C. S., H. Kobayashi, M. Sheffield-Moore, A. Aarsland, and R. R. Wolfe. 2006. A high proportion of leucine is required for optimal stimulation of the rate of muscle protein synthesis by essential amino acids in the elderly. *Am. J. Physiol. Endocrinol. Metab.* 291 (2): E381–E387.

Kaufman, B. A., C. Li, and S. A. Soleimanpour. 2015. Mitochondrial regulation of β-cell function: Maintaining the momentum for insulin release. *Mol. Aspects Med.* 42: 91–104.

Kelly-Hayes, M., A. Beiser, C. S. Kase, A. Scaramucci, R. B. D'Agostino, and P. A. Wolf. 2003. The influence of gender and age on disability following ischemic stroke: The Framingham Study. *J. Stroke Cerebrovasc. Dis.* 12 (3): 119–126.

Kent-Braun, J. A., A. V. Ng, M. Castro, M. W. Weiner, D. Gelinas, G. A. Dudley, and R. G. Miller. 1997. Strength, skeletal muscle composition, and enzyme activity in multiple sclerosis. *J. Appl. Physiol.(1985)* 83 (6): 1998–2004.

Khan, I., X. Perrard, G. Brunner, H. Lui, L. Sparks, S. Smith, X. Wang, Z. Shi, D. Lewis, and H. Wu. 2015. Intermuscular and perimuscular fat expansion in obesity correlates with skeletal muscle T cell and macrophage infiltration and insulin resistance. *Int. J.Obes.* 39: 1607.

Kim, Y., C. K. Kim, S. Jung, S. Ko, S. Lee, and B. Yoon. 2015. Prognostic importance of weight change on short-term functional outcome in acute ischemic stroke. *Int. J. Stroke* 10 (A100): 62–68.

King, D., M. L. Smith, and M. Lye. 1996. Gastro-intestinal protein loss in elderly patients with cardiac cachexia. *Age Ageing* 25 (3): 221–223.

King, D., M. L. Smith, T. J. Chapman, H. R. Stockdale, and M. Lye. 1996. Fat malabsorption in elderly patients with cardiac cachexia. *Age Ageing* 25 (2): 144–149.

Knights, D., K. G. Lassen, and R. J. Xavier. 2013. Advances in inflammatory bowel disease pathogenesis: Linking host genetics and the microbiome. *Gut* 62 (10): 1505–1510.

Kob, R., L. C. Bollheimer, T. Bertsch, C. Fellner, M. Djukic, C. C. Sieber, and B. E. Fischer. 2015. Sarcopenic obesity: Molecular clues to a better understanding of its pathogenesis? *Biogerontology* 16 (1): 15–29.

Kolterman, O. G., J. Insel, M. Saekow, and J. M. Olefsky. 1980. Mechanisms of insulin resistance in human obesity: Evidence for receptor and postreceptor defects. *J. Clin. Invest.* 65 (6): 1272–1284.

Kooman, J. P., P. Kotanko, A. M. Schols, P. G. Shiels, and P. Stenvinkel. 2014. Chronic kidney disease and premature ageing. *Nat. Rev. Nephrol.* 10 (12): 732–742.

Kosek, D. J., J. S. Kim, J. K. Petrella, J. M. Cross, and M. M. Bamman. 2006. Efficacy of 3 days/wk resistance training on myofiber hypertrophy and myogenic mechanisms in young vs. older adults. *J. Appl. Physiol. (1985)* 101 (2): 531–544.

Koves, T. R., R. C. Noland, A. L. Bates, S. T. Henes, D. M. Muoio, and R. N. Cortright. 2005. Subsarcolemmal and intermyofibrillar mitochondria play distinct roles in regulating skeletal muscle fatty acid metabolism. *Am. J. Physiol. Cell. Physiol.* 288 (5): C1074–C1082.

Kuo, H. K., S. G. Leveille, Y. H. Yu, and W. P. Milberg. 2007. Cognitive function, habitual gait speed, and late-life disability in the national health and nutrition examination survey (NHANES) 1999–2002. *Gerontology* 53 (2): 102–110.

Landi, F., E. Marzetti, A. M. Martone, R. Bernabei, and G. Onder. 2014. Exercise as a remedy for sarcopenia. *Curr. Opin. Clin. Nutr. Metab. Care* 17 (1): 25–31.

Langen, R. C., H. R. Gosker, A. H. Remels, and A. M. Schols. 2013. Triggers and mechanisms of skeletal muscle wasting in chronic obstructive pulmonary disease. *Int. J. Biochem. Cell Biol.* 45 (10): 2245–2256.

Larsson, L., B. Sjödin, and J. Karlsson. 1978. Histochemical and biochemical changes in human skeletal muscle with age in sedentary males, age 22–65 years. *Acta Physiol. Scand.* 103 (1): 31–39.

Lauretani, F., C. R. Russo, S. Bandinelli, B. Bartali, C. Cavazzini, A. Di Iorio, A. M. Corsi, T. Rantanen, J. M. Guralnik, and L. Ferrucci. 2003. Age-associated changes in skeletal muscles and their effect on mobility: An operational diagnosis of sarcopenia. *J. Appl. Physiol. (1985)* 95 (5): 1851–1860.

Learmonth, Y. C. and R. W. Motl. 2016. Physical activity and exercise training in multiple sclerosis: A review and content analysis of qualitative research identifying perceived determinants and consequences. *Disabil. Rehabil.* 38 (13): 1227–1242.

Lee, H., J. Song, C. Shin, D. Park, K. Park, K. Lee, and C. Koh. 1998. Decreased mitochondrial DNA content in peripheral blood precedes the development of non-insulin-dependent diabetes mellitus. *Diabetes Res. Clin. Pract.* 42 (3): 161–167.

Leveille, S. G., J. M. Guralnik, L. Ferrucci, and J. A. Langlois. 1999. Aging successfully until death in old age: Opportunities for increasing active life expectancy. *Am. J. Epidemiol.* 149 (7): 654–664.

Li, Y. P. and M. B. Reid. 2000. NF-kappaB mediates the protein loss induced by TNF-alpha in differentiated skeletal muscle myotubes. *Am. J. Physiol. Regul. Integr. Comp. Physiol.* 279 (4): R1165–R1170.

Li, Y. P., Y. Chen, A. S. Li, and M. B. Reid. 2003. Hydrogen peroxide stimulates ubiquitin-conjugating activity and expression of genes for specific E2 and E3 proteins in skeletal muscle myotubes. *Am. J. Physiol. Cell. Physiol.* 285 (4): C806–C812.

Licastro, F., S. Pedrini, L. Caputo, G. Annoni, L. J. Davis, C. Ferri, V. Casadei, and L. M. E. Grimaldi. 2000. Increased plasma levels of interleukin-1, interleukin-6 and α-1-antichymotrypsin in patients with Alzheimer's disease: Peripheral inflammation or signals from the brain? *J. Neuroimmunol.* 103 (1): 97–102.

Lim, P., Y. Cheng, and Y. Wei. 2000. Large-scale mitochondrial DNA deletions in skeletal muscle of patients with end-stage renal disease. *Free Radical Bio. Med.* 29 (5): 454–463.

Lindenbach, D. and C. Bishop. 2013. Critical involvement of the motor cortex in the pathophysiology and treatment of Parkinson's disease. *Neurosci. Biobehav. Rev.* 37 (10): 2737–2750.

Lin, S. Y., W. Y. Chen, F. Y. Lee, C. J. Huang, and W. H. Sheu. 2005. Activation of ubiquitin-proteasome pathway is involved in skeletal muscle wasting in a rat model with biliary cirrhosis: Potential role of TNF-alpha. *Am. J. Physiol. Endocrinol. Metab.* 288 (3): E493–E501.

Llovera, M., C. Garcí, N. Agell, F. J. López-Soriano, and J. M. Argilés. 1997. TNF can directly induce the expression of ubiquitin-dependent proteolytic system in rat soleus muscles. *Biochem. Biophys. Res. Commun.* 230 (2): 238–241.

Look AHEAD Research Group, X. Pi-Sunyer, G. Blackburn et al. 2007. Reduction in weight and cardiovascular disease risk factors in individuals with type 2 diabetes: One-year results of the look AHEAD trial. *Diabetes Care* 30 (6): 1374–1383.

Lorefält, B., A. Granérus, and M. Unosson. 2006. Avoidance of solid food in weight losing older patients with Parkinson's disease. *J. Clin. Nurs.* 15 (11): 1404–1412.

Lundholm, K., A. C. Bylund, J. Holm, and T. Schersten. 1976. Skeletal muscle metabolism in patients with malignant tumor. *Eur. J. Cancer* 12 (6): 465–473.

Ma, K., C. Mallidis, S. Bhasin, V. Mahabadi, J. Artaza, N. Gonzalez-Cadavid, J. Arias, and B. Salehian. 2003. Glucocorticoid-induced skeletal muscle atrophy is associated with upregulation of myostatin gene expression. *Am. J. Physiol. Endocrinol. Metab.* 285 (2): E363–E371.

Ma, Z. A., Z. Zhao, and J. Turk. 2012. Mitochondrial dysfunction and beta-cell failure in type 2 diabetes mellitus. *Exp. Diabetes Res.* 2012: 703538.

Machida, N., H. Tohara, K. Hara, A. Kumakura, Y. Wakasugi, A. Nakane, and S. Minakuchi. 2016. Effects of aging and sarcopenia on tongue pressure and jaw-opening force. *Geriatr. Gerontol. Int.* 17: 295–301.

Mackey, A., B. Esmarck, F. Kadi, S. Koskinen, M. Kongsgaard, A. Sylvestersen, J. Hansen, G. Larsen, and M. Kjaer. 2007. Enhanced satellite cell proliferation with resistance training in elderly men and women. *Scand. J. Med. Sci. Sports* 17 (1): 34–42.

Madaro, L. and M. Bouche. 2014. From innate to adaptive immune response in muscular dystrophies and skeletal muscle regeneration: The role of lymphocytes. *Biomed. Res. Int.* 2014: 438675.

Maeda, K. and J. Akagi. 2015. Decreased tongue pressure is associated with sarcopenia and sarcopenic dysphagia in the elderly. *Dysphagia* 30 (1): 80–87.

Manini, T. M. and M. Pahor. 2009. Physical activity and maintaining physical function in older adults. *Br. J. Sports Med.* 43 (1): 28–31.

Manini, T. M., B. C. Clark, M. A. Nalls, B. H. Goodpaster, L. L. Ploutz-Snyder, and T. B. Harris. 2007. Reduced physical activity increases intermuscular adipose tissue in healthy young adults. *Am. J. Clin. Nutr.* 85 (2): 377–384.

Manini, T. M., J. E. Everhart, K. V. Patel, et al. 2009. Activity energy expenditure and mobility limitation in older adults: Differential associations by sex. *Am. J. Epidemiol.* 169 (12): 1507–1516.

Mantovani, G., A. Macciò, L. Mura, E. Massa, M. C. Mudu, C. Mulas, M. R. Lusso, C. Madeddu, and A. Dessì. 2000. Serum levels of leptin and proinflammatory cytokines in patients with advanced-stage cancer at different sites. *J. Mol. Med.* 78 (10): 554–561.

Mantovani, G., A. Maccio, E. Massa, and C. Madeddu. 2001. Managing cancer-related anorexia/cachexia. *Drugs* 61 (4): 499–514.

Marcell, T. J. 2003. Sarcopenia: Causes, consequences, and preventions. *J. Gerontol. A Biol. Sci. Med. Sci.* 58 (10): M911–M916.

Marcell, T. J., S. A. Hawkins, and R. A. Wiswell. 2014. Leg strength declines with advancing age despite habitual endurance exercise in active older adults. *J. Strength Cond. Res.* 28 (2): 504–513.

Marquis, K., R. Debigaré, Y. Lacasse, P. LeBlanc, J. Jobin, G. Carrier, and F. Maltais. 2002. Midthigh muscle cross-sectional area is a better predictor of mortality than body mass index in patients with chronic obstructive pulmonary disease. *Am. J. Resp. Crit. Care Med.* 166 (6): 809–813.

Marzetti, E., J. M. Lawler, A. Hiona, T. Manini, A. Y. Seo, and C. Leeuwenburgh. 2008. Modulation of age-induced apoptotic signaling and cellular remodeling by exercise and calorie restriction in skeletal muscle. *Free Radical Bio. Med.* 44 (2): 160–168.

Massironi, S., R. E. Rossi, F. A. Cavalcoli, S. Della Valle, M. Fraquelli, and D. Conte. 2013. Nutritional deficiencies in inflammatory bowel disease: Therapeutic approaches. *Clin. Nutr.* 32 (6): 904–910.

Mastrocola, R., M. Collino, D. Nigro, F. Chiazza, G. D'Antona, M. Aragno, and M. A. Minetto. 2015. Accumulation of advanced glycation end-products and activation of the SCAP/SREBP lipogenetic pathway occur in diet-induced obese mouse skeletal muscle. *PLoS One* 10 (3): e0119587.

McDermott, M. M., F. Hoff, L. Ferrucci, W. H. Pearce, J. M. Guralnik, L. Tian, K. Liu, J. R. Schneider, L. Sharma, and J. Tan. 2007. Lower extremity ischemia, calf skeletal muscle characteristics, and functional impairment in peripheral arterial disease. *J. Am. Geriatr. Soc.* 55 (3): 400–406.

McIntire, K. L., Y. Chen, S. Sood, and R. Rabkin. 2014. Acute uremia suppresses leucine-induced signal transduction in skeletal muscle. *Kidney Int.* 85 (2): 374–382.

McNeil, C. J., T. J. Doherty, D. W. Stashuk, and C. L. Rice. 2005. Motor unit number estimates in the tibialis anterior muscle of young, old, and very old men. *Muscle Nerve* 31 (4): 461–467.

Melov, S., M. A. Tarnopolsky, K. Beckman, K. Felkey, and A. Hubbard. 2007. Resistance exercise reverses aging in human skeletal muscle. *PLoS One* 2 (5): e465.

Mendes, M. C., G. D. Pimentel, F. O. Costa, and J. B. Carvalheira. 2015. Molecular and neuroendocrine mechanisms of cancer cachexia. *J. Endocrinol.* 226 (3): R29–R43.

Meng, N., T. Wang, and I. Lien. 2000. Dysphagia in patients with brainstem stroke: Incidence and outcome. *Am. J. Phys. Med. Rehabil.* 79 (2): 170–175.

Metsios, G. S., A. Stavropoulos-Kalinoglou, and G. D. Kitas. 2015. The role of exercise in the management of rheumatoid arthritis. *Expert Rev. Clin. Immu.* 11 (10): 1121–1130.

Mijač, D. D., G. L. Janković, J. Jorga, and M. N. Krstić. 2010. Nutritional status in patients with active inflammatory bowel disease: Prevalence of malnutrition and methods for routine nutritional assessment. *Eur. J. Intern. Med.* 21 (4): 315–319.

Mijnarends, D. M., A. Koster, J. M. Schols et al. 2016. Physical activity and incidence of sarcopenia: The population-based AGES-reykjavik study. *Age Ageing* 45 (5): 614–620.

Moller, S., and U. Becker. 1992. Insulin-like growth factor 1 and growth hormone in chronic liver disease. *Dig. Dis.* 10 (4): 239–248.

Morais, J., S. Chevalier, and R. Gougeon. 2006. Protein turnover and requirements in the healthy and frail elderly. *J. Nutr. Health Aging* 10 (4): 272.

Moreno Catala, M., D. Woitalla, and A. Arampatzis. 2013. Central factors explain muscle weakness in young fallers with Parkinson's disease. *Neurorehabil. Neural. Repair* 27 (8): 753–759.

Morley, J. E. 2000. Diabetes mellitus: A major disease of older persons. *J. Gerontol. A Biol. Sci. Med. Sci.* 55 (5): M255–M256.

Morley, J. E. 2008. Diabetes, sarcopenia, and frailty. *Clin. Geriatr. Med.* 24 (3): 455–469.

Morley, J. E., A. M. Abbatecola, J. M. Argiles, V. Baracos, J. Bauer, S. Bhasin, T. Cederholm, A. J. S. Coats, S. R. Cummings, and W. J. Evans. 2011. Sarcopenia with limited mobility: An international consensus. *J. Am. Med. Dir. Assoc.* 12 (6): 403–409.

Morley, J., M. Kim, M. Haren, R. Kevorkian, and W. Banks. 2005. Frailty and the aging male. *Aging Male* 8 (3–4): 135–140.

Motl, R., E. McAuley, B. Sandroff, and E. Hubbard. 2015. Descriptive epidemiology of physical activity rates in multiple sclerosis. *Acta Neurol. Scand.* 131 (6): 422–425.

Mühlberg, W. and C. Sieber. 2004. Sarcopenia and frailty in geriatric patients: Implications for training and prevention. *Zeitschrift Für Gerontologie Und Geriatrie* 37 (1): 2–8.

Mul, J. D., R. J. Seeley, S. C. Woods, and D. P. Begg. 2013. Angiotensin-converting enzyme inhibition reduces food intake and weight gain and improves glucose tolerance in melanocortin-4 receptor deficient female rats. *Physiol. Behav.* 121: 43–48.

Muller, M. J. 2007. Malnutrition and hypermetabolism in patients with liver cirrhosis. *Am. J. Clin. Nutr.* 85 (5): 1167–1168.

Muscaritoli, M., S. Anker, J. Argiles, Z. Aversa, J. Bauer, G. Biolo, Y. Boirie, I. Bosaeus, T. Cederholm, and P. Costelli. 2010. Consensus definition of sarcopenia, cachexia and pre-cachexia: Joint document elaborated by special interest groups (SIG) "cachexia-anorexia in chronic wasting diseases" and "nutrition in geriatrics". *Clin. Nutr.* 29 (2): 154–159.

Naclerio, F. and E. Larumbe-Zabala. 2016. Effects of whey protein alone or as part of a multi-ingredient formulation on strength, fat-free mass, or lean body mass in resistance-trained individuals: A meta-analysis. *Sports Med.* 46 (1): 125–137.

Narici, M. V., N. D. Reeves, C. I. Morse, and C. N. Maganaris. 2004. Muscular adaptations to resistance exercise in the elderly. *J Musculoskelet. Neuronal. Interact.* 4 (2): 161–164.

Nathan, I., C. Norton, W. Czuber-Dochan, and A. Forbes. 2013. Exercise in individuals with inflammatory bowel disease. *Gastroenterol. Nurs.* 36 (6): 437–442.

Naudin, M. and B. Atanasova. 2014. Olfactory markers of depression and Alzheimer's disease. *Neurosci. Biobehav. Rev.* 45: 262–270.

Ness, K. K., K. S. Baker, D. R. Dengel, et al. 2007. Body composition, muscle strength deficits and mobility limitations in adult survivors of childhood acute lymphoblastic leukemia. *Pediatr. Blood Cancer* 49 (7): 975–981.

Ness, K. K., K. R. Krull, K. E. Jones et al. 2013. Physiologic frailty as a sign of accelerated aging among adult survivors of childhood cancer: A report from the St Jude Lifetime cohort study. *J. Clin. Oncol.* 31 (36): 4496–4503.

Nicolini, A., P. Ferrari, M. C. Masoni, M. Fini, S. Pagani, O. Giampietro, and A. Carpi. 2013. Malnutrition, anorexia and cachexia in cancer patients: A mini-review on pathogenesis and treatment. *Biomed. Pharmacother.* 67 (8): 807–817.

O'Keefe, J. H. and D. S. Bell. 2007. Postprandial hyperglycemia/hyperlipidemia (postprandial dysmetabolism) is a cardiovascular risk factor. *Am. J. Cardiol.* 100 (5): 899–904.

Oeffinger, K. C., A. C. Mertens, C. A. Sklar, et al. 2006. Chronic health conditions in adult survivors of childhood cancer. *N. Engl. J. Med.* 355 (15): 1572–1582.

Offner, H., S. Subramanian, S. M. Parker, M. E. Afentoulis, A. A. Vandenbark, and P. D. Hurn. 2006. Experimental stroke induces massive, rapid activation of the peripheral immune system. *J. Cereb. Blood Flow & Metab.* 26 (5): 654–665.

Oliviero, A., P. Profice, P. Tonali, F. Pilato, E. Saturno, M. Dileone, F. Ranieri, and V. Di Lazzaro. 2006. Effects of aging on motor cortex excitability. *Neurosci. Res.* 55 (1): 74–77.

Packer, N., L. Hoffman-Goetz, and G. Ward. 2010. Does physical activity affect quality of life, disease symptoms and immune measures in patients with inflammatory bowel disease? A systematic review. *J. Sports Med. Phys. Fitness* 50 (1): 1.

Paddon-Jones, D. and B. B. Rasmussen. 2009. Dietary protein recommendations and the prevention of sarcopenia. *Curr. Opin. Clin. Nutr. Metab. Care* 12 (1): 86–90.

Palmer, B. F. and D. J. Clegg. 2014. Ascent to altitude as a weight loss method: The good and bad of hypoxia inducible factor activation. *Obesity* 22 (2): 311–317.

Patra, S. K. and S. Arora. 2012. Integrative role of neuropeptides and cytokines in cancer anorexia–cachexia syndrome. *Clin. Chimica Acta* 413 (13): 1025–1034.

Pearson, S. J., A. Young, A. Macaluso, G. Devito, M. A. Nimmo, M. Cobbold, and S. D. Harridge. 2002. Muscle function in elite master weightlifters. *Med. Sci. Sports Exerc.* 34 (7): 1199–1206.

Peluso, I. and M. Palmery. 2016. The relationship between body weight and inflammation: Lesson from anti-TNF-α antibody therapy. *Hum. Immunol.* 77 (1): 47–53.

Pennisi, G., G. Alagona, R. Ferri, S. Greco, D. Santonocito, A. Pappalardo, and R. Bella. 2002. Motor cortex excitability in Alzheimer disease: One year follow-up study. *Neurosci. Lett.* 329 (3): 293–296.

Perandini, L. A., A. L. de Sa-Pinto, H. Roschel, F. B. Benatti, F. R. Lima, E. Bonfa, and B. Gualano. 2012. Exercise as a therapeutic tool to counteract inflammation and clinical symptoms in autoimmune rheumatic diseases. *Autoimmun. Rev.* 12 (2): 218–224.

Pesce, V., A. Cormio, L. C. Marangi, F. W. Guglielmi, A. M. Lezza, A. Francavilla, P. Cantatore, and M. N. Gadaleta. 2002. Depletion of mitochondrial DNA in the skeletal muscle of two cirrhotic patients with severe asthenia. *Gene* 286 (1): 143–148.

Petersen, K. F., D. Befroy, S. Dufour, J. Dziura, C. Ariyan, D. L. Rothman, L. DiPietro, G. W. Cline, and G. I. Shulman. 2003. Mitochondrial dysfunction in the elderly: Possible role in insulin resistance. *Science* 300 (5622): 1140–1142.

Peterson, J. M., Y. Wang, R. W. Bryner, D. L. Williamson, and S. E. Alway. 2008. Bax signaling regulates palmitate-mediated apoptosis in C(2)C(12) myotubes. *Am. J. Physiol. Endocrinol. Metab.* 295 (6): E1307–E1314.

Pipinos, I. I., A. R. Judge, Z. Zhu, J. T. Selsby, S. A. Swanson, J. M. Johanning, B. T. Baxter, T. G. Lynch, and S. L. Dodd. 2006. Mitochondrial defects and oxidative damage in patients with peripheral arterial disease. *Free Radical Bio. Med.* 41 (2): 262–269.

Plaza, S. M. 2002. Mitochondrial factors in the pathogenesis of diabetes: A hypothesis for treatment. *Altern. Med. Rev.* 7 (2): 94–111.

Plow, E. B., D. A. Cunningham, C. Bonnett et al. 2013. Neurophysiological correlates of aging-related muscle weakness. *J. Neurophysiol.* 110 (11): 2563–2573.

Porte, D. and S. E. Kahn. 1991. Mechanisms for hyperglycemia in type II diabetes mellitus: Therapeutic implications for sulfonylurea treatment—An update. *Am. J. Med.* 90 (6): S8–S14.

Pott Godoy, M. C., R. Tarelli, C. C. Ferrari, M. I. Sarchi, and F. J. Pitossi. 2008. Central and systemic IL-1 exacerbates neurodegeneration and motor symptoms in a model of Parkinson's disease. *Brain* 131 (Pt 7): 1880–1894.

Prado, C. M., V. E. Baracos, L. J. McCargar, T. Reiman, M. Mourtzakis, K. Tonkin, J. R. Mackey, S. Koski, E. Pituskin, and M. B. Sawyer. 2009. Sarcopenia as a determinant of chemotherapy toxicity and time to tumor progression in metastatic breast cancer patients receiving capecitabine treatment. *Clin. Cancer Res.* 15 (8): 2920–2926.

Preston, T., C. Slater, D. C. McMillan, J. S. Falconer, A. Shenkin, and K. C. Fearon. 1998. Fibrinogen synthesis is elevated in fasting cancer patients with an acute phase response. *J. Nutr.* 128 (8): 1355–1360.

Preston T., K. C. H. Fearon, I. Robertson, B. W. East and K. C. Calman. 1987. Tissue loss during severe wasting in lung cancer patients. In K. J. Ellis, S. Yasumura, W. D. Morgan (Eds.), *In Vivo Body Composition Studies*, pp. 60–69. London: Institute of Physical Sciences in Medicine.

Rahman, A., S. Jafry, K. Jeejeebhoy, A. D. Nagpal, B. Pisani, and R. Agarwala. 2016. Malnutrition and cachexia in heart failure. *JPEN J. Parenter. Enteral. Nutr.* 40 (4): 475–486.

Raj, D. S., T. Welbourne, E. A. Dominic, D. Waters, R. Wolfe, and A. Ferrando. 2005. Glutamine kinetics and protein turnover in end-stage renal disease. *Am. J. Physiol. Endocrinol. Metab.* 288 (1): E37–E46.

Ramos, E. J., S. Suzuki, D. Marks, A. Inui, A. Asakawa, and M. M. Meguid. 2004. Cancer anorexia-cachexia syndrome: Cytokines and neuropeptides. *Curr. Opin. Clin. Nutr. Metab. Care* 7 (4): 427–434.

Reale, M., C. Iarlori, A. Thomas, D. Gambi, B. Perfetti, M. Di Nicola, and M. Onofrj. 2009. Peripheral cytokines profile in Parkinson's disease. *Brain Behav. Immun.* 23 (1): 55–63.

Reeds, P. J., C. R. Fjeld, and F. Jahoor. 1994. Do the differences between the amino acid compositions of acute-phase and muscle proteins have a bearing on nitrogen loss in traumatic states? *J. Nutr.* 124 (6): 906–910.

Regensteiner, J. G., E. E. Wolfel, E. P. Brass, M. R. Carry, S. P. Ringel, M. E. Hargarten, E. R. Stamm, and W. R. Hiatt. 1993. Chronic changes in skeletal muscle histology and function in peripheral arterial disease. *Circulation* 87 (2): 413–421.

Rehm, K. E. and D. Konkle-Parker. 2016. Physical activity levels and perceived benefits and barriers to physical activity in HIV-infected women living in the deep south of the united states. *AIDS Care*: 1–6.

Rivas, D. A., D. J. McDonald, N. P. Rice, P. H. Haran, G. G. Dolnikowski, and R. A. Fielding. 2016. Diminished anabolic signaling response to insulin induced by intramuscular lipid accumulation is associated with inflammation in aging but not obesity. *Am. J. Physiol. Regul. Integr. Comp. Physiol.* 310 (7): R561–R569.

Roglic, Gojka. 2016. World Health Organization Global Report on Diabetes 2014. Accessed November 10, 2016, http://www.who.int/diabetes/global-report/en/.

Rolland, Y., G. Abellan van Kan, A. Benetos, H. Blain, M. Bonnefoy, P. Chassagne, C. Jeandel, et al. 2008. Frailty, osteoporosis and hip fracture: Causes, consequences and therapeutic perspectives. *J. Nutr. Health Aging* 12 (5): 335–346.

Romero-Gómez, M., R. Ramos-Guerrero, L. Grande, L. C. de Terán, R. Corpas, I. Camacho, and J. D. Bautista. 2004. Intestinal glutaminase activity is increased in liver cirrhosis and correlates with minimal hepatic encephalopathy. *J. Hepatol.* 41 (1): 49–54.

Roubenoff, R. and V. A. Hughes. 2000. Sarcopenia: Current concepts. *J. Gerontol. A Biol. Sci. Med. Sci.* 55 (12): M716–M724.

Rubin, H. 2003. Cancer cachexia: Its correlations and causes. *Proc. Natl. Acad. Sci. U.S.A.* 100 (9): 5384–5389.

Russ, D. W., K. Gregg-Cornell, M. J. Conaway, and B. C. Clark. 2012. Evolving concepts on the age-related changes in "muscle quality". *J. Cachexia, Sarcopenia and Muscle* 3 (2): 95–109.

Rutan, R. L. and D. N. Herndon. 1990. Growth delay in postburn pediatric patients. *Arch. Surg.* 125 (3): 392–395.

Ryan, A. S., A. Buscemi, L. Forrester, C. E. Hafer-Macko, and F. M. Ivey. 2011. Atrophy and intramuscular fat in specific muscles of the thigh: Associated weakness and hyperinsulinemia in stroke survivors. *Neurorehabil. Neural. Repair* 25 (9): 865–872.

Ryan, A. S., C. L. Dobrovolny, G. V. Smith, K. H. Silver, and R. F. Macko. 2002. Hemiparetic muscle atrophy and increased intramuscular fat in stroke patients. *Arch. Phys. Med. Rehabil.* 83 (12): 1703–1707.

Sacheck, J. M., J. P. Hyatt, A. Raffaello, R. T. Jagoe, R. R. Roy, V. R. Edgerton, S. H. Lecker, and A. L. Goldberg. 2007. Rapid disuse and denervation atrophy involve transcriptional changes similar to those of muscle wasting during systemic diseases. *FASEB J.* 21 (1): 140–155.

Saitoh, M., M. R. dos Santos, and S. von Haehling. 2016. Muscle wasting in heart failure. *Wien. Klin. Wochenschr.* 128 (7): 1–11.

Saitoh, M., M. R. dos Santos, N. Ebner, A. Emami, M. Konishi, J. Ishida, M. Valentova, A. Sandek, W. Doehner, and S. D. Anker. 2016. Nutritional status and its effects on muscle wasting in patients with chronic heart failure: Insights from studies investigating co-morbidities aggravating heart failure. *Wien. Klin. Wochenschr.* 128 (7): 1–8.

Sakkas, G. K., D. Ball, T. H. Mercer, A. J. Sargeant, K. Tolfrey, and P. F. Naish. 2003. Atrophy of non-locomotor muscle in patients with end-stage renal failure. *Nephrol. Dial. Transplant.* 18 (10): 2074–2081.

Sakkas, G. K., J. A. Kent-Braun, J. W. Doyle, T. Shubert, P. Gordon, and K. L. Johansen. 2006. Effect of diabetes mellitus on muscle size and strength in patients receiving dialysis therapy. *Am. J. Kidney Dis.* 47 (5): 862–869.

Sanchís, D., S. Busquets, B. Alvarez, D. Ricquier, F. J. López-Soriano, and J. M. Argilés. 1998. Skeletal muscle UCP2 and UCP3 gene expression in a rat cancer cachexia model. *FEBS Lett.* 436 (3): 415–418.

Sandberg, M. E., S. Wedren, L. Klareskog, I. E. Lundberg, C. H. Opava, L. Alfredsson, and S. Saevarsdottir. 2014. Patients with regular physical activity before onset of rheumatoid arthritis present with milder disease. *Ann. Rheum. Dis.* 73 (8): 1541–1544.

Sandek, A., J. Bauditz, A. Swidsinski, S. Buhner, J. Weber-Eibel, S. von Haehling, W. Schroedl, T. Karhausen, W. Doehner, and M. Rauchhaus. 2007. Altered intestinal function in patients with chronic heart failure. *J. Am. Coll. Cardiol.* 50 (16): 1561–1569.

Santos, E. L., K. de Picoli Souza, P. B. Guimarães, F. C. G. Reis, S. M. A. Silva, C. M. Costa-Neto, J. Luz, and J. B. Pesquero. 2008. Effect of angiotensin converting enzyme inhibitor enalapril on body weight and composition in young rats. *Int. Immunopharmacol.* 8 (2): 247–253.

Schellenbaum, G. D., N. L. Smith, S. R. Heckbert, T. Lumley, T. D. Rea, C. D. Furberg, M. F. Lyles, and B. M. Psaty. 2005. Weight loss, muscle strength, and Angiotensin-Converting enzyme inhibitors in older adults with congestive heart failure or hypertension. *J. Am. Geriatr. Soc.* 53 (11): 1996–2000.

Scherzer, R., S. B. Heymsfield, D. Lee, W. G. Powderly, P. C. Tien, P. Bacchetti, M. G. Shlipak, C. Grunfeld, and Study of Fat Redistribution and Metabolic Change in HIV Infection (FRAM). 2011. Decreased limb muscle and increased central adiposity are associated with 5-year all-cause mortality in HIV infection. *AIDS* 25 (11): 1405–1414.

Schilling, B. K., R. E. Karlage, M. S. LeDoux, R. F. Pfeiffer, L. W. Weiss, and M. J. Falvo. 2009. Impaired leg extensor strength in individuals with Parkinson disease and relatedness to functional mobility. *Parkinsonism Relat. Disord.* 15 (10): 776–780.

Schneider, J. A., J. Li, Y. Li, R. S. Wilson, J. H. Kordower, and D. A. Bennett. 2006. Substantia nigra tangles are related to gait impairment in older persons. *Ann. Neurol.* 59 (1): 166–173.

Schneider, S. M., R. Al-Jaouni, J. Filippi, J. B. Wiroth, G. Zeanandin, K. Arab, and X. Hebuterne. 2008. Sarcopenia is prevalent in patients with crohn's disease in clinical remission. *Inflamm. Bowel. Dis.* 14 (11): 1562–1568.

Schulze, P. C., A. Linke, N. Schoene, S. M. Winkler, V. Adams, S. Conradi, M. Busse, G. Schuler, and R. Hambrecht. 2004. Functional and morphological skeletal muscle abnormalities correlate with reduced electromyographic activity in chronic heart failure. *Eur. J. Cardiovasc. Prev. Rehabil.* 11 (2): 155–161.

Shachar, S. S., G. R. Williams, H. B. Muss, and T. F. Nishijima. 2016. Prognostic value of sarcopenia in adults with solid tumours: A meta-analysis and systematic review. *Eur. J. Cancer* 57: 58–67.

Shephard, R. J., H. Park, S. Park, and Y. Aoyagi. 2013. Objectively measured physical activity and progressive loss of lean tissue in older Japanese adults: Longitudinal data from the Nakanojo study. *J. Am. Geriatr. Soc.* 61 (11): 1887–1893.

Shimoyama, T., Y. Tamura, T. Sakamoto, and K. Inoue. 2009. Immune-mediated myositis in crohn's disease. *Muscle Nerve.* 39 (1): 101–105.

Shishikura, K., K. Tanimoto, S. Sakai, Y. Tanimoto, J. Terasaki, and T. Hanafusa. 2014. Association between skeletal muscle mass and insulin secretion in patients with type 2 diabetes mellitus. *Endocr. J.* 61 (3): 281–287.

Shrikrishna, D., R. Astin, P. R. Kemp, and N. S. Hopkinson. 2012. Renin-angiotensin system blockade: A novel therapeutic approach in chronic obstructive pulmonary disease. *Clin. Sci.(Lond)* 123 (8): 487–498.

Sietsema, K. E., A. Amato, S. G. Adler, and E. P. Brass. 2004. Exercise capacity as a predictor of survival among ambulatory patients with end-stage renal disease. *Kidney Int.* 65 (2): 719–724.

Signorelli, S. S., M. C. Mazzarino, D. A. Spandidos, and G. Malaponte. 2007. Proinflammatory circulating molecules in peripheral arterial disease (review). *Int. J. Mol. Med.* 20 (3): 279.

Signorelli, S. S., M. C. Mazzarino, L. Di Pino, G. Malaponte, C. Porto, G. Pennisi, G. Marchese, M. Pia Costa, D. Digrandi, and G. Celotta. 2003. High circulating levels of cytokines (IL-6 and TNFα), adhesion molecules (VCAM-1 and ICAM-1) and selectins in patients with peripheral arterial disease at rest and after a treadmill test. *Vascular Medicine* 8 (1): 15.

Sinclair, A. J. and L. Rodriguez-Mañas. 2016. Diabetes and frailty: Two converging conditions? *Can. J. Diabetes* 40 (1): 77–83.

Sinclair, M., P. J. Gow, M. Grossmann, and P. W. Angus. 2016. Review article: Sarcopenia in cirrhosis—Aetiology, implications and potential therapeutic interventions. *Aliment. Pharmacol. Ther.* 43 (7): 765–777.

Singh, V. K. and P. Guthikonda. 1997. Circulating cytokines in Alzheimer's disease. *J. Psychiatr. Res.* 31 (6): 657–660.

Slager, U. T., J. D. Hsu, and C. Jordan. 1985. Histochemical and morphometric changes in muscles of stroke patients. *Clin. Orthop.* 199: 159–168.

Smith, G. I., S. Julliand, D. N. Reeds, D. R. Sinacore, S. Klein, and B. Mittendorfer. 2015. Fish oil-derived n-3 PUFA therapy increases muscle mass and function in healthy older adults. *Am. J. Clin. Nutr.* 102 (1): 115–122.

Smolen, J. S., D. Aletaha, and I. B. McInnes. 2016. Rheumatoid arthritis. *Lancet* 388: 2023–2038.

Song, Y. H., Y. Li, J. Du, W. E. Mitch, N. Rosenthal, and P. Delafontaine. 2005. Muscle-specific expression of IGF-1 blocks angiotensin II-induced skeletal muscle wasting. *J. Clin. Invest.* 115 (2): 451–458.

Sood, N., W. L. Baker, and C. I. Coleman. 2008. Effect of glucomannan on plasma lipid and glucose concentrations, body weight, and blood pressure: Systematic review and meta-analysis. *Am. J. Clin. Nutr.* 88 (4): 1167–1175.

Souba, W. W., R. J. Smith, and D. W. Wilmore. 1985. Glutamine metabolism by the intestinal tract. *JPEN J. Parenter. Enteral. Nutr.* 9 (5): 608–617.

Springer, J., S. Schust, K. Peske et al. 2014. Catabolic signaling and muscle wasting after acute ischemic stroke in mice: Indication for a stroke-specific sarcopenia. *Stroke* 45 (12): 3675–3683.

Steves, C. J., S. Bird, F. M. Williams, and T. D. Spector. 2016. The microbiome and musculoskeletal conditions of aging: A review of evidence for impact and potential therapeutics. *J. Bone Miner. Res.* 31 (2): 261–269.

Strober, W., F. Zhang, A. Kitani, I. Fuss, and S. Fichtner-Feigl. 2010. Proinflammatory cytokines underlying the inflammation of Crohn's disease. *Curr. Opin. Gastroenterol.* 26 (4): 310–317.

Studenski, S. A., K. W. Peters, D. E. Alley et al. 2014. The FNIH sarcopenia project: Rationale, study description, conference recommendations, and final estimates. *J. Gerontol. A Biol. Sci. Med. Sci.* 69 (5): 547–558.

Subramaniam, K., K. Fallon, T. Ruut, D. Lane, R. McKay, B. Shadbolt, S. Ang, M. Cook, J. Platten, and P. Pavli. 2015. Infliximab reverses inflammatory muscle wasting (sarcopenia) in Crohn's disease. *Aliment. Pharmacol. Ther.* 41 (5): 419–428.

Sugawara, J., M. Miyachi, K. L. Moreau, F. A. Dinenno, C. A. DeSouza, and H. Tanaka. 2002. Age-related reductions in appendicular skeletal muscle mass: Association with habitual aerobic exercise status. *Clin. Physiol. Funct. Imaging* 22 (3): 169–172.

Suntrup, S., I. Teismann, J. Bejer, I. Suttrup, M. Winkels, D. Mehler, C. Pantev, R. Dziewas, and T. Warnecke. 2013. Evidence for adaptive cortical changes in swallowing in Parkinson's disease. *Brain* 136 (Pt 3): 726–738.

Sura, L., A. Madhavan, G. Carnaby, and M. A. Crary. 2012. Dysphagia in the elderly: Management and nutritional considerations. *Clin. Interv. Aging* 7 (287): 98.

Suva, D., I. Favre, R. Kraftsik, M. Esteban, A. Lobrinus, and J. Miklossy. 1999. Primary motor cortex involvement in Alzheimer disease. *J. Neuropathol. Exp. Neurol.* 58 (11): 1125–1134.

Symons, T. B., M. Sheffield-Moore, R. R. Wolfe, and D. Paddon-Jones. 2009. A moderate serving of high-quality protein maximally stimulates skeletal muscle protein synthesis in young and elderly subjects. *J. Am. Diet. Assoc.* 109 (9): 1582–1586.

Szabo, N., S. Lukacs, I. Kulcsar, W. Gunasekera, A. Nagy-Toldi, B. Dezso, and K. Danko. 2009. Association of idiopathic inflammatory myopathy and Crohn's disease. *Clin. Rheumatol.* 28 (1): 99–101.

Szulc, P., F. Munoz, F. Marchand, R. Chapurlat, and P. D. Delmas. 2010. Rapid loss of appendicular skeletal muscle mass is associated with higher all-cause mortality in older men: The prospective MINOS study. *Am. J. Clin. Nutr.* 91 (5): 1227–1236.

Tamura, F., T. Kikutani, T. Tohara, M. Yoshida, and K. Yaegaki. 2012. Tongue thickness relates to nutritional status in the elderly. *Dysphagia* 27 (4): 556–561.

Tessari, P. 2003. Protein metabolism in liver cirrhosis: From albumin to muscle myofibrils. *Curr. Opin. Clin. Nutr. Metab. Care* 6 (1): 79–85.

Tew, G. A., K. Jones, and A. Mikocka-Walus. 2016. Physical activity habits, limitations, and predictors in people with inflammatory bowel disease: A large cross-sectional online survey. *Inflamm. Bowel Dis.* 22 (12): 2933–2942.

Tisdale, M. J. 2003. Pathogenesis of cancer cachexia. *J. Support. Oncol.* 1 (3):159–168.

Tisdale, M. J. 2009. Mechanisms of cancer cachexia. *Physiol. Rev.* 89 (2): 381–410.

Tomlinson, B., D. Irving, and J. Rebeiz. 1973. Total numbers of limb motor neurones in the human lumbosacral cord and an analysis of the accuracy of various sampling procedures. *J. Neurol. Sci.* 20 (3): 313–327.

Toth, M. J., D. E. Matthews, R. P. Tracy, and M. J. Previs. 2005. Age-related differences in skeletal muscle protein synthesis: Relation to markers of immune activation. *Am. J. Physiol. Endocrinol. Metab.* 288 (5): E883–E891.

Toth, M. J., M. M. LeWinter, P. A. Ades, and D. E. Matthews. 2010. Impaired muscle protein anabolic response to insulin and amino acids in heart failure patients: Relationship with markers of immune activation. *Clin. Sci.(Lond)* 119 (11): 467–476.

Trappe, S., D. Williamson, and M. Godard. 2002. Maintenance of whole muscle strength and size following resistance training in older men. *J. Gerontol. A Biol. Sci. Med. Sci.* 57 (4): B138–B143.

Tsujinaka, T., J. Fujita, C. Ebisui et al. 1996. Interleukin 6 receptor antibody inhibits muscle atrophy and modulates proteolytic systems in interleukin 6 transgenic mice. *J. Clin. Invest.* 97 (1): 244–249.

Umegaki, H. 2016. Sarcopenia and frailty in older patients with diabetes mellitus. *Geriatr. Gerontol. Int.* 16 (3): 293–299.

Van Helvoort, H. A., Y. F. Heijdra, H. M. Thijs, J. Viña, G. J. Wanten, and P. R. Dekhuijzen. 2006. Exercise-induced systemic effects in muscle-wasted patients with COPD. *Med. Sci. Sports Exerc.* 38 (9): 1543.

van Nimwegen, M., A. D. Speelman, E. J. Hofman-van Rossum, S. Overeem, D. J. Deeg, G. F. Borm, van der Horst, Marleen HL, B. R. Bloem, and M. Munneke. 2011. Physical inactivity in Parkinson's disease. *J. Neurol.* 258 (12): 2214–2221.

van Zanten, J. J., P. C. Rouse, E. D. Hale, N. Ntoumanis, G. S. Metsios, J. L. Duda, and G. D. Kitas. 2015. Perceived barriers, facilitators and benefits for regular physical activity and exercise in patients with rheumatoid arthritis: A review of the literature. *Sports Med.* 45 (10): 1401–1412.

Vancampfort, D., J. Mugisha, M. De Hert, M. Probst, P. Firth, P. Gorczynski, and B. Stubbs. 2016. Global physical activity levels among people living with HIV: A systematic review and meta-analysis. *Disabil. Rehabil.*: 1–10.

Venables, M. C. and A. E. Jeukendrup. 2009. Physical inactivity and obesity: Links with insulin resistance and type 2 diabetes mellitus. *Diabetes. Metab. Res.* 25 (S1): S18–S23.

Verdijk, L. B., T. Snijders, M. Drost, T. Delhaas, F. Kadi, and L. J. van Loon. 2014. Satellite cells in human skeletal muscle; from birth to old age. *Age* 36 (2): 545–557.

Verhoeven, F., N. Tordi, C. Prati, C. Demougeot, F. Mougin, and D. Wendling. 2016. Physical activity in patients with rheumatoid arthritis. *Joint Bone Spine* 83 (3): 265–270.

Verhoeven, S., K. Vanschoonbeek, L. B. Verdijk, R. Koopman, W. K. Wodzig, P. Dendale, and L. J. van Loon. 2009. Long-term leucine supplementation does not increase muscle mass or strength in healthy elderly men. *Am. J. Clin. Nutr.* 89 (5): 1468–1475.

Vetta, F., S. Ronzoni, G. Taglieri, and M. R. Bollea. 1999. The impact of malnutrition on the quality of life in the elderly. *Clin. Nutr.* 18 (5): 259–267.

Victor, M. and A. Ropper. 2002. Degenerative diseases of the nervous system. *Adams and Victor's Principles of Neurology*, pp. 11–18. New York: Mc Graw-Hill Company.

Vidoni, E. D., G. P. Thomas, R. A. Honea, N. Loskutova, and J. M. Burns. 2012. Evidence of altered cortico-motor system connectivity in early-stage Alzheimer's disease. *J. Neurol. Phys. Ther.* 36 (1): 8–16.

Villareal, D. T., M. Banks, C. Siener, D. R. Sinacore, and S. Klein. 2004. Physical frailty and body composition in obese elderly men and women. *Obes. Res.* 12 (6): 913–920.

Visser, M., B. H. Goodpaster, S. B. Kritchevsky, A. B. Newman, M. Nevitt, S. M. Rubin, E. M. Simonsick, and T. B. Harris. 2005. Muscle mass, muscle strength, and muscle fat infiltration as predictors of incident mobility limitations in well-functioning older persons. *J. Gerontol. A Biol. Sci. Med. Sci.* 60 (3): 324–333.

Visser, M., T. B. Harris, J. Langlois, M. T. Hannan, R. Roubenoff, D. T. Felson, P. W. Wilson, and D. P. Kiel. 1998. Body fat and skeletal muscle mass in relation to physical disability in very old men and women of the Framingham Heart Study. *J. Gerontol. A Biol. Sci. Med. Sci.* 53 (3): M214–M221.

Visvanathan, R. and I. M. Chapman. 2009. Undernutrition and anorexia in the older person. *Gastroenterol. Clin. North Am.* 38 (3): 393–409.

Voigt, E., T. Griga, A. Tromm, M. Henschel, M. Vorgerd, and B. May. 1999. Polymyositis of the skeletal muscles as an extraintestinal complication in quiescent ulcerative colitis. *Int. J. Colorectal Dis.* 14 (6): 304–307.

Volpi, E., B. Mittendorfer, B. B. Rasmussen, and R. R. Wolfe. 2000. The response of muscle protein anabolism to combined hyperaminoacidemia and glucose-induced hyperinsulinemia is impaired in the elderly 1. *J. Clin. Endocrinol. Metab.* 85 (12): 4481–4490.

Volpi, E., B. Mittendorfer, S. E. Wolf, and R. R. Wolfe. 1999. Oral amino acids stimulate muscle protein anabolism in the elderly despite higher first-pass splanchnic extraction. *Am. J. Physiol.* 277 (3 Pt 1): E513–E520.

Volpi, E., M. Sheffield-Moore, B. B. Rasmussen, and R. R. Wolfe. 2001. Basal muscle amino acid kinetics and protein synthesis in healthy young and older men. *JAMA* 286 (10): 1206–1212.

Volpi, E., W. W. Campbell, J. T. Dwyer, M. A. Johnson, G. L. Jensen, J. E. Morley, and R. R. Wolfe. 2013. Is the optimal level of protein intake for older adults greater than the recommended dietary allowance? *J. Gerontol. A Biol. Sci. Med. Sci.* 68 (6): 677–681.

von Zglinicki, T., G. Saretzki, W. Döcke, and C. Lotze. 1995. Mild hyperoxia shortens telomeres and inhibits proliferation of fibroblasts: A model for senescence? *Exp. Cell Res.* 220 (1): 186–193.

Wagatsuma, A. and K. Sakuma. 2014. Vitamin D signaling in myogenesis: Potential for treatment of sarcopenia. *Biomed. Res. Int.* 2014: 121254.

Wakabayashi, H. and K. Sakuma. 2014. Comprehensive approach to sarcopenia treatment. *Current Clin. Pharmacol.* 9 (2): 171–180.

Warner, J. T. 2008. Body composition, exercise and energy expenditure in survivors of acute lymphoblastic leukaemia. *Pediatr. Blood Cancer* 50 (2 Suppl): 456–461; discussion 468.

Waters, D. L., W. M. Brooks, C. R. Qualls, and R. N. Baumgartner. 2003. Skeletal muscle mitochondrial function and lean body mass in healthy exercising elderly. *Mech. Ageing Dev.* 124 (3): 301–309.

Watters, J. M., S. M. Clancey, S. B. Moulton, K. M. Briere, and J. M. Zhu. 1993. Impaired recovery of strength in older patients after major abdominal surgery. *Ann. Surg.* 218 (3): 380–390; discussion 390–393.

Welle, S., C. Thornton, and M. Statt. 1995. Myofibrillar protein synthesis in young and old human subjects after three months of resistance training. *Am. J. Physiol.* 268 (3 Pt 1): E422–E427.

Welle, S., C. Thornton, M. Statt, and B. McHenry. 1994. Postprandial myofibrillar and whole body protein synthesis in young and old human subjects. *Am. J. Physiol.* 267 (4 Pt 1): E599–E604.

Welle, S., C. Thornton, R. Jozefowicz, and M. Statt. 1993. Myofibrillar protein synthesis in young and old men. *Am. J. Physiol.* 264 (5 Pt 1): E693–E698.

Wens, I., U. Dalgas, F. Vandenabeele, M. Krekels, L. Grevendonk, and B. O. Eijnde. 2014. Multiple sclerosis affects skeletal muscle characteristics. *PloS One* 9 (9): e108158.

White, S. L., D. W. Dunstan, K. R. Polkinghorne, R. C. Atkins, A. Cass, and S. J. Chadban. 2011. Physical inactivity and chronic kidney disease in Australian adults: The AusDiab study. *Nutr. Metab. Cardiovasc. Dis.* 21 (2): 104–112.

Wiedenmann, B., P. Malfertheiner, H. Friess et al. 2008. A multicenter, phase II study of infliximab plus gemcitabine in pancreatic cancer cachexia. *J. Support. Oncol.* 6 (1): 18–25.

Wingo, P. A., L. A. Ries, H. M. Rosenberg, D. S. Miller, and B. K. Edwards. 1998. Cancer incidence and mortality, 1973–1995: A report card for the U.S. *Cancer* 82 (6): 1197–1207.

Wust, R. C., R. T. Jaspers, A. F. van Heijst, M. T. Hopman, L. J. Hoofd, W. J. van der Laarse, and H. Degens. 2009. Region-specific adaptations in determinants of rat skeletal muscle oxygenation to chronic hypoxia. *Am. J. Physiol. Heart Circ. Physiol.* 297 (1): H364–H374.

Wysokiński, A., T. Sobów, I. Kłoszewska, and T. Kostka. 2015. Mechanisms of the anorexia of aging—A review. *Age* 37 (4): 1–14.

Xiao, D. Y., S. Luo, K. O'Brian et al. 2016. Longitudinal body composition changes in diffuse large B-cell lymphoma survivors: A retrospective cohort study of united states veterans. *J. Natl. Cancer Inst.* 108 (11). 10.1093/jnci/djw145. Print 2016 Nov.

Yan, X., J. Zhang, Q. Gong, and X. Weng. 2011. Appetite at high altitude: An fMRI study on the impact of prolonged high-altitude residence on gustatory neural processing. *Exp. Brain Res.* 209 (4): 495–499.

Yarasheski, K. E., J. J. Zachwieja, and D. M. Bier. 1993. Acute effects of resistance exercise on muscle protein synthesis rate in young and elderly men and women. *Am. J. Physiol.* 265 (2 Pt 1): E210–E214.

Yarasheski, K. E., J. Pak-Loduca, D. L. Hasten, K. A. Obert, M. B. Brown, and D. R. Sinacore. 1999. Resistance exercise training increases mixed muscle protein synthesis rate in frail women and men >/=76 yr old. *Am. J. Physiol.* 277 (1 Pt 1): E118–E125.

Zanetti, M., R. Barazzoni, and G. Guarnieri. 2008. Inflammation and insulin resistance in uremia. *J. Ren. Nutr.* 18 (1): 70–75.

Zanetti, M., S. S. Harris, and B. Dawson-Hughes. 2014. Ability of vitamin D to reduce inflammation in adults without acute illness. *Nutr. Rev.* 72 (2): 95–98.

Zhang, T., C. Ding, T. Xie et al. 2016. Skeletal muscle depletion correlates with disease activity in ulcerative colitis and is reversed after colectomy. *Clin. Nutr.* Vol. 36 (6): 1586–1592.

Zhang, T., L. Cao, T. Cao, J. Yang, J. Gong, W. Zhu, N. Li, and J. Li. 2015. Prevalence of sarcopenia and its impact on postoperative outcome in patients with Crohn's disease undergoing bowel resection. *JPEN J. Parenter. Enteral Nutr.* 41 (4): 592–600.

Zhou, J., B. Liu, C. Liang, Y. Li, and Y. Song. 2016. Cytokine signaling in skeletal muscle wasting. *Trends Endocrin. Met.* 27 (5): 335–347.

3 Sarcopenia in Physical Frailty

Maturin Tabue-Teguo, Emanuele Marzetti,
Riccardo Calvani, Bruno Vellas, and Matteo Cesari

CONTENTS

3.1   INTRODUCTION

A context that is overall featuring social and economic changes, improvement in technology, progress in medicine, and better public health care is favorable to a worldwide increase in the elderly population aged over 65 years (Christensen et al., 2009). Prevention in the case of disability proves essential as aging could lead to permanent incapacities (Morley, 2012). Medical care, in particular, and healthcare systems at large are more and more on the alert: they are requested to find the proper solutions and decide on the most efficient answers to the damaging effects of aging. Both sarcopenia (Janssen et al., 2000) and frailty (Clegg et al., 2013), which are quite frequent among the elderly, require specific attention (Cesari et al., 2014). Both are at times linked to health problems (Vetrano et al., 2014) that can possibly be reversible and dealt with via clinical practices (Vellas et al., 2013). Hence, this review will focus on the characterization of sarcopenia in the context of physical frailty (PF).

3.2   PF: A DEFINITION

PF is a syndrome relating to old age. It shows through diminished homeostatic reserves, thus increasing the risk of health problems ensuing, and then requiring hospitalization or even institutionalization. Frailty is a geriatric syndrome caused by the loss of reserve capacity in multiple physiological systems making older adults more vulnerable to adverse environment. Frailty has been associated with the risk of negative health-related outcomes including falls, hospitalizations, disability, institutionalization, and mortality (Clegg et al., 2013; Rodríguez-Mañas et al., 2013). Definitions for PF are several, depending on the syndrome that is examined and the risk profiles that are observed (Theou et al., 2013). With respect to aging, it concerns disability, comorbidity, or advanced old age (Corti et al., 1994; Ensrud et al., 2008; Fried et al., 2001; Klein et al., 2005; Rockwood et al., 1999; Theou et al., 2013; Winograd, 1991). Even if the concept is far reaching, it relies on the two key definitions: the frailty phenotype model and the frailty index (FI). Those are based on the clusters of components that are reliable as to characterize the elderly patient with negative health issues and mortality (García-González et al., 2009). The Frailty Tilburg indicator (Gobbens et al., 2010) will describe sociodemographic and economic features, education, family history, and psychological factors besides the physical condition of the patient (Gobbens et al., 2010). The Rockwood definition (Rockwood et al., 2005) shows all the effects of frailty on the physical, the psychological, the mental, and the cognitive. Fried et al. determine the patient's physical health difficulties through the evaluation of gait speed, grip strength, and sedentariness.

3.2.1 FRAILTY PREVALENCE

Reports on frailty prevalence are several in epidemiology research (Clegg et al., 2013). However, the results, which are mainly obtained through the frailty phenotype model, depend on many factors and may possibly show some variations, if not inconsistencies. This is due to compliance with specific criteria definition and characterization. The definition of frailty prevalence will evolve considering the resources available for each case study and the data inferred via strict observation in each environment (Ravindrarajah et al., 2013). If such changes occur, they do not alter the results of instrument measurements substantially, which would not be the case in using other tools (Cesari et al., 2014). Yet, some flexibility is required comparing the results, and comments on them may be necessary. The two following fundamentals will remain true: many among the elderly show high risk factors relating to negative health events, such as disability, and do not benefit from clinical care; frailty prevalence and incidence are altogether confirmed via our instruments, the targeted population, and the settings. The results show imperative necessity for health prevention.

3.2.2 FRAILTY MEASUREMENTS

Quite lately, progress has been made in instruments that could measure frailty with even more accuracy. Fried and colleagues have confirmed their hypotheses on frailty through essential clinical presentations; the operations were carried out via the frailty phenotype model and validated in the Cardiovascular Health Study (Fried et al., 2001). Frailty measurement is through three criteria out of the following ones: weight loss, exhaustion, weakness, slowness, and low energy.

Rockwood et al. have based their research on the Canadian Study of Health and Aging to develop and validate the frailty index (Rockwood et al., 2005). Recently, several other instruments, mostly based on those two models, have been recommended for frailty measuring. It is to be noticed that those instruments show a valuable prediction for negative outcomes (Clegg et al., 2013), even if their results are not wholly significant. Some analyses have been led by Magali Gonzales and her colleagues (Gonzalez-Colaço Harmand et al., 2017) in the 3C-Bordeaux study. The three definitions of frailty via those three different defining tools—Frailty Phenotype, Frailty Index, and Tilburg Frailty Indicator—are compared with their predictive value. The results show the fact that frailty consequences appear different depending on the specific measurement criteria; as each tool used for its specificities will measure a targeted population, definitions of frailty do not fully coincide. For example, Rockwood's FI is a statistically significant predictor of falls, incident disability, hospitalization, and mortality. If the nature of the syndrome reflects in the measurement, the operational definition chosen to implement it should be fully taken into account. Thus, the following criteria chosen in order to select the most convenient measuring instrument for frailty should be validated: the outcome, the definition, the tool, the targeted population, and the settings. To opt for one specific instrument at the expense of the others will automatically penalize the possibilities available. The more recent approach is in favor of focusing on the frailty syndrome specificities within the elderly population with physical disability, that is in direct connection to geriatric medicine (Studenski et al., 2003). Then, all will be centered on frailty epidemiology: its scope, area, and monitoring means. As a whole, epidemiological prognostic will take the following criteria into account: description, prediction of health events, causes and consequences, and remediation. Most scientists agree to the fact that loss of physical ability is the major criteria for the frailty syndrome (Ferrucci et al., 2004). Physical performance is measured to determine the level of frailty of the elderly, mainly gait speed and Short Physical Performance Battery (SPPB) (Guralnik et al., 1995). Today, the predictive value of physical performance measures for negative event outcomes in the older population is actually recognized (Gonzalez-Colaço Harmand et al., 2017; Studenski et al., 2011). The "physical parameters" criteria for frailty assessment will support the clinical response to the syndrome more easily: a more economical, objective, reliable process, and a rapid detection carried out among hospital settings and institutions, whether locally or worldwide, and of a real interest in research at an international level. An objective test of physical performance may be part of the surveys that are necessary among the aging population to help solve possible social, cultural, clinical, and environmental issues.

3.2.3 THE SHORT PHYSICAL PERFORMANCE BATTERY

The SPPB is the process for assessing physical performance. It was developed by Guralnik and colleagues to measure the lower-extremity function via three subtests: balance, 4-m walking test, and chair stand tests (Guralnik et al., 1995). That is, with feet side by side, semi tandem, and tandem balance positions for 10 s each, then, walk a distance of 4 m at usual pace. The ratio between distance and time (i.e., the gait speed) is measured in meter per second (m/s); finally, rise from a chair and return to the seated position five times, as quickly as possible, while keeping arms folded, each movement scored 0–4 points out of a total score from 0 to 12; higher scores show better physical performance (Guralnik et al., 1995).

3.2.4 USUAL GAIT SPEED

As part of the SPPB, the usual gait speed can be used as a fundamental parameter for measuring the increased vulnerability of the older patient (Abellan van Kan et al., 2009). Guralnik and colleagues have shown its predictive value capacity in major health-related outcomes such as disability and mortality. It can prove sufficient as to show the complete SPPB (Guralnik et al., 1995). It can accurately predict hospitalization, institutionalization, disability, and mortality in the elderly (Cesari et al., 2009, 2005; Guralnik et al., 1995). Studenski and his colleagues (2011) show that we can infer with certainty life expectancy of older people by knowing their age, gender, and gait speed, which correspond to a "vital sign" for elders (Cesari, 2011; Goodwin, 2012; Studenski et al., 2003). Yet, if it is wholly functional within the frailty area considered as a heterogeneous syndrome, a number of specific parameters lack to decide on a functional preventive action or a medical intervention. Those physical performance measures constitute a starting point for the identification of people at higher risk. At the origin, SPPB and the gait speed test were implemented to measure frailty. Now, they prove adequate to detect risks for older individuals, their vulnerability to stressors, and to negative health-related events, including disability. The concepts that define frailty can apply to physical performance measures. Besides geriatrics, some other medical specialties take the role of physical performance measures into account in the field of frailty, such as cardio surgery (Afilalo et al., 2010; Lilamand et al., 2014), cardiology (Odden et al., 2012), and respiratory medicine (Kon et al., 2013).

3.3 SARCOPENIA: A DEFINITION

Sarcopenia is a decline related to age in the skeletal muscle mass with a decrease in strength and/or function, as described by Irwin Rosenberg (1997). Sarcopenia concept and study appear more and more frequently in clinical practice and research. It is part of geriatric medicine and also many more medical specialties (Cruz-Jentoft et al., 2010; Cruz-Jentoft and Landi, 2014). Yet, the concept is not wholly operational, and the healthcare induces huge personal and social security costs.

Muscle strength and physical performance are related to muscle mass, but the latter cannot wholly account for disability or other issues (Guralnik et al., 1995). Today, a few definitions of sarcopenia the scientists agree to comprise measurements of the link between the loss of skeletal muscle mass and the loss of strength in advanced age (Cruz-Jentoft et al., 2010).

A clinical definition of sarcopenia has been given by the European Working Group on Sarcopenia in Older People (EWGSOP): "a syndrome characterized by progressive and generalized loss of skeletal muscle mass and strength (Goodpaster et al., 2006) with a risk of adverse outcomes such as physical disability, poor quality of life and death." Its identification, as recommended by the EWGSOP, must be based on the correlation between low muscle mass and low muscle function, either both in strength or in performance, because their physical decline does not coincide (Cruz-Jentoft et al., 2019). The code attributed to sarcopenia in the 10th revision of the *International Statistical Classification of Diseases and Related Health Problems* (ICD-10) was validated on October 1, 2016. The classification provides specific standards as to screening, diagnosing, and monitoring the decline in muscle strength that is greater than expected through the decrease in mass.

The term "dynapenia" is better to describe the loss of muscle strength and function in aging (Clark and Manini, 2012).

3.3.1 SARCOPENIA RISK FACTORS

Sarcopenia will show signs of reduced muscle activity, chronic inflammation, and metabolic derangements and malnutrition. The risk factors can either be personal and social (lifestyle) or be physical (hormonal, inflammation, chronic health problems). This is what Rosenberg says about this condition: "There is probably no decline in structure and function more dramatic than the decline in lean body mass or muscle mass over the decades of life" (Rosenberg, 1997). Aging shows through the loss in muscle and skeleton mass, which induces severe health condition. Both endogenous and exogenous occurrences in the aging process will affect the muscle quality and quantity.

3.3.2 SARCOPENIA PREVALENCE

EWGSOP criteria (Cruz-Jentoft et al., 2010) show the following figures: 1%–29% for sarcopenia prevalence, 14%–33% in the case of long-term hospitalization, 10% in acute cases. The review was carried out systematically, and it was proved a number of disparities liable to age differences and geographic areas (Cruz-Jentoft and Landi, 2014).

3.3.3 SARCOPENIA MEASURE

To measure muscle or body mass, modern or more classic methods are available: (1) Costly equipment operated by specialized staff proves reliable in their methods for measuring the body composition (Janssen et al., 2000) such as magnetic resonance imaging (MRI), computed tomography (CT), dual energy X-ray absorptiometry (DXA), and ultrasonography; (2) bioelectric impedance analysis (BIA), electrical impedance myography, neutron activation, and anthropometry (parameters for calf or mid-arm muscle circumference) are not wholly reliable for diagnosis; and (3) biochemical markers such as potassium, serum and urinary creatinine, and deuterated creatinine dilution method (Marzetti, 2012) are not sufficiently standardized for diagnosis.

3.4 FRAILTY AND SARCOPENIA: CONCEPTS

Both frailty and sarcopenia refer to the aging process. The two concepts have to be clearly defined as to determine the best prevention against physical dependence. The role of geriatricians is to identify the early stages of disability. The definitions of both concepts have to be stated separately.

Sarcopenia is an element of frailty as it comprises low physical performance, but both sarcopenia and frailty interact in practice and have been studied side by side over the last two decades. The difference is often marked as such: sarcopenia is part of research in the basic science of organs, while frailty applies to diagnosis in clinical settings (Bauer and Sieber, 2008). Yet, links are numerous, such as the malnutrition part in the pathogenesis of both frailty and sarcopenia inducing nutritional interventions. Sarcopenia is the biological basis of the physical function impairment that characterizes PF. The tools used for assessment are the following ones: muscle strength and function as a prerequisite for identifying PF and sarcopenia and also handgrip strength test (Alley et al., 2014), SPPB (Guralnik et al., 1994) in balance, walking speed and strength (SPPB), gait speed (Studenski et al., 2011), and lower extremity muscle power. All measurements are easily implemented in research and clinical settings. There is no point in debating whether frailty is due to sarcopenia, or sarcopenia is a clinical manifestation of frailty as both refer to physical function deficiency. Aging disability is the key problem requiring medical solutions. Both sarcopenia and frailty can be viewed as the consequences of a disruption of the organism with extremely limited chances of reversibility. Sarcopenia will feature signs of cachexia (Rolland et al., 2003) when frailty shows disabling condition (Fried et al., 2004). A combination of both will also be different in case of the presence or not of disability. The treatment of disability implies additional care because of the major risks incurred, like some more severe functional losses and incapacities. In the preclinical phase of the illness, action can be taken against disability at a secondary prevention level. However, it proves impossible when sarcopenia, frailty, and disability are entangled and interact negatively, complicating the condition exponentially with higher risks for the patient's health. If disability is absent—thus eliminating it from the condition—it becomes easier to implement medical care. As such, a main clinical care procedure will aim at preventing disability linked to incident (Subra et al., 2012). The fact that the elderly may present characteristics of both sarcopenia and frailty makes it difficult to connect theory and practice. However, here is the challenge in medical research.

Several consensus papers have provided recommendations on how to identify patients with sarcopenia (Cruz-Jentoft et al., 2010). For example, articles have been published by the Foundation for the National Institutes of Health (FNIH) (Alley et al., 2014; Cawthon et al., 2014; Dam et al., 2014; McLean et al., 2014). A major contribution has been made about a clearer distinction between the effect of skeletal muscle mass and function in the development of negative effects on health, like in the case of mobility disability. Even if scientists have diverging opinions about the frailty concept, there seems to be a community of view, nonetheless. An international board of experts (Morley, 2012) recently reported about the topic, and although they were from different schools and opinions, they agreed on taking a step forward in the study, both clinically and through research. Too many ways for assessing and measuring sarcopenia and frailty would slow down the interest of the scientific community and the possible initiatives of the public health authorities. It would compromise research, further developments, and clinical breakthroughs.

An accurate measuring tool is required in our field when assessing and replicating the conditions measured. If measurement is limited and arbitrary, it proves difficult to give an account of a health condition, especially when some of the pathophysiological features are not well documented, and/or when important details remain unavailable. This is what geriatrics and gerontology specialists may experiment at times. For instance, the effects of aging may blur fundamental data at clinical and subclinical levels (Cesari et al., 2014). Yet, those limits—or "evidence-based" issues in geriatric medicine (Straus and McAlister, 2000)—admit a fundamental basis for both sarcopenia and frailty as being connected to the physical functioning of muscles, coordination, and balance, which are altogether part of mobility. This is part and parcel of every living being. Then, animal models may be of some help on intervening against disability, specifically providing important vital information at a prior stage. Lack of mobility is a fundamental feature showing aging condition, and it is a major source of negative events in life. Physical functioning and skeletal muscle production can be measured in an objective way (Studenski et al., 2003). The measurements will display the risks at stake and help anticipate problematic issues. Considering sarcopenia and frailty as such, independently from their relation to skeletal muscle, for example, can be a way to include them under a concept both standardized and clinically relevant. Therefore, implementation in clinical and research settings can be viewed through a number of recent scientific developments in literatures about physical deficiency. Moreover, instruments are available, such as SPPB (Guralnik et al., 1994), usual gait speed (Studenski et al., 2011), and handgrip strength (Rantanen et al., 1999). Reduced muscle mass and impaired muscle performance that characterize sarcopenia prove a risk factor of physical limitation, chronic diseases, and clinical frailty in the elderly population.

Quality of life (QoL) is measured through generic instruments such as the SF-36. These measures are specific to elderly decline. Yet, there should be a better concept of disability and a disease-specific QoL instrument for sarcopenia/frailty. This could be part of a range of future more appropriate measurements, including psychometric properties. The report of an expert meeting (Reginster et al., 2016) clearly shows sarcopenia and frailty geriatric syndromes and recommends drug treatment under clinical tests against sarcopenia.

If frailty and sarcopenia are linked, they are also distinct. Both are related to the aging of the muscle and the skeleton. The causes are multiple: changes in body composition, inflammation, and hormonal imbalance. Fashionable therapies are inventive, suggesting ways of delaying the loss of muscle mass. Medically, it is fundamental and a priority to work at a consensus about an accurate definition of sarcopenia and PF, taking into account the loss of muscle mass, the decrease in muscle strength, and physical activity. The definition has to be valid and clinically operational. It may then induce an opportunity for actual clinical tests and treatments. Improvement in measuring muscle strength, physical activity, and clinical outcomes is imperative to initiate a better implementation in medical interventions and an actual improvement in the clinical issues for frail older patients (Cooper et al., 2012).

To better understand physiopathology and promote therapeutic strategies, the "Sarcopenia and physical frailty in older people: multi-component treatment strategies" project (a grant of European Union) has operationalized a specific condition, named PF and sarcopenia. It is a randomized controlled trial will be conducted to evaluate the efficacy of a multicomponent intervention for preventing mobility disability and other adverse health outcomes in older adults with PF and sarcopenia (Marzetti et al., 2017).

3.5 CONCLUSION

To conclude, sarcopenia and frailty should be assessed according to a new definition of their conceptual and clinical realities. Impairment of the physical function in the absence of disability is definitively the link between both conditions. Thus, we can first opt for targeted clinical interventions against disability; then, make both conditions not just concepts but part of our clinical implementations; finally, make both health conditions depend on an objective, clinically assessed and relevant standard that would be definitely part and parcel of public health protocols. To reach this goal, it is necessary to include sarcopenia and frailty into a unique and wholly effective operational concept.

REFERENCES

Abellan van Kan, G., Rolland, Y., Andrieu, et al. 2009. Gait speed at usual pace as a predictor of adverse outcomes in community-dwelling older people an international academy on nutrition and aging (IANA) task force. *J. Nutr. Health Aging* 13, 881–889.

Afilalo, J., Eisenberg, M.J., Morin, J.-F., et al. 2010. Gait speed as an incremental predictor of mortality and major morbidity in elderly patients undergoing cardiac surgery. *J. Am. Coll. Cardiol.* 56, 1668–1676.

Alley, D.E., Shardell, M.D., Peters, K.W., et al. 2014. Grip strength cutpoints for the identification of clinically relevant weakness. *J. Gerontol. A. Biol. Sci. Med. Sci.* 69, 559–566.

Bauer, J.M., Sieber, C.C., 2008. Sarcopenia and frailty: A clinician's controversial point of view. *Exp. Gerontol.* 43, 674–678.

Cawthon, P.M., Peters, K.W., Shardell, M.D., et al. 2014. Cutpoints for low appendicular lean mass that identify older adults with clinically significant weakness. *J. Gerontol. A. Biol. Sci. Med. Sci.* 69, 567–575.

Cesari, M., 2011. Role of gait speed in the assessment of older patients. *JAMA* 305, 93–94.

Cesari, M., Kritchevsky, S.B., Newman, A.B., et al. 2009, Added value of physical performance measures in predicting adverse health-related events: Results from the health, aging and body composition study. *J. Am. Geriatr. Soc.* 57, 251–259.

Cesari, M., Kritchevsky, S.B., Penninx, B.W.H.J., et al. 2005. Prognostic value of usual gait speed in well-functioning older people--results from the health, aging and body composition study. *J. Am. Geriatr. Soc.* 53, 1675–1680.

Cesari, M., Landi, F., Vellas, B., Bernabei, R., Marzetti, E. 2014. Sarcopenia and physical frailty: Two sides of the same coin. *Front. Aging Neurosci.* 6, 192.

Christensen, K., Doblhammer, G., Rau, R., Vaupel, J.W. 2009. Ageing populations: The challenges ahead. *Lancet Lond. Engl.* 374, 1196–1208.

Clark, B.C., Manini, T.M., 2012. What is dynapenia? *Nutr.* 28, 495–503.

Clegg, A., Young, J., Iliffe, S., Rikkert, M.O., Rockwood, K., 2013. Frailty in elderly people. *Lancet Lond. Engl.* 381, 752–762.

Cooper, C., Dere, W., Evans, W., et al. 2012. Frailty and sarcopenia: Definitions and outcome parameters. *Osteoporos. Int. J. Establ. Result Coop. Eur. Found. Osteoporos. Natl. Osteoporos. Found.* 23, 1839–1848.

Corti, M.C., Guralnik, J.M., Salive, M.E., Sorkin, J.D., 1994. Serum albumin level and physical disability as predictors of mortality in older persons. *JAMA* 272, 1036–1042.

Cruz-Jentoft, A.J., Bahat G., Bauer J., Bruyère O., 2019. Sarcopenia: Revised European consensus on definition and diagnosis. *Age Ageing.* 2019 1;48(1):16–31.

Cruz-Jentoft, A.J., Landi, F., 2014. Sarcopenia. *Clin. Med. Lond. Engl.* 14, 183–186.

Cruz-Jentoft, A.J., Landi, F., Topinková, E., Michel, J.-P., 2010. Understanding sarcopenia as a geriatric syndrome. *Curr. Opin. Clin. Nutr. Metab. Care* 13, 1–7.

Dam, T.-T., Peters, K.W., Fragala, M., et al. 2014. An evidence-based comparison of operational criteria for the presence of sarcopenia. *J. Gerontol. A. Biol. Sci. Med. Sci.* 69, 584–590.

Ensrud, K.E., Ewing, S.K., Taylor, B.C. et al. 2008. Comparison of 2 frailty indexes for prediction of falls, disability, fractures, and death in older women. *Arch. Intern. Med.* 168, 382–389.

Ferrucci, L., Guralnik, J.M., Studenski, S., Fried, L.P. et al. 2004. Interventions on frailty working group, designing randomized, controlled trials aimed at preventing or delaying functional decline and disability in frail, older persons: A consensus report. *J. Am. Geriatr. Soc.* 52, 625–634.

Fried, L.P., Ferrucci, L., Darer, J., Williamson, J.D., Anderson, G., 2004. Untangling the concepts of disability, frailty, and comorbidity: Implications for improved targeting and care. *J. Gerontol. A. Biol. Sci. Med. Sci.* 59, 255–263.

Fried, L.P., Tangen, C.M., Walston, J., Newman, A.B. et al. 2001. Cardiovascular health study collaborative research group. Frailty in older adults: Evidence for a phenotype. *J. Gerontol. A. Biol. Sci. Med. Sci.* 56, M146–M156.

García-González, J.J., García-Peña, C., Franco-Marina, F., Gutiérrez-Robledo, L.M., 2009. A frailty index to predict the mortality risk in a population of senior Mexican adults. *BMC Geriatr.* 9, 47.

Gobbens, R.J.J., van Assen, M.A.L.M., Luijkx, K.G. et al. 2010. The Tilburg Frailty Indicator: Psychometric properties. *J. Am. Med. Dir. Assoc.* 11, 344–355.

Gonzalez-Colaço Harmand, M., Meillon, C., Bergua, V. et al. 2017. Comparing the predictive value of three definitions of frailty: Results from the Three-City study. *Arch. Gerontol. Geriatr.* 72, 153–163.

Goodpaster, B.H., Park, S.W., Harris, T.B., et al. 2006. The loss of skeletal muscle strength, mass, and quality in older adults: The health, aging and body composition study. *J. Gerontol. A. Biol. Sci. Med. Sci.* 61, 1059–1064.

Goodwin, J.S., 2012. Gait speed: Comment on "rethinking the association of high blood pressure with mortality in elderly adults." *Arch. Intern. Med.* 172, 1168–1169.

Guralnik, J.M., Ferrucci, L., Simonsick, E.M., Salive, M.E., Wallace, R.B., 1995. Lower-extremity function in persons over the age of 70 years as a predictor of subsequent disability. *N. Engl. J. Med.* 332, 556–561.

Guralnik, J.M., Simonsick, E.M., Ferrucci, L., et al. 1994. A short physical performance battery assessing lower extremity function: Association with self-reported disability and prediction of mortality and nursing home admission. *J. Gerontol.* 49, M85–M94.

Janssen, I., Heymsfield, S.B., Baumgartner, R.N., Ross, R., 2000. Estimation of skeletal muscle mass by bioelectrical impedance analysis. *J. Appl. Physiol. Bethesda Md* 1985 89, 465–471.

Klein, B.E.K., Klein, R., Knudtson, M.D., Lee, K.E., 2005. Frailty, morbidity and survival. *Arch. Gerontol. Geriatr.* 41, 141–149.

Kon, S.S.C., Patel, M.S., Canavan, J.L. et al. 2013. Reliability and validity of 4-metre gait speed in COPD. *Eur. Respir. J.* 42, 333–340.

Lilamand, M., Dumonteil, N., Nourhashémi, F., et al. 2014. Gait speed and comprehensive geriatric assessment: Two keys to improve the management of older persons with aortic stenosis. *Int. J. Cardiol.* 173, 580–582.

Marzetti, E., 2012. Editorial: Imaging, functional and biological markers for sarcopenia: The pursuit of the golden ratio. *J. Frailty Aging* 1, 97–98.

Marzetti, E., Calvani R., Tosato M., Cesari M., Di Bari M., et al. 2017. Sarcopenia: An overview. *Aging Clin. Exp. Res.* 29 (1): 11–17.

McLean, R.R., Shardell, M.D., Alley, D.E., et al. 2014. Criteria for clinically relevant weakness and low lean mass and their longitudinal association with incident mobility impairment and mortality: The foundation for the National Institutes of Health (FNIH) sarcopenia project. *J. Gerontol. A. Biol. Sci. Med. Sci.* 69, 576–583.

Morley, J.E., 2012. Aging in place. *J. Am. Med. Dir. Assoc.* 13, 489–492.

Odden, M.C., Peralta, C.A., Haan, M.N., Covinsky, K.E., 2012. Rethinking the association of high blood pressure with mortality in elderly adults: The impact of frailty. *Arch. Intern. Med.* 172, 1162–1168.

Rantanen, T., Guralnik, J.M., Ferrucci, L., Leveille, S., Fried, L.P., 1999. Coimpairments: Strength and balance as predictors of severe walking disability. *J. Gerontol. A. Biol. Sci. Med. Sci.* 54, M172–M176.

Ravindrarajah, R., Lee, D.M., Pye, S.R., et al. 2013. The ability of three different models of frailty to predict all-cause mortality: Results from the European Male Aging Study (EMAS). *Arch. Gerontol. Geriatr.* 57, 360–368.

Reginster, J.-Y., Cooper, C., Rizzoli, R., et al. 2016. Recommendations for the conduct of clinical trials for drugs to treat or prevent sarcopenia. *Aging Clin. Exp. Res.* 28, 47–58.

Rockwood, K., Song, X., MacKnight, C., et al. 2005. A global clinical measure of fitness and frailty in elderly people. *CMAJ Can. Med. Assoc. J. J. Assoc. Medicale Can.* 173, 489–495.

Rockwood, K., Stadnyk, K., MacKnight, C., et al. 1999. A brief clinical instrument to classify frailty in elderly people. *Lancet Lond. Engl.* 353, 205–206.

Rodríguez-Mañas, L., Féart, C., Mann, G., et al. 2013. Searching for an operational definition of frailty: A Delphi method based consensus statement: The frailty operative definition-consensus conference project. *J. Gerontol. A. Biol. Sci. Med. Sci.* 68, 62–67.

Rolland, Y., Lauwers-Cances, V., Cournot, M., et al. 2003. Sarcopenia, calf circumference, and physical function of elderly women: A cross-sectional study. *J. Am. Geriatr. Soc.* 51, 1120–1124.

Rosenberg, I.H., 1997. Sarcopenia: Origins and clinical relevance. *J. Nutr.* 127, 990S–991S.

Straus, S.E., McAlister, F.A., 2000. Evidence-based medicine: A commentary on common criticisms. *CMAJ Can. Med. Assoc. J. J. Assoc. Medicale Can.* 163, 837–841.

Studenski, S., Perera, S., Patel, K., et al. 2011. Gait speed and survival in older adults. *JAMA* 305, 50–58.

Studenski, S., Perera, S., Wallace, D., et al. 2003. Physical performance measures in the clinical setting. *J. Am. Geriatr. Soc.* 51, 314–322.

Subra, J., Gillette-Guyonnet, S., Cesari, M., Oustric, S., Vellas, B., Platform Team, 2012. The integration of frailty into clinical practice: Preliminary results from the Gérontopôle. *J. Nutr. Health Aging* 16, 714–720.

Theou, O., Brothers, T.D., Mitnitski, A., Rockwood, K., 2013. Operationalization of frailty using eight commonly used scales and comparison of their ability to predict all-cause mortality. *J. Am. Geriatr. Soc.* 61, 1537–1551.

Vellas, B., Balardy, L., Gillette-Guyonnet, S., et al. 2013. Looking for frailty in community-dwelling older persons: The Gérontopôle Frailty Screening Tool (GFST). *J. Nutr. Health Aging* 17, 629–631.

Vetrano, D.L., Landi, F., Volpato, S., et al. 2014. Association of sarcopenia with short- and long-term mortality in older adults admitted to acute care wards: Results from the CRIME study. *J. Gerontol. A. Biol. Sci. Med. Sci.* 69, 1154–1161.

Winograd, C.H., 1991. Targeting strategies: An overview of criteria and outcomes. *J. Am. Geriatr. Soc.* 39, 25S–35S.

Section III

Molecular and Cellular Aspects
of Sarcopenia

4 The Role of Imaging Techniques in the Diagnosis of Sarcopenia

Thiago Gonzalez Barbosa-Silva and Carla M.M. Prado

CONTENTS

4.1 INTRODUCTION

The composition of the human body can be described in levels of complexity ranging from atoms to molecules, to cells, and, finally, to tissues. At the atomic level, elements are the building blocks that form the whole body, and they are mainly represented by oxygen, carbon, hydrogen, nitrogen, calcium, and phosphorus, representing 98% of the atoms in the body (Wang et al. 1992). At the molecular level, body composition can be divided into six major compartments: water, protein, lipid, carbohydrates, bone minerals, and soft tissue minerals (Shen et al. 2005). The cellular level of body composition is composed of cells, extracellular fluid, and extracellular solids. Adipose tissue, skeletal muscle, bone, visceral organs, brain, and heart form the tissue-organ level of body composition. Adipose tissue, formed by adipocytes, collagenous and elastic fibers, fibroblasts, and capillaries can be further subdivided into subcutaneous, visceral, interstitial, and yellow marrow (Wang et al. 1992). The five-level organizational model of body composition (i.e., atoms, molecules, cells, tissues, and whole body) provides the framework for understanding the different methodologies available to assess abnormalities in body composition (Wang et al. 1992) including sarcopenia.

Semantically, the word "sarcopenia" means depletion of flesh (muscle). As such, an essential aspect of the sarcopenia diagnosis is the assessment of muscle mass or the compartments including it, such as lean soft tissue and fat-free mass. Lean soft tissue is the compartment of the body that includes intra and extracellular water, total body protein (including skeletal muscle), carbohydrates, and soft tissue minerals (Wang et al. 1992). Fat-free mass includes lean soft tissue plus bone minerals.

Establishing the prevalence of sarcopenia is limited due to the lack of standardized methods to assess/define this condition and the lack of a standard reference population to differentiate between normal and abnormal levels of muscle mass (Melton et al. 2000). As such, a variety of cutoffs derived from different populations have been published utilizing several different techniques (Dam et al. 2014; Barbosa-Silva et al. 2016; Gonzalez et al. 2018).

Sarcopenia can also occur in the context of obesity (sarcopenic obesity), an emerging health problem due to the combined impact of both conditions. As no standard definitions are available, defining sarcopenic obesity is as limiting as defining either sarcopenia or obesity alone. The cutoffs used are somewhat arbitrary and may also ignore important aspects such as muscle quality and intramuscular fat (Zamboni et al. 2008), which have important clinical implications (Brown et al. 2018; Xiao et al. 2018). Recent studies have shown discrepant prevalence ranging from less than 5% to 100%, depending on the criterion used (Batsis et al. 2013; Johnson Stoklossa et al. 2017). This wide variability precludes a full understanding of the prevalence and impact of this condition across the continuum of care, as discussed extensively in other chapters of this book. Moreover, although this chapter focuses on the measurement of body composition as absolute/relative amounts of muscle mass, measurements of strength, and performance are also an important component of sarcopenia diagnosis and discussed elsewhere in this book.

4.2 METHODS

When choosing a body composition assessment method, certain aspects should be considered. Each method has its own accuracy, some being more precise and reliable than others. Also, one should evaluate the cost of the exam to be performed. The method's availability is something else to be considered, as some devices are "omnipresent" in clinical settings, while a few are restricted to reference hospitals or research centers (as they demand a large physical space in the institutions, require trained personnel, and are more expensive). Furthermore, some methods involve considerable doses of radiation, which limits the frequency of examinations to be performed and limit their indication for body composition purposes as "convenience exams"—in other words, exams performed for the specific clinical purposes other than body composition but allow the examiner to analyze the images retrospectively. At last, the amount of training to perform and interpret the exams should also be taken into account, as some methods require specific skills and practice time, while others provide the desired information in an automated manner.

4.2.1 Ultrasound

Ultrasound (US) is a method based on the conversion of electrical energy into sound waves, by the passage through a piezoelectric crystal in the US probe. The sound waves are emitted by the probe in the direction of the tissue to be evaluated. Some of them are absorbed by the tissue and some are reflected and captured by the probe. The device calculates the difference between the emitted and received waves and is able to convert this into images.

The most conventional type used for the body composition purpose is the B-mode US (where "B" stands for "bright"; Figure 4.1). The two-dimensional image generated by the device is represented in a grayscale, which is related to the ability of the analyzed tissues to reflect or absorb the sound waves emitted by the probe. In that manner, darker images on the US screen represent more absorbent surfaces, while whiter images stand for reflective tissues. This absorptive/reflective tissue's properties allow real-time visual differentiation (i.e., bone, muscle, and fat) between different body layers, considering the contrast generated between the sound wave visual representation while it travels through different conductive means.

In the US screen, bone usually will appear as white surface over a black background, as a result of the reflection of the sound waves in the bone surface and the "absence" of sound waves traveling further. Considering bone as the "most reflective" structure in the evaluated sites, fat should appear as a "less white" tissue, but still whiter than muscle—in normal conditions. Another important matter is taking into consideration the anatomical distribution of the tissue layers in each body region—as muscle fascia may appear "whiter" than regular muscle, considering its composition and, therefore, unique reflective properties.

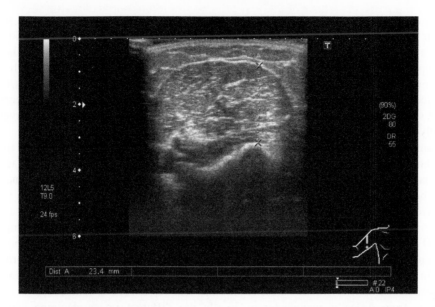

FIGURE 4.1 Right anterior arm's transversal image of a 71-year-old man by a US B-mode device. The muscle layer's thickness is indicated by the blue dotted line.

In the sarcopenia assessment context, US can be a useful tool for the evaluation of muscle. Given the device's capability of identifying different structures in a measurable fashion (with considerable detail precision) and quantifying their density through grayscale-estimated echogenicity, the range of available US evaluation methods is quite diverse. One can measure muscle thickness, or even the cross-sectional area of a given anatomical region, in order to evaluate the muscle quantity in a single body site or to predict the total muscle mass. Echo intensity, or even a given muscular group's inhomogeneity, can be useful to predict its fat infiltration—which is known to affect muscular contractile mechanics and, therefore, represents its quality. At last, muscle architectural analysis is also possible by measuring the muscle fascicle's length and pennation angles between fibers and their respective bone insertions.

If, on the one hand, there are many methods to use US in the body composition context; on the other hand, there is no consensus concerning the best use of US in this setting. To date, it seems that muscle thickness and cross-sectional area evaluation are equivalent methods to evaluate the muscle quantity. Muscle echogenicity analysis usually reflects muscle quality. Moreover, concerning sarcopenia, both these parameters may be useful—however, how well they complement each other is currently unknown.

US is usually referred as an "operator-dependent" method. To a certain point, this statement is true—however, there is plentiful evidence in the literature that suggest that US may be a highly intra- and inter-observer reproducible and accurate method, given the adequate training and the establishment of feasible evaluation protocols.

Research protocols for US usually differ in methodological aspects, which, sometimes, are not irrelevant and should be adequately discussed. The number and position of evaluated sites, the patient's positioning, probe compression, device's image-obtaining settings, and even the region of interest (ROI) to be analyzed in the obtained image may pose as details that can influence the results obtained.

The patient's inability to stand up may be a serious limitation to access certain body sites—specially the posterior ones. Abe's nine-site protocol (Abe et al. 1994) has been exhaustively studied and present very satisfying results in comparison with other imaging methods in terms of muscle mass estimation, but the patient is required to stand up—which is not always feasible. One-site protocols are available and usually enable the patients' evaluation in intensive care unit scenarios (Seymour et al. 2009) but may somewhat lack some accuracy. The middle ground should perhaps come as a somehow "universal" solution in the future, but, to the moment, such protocols are still in development or in need for validation. It is known that the anterior aspect of the thigh presents the better correlation with whole body muscle mass and is very susceptible to fluctuations, enabling early detection of muscle mass loss in acute clinical scenarios (Gruther et al. 2008). Recent data, however, suggest that complementing such a measure with other body sites—or, perhaps, measuring bilaterally—might improve muscle evaluation results (Paris et al. 2017).

The amount of compression to be applied in the probe is an important consideration to be discussed. The pressure should always be minimal, just enough to adequately visualize subjacent anatomical landmarks without deforming underlying structures, such as adipose tissue and muscle layers. That may require some considerable training, and the lack of adequate pressure is known to negatively influence the estimation of muscle mass. One must always remember to use a generous amount of conductive gel, in order to obtain better visualization with minimal compression. The ideal amount of compression usually produces the image of an arch in the upper pole of the image, representing the uncompressed skin on the borders of the image and the minimal probe–skin contact in the middle point. Protocols using maximal compression presented unsatisfying results and proved to be less effective than those of minimal compression (Paris et al. 2016).

Devices from different manufacturers usually use different image-obtaining and image-interpreting methods/equations, and these data are not routinely available for users. Musculoskeletal presets available in the devices also differ in settings (i.e., time-gain compensation, gain, dynamic range) between manufacturers. At last, the "raw" imaging data is difficult to obtain. These factors make echogenicity somehow incomparable between devices—unless conversion equations established from common phantoms or special (and, yet, complicated) backscatter data analysis are performed (Pillen et al. 2009; Zaidman et al. 2012). Different devices, however, do not seem to influence thickness measurements (Legerlotz et al. 2010).

The acquired images can be analyzed posteriorly through different softwares (i.e., ImageJ [NIH, Bethesda, MD, USA], Photoshop [Adobe Systems, San Jose, CA, USA]). The chosen software does not seem to influence the results (Harris-Love et al. 2016). However, concerning echogenicity evaluation, the area and depth of the selected ROI do seem to matter (Caresio et al. 2015). Careful analytic methodology concerning these aspects should be standardized, as the ideal analysis protocol is still uncertain. Muscle thickness evaluation, though, can be obtained in real time by the US device while performing the exam, which somehow makes it more practical.

A "novel" US variation is the A-mode device based on a single-point evaluation through time. Such devices (i.e., Bodymetrix [Intelametrix, California, USA]), commercially available as portable US probes attachable to a computer, provide depth estimations that are useful for adipose tissue and muscle thickness evaluation. These values are, then, able to be applied to predictive formulas inserted in the device's software and extrapolated to whole-body estimates. Alternatively, the so-called scan mode is able to provide real-time images of the evaluated tissues and can also be used in the body composition evaluation (Figure 4.2).

Currently, the A-mode US role in the total fat tissue estimation is more well consolidated than the muscle tissue. Still, the method does seem promising for the muscle evaluation, considering its low cost and simplicity of use, and its role in the body composition evaluation tools range should be better defined in the next years, with further research.

The use of the US in the body composition scenario is still a matter of discussion. If there were reference values available for different populations, a single measurement would enable to provide values, which could then be compared with cutoff values. However, this is not a reality for most populations. So, nowadays, US has been used as a seriated evaluation tool, comparing the patient's evolution with its own basal values (Puthucheary et al. 2013). This method allows for the identification of muscle gain/loss in specific body sites, which allow inferences concerning its prognosis and adequacy of treatment.

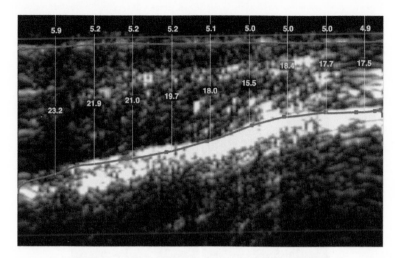

FIGURE 4.2 Left anterior thigh's scan of a 62-year-old man by an A-mode US device. The upper layer represents the subcutaneous fat tissue, and the middle layer represents the muscle tissue. The brighter line in the image stands for the bone surface of the femur.

US may be a valuable method in specific contexts in which other methods are not adequate, such as intensive care unit patients or large epidemiological studies. It is a relatively cheap and widely available device, which does not involve patient transportation and/or ability to stand still. It does not involve radiation, metallic or silicon prosthetics are not a limitation, and it is relatively simple to be applied—given the adequate protocols and training. Its precision, once described as unsatisfactory, is at least adequate—its accuracy is described as similar to dual-energy X-ray absorptiometry (DXA) and superior to anthropometry and electrical bioimpedance, considering magnetic resonance imaging (MRI) as a gold standard (Sanada et al. 2006). As limitations, US poses as a method that requires specific training; to the moment, there are not cutoff values available for all populations concerning adequacy of muscle mass; at last, there is the need for further discussion concerning application protocols and which parameters to consider.

4.2.2 DUAL-ENERGY X-RAY ABSORPTIOMETRY

DXA is a method based on the use of different frequencies' low-radiation X-ray. The device emits two different frequencies of photonic energy and captures them after they pass through a specific mean, therefore being able to estimate how much energy was attenuated in the process. In this fashion, DXA is able to access tissue density, quantify tissues of different densities, and depict a visual representation of the analysis (Prado et al. 2014).

The two different frequencies emitted by the DXA device allows differentiation between bone mineral mass vs. soft tissues and fat vs. fat-free tissues (Figure 4.3). Fat-free tissues are a good surrogate for lean soft tissue (since the nonmuscle part, constituted by connective tissue and skin, is negligible in practical means). Therefore, DXA allows the estimation of the bone mass (and its density, which may be useful for the osteopenia/osteoporosis diagnosis), fat, and "muscle."

Concerning bone density, the sites usually analyzed are the lumbar spine and the hip/femur neck. However, for fat-free tissue, the most common protocols are the whole-body and the appendicular analysis (by summing the fat-free tissue of four limbs). The latter allows for a very useful parameter in body composition: the estimation of the appendicular lean soft tissue (ALST).

The device requires constant calibration in variable periods (depending on the model and manufacturer), with factory-provided phantoms, which should be considered and adequately incorporated to the exam routine. Despite that, it is a relatively simple and quick exam, usually taking less than

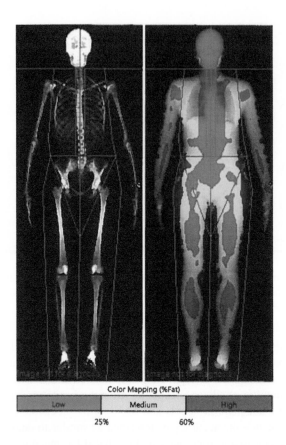

FIGURE 4.3 DXA evaluation screen (left: bone mineral mass vs. soft tissues; left: fat vs. fat-free tissues). The sample image is only a visual representation, as the DXA software usually also presents the respective bone and fat density values for each evaluated body segment.

15 min to perform (considering the lumbar, hip, and whole-body evaluations together). It requires the patient to lay still on the devices' table with legs and arms spread apart. Silicone implants and metallic prosthetics may influence the results, depending of the areas of interest, but are not a formal contraindication to the method. However, the height, width, or weight of the subject may pose as a limitation—as, if the subject does not fit in the examination table or if the table is not able to hold the subject's weight, the exam should not be performed. Different models may provide bigger and more resistant tables, but this should be checked before indicating the method. Also concerning wide subjects, an alternative is to "divide" the evaluation by measuring only half the body and multiplying the values by two or summing both halves individually measured at a time.

DXA is considered a very accurate method, being even described as a "gold standard" method by the scientific literature. Its ability to identify ALST with considerable precision and repeatability in a very feasible and low-radiation manner allows DXA to be considered one of the best choices for prospective studies and seriated evaluations in the research scenario. However, although very low, the radiation dose associated with the exam is not negligible, which somehow poses as a limitation to short-interval re-evaluations. It also provides the advantage of being able to inform bone density, which may be useful for the geriatric patient general health's evaluation—and some of the patients may already have the densitometry exam available, for other reasons, than body composition assessment. When soliciting the exam, however, the responsible assistant should also request for the whole-body evaluation, as the bone density evaluation usually only focuses on the hip and lumbar spine sites.

4.2.3 Computerized Tomography

The use of computerized tomography (CT) scans has spread widely lately as these images may be available in the medical records of patients with certain clinical conditions such as cancer, chronic obstructive pulmonary disease, kidney disease, cirrhosis, and human immunodeficiency virus. Likewise, patients requiring emergency care may also require a CT scan, which can be retrieved for the additional purpose of body composition analysis.

With this technique, the X-ray attenuation passing through tissues is detected, and a graphic image is reconstructed, assigning a Hounsfield unit (HU) number to represent the radio density at each point. The number for each point in the image (or pixel) is shown as a level of gray from black to white. As such, adipose tissue, skeletal muscle, bone, visceral organs, and brain can be identified by the different X-ray attenuation. The generated high-image resolutions can be retrieved and exported to a software capable of quantifying tissues (i.e., skeletal muscle and adipose tissue) based on the pre-established HU ranges as fully described elsewhere (Prado et al. 2016) (Figure 4.4). This identification can be manually or automated. A brief illustration of manual identification of the tissue can be seen here: https://www.youtube.com/watch?v=KJrsQ_dg5mM. Technological developments such as the use of automated software for segmenting skeletal muscle, visceral, and subcutaneous adipose tissue are available. Using the shape-modeling approach, they can promptly quantify body composition with a 2% coefficient of variation from manual segmentation (Popuri et al. 2016). An advantage of the CT scan is the use of one cross-sectional image (the third lumbar vertebrae) to estimate the whole body composition for specific tissues from these images, which are strong correlates of whole body fat and muscle masses (Shen et al. 2004a,b).

The high-resolution images make CT scan the most accurate method to determine the body composition at the tissue-organ level, specifically total and regional adipose tissue and skeletal muscle tissue. Furthermore, this technique is able to quantify intramuscular adipose tissue, a depot of emerging importance as a prognostic marker (Prado et al. 2016; Brown et al. 2018; Xiao et al. 2018). However, the large radiation exposure precludes its widespread, making it unethical to use CT imaging solely for the purpose of body composition analysis.

In the context of sarcopenia research, several cutoffs have been proposed for the diagnosis of muscle depletion using CT scans, particularly in the context of cancer (summarized in Prado et al. 2016). However, the use of medians and cohort-specific distributions such as lowest tercile or quartile has also been reported (Prado et al. 2016). The use of CT imaging in clinical population where

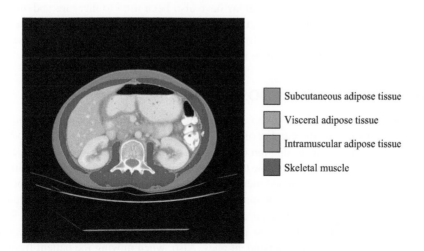

Subcutaneous adipose tissue

Visceral adipose tissue

Intramuscular adipose tissue

Skeletal muscle

FIGURE 4.4 Cross-sectional area of the third lumbar vertebrae evaluated for body composition using the SliceOmatic software, version 5.3 (Tomovision, Montreal, Canada).

these images are widely available has transformed our understanding of the value of body composition in patient prognostication, also highlighting sarcopenia as a hidden condition, present in patients across the body mass index spectrum (Prado et al. 2016).

4.2.4 MAGNETIC RESONANCE IMAGING

As with the CT imaging methodology, MRI measures body composition at the tissue-organ level. Cross-sectional areas are evaluated, and tissue volumes are obtained and then converted to mass units by multiplying the volume by the specific tissue density (Ross and Janssen 2005). This is possible due to the small intra-individual variability of tissue densities for skeletal muscle (0.92 g/cm^3) and adipose tissue (1.04 g/cm^3) (Ross and Janssen 2005).

The principle of the MRI technique is the use of a magnetic field to align hydrogen nuclei (as atomic protons), which behave like magnetics. These protons are then activated by a radio frequency wave, absorbing energy. The signal generated is then used to develop regional and whole-body cross-sectional images. As such, the MRI technique quantifies skeletal muscle, adipose tissue, and organ mass (Heymsfield et al. 1997). For the specific purpose of body composition analysis, once the data are obtained, it should be transferred to a computer with available image analysis software and subsequent procedures similar to that described above under CT image analysis. Once the area of interest has been manually delineated, its area (either muscle, adipose tissue, or organ mass) can be calculated by the number of pixels multiplied by the defined area of interest (Ross and Janssen 2005).

Together with CT scan, MRI is considered the most accurate method for the measurement of *in vivo* body composition at the tissue-organ level. Its advantages include the lack of exposure to ionizing radiation and therefore safety for use in children and pregnant women. It is the best method for multiple-image protocols and whole body and serial measurements. However, its high costs and need for high technical skill precludes widespread use in research or clinical settings. Image quality is also affected by respiratory motion, which becomes an issue for participants unable to hold their breath for a long period. An additional limitation includes the size of the participant as those with obesity (usually body-mass index [BMI] > 35 kg/m^2) may not fit in the scan.

Due to the limited availability of MRI scans, its use in sarcopenia research has been somewhat limited and mostly related to the evaluation of specific muscle groups in the context of changes over time (Lustgarten and Fielding 2011). Commonly measured muscles include quadriceps muscle mass (Gray et al. 2010), leg muscle area (Kent-Braun et al. 2000), and hamstrings (Macaluso et al. 2002). Similar to CT images, cross-sectional regions have also been used to measure abdominal muscle and adipose tissue volumes (Orsso et al. 2017). The use of MRI to assess adiposity has been more extensive, and the recent development of fully automated algorithms to quantify adipose tissue precluding human intervention may facilitate its use in research settings (Sun et al. 2016).

4.3 WHICH METHOD TO CHOOSE?

When choosing between one of the aforementioned methods, the main aspects to be considered are the context of the evaluation, the number of repetitions required, the cost and the method's safety to the patient. The scenario of the evaluation is a very important variable. Clinical outpatients usually have their mobility preserved and are able to stay immobile when requested—which allows them to be evaluated by all the aforementioned methods. However, hospitalized patients may present some form of mobility limitation, or even the inability to lay still. This may come as a problem, particularly concerning long exams, which require the patient to be collaborative—such as MRI. Another limitation of the MRI evaluation may be patients with metallic catheters, prosthetics, orthopedic traction devices, or even ventilator-dependent ones. Also, when moving the patient from its bed to the device is challenging, particularly in the intensive care unit setting, one of the best options available is the bedside US evaluation. This also applies for large epidemiological studies, where portable

US devices may be an alternative to assess body composition in the patient's own home—given that its practicality may overcome the relative loss of accuracy against other imaging methods.

Also, the number of assessments should be considered. As CT and DXA comprise radiation, the exams should not be repeated in short periods. Therefore, although these may be of choice for cross-sectional assessments, they are unfeasible in the context of longitudinal evaluations of serial measurements of patient's clinical evolution (i.e., over days or weeks). As US does not emit radiation, and given its low cost, it should be the method of choice for short-period longitudinal/serial evaluations. When the time frame between evaluations is months, or years, any of the methods should be adequate. MRI does not emit radiation, but each exam can be quite expensive, restricting its use in clinical settings.

Another important aspect to be noted is that patients may have performed the aforementioned exams for reasons other than of body composition assessment. If the exams evaluated the adequate sites, or the correct protocols were used, the so-called opportunistic evaluation is feasible. For instance, a hospitalized patient whose abdominal CT scan was requested for clinical reasons may have its body composition evaluated posteriorly (allowing, even, for retrospective studies to be performed, if the images were adequately stored). Also, an elderly performing osteoporosis preventive exams may be able to provide its DXA scans to the assistant doctor, allowing for retrospective body composition assessment. Those are quite relevant information, given the amount of radiation involved in a CT scan, as, for instance, its use for "body composition purposes only" may be considered unethical, as discussed above.

4.4 FINAL CONSIDERATIONS

While advances in imaging techniques to assess human body composition have greatly increased our understanding of the prevalence and significance of sarcopenia, to date no method is widely accepted for the diagnosis of low muscularity. The use of MRI is restricted to research settings and the use of CT scans to populations in which these images are already available in the medical records. When available, CT scans can be the method of choice due to its high accuracy and reliability. Although the use of DXA scans in clinical settings is likely to increase in the future, US may become the most readily available tool due to its cost-effectiveness.

In spite of the available technique/expertise, no consensus cutoffs for each of these techniques are available. In fact, it is unlikely that a sex-specific cutoff will be agreed upon for the identification of low muscle mass across aging and chronic conditions. Instead, cutoffs should likely be population specific and based on a specific outcome of interest (i.e., mobility vs. survival).

The advantages and limitations listed under each technique hereby discussed should be weighed in when deciding which method to implement in the research or clinical setting. The American Society for Parenteral and Enteral Nutrition will soon release body composition clinical guidelines to help researchers and professionals with the technique of choice in diverse clinical cohorts. There is momentum for further understanding and use of body composition in the prevention, diagnosis, and treatment of sarcopenia and we expect the literature to increase exponentially in the upcoming years.

REFERENCES

Abe, T., M. Kondo, Y. Kawakami, and T. Fukunaga. 1994. Prediction equations for body composition of Japanese adults by B-mode ultrasound. *Am J Hum Biol.* 6:161–170.

Barbosa-Silva, T. G., R. M. Bielemann, M. C. Gonzalez, and A. M. B. Menezes. 2016. Prevalence of sarcopenia among community-dwelling elderly of a medium-sized South American city: Results of the COMO VAI? study. *J Cachexia Sarcopenia Muscle* 7 (2):136–143.

Batsis, J. A., L. K. Barre, T. A. Mackenzie, S. I. Pratt, F. Lopez-Jimenez, and S. J. Bartels. 2013. Variation in the prevalence of sarcopenia and sarcopenic obesity in older adults associated with different research definitions: Dual-energy X-ray absorptiometry data from the National Health and Nutrition Examination Survey 1999–2004. *J Am Geriatr Soc* 61 (6):974–980.

Brown, J. C., B. J. Caan, J. A. Meyerhardt, E. Weltzien, J. Xiao, E. M. Cespedes Feliciano, et al. 2018. The deterioration of muscle mass and radiodensity is prognostic of poor survival in stage I-III colorectal cancer: A population-based cohort study (C-SCANS). *J Cachexia Sarcopenia Muscle* 9 (4):664–672.

Caresio, C., F. Molinari, G. Emanuel, and M. A. Minetto. 2015. Muscle echo intensity: Reliability and conditioning factors. *Clin Physiol Funct Imaging* 35:393–403.

Dam, T. T., K. W. Peters, M. Fragala, P. M. Cawthon, T. B. Harris, R. McLean, et al. 2014. An evidence-based comparison of operational criteria for the presence of sarcopenia. *J Gerontol A Biol Sci Med Sci* 69 (5):584–590.

Gonzalez, M. C., T. G. Barbosa-Silva, and S. B. Heymsfield. 2018. Bioelectrical impedance analysis in the assessment of sarcopenia. *Curr Opin Clin Nutr Metab Care* 21 (5):366–374.

Gray, C., T. J. Macgillivray, C. Eeley, N. A. Stephens, I. Beggs, K. C. Fearon, and C. A. Greig. 2010. Magnetic resonance imaging with k-means clustering objectively measures whole muscle volume compartments in sarcopenia/cancer cachexia. *Clin Nutr* 30 (1):106–111.

Gruther, W., T. Benesch, C. Zorn, T. Paternostro-Sluga, M. Quittan, V. Fialka-Moser, et al. 2008. Muscle wasting in intensive care patients: Ultrasound observation of the m. quadriceps femoris muscle layer. *J Rehabil Med.* 40:185.

Harris-Love, M. O., B. A. Seamon, C. Teixeira, and C. Ismail. 2016. Ultrasound estimates of muscle quality in older adults: Reliability and comparison of Photoshop and ImageJ for the grayscale analysis of muscle echogenicity. *PeerJ* 4 (e1721):1–23.

Heymsfield, S., R. Ross, Z. Wang, and D. Frager. 1997. Imaging techniques of body composition: Advantages of measurement and new uses. In *Emerging Technologies for Nutrition Research*, 127–150. National Academy Press, Washington, DC.

Johnson Stoklossa, C. A., A. M. Sharma, M. Forhan, M. Siervo, R. S. Padwal, and C. M. Prado. 2017. Prevalence of sarcopenic obesity in adults with class II/III obesity using different diagnostic criteria. *J Nutr Metab* 2017:7307618.

Kent-Braun, J. A., A. V. Ng, and K. Young. 2000. Skeletal muscle contractile and non-contractile components in young and older women and men. *J Appl Physiol* 88 (2):662–668.

Legerlotz, K., H. K. Smith, and W. A. Hing. 2010. Variation and reliability of ultrasonographic quantification of the architecture of the medial gastrocnemius muscle in young children. *Clin Physiol Funct Imaging* 30:198–205.

Lustgarten, M. S., and R. A. Fielding. 2011. Assessment of analytical methods used to measure changes in body composition in the elderly and recommendations for their use in phase II clinical trials. *J Nutr Health Aging* 15 (5):368–375.

Macaluso, A., M. Nimmo, J. Foster, M. Cockburn, N. McMillan, and G. De Vito. 2002. Contractile muscle volume and agonist-antagonist co activation account for differences in torque between young and older women. *Muscle Nerve* 25 (6):858–863.

Melton, L. J., 3rd, S. Khosla, and B. L. Riggs. 2000. Epidemiology of sarcopenia. *Mayo Clin Proc* 75 (Suppl):10.

Orsso, C. E., M. Mackenzie, A. S. Alberga, A. M. Sharma, L. Richer, D. A. Rubin, C. M. Prado, and A. M. Haqq. 2017. The use of magnetic resonance imaging to characterize abnormal body composition phenotypes in youth with Prader-Willi syndrome. *Metabolism* 69 (67–75):67.

Paris, M. T., B. Lafleur, J. A. Dubin, and M. Mourtzakis. 2017. Development of a bedside viable ultrasound protocol to quantify appendicular lean tissue mass. *J Cachexia Sarcopenia Muscle* 8 (5):713–726.

Paris, M. T., M. Mourtzakis, A. Day, R. Leung, S. Watharkar, R. Kozar, et al. 2016. Validation of bedside ultrasound of muscle layer thickness of the quadriceps in the critically ill patient (VALIDUM study). *J Parenter Enteral Nutr* 41 (2):171–180.

Pillen S., J. P. van Dijk, G. Weijers, W. Raijmann, C. L. de Korte, and M. J. Zwarts. 2009. Quantitative grayscale analysis in skeletal muscle ultrasound: A comparison study of two ultrasound devices. *Muscle Nerve* 39 (6):781–786.

Popuri, K., D. Cobzas, N. Esfandiari, V. Baracos, and M. Jägersand. 2016. Body composition assessment in axial CT images using FEM-based automatic segmentation of skeletal muscle. *IEEE Trans Med Imaging* 35:512–520.

Prado, C. M., and S. B. Heymsfield. 2014. Lean tissue imaging: A new era for nutritional assessment and intervention. *J Parenter Enteral Nutr* 38 (8):940–953.

Prado, C. M., S. J. Cushen, C. E. Orsso, and A. M. Ryan. 2016. Sarcopenia and cachexia in the era of obesity: Clinical and nutritional impact. *Proc Nutr Soc* 75 (2):188–198.

Puthucheary, Z. A., J. Rawal, M. McPhail, B. Connolly, G. Ratnayake, P. Chan, et al. 2013. Acute skeletal muscle wasting in critical illness. *JAMA* 310 (15):1591–1600.

Ross, R., and I. Janssen. 2005. Computed tomography and magnetic resonance imaging. In *Human Body Composition*, ed. S. B. Heymsfield, T. Lohman, Z. Wang, and S. Going, 89–108: Champaign, IL: Human Kinetics.

Sanada, K., C. F. Kearns, T. Midorikawa, and T. Abe. 2006. Prediction and validation of total and regional skeletal muscle mass by ultrasound in Japanese adults. *Eur J Appl Physiol* 96 (1):24–31.

Seymour, J. M., K. Ward, P. S. Sidhu, Z. A. Puthucheary, J. Steier, C. J. Jolley, et al. 2009. Ultrasound measurement of rectus femoris crosssectional area and the relationship with quadriceps strength in COPD. *Thorax* 64 (5):418–423.

Shen, W., M. P. St-Onge, Z. Wang, and S. B. Heymsfield. 2005. Study of body composition: An overview. In *Human body composition*, ed. S. B. Heymsfield, T. Lohman, Z. Wang, and S. Going, 3–14. Champaign, IL: Human Kinetics.

Shen, W., M. Punyanitya, Z. Wang, D. Gallagher, M. P. St-Onge, J. Albu, et al. 2004a. Total body skeletal muscle and adipose tissue volumes: Estimation from a single abdominal cross-sectional image. *J Appl Physiol* 97 (6):2333–2338.

Shen, W., M. Punyanitya, Z. Wang, D. Gallagher, M. P. St-Onge, J. Albu, and et al. 2004b. Visceral adipose tissue: Relations between single-slice areas and total volume. *Am J Clin Nutr* 80 (2):271–278.

Sun, J., B. Xu, and J. Freeland-Graves. 2016. Automated quantification of abdominal adiposity by magnetic resonance imaging. *Am J Hum Biol.* 28 (6):757–766.

Wang, Z. M., R. N. Pierson, Jr., and S. B. Heymsfield. 1992. The five- level model: A new approach to organising body-composition research. *Am J Clin Nutr* 56 (1):19–28.

Xiao, J., B. J. Caan, E. Weltzien, E. M. Cespedes Feliciano, C. H. Kroenke, J. A. Meyerhardt, et al. 2018. Associations of pre-existing co-morbidities with skeletal muscle mass and radiodensity in patients with non-metastatic colorectal cancer. *J Cachexia Sarcopenia Muscle* 9 (4):654–663.

Zaidman, C. M., Holland, M.R., Hughes, M.S. 2012. Quantitative ultrasound of skeletal muscle: Reliable measurements of calibrated muscle backscatter from different ultrasound systems. *Ultrasound Med Biol* 38 (9):1616–1625.

Zamboni, M., G. Mazzali, F. Fantin, A. Rossi, and V. Di Francesco. 2008. Sarcopenic obesity: A new category of obesity in the elderly. *Nutr Metab Cardiovasc Dis* 18 (5):388–395.

5 Nutrient Sensing and mTORC1 Regulation in Sarcopenia

Ted G. Graber and Blake B. Rasmussen

CONTENTS

5.1 INTRODUCTION

In 1950, 205 million people were over the age of 60 years, 810 million in 2010, and by 2050, the number is estimated to be over 2 billion (UNDESA 2013). With increasing age comes the risk of sarcopenia (Marzetti and Leeuwenburgh 2006, Sundell 2011), which is the age-related loss of muscle mass and strength. Sarcopenia leads to falls, deteriorating quality of life, difficulty performing activities of daily living, eventual loss of independence, and which is a major predictor of poor prognoses after procedures and increased mortality (Landi et al. 2012, Rolland et al. 2008, Breen and Phillips 2011, Janssen et al. 2002). Muscle mass typically peaks by the mid-20s to 30s and then declines steadily after the mid-40s (Janssen, Heymsfield, and Ross 2002); by the age of 80 years, people have lost about 40% of their peak muscle mass (Deschenes 2004), with about 8% of older adults having the most serious cases of sarcopenia, causing functional disability (Janssen, Heymsfield, and Ross 2002). Despite a lack of a cure, many individual treatments do show some promise (exercise, nutrition, hormonal). The multifactorial etiology of sarcopenia suggests the benefit of combinatorial treatment strategies (Narici and Maffulli 2010).

Skeletal muscle is a dynamic plastic organ, constantly being remodeled. This remodeling involves a balance of catabolism (muscle protein breakdown [MPB]) with anabolism (muscle protein synthesis [MPS]). Even a small imbalance in this system favoring MPB would result in a dramatic loss of muscle mass over decades. Thus, any resistance to signaling pathways that control these two cellular mechanisms may be an underlying contributor to sarcopenia. The mechanistic target of rapamycin complex 1 (mTORC1) is a major regulator of protein synthesis and is influenced by diverse cell signals, such as insulin, amino acids, growth factors, exercise, energy balance, and oxygen levels (see Figure 5.1).

The following sections in this chapter will explore the relation among mTORC1 signaling, protein synthesis, and protein anabolic resistance. First, we will discuss the regulation of mTORC1 and its downstream effectors, focusing primarily on the control of nutrient signaling. Then, we will discuss the potential etiology for protein anabolic resistance and possible mechanisms.

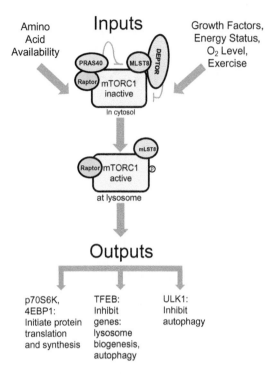

Inputs

Amino Acid Availability

Growth Factors, Energy Status, O$_2$ Level, Exercise

Outputs

p70S6K, 4EBP1: Initiate protein translation and synthesis

TFEB: Inhibit genes: lysosome biogenesis, autophagy

ULK1: Inhibit autophagy

FIGURE 5.1 mTORC1 is the master regulator of protein synthesis. The mechanistic target of rapamycin complex 1 (mTORC1) receives inputs from numerous sources and when conditions are favorable, it outputs to number of downstream proteins, including ones that control protein translation and autophagy.

5.2 mTORC1: MASTER REGULATOR OF PROTEIN SYNTHESIS

mTORC1 is a large dimeric protein complex receiving inputs from many sources (see Figure 5.1). The three core subunits are mTOR (the serine/threonine kinase core), mLST8 (mammalian lethal with SEC13 protein), and raptor (regulatory-associated protein of mTOR, raptor binds to rapamycin via FKBP12—FK506-binding protein—to shut down mTORC1 activity). The regulatory-associated proteins are deptor (DEP domain-containing mTOR-interacting protein), pras40 (proline-rich Akt substrate 40), and fkBP38 (Yoon and Choi 2016).

mTORC1 kinase activity affects downstream activation of anti-autophagy pathways via phosphorylation of transcription factor EB (TFEB) and uncoordinated movement-51-like kinase 1 (ULK1) (Lin and Hurley 2016, Sardiello et al. 2009). Activation of mTORC1 also results in the phosphorylation of insulin receptor substrate 1 (IRS-1), which promotes degradation of IRS-1 by the proteasome, ultimately resulting in a negative feedback loop that helps to inhibit mTORC1 activation by inducing insulin resistance (Magnuson, Ekim and Fingar 2012, Dibble and Cantley 2015). However, more important to the question of sarcopenia regulation is a third function of mTORC1, enabling messenger RNA translation initiation and polypeptide elongation resulting in increased protein synthesis. The rest of this section will focus on signaling cascades that regulate protein synthesis rates.

The mTORC1 serine/threonine kinase phosphorylates S6K1 and S6K2 (ribosomal protein S6 kinase). S6K has at least two isoforms: 70-kDa ribosomal protein S6 kinase (p70S6K) and 85-kDa ribosomal protein S6 kinase (p85S6K). S6K1 also has multiple phosphorylation sites (eight discovered) regulating its activity, with three of the sites being required for activation, which include the position affected by mTORC1 (on p70S6K = threonine 389; on p85S6K = threonine 412) (Weng et al. 1998). S6K1 in turn phosphorylates ribosomal protein S6 (rpS6) and eukaryotic elongation factor 2 (eEF2) (Glass 2010). rpS6 is one of the 33 proteins comprising the 40S ribonucleoprotein complex (which together with the 60S subunit comprise the eukaryotic protein translation engine 80S ribosome) and is phosphorylated at five positions (serine 235, 236, 240, 244, and 247); however, the exact role of rpS6 in protein synthesis has yet to be determined (Roux et al. 2007, Magnuson, Ekim, and Fingar 2012). eEF2 is phosphorylated by S6K at position serine 366 and once activated has GTPase activity necessary in the elongation step of protein synthesis (Kaul, Pattan, and Rafeequi 2011, Gingras, Raught, and Sonenberg 2004).

The kinase activity of mTORC1 also phosphorylates eukaryotic initiation factor 4E binding protein 1 (4E-BP1) at four (threonine 37, 46, and 70 and serine 65) of its known six sites (Schalm et al. 2003). 4E-BP1 is recruited by raptor to an active mTORC1 where it is then phosphorylated (Schalm et al. 2003). Phosphorylation of 4E-BP1 causes its dissociation from eukaryotic initiation factor 4E (eIF4E). Once free from 4E-BP1, eIF4E is vital to forming the translation initiation complex (with other proteins including eIF4G, eukaryotic initiation factor 4G, serving as the backbone of the complex, which is attached to the ribosome by eIF3, eukaryotic initiation factor 3) as it binds the cap structure on the 5′ end of mRNAs (Richter and Sonenburg 2005).

In this chapter, we will primarily focus on the action of upstream activators (mainly nutritional regulation) of mTORC1 that act to turn on protein synthesis. Regulation of mTORC1 is convoluted, with many branching signaling pathways. However, common protein complex effectors transmit these numerous signal inputs to activate mTORC1. Figure 5.2 is a simplified depiction of the conditions in the basal, fasted state with mTORC1 inactive, without sufficient amino acids, growth factors, or other signals to activate the protein complex. In Figure 5.3, signals derived from growth

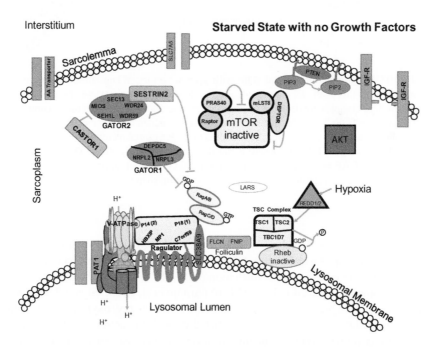

FIGURE 5.2 mTORC1 regulation: Fasted state without growth factors. When insufficient oxygen levels, no growth factors (IGF-1 in this example), and low nutrient states are present, AKT is inactivated; Redd and the TSC complex are activated, leading to the inactivation of Rheb at the lysosomal membrane (active Rheb activates mTORC1). In the absence of growth factors, the subunits Pras40 and DEPTOR act to inhibit mTORC1 activation as well. In the starved state (without sufficient amino acids present), mTORC1 in the sarcoplasm of the muscle fiber is not able to be recruited to the lysosomal membrane by the Rag heterodimer (Rag A/B in the high-energy GTP state) where it can then be activated by Rheb. Thus, protein translation/synthesis is turned off. GAP = GTPase-activating protein, GTP = guanine triphosphate, GDP = guanine diphosphate, H⁺ = hydrogen ion (proton), O_2 = oxygen. Text within shapes = protein or subunit names.

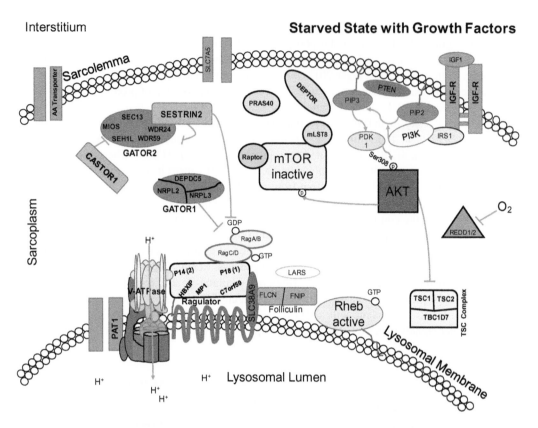

FIGURE 5.3 mTORC1 regulation: Fasted state with growth factors. When sufficient oxygen levels and growth factors (IGF-1 in this example) are present, AKT is activated (phosphorylated by PDK1 at ser [serine] 308); Redd and the TSC complex are deactivated, leading to the activation of Rheb at the lysosomal membrane (active Rheb activates mTORC1). In the starved state without sufficient amino acids present, mTORC1 is in the sarcoplasm of the muscle fiber and cannot be recruited to the lysosomal membrane by the Rag heterodimer (Rag A/B in the high-energy GTP state) where it can then be activated by Rheb. Thus, protein translation/synthesis is turned off. GAP = GTPase-activating protein, GTP = guanine triphosphate, GDP = guanine diphosphate. H$^+$ = hydrogen ion (proton), O$_2$ = oxygen. Text within shapes = protein or subunit names.

factors or insulin (in this case, insulin-like growth factor-1 [IGF-1]), oxygen (via REDD, DNA-damage-inducible transcript 4), and energy levels (via adenosine monophosphate-activated protein kinase [AMPK]) are filtered through the mediation of the TSC1/TSC2 (tuberous sclerosis) protein complex on Ras homolog enriched in brain (Rheb) (Armijo et al. 2016). Figure 5.4 shows the activation of mTORC1 in the presence of sufficient amino acids, energy, oxygen, and growth factors. Amino acid-sensing signaling pathways are mediated through the actions of the small Rag (Ras-related GTP-binding protein; GTP = guanine triphosphate) GTPase heterodimer RAG A (or B)/RAG C (or D). We will first detail amino acid sensing; much of this research being relatively recent, followed by a description of the well-established signaling pathways initiated by insulin/growth factors, energy, and oxygen levels in the cell. We will briefly discuss the recent initial characterization of independent pathways deriving from mechanical muscle cell stimulation at the end of this section. Of note, much of the work detailing mTORC1 signaling, and in particular amino acid sensing, was performed using HEK-293T kidney cells in culture and not in muscle tissue (or cells). We assume similar signaling to occur within different mammalian cell types, particularly as the basic pathways are conserved in all eukaryotes, from yeast to humans.

Fed State with Growth Factors

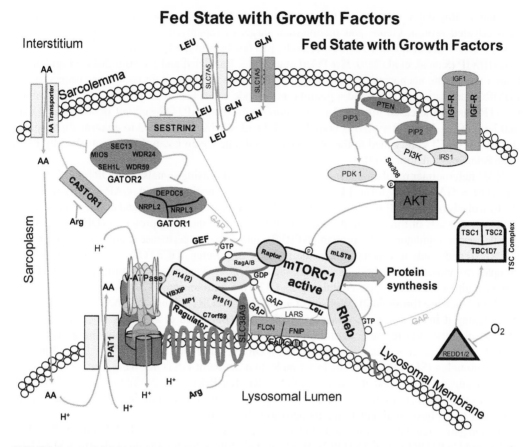

FIGURE 5.4 mTORC1 regulation: Fed state with growth factors. When sufficient oxygen levels and growth factors (IGF-1 in this example) are present, AKT is activated (phosphorylated by PDK1 at ser [serine] 308); Redd and the TSC complex are deactivated, leading to the activation of Rheb at the lysosomal membrane (active Rheb activates mTORC1). In the fed state with sufficient amino acids present, mTORC1 is recruited to the lysosomal membrane by the Rag heterodimer (Rag A/B in the high-energy GTP state) where it can then be activated by Rheb. mTORC1 then phosphorylates its downstream substrates such as S6K1, which turns on protein translation. AA = amino acids, LEU = leucine, ARG = arginine, GLN = glutamine, GAP = GTPase-activating protein, GEF = guanine exchange factor, GTP = guanine triphosphate, GDP = guanine diphosphate. H$^+$ = hydrogen ion (proton), O$_2$ = oxygen. Text within shapes = protein or subunit names. Dotted lines and arrows indicate repressed functions under the stated conditions.

5.3 mTORC1 REGULATION AND AMINO ACID SENSING

mTORC1 must translocate from the cytosol to the lysosomal membrane to stimulate its kinase activity. Rheb is also located at the lysosomal membrane and in its GTP state will promote mTORC1 activation. The signaling activation of Rheb is shown in Figure 5.3 and described in detail in the next section; however, the precise mechanism by which Rheb activates mTORC1 is unknown. We are only now deciphering the molecular mechanisms of amino acid sensing in the control of mTORC1 signaling.

In the presence of sufficient amino acids, a complex signaling cascade is activated (Figure 5.4) that ultimately converts its signal through a single protein complex, the small Rag GTPase heterodimer (consisting of either Rag A or Rag B and either Rag C or Rag D). The Rag heterodimer interacts with, and is tethered at the lysosomal membrane by, the scaffolding protein complex

Ragulator. Ragulator consists of five subunits, LAMTOR1–5—lysosomal adaptor and mitogen-activated protein kinase and mammalian target of rapamycin (mTOR) activator/regulator. LAMTOR1–5 are also known as, respectively, p18 (anchors to the lysosome), p14, MP1, C7orf59, and HBXIP (Sancak et al. 2010, Bar-Peled et al. 2012, Bar-Peled and Sabatini 2014). Ragulator is both a membrane anchoring scaffold complex that holds all the protein complexes necessary for mTORC1 activation in place and a guanine nucleotide exchange factor (GEF). RAG A/B bound to GTP has an increased affinity for interaction with the RAPTOR subunit of mTORC1 (Kim and Kim 2016). When Rag A/B is bound to GTP and RAG C/D is bound to guanine diphosphate (GDP), mTORC1 is recruited to the lysosomal membrane where it can then be activated by Rheb (Bar-Peled and Sabatini 2014).

RAG heterodimer regulation is initiated by amino acid-sensing signals. GATOR1 and GATOR2 (GATOR = GAP activity toward Rag; GAP = GTPase-activating protein; GAP activity increases the hydrolysis rate of GTP, reducing GTP to GDP) act as sensors of leucine and arginine sufficiency. GATOR1 consists of three subunits, such as DEP domain-containing protein 5 (DEPDC5), nitrogen permease regulator 2-like protein (NPRL2), and nitrogen permease regulator 3-like protein (NPRL3), and binds to a heterodimer, the recently discovered KICSTOR, a scaffolding protein complex composed of four protein subunits such as KPTN (kaptin, actin-binding protein), ITFG2 (integrin alpha FG-GAP repeat containing 2), C12orf66 (chromosome 12 open reading frame 66), and SZT2 (seizure threshold 2) (Wolfson et al. 2017). GATOR1 inhibits Rag A/B with GAP activity (keeping it in the low-energy inactive GDP state) (Bar-Peled et al. 2013). GATOR2 consists of five subunits: WD repeat-containing protein mio (MIOS), WD repeat-containing protein 24 (WDR24), WD repeat-containing protein 59 (WDR59), nucleoporin SEH1 (SEH1L), and protein SEC13 homolog (SEC13). While the exact molecular action of GATOR2 has yet to be elucidated, evidence suggests it is a potent inhibitor of GATOR1 (Bar-Peled et al. 2013); thus, active GATOR2 represses the GAP activity of GATOR1, allowing RAG A/B to remain in the GTP state and recruit mTORC1 to the lysosomal membrane for activation by Rheb.

Two negative regulatory inputs to GATOR2 have been discovered: SESTRIN2 (and SESTRIN1 and 3) and CASTOR1 (and CASTOR2) (Wolfson et al. 2016, Chantranupong et al. 2016). SESTRIN may function as a guanine nucleotide dissociation inhibitor toward Rag A/B (Peng, Ying, and Li 2014), but its primary role in mTORC1 activation may involve leucine sensing and control of GATOR2 activation (Chantranupong et al. 2014). SESTRIN2 (and SESTRIN1 and 3) is thought to control GATOR2, with sufficient leucine in the cell promoting dephosphorylation and inactivation of SESTRIN2, whereby it disassociates from GATOR2 (Kimball et al. 2016). CASTOR1 (or 2) binds to a different site on GATOR2 than SESTRIN. Similar to the leucine-sensing ability of SESTRIN, sufficient arginine in the cell causes the inhibition of the CASTOR1 dimer (homodimer or heterodimer with CASTOR2), which then also allows GATOR2 to repress the GAP activity of GATOR1 (Chantranupong et al. 2016). The CASTOR1 protein is thought to be inhibited from interacting with GATOR when arginine binds to the ACT domain in CASTOR1, causing a disassociation of the castor dimer. CASTOR2 binds with more affinity to GATOR2 and may be arginine insensitive, but its mRNA expression may be regulated upstream by stress-responsive transcription factors such as FOXO1 (Ouyang et al. 2012). The exact mechanism of how the CASTOR dimer regulates GATOR2 is currently unknown.

The other half of the Rag heterodimer, RAG C/D, is hydrolyzed from the inactive GTP state to its active GDP state through the leucine-sufficient sensitive GAP activity of both leucine–tRNA ligase (LARS) and the amino acid-sensitive folliculin (FLCN) complex consisting of FLCN and folliculin-interacting protein 1 (FNIP) (Dibble and Cantley 2015).

At the lysosomal membrane, other protein–protein complexes also sense amino acid sufficiency. Arginine is sensed by solute carrier family 38 member 9 (SLC38A9) (Wang et al. 2015), which is necessary with vacuolar proton-ATPase (v-ATPase) (Zhang et al. 2014), and possibly with proton-assisted amino acid transporter (PAT1) (Ögmundsdóttir et al. 2012), to generate a conformational shift in Ragulator that initiates GEF activity toward RAG A/B (Rebsamen et al. 2015,

Wang et al. 2015). The GEF activation of Ragulator exchanges a GDP on RAG A/B for a GTP, thus enabling the Rag heterodimer to initiate translocation of mTORC1 from the cytosol to the lysosome surface. PAT1 symports amino acids out of the lysosome with protons (Ögmundsdóttir et al. 2012). The v-ATPase pumps protons back into the lysosome against the concentration gradient, using ATP for power. SLC38A9 is a neutral amino acid transporter, with a weak affinity that acts less as a transporter and more as a sensor of leucine that physically interacts with Ragulator with v-ATPase acting as the intermediary (Wang et al. 2015).

Recruitment of mTORC1 to the lysosomal membrane puts mTORC1 in the vicinity of Rheb. Rheb is a GTPase that has to be in the high-energy GTP bound state in order to activate mTORC1. The amino acid-sensing mechanisms outlined above are currently being elucidated, and we expect more discoveries will soon add more information to our evolving understanding of how cells sense amino acid availability. The next section will briefly explain how the various signaling pathways of energy sufficiency, oxygen levels, and growth factor availability converge to activate Rheb.

5.4 mTORC1 REGULATION BY INSULIN, HYPOXIA, AND CELLULAR ENERGY LEVELS

Postprandial metabolic control of mTORC1, from a nutritional standpoint, originates from both protein and carbohydrates. Both macronutrients elicit an increase in circulating insulin concentrations. Growth factors such as insulin and IGF-1 promote activation of mTORC1. Insulin signaling within muscle cells is generated when insulin binds to insulin receptors (tetramer receptor tyrosine kinases) located on the sarcolemma, which then form enzymatically active dimers that move the kinase domains closer to each other, creating self-phosphorylating tyrosine docking sites. IRS-1 docks at and is phosphorylated at these phosphotyrosine sites. IRS-1 docking to the insulin receptor provides multiple sites for the binding and phosphorylation of phosphatidylinositol-3-kinase (PI3K). Activation of PI3K by IRS1 in turn results in PI3K phosphorylating phosphoinositide diphosphate (PIP2), thus converting PIP2 to phosphoinositide triphosphate (PIP3). Akt has a high affinity for binding to PIP3. Phosphoinositide-dependent protein kinase 1 (PDK1) is also recruited at the sarcolemma by PIP3. PDK1 phosphorylates Akt at the threonine308 site (and the serine473 site is phosphorylated as well by mTORC2—mTORC2 being relatively insensitive to rapamycin and having the RICTOR, rather than RAPTOR, subunit) (Glass 2010). Akt, now an active kinase, phosphorylates mTORC1 on the mTOR subunit at the serine2443 site and the PRAS40 inhibitory subunit, allowing mTOR to self-phosphorylate, resulting in disassociation of the inhibitory subunits (Egerman and Glass 2014). Akt also phosphorylates and inactivates the tuberous sclerosis 1 protein (TSC) complex.

Figure 5.2 shows the status of mTORC1 signaling without growth factors or insulin and under low energy/hypoxic conditions. Low oxygen levels downregulate mTORC1 activity through DNA-damage-inducible transcript 4 (REDD1) or DNA-damage-inducible transcript 4-like protein (REDD2). REDD1 is also implicated as important in the nutritional regulation of mTORC1 during atrophic conditions such as glucose starvation, acute alcohol intoxication, and sepsis (Gordon et al. 2015). The REDD proteins are thought to activate the TSC complex via an unknown mechanism although recent studies indicate that the mechanism is through promotion of the dephosphorylation of Akt at the threonine308 site mediated by protein phosphatase 2A (Dennis et al. 2014).

The TSC complex (consisting of three subunits: TSC1, TSC2, and TBC1 domain family member 7 [TBC1D7]) inhibits Rheb in part via GAP activity to hydrolyze Rheb-GTP to Rheb-GDP, Rheb being active only in the high-energy GTP state. Rheb is necessary for activation of mTORC1 (Brugarolas et al. 2004). Moreover, evidence suggests that the TSC complex also acts to sterically hinder the Rheb kinase site and that it may be recruited to the lysosome by Rag A/B during times of low amino acid availability in the cell (Carroll et al. 2016).

Energy level in the cell (the ratio of adenosine monophosphate [AMP] to adenosine triphosphate [ATP]) is mediated through 5′ AMP-activated protein kinase (AMPK). AMPK has numerous targets but negatively controls mTORC1 although activation of the TSC complex by direct phosphorylation of TSC2 on serine1387 and phosphorylation of the mTORC1 subunit RAPTOR on serine792 (Mihaylova and Shaw 2011, Dibble and Manning 2013). Sufficient energy and oxygen within the cell limit the activity of REDD and AMPK, thus allowing for increased protein synthesis.

In Figure 5.3, growth factors (in this case, IGF-1, but could also be others such as insulin), sufficient oxygen, and energy are present. The presence of growth factors causes a signaling cascade (see above) resulting in the Akt activating phosphorylation of threonine308 by PDK1. Akt, in turn, phosphorylates mTOR, PRAS40, and TSC2 (removing TSC complex inhibition on Rheb) (Dibble and Cantley 2015, Glass 2010). Thus, as shown, oxygen levels, growth factors, and energy levels all convert their signals through the TSC complex to inhibit or activate Rheb. Rheb is located at the lysosomal membrane. TSC complex is a GAP on Rheb and additionally acts to sterically hinder Rheb's kinase sites. As outlined above, Rheb must be in the high-energy GTP state to activate mTORC1 (Armijo et al. 2016) when mTORC1 translocates to the lysosomal membrane (see Figure 5.2).

5.5 mTORC1 REGULATION BY EXERCISE/MECHANICAL STIMULATION

Mechanosensitive signaling provides an additional and separate level of mTORC1 regulation. Previously, exercise-mediated increases in mTORC1 activity were assumed to be induced by the paracrine/autocrine action of locally muscle-produced IGF-1. Intramuscular-produced IGF-1 is increased post-exercise, but it has been shown that mechanically induced hypertrophy can be achieved in the absence of IGF-1 receptors (Spangenburg et al. 2008), and other studies have shown that hypertrophy can follow loading even with genetic or inhibitor-based ablation of the PI3-Akt pathway (Hornberger et al. 2004, 2007). Thus, other mechanisms beyond IGF-1 must be involved in mechanical-induced mTORC1 activation.

While the exact mechanisms are currently unknown, at least two models show how exercise, passive stretch, or contraction influence mTORC1 activity. First, mechanical deformation of the extracellular matrix linking trans-sarcolemma proteins during contraction may impinge on the TSC complex via a focal adhesion kinase (FAK) signaling cascade (Crossland et al. 2013). Second, muscle fiber mechanical loading acts to upregulate diaglycerol kinase (DGKζ), which facilitates binding of phosphatidic acid (PA) to the mTORC1 complex (You et al. 2014, Zanchi and Lancha 2008, Hornberger 2011).

Integrins are an example of proteins that bridge the extracellular matrix to the interior of the muscle cell via the multiprotein focal adhesion complex. Integrins and the focal adhesion complex are highly concentrated in the parts of the muscle cell that transmit force, both longitudinally at the myotendinous junction and laterally at the costamere (Anastasi et al. 2008, Burkin and Kaufman 1999). Tensile stress during contractile events may deform these types of proteins, resulting in downstream phosphorylation and inactivation of the TSC complex (Jacobs et al. 2013). A primary enzyme within the focal adhesion complex is focal adhesion kinase (FAK). FAK has numerous pleiotropic effects, ranging from modulation of myoblast development to involvement in growth factor-mediated mTORC1 regulation. FAK has increased autophosphorylation postloading (Klossner et al. 2013). Interestingly, the phosphorylation sites of TSC2 after mechanical loading appear to differ from the threonine 1462 site that is phosphorylated by Akt (Inoki et al. 2002). However, the exact identity and specific mechanism of the mechanical sensor in this model has yet to be elucidated.

The other widely touted model involves PA. PA is synthesized by phospholipase D (PLD) from phosphatidylcholine (PC) and by diaglycerol kinase zeta isoform (DGK-zeta) from diacylglycerol (DAG). PA directly binds to mTOR and is sufficient to stimulate mTOR kinase activity *in vitro* (Avila-Flores et al. 2005). PA has been shown to increase postmechanical loading, an effect initially attributed to PLD (Fang et al. 2001). However, recent research suggests that PLD is not necessary, but DGK-zeta is necessary, for PA to activate mTORC1 (You et al. 2014). The current model is that through an unknown mechanical sensor, DGK-zeta synthesizes increased levels of PA to stimulate mTORC1 signaling.

5.6 AGE, mTORC1 SIGNALING, AND PROTEIN ANABOLIC RESISTANCE

While many specific details of mTORC1 regulation have only recently been determined, other aspects are well established. Anabolic signaling pathways, for example, mTORC1-associated signaling cascades, are stimulated by growth factors (i.e., insulin or IGF-1) or nutrients (i.e., the branch chain amino acid leucine) but may not be as responsive in the elderly as in younger individuals (Burd, Gorissen, and Luc van Loon 2013). Numerous studies have shown a disconnection between anabolic stimuli and mTORC1 signaling/anabolic response in older subjects (Francaux et al. 2016, Drummond et al. 2008, Guillet et al. 2004, Fry et al. 2011b, Cuthbertson et al. 2005), although some research has shown little or no change (Chevalier et al. 2011, Symons et al. 2011). A recent study in human skeletal muscle showed that the gene expression pattern of genes that code for proteins involved in mTORC1 signaling and amino acid transport change significantly in response to amino acid feeding (Graber et al. 2017). Therefore, it will be interesting to observe in future follow-up studies if this pattern of gene expression is altered in older adults. Older mice-fed high-fat diets have increased intramuscular lipid accumulation, particularly the bioactive ceramides and diaglycerols, and impaired mTORC1 signaling via enhanced insulin resistance, which leads to loss of muscle mass and strength (Rivas et al. 2016). The inability of anabolic stimuli to activate mTORC1 signaling and protein synthesis has been termed protein anabolic resistance. Protein anabolic resistance may be directly linked to the development of sarcopenia in that even a small decrease in protein synthetic capacity could result in a dramatic loss of muscle mass over decades.

An overall reduced capacity in older adults to respond to anabolic stimuli has long been presumed for protein synthesis (Fry and Rasmussen 2011a, Koopman and Luc van Loon 2009). Aging results in reduced nutrient signaling and a reduced anabolic response to exercise (Kumar et al. 2009, Degens and Alway 2003). However, older individuals may have the same capacity for protein synthesis after feeding. For example, older and younger adults both increased protein synthesis after an acute bout of resistance training followed by a high-protein meal (Symons et al. 2011). There is some evidence that older individuals may require larger amount of proteins or amino acids in order to initiate transcription and translation at the same level as younger adults (Walker et al. 2011, Moore et al. 2015, Cuthbertson et al. 2005). The inability of insulin to produce a protein anabolic response has also been shown in older adults (Guillet et al. 2004, Rasmussen et al. 2006). However, when a sufficient quantity of protein or essential amino acids is consumed, older adults can achieve similar rates of MPS (Moore et al. 2015, Paddon-Jones et al. 2004). Likewise, high levels of insulin can also overcome the protein anabolic resistance in older adults (Fujita et al. 2009). The amount of essential amino acids, especially the branched chain amino acid leucine, seems to determine the level of response. Specific examples include a study in which 7 g of essential amino acids was sufficient to increase protein synthesis in younger, but not older adults (Katsanos et al. 2005); however, when the percentage of leucine in 7 g was increased (from 26% to 46%), protein synthesis was stimulated (Katsanos et al. 2006). However, these studies were performed in a small number of subjects, and the literature is inconclusive about whether protein anabolic resistance to amino acids is a real phenomenon in older adults. Recently, it was shown that resistance exercise training does not improve the muscle protein anabolic response to amino acid feeding in older adults, suggesting that anabolic resistance to amino acids is not a significant problem in healthy older adults (Moro et al. 2018). Furthermore, most studies in humans also show that *fasting* protein synthesis rates are similar in both younger and older adults (Volpi et al. 2001, Dickinson et al. 2014, Moore et al. 2015, Markofski et al. 2015). However, such measurements may not be sufficiently sensitive to detect the tiny changes in daily protein synthesis rates that would, over decades, lead to sarcopenia.

Exercise-induced protein synthesis may also be influenced by age. Exercise is, of course, an accepted intervention for sarcopenia although not a cure. Progressive resistance exercise training is the gold standard intervention to increase muscle strength, function, and mass in older adults (Marini et al. 2008, Fiatarone et al. 1990, Pillard et al. 2011). Resistance exercise training in older adults reduces age-related myofiber apoptosis (Marzetti and Leeuwenburgh 2006), induces gains in strength and muscle mass (Fiatarone et al. 1990, Latham 2004), improves functional outcome measurements (Marini et al. 2008, Capodaglio et al. 2007, Fiatarone et al. 1990), and confers positive metabolic changes (Sundell 2011). However, the protein anabolic response to chronic exercise training is reduced in older adults compared with their younger counterparts (Welle, Thornton, and Statt 1995). The protein anabolic response to an acute bout of resistance exercise is also reduced in older adults (Fry et al. 2011b, Mayhew et al. 2009, Walker et al. 2011). In a study of maintaining training adaptations, younger adults were found to maintain benefits from exercise with lower doses of maintenance exercise than that needed by older adults (Bickel et al. 2011). Similarly, the genetic response to resistance exercise training was altered in older adults compared with younger adults (Thalacker-Mercer et al. 2009). A study combining resistance exercise training and post-exercise protein feeding found that phosphorylated S6K1 (a marker of mTORC1 activation) was elevated in both the older and younger cohorts at the acute level but was blunted in the older group after a training program (Farnfield et al. 2012). This result contrasts with previous research showing no added benefit of protein ingestion combined with resistance exercise training in the elderly (Verdijk et al. 2009).

Protein anabolic resistance is clearly a problem in older adults performing acute or chronic resistance exercise. The mechanisms responsible for why older adults have difficulty adding muscle mass in response to resistance exercise training are unknown. A couple potential mechanism may be (1) dysfunction in the mechanically stimulated activation of mTORC1 (Hornberger 2011) and/or (2) the systemic physiological environment in the older (particularly the less-healthy) population may be less conducive to muscle hypertrophy. For example, global inflammation, hormonal alterations, growth factor reduction, reduced satellite cell proliferative capacity, amino acid transporter expression, and endothelial dysfunction may all contribute to reduced training response in older adults (Burd, Gorissen, and Luc van Loon 2013).

Age-associated global inflammation has been linked to reduced capacity to respond to exercise (Merritt et al. 2013). Many of the repair mechanisms in muscle activated after a vigorous exercise routine rely upon inflammatory signaling to activate response (Evans and Cannon 1991). Thus, a high level of background noise in the form of global elevation of inflammatory signaling may obscure the normal repair signaling response, limiting adaptation (Jo et al. 2012).

Anabolic hormone production (i.e., testosterone and growth hormone [GH]) declines with age and may be another potential cause of sarcopenia (Sipilä et al. 2013, Valenti 2010). The production of testosterone declines 1% per year after age of 40 years in men (Gray et al. 1991) and declines up to 64% and 28%, in men and women, respectively (Van der Beld et al. 2000). Growth hormone (GH) declines steadily with age (Sipilä et al. 2013, Valenti 2010, Gray et al. 1991, van den Beld et al. 2000), starting in the early 30s in men (Sattler 2013). Testosterone is not bioactive when bound to sex hormone-binding globulin (SHBG). SHBG increases more than twofold in quantity with age (van den Beld et al. 2000). Meta-analysis of testosterone replacement therapy has demonstrated increasing strength and mass with treatment (Ottenbacher et al. 2006, Neto et al. 2015). GH has numerous metabolic functions including mitochondrial efficiency, improving overall body composition by increasing fat metabolism, and increasing resistance to fatigue (Ryall, Schertzer, and Lynch 2008, Sattler 2013). However, to date, GH has not been proven effective to stimulate muscle mass increase (Lynch 2004) or strength (Borst 2004), although GH has been shown to improve body composition and general well-being (Sattler 2013).

Insulin and IGF-1 and 2 are also powerful anabolic agents (Cushman and Wardzala 1980, Klip et al. 2014, Glass 2010). The insulin resistance often seen in older individuals may contribute to deviations from normal postprandial activation of protein anabolism (Fujita et al. 2009). IGF-1 is important in regulating apoptosis and denervation (Messi and Delbono 2007), as well as in satellite cell activation and differentiation (Snijders et al. 2015). Systemic IGF is produced in the liver, controlled by growth hormone, but other IGF-1 is produced inside the muscles and used locally. IGF-1 is reduced by a third in older adults with sarcopenia (Benbassat, Maki, and Unterman 1997), and the IGF-1 muscle-derived splice variants are reduced by 45% in older versus younger adults (Léger et al. 2008). The regulation of IGF-1 by binding proteins and regulation of intramuscular production has not been fully elucidated. Some forms of IGF-1-binding proteins are more prevalent in older individuals than in younger (Ryall, Schertzer, and Lynch 2008).

Satellite cells perform a critical function in repair of damaged muscle fibers. Older subjects have a reduced satellite cell functionality (Conboy et al. 2005, Gopinath and Rando 2008). Although recent data show that satellite cell function may not be necessary for muscle hypertrophy, at least as far as maintaining the myonuclear domain or affecting sarcopenia in mice (Jackson et al. 2012, Fry et al. 2015), older subjects have an undisputed decline in proliferative/differential capacity, which could certainly hinder injury recovery, possibly contributing to increased fibrosis and potentially hindering the exercise response.

Amino acid transporters also may play a role in anabolic response to protein feeding (Dickinson et al. 2013) and have age-associated difference in response to exercise (Drummond et al. 2011). We have already mentioned the contribution of SLC38A9 as an arginine sensor, and of PAT1, but the transport of amino acids into the cell by other transporters such as L-type amino acid transporter (LAT1) and sodium-coupled neutral amino acid transporter (SNAT2) also plays a critical role in producing a protein anabolic response. Nutrients must be absorbed in the gut and transported through the blood to the muscle before they can be utilized. Insulin induces vasodilation (Fujita et al. 2006). With age, there is a reduction in induced vasodilation following feeding or exercise. Pharmacologically restoring vasodilation restores, to youthful levels, amino acid flux from blood to muscle (Timmerman et al. 2010). Endothelial dysfunction with aging is a likely contributor to protein anabolic resistance. We point to a recent review that goes into much greater detail of the potential etiology and mechanisms for the role of endothelial dysfunction to alter amino acid sensing with age (Moro et al. 2016).

In conclusion, we have provided evidence that mTORC1 signaling is a vital process required for maintaining muscle mass, strength, and function. Dysregulation of mTORC1 is clear during aging and may contribute to sarcopenia or the ability to maintain muscle mass. However, we readily acknowledge the negative effects of chronic mTORC1 activation on metabolic abnormalities, myopathies, and reduced longevity. We propose, in contrast, that *periodic and transient* increases in mTORC1 activity in response to anabolic stimuli such as feeding, growth factors, and exercise are a necessary process for maintaining muscle mass, quality, and function, and the utilization of interventions designed to enhance the protein anabolic response in older adults will be essential in counteracting sarcopenia.

REFERENCES

Anastasi, G., G. Cutroneo, G. Santor, A. Arco, G. Rizzo, P. Bramanti, C. Rinaldi, A. Sidoti, A. Amato, and A. Favaloro. 2008. Costameric proteins in human skeletal muscle during muscular inactivity. *Journal of Anatomy* 213 (3):284–295.

Armijo, M.E., T. Campos, F. Fuentes-Villalobos, M.E. Palma, R. Pincheira, and A.F. Castro. 2016. Rheb signaling and tumorigenesis: mTORC1 and new horizons. *International Journal of Cancer* 138 (8):1815–1823.

Avila-Flores, A., T. Santos, E. Rincón, and I. Mérida. 2005. Modulation of the mammalian target of rapamycin pathway by diacylglycerol kinase-produced phosphatidic acid. *Journal of Biological Chemistry* 280 (11):10091–10099.

Bar-Peled, L., L. Chantranupong, A.D. Cherniack, W.W. Chen, K.A. Ottina, B.C. Grabiner, E.D. Spear, S.L. Carter, M. Meyerson, and D.M. Sabitini. 2013. A tumor suppressor complex with GAP activity for the Rag GTPases that signal amino acid sufficiency to mTORC1. *Science* 340 (6136):1100–1106.

Bar-Peled, L., and D.M. Sabatini. 2014. Regulation of mTORC1 by amino acids. *Trends in Cell Biology* 24 (7):400–406.

Bar-Peled, L., L.D. Schweitzer, R. Zoncu, and D.M. Sabatini. 2012. An expanded Ragulator is a GEF for the Rag GTPases that signal amino acid levels to mTORC1. *Cell* 150 (6):1196–1208.

Benbassat, C.A., K.C. Maki, and T.G. Unterman. 1997. Circulating levels of insulin-like growth factor (IGF) binding protein-1 and -3 in aging men: relationships to insulin, glucose, IGF, and dehydroepiandrosterone sulfate levels and anthropometric measures. *The Journal of Clinical Endocrinology and Metabolism* 82 (5):1484–1491.

Bickel, C.S., J. Cross, and M. Bamman. 2011. Exercise dosing to retain resistance training adaptations in young and older adults. *Medicine Science Sports Exercise* 43 (7):1177–1187.

Borst, S. 2004. Interventions for sarcopenia and muscle weakness in older people. *Age Ageing* 33 (6):548–555.

Breen, L., and S. Phillips. 2011. Skeletal muscle protein metabolism in the elderly: Interventions to counteract the "anabolic resistance" of ageing. *Nutrition & Metabolism* 8:68.

Brugarolas, J., K. Lei, R.L. Hurley, B.D. Manning, J.H. Reiling, E. Hafen, L.A. Witters, L.W. Ellisen, and W.G. Kaelin Jr. 2004. Regulation of mTOR function in response to hypoxia by REDD1 and the TSC1/TSC2 tumor suppressor complex. *Genes & Development* 18 (23):2893–2904.

Burd, N., S. Gorissen, and J.C. Luc van Loon. 2013. Anabolic resistance of muscle protein synthesis with aging. *Exercise and Sport Sciences Reviews* 41 (3):169–173.

Burkin, D.J., and S.J. Kaufman.1999. The alpha7beta1 integrin in muscle development and disease. *Cell and Tissue Research* 296 (1):183–190.

Capodaglio, P., M. Capodaglio Edda, M. Facioli, and F. Saibene. 2007. Long-term strength training for community-dwelling people over 75: Impact on muscle function, functional ability and life style. *European Journal of Applied Physiology* 100 (5):535–542.

Carroll, B., D. Maetzel, O.D. Maddocks et al. 2016. Control of TSC2-Rheb signaling axis by arginine regulates mTORC1 activity. Mizushima N, ed. *eLife* 5:e11058.

Chantranupong, L., S.M. Scaria, R.A. Saxton, M.P. Gygi, K. Shen, G.A. Wyant, T. Wang, J.W. Harper, S.P. Gygi, and D.M. Sabitini. 2016. The CASTOR proteins are arginine sensors for the mTORC1 pathway. *Cell* 165 (1):153–164.

Chantranupong, L., R.L. Wolfson, J.M. Orozco, R.A. Saxton, S.M. Scaria, L. Bar-Peled, E. Spooner, M. Isasa, S.P. Gygi, and D.M. Sabatini. 2014. The Sestrins interact with GATOR2 to negatively regulate the amino-acid-sensing pathway upstream of mTORC1. *Cell Reports* 9:1–8.

Chevalier, S., E.D. Goulet, S.A. Burgos, L.J. Wykes, and J.A. Morais. 2011. Protein anabolic responses to a fed steady state in healthy aging. *The Journals of Gerontology: Series A, Biological Sciences and Medical Sciences* 66:681–688.

Conboy, I., M. Conboy, A. Wagers, E. Girma, I. Weissman, and T. Rando. 2005. Rejuvenation of aged progenitor cells by exposure to a young systemic environment. *Nature* 433 (7027):760–764.

Crossland, H., A.A. Kazi, C.H. Lang, J.A. Timmons, P. Pierre, D.J. Wilkinson, K. Smith, N.J. Szewczyk, and P.J. Atherton. 2013. Focal adhesion kinase is required for IGF-I-mediated growth of skeletal muscle cells via a TSC2/mTOR/S6K1-associated pathway. *American Journal of Physiology. Endocrinology and Metabolism* 305 (2):E183–E193.

Cushman, S.W., and L.J. Wardzala. 1980. Potential mechanism of insulin action on glucose transport in the isolated rat adipose cell: Apparent translocation of intracellular transport systems to the plasma membrane. *Journal of Biological Chemistry* 255 (10):4758–4762.

Cuthbertson, D., K. Smith, J. Babraj, G. Leese, T. Waddell, P. Atherton, H. Wackerhage, P.M. Taylor, and M.J. Rennie. 2005. Anabolic signaling deficits underlie amino acid resistance of wasting, aging muscle. *The FASEB Journal* 19 (3):422–424.

Degens, H., and S.E. Alway. 2003. Skeletal muscle function and hypertrophy are diminished in old age. *Muscle and Nerve* 27 (3):339–347.

Dennis, M.D., C.S. Coleman, A. Berg, L.S. Jefferson, and S.R. Kimball. 2014. REDD1 enhances protein phosphatase 2A-mediated dephosphorylation of Akt to repress mTORC1 signaling. *Science Signaling* 7 (335):ra68.

Deschenes, M.R. 2004. Effects of aging on muscle fibre type and size. *Sports Medicine* 34 (12):809–824.

Dibble, C.C., and L.C. Cantley. 2015. Regulation of mTORC1 by PI3K signaling. *Trends in Cell Biology* 25 (9):545–555.

Dibble, C.C., and B.D. Manning. 2013. Signal integration by mTORC1 coordinates nutrient input with biosynthetic output. *Nature Cell Biology* 15 (6):555–564.

Dickinson, J., M.J. Drummond, J. Coben, E. Volpi, and B. Rasmussen. 2013. Aging differentially affects human skeletal muscle amino acid transporter expression when essential amino acids are ingested after exercise. *Clinical Nutrition* 32 (2):273–280.

Dickinson, J.M., D.M. Gundermann, D.K. Walker, P.T. Reidy, M.S. Borack, M.J. Drummond, M. Arora, E. Volpi, and B.B. Rasmussen. 2014. Leucine-enriched amino acid ingestion after resistance exercise prolongs myofibrillar protein synthesis and amino acid transporter expression in older men. *The Journal of Nutrition* 144 (11):1694–1702.

Drummond, M.J., H.C. Dreyer, B. Pennings, C.S. Fry, S. Dhanani, E.L. Dillon, M. Sheffield-Moore, E. Volpi, and B.B. Rasmussen. 2008. Skeletal muscle protein anabolic response to resistance exercise and essential amino acids is delayed with aging. *Journal of Applied Physiology* 104:1452–1461.

Drummond, M.J., C.S. Fry, E.L. Glynn, K.L. Timmerman, J.M. Dickinson, D.K. Walker, D.M. Gundermann, E. Volpi, and B.B. Rasmussen. 2011. Skeletal muscle amino acid transporter expression is increased in young and older adults following resistance exercise. *Journal of Applied Physiology* 111 (1):135–142.

Egerman, M.A., and D.J. Glass. 2014. Signaling pathways controlling skeletal muscle mass. *Critical Reviews in Biochemistry and Molecular Biology* 49 (1):59–68.

Evans, W.J., and J.G. Cannon. 1991. The metabolic effects of exercise-induced muscle damage. *Exercise Sport Science Reviews* 19:99–125.

Fang, Y., M. Vilella-Bach, R. Bachmann, A. Flanigan, and J. Chen. 2001. Phosphatidic acid-mediated mitogenic activation of mTOR signaling. *Science* 294 (5548):1942–1945.

Farnfield, M., L. Breen, K. Carey, A. Garnham, and D. Cameron Smith. 2012. Activation of mTOR signalling in young and old human skeletal muscle in response to combined resistance exercise and whey protein ingestion. *Applied Physiology, Nutrition, and Metabolism* 37 (1):21–30.

Fiatarone, M.A., E.C. Marks, N.D. Ryan, C.N. Meredith, L.A. Lisitz, and W.J. Evans. 1990. High-intensity strength training in nonagenarians. Effects on skeletal muscle. *JAMA: the Journal of the American Medical Association* 263 (22):3029–3034.

Francaux, M., B. Demeulder, D. Naslain, R. Fortin, O. Lutz, G. Caty, and L. Deldicque. 2016. Aging reduces the activation of the mTORC1 pathway after resistance exercise and protein intake in human skeletal muscle: Potential role of REDD1 and impaired anabolic sensitivity. *Nutrients* 8 (1):47.

Fry, C.S., M.J. Drummond, E.L. Glynn, J.M. Dickinson, D.M. Gundermann, K.L. Timmerman, D.K. Walker, S. Dhanani, E. Volpi, and B.B. Rasmussen. 2011b. Aging impairs contraction-induced human skeletal muscle mTORC1 signaling and protein synthesis. *Skeletal Muscle B* 1:47.

Fry, C.S., J.D. Lee, J. Mula et al. 2015. Inducible depletion of satellite cells in adult, sedentary mice impairs muscle regenerative capacity but does not contribute to sarcopenia. *Nature Medicine* 21 (1):76–80.

Fry, C.S., and B. Rasmussen. 2011a. Skeletal muscle protein balance and metabolism in the elderly. *Current Aging Science* 4 (3):260–268.

Fujita, S., E.L. Glynn, K.L. Timmerman, B.B. Rasmussen, and E. Volpi. 2009. Supraphysiological hyperinsulinaemia is necessary to stimulate skeletal muscle protein anabolism in older adults: Evidence of a true age-related insulin resistance of muscle protein metabolism. *Diabetologia* 52: 1889–1898.

Fujita S., B.B. Rasmussen, J.G. Cadenas, J.J. Grady, and E. Volpi. 2006. Effect of insulin on human skeletal muscle protein synthesis is modulated by insulin-induced changes in muscle blood flow and amino acid availability. *American Journal of Physiology: Endocrinology and Metabolism* 291 (4):E745–E754.

Gingras, A.C., B. Raught, and N. Sonenberg. 2004. mTOR signaling to translation. *Current Topics in Microbiology and Immunology* 279:169–197.

Glass, D.J. 2010. Pi3 kinase regulation of skeletal muscle hypertrophy and atrophy. *Current Topics in Microbiology and Immunology* 346:267–278.

Gopinath, S.D., and T.A. Rando. 2008. Stem cell review series: Aging of the skeletal muscle stem cell niche. *Aging Cell* 7 (4):590–598.

Gordon, B.S., D.L. Williamson, C.H. Lang, L.S. Jefferson, and S.R. Kimball. 2015. Nutrient induced stimulation of protein synthesis in mouse skeletal muscle is limited by the mTORC1 repressor REDD1. *The Journal of Nutrition* 145 (4):708–713.

Graber, T.G., M.S. Borack, P.T. Reidy, E. Volpi, B.B. Rasmussen. 2017. Essential amino acid ingestion alters expression of genes associated with amino acid sensing, transport, and mTORC1 regulation in human skeletal muscle. *Nutrition & Metabolism* 14:35.

Gray, A., H.A. Feldman, J.B. McKinlay, and C. Longcope. 1991. Age, disease, and changing sex hormone levels in middle-aged men: results of the Massachusetts Male Aging Study. *Journal of Clinical Endocrinology and Metabolism* 73:1016–1025.

Guillet, C., M. Prod'homme, M. Balage, P. Gachon, C. Giraudet, L. Morin, J. Grizard, and Y. Boirie. 2004. Impaired anabolic response of muscle protein synthesis is associated with S6K1 dysregulation in elderly humans. *The FASEB Journal* 18:1586–1587.

Hornberger, T.A. 2011. Mechanotransduction and the regulation of mtorc1 signaling in skeletal muscle. *International Journal of Biochemistry & Cell Biology* 43 (9):1267–1276.

Hornberger, T.A., R. Stuppard, K.E. Conley, M.J. Fedele, M.L. Fiorotto, E.R. Chin, and K.A Esser. 2004. Mechanical stimuli regulate rapamycin-sensitive signalling by a phosphoinositide 3-kinase-, protein kinase B- and growth factor-independent mechanism. *Biochemical Journal*, 380 (Pt 3):795–804.

Hornberger, T.A., K.B. Sukhija, X.-R. Wang, and S. Chien. 2007. mTOR is the Rapamycin-sensitive Kinase that confers mechanically-induced Phosphorylation of the Hydrophobic Motif Site Thr(389) in p70S6k. *FEBS Letters* 581 (24):4562–4566.

Inoki, K., Y. Li, T. Zhu, J. Wu, and K.L. Guan. 2002. TSC2 is phosphorylated and inhibited by Akt and suppresses mTOR signalling. *Nature Cell Biology* 4 (9):648–657.

Jackson, J., J. Mula, T. Kirby, C.S. Fry, J.D. Lee, M.F. Ublele, K.S. Campbell, C.A. Peterson, and E.E. Dupont-Versteeqden. 2012. Satellite cell depletion does not inhibit adult skeletal muscle regrowth following unloading-induced atrophy. *American Journal of Physiology: Cell Physiology* 303 (8):C854–C861.

Jacobs, B.L., J.-S. You, J.W. Frey, C.A. Goodman, D.M. Gundermann, and T.A. Hornberger. 2013. Eccentric contractions increase the phosphorylation of tuberous sclerosis complex-2 (TSC2) and alter the targeting of TSC2 and the mechanistic target of rapamycin to the lysosome. *The Journal of Physiology* 591 (Pt 18):4611–4620.

Janssen, I., S.B. Heymsfield, and R. Ross. 2002. Low relative skeletal muscle mass (sarcopenia) in older persons is associated with functional impairment and physical disability. *Journal of the American Geriatrics Society* 50 (5):889–896.

Jo, E., S. Lee, B. Park, and J. Kim. 2012. Potential mechanisms underlying the role of chronic inflammation in age-related muscle wasting. *Aging Clinical and Experimental Research* 24 (5):412–422.

Katsanos C.S., H. Kobayashi, M. Sheffield-Moore, A. Aarsland, and R.R. Wolfe. 2005. Aging is associated with diminished accretion of muscle proteins after the ingestion of a small bolus of essential amino acids. *The American Journal of Clinical Nutrition* 82 (5):1065–1073.

Katsanos C.S., H. Kobayashi, M. Sheffield-Moore, A. Aarsland, and R.R. Wolfe. 2006. A high proportion of leucine is required for optimal stimulation of the rate of muscle protein synthesis by essential amino acids in the elderly. *The American Journal of Physiology-Endocrinology and Metabolism* 291 (2):E381–E387.

Kaul, G., G. Pattan, and T. Rafeequi. 2011. Eukaryotic elongation factor-2 (eEF2): Its regulation and peptide chain elongation. *Cell Biochemistry & Function* 29 (3):227–234.

Kim, J., and E. Kim. 2016. Rag GTPase in amino acid signaling. *Amino Acids* 48 (4):915–928.

Kimball, S.R., B.S. Gordon, J.E. Moyer, M.D. Dennis, and L.S. Jefferson. 2016. Leucine induced dephosphorylation of Sestrin2 promotes mTORC1 activation. *Cell Signal* 28 (8):896–906.

Klossner, S., R. Li, S. Ruoss, A.C. Durieux, and M. Flück. 2013. Quantitative changes in focal adhesion kinase and its inhibitor, FRNK, drive load-dependent expression of costamere components. *The American Journal of Physiology—Regulatory, Integrative and Comparative Physiology* 305 (6):R647–R657.

Klip, A., Y. Sun, T.T. Chui, and K.P. Foley. 2014. Signal transduction meets vesicle traffic: The software and hardware of GLUT4 translocation. *AJP-Cell Physiology* 306 (10):C879–C886.

Koopman, R., and Luc van Loon. 2009. Aging, exercise, and muscle protein metabolism. *Journal of Applied Physiology* 106 (6):2040–2048.

Kumar, V., A. Selby, D. Rankin, R. Patel, P. Atherton, J. Williams, K. Smith, O. Seynnes, N. Hicock, and M.J. Rennie. 2009. Age-related differences in the dose-response relationship of muscle protein synthesis to resistance exercise in young and old men. *Journal of Physiology* 587 (1):211–217.

Landi, F., R. Liperoti, A. Russo et al. 2012. Sarcopenia as a risk factor for falls in elderly individuals: Results from the ilSIRENTE study. *Clinical Nutrition* 31 (5):652–658.

Latham, N. 2004. Physiotherapy to treat sarcopenia in older adults. *New Zealand Journal of Physiotherapy* 32 (1):16–21.

Léger, B., W. Derave, K. De Bock, P. Hespel, and A.P. Russell. 2008. Human sarcopenia reveals an increase in SOCS-3 and myostatin and a reduced efficiency of Akt phosphorylation. *Rejuvenation Research* 11 (1):163–175B.

Lin, M.G., and J.H. Hurley. 2016. Structure and function of the ULK1 complex in autophagy. *Current Opinion in Cell Biology* 39:61–68.

Lynch, G.S. 2004. Emerging drugs for sarcopenia: Age-related muscle wasting. *Expert Opinion on Emerging Drugs* 9 (2):345–361.

Magnuson, B., B. Ekim, and D.C. Fingar. 2012. Regulation and function of ribosomal protein S6 kinase (S6K) within mTOR signalling networks. *Biochemical Journal* 441 (1):1–21.

Marini, M., E. Sarchielli, L. Brogi, R. Lazzeri, R. Salerno, E. Sgambati, and M. Monaci. 2008. Role of adapted physical activity to prevent the adverse effects of the sarcopenia: A pilot study. *Italian Journal of Anatomy and Embryology* 113 (4):217–225.

Markofski, M.M., J.M. Dickinson, M.J. Drummond et al. 2015. Effect of age on basal muscle protein synthesis and mTORC1 signaling in a large cohort of young and older men and women. *Experimental Gerontology* 65:1–7.

Marzetti, E., and C. Leeuwenburgh. 2006. Skeletal muscle apoptosis, sarcopenia and frailty at old age. *Experimental Gerontology* 41 (12):1234–1238.

Mayhew, D.L., J. Kim, J.M. Cross, A.A. Ferrando, and M.M. Bamman. 2009. Translational signaling responses preceding resistance training-mediated myofiber hypertrophy in young and old humans. *Journal of Applied Physiology* 107 (5):1655–1662.

Merritt, E., M. Stec, A. Thalacker-Mercer, S. Windham, J. Cross, D. Shelley, C. Tuggle, S. Kosek, J-S. Kim, and M.M. Bamman. 2013. Heightened muscle inflammation susceptibility may impair regenerative capacity in aging humans. *Journal of Applied Physiology* 115 (6):937–948.

Messi, M.L., and O. Delbono. 2007. Target-derived trophic effect on skeletal muscle innervations in senescent mice. *Journal Neuroscience* 23:1351–1359.

Mihaylova, M.M., and R.J. Shaw. 2011. The AMP-activated protein kinase (AMPK) signaling pathway coordinates cell growth, autophagy, & metabolism. *Nature Cell Biology* 13 (9):1016–1023.

Moore, D., T. Churchward Venne, O. Witard, L. Breen, N.A. Burd, K.D. Tipton, and S.M. Phillips. 2015. Protein ingestion to stimulate myofibrillar protein synthesis requires greater relative protein intakes in healthy older versus younger men. *The Journals of Gerontology Series A, Biological Sciences and Medical Sciences* 70 (1):57–62.

Moro, T., C.R. Brightwell, R.R. Deer, T.G. Graber, E. Galvan, C.S. Fry, E. Volpi, and B.B. Rasmussen. 2018. Muscle protein anabolic resistance to essential amino acids does not occur in healthy older adults before or after resistance exercise training. *Journal of Nutrition* 148 (6):900–909.

Moro, T., S.M. Ebert, C.M. Adams, and B.B. Rasmussen. 2016. Amino Acid Sensing in Skeletal Muscle. *Trends in Endocrinology & Metabolism* 27 (11):796–806.

Narici, M., and N. Maffulli. 2010. Sarcopenia: Characteristics, mechanisms and functional significance. *British Medical Bulletin* 95:139–159.

Neto, W., E. Gama, L. Rocha, C.C. Ramos, W. Taets, K.B. Scapini, J.B. Ferreira, B. Rodrigues, and E. Casperuto. 2015. Effects of testosterone on lean mass gain in elderly men: Systematic review with meta-analysis of controlled and randomized studies. *Age* 37 (1):9742.

Ögmundsdóttir, M.H., S. Heublein, S. Kazi, B. Reynolds, S.M. Visvalingam, M.K. Shaw, and D.C.I. Goberdhan. 2012. Proton-assisted amino acid transporter PAT1 complexes with Rag GTPases and activates TORC1 on late endosomal and lysosomal membranes. *PLoS One* 7 (5):e36616.

Ottenbacher, K., M. Ottenbacher, A. Ottenbacher, A. Acha, and G. Ostir. 2006. Androgen treatment and muscle strength in elderly men: A meta-analysis. *Journal of the American Geriatric Society* 54 (11):1666–1673.

Ouyang, W., W. Liao, C.T. Luo et al. 2012. Novel Foxo1-dependent transcriptional programs control T(reg) cell function. *Nature* 491 (7425):554–559.

Paddon-Jones, D., M. Sheffield-Moore, X.J. Zhang, E. Volpi, S.E. Wolf, A. Aarsland, A.A. Ferrando, and R.R. Wolfe. 2004. Amino acid ingestion improves muscle protein synthesis in the young and elderly. *The American Journal of Physiology—Endocrinology and Metabolism* 286 (3):E321–E328.

Peng, M., N. Yin, and M.O. Li. 2014. Sestrins function as guanine nucleotide dissociation inhibitors for Rag GTPases to control mTORC1 signaling. *Cell* 159:122–133.

Pillard, F., D. Laoudj Chenivesse, G. Carnac, J. Mercier, J. Rami, D. Riviere, and Y. Rolland. 2011. Physical activity and sarcopenia. *Clinics in Geriatric Medicine* 27 (3):449–470.

Rasmussen, B.B., S. Fujita, R.R. Wolfe, B. Mittendorfer, M. Roy, V.L. Rowe, and E. Volpi. 2006. Insulin resistance of muscle protein metabolism in aging. *The FASEB Journal* 20: 768–769.

Rebsamen, M, L. Pochini, T. Stasyk et al. 2015. SLC38A9 is a component of the lysosomal amino acid sensing machinery that controls mTORC1. *Nature* 519:477–481.

Richter, J.D., and N. Sonenberg. 2005. Regulation of cap-dependent translation by eIF4E inhibitory proteins. *Nature* 433 (7025):477–480.

Rivas, D.A., D.J. McDonald, N.P. Rice, P.H. Haran, G.G. Dolnikowski, and R.A. Fielding. 2016. Diminished anabolic signaling response to insulin induced by intramuscular lipid accumulation is associated with inflammation in aging but not obesity. *The American Journal of Physiology—Regulatory, Integrative and Comparative Physiology* 310 (7):R561–R569.

Rolland, Y., S. Czerwinski, G. Abellan Van Kan et al. 2008. Sarcopenia: Its assessment, etiology, pathogenesis, consequences and future perspectives. *Journal of Nutrition Health Aging* 12:433–450.

Roux, P.P., D. Shahbazian, H. Vu, M.K. Holz, M.S. Cohen, J. Taunton, N. Sonenberg, and J. Blenis. 2007. RAS/ERK signaling promotes site-specific ribosomal protein S6 phosphorylation via RSK and stimulates cap-dependent translation. *The Journal of Biological Chemistry* 282 (19):14056–14064.

Ryall, J., J. Schertzer, and G. Lynch. 2008. Cellular and molecular mechanisms underlying age related skeletal muscle wasting and weakness. *Biogerontology* 9 (4):213–228.

Sancak, Y., L. Bar-Peled, R. Zoncu, A.L. Markhard, S. Nada, and D.M. Sabatini. 2010. Ragulator-Rag complex targets mTORC1 to the lysosomal surface and is necessary for its activation by amino acids. *Cell* 141 (2):290–303.

Sardiello, M., M. Palmieri, A. di Ronza et al. 2009. A gene network regulating lysosomal biogenesis and function. *Science* 325 (5939):473–477.

Sattler, F. 2013. Growth hormone in the aging male. *Best Practice & Research Clinical Endocrinology & Metabolism* 27 (4):541–555.

Schalm, S.S., D.C. Fingar, D.M. Sabatini, and J. Blenis. 2003. TOS motif-mediated raptor binding regulates 4E-BP1 multisite phosphorylation and function. *Current Biology* 13 (10):797–806.

Sipilä, S., M. Narici, M. Kjaer, E. Pollanen, R.A. Atkinson, M. Hansen, and V. Kovanen. 2013. Sex hormones and skeletal muscle weakness. *Biogerontology* 14 (3):231–245.

Snijders, T., J.P. Nederveen, B.R. McKay, S. Joanisse, L.B. Verdijk, L.J.C. van Loon, and G. Parise. 2015. Satellite cells in human skeletal muscle plasticity. *Frontiers in Physiology* 6:283.

Spangenburg, E.E., D. Le Roith, C.W. Ward, and S.C. Bodine. 2008. A functional insulin-like growth factor receptor is not necessary for load-induced skeletal muscle hypertrophy. *The Journal of Physiology* 586 (Pt 1):283–291.

Sundell, J. 2011. Resistance training is an effective tool against metabolic and frailty syndromes. *Advances in Preventive Medicine* 984683–984683.

Symons, T.B., M. Sheffield Moore, M.M. Mamerow, R.R. Wolfe, and D. Paddon Jones. 2011. The anabolic response to resistance exercise and a protein-rich meal is not diminished by age. *Journal Nutrition Health Aging* 15 (5):376–381.

Thalacker-Mercer, A., L.J. Dell'italia, X. Cui, J.M. Cross, and M.M. Bamman. 2009. Differential genomic responses in old vs. young humans despite similar levels of modest muscle damage after resistance loading. *Physiological Genomics* 40:141–149.

Timmerman, K.L., J.L. Lee, S. Fujita, S. Dhanani, H.C. Dreyer, C.S. Fry, M.J. Drummond, M. Sheffield-Moore, B.B. Rasmussen, and E. Volpi. 2010. Pharmacological vasodilation improves insulin-stimulated muscle protein anabolism but not glucose utilization in older adults. *Diabetes* 59 (11):2764–2771.

UNDESA. 2013. United Nations Department of Economic and Social Affairs Ageing Social Policy and Development Division. World Population Aging Report. Accessed via WWW on 02/03/2016: http://undesadspd.org/ageing.aspx

Valenti, G. 2010. Aging as an allostasis condition of hormones secretion: Summing up the endocrine data from the inChianti study. *Acta Bio-Medica de L'Ateneo Parmense* 81 (Suppl 1):9–14.

van den Beld, A.W., F.H. de Jong, D.E. Grobbee, H.A. Pols, and S.W. Lamberts. 2000. Measures of bioavailable serum testosterone and estradiol and their relationships with muscle strength, bone density, and body composition in elderly men. *Journal of Clinical Endocrinology Metabolism* 85 (9):3276–3282.

Verdijk, L.B., R.A.M. Jonkers, B.G. Gleeson, M. Beelen, K. Meijer, H.H.C.M. Savelberg, W.K.W.H. Wodzig, P. Dendale, and L.J.C. van Loon. 2009. Protein supplementation before and after exercise does not further augment skeletal muscle hypertrophy after resistance training in elderly men. *The American Journal of Clinical Nutrition* 89 (2):608–616.

Volpi, E., M. Sheffield-Moore, B.B. Rasmussen, and R.R. Wolfe. 2001. Basal muscle amino acid kinetics and protein synthesis in healthy young and older men. *JAMA: The Journal of the American Medical Association* 286 (10):1206–1212.

Walker, D.K., J.M. Dickinson, K.L. Timmerman, M.J. Drummond, P.T. Reidy, C.S. Fry, D.M. Gundermann, and B.B. Rasmussen. 2011. Exercise, amino acids, and aging in the control of human muscle protein synthesis. *Medicine and Science in Sports and Exercise* 43 (12):2249–2258.

Wang, S., Z.Y. Tsun, R.L. Wolfson et al. 2015. Metabolism. Lysosomal amino acid transporter SLC38A9 signals arginine sufficiency to mTORC1. *Science* 347:188–194.

Welle S., C. Thornton, and M. Statt. 1995. Myofibrillar protein synthesis in young and old human subjects after three months of resistance training. *American Journal of Physiology* 268:E422–E427.

Weng, Q.P., M. Kozlowski, C. Belham, A. Zhang, M.J. Comb, and J.J. Avruch. 1998. Regulation of the p70 S6 kinase by phosphorylation in vivo. Analysis using site-specific anti-phosphopeptide antibodies. *Journal of Biological Chemistry* 273 (26):16621–16629.

Wolfson, R.L., L. Chantranupong, R.A. Saxton, K. Shen, S.M. Scaria, J.R. Cantor, and D.M. Sabatini. 2016. Sestrin2 is a leucine sensor for the mTORC1 pathway. *Science* 351 (6268):43–48.

Wolfson, R.L., L. Chantranupong, G.A. Wyant et al. 2017. KICSTOR recruits GATOR1 to the lysosome and is necessary for nutrients to regulate mTORC1. *Nature* 543 (7645):438–442.

Yoon, M.-S., and C.S. Choi. 2016. The role of amino acid-induced mammalian target of rapamycin complex 1(mTORC1) signaling in insulin resistance. *Experimental & Molecular Medicine* 48 (1):e201.

You, J.S., H.C. Lincoln, C.R. Kim, J.W. Frey, C.A. Goodman, X.P. Zhong, and T.A. Hornberger. 2014. The role of diacylglycerol kinase ζ and phosphatidic acid in the mechanical activation of mammalian target of rapamycin (mtor) signaling and skeletal muscle hypertrophy. *Journal of Biological Chemistry* 289 (3):1551–1563.

Zanchi, N.E., and A.H. Lancha Jr. 2008. Mechanical stimuli of skeletal muscle: Implications on mTOR/p70s6k and protein synthesis. *European Journal of Applied Physiology* 102 (3):253–263.

Zhang, C.S., B. Jiang, M. Li et al. 2014. The lysosomal v-ATPase-Ragulator complex is a common activator for AMPK and mTORC1, acting as a switch between catabolism and anabolism. *Cell Metabolism* 20 (3):526–540.

6 Different Adaptation of Ubiquitin-Proteasome and Lysosome-Autophagy Signaling in Sarcopenic Muscle

Kunihiro Sakuma, Hidetaka Wakabayashi,
and Akihiko Yamaguchi

CONTENTS

6.1 INTRODUCTION

In humans, skeletal muscle is the most abundant tissue in the body, comprising 40%–50% of the body mass and playing vital roles in locomotion, heat production during periods of cold stress, and overall metabolism. It makes up the largest pool of proteins in the body, highlighting why this specific tissue is highly sensitive to conditions that act to alter the balance between protein synthesis and degradation. Loss of muscle is a serious consequence of many chronic diseases and of aging itself because it leads to weakness, loss of independence, and an increased risk of death.

Previous studies using animal models demonstrated that muscle atrophy caused by various catabolic stimuli lead to the activation of protein degradation similar to that caused by both the ubiquitin–proteasome system (UPS) and autophagy. Most muscle proteins, particularly myofibrillar components, are considered to be degraded by the UPS. Two muscle-specific ubiquitin ligases, muscle RING finger-1 (MuRF-1) and atrophy gene-1 (atrogin-1), are markedly induced in a wide range of *in vivo* models of skeletal muscle atrophy, including diabetes, cancer, denervation, unweighting, and glucocorticoid treatment (Bodine et al., 2001; Lecker et al., 2004). The importance of these atrophy-regulated genes in muscle wasting was confirmed through studies in these knockout mice by attenuating denervation-, fasting-, and dexamethasone-induced muscle atrophy (Drummond et al., 2008; Baehr et al., 2011; Cong et al., 2011). Interestingly, recent findings indicated that atrogin-1-knockout mice are short lived and experience greater loss of muscle mass during aging than the control mice (Sandri et al., 2013), indicating that the chronic inhibition of these atrogenes should not be considered as a therapeutic target to counteract sarcopenia (Sakuma and Yamaguchi, 2012a, Sakuma et al. 2015a, 2017).

Autophagy occurs in all eukaryotic cells and is evolutionarily conserved from yeast to humans (Levine and Klionsky, 2004). The turnover of most long-lived proteins, macromolecules, biological membranes, and whole organelles, including mitochondria, ribosomes, the endoplasmic reticulum, and peroxisomes, is mediated by autophagy (Cuervo, 2004). Three major mechanisms of autophagy have been described. Microautophagy is when lysosomes directly take up cytosol, inclusion bodies, and organelles for degradation. Chaperone-mediated autophagy is when soluble proteins with a particular pentapeptide motif are recognized and transported across the lysosomal membrane for degradation. Macroautophagy (herein described as autophagy) is a ubiquitous catabolic process that involves the bulk degradation of cytoplasmic components by interacting lysosomes (Neel et al., 2013). This process is characterized by the engulfment of part of the cytoplasm inside double-membrane vesicles (autophagosomes). Autophagosomes subsequently fuse with lysosomes to form autophagolysosomes, in which the cytoplasmic cargo is degraded.

The autophagy machinery, a critical pathway for cell homeostasis that had long been forgotten in skeletal muscle, has been intensively studied in the past few years. Particular emphasis has been placed on the role played by autophagic defects in disease pathogenesis, its involvement in atrophy, and the possible effects of exercise as a countermeasure (Ferraro et al., 2014; Vainshtein et al., 2014; Sanchez et al., 2014). Indeed, sarcopenic muscle of humans and rodents exhibits a marked autophagic defect (Carnio et al., 2014; Sakuma et al., 2016), whereby it cannot degrade the accumulated denatured proteins, abnormal mitochondria, and sarcoplasmic reticulum. This chapter outlines the UPS-dependent signaling and autophagy-dependent system and these adaptations in sarcopenic muscle.

6.2   SARCOPENIA

Sarcopenia is widely considered to be the reason for the age-related decline of muscle mass, quality, and strength. Moreover, it is often used to describe both the cellular processes (denervation, mitochondrial dysfunction, inflammatory, and hormonal changes) and the outcomes such as decreased muscle strength, mobility, and function, a greater risk of falls, and reduced energy needs. Sarcopenia can be considered "primary" (or age related) when no other cause is evident but aging itself (Cruz-Jentoft et al., 2010). Primary sarcopenia is especially associated with physical inactivity, derived from a reduction of physical exercise during leisure time or work related. Secondary sarcopenia usually occurs when one or more identifiable causes coexist (Cruz-Jentoft et al., 2010). This condition is a proxy of chronic or acute diseases that are highly prevalent in older persons, such as Parkinson's disease, diabetes, chronic heart failure, chronic obstructive pulmonary disease, stroke, and hip fracture. von Haehling et al. (2010) estimated its prevalence at 5%–13% for elderly people aged 60–70 years and 11%–50% for those aged 80 years or older. The lean muscle mass generally contributes up to ~50% of the total body weight in young adults but declines with aging to 25% at 75–80 years old. The loss of muscle mass is most notable in the lower-limb muscle groups, with the cross-sectional area of the vastus lateralis being reduced by as much as 40% between the age of 20 and 80 years old. At the muscle fiber level, sarcopenia is characterized by specific type II muscle fiber atrophy, fiber necrosis, and fiber-type grouping.

Several possible mechanisms of age-related muscle atrophy have been described. Age-related muscle loss is a result of reductions in the size and number of muscle fibers, possibly due to a multifactorial process that involves physical activity, nutritional metabolic homeostasis, oxidative stress, hormonal changes, and the lifespan. The specific contribution of each of these factors is unknown, but there is emerging evidence that the disruption of several positive regulators (Akt and serum response factor) of muscle hypertrophy with age is an important feature in the progression of sarcopenia (Sakuma and Yamaguchi, 2010). Very intriguingly, more recent studies indicated an apparent functional defect in autophagy- and myostatin-dependent signaling in sarcopenic muscle (McKay et al., 2012; Wohlgemuth et al., 2010; Zhou et al., 2013). In contrast, many investigators have failed to demonstrate age-related enhancement in the levels of common negative regulators (atrogin-1, nuclear factor-kappaB [NF-κB], and calpain) in senescent mammalian muscles (Sakuma and Yamaguchi, 2010, Sakuma et al. 2015a).

The progress of age-related muscle wasting and weakness is effectively prevented by the combination of resistance training and amino acid-containing supplements (Drummond et al., 2008; Sakuma and Yamaguchi, 2010, 2018). In contrast, sarcopenia has been most markedly attenuated by mild caloric restriction in all of the mammals tested (McKiernan et al., 2011, 2012; Wohlgemuth et al., 2010). Moreover, many researchers have considered the strategy of inhibiting myostatin to treat various muscle disorders such as muscular dystrophy, cachexia, and sarcopenia (Sakuma and Yamaguchi, 2011b). Furthermore, more recent studies indicated the possible application of new supplements (e.g., ursolic acid) to prevent muscle atrophy.

6.3 UPS IN SKELETAL MUSCLE

The UPS is essential for protein degradation. The degradation of a protein via the UPS involves two steps: (1) tagging of the substrate by covalent attachment of multiple ubiquitin molecules and (2) degradation of the tagged protein by the 26S proteasome (Figure 6.1). The ubiquitination of proteins is regulated by at least three enzymes: ubiquitin-activating enzyme (E1), ubiquitin-conjugating enzyme (E2), and ubiquitin ligase (E3). Consistent increases in the gene expression of two important E3 ubiquitin ligases (atrogin-1 and MuRF-1) have been observed in a wide range of *in vivo* models of skeletal muscle atrophy including diabetes, cancer, renal failure, denervation, unweighting, and glucocorticoid or cytokine treatment (Bodine et al., 2013).

Overexpression of atrogin-1 results in the polyubiquitination of MyoD and an inhibition of MyoD-induced myotube formation (Lagirand-Cantaloube et al., 2009). In contrast, the knockdown of atrogin-1 reversed endogenous MyoD proteolysis and the overexpression of a mutant MyoD, unable to be ubiquitinated, prevented muscle atrophy *in vivo* (Lagirand-Cantaloube et al., 2009). These results confirmed MyoD as a substrate of atrogin-1 during dexamethasone-induced myotube atrophy (Jogo et al., 2009). In recent work, atrogin-1 was found to interact with sarcomeric proteins, including myosins, desmin, and vimentin, as well as transcription factors, components of the translational machinery, enzymes involved in glycolysis and gluconeogenesis, and mitochondrial proteins (Lokireddy et al., 2011). In contrast to atrogin-1, it appears that MuRF-1 mainly interacts with structural proteins. MuRF-1 binds to titin and potentially affects titin signaling. It also binds to and degrades myosin heavy chain proteins following the treatment of skeletal muscle with dexamethasone. Moreover, MuRF-1 degrades myosin light chains (LCs) 1 and 2 during denervation and fasting conditions (Cohen et al., 2009). These studies suggest that, while numerous stimuli can activate both atrogin-1 and MuRF-1, the downstream pathways affected may be separate for each protein.

Well-known factors regulating the transcription of both E3 ligases are the class O-type forkhead transcription factors (FOXOs), which include FOXO1, FOXO31, and FOXO4. All FOXO are expressed in skeletal muscle, and several forms of atrophy show the upregulation of FOXO1 and FOXO3a expressions (Baehr LM, 2011; Furuyama et al., 2003; Lecker et al., 2004; Sandri et al., 2004). Sandri et al. (2004) demonstrated that constitutively activated FOXO3a activates the MAFbx promoter *in vitro* and *in vivo* during muscle atrophy. In contrast, Stitt et al. (2004) indicated that activated FOXO1 was necessary but not sufficient to increase both gene expressions in cultured myotubes. Moreover, Waddell et al. (2008) suggested that FOXO transcription factors can directly bind to the atrogin-1 and MuRF-1 promoters, although not all FOXO family members equally

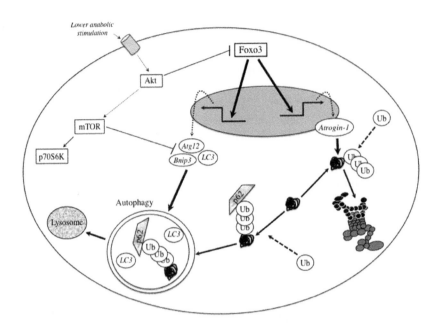

FIGURE 6.1 Contribution of the proteolytic pathways to muscle atrophy during catabolic conditions. In catabolic conditions such as denervation, cancer, and fasting, an atrophy program is induced to degrade muscle proteins and organelles. Proteins can have a double fate, being recognized and removed by the proteasome or docked to the autophagosome. In the latter case, the chains of polyubiquitins are interacting with the p62. These proteins have also a domain for the interaction with LC3, therefore bringing the ubiquitinated proteins to the growing autophagosome. Less anabolic stimulation (IGF-I, mechanical loading, amino acids, etc.) reduces the amount of activated Akt, not promoting protein synthesis by activating the mTOR/p70S6K pathway. Lower Akt activity also does not block the nuclear translocation of FOXO3 to enhance the expression of autophagy-related genes (Bnip3, LC3, Atg12) and Atrogin-1 and the consequent protein degradation. FOXO, forkhead box O; IGF-I, insulin-like growth factor-I; LC3, microtubule-associated protein light chain; mTOR, mammalian target of rapamycin; p70S6K, 70 kDa ribosomal protein S6 kinase; Ub, ubiquitin. (Data from Sakuma, K. and Yamaguchi, A., Cellular and molecular mechanism regulating hypertrophy and atrophy of skeletal muscle, in Willems, M., Ed., *Skeletal Muscle: Physiology, Classification and Disease*, Nova Science Publishers, New York, pp. 141–93, 2012b.)

activate them. FOXO family members may regulate the transcription of atrogin-1 and MuRF-1, but this is highly controversial, and it remains to be elucidated by further studies.

The transcription factor KLF-15 is a glucocorticoid-responsive gene that is upregulated in skeletal muscle by dexamethasone. Atrogin-1 and MuRF-1 have multiple KLF-15-binding sites in the promoter region; some of which are in close proximity to the FOXO-binding sites and glucocorticoid response element (Shimizu et al., 2011). Intriguingly, KLF-15 can activate the transcription of atrogin-1 and MuRF-1 directly and through the induction of FOXO transcription factors. Furthermore, atrogin-1 and MuRF-1 promoters have putative binding sites for other transcription factors such as NF-κB transcription factors VCCAAT/enhancer-binding protein-β (C/EBP-β) and Smad 3. Numerous studies demonstrated that NF-κB transcription factors (p65, c-Rel, RelB, p52, and p50) are induced in atrophied muscles caused by disuse (Van Gammeren et al., 2009), denervation (Mourkioti et al., 2006), and cachexia (Rhoads et al., 2010). Recent studies indicated that both genes are direct targets of p50 (Wu et al., 2011). More recently, it was shown that the overexpression of Smad3, a downstream regulator of myostatin, in myofibers resulted in a 1.8-fold induction of atrogin-1 promoter activity and no change in MuRF-1 activity (Goodman et al., 2013). On considering all of these findings, the transcription of atrogin-1 and MuRF-1 may be complicatedly regulated by many factors.

6.3.1 THE ADAPTATION OF UPS SIGNALING IN ATROPHIED MUSCLE

In 2001, the group of Bodine et al. (2001) discovered two E3 ubiquitin ligases (atrogin-1 and MuRF-1) to critically regulate muscle atrophy (protein degradation) using immobilization, hindlimb suspension, and denervation with both mRNA expression screening (differential display analysis) and Northern blot analysis. Moreover, mice with a null deletion of either atrogin-1 or MuRF-1 (TRIM63) showed a significant sparing of the muscle mass in both the tibialis anterior and gastroc-nemius muscles: atrogin-1-KO showed sparing at both 7 and 14 days, and MuRF-1-KO showed sparing at 14 days. In young female Sprague-Dawley rats, Haddad et al. (2006) indicated the upregu-lation of atrogin-1 (230% of control levels) and MuRF-1 (140% of control levels) mRNA levels after 5 days of unloading (hindlimb suspension). Andrianjafiniony et al. (2010) also demonstrated that 2 weeks of hindlimb suspension induced the upregulation of these two genes (57%–73% increase). However, the adaptive response of atrogin-1 and MuRF-1 after unloading is highly contradictory in humans. For example, bed rest for 20 days induced a marked increase in atrogin-1 mRNA in humans (healthy volunteers) (Ogawa et al., 2006), whereas lower limb unloading for 20 days did not lead to a significant increase in the vastus lateralis muscle of young healthy untrained men (Sakuma et al., 2009). Furthermore, Gustafsson et al. (2010) reported that 3 days of unilateral lower limb suspension upregulated atrogin-1 and MuRF-1 gene expression in the vastus lateralis but not soleus muscle. These studies dealing with two atrogenes in human muscle after unloading were designed to investigate the expression levels of only mRNA and not protein. mRNA adaptation of these atro-genes in muscle shows marked differences in time dependence after unloading. For example, Caron et al. (2009) demonstrated that hindlimb immobilization of mice induced a significant increase in atrogin-1 and MuRF-1 mRNA at 3.5 and 7 days but not 14 days (atrogin-1 mRNA level was significantly lower than in control). More descriptively, Pond et al. (2014) indicated that hindlimb suspension induced a significant increase in the mRNA level of these atrogenes at 4, 5, and 7 days but not before or after (1, 2, 3, 10, and 14 days). Intriguingly, significant increases in atrogin-1 and MuRF-1 proteins were observed in the unloaded mouse muscle by hindlimb suspension (Dong et al., 2009; Pond et al., 2014) and immobilization (Bae et al., 2012; Kang and Ji, 2013) after 1–2 weeks. The adaptation of ubiquitin ligases to various atrophy-promoting stimuli may more readily occur in rodents (mice and rats) than humans. As almost all human studies concerning ubiquitin ligase in various atrophy models (unloading, cachexia, and neuromuscular disease) have been conducted to determine the level of mRNA only, the adaptive response of ubiquitin ligases at protein levels should be investigated more extensively.

6.3.2 UPS CANNOT MODULATE SARCOPENIA

Atrogin-1 and/or MuRF-1 mRNA levels in aged muscle reportedly increase (Clavel et al., 2006) or remain unchanged (Welle et al., 2003; Whitman et al., 2005) in humans and rats or decrease in rats (DeRuisseau et al., 2005; Edström et al., 2006). Even when the mRNA expression of these atrogenes was reported to increase in sarcopenic muscles, this was very limited (1.5- to 2.5-fold) compared with those in other catabolic conditions (10-fold). Although various findings have been reported regarding the mRNA levels of both ubiquitin ligases in aged mammalian muscle, the examina-tion of protein levels in sarcopenic muscles did not support age-related increases in the mRNA of several ubiquitin ligases. For example, a descriptive analysis using muscle samples ($n = 10$) by Edström et al. (2006) indicated the marked upregulation of phosphorylated Akt and FOXO4 in the gastrocnemius muscle of aged female rats, probably contributing to the downregulation of atrogin-1 and MuRF-1 mRNA. This result is further supported by the more recent finding of Léger et al. (2008) who, using human subjects aged 70 years old, demonstrated decreases in nuclear FOXO1 and FOXO3a by 73% and 50%, respectively, although they did not recognize significant age-dependent changes in the expressions of atrogin-1 and MuRF-1 mRNA. The major peptidase activities of the proteasome (i.e., the chymotrypsin-like, trypsin-like, and caspase-like activities) were always reduced (as reported in other tissues) or remained unchanged with aging (Combaret et al., 2009; Sakuma and Yamaguchi, 2011a). In contrast, Altun et al. (2010) recently found that the hindlimb muscles of (30-month-old) rats contained two- to threefold more 26S proteasomes than the muscles of aged rats, and adult (control) rats showed a similar capacity to degrade peptides, proteins, and a ubiquitinated substrate but differed in the levels of proteasome-associated proteins (e.g., the deubiquitinating enzyme USP14). Although the activities of many other deubiquitinating enzymes were markedly enhanced in aged muscles, the levels of polyubiquitinated proteins were higher than in the adult animals. Interestingly, recent findings indicated that atrogin-1-knockout mice are short lived and experience a greater loss of muscle mass during aging than control mice (Sandri et al., 2013), indicating that the activity of this E3 ubiquitin ligase is required to preserve the muscle mass during aging in mice. Similarly, recent studies demonstrated that the inhibition of atrogin-1 did not block other forms of muscular atrophy (caused by denervation and treatment with glucocorticoids) (Baehr et al., 2011; Gomes et al., 2012). Moreover, MuRF-1-null mice experience a higher rate of muscle strength loss during aging than controls, although the muscle mass is at least partly preserved in these mice (Hwee et al., 2014). As indicated by Sandri et al. (2013), the chronic inhibition of these atrogenes should not be considered as a therapeutic target to counteract sarcopenia because this does not prevent muscle loss and exacerbates weakness.

6.4 AUTOPHAGY-DEPENDENT SYSTEM

Autophagy represents an extremely refined collection of altered organelles, abnormal protein aggregates, and pathogens, similar to a selective recycling center (Park and Cuervo, 2013). The selectivity of the autophagy process is conferred by a growing number of specific cargo receptors such as p62/SQSTM1, Nbr1, and Nix (B-cell lymphoma 2 [BCL2]/adenovirus E1B 19 kd-interacting protein [BNIP] 3L) (Shaid et al., 2013). These adaptor proteins are equipped with a cargo-binding domain, with the capability to recognize and attach directly to molecular tags on organelles. At the same time, these adaptor proteins bind to essential autophagosome membrane proteins. Three molecular complexes mainly regulate the formation of autophagosomes: the microtubule-associated protein LC3 conjugation system and the regulatory complexes governed by unc51-like kinase-1 (ULK1) and Beclin-1. The conjugation complex is composed of different proteins encoded by autophagy-related genes (Atg) (Mizushima and Komatsu, 2011). The Atg12-Atg5-Atg16L1 complex, along with Atg7, plays an essential role in the conjugation of LC3 to phosphatidylethanolamine, which is required for the elongation and closure of the isolation membrane (Mizushima and Komatsu, 2011). This system is under the regulation of at least two major cellular energy-sensing complexes. Under basal conditions, the ULK1 complex is inactivated by phosphorylation through mammalian target of rapamycin (mTOR) signaling complex1 (mTORC1), whereas during autophagy induction mTORC1 is inhibited, thus enhancing the formation of a complex involving ULK1, Atg13, and FIP200. Moreover, mTORC1 can also be negatively regulated independently of Akt by energy stress sensors such as adenosine monophosphate (AMP)-activated protein kinase (AMPK) and, in a mechanical-activity-dependent manner, through tuberous sclerosis complex 1/2. Moreover, AMPK can also directly phosphorylate ULK1 and Beclin-1 (Kim et al., 2013). During autophagy, the ULK1 complex is localized to the isolation membrane, where it facilitates the formation of autophagosomes through interaction with the Beclin-1 complex.

The UPS and lysosomal-autophagy system in skeletal muscle are interconnected (Mammucari et al., 2007; Zhao et al., 2007). Both of these studies identified FOXO3 as a regulator of these two pathways in muscle wasting (Figure 6.1). FOXO3 is a transcriptional regulator of atrogin-1 and MuRF-1. FOXO3 modulates the expression of Atg in mammalian skeletal muscle and C2C12 myotubes (Zhao et al., 2007). Masiero et al. (2009) found an intriguing characteristic using muscle-specific autophagy knockout mice, which exhibit fiber atrophy, weakness, and mitochondrial abnormalities. Autophagy-dependent protein degradation may also be modulated by tumor necrosis factor receptor-associated factor 6 and peroxisome proliferator-activated receptor γ coactivator 1α (PGC-1α) (Paul et al., 2010). Wenz et al. (2009) reported an age-related increase in the ratio of LC3-II to LC3-I in MCK-PGC-1α mice. Therefore, PGC-1α would attenuate the autophagic process probably through increased antioxidant defense and mitochondrial biogenesis.

6.4.1 The Adaptation of Autophagy-Dependent Signaling in Atrophied Muscle

6.4.1.1 Denervation and Autophagy-Dependent System

Denervation of skeletal muscle is a very popular model of atrophy because of the loss of neuromuscular activity and neurotrophic factors. Denervation induced a marked increase in all autophagic markers such as LC3-II, p62/SQSTM1, Beclin-1, Atg7, and ULK1 proteins (O'Leary et al., 2013). Intriguingly, the amount of LC3-II and p62 in the mitochondria also increases probably due to activation of mitophagy. Such a marked upregulation of the autophagy system is modulated by various upstream mediators. For example, Vainshtein et al. (2015) demonstrated that animals lacking PGC-1α exhibited a reduced mitochondrial density alongside myopathic characteristics similar to autophagy-deficient muscle. Furthermore, their group indicated that PGC-1α overexpression led to an increase in the lysosomal capacity as well as autophagy flux but led to the reduced localization of LC3-II and p62/SQSTM1 to mitochondria. Thus, they hypothesized that PGC-1α regulates mitochondrial turnover, not only through biogenesis but also via degradation using the autophagy-lysosome machinery. In contrast, Quy et al. (2013) indicated that denervated muscle during the short-term (24 and 48 h) exhibited autophagic suppression through the proteasome-dependent activation of mTORC1. Autophagic activating patterns may be different in early (1–2 days) and late (14 days) phase in the denervated muscle.

TRIM63 (MuRF-1) would also modulate the autophagic process of skeletal muscle by denervation. Recently, Rudolf et al. (2013) demonstrated that TRIM63, an E3 ubiquitin ligase, plays an important role in the turnover of pentameric transmembrane protein, cholinergic receptor, nicotinic (CHRN), which is the major postsynaptic ion channel of the neuromuscular junction (NMJ). Moreover, most SH3GLB1 (SH3-domain GRB2-like endophilin B1) and CHRN-positive puncta were exactly overlaid with TRIM63 and accompanied by LC3 (Khan et al., 2014). Denervation led to a marked increase of SH3GLB1 and CHRN double-positive puncta, which was almost completely blocked in TRIM63 knockout mice. TRIM63 knockout mice exhibited no significant muscle atrophy after denervation. Therefore, TRIM63 plays a crucial role in the atrophy-induced endocytic retrieval of CHRN and its subsequent autophagic processing by cooperation with SH3GLB1, LC3, and p62/SQSTM1.

Such an upstream modulator may differ between slow-twitch and fast-twitch muscle fibers. Mitochondria have been postulated to play an important role in triggering signals that contribute to muscle atrophy. PARK-2 participates in mitochondrial quality control by mitophagy. In normal mice, the denervation of slow-twitch muscle activated mitophagy by the accumulation of PARK-2 and then activated the UPS system through the translocation of nuclear factor erythroid 2-related factor 1 (NFE2L1) from the cytosol to nucleus. In contrast, PARK-2-deficient mice exhibited an atrophic delay of the soleus muscle through the inactivation of UPS by the inability to carry out nuclear translocation of NFE2L1. Autophagy deficiency in denervated soleus muscle delayed skeletal muscle atrophy, reduced mitochondrial activity, and induced oxidative stress and the accumulation of PARK2/Parkin (Furuya et al., 2014).

6.4.1.2 Unloading and Autophagy-Dependent System

Several studies investigated the changes in autophagy-linked molecules in the unloaded mammalian muscle. Recently, Smith et al. (2014) demonstrated that hindlimb unweighting led to a marked increase in mRNA expressions of LC3B, Gabarapl1, and Atg4b mRNA in the quadriceps femoris muscle of mice (16–18 weeks of age) in earlier (2 days after operation) but not later (7 days after this) periods. Using 6-month-old male mice, hindlimb unweighting for 3 but not 7 days induced the significant upregulation of p62 p62/SQSTM1 mRNA in the soleus muscle, in spite of there being no change in Beclin-1 mRNA in either period. Intriguingly, Cannavino et al. (2014) demonstrated that the significant induction of p62/SQSTM1 mRNA in unloaded muscle is prevented by PGC-1α overexpression but not Trolox (antioxidant treatment). Thus, the decrease in PGC-1α expression recognized in the unloaded muscle would be attributable to the increase in p62/SQSTM1 expression. In contrast, Dupré-Aucouturier et al. (2015) showed that there was no significant change in mRNA levels of Gabarapl1, LC3B, or ULK1 in the rat soleus muscle after 2 weeks of hindlimb suspension. Although some researchers reported the mRNA induction of autophagy-related molecules after hindlimb unweighting, many researchers demonstrated no significant change in the protein level of autophagy-linked molecules. For example, the Western blot analysis conducted by Baehr et al. (2016) demonstrated that the levels in p62/SQSTM1, Beclin-1, and Atg7 proteins did not significantly change in the unloaded (2 weeks) soleus or tibialis anterior muscles of male F344BN rats (9 months). Moreover, hindlimb unloading elicits no significant change in the amounts of Beclin-1 protein (Andrianjafiniony et al., 2010) and LC3 (Liu et al., 2013). In contrast, some researchers indicated the elevations of p62/SQSTM1 (Liu et al., 2013) and LC3-II/I (Dupré-Aucouturier et al., 2015) at the protein levels. As the adaptive changes in autophagy-related molecules have only been elucidated in the muscle of rodents, the manner of adaptation should be investigated using human muscle samples with unloading models. Table 6.1 is a summary of the adaptation of autophagy-linked molecules in muscle caused by hindlimb unloading.

6.4.1.3 Autophagic Adaptation in Muscular Dystrophy

A finely tuned system for protein degradation and organelle removal is required for the proper functioning and contractility of skeletal muscle (Vainshtein et al., 2014). Inhibition/alteration of autophagy contributes to myofiber degeneration leading to the accumulation of abnormal (dysfunctional) organelles and of unfolded and aggregation-prone proteins (Masiero et al., 2009; Irwin et al., 2003), which are typical features of several myopathies (Grumati et al., 2010; Vergne et al., 2009).

TABLE 6.1
The Adaptive Response in Autophagy-Linked Molecules in the Unloaded Muscle

Author/Year	Journal/Volume/Pages	Manner of Analysis	Results	Species
Andrianjafiniony et al. (2010)	*Am J Physiol Cell Physiol* 299: C307–15	Western blot	Beclin-1 protein ⇔	Mouse
Liu et al. (2013)	*IUBMB Life* 64: 393–402	Real-time PCR Western blot	Beclin-1 ↓ LC3 protein ⇔ p62/SQSTM1 mRNA ⇔ p62/SQSTM1 protein ↑	Mouse
Cannavino et al. (2014)	*J Physiol* 592: 901–10	Real-time PCR	Beclin-1 mRNA ⇔ P62/SQSTM1 mRNA ↑	Mouse
Dupré-Aucouturier et al. (2015)	*J Appl Physiol* 119: 342–51	Real-time PCR Western blot	LC3B mRNA ⇔ LC3II/I protein ratio ↑ Gabalapl1 mRNA ⇔	Rats
Smith et al. (2014)	*PLoS One* 9: e94356	Real-time PCR	LC3B mRNA ↑ Gabarapl1 mRNA ↑ Atg4b mRNA ↑	Mouse
Baehr et al. (2016)	*Aging* 8: 127–46	Western blot	Beclin-1 protein ⇔ P62/SQSTM1 protein ⇔ Atg7 protein ⇔	Mouse

The generation of Atg5 and Atg7 muscle-specific knockout mice confirmed the physiological importance of the autophagy system in muscle mass maintenance (Masiero et al., 2009; Raben et al., 2008). The muscle-specific Atg7 knockout mice are characterized by the presence of abnormal mitochondria, oxidative stress, accumulation of polyubiquitinated proteins, and consequent sarcomere disorganization (Masiero et al., 2009). Moreover, the central role of the autophagy-lysosome system in muscle homeostasis is highlighted by lysosomal storage diseases (Pompe disease, Danon disease, and X-linked myopathy), a group of debilitating muscle disorders characterized by alterations in lysosomal proteins and autophagosome buildup (Tardif et al., 2013). Interestingly, all of these myopathies exhibit the accumulation of autophagic vacuoles inside myofibers due to defects in their clearance.

An apparent defect of autophagy-dependent signaling is also observed in various muscular dystrophies. The first evidence of impaired autophagy in these models was provided by studies in mice and patients with mutations in collagen VI (Irwin et al., 2003). Mutations that inactivate Jumpy, a phosphatase that counteracts the activation of VPS34 for autophagosome formation and reduces autophagy, are associated with centronuclear myopathy (Vergne et al., 2009). De Palma et al. (2012) reported the marked defect of autophagy in dystrophin-deficient mdx mice and Duchenne muscular dystrophy (DMD) patients. This evidence included the electron microscopic evaluation of muscle tissue morphology as well as the decreased expression of autophagic regulator proteins (i.e., LC3-II, Atg12, Gabarapl1, BNIP3). Moreover, starvation and treatment with chloroquine, potent inducers of autophagy, did not activate autophagy-dependent signaling in either the tibialis anterior or diaphragm muscles of mdx mice (De Palma et al., 2012). Furthermore, mdx mice and DMD patients exhibited an unnecessary accumulation of p62/SQSTM1 protein, which was lost after prolonged autophagy induction by a low-protein diet (De Palma et al., 2012). A similar block in autophagy progression was described in lamin A/C null mice (Ramos et al., 2012). LGMD2A muscles showed the upregulations of p62/SQSTM1 (2.1-fold) and BNIP3 (3-fold) mRNA and a slightly increased LC3-II/LC3-I protein ratio and level of p62/SQSTM1 (Fanin et al., 2013). Conversely, laminin-mutated (dy/dy) animals displayed an excessive level of autophagy, which is equally detrimental (Carmignac et al., 2011). These findings suggest that the defect of autophagy signaling has a central role in the degenerative symptoms in various types of muscular dystrophy.

6.4.2 Autophagic Defect in Sarcopenic Muscle

A decline in autophagy during normal aging has been described for invertebrates and higher organisms (Cuervo et al., 2005). Inefficient autophagy has been considered to play a major role in the age-related accumulation of damaged cellular components, such as undergradable lysosome-bound lipofuscin, protein aggregates, and damaged mitochondria (Cuervo et al., 2005). Demontis and Perrimon (2010) showed that the function of the autophagy/lysosome system of protein degradation declined during aging in the skeletal muscle of *Drosophila*. This results in the progressive accumulation of polyubiquitin protein aggregates in senescent *Drosophila* muscle. Intriguingly, the overexpression of FOXO increases the expression of many autophagy genes, preserves the function of the autophagy pathway, and prevents the accumulation of polyubiquitin protein aggregates in sarcopenic *Drosophila* muscle (Demontis and Perrimon, 2009). Several investigators reported the autophagic changes in aged mammalian skeletal muscle (Gaugler et al., 2011; McMullen et al., 2009; Sakuma et al., 2016; Wenz et al., 2009; Wohlgemuth et al., 2010). Compared with those in young male Fischer 344 rats, amounts of Beclin-1 were significantly increased in the plantaris muscles of senescent rats (Wohlgemuth et al., 2010). Using Western blot analysis of fractionated homogenates and immunofluorescence microscopy, we recently demonstrated the selective induction of p62/SQSTM1 and Beclin-1 but not LC3 in the cytosol of sarcopenic muscle fibers in mice (Sakuma et al., 2016, Figure 6.2). Moreover, we observed significantly smaller p62/SQSTM1-positive muscle fibers in aged muscle compared with the surrounding p62/

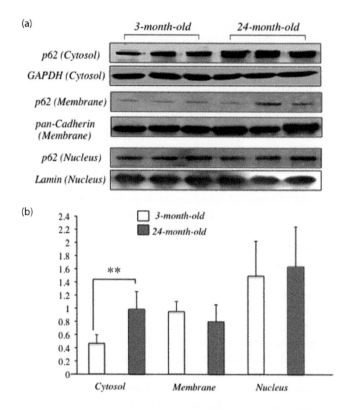

FIGURE 6.2 Western blot analysis showed that p62/SQSTM1 protein was more abundantly expressed in the cytosolic fraction of the quadriceps muscle of aged mice than that of young mice (a). No significant difference in the amount of p62/SQSTM1 in the nuclear and membrane fractions was observed in the quadriceps muscle between 3- and 24-month-old mice (b). The integrated optical density (IOD) of p62/SQSTM1 protein was normalized to the IOD of each internal control (GAPDH for cytosol fraction, pan-cadherin for membrane fraction, or lamin protein for nucleus fraction) (arbitrary units). Values are means ± SD (n = 6/group). (Data from Sakuma, K. et al., *J. Cachexia Sarcopenia Muscle*, 7, 204–12, 2016.)

SQSTM1-negative fibers (Sakuma et al., 2016; Figure 6.3). In contrast, aging did not influence the amount of Atg7 or Atg9 proteins in the rat plantaris muscle (Wohlgemuth et al., 2010). Western blot analysis by Wohlgemuth et al. (2010) clearly showed a marked increase in the amount of LC3 in muscle during aging. However, they could not demonstrate an aging-related increase of the ratio of LC3-II to LC3-I, a better biochemical marker to assess ongoing autophagy. Moreover, we failed to detect a marked increase in LC3-I and LC3-II (active form) proteins in aged quadriceps muscle (Sakuma et al., 2016). In contrast, Wenz et al. (2009) recognized a significant increase in the ratio of LC3-II to LC3-I during aging (3 vs. 22 months) in the biceps femoris muscle of wild-type mice. None of the studies determining the transcript level of autophagy-linked molecules found a significant increase with age (Gaugler et al., 2011; Sakuma et al., 2016; Wohlgemuth et al., 2010). Not all contributors to autophagy signaling seem to change similarly at both mRNA and protein levels in senescent skeletal muscle. Therefore, sarcopenia may include a partial defect of autophagy signaling although more exhaustive investigation is needed in this field. Intriguingly, a more recent study (Carnio et al., 2014) using biopsy samples of young and aged human volunteers clearly showed an age-dependent autophagic defect, such as decreases in the amount of Atg7 protein and the ratio of LC3-II/LC3-I protein.

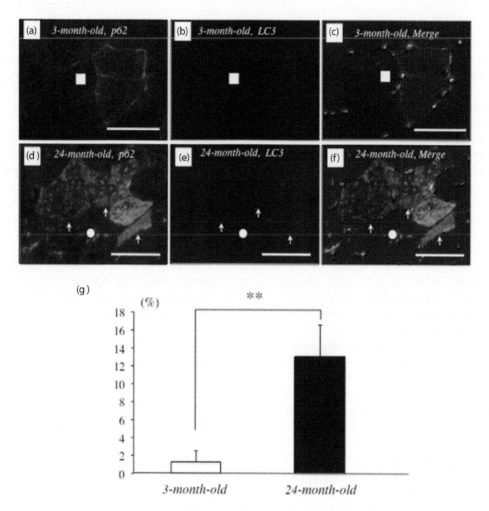

FIGURE 6.3 Serial cryosections of the quadriceps muscle of 3- and 24-month-old mice. p62/SQSTM1 and LC3 immunoreactivity was visualized using FITC- or rhodamine-conjugated antibodies. In young quadriceps muscle, immunofluorescence labeling showed that p62/SQSTM1 was present in the membrane and at a low level in the cytosol of several muscle fibers (a). Marked increases of p62/SQSTM1 immunoreactivity were observed in the membrane and the cytosol of aged muscle fibers (d). No apparent difference in LC3 immunoreactivity was observed in the muscle between 3- and 24-month-old mice (b and e). White circles and squares indicate the same fibers on different immuno images. White arrows denote the muscle fibers possessing p62/SQSTM1. Bar = 50 μm (c) and (f) are merge immunoimages among p62, LC3, and DAPI. and (g) is the percentage of p62-positive fibers. (Data from Sakuma, K. et al., *J. Cachexia Sarcopenia Muscle*, 7, 204–12, 2016.)

6.5 SEVERAL THERAPEUTIC STRATEGIES ATTENUATING SARCOPENIA DUE TO AUTOPHAGIC-ACTIVATION

Various hormonal, supplemental, and pharmacological approaches may attenuate muscle atrophy by activating autophagy. The most popular approach is resistance and endurance exercise. A recent systematic review of the international sarcopenia initiative (EWGSOP and IWGS) (Cruz-Jentoft et al., 2014) indicated that exercise interventions improve muscle strength and physical performance in sarcopenic patients. Nine weeks of resistance training prevented the loss of muscle mass and improved muscle strength in 18- to 20-month-old rats, accompanied by a reduced LC3-II/LC3-I ratio, reduced p62/SQSTM1 protein levels, and increased levels of autophagy regulatory proteins (Beclin-1, Atg5/12, Atg7) (Luo et al., 2013). Moreover, endurance exercise (training) activates autophagic flux probably by preventing binding between BCL2 and Beclin-1 (He et al., 2012) although recent findings demonstrated that exercise-induced autophagy does not have an impact on physical performance, the activation of a metabolic sensor (PRKAA1), or glucose homeostasis (Lo Verso et al., 2014). It is very reasonable for both exercises to prevent sarcopenia by inhibiting the autophagic defect and the denaturing and unfolded proteins or accumulated unfunctional mitochondrias and the other internal organs.

Calorie restriction (CR: typically 20%–40% fewer calories) activates autophagic flux, preserves mitochondrial health, and attenuates sarcopenia in mice, rats, and rhesus monkeys (Wohlgemuth et al., 2010; McKiernan et al., 2011, 2012). CR is recognized as the most robust intervention to retard both primary aging (natural age-related deterioration) and secondary aging (accelerated aging due to disease and negative lifestyle behaviors), thereby increasing the lifespan in many species. Yang et al. (2016) recently indicated that long-term CR for humans (3–15 years) also elicits the upregulation of several autophagic markers (LC3, Beclin-1) as well as that of a molecular chaperone (heat shock protein 70). A low-protein diet in spite of ad libitum feeding also attenuates muscular atrophy of DMD model mice (De Palma et al., 2012) and type 2 diabetic nephropathy (Huang et al., 2013). Moreover, a low-protein diet containing keto acids also inhibits muscle atrophy of nephrectomized rats (one of the CKD models) (Zhang et al., 2015). Therefore, dietary types and patterns are also important to inhibit muscular atrophy with autophagic defects such as sarcopenia.

Whether or not a supplemental approach (protein, amino acids, etc.) alleviates sarcopenia has been tested in many studies. It is widely accepted that the combination of resistance training and supplementation with amino acids has a promising role to attenuate sarcopenia (Sakuma and Yamaguchi, 2010; Wakabayashi et al., 2013). However, supplementation of amino acids or leucine without exercise frequently failed to prevent sarcopenia (Cruz-Jentoft et al., 2014; Sakuma and Yamaguchi, 2010), probably due to inhibiting autophagy. More recently, it was found that supplementation with dihydromyricetin, the main flavonoid component of *Ampelopsis grossedettata* (10, 50, and 100 mg/kg body weight), decreases the amount of p62/SQSTM1 proteins and increases the amount of Atg5 and Beclin-1 proteins and the ratio of LC3-II/LC3-I protein in male Sprague-Dawley rats (Shi et al., 2015). Moreover, dihydromyricetin supplementation induced the upregulation of autophagy-signaling inducers (p-AMPK/AMPK and p-ULK1/ULK1) and the downregulation of autophagy-signaling inhibitors (p-mTOR/mTOR) in both normal and autophagy-defective muscle of mice fed a high-fat diet (8 weeks). Treatment with beta-hydroxy-beta-methylbutyrate (HMB) alleviates muscle atrophy caused by dexamethasone in Sprague-Dawley rats. Girón et al. (2015) reported that supplementation with HMB (320 mg/kg in water) increased the grip strength and weight of the soleus and gastrocnemius muscles. Using the L6 myotube *in vitro*, they indirectly examined possible mechanisms of HMB in muscle fibers, demonstrating the marked autophagic inhibition of LC3-II induction and p62/SQSTM1 decrease after treatment with dexamethasone. Therefore, HMB supplementation would be a possible approach for attenuating sarcopenia (Brioche et al., 2016; Argilés et al., 2016) irrespective of avoiding an autophagic defect. In contrast, supplementation with alanine and citrulline (0.81 g/kg/d) for 14 days in mice cannot inhibit muscle atrophy under limb immobilization in spite of marked increase of a serum arginine in both cases and apparent autophagic activation in the former (Ham et al., 2015).

Similar to sarcopenic muscle, a marked autophagic defect is observed in collagen-VI null mice, which cannot induce any autophagic markers after the general stimulation of autophagy (24-h starvation), being different to normal mice. Chrisam et al. (2015) demonstrated that the intraperitoneal injection of spermidine (50 mg/kg) in this mutant mouse upregulates the levels of BNIP3 and LC3 mRNA and LC3B-II protein after 24-h starvation. Although the effect of CR on sarcopenia remains to further elucidate using humans, CR may exhibit a critically important role for alleviating sarcopenic disorders.

Several hormonal treatments, such as testosterone, growth hormone, and insulin-like growth factor-I, have been attempted to attenuate sarcopenia in rodents and humans for 20 years. However, many negative and some positive findings have been obtained (Sakuma and Yamaguchi, 2010, 2015b). More recently, some hormonal supplementations attenuated sarcopenia by autophagic activation. In normal male adult mice, the intraperitoneal injection of T3 (20 μg/100 g BW) induced the marked activation of autophagy regarding both protein (increase of LC3B and decrease of p62/SQSTM1) and mRNA expressions (LC3, p62/SQSTM1, Beclin-1, Atg5) (Lesmana et al., 2016). Furthermore, treatment with T3 further elicited the autophagy-upstream regulators, such as downregulation of the ratio of p-mTOR/mTOR, and upregulation of the ratio of pAMPK/AMP and pULK1Ser555 in skeletal muscle of normal mice. Circulating thyroid hormone levels decrease with normal aging in humans (Chakraborti et al., 1999; Hertoghe, 2005) and rodents (Cao et al., 2012), and treatment with T3 may attenuate sarcopenia by marked autophagic activation.

6.6 CONCLUDING REMARKS

Over the past decade, studies using rodent muscles have indicated that atrogin-1 and MuRF-1 contribute to the protein degradation in various muscular wasting (Bodine et al., 2014). However, recent studies using human muscle do not necessarily support such a role for these atrogenes (Foletta et al., 2011). In addition, chronic inhibition of these atrogenes should not be considered a therapeutic target to counteract sarcopenia (Sakuma et al., 2015a; Sandri et al., 2013). Particular emphasis has been placed on the role played by autophagic defects in disease pathogenesis, its involvement in atrophy, and the possible effects of exercise as a countermeasure. Indeed, endurance training possesses a positive effect for some disease model by modulating autophagy. The evidence described above supports that autophagy-dependent system regulates several atrophy model such as sarcopenia, cancer cachexia, and muscular dystrophy but not unloaded muscle. Intriguingly, the disorganization of the autophagy system seems to accelerate sarcopenic symptom in rodents and human because of no disposal of denaturing proteins and unfunctional mitochondria (Figure 6.4).

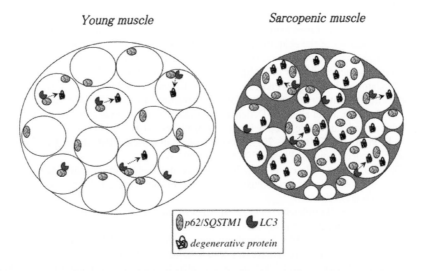

FIGURE 6.4 The comparison of an autophagy-dependent system between young and sarcopenic muscle. In contrast to young muscle, sarcopenic muscle exhibits abundant p62/SQSTM1 proteins with no activation of LC3, showing apparent autophagy defects, which cannot destroy the degenerative proteins. (Data from Sakuma, K. and Yamaguchi, A., Molecular mechanisms controlling skeletal muscle mass, in Sakuma, K., Ed., *Muscle Cell and Tissue*, InTech, Rijeka, Croatia, pp. 143–170, 2015c.)

ACKNOWLEDGMENTS

This work was supported by a research grant-in-aid for Scientific Research C (No. 26350815) from the Ministry of Education, Science, Culture, Sports, Science and Technology of Japan.

ABBREVIATIONS

AMP	Adenosine monophosphate
AMPK	AMP-activated protein kinase
Atg	Autophagy-related genes
Atrogin-1	Atrophy gene-1
BCL2	B-cell lymphoma 2
BNIP	BCL2/adenovirus E1B 19 kd-interacting protein
CHRN	Cholinergic receptor, nicotinic
CR	Caloric restriction
DMD	Duchenne muscular dystrophy
FOXO	Class O-type forkhead transcription factors
HMB	Beta-hydroxy-beta-methylbutyrate
LC3	Microtubule-associated protein light chain 3
mTOR	Mammalian target of rapamycin
mTORC1	mTOR signaling complex1
MuRF-1	Muscle RING finger-1
NFE2L1	Nuclear factor erythroid 2-related factor 1
NF-κB	Nuclear factor-κB
PGC-1α	Peroxisome proliferator-activated receptor γ coactivator 1α
ULK1	unc51-like kinase-1
UPS	Ubiquitin–proteasome system

REFERENCES

Altun, M., H. C. Besche, H. S. Overkleeft, et al. 2010. Muscle wasting in aged, sarcopenic rats is associated with enhanced activity of the ubiquitin proteasome pathway. *J Biol Chem* 285(51):39597–608.

Andrianjafiniony, T., S. Dupré-Aucouturier, D. Letexier, et al. 2010. Oxidative stress, apoptosis, and proteolysis in skeletal muscle repair after unloading. *Am J Physiol Cell Physiol* 299(2):C307–15.

Argilés, J. M., N. Campos, J. M. Lopez-Pedrosa, R. Rueda, L. Rodriguez-Manas. 2016. Skeletal muscle regulates metabolism via interorgan crosstalk: Roles in health and disease. *J Am Med Dir Assoc* 17(9):789–96.

Bae, S.-K., H.-N. Cha, T.-J. Ju, et al. 2012. Deficiency of inducible nitric oxide synthase attenuates immobilization-induced skeletal muscle atrophy in mice. *J Appl Physiol* 113(1):114–123.

Baehr, L. M., J. D. Furlow, S. C. Bodine. 2011. Muscle sparing in muscle RING finger 1 null mice: Response to synthetic glucocorticoids. *J Physiol* 589(Pt 19):4759–76.

Baehr, L. M., D. W. D. West, G. Marcotte, et al. 2016. Age-related deficits in skeletal muscle recovery following disuse are associated with neuromuscular junction instability and ER stress, not impaired protein synthesis. *Aging* 8(1):127–46.

Bodine, S. C., E. Latres, S. Baumhueter, et al. 2001. Identification of ubiquitin ligases required for skeletal muscle atrophy. *Science* 294(5547):1704–8.

Bodine, S. C., and L. M. Baehr. 2013. Skeletal muscle atrophy and the E3 ubiquitin ligases MuRF1 and MAFbx/atrogin-1. *Am J Physiol Endocrinol Metab* 307(6):E469–84.

Bodine, S. C., and L. M. Baehr. 2014. Skeletal muscle atrophy and the E3 ubiquitin ligases MuRF1 and MAFbx/atrogin-1. *Am J Physiol Endocrinol Metab* 307(6):E469–84.

Brioche, T., A. F. Pagano, G. Py, A. Chopard. 2016. Muscle wasting and aging: Experimental models, fatty infiltrations, and prevention. *Mol Aspects Med* 50:56–87.

Cannavino, J, L. Brocca, M. Sandri, et al. 2014. PGC1-α over-expression prevents metabolic alterations and soleus muscle atrophy in hindlimb unloaded mice. *J Physiol* 592(20):4575–89.

Cao, L., F. Wang, Q.-G.Yang, et al. 2012. Reduced thyroid hormones with increased hippocampal SNAP-25 and Munc18-1 might involve cognitive impairment during aging. *Behav Brain Res* 229(1):131–7.

Carmignac, V., M. Svensson, Z. Körner, et al. 2011. Autophagy is increased in laminin α2 chain-deficient muscle and its inhibition improves muscle morphology in a mouse model of MDC1A. *Hum Mol Genet* 20(24):4891–902.

Carnio, S., F. LoVerso, M. A. Baraibar, et al. 2014. Autophagy impairment in muscle induces neuromuscular junction degeneration and precocious aging. *Cell Reports* 8(5):1509–21.

Caron, A. Z., G. Drouin, J. Desrosiers, F. Trensz, G. Grenier. 2009. A novel hindlimb immobilization procedure for studying skeletal muscle atrophy and recovery in mouse. *J Appl Physiol* 106(6):2049–59.

Chakraborti, S., T. Chakraborti, M. Mandal, S. Das, S. K. Batabyal. 1999. Hypothalamic-pituitary-thyroid axis status of humans during development of ageing process. *Clin Chim Acta* 288(1–2):137–45.

Chrisam, M., M. Pirozzi, S. Castagnaro, et al. 2015. Reactivation of autophagy by spermidine ameriorates the myopathic defects of collagen VI-null mice. *Autophagy* 11(12):2142–52.

Clavel, S., A. S. Coldefy, E. Kurkdjian, J. Salles, I. Margaritis, B. Derijard. 2006. Atrophy-related ubiquitin ligases, atrogin-1 and MuRF1 are up-regulated in aged rat tibialis anterior muscle. *Mech Ageing Dev* 127(10):794–801.

Cohen, S., J. J. Brault, S. P. Gygi, et al. 2009. During muscle atrophy, thick, but not thin, filament components are degraded by MuRF1-dependent ubiquitylation. *J Cell Biol* 185(6):1083–95.

Combaret, L., D. Dardevet, D. Béchet, D. Taillandier, L. Mosoni, D. Attaix. 2009. Skeletal muscle proteolysis in aging. *Curr Opin Clin Nutr Metab Care* 12(1):37–41.

Cong, H., L. Sun, C. Liu, et al. 2011. Inhibition of atrogin-1/MAFbx expression by adenovirus-delivered small hairpin RNAs attenuates muscle atrophy in fasting mice. *Hum Gene Ther* 22(3):313–24.

Cruz-Jentoft, A. J., J. P. Baeyens, J. M. Bauer, et al. 2010. Sarcopenia: European consensus of definition and diagnosis: Report of the European Working Group on Sarcopenia in Older People. *Age Aging* 39(4):412–23.

Cruz-Jentoft, A., F. Landi, S. M. Scheider, et al. 2014. Prevalence of and interventions for sarcopenia in ageing adults: A systematic review—Report of the International Sarcopenia Initiative (EQGSOP and IWGS). *Age Aging* 43(6):748–59.

Cuervo, A. M. 2004. Autophagy: Many paths to the same end. *Mol Cell Biochem* 263(1–2):55–72.

Cuervo, A. M., E. Bergamini, U. T. Brunk, et al. 2005. Autophagy and aging: The importance of maintaining "clean" cells. *Autophagy* 1(3):131–40.

Demontis, F., and N. Perrimon. 2009. Integration of insulin receptor/Foxo signaling and dMyc activity during muscle growth regulates body size in *Drosophila*. *Development* 136(6):983–93.

Demontis, F., and N. Perrimon. 2010. FOXO/4E-BP signaling in *Drosophila* muscles regulates organism-wide proteostasis during aging. *Cell* 143(5):813–25.

De Palma, C., F. Morisi, S. Cheli, et al. 2012. Autophagy as a new therapeutic target in Duchenne muscular dystrophy. *Cell Death Dis* 3:e418.

DeRuisseau, K. C., A. N. Kavazis, S. K. Powers. 2005. Selective downregulation of ubiquitin conjugation cascade mRNA occurs in the senescent rat soleus muscle. *Exp Gerontol* 40(6):526–31.

Dong, F., Y. Hua, P. Zhao, J. Ren, M. Du, N. Sreejayan. 2009. Chromium supplement inhibits skeletal muscle atrophy in hindlimb-suspended mice. *J Nutr Biochem* 20(12):992–9.

Drummond, M. J., H. C. Dreyer, B. Pennings, et al. 2008. Skeletal muscle protein anabolic response to resistance exercise and essential amino acids is delayed with aging. *J Appl Physiol* 104(5):1452–61.

Dupré-Aucouturier, S., J. Castells, D. Freyssenet, et al. 2015. Trichostatin A, a histone deacetylase inhibitor, modulates unloaded-induced skeletal muscle atrophy. *J Appl Physiol* 119(4):342–51.

Edström, E., M. Altun, M. Hägglund, B. Ulfhake. 2006. Atrogin-1/MAFbx and MuRF1 are downregulated in ageing-related loss of skeletal muscle. *J Gerontol Series A Biol Sci Med Sci* 61(7):663–74.

Fanin, M., A. C. Nascimbeni, C. Angelini. 2013. Muscle atrophy in limb girdle muscular dystrophy 2A: A morphometric and molecular study. *Neuropath Appl Neurobiol* 39(7):762–71.

Ferraro, E., A. M. Giammarioli, S. Chiandotto, et al. 2014. Exercise-induced skeletal muscle remodeling and metabolic adaptation: Redox signaling and role of autophagy. *Antioxid Redox Signal* 21(1):154–76.

Foletta, V. C., L. J. White, A. E. Larsen, et al. 2011. The role and regulation of MAFbx/atrogin-1 and MuRF1 in skeletal muscle atrophy. *Pflügers Arch* 461(3):325–35.

Furuya, N., S. Ikeda, S. Sato, et al. 2014. PARK2/Parkin-mediated mitochondrial clearance contributes to proteasome activation during slow-twitch muscle atrophy via NFE2L1 nuclear translocation. *Autophagy* 10(4):631–41.

Furuyama, T., K. Kitayama, H. Yamashita, N. Mori. 2003. Forkhead transcription factor FOXO1 (FKHR)-dependent induction of PDK4 gene expression in skeletal muscle during energy deprivation. *Biochem J* 375(Pt 2):365–71.

Gaugler, M., A. Brown, E. Merrell, et al. 2011. PKB signaling and atrogene expression in skeletal muscle of aged mice. *J Appl Physiol* 111(1):192–9.

Girón, M. D., J. D. Vílchez, S. Shreeram, et al. 2015. β-hydroxyl-β-methylbutyrate (HMB) normalizes dexamethasone-induced autophagy-lysosomal pathway in skeletal muscle. *PLoS One* 10(2):e0117520.

Gomes, A. V., D. S. Waddell, R. Siu, et al. 2012. Upregulation of proteasome activity in muscle RING finger 1-null mice following denervation. *FASEB J* 26(7):2986–99.

Goodman, C. A., R. M. McNally, F. M. Hoffmann, T. A. Hornberger. 2013. Smad3 induced atrogin-1, inhibits mTOR and protein synthesis, and promotes muscle atrophy in vivo. *Mol Endocrinol* 27(11):1946–57.

Grumati, P., L. Coletto, P. Sabatelli, et al. 2010. Autophagy is defective in collagen VI muscular dystrophies, and its reactivation rescues myofiber degeneration. *Nat Med* 16(11):1313–20.

Gustafsson, T., T. Osterlund, J. N. Flanagan, et al. 2010. Effect of 3 days unloading on molecular regulators of muscle size in humans. *J Appl Physiol* 109(3):721–7.

Haddad, F., G. R. Adams, P. W. Bodell, K. M. Baldwin. 2006. Isometric resistance exercise fails to counteract skeletal muscle atrophy processes during the initial stages of unloading. *J Appl Physiol* 100(2):433–41.

Ham, D. J., T. L. Kennedy, M. K. Caldow, A. Chee, G. S. Lynch, R. Koopman. 2014. Citrulline does not prevent skeletal muscle wasting or weakness in limb-casted mice. *J Nutr* 145(5):900–6.

He, C., M. C. Bassik, V. Moresi, et al. 2012. Exercise-induced BCL2-regulated autophagy is required for muscle glucose homeostasis. *Nature* 481(7382):511–15.

Hertoghe, T. 2005. The "multiple hormone deficiency" theory of aging: Is human senescence caused mainly by multiple hormone deficiencies? *Ann NY Acad Sci* 1057:448–65.

Huang, J., J. Wang, L. Gu, et al. 2013. Effect of a low-protein diet supplemented with ketoacids on skeletal muscle atrophy and autophagy in rats with type 2 diabetic nephropathy. *PLoS One* 8(11):e81464.

Hwee, D. T., L. M. Baehr, A. Philp, K. Baar, S. C. Bodine. 2014. Maintenance of muscle mass and load-induced growth in Muscle RING Finger 1 null mice with age. *Aging Cell* 13(1):92–101.

Irwin, W. A., N. Bergamin, P. Sabatelli, et al. 2003. Mitochondrial dysfunction and apoptosis in myopathic mice with collagen VI deficiency. *Nat Genet* 35(4):361–71.

Jogo, M., S. Shiraishi, T. A. Tamura. 2009. Identification of MAFbx as a myogenin-engaged F-box protein in SCF ubiquitin ligase. *FEBS Lett* 583(17):2715–19.

Kang, C., and L. L. Ji. 2013. Muscle immobilization and remobilization downregulates PGC-1α signaling and the mitochondrial biogenesis pathway. *J Appl Physiol* 115 (11):1618–25.

Khan, M. M., S. Strack, F. Wild, et al. 2014. Role of autophagy, SQSTM1, SH3GLB1, and TRIM63 in the turnover of nicotinic acetylcholine receptors. *Autophagy* 10(1):123–36.

Kim, J., Y. C. Kim, C. Fang, et al. 2013. Differential regulation of distinct Vps34 complexes by AMPK in nutrient stress and autophagy. *Cell* 152(1–2):290–303.

Lagirand-Cantaloube, J., K. Cornille, A. Csibi, S. Batonet-Pichon, M. P. Leibovitch, S. A. Leibovitch. 2009. Inhibition of atrogin-1/MAFbx mediated MyoD proteolysis prevents skeletal muscle atrophy in vivo. *PLoS One* 4(3):e4973.

Lecker, S. H., R. T. Jagoe, A. Gilbert A, et al. 2004. Multiple types of skeletal muscle atrophy involve a common program of changes in gene expression. *FASEB J* 18(1):39–51.

Léger, B., W. Derave, K. De Bock, P. Hespel, A. P. Russell. 2008. Human sarcopenia reveals an increase in SOCS-3 and myostatin and a reduced efficiency of Akt phosphorylation. *Rejuvenation Res* 11(1):163–75B.

Lesmana, R., R. A. Sinha, B. K. Singh, et al. 2016. Thyroid hormone stimulation of autophagy is essential for mitochondrial biogenesis and activity in skeletal muscle. *Endocrinology* 157(1):23–38.

Levine, B., and D. J. Klionsky. 2004. Development of self-digestion: Molecular mechanisms and biological functions of autophagy. *Dev Cell* 6(4):463–77.

Liu, J., Y. Peng, Z. Cui, et al. 2013. Depressed mitochondrial biogenesis and dynamic remodeling in mouse tibialis anterior and gastrocnemius induced by 4-week hindlimb unloading. *IUBMB Life* 64(11):901–10.

Lokireddy, S., C. McFarlane, X. Ge, et al. 2011. Myostatin induces degradation of sarcomeric proteins through a Smad3 signaling mechanism during skeletal muscle wasting. *Mol Endocrinol* 25(11):1936–49.

Lo Verso, F., S. Carnio, A. Vainshtein, M. Sandri. 2014. Autophagy is not required to sustain exercise and PRKAA1/AMPK activity but is important to prevent mitochondrial damage during physical activity. *Autophagy* 10(11):1883–94.

Luo, L., A. M. Lu, Y. Wang, et al. 2013. Chronic resistance training activates autophagy and reduces apoptosis of muscle cells by modulating IGF-1 and its receptors, Akt/mTOR and Akt/FOXO3a signaling in aged rats. *Exp Gerontol* 48(4):427–36.

Mammucari, C., G. Milan, V. Romanello, et al. 2007. FoxO3 controls autophagy in skeletal muscle in vivo. *Cell Metab* 6 (6):458–71.

Masiero, E., L. Agatea, C. Mammucari, et al. 2009. Autophagy is required to maintain muscle mass. *Cell Metab* 10(6):507–15.

McKay, B. R., D. I. Ogborn, L. M. Bellamy, et al. 2012. Myostatin is associated with age-related human muscle stem cell dysfunction. *FASEB J* 26(6):2509–21.

McKiernan, S. H., R. J. Colman, M. Lopez, et al. 2011. Caloric restriction delays aging-induced cellular phenotypes in rhesus monkey skeletal muscle. *Exp Gerontol* 46(1):23–9.

McKiernan, S. H., R. J. Colman, E. Aiken, et al. 2012. Cellular adaptation contributes to calorie restriction-induced preservation of skeletal muscle in aged rhesus monkeys. *Exp Gerontol* 47(3):229–36.

McMullen, C. A., A. L. Ferry, J. L. Gamboa, et al. 2009. Age-related changes of cell death pathways in rat extraocular muscle. *Exp Gerontol* 44(6–7):420–5.

Mizushima, N., and M. Komatsu M. 2011. Autophagy: Renovation of cells and tissues. *Cell* 147(4):728–41.

Mourkioti, F., P. Kratsios, T. Luedde, et al. 2006. Targeted ablation of IKK2 improves skeletal muscle strength, maintains mass, and promotes regeneration. *J Clin Invest* 116(11):2945–54.

Neel, B. A., Y. Lin, and J. E. Pessin. 2013. Skeletal muscle autophagy: A new metabolic regulator. *Trends Endocrinol Metab* 24(12):635–43.

Ogawa, T., H. Furochi, M. Mameoka, et al. 2006. Ubiquitin ligase gene expression in healthy volunteers with 20-day bedrest. *Muscle Nerve* 34(4):463–9.

O'Leary, M. F., A. Vainshtein, S. Iqbal, O. Ostojic, D. A. Hood. 2013. Adaptive plasticity of autophagic proteins to denervation in aging skeletal muscle. *Am J Physiol Cell Physiol* 304(5):C422–30.

Park, C., and A. M. Cuervo. 2013. Selective autophagy: Talking with the UPS. *Cell Biochem Biophys* 67(1):3–13.

Paul, P. K., S. K. Gupta, S. Bhatnagar, et al. 2010. Targeted ablation of TRAF6 inhibits skeletal muscle wasting in mice. *J Cell Biol* 191(7):1395–411.

Pond, A. L., C. Nedele, W.-H. Wang, et al. 2014. The mERG1a channel modulates skeletal muscle MuRF1 but not MAFbx, expression. *Muscle Nerve* 49(3):378–88.

Quy, P. N., A. Kuma, P. Pierre, N. Mizushima. 2013. Proteasome-dependent activation of mammalian target of rapamycin complex1 (mTORC1) is essential for autophagy suppression and muscle remodeling following denervation. *J Biol Chem* 288(2):1125–34.

Raben, N., V. Hill, L. Shea, et al. 2008. Suppression of autophagy in skeletal muscle uncovers the accumulation of ubiquitinated proteins and their potential role in muscle damage in Pompe disease. *Hum Mol Genet* 17(24):3897–908.

Ramos, F. J., S. C. Chen, M. G. Garelick, et al. 2012. Rapamycin reverses elevated mTORC1 signaling in lamin A/C-deficient mice, rescues cardiac and skeletal muscle function, and extends survival. *Sci Transl Med* 4(144):144ra103.

Rhoads, M. G., S. C. Kandarian, F. Pacelli, G. B. Doglietto, M. Bossola. 2010. Expression of NF-kappaB and IkappaB proteins in skeletal muscle of gastric cancer patients. *Eur J Cancer* 46(1):191–7.

Rudolf, R., J. Bogomolovas, S. Strack, et al. 2013. Regulation of nicotinic acetylcholine receptor turnover by MuRF1 connects muscle activity to endo/lysosomal and atrophy pathways. *Age* 35(5):1663–74.

Sakuma, K., K. Watanabe, N. Hotta, et al. 2009. The adaptive responses in several mediators linked with hypertrophy and atrophy of skeletal muscle after lower limb unloading in humans. *Acta Physiol* 197(2):151–9.

Sakuma, K., and A. Yamaguchi. 2010. Molecular mechanisms in aging and current strategies to counteract sarcopenia. *Curr Aging Sci* 3(2):90–101.

Sakuma, K., and A. Yamaguchi. 2011a. Sarcopenia: Molecular mechanisms and current therapeutic strategy. In: Perloft, J. W. and Wong, A. H., Eds. *Cell Aging*. Nova Science Publishers, New York, pp. 93–152.

Sakuma, K., and A. Yamaguchi. 2011b. Inhibitors of myostatin- and proteasome-dependent signaling for attenuating muscle wasting. *Recent Pat Regen Med* 1(3):284–98.

Sakuma, K., and A. Yamaguchi. 2012a. Sarcopenia and cachexia: The adaptation of negative regulators of skeletal muscle mass. *J Cachexia Sarcopenia Muscle* 3(2):77–94.

Sakuma, K., and A. Yamaguchi. 2012b. Cellular and molecular mechanism regulating hypertrophy and atrophy of skeletal muscle. In: Willems, M., Ed. *Skeletal Muscle: Physiology, Classification and Disease*. Nova Science Publishers, New York, pp. 141–93.

Sakuma, K., W. Aoi, and A. Yamaguchi. 2015a. Current understanding of sarcopenia: Possible candidates modulating muscle mass. *Pflügers Arch* 467(2):213–29.

Sakuma, K., and A. Yamaguchi. 2015b. Sarcopenia and its intervention. In: Yu, B. P., Ed. *Nutrition, Exercise and Epigenetics: Ageing Interventions*. Springer, Cham, Switzerland, pp. 127–151.

Sakuma, K., and A. Yamaguchi. 2015c. Molecular mechanisms controlling skeletal muscle mass. In: Sakuma, K., Ed. *Muscle Cell and Tissue*. InTech, Rijeka, Croatia, pp. 143–170.

Sakuma, K., N. Kinoshita, Y. Ito, et al. 2016. p62/SQSTM1 but not LC3 is accumulated in sarcopenic muscle of mice. *J Cachexia Sarcopenia Muscle* 7(2):204–12.

Sakuma K., W. Aoi, and A. Yamaguchi. 2017. Molecular mechanism of sarcopenia and cachexia: Recent research advances. *Pflügers Arch* 469(5–6):573–91.

Sakuma K., and A. Yamaguchi. 2018. Recent advances in pharmacological, hormonal, and nutritional intervention for sarcopenia. *Pflügers Arch* 470(3):449–60.

Sanchez, A. M., H. Bernardi, G. Py, et al. 2014. Autophagy is essential to support skeletal muscle plasticity in response to endurance exercise. *Am J Physiol Regul Integr Comp Physiol* 307(8):R956–69.

Sandri, M., C. Sandri, A. Gilbert, et al. 2004. Foxo transcription factors induce the atrophy-related ubiquitin ligase atrogin-1 and cause skeletal muscle atrophy. *Cell* 117(3):399–412.

Sandri, M., L. Barberi, A. Y. Bijlsma, et al. 2013. Signaling pathways regulating muscle mass in ageing skeletal muscle. The role of IGF-1-Akt-mTOR-FoxO pathway. *Biogerontology* 14(3):303–23.

Shaid, S., C. H. Brandts, H. Serve, et al. 2013. Ubiquitination and selective autophagy. *Cell Death Differ* 20(1):21–30.

Shi, L., T. Zhang, X. Liang, et al. 2015. Dihyfromyricetin improves skeletal muscle insulin resistance by inducing autophagy via the AMPK signaling pathway. *Mol Cell Endocrinol* 409:92–102.

Shimizu, N., N. Yoshikawa, N. Ito, et al. 2011. Crosstalk between glucocorticoid receptor and nutritional sensor mTOR in skeletal muscle. *Cell Metab* 13(2):170–82.

Smith, H. K., K. G. Matthews, J. M. Oldham, et al. 2014. Translational signaling, atrogenic and myogenic gene expression during unloading and reloading of skeletal muscle in myostatin-deficient mice. *PLoS One* 9(4):e94356.

Stitt, T. N., D. Drujan, B. A. Clarke, et al. 2004. The IGF-1/PI3K/Akt pathway prevents expression of muscle atrophy-induced ubiquitin ligases by inhibiting FOXO transcription factors. *Mol Cell* 14(3):395–403.

Tardif, N., M. Klaude, L. Lundell, A. Thorell, O. Rooyackers. 2013. Autophagic-lysosomal pathway is the main proteolytic system modified in the skeletal muscle of esophageal cancer patients. *Am J Clin Nutr* 98(6):1485–92.

Vainshtein, A., P. Grumati, M. Sandri, et al. 2014. Skeletal muscle, autophagy, and physical activity: The ménage à trois of metabolic regulation in health and disease. *J Mol Med* 92(2):127–37.

Vainshtein A., E. M. A. Desjardins, A. Armani, M. Sandri, D. A. Hood. 2015. PGC-1α modulates denervation-induced mitophagy in skeletal muscle. *Skeletal Muscle* 5:9.

Van Gammeren, D., J. S. Damrauer, R. W. Jackman, S. C. Kandarian. 2009. The IκB kinases IKKα and IKKβ are necessary and sufficient for skeletal muscle atrophy. *FASEB J* 23(2):362–70.

Vergne, I., E. Roberts, R. A. Elmaoued, et al. 2009. Control of autophagy initiation by phosphoinositide 3-phosphatase Jumpy. *EMBO J* 28(15):2244–58.

von Haehling, S., J. E. Morley, and S. D. Anker. 2010. An overview of sarcopenia: Facts and numbers on prevalence and clinical impact. *J Cachexia Sarcopenia Muscle* 1(2):129–33.

Waddell, D. S., L. M. Baehr, J. van den Brandt, et al. 2008. The glucocorticoid receptor and FOXO1 synergistically activate the skeletal muscle atrophy-associated MuRF1 gene. *Am J Physiol Endocrinol Metab* 295(4):E785–97.

Wakabayashi, H., and K. Sakuma. 2014. Comprehensive approach to sarcopenia treatment. *Curr. Clin. Pharmacol.* 9(2):171-80.

Welle, S., A. I. Brooks, J. M. Delehanty, N. Needler, C. A. Thornton. 2003. Gene expression profile of aging in human muscle. *Physiol Genomics* 14(2):149–59.

Wenz, T., S. G. Rossi, R. L. Rotundo, et al. 2009. Increased muscle PGC-1alpha expression protects from sarcopenia and metabolic disease during aging. *Proc Natl Acad Sci USA* 106(48):20405–10.

Whitman, S. A., M. J. Wacker, S. R. Richmond, M. P. Godard. 2005. Contributions of the ubiquitin-proteasome pathway and apoptosis to human skeletal muscle wasting with age. *Pflügers Arch* 450(6):437–46.

Wohlgemuth, S. E., A. Y. Seo, E. Marzetti, et al. 2010. Skeletal muscle autophagy and apoptosis during aging: Effects of calorie restriction and life-long exercise. *Exp Gerontol* 45(2):138–48.

Wu, C. L., S. C. Kandarian, R. W. Jackman. 2011. Identification of genes that elicit disuse muscle atrophy via the transcription factors p50 and Bcl-3. *PLoS One* 6(1):e16171.

Yang, L., D. Licastro, E. Cava, et al. 2016. Long-term calorie restriction enhances cellular quality-control processes in human skeletal muscle. *Cell Reports* 14(3):422–8.

Zhang, Y.-Y., J. Huang, M. Yang, et al. 2015. Effect of a low-protein diet supplemented with keto-acids on autophagy and inflammation in 5/6 nephrectomized rats. *Biosci Rep* 35(5):e00263.

Zhao, J., J. J. Brault, A. Schild, et al. 2007. FoxO3 coordinately activates protein degradation by the autophagic/lysosomal and proteasomal pathways in atrophying muscle cells. *Cell Metab* 6(6):472–83.

Zhou, J., T. A. Freeman, F. Ahmad, et al. 2013. GSK-3α is a central regulator of age-related pathologies in mice. *J Clin Invest* 123(4):1821–32.

7 Myokines in Aging Muscle

Katie Brown, Aaron Persinger, and Melissa Puppa

CONTENTS

7.1 INTRODUCTION

Skeletal muscle comprises approximately 40% of the human body in healthy adults. In addition to facilitating movement and posture, skeletal muscle is an endocrine organ that secretes many cytokines and small molecules called myokines. Computational studies demonstrate that muscle makes hundreds of proteins that may be secreted (Catoire et al. 2014, Bortoluzzi et al. 2006, Hartwig et al. 2014). Myokines can elicit autocrine, paracrine, or endocrine effects allowing for cross-talk between the muscle and other organs to maintain whole body homeostasis (Figure 7.1). During aging, several myokines are disrupted and may contribute to age-related decline of skeletal muscle mass, quality, and strength termed sarcopenia.

The first myokine to be discovered was interleukin-6 (IL-6) in the late 1990s. IL-6 was found to be secreted from skeletal muscle during exercise (Ostrowski et al. 1998b). Researchers have been searching for an exercise factor that is responsible for many of the positive effects that are seen from exercise training. While there has not been one "exercise factor" found, several other myokines have been discovered that aid in systemic exercise adaptations and are also seen to be dysregulated with disease and aging. The growing availability of secretome analysis has aided in the discovery and confirmation of many myokines involved in the exercise response and in muscle atrophy and disease.

The exact mechanisms underlying sarcopenia are unclear; however, several myokines have been implicated in the development and progression of sarcopenia including IL-6, IL-8, IL-15, leukemia inhibitory factor (LIF), myostatin, brain-derived neurotrophic factor (BDNF), fibroblast growth factor 21 (FGF21), and irisin. Many of these myokines have similar signaling mechanisms and are known to be involved with regulation of metabolism, cell proliferation, and inflammatory signaling. We will discuss the role of several of these myokines on muscle homeostasis and aging.

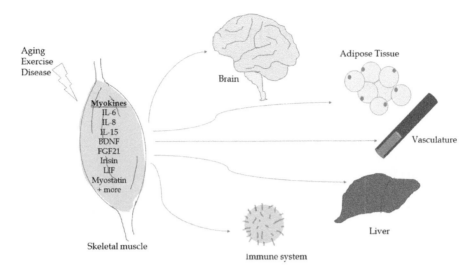

FIGURE 7.1 Schematic of skeletal muscle regulation by myokines.

7.2 INTERLEUKIN-6

IL-6 was first cloned in 1986 (Hirano 1998, Hirano et al. 1986). It is a member of the gp130 receptor family of cytokines, which includes IL-11, LIF, oncostatin M, and cardiotropin-1. IL-6 can signal either through the classical membrane-bound IL-6 receptors or through an alternative trans signaling pathway using the soluble IL-6 receptors (sIL-6Rs). Because not all cells have a membrane-bound IL-6 receptor, the trans signaling pathway allows IL-6 to have actions on all tissue types by binding to the sIL-6R, which can bind to ubiquitously expressed gp130r to initiate downstream signaling. Once bound to the IL-6 receptor, the complex binds with the gp130 receptors causing homodimerization and autophosphorylation of tyrosine residues on gp130. This allows for activation of the Janus kinase/signal transducer and activator of transcription (JAK/STAT), Ras/ERK, and phosphatidylinositol 3-kinase/protein kinase B (PI3K/Akt) signaling pathways (Ernst and Jenkins 2004, Heinrich et al. 1998).

IL-6 acts as both a pro-inflammatory and an anti-inflammatory cytokine. It is primarily secreted from T cells and macrophages to produce an immune response; however, many other tissues secrete this cytokine including muscle, adipocytes, tumors, hepatocytes, and osteoblasts (Akira, Taga, and Kishimoto 1993). Different tissues utilize IL-6 for various functions including differentiation, proliferation, growth, autoimmunity, hematopoiesis, acute phase response, and angiogenesis.

IL-6 was one of the first cytokines implicated in the aging process (Ershler 2003). The increases in IL-6 with age are independent of polymorphisms in the IL-6 gene (Walston et al. 2005). In addition to functioning as a regulator of aging, IL-6 has also been implicated in the development of sarcopenia. Several studies demonstrate elevated IL-6 in the risk and development of sarcopenia (Bian et al. 2017, Visser et al. 2002, Schaap et al. 2006, Payette et al. 2003). In addition to increasing the risk for sarcopenia, serum IL-6 levels correlate with disability and mortality (Schrager et al. 2007, Ferrucci et al. 2002) and may be a predictor of muscle strength reductions over time in elderly people (Miller et al. 2008). In addition to circulating IL-6 levels regulating the development of sarcopenia, Schaap et al. (2009) showed an association of high IL-6-soluble receptor levels with greater declines in grip strength. In otherwise healthy individuals, hand grip strength drops by approximately 50% from age 25 to 95 years. In individuals with an underlying inflammatory condition, the decline in grip strength is steeper after the age of 50 years (Beenakker et al. 2010). This suggests that inflammation may speed the natural declines in strength.

Studies have found that high plasma levels of IL-6 are associated with obesity and inactivity, and IL-6 has a role in the pathogenesis of many diseases including arthritis, asthma, and cancer, to name a few (Hamer et al. 2012, Colbert et al. 2004, Lin, Chen, and Wang 2016, Mauer, Denson, and Bruning 2015, Hunter and Jones 2017). Although IL-6 is implicated in many disease states, IL-6 is important for normal cellular function and healing in skeletal muscle. Mice lacking IL-6 demonstrate delayed recovery from overload-induced injury, which is associated with decreased Insulin-Like Growth Factor-1 (IGF-1) expression and Mammalian target of rapamycin (mTOR) activation (Washington et al. 2011). Moreover, IL-6 is involved in satellite cell proliferation through the regulation of cell cycle-associated genes, as well as differentiation and fusion of satellite cells to muscle fibers (Serrano et al. 2008, Baeza-Raja and Munoz-Canoves 2004, Hoene et al. 2013).

IL-6 signaling is accomplished through gp130r to activate JAK/STAT and downstream effectors. When phosphorylated by gp130, JAK/STAT can translocate to the nucleus of the cell to upregulate gene transcription. In muscle, STAT is activated promoting satellite cell proliferation and inhibiting myogenic genes such as MyoD, MEF2, and myogenin that would promote differentiation (Kami and Senba 2002, Sun et al. 2007). Once enough cells have proliferated, the cell proliferation pathway needs to be suppressed and the differentiation pathway activated for myogenesis to continue. Suppressor of cytokine signaling (SOCS) proteins inhibit gp130/STAT3 signaling allowing for the activation of myogenic differentiation from proliferated satellite cells. Interestingly, SOCS3 signaling increases during aging and may contribute to the development of sarcopenia (Leger et al. 2008, Trenerry et al. 2008).

IL-6 has been implicated in the loss of skeletal muscle. During cancer-induced muscle wasting IL-6 is increased (Carson and Baltgalvis 2010). Studies have shown that inhibition of IL-6 signaling in a cancer model can attenuate protein degradation; however, it is unable to attenuate the suppression of protein synthesis (White et al. 2011). Administration of high doses of IL-6 and chronic exposure to IL-6 increases proteolysis in rodents, which is reversed when IL-6 is inhibited (Goodman 1994, Tsujinaka et al. 1996). IL-6 work in conjunction with other factors to induce muscle breakdown as IL-6 knockout (KO) mice still displays muscle atrophy and protein degradation in a model of sepsis (Williams et al. 1998). IL-6 can also inhibit pathways promoting growth and protein synthesis as well as its effects on muscle protein breakdown. Reduced growth is observed in mouse models overexpressing IL-6 (De Benedetti et al. 1997). These reductions in growth are likely due to the suppression of IGF-1 and its downstream signaling (De Benedetti et al. 1997, Nemet et al. 2006). Neutralization of IL-6 signaling restores IGF-1 levels suggesting an interaction of IL-6 in the regulation of IGF-1 signaling (De Benedetti et al. 2001).

In addition to signaling the muscle for repair and protein turnover, IL-6 is important for the regulation of substrate utilization and may be an energy sensor for skeletal muscle. Several studies have demonstrated that IL-6 is increased when skeletal muscle glycogen content is low. Ingestion of glucose when exercising, to prevent glycogen depletion, attenuates the exercise-induced increase in IL-6 (Keller et al. 2005, Febbraio et al. 2003, Steensberg et al. 2001). Interestingly, glycogen stores are decreased in the muscle of untrained older humans (Cartee 1994). This may be one cause for the increase in IL-6 seen in sedentary older individuals. IL-6 is also a known stimulator of lipolysis and enhances fatty acid oxidation in skeletal muscle (Al-Khalili et al. 2006, Carey et al. 2006).

IL-6 is acutely elevated in skeletal muscle during contraction and may function to increase the availability of extracellular substrates for fuel during contraction (Febbraio and Pedersen 2005, Petersen and Pedersen 2005). As mentioned previously, IL-6 is sensitive to muscle glycogen concentrations (Steensberg et al. 2001, Keller et al. 2005). As glycogen stores are utilized, IL-6 is secreted from the muscle to mobilize fatty acids for beta-oxidation. Muscle-derived IL-6 is also involved in whole-body carbohydrate and lipid metabolism during exercise, decreasing the carbohydrate utilization during prolonged exercise (Gudiksen et al. 2016, Knudsen et al. 2017). Not only does the skeletal muscle release IL-6 after endurance type exercise but also after resistance exercise (Begue et al. 2013). This increase with resistance exercise is likely associated with activation of satellite cells for muscle repair rather than for the mediation of substrate utilization seen with endurance exercise. IL-6 production by the muscle in response to damage is smaller and occurs later than the IL-6 production that is elicited by muscle contraction (Pedersen, Steensberg, and Schjerling 2001). IL-6 release during endurance exercise is acute and typically returns to basal levels rapidly upon completion of the exercise (Ostrowski et al. 1998b, 1999, 1998a), whereas IL-6 release in response to muscle damage occurs later and remains elevated for a longer period (Bruunsgaard et al. 1997, Hellsten et al. 1997). The prolonged elevation of IL-6 in response to damage is necessary for normal recovery. Mice lacking IL-6 have a delayed recovery from muscle damage, demonstrating that it is necessary for recovery from eccentric exercise (Serrano et al. 2008, White et al. 2009, Washington et al. 2011).

IL-6 is released from a wide variety of cells; thus, the effects of systemic IL-6, which is released from other tissues, may have differential effects on the skeletal muscle. In muscle exposed to local IL-6, there is no effect on cytochrome c protein concentration; however, in muscle that is systemically exposed to IL-6, there is a decrease in cytochrome c (Puppa et al. 2012). Regardless of the exposure, muscle maintained the capacity to adapt to exercise (Puppa et al. 2012). In aging where there is an increase in systemic IL-6 concentrations, the capacity of exercising skeletal muscle to release IL-6 is maintained (Pedersen et al. 2004). Maintaining the capacity to acutely increase IL-6 concentrations may be important for the beneficial effects of exercise both on the muscle tissue itself and the body as a whole particularly in aging.

IL-6 has recently become a target for the treatment of inflammatory diseases. Blocking IL-6 signaling with neutralizing antibodies to either IL-6 or IL-6R has shown promise in decreasing disease severity in rheumatoid arthritis (Nishimoto et al. 2004, Yokota et al. 2008). Treatment of IL-6 inhibiting antibodies may also help to preserve muscle mass in cancer-induced muscle wasting (White et al. 2011, Narsale and Carson 2014, Puppa et al. 2014). While these therapies may be effective in treating pathological conditions, the long-term effects of these treatments associated with aging are unclear. More research needs to be done to understand the effects of IL-6 inhibition on aging and sarcopenia.

7.3 IL-8

IL-8, also known as CXCL8, is expressed by skeletal muscle after contraction (Frydelund-Larsen et al. 2007). IL-8 is a member of CXC family of chemokines and has two cysteine residues at the amino terminus. In addition to being secreted by skeletal muscle, IL-8 can be secreted by macrophages and endothelial cells. One potential benefit of this myokine is that it is thought to stimulate angiogenesis and thus increase capillary density in exercised muscle (Pedersen 2009).

IL-8 signals through the CXCR1 and CXCR2 receptors to elicit downstream signaling though PI3K/Akt, Mitogen Activated Protein Kinase (MAPK), and Focal Adhesion Kinase (FAK) pathways to induce cell migration, proliferation, and angiogenesis. Too much or too little IL-8 can alter signaling to inhibit these pathways. CXCR2 is expressed in the microvasculature and is responsible for IL-8-mediated increases in angiogenesis (Addison et al. 2000). In myotubes from diabetic mice, IL-8 is elevated and inhibits capillary growth potentially exacerbating the diabetic phenotype by minimizing the glucose availability to skeletal muscles (Amir Levy et al. 2015). Low levels of IL-8, similar to those seen during concentric exercise, appear to be angiogenic, whereas high concentrations of IL-8 inhibit angiogenesis. The transient increase in muscle IL-8 during concentric contraction may induce fiber capillary density increases seen with endurance training.

Similar to IL-6, IL-8 secretion by a muscle may be regulated in part by glycogen availability. Contraction induces the expression of both IL-6 and IL-8, and these cytokines increase further in low glycogen conditions (Chan et al. 2004, Nieman et al. 2015). Unlike IL-6, IL-8 does not appear to increase systemically during concentric exercise; however, there is still a transient increase within a muscle (Akerstrom et al. 2005). This suggests that IL-8 secreted by a muscle may be responsible for a paracrine or autocrine response to exercise and is not likely responsible for the systemic increase associated with the inflammatory response seen with eccentric exercise.

Elevations in circulating IL-8 are shown to be associated with decreased appendicular lean mass and increased risk of developing sarcopenia (Westbury et al. 2018). Like the diabetic phenotype, this observed association may be due to the inhibition of muscle angiogenesis and thus a limited supply of nutrients and oxygen to the muscle. More research is needed to fully understand the role of skeletal muscle IL-8 in aging and sarcopenia.

7.4 IL-15

IL-15 is a growth factor that is highly expressed in skeletal muscle. IL-15 is secreted by contraction of skeletal muscles as well as monocytes and macrophages from viral or bacterial infections (Molanouri Shamsi et al. 2015, Mak and Saunders 2006). It is theorized that IL-15 plays a role in muscle-fat cross-talk through an anabolic effect on skeletal muscle and a role in reduction of adipose tissue mass (Nielsen and Pedersen 2018). It seems that muscle cells can actively secrete IL-15 (Pistilli and Quinn 2013). IL-15 is similar to another member of the IL family, IL-2 but with more widely expressed receptors and a pro-inflammatory effect (Gravallese and Monach 2015). In healthy animals, IL-15 does not induce anabolic effects, however, in muscle wasting-diseased conditions, IL-15 acts to spare muscles (Pistilli and Quinn 2013). Moreover, IL-15 appears to be expressed more in atrophic conditions including aging (Pistilli and Quinn 2013).

Healthy adults that are 95 years old or older have higher IL-15 serum levels than control groups of 30–59 and 60–89 years old (Gangemi et al. 2005). In mice, muscle and serum IL-15 levels decline progressively with age (Quinn et al. 2010). With aging, IL-15 protein was lower in aged muscle. With a caloric restriction diet, these age-related reductions were partially attenuated (Pistilli and Quinn 2013). The exact mechanism of muscle IL-15 in the aging process needs to be further elucidated.

IL-15 signaling leads to the activation of the JAK3/STAT3 pathway (Krolopp, Thornton, and Abbott 2016). IL-15 can also induce Peroxisome proliferator-activated receptor gamma coactivator 1 alpha (PGC-1α) expression. This can lead to the expression of another myokine, irisin, and its genetic precursor, FNDC5. In mice, muscle FNDC5 mRNA expression and irisin release are not dependent on IL-15 (Quinn et al. 2015). Overexpressing IL-15 can lead to muscular hypertrophy in skeletal myogenic cultures (Schnyder and Handschin 2015, Quinn et al. 2015, Kim et al. 2013a). Researchers have proposed that IL-15 is an anabolic factor for skeletal muscle, with potential therapeutic use for muscle wasting conditions (Quinn, Haugk, and Grabstein 1995, Quinn, Haugk, and Damon 1997, Quinn et al. 2015, Pistilli and Quinn 2013). Overexpression of IL-15 was seen to both increase protein synthesis and inhibit myotube protein degradation without an effect on myoblasts (Pistilli and Quinn 2013, Quinn et al. 2002). In tumor bearing rodents, IL-15 attenuates apoptotic pathways downstream of tumor necrosis factor alpha (Carbo et al. 2000, Figueras et al. 2004).

In addition to its regulation of protein turnover, IL-15 may play a role in metabolism. Overexpression increases SIRT1 and uncoupling protein (UCP) expression in muscles (Quinn et al. 2011). Beyond its anabolic effects in skeletal muscle, it has been found that there are reductions in adipose tissue mass when skeletal muscles secrete IL-15 (Pierce, Maples, and Hickner 2015).

Exercise can increase IL-15. Following resistance exercise, IL-15 gene expression is increased, without any changes in plasma IL-15 or muscle protein IL-15 in type 2 muscle fibers (Nielsen et al. 2007). Researchers found that resistance exercise can activate the IL-15/IL-15R15Rα signaling pathway and that one bout of exercise can increase IL-15 mRNA (Perez-Lopez et al. 2018, Nielsen et al. 2007). Chronic aerobic exercise can also increase IL-15 levels (Duzova et al. 2018). The exact role that IL-15 plays in response to exercise is not fully understood; however, daily IL-15 therapy can mimic the anti-aging effects of exercise (Crane et al. 2015).

7.5 FGF21

FGF21 functions as a metabolic regulator. FGF21 was discovered to be produced and secreted by skeletal muscle in 2008 from a mouse model with constitutively active Akt1 in the skeletal muscle (Izumiya et al. 2008). It is primarily secreted from the liver; however, we now know that it is also released from muscle, adipose tissue, and the pancreas (Fisher and Maratos-Flier 2016).

To signal, FGF21 must bind to fibroblast growth factor receptors (FGFRs) and the co-receptor Klotho (KLB). FGF21 then signals through the Ras/Raf/MAPK and PI3K signaling pathways. In the skeletal muscle, FGF21 has been shown to regulate glucose uptake, but muscle has low levels of KLB indicating that skeletal muscle may not be very sensitive to FGF21. There is some evidence that KLB expression is increased in muscles with high FGF21 expression (Tezze et al. 2017). This suggests that FGF21 may have more of an endocrine role when released from skeletal muscle. Muscle FGF21 is regulated by insulin in both humans and rodents (Hojman et al. 2009, Izumiya et al. 2008). Overexpression of FGF21 in skeletal muscle leads to resistance of diet-induced obesity and browning of adipose tissue (Kim et al. 2013b). While the exact actions of muscle FGF21 are unknown, there is evidence that FGF21 is acting to regulate metabolism. Skeletal muscle with mitochondrial dysfunction increases FGF21 expression to activate the PGC-1α signaling pathway (Ji et al. 2015, Pereira et al. 2017, Vandanmagsar et al. 2016, Keipert et al. 2014). Additionally, endoplasmic reticulum stress triggered by sustained mTORC1 activation can also induce the production of FGF21 (Guridi et al. 2015).

During aging, there is a loss of functional mitochondria and decrease in mitochondrial turnover. FGF21 may be responsible for some of the age-associated metabolic dysfunction. FGF21 is increased in OPA1 KO mice, and OPA1 is known to decrease with age (Tezze et al. 2017). Additionally, deletion of FGF21 in skeletal muscle was able to reverse the aging phenotype seen in OPA1 KO mice supporting a role for FGF21 in aging and sarcopenia (Tezze et al. 2017). High levels of serum FGF21 are seen with aging (Hanks et al. 2015) and progeria (Suomalainen et al. 2011). As it stands, it appears that muscle-derived FGF21 plays a systemic role in aging with minimal effect on the muscle itself. More research is needed to fully understand the role of FGF21 on skeletal muscle during aging.

7.6 BDNF

BDNF has been associated with cognitive deficits including memory and is decreased in both the brain and serum of patients suffering from Alzheimer's disease (Connor et al. 1997, Laske et al. 2006, Komulainen et al. 2008). While the brain is thought to be the primary site for BDNF synthesis and is likely responsible for the exercise-induced increases in plasma BDNF, it is also produced in the skeletal muscle during contraction (Matthews et al. 2009).

BDNF signals through the tropomyosin-related kinase B (TrkB) receptor, which associates with p75NTR. BDNF released from the brain has been implicated in the regulation of food intake and body mass (Rothman et al. 2012, Pelleymounter, Cullen, and Wellman 1995). In skeletal muscle BDNF increases phosphorylation of AMP-activated protein kinase (AMPK) and acetyl CoA carboxylase (ACC), two energy sensors in the cell, as well as increased fatty acid oxidation. Interestingly, BDNF that is synthesized by the muscle during contraction does not seem to be released into circulation.

BDNF is an important regulator of skeletal muscle differentiation. Inhibition of BDNF in myoblasts enhances myogenic differentiation, and BDNF is expressed in satellite cells (Mousavi and Jasmin 2006). A decrease in Pax7 marker of satellite cells is found in mice with a muscle-specific deletion of BDNF and mice lacking muscle BDNF display a delay in fiber regeneration (Clow and Jasmin 2010). BDNF may also play a role in maintenance of the synapse through the regulation of neurotransmitter release from the neuron as well as regulating the postsynaptic region on the muscle fiber (Kalinkovich and Livshits 2015). Inhibition of the TrkB receptor impairs neuromuscular transmission in young but not old mice, suggesting that there is reduced TrkB activity in aged muscle and decreased BDNF with age (Greising et al. 2015). While the exact role of BDNF in the development and progression of sarcopenia is unknown, BDNF is correlated with strength in subjects with metabolic syndrome and diabetes (Tsai et al. 2015).

7.7 IRISIN

Irisin is a 112 amino acid peptide that is proteolytically cleaved from the fibronectin type 3 domain containing protein 5, predominately found in skeletal muscle (Chang et al. 2017). Recently, low levels of irisin have been correlated with aging; specifically, older individuals had significantly lower levels of circulating irisin compared with the young and middle-aged individuals (Chang et al. 2017). Healthy centenarians, defined as age > 100, have an increased relative level of serum irisin compared with age-matched controls, while younger patients with myocardial infarctions had significantly lower levels of serum irisin (Emmanuele et al. 2014). Moreover, irisin has been linked with sarcopenia as indicated by a study that showed that men with sarcopenia tended to have lower irisin level signaling than individuals without sarcopenia (Chang et al. 2017). Telomere length, specifically the shortening of telomeres, is an established genetic marker of aging. Plasma irisin levels can predict relative telomere length in healthy individuals, suggesting that irisin may be used as a predictor of sarcopenia development (Rana et al. 2014).

In the body, white adipose tissue (WAT) stores fat, while brown adipose tissue (BAT) is used for the energy expenditure (Timmons et al. 2007). Irisin beiges/browns the white adipocytes (Bostrom et al. 2012). This increases cellular mitochondrial density and expression of UCP1, which leads to increased energy expenditure through thermogenesis (Chang et al. 2017, Rana et al. 2014). PGC-1α expression in muscle stimulates increased expression of FNDC5, which is cleaved and secreted as irisin (Sundarrjan et al. 2018). Once in circulation, irisin acts on WAT to stimulate UCP1 expression (Bostrom et al. 2012). While largely thought to be secreted by skeletal muscle, adipocytes contribute about ¼ of secreted irisin to the body (Roca-Rivada et al. 2018).

In addition to its effects on adipose tissue, irisin has effects on skeletal muscle. Myostatin and troponin are targets of irisin (Sundarrjan et al. 2018). Knockdown of irisin upregulates myostatin mRNA expression in zebrafish heart and skeletal muscle tissues (Sundarrjan et al. 2018). Troponin, among other uses, plays a vital role in the normal functioning of skeletal muscle by binding to calcium and pulling tropomyosin off myosin-binding sites. In zebrafish, treatment with irisin increases both troponin and troponin C while knockdown has the opposite effects in skeletal and cardiac muscle (Sundarrjan et al. 2018).

Irisin is positively correlated with muscle mass, strength, and function (Huh et al. 2012, Kurdiova et al. 2014, Kim et al. 2015). Even accounting for the irisin secretion from fat mentioned above, there is a negative correlation between irisin and muscle atrophy and weakness (Chang et al. 2017). Irisin levels are closely related to IGF-1, which contributes to muscular hypertrophy (Huh et al. 2012, 2014). Evidence suggests that irisin may have a regulatory role in muscle metabolism (Pardo et al. 2014, Kurdiova et al. 2014). Irisin increases energy expenditure and muscle oxidative metabolism (Schnyder and Handschin 2015). Myocytes treated with irisin have increased PGC-1α, NRF-1, TFAM, GLUT4, and UCP3 expression suggesting a role for irisin in skeletal muscle metabolism and mitochondrial biogenesis (Vaughan et al. 2014).

During aging, sedentary behavior is associated with lower levels of FNDC5, BDNF, and PGC-1α than aging that is accompanied by exercise (Belviranli and Okudan 2018). Single sessions of intense endurance exercise and heavy strength training cause acute increases of irisin concentrations in blood. Further, results suggest that irisin response to heavy resistance exercise, but not endurance exercise, is higher in individuals with lower proportions of lean body mass (Nygaard et al. 2015). Intensity, regardless of energy consumption, affects irisin secretion after an acute running effort. In a study where subjects performed exercise trials at 40%, and 80% of VO2, immediately postexercise the higher intensity groups had significantly greater irisin concentrations compared with their pre-exercise levels (Tsuchiya et al. 2014).

Taken together, exercise seems to have caused an acute increase in irisin that may be responsible for some of the adaptations known to occur with exercise training. Long-term exposure to exercise has no effect on circulating irisin levels; however, the same study found increased circulating irisin immediately after exercise (Norheim et al. 2014). Another study found that exercise increased circulating irisin levels as well as two different strength measures without changing body composition (Kim et al. 2015). Additionally, endurance exercise increases systemic irisin levels through FNDC5 expression. Increases in hippocampal FNDC5, which is PGC-1α dependent, increases BDNF expression causing neurogenesis and may help to account for exercise-induced cognitive improvements (Wrann et al. 2013, Schnyder and Handschin 2015).

7.8 MYOSTATIN

Myostatin, or growth and differentiation factor-8 (GDF-8), is a highly regulated member of the transforming growth factor-β (TGF-β) superfamily. The TGF-β superfamily of genes is responsible for encoding secreted factors important in regulating embryonic development and tissue homeostasis in adults (Thomas et al. 2000). Myostatin is expressed and secreted mainly by the skeletal muscle; it is well established as the negative regulator of skeletal muscle growth, as it affects cell growth and differentiation as well as intracellular catabolic and anabolic signaling pathways. Myostatin expression inhibits protein synthesis and promotes protein degradation by binding to its receptor, activin type IIB (ActRIIB), and forming a heterodimer with activin-like kinase 4 (ALK4) or ALK5. The ALK4 and 5 domain phosphorylates Smad2 and 3, which then form a complex with Smad4. This complex translocates to the nucleus to regulate gene transcription involving skeletal muscle proliferation/differentiation and degradation processes. In the presence of growth signals (i.e., insulin and IGF-1), activation of Smad2 and 3 by myostatin inhibits the Akt/mTOR pathway to suppress protein synthesis (White and LeBrasseur 2014). Moreover, myostatin/Smad3 activation has been shown to inhibit myoblast differentiation in vitro by downregulating MyoD expression (Langley et al. 2002). Myostatin also affects the proteolytic systems by upregulating atrogenes in the ubiquitin–proteasome system and increasing autophagic enzymes.

Myostatin gene mutations and deletions exhibit remarkable phenotypical changes in musculature. One example is seen in children born with a myostatin gene mutation exhibiting profound muscle hypertrophy (Schuelke et al. 2009). Because of this evidence, myostatin inhibition has been of popular interest in the research field of muscle wasting; myostatin inhibitors seem to have a promising therapeutic effect on skeletal muscle loss. Yet, the literature on age-related physiological changes in myostatin abundance and activity remains inconsistent. This is partially due to difficulty in distinguishing myostatin from its homologous TGF-β family member, growth, and differentiation factor 11 (GDF11). Assays measuring these proteins, such as Enzyme-linked immunosorbent assay (ELISA), are unable to detect differences in the similar amino acid sequences and therefore yield varying and inconsistent results (Schafer et al. 2016). In an earlier cross-sectional study of young, middle-aged, and older men and women, serum myostatin levels were found to increase with age, be greatest in "physically frail" older women, and were inversely associated with skeletal muscle mass (Yarasheski et al. 2002). More recently, a specific and sensitive liquid chromatography combined with tandem mass spectrometry (LC–MS/MS) assay has been used to distinguish myostatin levels from GDF11. Studies using this approach confirm that serum myostatin concentrations increase with both age (>65 years) and sarcopenia in women; however, myostatin levels tend to decrease with age in men with and without sarcopenia compared with younger (<40 years) men. This suggests myostatin may be a key contributor to sarcopenia development in women but act as a key regulator of muscle mass in men (Schafer et al. 2016, Bergen et al. 2015). Moreover, myostatin function with aging has been investigated in mice. Both aged myostatin null and KO mice have been shown to exhibit minimal muscle atrophy, retain glycolytic fibers, and maintain muscle mass compared with wild-type (WT) controls (Siriett et al. 2006, Morissette et al. 2009).

Although skeletal muscle is the main secretory site of myostatin, adipose tissue and cardiac muscle express small amounts as well. Further, myostatin secretion affects a variety of tissues and metabolic processes including adipogenesis, insulin sensitivity, glucose metabolism, cardiac function, and bone density in normal physiological functioning and aging.

The role of myostatin in adipogenesis is controversial; it can both promote and inhibit adipogenesis depending on the parameters. For example, in preadipocytes, myostatin acts as an adipocyte inhibitor in cell differentiation. This inhibition is regulated through reduced expression of peroxisome proliferator-activated receptor γ (PPARγ) and CCAAT enhancer-binding protein α (C/EBPα) transcription factors, as well as adipogenic markers adipocyte protein 2 (aP2) and leptin (Schnyder and Handschin 2015, Deng et al. 2017). Similarly, myostatin inhibits BAT adipogenesis; however, this process is thought to be regulated by TGF-β/Smad3 signaling (Deng et al. 2017). Whereas in mesenchymal multipotent cells, myostatin promotes adipogenesis in early stages of differentiation. Most of the available evidence has shown myostatin deletion and inhibition in animals increases muscle mass and decreases fat mass. In myostatin null mice, total and intramuscular body fat was reduced compared with WT mice; additionally, myostatin deletion in muscle, but not adipose tissue, decreased fat mass with concomitant increases in muscle mass (McPherron and Lee 2002, Guo et al. 2009). Myostatin KO mice have been found to be resistant to diet-induced obesity, possibly driven by enhanced BAT formation and increased fatty acid oxidation. Moreover, myostatin KO mice allow BAT formation from WAT by activating the AMPK-PGC-1-Fndc5 pathway (Shan et al. 2013). The cross-talk between adipose tissue and skeletal muscle in myostatin signaling is complex. Studies show that both myogenesis-related factors (i.e., muscle-related transcription factor myogenic factor 5 (Myf5), myocyte enhancer factor-2 (MEF2), and MyoD) and adipogenesis-related factors (IBMX, C/EBPα and β) regulate myostatin in multiple animal models. Furthermore, myostatin can be upregulated by PPARγ and glucocorticoids (i.e., dexamethasone) (Deng et al. 2017). Yet more research is needed to determine exact mechanisms of myostatin signaling to adipose tissue.

Myostatin inhibition has also resulted in improved glucose metabolism and insulin sensitivity in mice-fed standard and high-fat diets. Myostatin null mice exhibit increased glucose utilization and insulin sensitivity as measured by indirect calorimetry, insulin and glucose tolerance test, and hyperinsulinemic-euglycemic clamps. Increased glucose utilization and insulin sensitivity were found to be tissue specific as myostatin deletion in muscle, not adipose tissue, produced these beneficial effects. This suggests that, like with fat mass reduction, the positive changes in glucose metabolism result from skeletal muscle alterations rather than a direct inhibition of myostatin in adipose tissue (Guo et al. 2009).

The evidence is promising for myostatin attenuation as a strategy against developing pathological conditions associated with age—including obesity, insulin resistance, and sarcopenia development. Aged myostatin null and KO mice, compared with their WT counterparts, exhibit the same effects mentioned previously on muscle mass function and adiposity: increased lean fat-free mass, improved insulin sensitivity, and reduced total adiposity. Moreover, enhanced cardiac stress response, measured by ejection fraction and fractional shortening, reduced bone mineralization, and increased bone density, is seen in the absence of myostatin in aged mice. Together these data confirm myostatin's role in exacerbating conditions associated with aging and sarcopenia (Morissette et al. 2009, Jackson et al. 2012).

The function of myostatin as a negative regulator of muscle growth was first determined by the phenotypic changes seen in myostatin-deficient animals. By knocking out the myostatin gene, these "mighty mice" exhibited muscle overgrowth due to hypertrophy and hyperplasia (McPherron, Lawler, and Lee 1997). Further studies assessing inhibition in mice show myostatin affects growth and repair (satellite cell activation, proliferation, and differentiation) and muscle size and mass. In satellite cell activation, myostatin has an antiproliferative property and is associated with upregulation of p21, a cyclin-dependent kinase inhibitor, and downregulation of Cdk2, Cdk4, and retinoblastoma protein Rb. Whereas in cell differentiation, myostatin downregulates Myfs such as

MyoD, Myf5, and myogenin (Carnac, Vernus, and Bonnieu 2007). Myostatin causes muscle atrophy by inhibiting protein synthesis and/or increasing protein degradation; it appears to do this by inhibiting the Akt-mTOR pathway, inhibiting satellite cell proliferation, and upregulating the ubiquitin–proteasome pathway (Elliot et al. 2012). Other ligands, in cooperation with myostatin, control muscle growth by negatively regulating its activation, secretion, and receptor binding; these antagonists include myostatin propeptide, FLRG, GASP1, hSGT, follistatin, decorin, and titan-cap. These are only a few of the known regulators, and other signaling pathways remain to be elucidated (Carnac, Vernus, and Bonnieu 2007). Because of its nature as a key catabolic mediator, myostatin function has been studied in a variety of muscle disuse models and wasting diseases including muscle dystrophies, glucocorticoid-induced atrophy, cachexia, and sarcopenia among others. Below we discuss the current findings of myostatin's role in a variety of atrophy models, as well as subsequent applications for aging and sarcopenia development.

In disuse atrophy in older patients, elevated myostatin is associated with muscle loss (Reardon et al. 2001). Likewise, increases in serum myostatin are observed in older sarcopenic men and women (Yarasheski et al. 2002). Human studies have reported mixed evidence for myostatin abundance in sarcopenia; some have observed a correlation between increased circulating myostatin levels and muscle mass loss, while others have found no relationship between muscle mass and myostatin mRNA abundance (Schulte and Yarasheski 2001, Marcell et al. 2001). However, positive results have been observed in aged sarcopenic mice without myostatin function. Myostatin null mice demonstrate that prolonged absence of myostatin reduces age-related sarcopenia and loss of muscle regenerative capacity; these mice exhibited minimal atrophy and no fiber-type switching compared with WT mice with significant atrophy and switching to oxidative fibers (Siriett et al. 2006). Furthermore, a short-term myostatin antagonist given to aged mice significantly improved muscle regeneration after injury and during sarcopenia by increasing MyoD and Pax7, markers of myogenesis, and satellite cell proliferation (Siriett et al. 2007).

Myostatin has been shown to induce cachexia *in vivo* and *in vitro* in a FoxO1-dependent mechanism (McFarlane et al. 2006). In contrast, inhibition of myostatin in cachectic mice-bearing tumors prevented atrophy (Murphy et al. 2011). Indeed, in other muscle-wasting diseases and conditions, inhibition of myostatin seems to rescue muscle atrophy. In glucocorticoid-induced atrophy, the gene deletion of myostatin blunts proteolysis by upregulating atrogenes MurF1 and Atrogin-1 and increasing lysosomal enzyme cathepsin L; moreover, inhibition prevents both body and muscle weight loss in myostatin-deficient mice (Gilson et al. 2007). In murine cancer and cachectic models, inhibition of myostatin increased body and muscle weights (Busquets et al. 2012). When assessing myostatin inhibition as a treatment for muscle dystrophy diseases, generally the evidence has shown increases in muscle mass; however, effects on strength and function remain unclear (Smith and Lin 2013). Overall, the evidence seems conclusive: treatments targeting myostatin inhibition seem effective for augmenting muscle atrophy in multiple atrophic conditions, especially sarcopenia. Early stages of clinical trials for myostatin inhibitors are currently under investigation in humans (Smith and Lin 2013). Although there is still much to be investigated, the evidence is exciting as a potential new therapy for treating age-related sarcopenia.

Adaptations of myostatin expression with both resistance training (RT) and endurance exercise generally demonstrate an inverse relationship. A majority of the evidence shows endurance training and aerobic exercise in humans and animals reduces myostatin expression in skeletal muscle, which could be beneficial in attenuating sarcopenia. Long-term endurance exercise training causes a reduction in myostatin expression in obese individuals, insulin resistant, and middle-aged men (Hoffman and Weigert 2018, Hittel et al. 2010). Similarly, hemodialysis patients performing endurance exercise for 9 weeks also show reduced myostatin mRNA levels. This response is also seen in short bouts of endurance exercise. Myostatin mRNA is reduced in response to 30 min of running in physically active men and women. Moreover, endurance exercise training in older mice has been investigated; inhibition of myostatin augmented exercise performance on treadmill running in aged mice compared with untrained controls and controls without anti-myostatin antibody injections.

Taken together, it appears that endurance exercise, both short term and long term, is indeed effective in reducing myostatin expression and mRNA in healthy and disease conditions.

Resistance exercise is known for effectively promoting muscle hypertrophy, strength, and performance, mainly by increasing protein synthesis. The effect of RT on myostatin expression seems to act in a dose-dependent manner. Light RT in healthy individuals does not alter myostatin mRNA, while heavy strength training decreases myostatin gene expression (Manini et al. 2011, Roth et al. 2003). Further, the effect on myostatin mRNA and serum protein levels seems to depend on overall duration and training regimen (Allen, Hittel, and McPherron 2011). Although sarcopenic individuals can reap the benefits of resistance exercise such as increased muscle mass, the mechanism behind this seems to be related to reduced oxidative stress, increased protein synthesis via activation of mTOR, and increased number of myofibrils rather than a myostatin-dependent mechanism. Recent data suggest RT-induced hypertrophy may be due to an increased PGC-1α isoform, which further increases IGF-1 expression while repressing myostatin. More research is needed in assessing age-related changes in myostatin in response to exercise.

7.9 LIF

LIF was first discovered in 1988 as a protein secreted from ascites tumor cells that stimulated differentiation of myeloid leukemic cells (hence the origin of its name). Since then, LIF has recently been discovered as a myokine known to be highly pleotropic and able to both stimulate and inhibit cell proliferation, differentiation, and survival (Nicola and Babon 2015, Broholm and Pedersen 2010). LIF is a highly glycosylated, long-chain four α-helix bundle; as a member of the IL-6 superfamily of cytokines, LIF signals through the shared receptor chain gp130, but in a heterodimer form as gp130 combines with the LIF receptor (known as LIFRβ) in order to initiate binding. Once LIF binds to its receptor, it phosphorylates and activates JAK and STAT; further, this activation initiates a cascade of tyrosine phosphorylation, which is able to activate the JAK/STAT, MAP kinase, and PI3K/Akt pathways to promote differentiation, survival, and self-renewal of cells (Nicola and Babon 2015). Interestingly, activation of JAK/STAT plays two roles in myogenic differentiation: it aids in myoblast proliferation via activation of JAK2/STAT2/STAT3 and suppresses myogenic differentiation through activation of JAK1/STAT1/STAT3 (Diao, Wang, and Wu 2009).

LIF is expressed in the muscle constitutively at low levels in type 1 muscle fibers. However, its secretion is regulated by certain factors influencing muscle activity; mechanical overloading and denervation in rat muscle have shown to rapidly increase LIF expression. Other external stressors such as exercise, injury, and inflammation lead to activation of LIF (Nylén et al. 2018). Taken together, LIF has been identified as a key contributor in promoting muscle regeneration and hypertrophy (Sakuma et al. 2000). However, evidence on age-related physiological changes in LIF abundance and activity in muscle is limited. LIF receptors are found in a variety of tissues including the central nervous system, bone, liver, uterus, and embryonic stem cells and seems to exert its regenerative effects systemically. Aging is associated with physiological decline and disease; LIF has been implicated in disease and injury (Kurek et al. 1996, Kandarian et al. 2018, Davis et al. 2018). With regard to sarcopenia, LIF appears to be a potential myokine in the regulation of muscle and adipose tissue metabolism; yet more research is necessary to further understand how LIF functions in both aging and sarcopenia (Aryana, Hapsari, and Kuswardhani 2018).

LIF receptors are found in a variety of tissues such as bone, liver, kidney, monocytes/macrophages, and the CNS to name a few (Gouin et al. 1999, Hilton, Nicola, and Metcalf 1991, Scott et al. 2000). The overall effect of LIF expression on these tissues seems to exert a systemic response to injury or tissue damage, especially in nerve and muscle. LIF has a role in neuromuscular development and regeneration. In cultured sensory neurons, LIF is activated in neuron development, which is augmented by co-treatment of FGF2 (Murphy et al. 1991). LIF expression is upregulated in response to neuron injury and further rescues motor neuron loss after intense nerve trauma (Curtis et al. 1994, Cheema et al. 1994). Moreover, LIF reduces atrophy in denervated rat muscle and stimulates muscle re-innervation (Finkelstein et al. 1996). There is some evidence that LIF expression increases glucose uptake in skeletal muscle of mice fed both a standard and a high-fat diet. These findings were associated with increases in Akt Ser473 phosphorylation, and this effect was found to be dependent on JAK/PI3K and mTORC2 signaling. However, in the same study, LIF stimulation did not alter palmitate oxidation (Brandt et al. 2015). LIF expression has shown to alter fat metabolism in other tissues by having a protective effect on diet-induced obesity, reducing body fat mass, and increasing lipolysis in pathological states; however, in skeletal muscle, the metabolic effects of LIF expression on fatty acid oxidation are not well characterized (Fioravante et al. 2017, Jansson et al. 2006, Marshall et al. 1994). Yet, LIF appears to be an important pleotropic myokine as it exhibits cross-talk with several tissues to regulate a variety of processes including inflammation, neuromuscular connectivity, and lipid metabolism—all of which are implicated in aging and sarcopenia. Yet further research is necessary to understand these relationships.

LIF secretion in skeletal muscle is a potent stimulator of myoblast growth and differentiation in mice and humans. As mentioned previously, LIF stimulates proliferation via activation of JAK1/STAT1/STAT3 in myoblasts but inhibits differentiation in myotubes (Austin and Burgess 1991). There are a few proposed mechanisms for this; however, LIF signaling seems context dependent and involves several pathways; LIF activates a variety of differentiation, survival, and renewal programs, thus determining the cell's fate (Nicola and Babon 2015). LIF mRNA is upregulated in several instances such as muscle injury and regeneration, disease, and dystrophy (Barnard et al. 1994). Following injury, LIF administration enhances muscle regeneration *in vivo* by increasing muscle fiber size rather than number (Barnard et al. 1994). Further, an increase in myoblast cell number is associated with PI3K inhibition of apoptosis; this suggests that LIF is a survival factor rather than a prompter of mitosis (Hunt, Tudor, and White 2010). This hypertrophic ability is also seen under mechanical loading; LIF KO mice fail to increase muscle mass compared with WT counterparts; however, delivery of LIF rescues hypertrophy (Spangenburg and Booth 2006). Whereas in differentiation, LIF inhibits early myogenic differentiation by activation of the extracellular signal-regulated kinase (ERK) pathway (Jo et al. 2005). In the context of disuse and muscle dystrophy, LIF mRNA is upregulated in human and mouse models denoting it as a trauma factor. Yet it exerts this regenerative effect locally in the muscle (Reardon et al. 2001, Hunt et al. 2011, Kurek et al. 1996). LIF has also been implicated in cancer cachexia acting as a regulator of this disease. Here, LIF promotes atrophy in myotubes treated with c26 cancer cell medium by activating JAK2/STAT3 and ERK signaling (Seto, Kandarian, and Jackman 2015). To date, the effects of LIF secretion in aging and sarcopenia remain elucidated; therefore more research is needed to further understand these mechanisms.

LIF has been termed an exercise-induced myokine and further a contraction-induced myokine; yet, the literature over exercise-induced LIF expression remains controversial (Broholm and Pedersen 2010, Broholm et al. 2011). One explanation for this could be that serum LIF has a short half-life making it hard to detect LIF levels in the circulation (Broholm et al. 2008). However, LIF seems to exert its affects locally rather than systemically as circulating levels of LIF are not increased following exercise (Broholm and Pedersen 2010). This suggests LIF signals to skeletal muscle in an autocrine and/or paracrine fashion (Broholm et al. 2011). LIF −/− mice reveal an important finding regarding LIF function in the development of hypertrophy; LIF KO are unable to increase muscle mass under muscle loading conditions. However, systemic delivery of LIF to these −/− mice rescues the hypertrophic response to the same extent seen in WT counterparts (Spangenburg and Booth 2006). One proposed mechanism for LIF-induced hypertrophy after exercise is the regulation of Ca^{2+} fluctuations. LIF mRNA levels are increased fourfold after a single bout of cycle ergometer exercise but gradually decline during the postexercise period. Moreover, it was demonstrated that concentric muscle contraction modulates LIF mRNA expression in human skeletal muscle. These increases in LIF mRNA and protein were associated with increases in the Ca^{2+} inophore, ionomycin, further demonstrating Ca^{2+} may be the mechanism implicated in LIF regulation after endurance exercise (Broholm et al. 2008). Whereas in heavy RT, LIF seems to be regulated by PI3K-Akt-mTOR signaling. Here, LIF mRNA was increased ninefold in healthy human quadriceps muscles but not detected in plasma. The PI3K-Akt-mTOR pathway is upregulated after exercise, and when PI3K, mTOR, and siRNA knockdown of Akt were inhibited, LIF was sufficiently downregulated. Moreover, LIF-induced transcription factors, JunB and c-Myc, potent promoters of myoblast proliferation in human myotubes, were induced by LIF expression. These data demonstrate both eccentric muscle contractions and heavy RT regulate LIF mRNA expression; moreover, LIF seems to promote satellite cell proliferation as these effects are only observed locally (Broholm et al. 2011).

Although LIF enhances survival of myoblast in dystrophic muscle, in another study investigating mdx mice, a model of muscle dystrophy, LIF mRNA levels were reduced after 2 weeks of voluntary wheel running. These data may be controversial; however, it does suggest LIF activity is implicated in dystrophic muscle following exercise (Hunt et al. 2011). Indeed, LIF is essential for muscle adaptation during exercise training; moreover, the effect of LIF-induced hypertrophy and muscle regeneration appears to be dependent on exercise type. These findings suggest that exercise-induced LIF may have potential in exerting an antiaging effect. It is well established that exercise is beneficial in sarcopenia, and because regeneration processes are impaired in sarcopenia, LIF may be a therapeutic target against disease progression (Arnold, Egger, and Handschin 2011).

7.10 CONCLUSION

Sarcopenia is associated with loss of both skeletal muscle mass and function. Several myokines may play a role in the aging process and development of sarcopenia. We have provided insights into several myokines that have been demonstrated to be altered during aging and sarcopenia. While several myokines are involved in sarcopenia and aging, it is likely a combination of dysregulation of several myokines, not just one that ultimately leads to sarcopenia. Exercise may be one method to offset the age-related dysregulation of myokines. More research is needed to fully understand the contribution of muscle released myokines in the aging process as many myokines are secreted not only from the skeletal muscle but also adipose tissue and the immune system; further, these systems work together to regulate whole-body homeostasis during aging and disease.

REFERENCES

Addison, C. L., T. O. Daniel, M. D. Burdick, H. Liu, J. E. Ehlert, Y. Y. Xue, L. Buechi, A. Walz, A. Richmond, and R. M. Strieter. 2000. "The CXC chemokine receptor 2, CXCR2, is the putative receptor for ELR+ CXC chemokine-induced angiogenic activity." *J Immunol* 165 (9):5269–77.

Akerstrom, T., A. Steensberg, P. Keller, C. Keller, M. Penkowa, and B. K. Pedersen. 2005. "Exercise induces interleukin-8 expression in human skeletal muscle." *J Physiol* 563 (Pt 2):507–16. doi:10.1113/jphysiol.2004.077610.

Akira, S., T. Taga, and T. Kishimoto. 1993. "Interleukin-6 in biology and medicine." *Adv Immunol* 54:1–78.

Al-Khalili, L., K. Bouzakri, S. Glund, F. Lonnqvist, H. A. Koistinen, and A. Krook. 2006. "Signaling specificity of interleukin-6 action on glucose and lipid metabolism in skeletal muscle." *Mol Endocrinol* 20 (12):3364–75. doi:10.1210/me.2005-0490.

Allen, D. L., D. S. Hittel, and A. C. McPherron. 2011. "Expression and function of myostatin in obesity, diabetes, and exercise adaptation." *Med Sci Sports Exerc* 43 (10):1828–35. doi:10.1249/MSS.0b013e3182178bb4.

Amir Levy, Y., T. P. Ciaraldi, S. R. Mudaliar, S. A. Phillips, and R. R. Henry. 2015. "Excessive secretion of IL-8 by skeletal muscle in type 2 diabetes impairs tube growth: Potential role of PI3K and the Tie2 receptor." *Am J Physiol Endocrinol Metab* 309 (1):E22–34. doi:10.1152/ajpendo.00513.2014.

Arnold, A. S., A. Egger, and C. Handschin. 2011. "PGC-1alpha and myokines in the aging muscle—A mini-review." *Gerontology* 57 (1):37–43. doi:10.1159/000281883.

Aryana, G. P. S., A. A. A. R. Hapsari, and R. A. T. Kuswardhani. 2018. "Myokine regulation as marker of sarcopenia in elderly." *Mol Cell Biomed Sci* 2 (2):38–47. doi:10.21705/mcbs.v2i2.32.

Austin, L., and A. W. Burgess. 1991. "Stimulation of myoblast proliferation in culture by leukaemia inhibitory factor and other cytokines." *J Neurol Sci* 101 (2):193–7.

Baeza-Raja, B., and P. Munoz-Canoves. 2004. "p38 MAPK-induced nuclear factor-kappaB activity is required for skeletal muscle differentiation: Role of interleukin-6." *Mol Biol Cell* 15 (4):2013–26. doi:10.1091/mbc.e03-08-0585.

Barnard, W., J. Bower, M. A. Brown, M. Murphy, and L. Austin. 1994. "Leukemia inhibitory factor (LIF) infusion stimulates skeletal muscle regeneration after injury: Injured muscle expresses lif mRNA." *J Neurol Sci* 123 (1–2):108–13.

Beenakker, K. G., C. H. Ling, C. G. Meskers, A. J. de Craen, T. Stijnen, R. G. Westendorp, and A. B. Maier. 2010. "Patterns of muscle strength loss with age in the general population and patients with a chronic inflammatory state." *Ageing Res Rev* 9 (4):431–6. doi:10.1016/j.arr.2010.05.005.

Begue, G., A. Douillard, O. Galbes, B. Rossano, B. Vernus, R. Candau, and G. Py. 2013. "Early activation of rat skeletal muscle IL-6/STAT1/STAT3 dependent gene expression in resistance exercise linked to hypertrophy." *PLoS One* 8 (2):e57141. doi:10.1371/journal.pone.0057141.

Belviranli, M., and N. Okudan. 2018. "Exercise training protects against aging-induced cognitive dysfunction via activation of the hippocampal PGC-1alpha/FNDC5/BDNF pathway." *Neuromolecular Med* 20 (3):386–400. doi:10.1007/s12017-018-8500-3.

Bergen, H. R., J. N. Farr, P. M. Vanderboom, E. J. Atkinson, T. A. White, R. J. Singh, S. Khosla, and N. K. LeBrasseur. 2015. "Myostatin as a mediator of sarcopenia versus homeostatic regulator of muscle mass: Insights using a new mass spectrometry-based assay." *Skelet Muscle* 5. doi:10.1186/s13395-015-0047-5.

Bian, A. L., H. Y. Hu, Y. D. Rong, J. Wang, J. X. Wang, and X. Z. Zhou. 2017. "A study on relationship between elderly sarcopenia and inflammatory factors IL-6 and TNF-alpha." *Eur J Med Res* 22 (1):25. doi:10.1186/s40001-017-0266-9.

Bortoluzzi, S., P. Scannapieco, A. Cestaro, G. A. Danieli, and S. Schiaffino. 2006. "Computational reconstruction of the human skeletal muscle secretome." *Proteins* 62 (3):776–92. doi:10.1002/prot.20803.

Bostrom, P., J. Wu, M. P. Jedrychowski, A. Korde, L. Ye, J. C. Lo, K. A. Rasbach et al. 2012. "A PGC1-alpha-dependent myokine that drives brown-fat-like development of white fat and thermogenesis." *Nature* 481 (7382):463–8. doi:10.1038/nature10777.

Brandt, N., H. M. O'Neill, M. Kleinert, P. Schjerling, E. Vernet, G. R. Steinberg, E. A. Richter, and S. B. Jørgensen. 2015. "Leukemia inhibitory factor increases glucose uptake in mouse skeletal muscle." 309 (2):E142–E153. doi:10.1152/ajpendo.00313.2014.

Broholm, C., M. J. Laye, C. Brandt, R. Vadalasetty, H. Pilegaard, B. K. Pedersen, and C. Scheele. 2011. "LIF is a contraction-induced myokine stimulating human myocyte proliferation." *J Appl Physiol* 111 (1):251–259. doi:10.1152/japplphysiol.01399.2010.

Broholm, C., O. H. Mortensen, S. Nielsen, T. Akerstrom, A. Zankari, B. Dahl, and B. K. Pedersen. 2008. "Exercise induces expression of leukaemia inhibitory factor in human skeletal muscle." *J Physiol* 586 (8):2195–201. doi:10.1113/jphysiol.2007.149781.

Broholm, C., and B. K. Pedersen. 2010. "Leukaemia inhibitory factor—An exercise-induced myokine." *Exerc Immunol Rev* 16:77–85.

Bruunsgaard, H., H. Galbo, J. Halkjaer-Kristensen, T. L. Johansen, D. A. MacLean, and B. K. Pedersen. 1997. "Exercise-induced increase in serum interleukin-6 in humans is related to muscle damage." *J Physiol* 499 (Pt 3):833–41.

Busquets, S., M. Toledo, M. Orpí, D. Massa, M. Porta, E. Capdevila, N. Padilla et al. 2012. "Myostatin blockage using actRIIB antagonism in mice bearing the Lewis lung carcinoma results in the improvement of muscle wasting and physical performance." *J Cachexia Sarcopenia Muscle* 3 (1):37–43. doi:10.1007/s13539-011-0049-z.

Carbo, N., J. Lopez-Soriano, P. Costelli, S. Busquets, B. Alvarez, F. M. Baccino, L. S. Quinn, F. J. Lopez-Soriano, and J. M. Argiles. 2000. "Interleukin-15 antagonizes muscle protein waste in tumour-bearing rats." *Br J Cancer* 83 (4):526–31. doi:10.1054/bjoc.2000.1299.

Carey, A. L., G. R. Steinberg, S. L. Macaulay, W. G. Thomas, A. G. Holmes, G. Ramm, O. Prelovsek et al. 2006. "Interleukin-6 increases insulin-stimulated glucose disposal in humans and glucose uptake and fatty acid oxidation in vitro via AMP-activated protein kinase." *Diabetes* 55 (10):2688–97. doi:10.2337/db05-1404.

Carnac, G., B. Vernus, and A. Bonnieu. 2007. "Myostatin in the pathophysiology of skeletal muscle." *Curr Genomics* 8 (7):415–22. doi:10.2174/138920207783591672.

Carson, J. A., and K. A. Baltgalvis. 2010. "Interleukin 6 as a key regulator of muscle mass during cachexia." *Exerc Sport Sci Rev* 38 (4):168–76. doi:10.1097/JES.0b013e3181f44f11.

Cartee, G. D. 1994. "Influence of age on skeletal muscle glucose transport and glycogen metabolism." *Med Sci Sports Exerc* 26 (5):577–85.

Catoire, M., M. Mensink, E. Kalkhoven, P. Schrauwen, and S. Kersten. 2014. "Identification of human exercise-induced myokines using secretome analysis." *Physiol Genomics* 46 (7):256–67. doi:10.1152/physiolgenomics.00174.2013.

Chan, M. H., A. L. Carey, M. J. Watt, and M. A. Febbraio. 2004. "Cytokine gene expression in human skeletal muscle during concentric contraction: Evidence that IL-8, like IL-6, is influenced by glycogen availability." *Am J Physiol Regul Integr Comp Physiol* 287 (2):R322–7. doi:10.1152/ajpregu.00030.2004.

Chang, J. S., T. H. Kim, T. T. Nguyen, K. S. Park, N. Kim, and I. D. Kong. 2017. "Circulating irisin levels as a predictive biomarker for sarcopenia: A cross-sectional community-based study." *Geriatr Gerontol Int* 17 (11):2266–2273. doi:10.1111/ggi.13030.

Cheema, S. S., L. J. Richards, M. Murphy, and P. F. Bartlett. 1994. "Leukaemia inhibitory factor rescues motoneurones from axotomy-induced cell death." *Neuroreport* 5 (8):989–92.

Clow, C., and B. J. Jasmin. 2010. "Brain-derived neurotrophic factor regulates satellite cell differentiation and skeletal muscle regeneration." *Mol Biol Cell* 21 (13):2182–90. doi:10.1091/mbc. E10-02-0154.

Colbert, L. H., M. Visser, E. M. Simonsick, R. P. Tracy, A. B. Newman, S. B. Kritchevsky, M. Pahor et al. 2004. "Physical activity, exercise, and inflammatory markers in older adults: Findings from the health, aging and body composition study." *J Am Geriatr Soc* 52 (7):1098–104. doi:10.1111/j.1532-5415.2004.52307.x.

Connor, B., D. Young, Q. Yan, R. L. Faull, B. Synek, and M. Dragunow. 1997. "Brain-derived neurotrophic factor is reduced in Alzheimer's disease." *Brain Res Mol Brain Res* 49 (1–2):71–81.

Crane, J. D., L. G. MacNeil, J. S. Lally, R. J. Ford, A. L. Bujak, I. K. Brar, B. E. Kemp, S. Raha, G. R. Steinberg, and M. A. Tarnopolsky. 2015. "Exercise-stimulated interleukin-15 is controlled by AMPK and regulates skin metabolism and aging." *Aging Cell* 14 (4):625–34. doi:10.1111/acel.12341.

Curtis, R., S. S. Scherer, R. Somogyi, K. M. Adryan, N. Y. Ip, Y. Zhu, R. M. Lindsay, and P. S. DiStefano. 1994. "Retrograde axonal transport of LIF is increased by peripheral nerve injury: Correlation with increased LIF expression in distal nerve." *Neuron* 12 (1):191–204.

Davis, S. M., L. A. Collier, S. Goodwin, D. E. Lukins, D. K. Powell, and K. R. Pennypacker. 2018. "Efficacy of leukemia inhibitory factor as a therapeutic for permanent large vessel stroke differs among aged male and female rats." *Brain Research* 1707 (2019):62–73. doi:10.1016/j.brainres.2018.11.017.

De Benedetti, F., T. Alonzi, A. Moretta, D. Lazzaro, P. Costa, V. Poli, A. Martini, G. Ciliberto, and E. Fattori. 1997. "Interleukin 6 causes growth impairment in transgenic mice through a decrease in insulin-like growth factor-I. A model for stunted growth in children with chronic inflammation." *J Clin Invest* 99 (4):643–50. doi:10.1172/JCI119207.

De Benedetti, F., P. Pignatti, R. Vivarelli, C. Meazza, G. Ciliberto, R. Savino, and A. Martini. 2001. "In vivo neutralization of human IL-6 (hIL-6) achieved by immunization of hIL-6-transgenic mice with a hIL-6 receptor antagonist." *J Immunol* 166 (7):4334–40.

Deng, B., F. Zhang, J. Wen, S. Ye, L. Wang, Y. Yang, P. Gong, and S. Jiang. 2017. "The function of myostatin in the regulation of fat mass in mammals." *Nutr Metab (Lond)* 14. doi:10.1186/s12986-017-0179-1.

Diao, Y., X. Wang, and Z. Wu. 2009. "SOCS1, SOCS3, and PIAS1 promote myogenic differentiation by inhibiting the leukemia inhibitory factor-induced JAK1/STAT1/STAT3 pathway." *Mol Cell Biol* 29 (18):5084–93. doi:10.1128/mcb.00267-09.

Duzova, H., E. Gullu, G. Cicek, B. K. Koksal, B. Kayhan, A. Gullu, and I. Sahin. 2018. "The effect of exercise induced weight-loss on myokines and adipokines in overweight sedentary females: Steps-aerobics vs. jogging-walking exercises." *J Sports Med Phys Fitness* 58 (3):295–308. doi:10.23736/s0022-4707.16.06565-8.

Elliot, B., D. Renshaw, S. Getting, and R. Mackenzie. 2012. "The central role of myostatin in skeletal muscle and whole body homeostasis—Elliott—2012—Acta Physiologica—Wiley Online Library." *Acta Physiologica* 205 (3):324–40. doi:10.1111/j.1748-1716.2012.02423.x.

Emanuele, E., P. Minoretti, H. Pareja-Galeano, F. Sanchis-Gomar, N. Garatachea, and A. Lucia. 2014. "Serum irisin levels, precocious myocardial infarction, and healthy exceptional longevity." *Am J Med* 127 (9):888–890. doi:10.1016/j.amjmed.2014.04.025.

Ernst, M., and B. J. Jenkins. 2004. "Acquiring signalling specificity from the cytokine receptor gp130." *Trends Genet* 20 (1):23–32.

Ershler, W. B. 2003. "Biological interactions of aging and anemia: A focus on cytokines." *J Am Geriatr Soc* 51 (3 Suppl):S18–21.

Febbraio, M. A., and B. K. Pedersen. 2005. "Contraction-induced myokine production and release: Is skeletal muscle an endocrine organ?" *Exerc Sport Sci Rev* 33 (3):114–19. doi:00003677-200507000-00003 [pii].

Febbraio, M. A., A. Steensberg, C. Keller, R. L. Starkie, H. B. Nielsen, P. Krustrup, P. Ott, N. H. Secher, and B. K. Pedersen. 2003. "Glucose ingestion attenuates interleukin-6 release from contracting skeletal muscle in humans." *J Physiol* 549 (Pt 2):607–12. doi:10.1113/jphysiol.2003.042374.

Ferrucci, L., B. W. Penninx, S. Volpato, T. B. Harris, K. Bandeen-Roche, J. Balfour, S. G. Leveille, L. P. Fried, and J. M. Md. 2002. "Change in muscle strength explains accelerated decline of physical function in older women with high interleukin-6 serum levels." *J Am Geriatr Soc* 50 (12):1947–54.

Figueras, M., S. Busquets, N. Carbo, E. Barreiro, V. Almendro, J. M. Argiles, and F. J. Lopez-Soriano. 2004. "Interleukin-15 is able to suppress the increased DNA fragmentation associated with muscle wasting in tumour-bearing rats." *FEBS Lett* 569 (1–3):201–6. doi:10.1016/j.febslet.2004.05.066.

Finkelstein, D. I., P. F. Bartlett, M. K. Horne, and S. S. Cheema. 1996. "Leukemia inhibitory factor is a myotrophic and neurotrophic agent that enhances the reinnervation of muscle in the rat." *J Neurosci Res* 46 (1):122–8.

Fioravante, M., B. Bombassaro, A. F. Ramalho, N. R. Dragano, J. Morari, C. Solon, N. Tobar, C. D. Ramos, and L. A. Velloso. 2017. "Inhibition of hypothalamic leukemia inhibitory factor exacerbates diet-induced obesity phenotype." *J Neuroinflammation* 14:1–12. doi:10.1186/s12974-017-0956-9.

Fisher, F. M., and E. Maratos-Flier. 2016. "Understanding the physiology of FGF21." *Annu Rev Physiol* 78:223–41. doi:10.1146/annurev-physiol-021115-105339.

Frydelund-Larsen, L., M. Penkowa, T. Akerstrom, A. Zankari, S. Nielsen, and B. K. Pedersen. 2007. "Exercise induces interleukin-8 receptor (CXCR2) expression in human skeletal muscle." *Exp Physiol* 92 (1):233–40. doi:10.1113/expphysiol.2006.034769.

Gangemi, S., G. Basile, D. Monti, R. Alba Merendino, G. Di Pasquale, U. Bisignano, V. Nicita-Mauro, and C. Franceschi. 2005. "Age-related modifications in circulating IL-15 levels in humans." *Mediators Inflamm* 2005 (4):245–7. doi:10.1155/mi.2005.245.

Gilson, H., O. Schakman, L. Combaret, P. Lause, L. Grobet, D. Attaix, J. M. Ketelslegers, and J. P. Thissen. 2007. "Myostatin gene deletion prevents glucocorticoid-induced muscle atrophy." *Endocrinology* 148 (1):452–60. doi:10.1210/en.2006-0539.

Goodman, M. N. 1994. "Interleukin-6 induces skeletal muscle protein breakdown in rats." *Proc Soc Exp Biol Med* 205 (2):182–5.

Gouin, F., S. Couillaud, M. Cottrel, A. Godard, N. Passuti, and D. Heymann. 1999. "Presence of leukaemia inhibitory factor (LIF) and LIF-receptor chain (gp190) in osteoclast-like cells cultured from human giant cell tumour of bone. Ultrastructural distribution." *Cytokine* 11 (4):282–9. doi:10.1006/cyto.1998.0429.

Gravallese, E. M, and P. A. Monach. 2015. *Rheumatology* (6th ed.). Edited by Mosby Philadelphia: Elsevier.

Greising, S. M., L. G. Ermilov, G. C. Sieck, and C. B. Mantilla. 2015. "Ageing and neurotrophic signalling effects on diaphragm neuromuscular function." *J Physiol* 593 (2):431–40. doi:10.1113/jphysiol.2014.282244.

Gudiksen, A., C. L. Schwartz, L. Bertholdt, E. Joensen, J. G. Knudsen, and H. Pilegaard. 2016. "Lack of skeletal muscle IL-6 affects pyruvate dehydrogenase activity at rest and during prolonged exercise." *PLoS One* 11 (6):e0156460. doi:10.1371/journal.pone.0156460.

Guo, T., W. Jou, T. Chanturiya, J. Portas, O. Gavrilova, and A. C. McPherron. 2009. "Myostatin inhibition in muscle, but not adipose tissue, decreases fat mass and improves insulin sensitivity." *PLoS One* 4 (3). doi:10.1371/journal.pone.0004937.

Guridi, M., L. A. Tintignac, S. Lin, B. Kupr, P. Castets, and M. A. Ruegg. 2015. "Activation of mTORC1 in skeletal muscle regulates whole-body metabolism through FGF21." *Sci Signal* 8 (402):ra113. doi:10.1126/scisignal.aab3715.

Hamer, M., S. Sabia, G. D. Batty, M. J. Shipley, A. G. Tabak, A. Singh-Manoux, and M. Kivimaki. 2012. "Physical activity and inflammatory markers over 10 years: Follow-up in men and women from the Whitehall II cohort study." *Circulation* 126 (8):928–33. doi:10.1161/CIRCULATIONAHA.112.103879.

Hanks, L. J., O. M. Gutierrez, M. M. Bamman, A. Ashraf, K. L. McCormick, and K. Casazza. 2015. "Circulating levels of fibroblast growth factor-21 increase with age independently of body composition indices among healthy individuals." *J Clin Transl Endocrinol* 2 (2):77–82. doi:10.1016/j.jcte.2015.02.001.

Hartwig, S., S. Raschke, B. Knebel, M. Scheler, M. Irmler, W. Passlack, S. Muller et al. 2014. "Secretome profiling of primary human skeletal muscle cells." *Biochim Biophys Acta* 1844 (5):1011–17. doi:10.1016/j.bbapap.2013.08.004.

Heinrich, P. C., I. Behrmann, G. Muller-Newen, F. Schaper, and L. Graeve. 1998. "Interleukin-6-type cytokine signalling through the gp130/Jak/STAT pathway." *Biochem J* 334 (Pt 2):297–314.

Hellsten, Y., U. Frandsen, N. Orthenblad, B. Sjodin, and E. A. Richter. 1997. "Xanthine oxidase in human skeletal muscle following eccentric exercise: A role in inflammation." *J Physiol* 498 (Pt 1):239–48.

Hilton, D. J., N. A. Nicola, and D. Metcalf. 1991. "Distribution and comparison of receptors for leukemia inhibitory factor on murine hemopoietic and hepatic cells." *J Cell Physiol* 146 (2):207–15. doi:10.1002/jcp.1041460204.

Hirano, T. 1998. "Interleukin 6 and its receptor: Ten years later." *Int Rev Immunol* 16 (3–4):249–84. doi:10.3109/08830189809042997.

Hirano, T., K. Yasukawa, H. Harada, T. Taga, Y. Watanabe, T. Matsuda, S. Kashiwamura et al. 1986. "Complementary DNA for a novel human interleukin (BSF-2) that induces B lymphocytes to produce immunoglobulin." *Nature* 324 (6092):73–6. doi:10.1038/324073a0.

Hittel, D. S., M. Axelson, N. Sarna, J. Shearer, K. M. Huffman, and W. E. Kraus. 2010. "Myostatin decreases with aerobic exercise and associates with insulin resistance." *Med Sci Sports Exerc* 42 (11):2023–9. doi:10.1249/MSS.0b013e3181e0b9a8.

Hoene, M., H. Runge, H. U. Haring, E. D. Schleicher, and C. Weigert. 2013. "Interleukin-6 promotes myogenic differentiation of mouse skeletal muscle cells: Role of the STAT3 pathway." *Am J Physiol Cell Physiol* 304 (2):C128–36. doi:10.1152/ajpcell.00025.2012.

Hoffman, C., and C. Weigert. 2018. "Skeletal muscle as an endocrine organ: The role of myokines in exercise adaptations." *Cold Spring Harb Perspect Med* 8 (11). doi:10.1101/cshperspec.a029793.

Hojman, P., M. Pedersen, A. R. Nielsen, R. Krogh-Madsen, C. Yfanti, T. Akerstrom, S. Nielsen, and B. K. Pedersen. 2009. "Fibroblast growth factor-21 is induced in human skeletal muscles by hyperinsulinemia." *Diabetes* 58 (12):2797–801. doi:10.2337/db09-0713.

Huh, J. Y., F. Dincer, E. Mesfum, and C. S. Mantzoros. 2014. "Irisin stimulates muscle growth-related genes and regulates adipocyte differentiation and metabolism in humans." *Int J Obes (Lond)* 38 (12):1538–44. doi:10.1038/ijo.2014.42.

Huh, J. Y., G. Panagiotou, V. Mougios, M. Brinkoetter, M. T. Vamvini, B. E. Schneider, and C. S. Mantzoros. 2012. "FNDC5 and irisin in humans: I. Predictors of circulating concentrations in serum and plasma and II. mRNA expression and circulating concentrations in response to weight loss and exercise." *Metabolism* 61 (12):1725–38. doi:10.1016/j.metabol.2012.09.002.

Hunt, L. C., A. C. Coles, C. M. Gorman, E. M. Tudor, G. M. Smythe, and J. D. White. 2011. "Alterations in the expression of leukemia inhibitory factor following exercise: Comparisons between wild-type and mdx muscles." *PLoS Currents* 3. doi:10.1371/currents. RRN1277.

Hunt, L. C., E. M. Tudor, and J. D. White. 2010. "Leukemia inhibitory factor-dependent increase in myoblast cell number is associated with phosphotidylinositol 3-kinase-mediated inhibition of apoptosis adn not mitosis." *Exp Cell Res* 316 (6). doi:10.1016/j.yexcr.2009.11.022.

Hunter, C. A., and S. A. Jones. 2017. "Corrigendum: IL-6 as a keystone cytokine in health and disease." *Nat Immunol* 18 (11):1271. doi:10.1038/ni1117-1271b.

Izumiya, Y., H. A. Bina, N. Ouchi, Y. Akasaki, A. Kharitonenkov, and K. Walsh. 2008. "FGF21 is an Akt-regulated myokine." *FEBS Lett* 582 (27):3805–10. doi:10.1016/j.febslet.2008.10.021.

Jackson, M. F., D. Luong, D. D. Vang, D. K. Garikipati, J. B. Stanton, O. L. Nelson, and B. D. Rodgers. 2012. "The aging myostatin null phenotype: Reduced adiposity, cardiac hypertrophy, enhanced cardiac stress response, and sexual dimorphism." *Journal of Endocrinology* 213 (3):263–275. doi:10.1530/JOE-11-0455.

Jansson, J. O., S. Moverare-Skrtic, A. Berndtsson, I. Wernstedt, H. Carlsten, and C. Ohlsson. 2006. "Leukemia inhibitory factor reduces body fat mass in ovariectomized mice." *Eur J Endocrinol* 154 (2):349–54. doi:10.1530/eje.1.02082.

Ji, K., J. Zheng, J. Lv, J. Xu, X. Ji, Y. B. Luo, W. Li, Y. Zhao, and C. Yan. 2015. "Skeletal muscle increases FGF21 expression in mitochondrial disorders to compensate for energy metabolic insufficiency by activating the mTOR-YY1-PGC1alpha pathway." *Free Radic Biol Med* 84:161–70. doi:10.1016/j.freeradbiomed.2015.03.020.

Jo, C., H. Kim, I. Jo, I. Choi, S. C. Jung, J. Kim, S. S. Kim, and S. A. Jo. 2005. "Leukemia inhibitory factor blocks early differentiation of skeletal muscle cells by activating ERK." *Biochim Biophys Acta* 1743 (3):187–97. doi:10.1016/j.bbamcr.2004.11.002.

Kalinkovich, A., and G. Livshits. 2015. "Sarcopenia—The search for emerging biomarkers." *Ageing Res Rev* 22:58–71. doi:10.1016/j.arr.2015.05.001.

Kami, K., and E. Senba. 2002. "In vivo activation of STAT3 signaling in satellite cells and myofibers in regenerating rat skeletal muscles." *J Histochem Cytochem* 50 (12):1579–89. doi:10.1177/002215540205001202.

Kandarian, S. C., R. L. Nosacka, A. E. Delitto, A. R. Judge, S. M. Judge, J. D. Ganey, J. D. Moreira, and R. W. Jackman. 2018. "Tumour-derived leukaemia inhibitory factor is a major driver of cancer cachexia and morbidity in C26 tumour-bearing mice." *J Cachexia Sarcopenia Muscle* 9 (6):1109–120. doi:10.1002/jcsm.12346.

Keipert, S., M. Ost, K. Johann, F. Imber, M. Jastroch, E. M. van Schothorst, J. Keijer, and S. Klaus. 2014. "Skeletal muscle mitochondrial uncoupling drives endocrine cross-talk through the induction of FGF21 as a myokine." *Am J Physiol Endocrinol Metab* 306 (5):E469–82. doi:10.1152/ajpendo.00330.2013.

Keller, C., A. Steensberg, A. K. Hansen, C. P. Fischer, P. Plomgaard, and B. K. Pedersen. 2005. "Effect of exercise, training, and glycogen availability on IL-6 receptor expression in human skeletal muscle." *J Appl Physiol (1985)* 99 (6):2075–9. doi:10.1152/japplphysiol.00590.2005.

Kim, H. J., J. Y. Park, S. L. Oh, Y. A. Kim, B. So, J. K. Seong, and W. Song. 2013a. "Effect of treadmill exercise on interleukin-15 expression and glucose tolerance in Zucker diabetic fatty rats." *Diabetes Metab J* 37 (5):358–64. doi:10.4093/dmj.2013.37.5.358.

Kim, H. J., B. So, M. Choi, D. Kang, and W. Song. 2015. "Resistance exercise training increases the expression of irisin concomitant with improvement of muscle function in aging mice and humans." *Exp Gerontol* 70:11–17. doi:10.1016/j.exger.2015.07.006.

Kim, K. H., Y. T. Jeong, H. Oh, S. H. Kim, J. M. Cho, Y. N. Kim, S. S. Kim et al. 2013b. "Autophagy deficiency leads to protection from obesity and insulin resistance by inducing Fgf21 as a mitokine." *Nat Med* 19 (1):83–92. doi:10.1038/nm.3014.

Knudsen, J. G., A. Gudiksen, L. Bertholdt, P. Overby, I. Villesen, C. L. Schwartz, and H. Pilegaard. 2017. "Skeletal muscle IL-6 regulates muscle substrate utilization and adipose tissue metabolism during recovery from an acute bout of exercise." *PLoS One* 12 (12):e0189301. doi:10.1371/journal.pone.0189301.

Komulainen, P., M. Pedersen, T. Hanninen, H. Bruunsgaard, T. A. Lakka, M. Kivipelto, M. Hassinen, T. H. Rauramaa, B. K. Pedersen, and R. Rauramaa. 2008. "BDNF is a novel marker of cognitive function in ageing women: The DR's EXTRA study." *Neurobiol Learn Mem* 90 (4):596–603. doi:10.1016/j.nlm.2008.07.014.

Krolopp, J. E., S. M. Thornton, and M. J. Abbott. 2016. "IL-15 activates the Jak3/STAT3 signaling pathway to mediate glucose uptake in skeletal muscle cells." *Front Physiol* 7:626. doi:10.3389/fphys.2016.00626.

Kurdiova, T., M. Balaz, M. Vician, D. Maderova, M. Vlcek, L. Valkovic, M. Srbecky et al. 2014. "Effects of obesity, diabetes and exercise on Fndc5 gene expression and irisin release in human skeletal muscle and adipose tissue: In vivo and in vitro studies." *J Physiol* 592 (5):1091–107. doi:10.1113/jphysiol.2013.264655.

Kurek, J. B., S. Nouri, G. Kannourakis, M. Murphy, and L. Austin. 1996. "Leukemia inhibitory factor and interleukin-6 are produced by diseased and regenerating skeletal muscle—Kurek—1996—Muscle & Nerve—Wiley Online Library." *Muscle & Nerve* 19 (10).

Langley, B., M. Thomas, A. Bishop, M. Sharma, S. Gilmour, and R. Kambadur. 2002. "Myostatin inhibits myoblast differentiation by down-regulating myod expression." *Journal of Biological Chemistry.* doi:10.1074/jbc. M204291200.

Laske, C., E. Stransky, T. Leyhe, G. W. Eschweiler, A. Wittorf, E. Richartz, M. Bartels, G. Buchkremer, and K. Schott. 2006. "Stage-dependent BDNF serum concentrations in Alzheimer's disease." *J Neural Transm (Vienna)* 113 (9):1217–24. doi:10.1007/s00702-005-0397-y.

Leger, B., W. Derave, K. De Bock, P. Hespel, and A. P. Russell. 2008. "Human sarcopenia reveals an increase in SOCS-3 and myostatin and a reduced efficiency of Akt phosphorylation." *Rejuvenation Res* 11 (1):163–175B. doi:10.1089/rej.2007.0588.

Lin, Y. L., S. H. Chen, and J. Y. Wang. 2016. "Critical role of IL-6 in dendritic cell-induced allergic inflammation of asthma." *J Mol Med (Berl)* 94 (1):51–9. doi:10.1007/s00109-015-1325-8.

Mak, T. W., and M. E. Saunders. 2006. *The Immune Response Basic and Clinical Principles.* New York: Academic Press.

Manini, T. M., K. R. Vincent, C. L. Leeuwenburgh, H. A. Lees, A. N. Kavazis, S. E. Borst, and B. C. Clark. 2011. "Myogenic and proteolytic mRNA expression following blood flow restricted exercise." *Acta Physiol (Oxf)* 201 (2):255–63. doi:10.1111/j.1748-1716.2010.02172.x.

Marcell, T. J., S. M. Harman, R. J. Urban, D. D. Metz, B. D. Rodgers, and M. R. Blackman. 2001. "Comparison of GH, IGF-1 and testosterone with mRNA of receptors and myostatin in skeletal muscle in older men." *Am J Physiol* 281 (6):E1159–64. doi:10.1152/ajpendo.2001.281.6.E1159.

Marshall, M. K., W. Doerrler, K. R. Feingold, and C. Grunfeld. 1994. "Leukemia inhibitory factor induces changes in lipid metabolism in cultured adipocytes." *Endocrinology* 135 (1):141–7. doi:10.1210/en.135.1.141.

Matthews, V. B., M. B. Astrom, M. H. Chan, C. R. Bruce, K. S. Krabbe, O. Prelovsek, T. Akerstrom et al. 2009. "Brain-derived neurotrophic factor is produced by skeletal muscle cells in response to contraction and enhances fat oxidation via activation of AMP-activated protein kinase." *Diabetologia* 52 (7):1409–18. doi:10.1007/s00125-009-1364-1.

Mauer, J., J. L. Denson, and J. C. Bruning. 2015. "Versatile functions for IL-6 in metabolism and cancer." *Trends Immunol* 36 (2):92–101. doi:10.1016/j.it.2014.12.008.

McFarlane, C., E. Plummer, M. Thomas, A. Hennebry, M. Ashby, N. Ling, H. Smith, M. Sharma, and R. Kambadur. 2006. "Myostatin induces cachexia by activating the ubiquitin proteolytic system through an NF-kappaB-independent, FoxO1-dependent mechanism." *J Cell Physiol* 209 (2):501–14. doi:10.1002/jcp.20757.

McPherron, A. C., A. M. Lawler, and S. J. Lee. 1997. "Regulation of skeletal muscle mass in mice by a new TGF-beta superfamily member." *Nature* 387 (6628):83–90. doi:10.1038/387083a0.

McPherron, A. C., and S. J. Lee. 2002. "Suppression of body fat accumulation in myostatin-deficient mice." *J Clin Invest* 109 (5):595–601. doi:10.1172/jci13562.

Miller, R. R., M. D. Shardell, G. E. Hicks, A. R. Cappola, W. G. Hawkes, J. A. Yu-Yahiro, and J. Magaziner. 2008. "Association between interleukin-6 and lower extremity function after hip fracture—The role of muscle mass and strength." *J Am Geriatr Soc* 56 (6):1050–6. doi:10.1111/j.1532-5415.2008.01708.x.

Molanouri Shamsi, M., Z. M. Hassan, L. S. Quinn, R. Gharakhanlou, L. Baghersad, and M. Mahdavi. 2015. "Time course of IL-15 expression after acute resistance exercise in trained rats: Effect of diabetes and skeletal muscle phenotype." *Endocrine* 49 (2):396–403. doi:10.1007/s12020-014-0501-x.

Morissette, M. R., J. C. Stricker, M. A. Rosenberg, C. Buranasombati, E. B. Levitan, M. A. Mittleman, and A. Rosenzweig. 2009. "Effects of myostatin deletion in aging mice." *Aging Cell* 8 (5):573–83. doi:10.1111/j.1474-9726.2009.00508.x.

Mousavi, K., and B. J. Jasmin. 2006. "BDNF is expressed in skeletal muscle satellite cells and inhibits myogenic differentiation." *J Neurosci* 26 (21):5739–49. doi:10.1523/JNEUROSCI.5398-05.2006.

Murphy, K. T., A. Chee, B. G. Gleeson, T. Naim, K. Swiderski, R. Koopman, and G. S. Lynch. 2011. "Antibody-directed myostatin inhibition enhances muscle mass and function in tumor-bearing mice." *Am J Physiol Regul Integr Comp Physiol* 301 (3):R716–26. doi:10.1152/ajpregu.00121.2011.

Murphy, M., K. Reid, D. J. Hilton, and P. F. Bartlett. 1991. "Generation of sensory neurons is stimulated by leukemia inhibitory factor." *Proc Natl Acad Sci USA* 88 (8):3498–501.

Narsale, A. A., and J. A. Carson. 2014. "Role of interleukin-6 in cachexia: Therapeutic implications." *Curr Opin Support Palliat Care* 8 (4):321–7. doi:10.1097/SPC.0000000000000091.

Nemet, D., A. Eliakim, F. Zaldivar, and D. M. Cooper. 2006. "Effect of rhIL-6 infusion on GH→IGF-I axis mediators in humans." *Am J Physiol Regul Integr Comp Physiol* 291 (6):R1663–8. doi:10.1152/ajpregu.00053.2006.

Nicola, N. A., and J. J. Babon. 2015. "Leukemia inhibitory factor (LIF)." *Cytokine Growth Factor Rev* 26 (5):533–44. doi:10.1016/j.cytogfr.2015.07.001.

Nielsen, A. R., R. Mounier, P. Plomgaard, O. H. Mortensen, M. Penkowa, T. Speerschneider, H. Pilegaard, and B. K. Pedersen. 2007. "Expression of interleukin-15 in human skeletal muscle—Effect of exercise and muscle fibre type composition." *J Physiol* 584 (Pt 1):305–12. doi:10.1113/jphysiol.2007.139618.

Nielsen, A. R., and B. K. Pedersen. 2018. "The biological roles of exercise-induced cytokines: IL-6, IL-8, and IL-15." *Applied Physiology, Nutrition, and Metabolism* 32 (5):833–9. doi:10.1139/H07-054.

Nieman, D. C., K. A. Zwetsloot, M. P. Meaney, D. D. Lomiwes, S. M. Hurst, and R. D. Hurst. 2015. "Post-exercise skeletal muscle glycogen related to plasma cytokines and muscle IL-6 protein content, but not muscle cytokine mRNA expression." *Front Nutr* 2:27. doi:10.3389/fnut.2015.00027.

Nishimoto, N., K. Yoshizaki, N. Miyasaka, K. Yamamoto, S. Kawai, T. Takeuchi, J. Hashimoto, J. Azuma, and T. Kishimoto. 2004. "Treatment of rheumatoid arthritis with humanized anti-interleukin-6 receptor antibody: A multicenter, double-blind, placebo-controlled trial." *Arthritis Rheum* 50 (6):1761–9. doi:10.1002/art.20303.

Norheim, F., T. M. Langleite, M. Hjorth, T. Holen, A. Kielland, H. K. Stadheim, H. L. Gulseth, K. I. Birkeland, J. Jensen, and C. A. Drevon. 2014. "The effects of acute and chronic exercise on PGC-1alpha, irisin and browning of subcutaneous adipose tissue in humans." *FEBS J* 281 (3):739–49. doi:10.1111/febs.12619.

Nygaard, H., G. Slettalokken, G. Vegge, I. Hollan, J. E. Whist, T. Strand, B. R. Ronnestad, and S. Ellefsen. 2015. "Irisin in blood increases transiently after single sessions of intense endurance exercise and heavy strength training." *PLoS One* 10 (3):e0121367. doi:10.1371/journal.pone.0121367.

Nylén, C., W. Aoi, A. M. Abdelmoez, D. G. Lassiter, L. S. Lundell, H. Wallberg-Henriksson, E. Näslund, N. J. Pillon, and A. Krook. 2018. "IL6 and LIF mRNA expression in skeletal muscle is regulated by AMPK and the transcription factors NFYC, ZBTB14, and SP1." *Am J Physiol Endocrinol Metab* 315 (5):E995–1004. doi:10.1152/ajpendo.00398.2017.

Ostrowski, K., C. Hermann, A. Bangash, P. Schjerling, J. N. Nielsen, and B. K. Pedersen. 1998a. "A trauma-like elevation of plasma cytokines in humans in response to treadmill running." *J Physiol* 513 (Pt 3):889–94.

Ostrowski, K., T. Rohde, S. Asp, P. Schjerling, and B. K. Pedersen. 1999. "Pro- and anti-inflammatory cytokine balance in strenuous exercise in humans." *J Physiol* 515 (Pt 1):287–91.

Ostrowski, K., T. Rohde, M. Zacho, S. Asp, and B. K. Pedersen. 1998b. "Evidence that interleukin-6 is produced in human skeletal muscle during prolonged running." *J Physiol* 508 (Pt 3):949–53.

Pardo, M., A. B. Crujeiras, M. Amil, Z. Aguera, S. Jimenez-Murcia, R. Banos, C. Botella et al. 2014.
"Association of irisin with fat mass, resting energy expenditure, and daily activity in conditions of
extreme body mass index." *Int J Endocrinol* 2014:857270. doi:10.1155/2014/857270.

Payette, H., R. Roubenoff, P. F. Jacques, C. A. Dinarello, P. W. Wilson, L. W. Abad, and T. Harris. 2003.
"Insulin-like growth factor-1 and interleukin 6 predict sarcopenia in very old community-living men
and women: The Framingham heart study." *J Am Geriatr Soc* 51 (9):1237–43.

Pedersen, B. K. 2009. "Edward F. Adolph distinguished lecture: Muscle as an endocrine organ: IL-6 and other
myokines." *J Appl Physiol (1985)* 107 (4):1006–14. doi:10.1152/japplphysiol.00734.2009.

Pedersen, B. K., A. Steensberg, and P. Schjerling. 2001. "Muscle-derived interleukin-6: Possible biological
effects." *J Physiol* 536 (Pt 2):329–37.

Pedersen, M., A. Steensberg, C. Keller, T. Osada, M. Zacho, B. Saltin, M. A. Febbraio, and B. K. Pedersen.
2004. "Does the aging skeletal muscle maintain its endocrine function?" *Exerc Immunol Rev* 10:42–55.

Pelleymounter, M. A., M. J. Cullen, and C. L. Wellman. 1995. "Characteristics of BDNF-induced weight loss."
Exp Neurol 131 (2):229–38.

Pereira, R. O., S. M. Tadinada, F. M. Zasadny, K. J. Oliveira, K. M. P. Pires, A. Olvera, J. Jeffers, R. Souvenir,
R. McGlauflin, A. Seei, T. Funari, H. Sesaki, M. J. Potthoff, C. M. Adams, E. J. Anderson, and E. D.
Abel. 2017. "OPA1 deficiency promotes secretion of FGF21 from muscle that prevents obesity and insu-
lin resistance." *EMBO J* 36 (14):2126–45. doi:10.15252/embj.201696179.

Perez-Lopez, A., J. McKendry, M. Martin-Rincon, D. Morales-Alamo, B. Perez-Kohler, D. Valades, J. Bujan,
J. A. L. Calbet, and L. Breen. 2018. "Skeletal muscle IL-15/IL-15Ralpha and myofibrillar protein synthesis after resistance exercise." *Scand J Med Sci Sports* 28 (1):116–125. doi:10.1111/sms.12901.

Petersen, A. M., and B. K. Pedersen. 2005. "The anti-inflammatory effect of exercise." *J Appl Physiol*
98 (4):1154–62.

Pierce, J. R., J. M. Maples, and R. C. Hickner. 2015. "IL-15 concentrations in skeletal muscle and subcutaneous adipose tissue in lean and obese humans: Local effects of IL-15 on adipose tissue lipolysis."
Am J Physiol Endocrinol Metab 308 (12):E1131–9. doi:10.1152/ajpendo.00575.2014.

Pistilli, E. E., and L. S. Quinn. 2013. "From anabolic to oxidative: Reconsidering the roles of IL-15 and
IL-15Rα in skeletal muscle." *Exerc Sport Sci Rev* 41 (2):100–6. doi:10.1097/JES.0b013e318275d230.

Puppa, M. J., S. Gao, A. A. Narsale, and J. A. Carson. 2014. "Skeletal muscle glycoprotein 130's role in Lewis
lung carcinoma-induced cachexia." *FASEB J* 28 (2):998–1009. doi:10.1096/fj.13-240580.

Puppa, M. J., J. P. White, K. T. Velazquez, K. A. Baltgalvis, S. Sato, J. W. Baynes, and J. A. Carson. 2012.
"The effect of exercise on IL-6-induced cachexia in the Apc (Min/+) mouse." *J Cachexia Sarcopenia
Muscle* 3 (2):117–37. doi:10.1007/s13539-011-0047-1.

Quinn, L. S., B. G. Anderson, J. D. Conner, E. E. Pistilli, and T. Wolden-Hanson. 2011. "Overexpression of
interleukin-15 in mice promotes resistance to diet-induced obesity, increased insulin sensitivity, and
markers of oxidative skeletal muscle metabolism." *Int J Interferon Cytokine Mediat Res* 3:29–42.
doi:10.2147/IJICMR.S19007.

Quinn, L. S., B. G. Anderson, J. D. Conner, and T. Wolden-Hanson. 2015. "Circulating irisin levels and
muscle FNDC5 mRNA expression are independent of IL-15 levels in mice." *Endocrine* 50 (2):368–77.
doi:10.1007/s12020-015-0607-9.

Quinn, L. S., B. G. Anderson, R. H. Drivdahl, B. Alvarez, and J. M. Argiles. 2002. "Overexpression of inter-
leukin-15 induces skeletal muscle hypertrophy in vitro: Implications for treatment of muscle wasting
disorders." *Exp Cell Res* 280 (1):55–63.

Quinn, L. S., B. G. Anderson, L. Strait-Bodey, and T. Wolden-Hanson. 2010. "Serum and muscle interleu-
kin-15 levels decrease in aging mice: Correlation with declines in soluble interleukin-15 receptor alpha
expression." *Exp Gerontol* 45 (2):106–12. doi:10.1016/j.exger.2009.10.012.

Quinn, L. S., K. L. Haugk, and S. E. Damon. 1997. "Interleukin-15 stimulates C2 skeletal myoblast differentiation." *Biochem Biophys Res Commun* 239 (1):6–10. doi:10.1006/bbrc.1997.7414.

Quinn, L. S., K. L. Haugk, and K. H. Grabstein. 1995. "Interleukin-15: A novel anabolic cytokine for skeletal
muscle." *Endocrinology* 136 (8):3669–72. doi:10.1210/endo.136.8.7628408.

Rana, K. S., M. Arif, E. J. Hill, S. Aldred, D. A. Nagel, A. Nevill, H. S. Randeva, C. J. Bailey, S. Bellary,
and J. E. Brown. 2014. "Plasma irisin levels predict telomere length in healthy adults." *Age (Dordr)*,
36, 995–1001.

Reardon, K. A., J. Davis, R. M. Kapsa, P. Choong, and E. Byrne. 2001. "Myostatin, insulin-like growth factor-1, and leukemia inhibitory factor mRNAs are upregulated in chronic human disuse muscle atrophy." *Muscle Nerve* 24 (7):893–9.

Roca-Rivada, A., C. Castelao, L. Senin, M. Landrove, J. Baltar, A. Crujeiras, L. Seoane, F. Casanueva, and M. Pardo. 2018. "FNDC5/Irisin is not only a myokine but also an adipokine." doi:10.1371/journal.pone.0060563.

Roth, S. M., G. F. Martel, R. E. Ferrell, E. J. Metter, B. F. Hurley, and M. A. Rogers. 2003. "Myostatin gene expression is reduced in humans with heavy-resistance strength training: A brief communication." *Exp Biol Med (Maywood)* 228 (6):706–9.

Rothman, S. M., K. J. Griffioen, R. Wan, and M. P. Mattson. 2012. "Brain-derived neurotrophic factor as a regulator of systemic and brain energy metabolism and cardiovascular health." *Ann N Y Acad Sci* 1264:49–63. doi:10.1111/j.1749-6632.2012.06525.x.

Sakuma, K., K. Watanabe, M. Sano, I. Uramoto, and T. Totsuka. 2000. "Differential adaptation of growth and differentiation factor 8/myosatatin, fibroblast growth factor 6, and leukemia inhibitory factor in overloaded, regenerating and denervated rat muscles." *Biochemica et Biophysica Acta (BBA)-Molecular Cell Research* 1497 (1):77–88. doi:10.1016/S0167-4889(00)00044-6.

Schaap, L. A., S. M. Pluijm, D. J. Deeg, T. B. Harris, S. B. Kritchevsky, A. B. Newman, L. H. Colbert et al. 2009. "Higher inflammatory marker levels in older persons: Associations with 5-year change in muscle mass and muscle strength." *J Gerontol A Biol Sci Med Sci* 64 (11):1183–9. doi:10.1093/gerona/glp097.

Schaap, L. A., S. M. Pluijm, D. J. Deeg, and M. Visser. 2006. "Inflammatory markers and loss of muscle mass (sarcopenia) and strength." *Am J Med* 119 (6):526 e9–17. doi:10.1016/j.amjmed.2005.10.049.

Schafer, M. J., E. J. Atkinson, P. M. Vanderboom, B. Kotajarvi, T. A. White, M. M. Moore, C. J. Bruce et al. 2016. "Quantification of GDF11 and myostatin in human aging and cardiovascular disease." *Cell Metab* 23 (6):1207–15. doi:10.1016/j.cmet.2016.05.023.

Schnyder, S., and C. Handschin. 2015. "Skeletal muscle as an endocrine organ: PGC-1α, myokines and exercise." *Bone* 80:115–25. doi:10.1016/j.bone.2015.02.008.

Schrager, M. A., E. J. Metter, E. Simonsick, A. Ble, S. Bandinelli, F. Lauretani, and L. Ferrucci. 2007. "Sarcopenic obesity and inflammation in the InCHIANTI study." *J Appl Physiol (1985)* 102 (3):919–25. doi:10.1152/japplphysiol.00627.2006.

Schuelke, M., K. R. Wagner, L. E. Stolz, C. Hübner, T. Riebel, W. Kömen, T. Braun, J. F. Tobin, and S.-J. Lee. 2009. "Myostatin mutation associated with gross muscle hypertrophy in a child." doi:10.1056/NEJMoa040933.

Schulte, J. N., and K. E. Yarasheski. 2001. "Effects of resistance training on the rate of muscle protein synthesis in frail elderly people." *Int J Sport Nutr Exerc Metab* 11 Suppl:S111–18.

Scott, R. L., A. D. Gurusinghe, A. A. Rudvosky, V. Kozlakivsky, S. S. Murray, M. Satoh, and S. S. Cheema. 2000. "Expression of leukemia inhibitory factor receptor mRNA in sensory dorsal root ganglion and spinal motor neurons of the neonatal rat." *Neurosci Lett* 295 (1–2):49–53.

Serrano, A. L., B. Baeza-Raja, E. Perdiguero, M. Jardi, and P. Munoz-Canoves. 2008. "Interleukin-6 is an essential regulator of satellite cell-mediated skeletal muscle hypertrophy." *Cell Metab* 7 (1):33–44. doi:10.1016/j.cmet.2007.11.011.

Seto, D. N., S. C. Kandarian, and R. W. Jackman. 2015. "A key role for leukemia inhibitory factor in C26 cancer cachexia." *J Biol Chem* 290 (32):19976–86. doi:10.1074/jbc. M115.638411.

Shan, T., X. Liang, P. Bi, and S. Kuang. 2013. "Myostatin knockout drives browning of white adipose tissue through activating the AMPK-PGC1α-Fndc5 pathway in muscle." *FASEB J* 27 (5):1981–9. doi:10.1096/fj.12-225755.

Siriett, V., L. Platt, M. S. Salernoa, N. Ling, R. Kambadur, and M. Sharma. 2006. "Prolonged absence of myostatin reduces sarcopenia—Siriett—2006—Journal of cellular physiology—Wiley Online Library." *J Cell Physiol* 209 (3):866–73. doi:10.1002/jcp.20778.

Siriett, V., M. S. Salerno, C. Berry, G. Nicholas, R. Bower, R. Kambadur, and M. Sharma. 2007. "Antagonism of myostatin enhances muscle regeneration during sarcopenia." *Mol Ther* 15 (8):1463–70.

Smith, R. C., and B. K. Lin. 2013. "Myostatin inhibitors as therapies for muscle wasting associated with cancer and other disorders." *Curr Opin Support Palliat Care* 7 (4):352–60. doi:10.1097/spc.0000000000000013.

Spangenburg, E. E., and F. W. Booth. 2006. "Leukemia inhibitory factor restores the hypertrophic response to increased loading in the LIF(-/-) mouse." *Cytokine* 34 (3–4):125–30. doi:10.1016/j.cyto.2006.05.001.

Steensberg, A., M. A. Febbraio, T. Osada, P. Schjerling, G. van Hall, B. Saltin, and B. K. Pedersen. 2001. "Interleukin-6 production in contracting human skeletal muscle is influenced by pre-exercise muscle glycogen content." *J Physiol* 537 (Pt 2):633–9.

Sun, L., K. Ma, H. Wang, F. Xiao, Y. Gao, W. Zhang, K. Wang, X. Gao, N. Ip, and Z. Wu. 2007. "JAK1-STAT1-STAT3, a key pathway promoting proliferation and preventing premature differentiation of myoblasts." *J Cell Biol* 179 (1):129–38. doi:10.1083/jcb.200703184.

Sundarrjan, L., C. Yeung, L. Hahn, L. Weber, and S. Unniappan. 2018. "Irisin regulates cardiac physiology in zebrafish." doi:10.1371/journal.pone.0181461.

Suomalainen, A., J. M. Elo, K. H. Pietilainen, A. H. Hakonen, K. Sevastianova, M. Korpela, P. Isohanni et al. 2011. "FGF-21 as a biomarker for muscle-manifesting mitochondrial respiratory chain deficiencies: A diagnostic study." *Lancet Neurol* 10 (9):806–18. doi:10.1016/S1474-4422(11)70155-7.

Tezze, C., V. Romanello, M. A. Desbats, G. P. Fadini, M. Albiero, G. Favaro, S. Ciciliot et al. 2017. "Age-associated loss of OPA1 in muscle impacts muscle mass, metabolic homeostasis, systemic inflammation, and epithelial senescence." *Cell Metab* 25 (6):1374–89 e6. doi:10.1016/j.cmet.2017.04.021.

Thomas, M., B. Langley, C. Berry, M. Sharma, S. Kirk, J. Bass, and R. Kambadur. 2000. "Myostatin, a negative regulator of muscle growth, functions by inhibiting myoblast proliferation." *J Biol Chem.* doi:10.1074/jbc. M004356200.

Timmons, J. A., K. Wennmalm, O. Larsson, T. B. Walden, T. Lassmann, N. Petrovic, D. L. Hamilton et al. 2007. "Myogenic gene expression signature establishes that brown and white adipocytes originate from distinct cell lineages." *Proc Natl Acad Sci USA* 104:4401–6.

Trenerry, M. K., K. A. Carey, A. C. Ward, M. M. Farnfield, and D. Cameron-Smith. 2008. "Exercise-induced activation of STAT3 signaling is increased with age." *Rejuvenation Res* 11 (4):717–24. doi:10.1089/rej.2007.0643.

Tsai, S. W., Y. C. Chan, F. Liang, C. Y. Hsu, and I. T. Lee. 2015. "Brain-derived neurotrophic factor correlated with muscle strength in subjects undergoing stationary bicycle exercise training." *J Diabetes Complicat* 29 (3):367–71. doi:10.1016/j.jdiacomp.2015.01.014.

Tsuchiya, Y., D. Ando, K. Goto, M. Kiuchi, M. Yamakita, and K. Koyama. 2014. "High-intensity exercise causes greater irisin response compared with low-intensity exercise under similar energy consumption." *Tohoku J Exp Med* 233 (2):135–40.

Tsujinaka, T., J. Fujita, C. Ebisui, M. Yano, E. Kominami, K. Suzuki, K. Tanaka et al. 1996. "Interleukin 6 receptor antibody inhibits muscle atrophy and modulates proteolytic systems in interleukin 6 transgenic mice." *J Clin Invest* 97 (1):244–9. doi:10.1172/JCI118398.

Vandanmagsar, B., J. D. Warfel, S. E. Wicks, S. Ghosh, J. M. Salbaum, D. Burk, O. S. Dubuisson et al. 2016. "Impaired mitochondrial fat oxidation induces FGF21 in muscle." *Cell Rep* 15 (8):1686–99. doi:10.1016/j.celrep.2016.04.057.

Vaughan, R. A., N. P. Gannon, M. A. Barberena, R. Garcia-Smith, M. Bisoffi, C. M. Mermier, C. A. Conn, and K. A. Trujillo. 2014. "Characterization of the metabolic effects of irisin on skeletal muscle in vitro." *Diabetes Obes Metab* 16 (8):711–18. doi:10.1111/dom.12268.

Visser, M., M. Pahor, D. R. Taaffe, B. H. Goodpaster, E. M. Simonsick, A. B. Newman, M. Nevitt, and T. B. Harris. 2002. "Relationship of interleukin-6 and tumor necrosis factor-alpha with muscle mass and muscle strength in elderly men and women: The health ABC study." *J Gerontol A Biol Sci Med Sci* 57 (5):M326–32.

Walston, J., D. E. Arking, D. Fallin, T. Li, B. Beamer, Q. Xue, L. Ferrucci, L. P. Fried, and A. Chakravarti. 2005. "IL-6 gene variation is not associated with increased serum levels of IL-6, muscle, weakness, or frailty in older women." *Exp Gerontol* 40 (4):344–52. doi:10.1016/j.exger.2005.01.012.

Washington, T. A., J. P. White, J. M. Davis, L. B. Wilson, L. L. Lowe, S. Sato, and J. A. Carson. 2011. "Skeletal muscle mass recovery from atrophy in IL-6 knockout mice." *Acta Physiol* 202 (4):657–69. doi:10.1111/j.1748-1716.2011.02281.x.

Westbury, L. D., N. R. Fuggle, H. E. Syddall, N. A. Duggal, S. C. Shaw, K. Maslin, E. M. Dennison, J. M. Lord, and C. Cooper. 2018. "Relationships between markers of inflammation and muscle mass, strength and function: Findings from the Hertfordshire Cohort Study." *Calcif Tissue Int* 102 (3):287–95. doi:10.1007/s00223-017-0354-4.

White, J. P., J. W. Baynes, S. L. Welle, M. C. Kostek, L. E. Matesic, S. Sato, and J. A. Carson. 2011. "The regulation of skeletal muscle protein turnover during the progression of cancer cachexia in the Apc(Min/+) mouse." *PLoS One* 6 (9):e24650. doi:10.1371/journal.pone.0024650.

White, J. P., J. M. Reecy, T. A. Washington, S. Sato, M. E. Le, J. M. Davis, L. B. Wilson, and J. A. Carson. 2009. "Overload-induced skeletal muscle extracellular matrix remodelling and myofibre growth in mice lacking IL-6." *Acta Physiol* 197 (4):321–32. doi:10.1111/j.1748-1716.2009.02029.x.

White, T. A., and N. K. LeBrasseur. 2014. "Myostatin and sarcopenia: Opportunities and challenges—A mini-review." *Gerentology* 60. doi:10.1159/000356740.

Williams, A., J. J. Wang, L. Wang, X. Sun, J. E. Fischer, and P. O. Hasselgren. 1998. "Sepsis in mice stimulates muscle proteolysis in the absence of IL-6." *Am J Physiol* 275 (6 Pt 2):R1983–91.

Wrann, C., J. White, J. Salogiannis, D. Laznik-Bogoslavski, J. Wu, D. Ma, J. Lin, M. Greenberg, and B. Spiegelman. 2013. "Exercise induces hippocampal BDNF through a PGC-1α/FNDC5 pathway." *Cell Metab* 18 (5):649–59. doi:10.1016/j.cmet.2013.09.008.

Yarasheski, K. E., S. Bhasin, I. Sinha-Hikim, J. Pak-Loduca, and N. F. Gonzalez-Cadavid. 2002. "Serum myostatin-immunoreactive protein is increased in 60–92 year old women and men with muscle wasting." *J Nutr Health Aging* 6 (5):343–8.

Yokota, S., T. Imagawa, M. Mori, T. Miyamae, Y. Aihara, S. Takei, N. Iwata et al. 2008. "Efficacy and safety of tocilizumab in patients with systemic-onset juvenile idiopathic arthritis: A randomised, double-blind, placebo-controlled, withdrawal phase III trial." *Lancet* 371 (9617):998–1006. doi:10.1016/S0140-6736(08)60454-7.

8 The Contribution of Satellite Cells to Skeletal Muscle Aging

Christopher S. Fry

CONTENTS

8.1 INTRODUCTION

The maintenance and repair of many adult tissues are dependent on the tissue-resident stem cells. In the most basic sense, stem cells can pursue two distinct paths: differentiation to repair damaged tissue or self-renewal to repopulate the existing pool. Skeletal muscle possesses a robust ability to respond to a variety of environmental stimuli. The inherent plasticity of skeletal muscle is largely attributable to the tissue-resident stem cell known as the satellite cell. Satellite cells are often long lived; however, they do not retain their capacity and function indefinitely. Muscle dependency on satellite cell regenerative capacity, and satellite cells' unique longevity requirement may explain why age-related decline has such a potentially detrimental effect on healthy skeletal muscle aging. Hence, this chapter will focus on the age-related adaptations to muscle stem cells.

8.2 HISTORICAL PERSPECTIVES AND IDENTIFICATION

Satellite cells are relatively few in number, typically accounting for ~4% of muscle fiber nuclei (Venable 1966), but confer a remarkable adaptive capacity to muscle. The functional unit of skeletal muscle, the myofiber, is a contractile multinucleated cell, the result of embryonic fusion of mesodermal mononucleated myoblasts, the daughters of activated satellite cells (Figure 8.1). Early research deemed myonuclei as postmitotic, demonstrating that the cellular replication observed in skeletal muscle is the result of single-cell proliferation (Cooper and Konigsberg 1961, Capers 1960). The source of this single-cell replicative capacity would soon be determined.

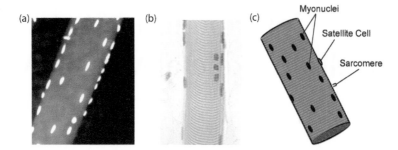

FIGURE 8.1   Multinucleated skeletal muscle fiber. (a) Representative image depicting several myonuclei (white) along the length of a skeletal muscle myofiber, stained with 4′,6-diamidino-2-phenylindole (DAPI), a fluorescent stain that binds strongly to adenine–thymine-rich regions in DNA. (b) Representative image of a single myofiber stained with hematoxylin and eosin. Multiple myonuclei (stained purple) can be seen along the length of the myofiber, and striations representing sarcomeres can also be seen. (c) An illustrated representation of a myofiber with myonuclei, a satellite cell, and sarcomeres depicted. Skeletal muscle myofibers represent a unique multinucleated cell that serves as the basic contractile unit within muscle. Early research found myonuclei to be postmitotic, proposing the existence of an alternate mononuclear cell responsible for myogenic proliferation, which would lead to the discovery of the satellite cell. (Images courtesy of the Fry Laboratory.)

Satellite cells were initially identified in 1961; they were observed as a mononucleated cell occupying a space that was described as "wedged between the plasma membrane of the muscle fiber and basement membrane" (Mauro 1961). It was the intimate association of these cells along the length of the myofiber that prompted their name, referencing the satellite position these cells occupied and forever linking the anatomical location with the name of the cell (Mauro 1961). Figure 8.2 shows the location of a satellite cell, residing within the basement membrane/basal lamina (illustrated by laminin) of the myofiber but outside of the sarcolemma (illustrated by dystrophin). The anatomical location of a satellite cell is distinguished from that of a myonucleus by the sublaminal and sub-sarcolemmal location of the myonucleus in relation to the sublaminal position of the satellite cell (Figure 8.2).

The regenerative potential of satellite cells and their role as a precursor to myogenic nuclei were even postulated by Mauro, who stated "satellite cells are merely dormant myoblasts that failed to fuse with other myoblasts and are ready to recapitulate the embryonic development of skeletal muscle fiber when the main multinucleate cell is damaged" (Mauro 1961).

While the discovery of satellite cells provided a viable candidate for the source of myoblasts capable of fusing into myofibers during growth and repair, direct evidence linking satellite cells as the myogenic precursor was lacking (Carlson 1973). The culturing of isolated myofibers provided evidence that mononucleated cells along the myofiber give rise to myoblasts that are capable of producing multinucleated myotubes in culture (Konigsberg et al. 1975). These observations were instrumental in the identification of myoblasts as capable of fusing together to generate a new myotube or *de novo* fiber formation. Additional evidence for myogenic replication came from studies utilizing radiolabeled nucleotide localization in damaged myofibers (Moss and Leblond 1971). Confirmation that satellite cells function as stem cells, capable of differentiating into multinucleated myofibers in addition to maintaining a quiescent pool through self-renewal, would come many decades later. The true stem cell nature of satellite cells was keenly observed through the use of myofiber transplantation (Collins et al. 2005).

More recent genetic studies have demonstrated that the expression of *paired box 7* (*Pax7*) is restricted to satellite cells within the skeletal muscle (Seale et al. 2000), and its expression is now universally considered the best marker of quiescent satellite cells. Indeed, *Pax7* knock out

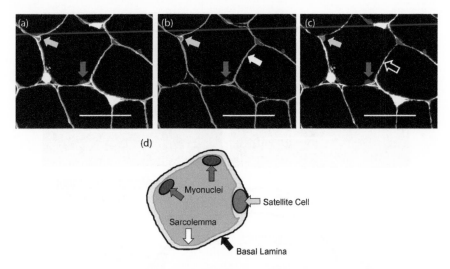

FIGURE 8.2 Anatomical location of satellite cells. (a–c) Immunohistochemical images of human skeletal muscle cross-sections depicting satellite cells (Pax7+, pink), sarcolemma (dystrophin, green), basal lamina (laminin, white), and nuclear content (DAPI, blue). A satellite cell is indicated with a yellow arrow, a myonucleus is indicated with a red arrow, the sarcolemma is indicated with a white arrow, and the basal lamina is indicated with a black arrow. Panel (a) is a merged image, depicting the satellite cell's location outside of dystrophin (sarcolemma) but inside laminin (basal lamina), and a myonucleus residing within both dystrophin and laminin. Panel (b) depicts the subsarcolemmal location of the myonucleus, and while the satellite cell is located outside of the sarcolemma. Panel (c) depicts the sublaminal location of both the myonucleus and satellite cell. Scale bar = 50 μm. Panel (d) is an illustrated schematic depicting the location of both myonuclei (blue, red arrow) and a satellite cell (pink, yellow arrow) in relation to the sarcolemma (green, white arrow) and basal lamina (black, black arrow). While the images above utilize modern immunohistochemical imaging techniques to visualize the location of satellite cells adjacent to muscle fibers, the unique location of satellite cells along the periphery of the fibers, within a "satellite" position, enabled their discovery over 50 years ago. (Images courtesy of the Fry Laboratory.)

mice ($Pax7^{-/-}$) are viable, but display reduced muscle mass, are unable to mount a regenerative response to muscle injury and lack any functional satellite cells (Kuang et al. 2006, Seale et al. 2004). In addition to anatomical location, identification of satellite cells can also be accomplished through antibody-mediated labeling of the Pax7 protein within satellite cells. The satellite cell shown in Figure 8.2 is labeled immunohistochemically with an antibody against the Pax7 protein, demonstrating the nuclear location of the Pax7 transcription factor in addition to its location between the sarcolemma and the basal lamina. The identification of unique molecular targets within satellite cells allowed for eloquent lineage tracing studies in rodent muscle, whereby satellite cells are genetically labeled to trace their proliferation and fusion into damaged myofibers (Lepper et al. 2009). Satellite cells can be labeled with various molecular tags, including fluorescent markers, providing a visual method of detection for satellite cells within the mouse skeletal muscle.

In the human skeletal muscle, the Pax7 transcription factor also serves as a useful tool for the detection of satellite cells (Arentson-Lantz et al. 2016, Fry, Noehren et al. 2014, Mackey et al. 2009). Moreover, cluster of differentiation 56 (CD56)/neural cell adhesion molecule (NCAM) can also label satellite cells in human skeletal muscle (Illa et al. 1992, Mackey et al. 2009), allowing for identification in clinical muscle samples. Some studies have shown slight deviations in satellite

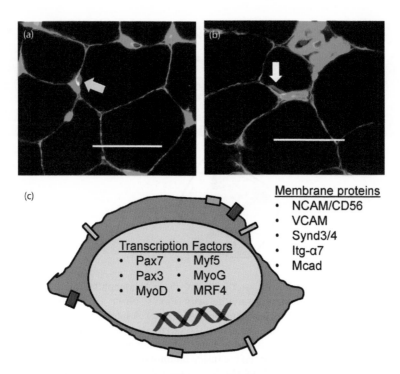

FIGURE 8.3 Immunohistochemical identification of satellite cells in human skeletal muscle. (a) Representative immunohistochemical image of a human skeletal muscle cross-section depicting a satellite cell labeled with a nuclear marker (yellow arrow, identified by its expression of the Pax7 transcription factor) located within the basal lamina (laminin, green). The nuclear location of the Pax7 transcription factor is visualized with the overlap of the Pax7 signal with a DAPI-labeled nucleus. (b) Representative immunohistochemical image of a human skeletal muscle cross-section depicting a satellite cell labeled with a cell-surface marker (white arrow, identified by its expression of the neural cell adhesion molecule [NCAM]) located within the basal lamina (laminin, green). With the location of NCAM on the surface of the satellite cell membrane, it appears that the NCAM label is "surrounding" the DAPI-labeled nucleus. Both Pax7 and NCAM are commonly used for the identification of satellite cells within human skeletal muscle. Similar results from the quantification of satellite cells using both methods have been reported, and the use of multiple markers allows for confirmation of results obtained from clinical samples. (c) An illustrated schematic depicting a satellite cell with various proteins that allow for identification of satellite cells in human and animal muscle. Specific transcription factors present in satellite cells include paired box 7 and 3 (Pax7 and Pax3), myogenic differentiation (MyoD), myogenic factor 5 (Myf5), myogenin (MyoG), and myogenic regulatory factor 4 (MRF4). Various proteins found along the surface of satellite cells also allow for their identification and include neural cell adhesion molecule/cluster of differentiation 56 (NCAM/CD56), vascular cell adhesion molecule/cluster of differentiation 106 (VCAM/CD106), syndecan 3/4 (Synd3/4), integrin alpha 7 (Itg-α7), and M-cadherin (Mcad). These lists are not exhaustive but provide more commonly reported markers used to identify satellite cells. Scale bar = 50 μm. (Images courtesy of the Fry Laboratory.)

cell abundance when assessed by Pax7 and CD56 (Mackey et al. 2009), where other studies report strongly correlated satellite cell content when measured using both markers (Arentson-Lantz et al. 2016). Figure 8.3 shows satellite cells in human skeletal muscle, labeled with Pax7 and CD56, showing the different staining patterns of transcription factors versus cell surface markers. Several molecular targets have been used to identify satellite cells in both animal and human skeletal muscle, and a brief list is seen in Figure 8.3.

8.3 SATELLITE CELL ACTIVITY AND MAINTENANCE

Satellite cells are quiescent under resting or basal conditions within the muscle, residing in a dormant state adjacent to the myofiber in their niche (Schultz et al. 1978). Quiescent satellite cells largely express *Pax7*, with expression of other myogenic regulatory factors (MRFs) uncommon in this state. In their quiescent state, satellite cells display a reduced metabolism and are more resistant to DNA damage, which both provide stability and long-term maintenance of the satellite cell pool. The combined high expression of cell cycle inhibitors and low expression of cyclins and cyclin-dependent kinases (CDKs) prevent cell cycle progression and promote maintenance of the quiescent state (Fukada et al. 2007). When the capacity of the quiescent state is challenged, the sporadic differentiation of satellite cells can occur, leading to reduced abundance over time (Chakkalakal et al. 2014).

Myogenic lineage specificity is conferred to satellite cells through transcriptional regulation by MRFs, which include myogenic differentiation 1 protein (MyoD), myogenic factor 5 (Myf5), myogenin (MyoG), and muscle-specific regulatory factor 4 (Mrf4). When myofibers are exposed to an injury, insult, or other stimulus (loading, exercise-induced stress, etc.), satellite cells become activated and drastically upregulate their expression of *MyoD* and/or *Myf5*, which are transcriptionally regulated by Pax7 (Cornelison and Wold 1997, Duprey and Lesens 1994). These Pax7+/MyoD+ satellite cells undergo rapid expansion and are capable of proliferating several hundred percent in a few days' time (McCarthy et al. 2011, Murphy et al. 2011). While Pax7 is integral in the upregulation of MyoD and other MRFs, it must be downregulated in order for activated satellite cells to proceed to terminal differentiation (Olguin and Olwin 2004, Olguin et al. 2007). MyoD works to accomplish this task, functioning as a transcriptional co-activator and regulating expression of both proliferation and differentiation genes, including *Pax7* and MyoG (Soleimani et al. 2012).

A decline in *Pax7* expression along with the concomitant induction of MyoG is indicative of myoblasts that have entered terminal differentiation and initiated withdrawal from the cell cycle. Concurrent with the increase in MyoG expression, differentiating myoblasts also initiate transcription of various structural proteins, including actin and myosin, which facilitates the fusion of myoblasts into *de novo* myotubes or into existing myofibers (Smith et al. 1994, Yablonka-Reuveni et al. 1999, Andres and Walsh 1996).

Routine repair of skeletal muscle and satellite cell activation would in theory lead to the depletion of the satellite cell pool. Fortunately, satellite cells possess robust self-renewal ability. Studies of isolated satellite cells have shown that, in addition to myoblast differentiation and fusion, a subpopulation of activated satellite cells can subsequently downregulate *MyoD* expression, exit the cell cycle, maintain *Pax7* expression, and re-enter quiescence (Halevy et al. 2004, Zammit et al. 2004). This allows for the homeostatic maintenance of the total satellite cell pool within skeletal muscle and is facilitated by the ability of satellite cells to undergo asymmetric cell division to generate committed progeny without losing their own stem cell identity (Kuang et al. 2007, Shinin et al. 2006). Moreover, satellite cells can also undergo symmetrical expansion to replenish and maintain the stem cell pool (Bentzinger et al. 2013).

These events are shown in Figure 8.4. The myofiber at the top of the figure shows satellite cells in the quiescent state, depicted as Pax7+/MyoD−/MyoG−. Following a stimulus or insult to the myofiber (injury or loading), satellite cells become activated and begin to express MyoD while undergoing rapid proliferation, increasing several fold in less than a week. These cells will express both *Pax7* and *MyoD* in the activated and proliferative state. As the repair/adaptation process continues, satellite cells will continue to express *MyoD* while downregulating the expression of *Pax7*. Expression of *MyoG* (blue nuclei, bottom fiber) is indicative of MyoD commitment, and these myoblasts will ultimately fuse into existing fibers or fuse together to form de novo myotubes. Furthermore, through asymmetric cell division, a subset of Pax7+/MyoD+ satellite cells will downregulate *MyoD* expression and return to quiescence.

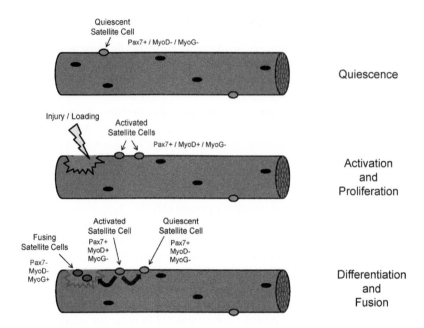

FIGURE 8.4 Illustration of satellite cell activation and fusion. Satellite cells reside in a niche located between the basal lamina and sarcolemma (Figure 8.2) and express the transcription factor Pax7 in their quiescent state. Following an insult (injury, loading, etc.) to nearby myofibers, satellite cells are activated and proliferated, expressing both Pax7 and MyoD. Following activation and expansion, activated satellite cells will transition to the site of the insult, passing through the sarcolemma. These cells begin expressing myogenin (MyoG), which will be followed by a structural myogenic differentiation program that will facilitate the fusion of the myoblasts to the myofibers to repair the injury or contribute to myofiber hypertrophy. A subset of activated satellite cells will return to a quiescent state, expressing Pax7 within the niche adjacent to myofibers, maintaining the pool of satellite cells.

8.4 SATELLITE CELL REQUIREMENTS THROUGHOUT THE LIFESPAN

Postnatal muscle growth, myogenesis, involves significant myofiber hypertrophy with a concomitant increase in myonuclear content (Oertel 1988, Enesco and Puddy 1964). These observations provide support for an active role of satellite cells during childhood muscle development. Indeed, satellite cells in developing mammalian muscles appear more "active" than those found in mature muscles, through observation of a well-developed granular endoplasmic reticulum, indicative of enhanced protein synthesis (Ishikawa 1966). Fewer ribosomes are present in the satellite cells of adult mammals, consistent with the reduced metabolic activity (Schultz 1976). With the significant contribution of satellite cells to developing skeletal muscle, it is unsurprising that the relative number of satellite cells declines during maturation. In rodent models, perinatal skeletal muscle nuclei are composed of approximately 30% satellite cells, which declines to approximately 5% in adults (Allbrook et al. 1971). This observation is confirmed in humans, where children aged 0–18 years exhibit a precipitous drop in satellite cell content normalized to the muscle area (Verdijk et al. 2014). As maturity nears completion, satellite cell content closely resembles that of adults, characterized by lower metabolic and proliferative activity (Schultz et al. 1978).

Skeletal muscle mass often peaks around the third decade of life and then begins to decline, with greater deficits in lean muscle mass and strength apparent in older adults (Janssen et al. 2000). Sarcopenia can lead to frailty, impeding the functional capacity of older adults and contributing to their loss of independence (Evans 1997). Satellite cells directly contribute to myofiber maintenance, and previous research has shown age-related declines in satellite cell abundance that mirror the loss of skeletal muscle mass that occurs during older age (Renault et al. 2002, Kadi et al. 2004). However, the age-related decline in satellite cell content has not been universally confirmed, as several studies have shown no decrement in satellite cell abundance in older adults (Roth et al. 2000, Dreyer et al. 2006). The conflicting nature of these studies could be due to the relatively small number of subjects sampled in addition to methodological variations. More recently, larger studies, including many more human subjects, have sought to clarify previous discrepancies. Skeletal muscle biopsies from 150 adults (age 18–86 years) demonstrate a negative linear association between satellite cell content and age (Verdijk et al. 2014), and data of 70 subjects (age 18–83 years) presented in Figure 8.5 confirms these findings, providing evidence for the age-related decline in satellite cell abundance. While far from undisputed, these studies of relatively large cohorts of human subjects support the notion that satellite cell content is reduced with the advancing age. Whether the decline in satellite cell abundance is a product of age-related muscle loss or contributory to sarcopenia remains a fiercely contested topic.

While recent clinical evidence provides support for reduced satellite cell content in muscle from older adults, these data are correlative in nature and do not ascertain the contribution of satellite cells (or lack thereof) to sarcopenia. Studies in rodents have confirmed the decrement in satellite cell abundance in aging muscle (Keefe et al. 2015, Fry et al. 2015, Shefer et al. 2006) and have offered support for a direct role of satellite cells in restoring age-related decrements in muscle function and adaptation (Sinha et al. 2014, Cosgrove et al. 2014). Figure 8.6 shows the relative satellite cell content in various skeletal muscles in mice throughout the lifespan. The age-related decline in satellite cell abundance displays heterogeneity among the various muscle groups profiled, but the decline typically does not become apparent until approximately 20 months of age, which corresponds to an approximate human age of 65 years, supportive of clinical data on satellite cell abundance in older adults.

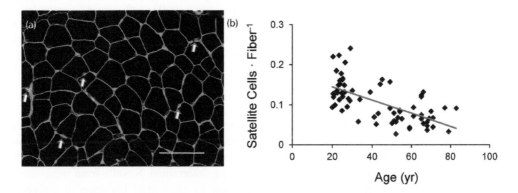

FIGURE 8.5 Aged-related decline in satellite cell abundance in human skeletal muscle. (a) Representative immunohistochemical image of a human skeletal muscle cross-section depicting satellite cells (expressing Pax7 [pink], denoted by red and white arrows), laminin (green), and nuclear content (DAPI, blue). Scale bar = 50 μm. (b) Satellite cells are quantified and expressed relative to myofiber number in 70 subjects. Satellite cell content is plotted against the age (in years) of various human subject volunteers and depicts a negative association (red line), with reduced satellite cell abundance in older adults. This association illustrates the nature of satellite cell abundance within the adult lifespan in a group of human subjects. Skeletal muscle mass is also known to decline with advancing age, and the decline in satellite cell abundance in older follows a similar pattern. This observation has led to the hypothesis that the age-related decline in satellite cell content may be related to, or contribute to, sarcopenia, but the correlative nature of this relationship does not provide evidence of a direct relationship, only a similar trend in aging skeletal muscle. (Images and data courtesy of the Fry Laboratory.)

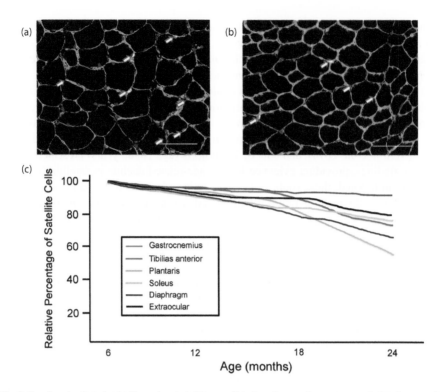

FIGURE 8.6 Aged-related decline in satellite cell abundance in mouse skeletal muscle. (a–b) Representative immunohistochemical images of skeletal muscle cross-sections in young (a, 4 months) and older (b, 24 months) mice. The images depict satellite cells (expressing Pax7 [pink], denoted by red and white arrows), laminin (green), and nuclear content (DAPI, blue). Scale bar = 100 μm. In addition to fewer satellite cells in the older mouse, the size of the myofibers (outlined in green) is also reduced, illustrating the age-related atrophy of muscle (sarcopenia). Several studies have assessed satellite cell content in various mouse muscles across the lifespan, and results of relative satellite cell abundance are summarized in (c). Both limb muscles (gastrocnemius, tibialias anterior, plantaris, and soleus) and diaphragm and extraocular muscles were sampled. These data support findings in human skeletal muscle that satellite cells exhibit a decline in content with advancing age. (Adapted from Fry, C.S. et al., *Nat. Med.*, 21, 76–80, 2015; Keefe, A.C. et al., *Nat. Commun.*, 6, 7087, 2015.)

To directly determine whether satellite cell abundance is responsible for the promotion or exacerbation of sarcopenia, genetic studies in mice were conducted. Young mice underwent conditional genetic depletion of their satellite cells and were allowed to age, so that the effect of total satellite cell loss on muscle mass and function with age could be determined. Results from these studies showed that depletion of satellite cells does not promote or exacerbate the onset of sarcopenia (Keefe et al. 2015, Fry et al. 2015). These studies were the first to determine that the absence of satellite cells, which was far more severe than typical age-related decline, does not quicken or worsen the onset of sarcopenia. While these results argue against a causative role for satellite cells in the promotion of age-related muscle loss, limitations need to be considered. Mice in these studies led relatively sedentary lives, and the introduction of physical activity or other stimuli that would activate satellite cells may demonstrate satellite cell-dependent muscle atrophy or dysfunction during aging.

Intriguingly, the genetic labeling of satellite cells with fluorescent markers demonstrates fusion of satellite cells into uninured myofibers (Keefe et al. 2015, Pawlikowski et al. 2015). The fusion of satellite cells occurred both with and without satellite cell division, highlighting the replicative potential of satellite cells needed to maintain an adequate pool size (Pawlikowski et al. 2015). Hind limb muscles demonstrate minimal fusion of satellite cells during adulthood, with fusion largely limited to the postnatal muscle growth period. This contrasts with extraocular and diaphragm muscles that display elevated fusion of satellite cells into adulthood, supporting differential satellite cell requirements that are muscle dependent (Pawlikowski et al. 2015, McLoon et al. 2004). These lineage tracing studies would indicate that satellite cell fusion into myofibers occurs more extensively than previously thought. While most research would support satellite cell quiescence in uninjured muscle at rest, satellite cells may have a more active role in skeletal muscle homeostasis, requiring greater basal satellite cell activity. The molecular signal(s) that spur the fusion of satellite cells into uninjured myofibers are currently unknown, and subsequent sections will discuss satellite cell activity during muscle adaptation and age-related decline.

8.5 DECLINE IN SATELLITE CELL ACTIVITY IN AGING MUSCLE

Previous sections have discussed satellite cell activation and requirements throughout the lifespan; here we will focus on perturbations in satellite cell function that occur in aging skeletal muscle. Various stimuli can activate satellite cells, prompting their exit from quiescence and entrance into the cell cycle, including injury to myofibers and exercise/loading (Figure 8.4). In addition to the previously discussed age-related decline in satellite cell content, there also exists a deficit in the functional capacity of aged satellite cells (Sousa-Victor et al. 2014, Zwetsloot et al. 2013). The decline in satellite cell functionality with age has been attributed to a combination of factors, including defects in self-renewing mechanisms, exhaustion by forced differentiation as well as apoptosis and senescence. Aging satellite cells also exhibit deficits in their ability to properly activate and produce the necessary myogenic lineage following damage to nearby myofibers. The following subsections will explore the molecular changes that satellite cells undergo during aging, as well as both extrinsic and intrinsic factors contributing to these changes.

8.5.1 EXTERNAL ALTERATIONS CONTRIBUTING TO FUNCTIONAL DECLINE

Some of the most conclusive evidence supporting extrinsic alterations to aging satellite cells can be seen in the results of heterochronic tissue transplant and parabiosis experiments. These experiments involve the transplantation of tissue from a young mouse to an older mouse or the joining of the circulatory system between a young and older mouse. Early studies involving transplantation and engraftment of muscle groups between young and older animals found that the aged environment was key in conferring regenerative potential to transplanted tissue (Carlson and Faulkner 1989, Gutmann and Carlson 1976). These studies provided early clues that alterations in the local environment surrounding satellite cells, termed the niche, greatly affect regenerative capacity by altering the satellite cell function. Later studies would confirm these initial observations through the use of shared circulatory systems, parabiosis, between younger and older animals. The exchange of systemic factors from a younger to an older mouse rejuvenated many age-related physiological adaptations, including enhancing the regenerative capacity of skeletal muscle in older animals (Brack and Rando 2007, Conboy et al. 2005). Intriguingly, as aged muscle regenerative ability was restored, skeletal muscle from the younger counterparts exhibited a decline in regenerative potential, supportive of the presence of circulating factors that change with age and influence satellite cell activity. Subsequent research was undertaken with the purpose of identifying specific factors derived from the young/old environment that were responsible for mediating the effects of age on satellite cell function. The successful identification of such factors would provide defined therapeutic targets against which to develop strategies aimed at restoring the regenerative capacity of sarcopenic muscle.

Early successes have recently been met with controversy, as studies with conflicting results demonstrate complexity that is just now being appreciated (Sinha et al. 2014, Egerman et al. 2015). One factor in question is growth differentiation factor 11 (GDF11), originally reported to be present in the circulating environment of young mice but undergoing a steep decline in older animals (Sinha et al. 2014). Administration of GDF11 to older animals attenuated the age-related decline in satellite cell activity and muscle regeneration. A more recent study has directly challenged those findings with results indicating *elevated* levels of GDF11 levels in the plasma of older animals and humans (Egerman et al. 2015). It is likely that GDF11 associates with myostatin, a negative regulator of skeletal muscle size, to reduce the proliferative potential of satellite cells in older animals. Further research is needed to better elucidate the downstream action of GDF11 and myostatin on satellite cell activity and skeletal muscle homoeostasis during aging to determine the viability of these targets in ameliorating sarcopenia and the age-related impairment in regenerative capacity.

Moreover, transforming growth factor (TGF) β influences satellite cell activity and undergoes age-related adaptations, contributing to depressed satellite cell function in aging muscle. Increased expression of TGF-β within aging skeletal muscle negatively influences satellite cell proliferation through interaction with CDK inhibitors, leading to impairments in the regenerative capacity of aged satellite cells (Carlson et al. 2008). TGF-β may also promote a fibrotic muscle environment that is not conducive to proper satellite cell activity (Biressi et al. 2014). Age-related adaptations in local TGF-β signaling can alter the fate of satellite cells, diminishing their myogenic potential and converting them toward a fibrogenic lineage (Brack et al. 2007). This extrinsic regulation of satellite cells then promotes a fibrotic muscle environment, which further reduces satellite cell activity and ability to self-renew through alterations in the composition and stiffness of the satellite cell niche (Gilbert et al. 2010, Urciuolo et al. 2013). Attenuation of downstream effectors of TGF-β can restore the proper activity of satellite cells and regenerative capacity of aged skeletal muscle (Carlson et al. 2008). Taken together, maladaptations in both systemic and local environments attenuate innate satellite cell function in aged muscle, contributing to age-related dysfunction.

8.5.2 INTERNAL ALTERATIONS DURING AGING IN SATELLITE CELLS

While some evidence exists to the contrary (Pawlikowski et al. 2015), satellite cells largely exist within a quiescent state throughout an organism's lifespan but must maintain the ability to rapidly activate under certain external cues. Loss of this intrinsic ability to undergo robust proliferation and differentiation drastically diminishes skeletal muscle plasticity and places muscle at an increased risk of irreparable damage. These changes are manifested through alterations in gene expression patterns, with aging satellite cells showing increased expression of genes associated with inflammation and fibrosis, which alters the tissue microenvironment and leads to a decline in satellite cell functional and regenerative capacity (Budai et al. 2018). Moreover, satellite cell proliferative capacity has been recently shown to be challenged during aging through impairments in Notch signaling that leads to activation of p53, attenuating the proliferative response of aged satellite cells (Liu et al. 2018). During the aging process, satellite cells are exposed to greater stress from a variety of sources, which can lead to the transition into cellular senescence. Cellular senescence is characterized by the irreversible arrest of the cell cycle, preventing both further proliferation and apoptosis. Cellular senescence likely contributes to many age-related pathologies (Campisi and Robert 2014), including reduced satellite cell abundance and sarcopenia (Sousa-Victor et al. 2014). Induction of senescence from oxidative damage impairs aging satellite cell activity but can be rescued (Garcia-Prat et al. 2016). Much of the work underlying satellite cell senescence during aging has utilized rodent models, and the ubiquity of these models to define the human condition may not be wholly appropriate (Bigot et al. 2015), demonstrating the need for further investigation into differences between age-related satellite cell dysfunction between rodents and humans.

8.6 SATELLITE CELL RESPONSE TO RESISTANCE EXERCISE/ MECHANICAL LOADING DURING AGING

Skeletal muscle displays remarkable plasticity, and the inherent stem cell quality of satellite cells is a key component in the adaptation of muscle. Exercise confers a number of beneficial adaptations to muscle, and the nature of the exercise stimulus endows specific alterations to muscle. This section will focus primarily on resistance exercise training, which is characterized by increases in muscle strength and myofiber cross-sectional area. To attenuate or counteract the deleterious effects of sarcopenia, resistance exercise training is the most effective countermeasure, as training can preserve and, in some cases, enhance muscle mass in older adults. In response to resistance exercise, satellite cells are activated and fuse into myofibers, increasing myonuclear number as myofiber cytoplasmic volume increases. Satellite cell abundance and myonuclear accretion are strongly associated with myofiber hypertrophy in response to resistance exercise training in humans (Petrella et al. 2008) and mechanical overload in rodents (Ishido et al. 2009). The adaptive response of aged muscle to mechanical overload is attenuated in comparison with younger controls (Blough and Linderman 2000), and older adults demonstrates diminished improvements following resistance exercise training than their younger counterparts (Kosek et al. 2006). The age-related decline in satellite cell abundance and function likely mitigate muscle adaptation during exercise and overload, and experimental evidence in rodents supports this notion (Ballak et al. 2015).

The study of rodents has greatly furthered knowledge regarding satellite cell contributions to skeletal muscle hypertrophy; however, resistance exercise modalities are difficult to establish in animal research. To work around this, hypertrophy in rodents is often induced through the surgical removal of the gastrocnemius and soleus, placing a functional/mechanical overload on the remaining synergistic plantaris (Fry, Lee et al. 2014, Kirby et al. 2016, Fry et al. 2017). Mechanical overload induces satellite cell activation and fusion into myofibers, in addition to significantly increasing myofiber size (Timson 1990, Ishido et al. 2009). While mechanical overload in rodents does not offer direct comparison with resistance training in humans, it is useful to elucidate regulatory processes during myofiber hypertrophy, such as satellite cell activation. Figure 8.7a offers a time course of satellite cell proliferative capacity during mechanical overload. Dramatic increases in satellite cell abundance are evident in the first 2 weeks, and the population of satellite cells remains elevated through 8 weeks. These correlative findings would indicate a likely role for satellite cells in mediating muscle hypertrophy and reduced satellite cell function with age may contribute to diminished hypertrophic gains in aged muscle.

The genetic depletion of satellite cells from adult muscle allowed for the necessity of satellite cells during muscle hypertrophy to be directly assessed, and satellite cells were found to be dispensable during muscle growth in younger rodents (McCarthy et al. 2011). A similar pattern was observed in older animals, which also displayed minimal muscle growth as previously observed (Lee et al. 2015). Interestingly, despite negligible myofiber growth, myonuclear content increased (Lee et al. 2015), indicative of satellite cell fusion, and supportive of an active role for satellite cells even in the absence of clear muscle adaptation. While the presence of satellite cells did not appear needed during overload-induced muscle adaptation, mechanical overload induced a similar increase in satellite cell content in older rodent muscle compared with younger animals (Lee et al. 2015). Age-related declines in satellite cell abundance may be responsive to increased mechanical load, which could then potentially restore regenerative capacity to aged muscle. Figure 8.7b shows satellite cell content in the plantaris at rest and following 14 days of mechanical overload. A similar increase is seen in both young and older animals, demonstrating the restorative potential of mechanical loading on satellite cell function in aging muscle. This provides enthusiasm that satellite cell function and proliferative capacity can be rescued and highlights the therapeutic potential of increased loading on the satellite cell activity.

(a)

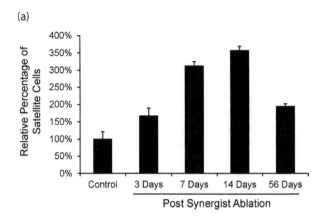

(b)

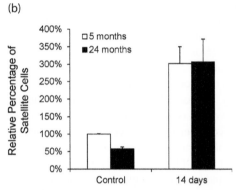

FIGURE 8.7 Satellite cell proliferation following overload in mice. Mechanical overload is a commonly used resistance exercise memetic in rodents to induce myofiber hypertrophy. The functional overload of rodent hind limb muscle induces significant satellite cell activation (a). Satellite cell abundance was assessed in the plantaris following the surgical removal of the synergist gastrocnemius and soleus, placing a functional overload on the remaining plantaris in a young male mice. Satellite cell content increases by several hundred percent within 14 days and remains elevated for up to 56 days following the surgery, which is accompanied by robust hypertrophy of myofibers within the plantaris. (b) A similar assessment of satellite cell abundance from the plantaris in young (5-month) and older (24-month) male mice following the functional overload. Satellite cell content increases to a similar degree in both young and older mice by 14 days. The results of these studies demonstrate that mechanical overload is capable of stimulating satellite cell proliferation and expansion to a similar extent in both young and older rodents, which may provide a means to promote improvement in regenerative capacity in aging skeletal muscle. ([a] Adapted from McCarthy, J.J. et al., *Development*, 138, 3657–3666, 2011; Fry, C.S. et al., *FASEB J.*, 28, 1654–1665, 2014a; [b] Adapted from McCarthy, J.J. et al., *Development*, 138, 3657–3666, 2011; Lee, J.D. et al., *J. Gerontol. A Biol. Sci. Med. Sci.*, 2015.)

A similar increase in satellite cell abundance following traditional resistance exercise in humans is seen. The acute effects of resistance exercise are often measured following the completion of a single unaccustomed bout of exercise engaging the leg extensor (i.e., quadriceps) muscles. Exercises often include the leg press, extension, and, in some instance, the squat or lunge. Biopsies of the *muscle vastus lateralis* are sampled in the hours and days following completion of acute exercise to study effects at the myofiber level. In the first few days following performance of a single bout of resistance exercise, robust increases in satellite cell number are observed (Dreyer et al. 2006, Walker et al. 2012). Figure 8.8 shows the summation of several recent clinical studies illustrating the differential proliferative ability of satellite cells in older and younger adults following acute resistance exercise. In younger subjects, the increase in satellite cells becomes apparent at 24 h and remains elevated through 72 h of recovery. Following the completion of a matched exercise bout,

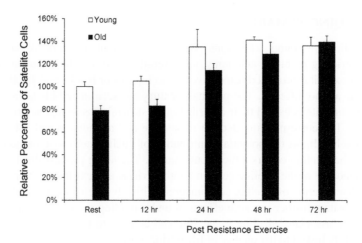

FIGURE 8.8 Delayed satellite cell response following resistance exercise in older adults. Resistance exercise is a potent stimulus for myofiber hypertrophy and concurrently activates satellite cells to induce fusion into existing myofibers. Several studies have explored the effects of an acute resistance exercise bout on satellite cell dynamics. A summary of several studies quantifying satellite cell abundance in the *muscle vastus lateralis* in young (<40 years) and old (>65 years) adult men following a single, comparable bout of resistance exercise involving the leg extensor muscles. The increase in satellite cell content is over 40% at 48 h following completion of exercise and remains elevated through 72 h postexercise in young men. The older adults display a slight delay in satellite cell activation and proliferation at 24 h postexercise but demonstrate a similar percent increase in satellite cell content at 48 and 72 h postexercise. Resistance exercise induces robust expansion of the satellite cell pool within the first few days after exercise. The results of these studies demonstrate that resistance exercise results in a similar increase in satellite cell content in both young and older adults, providing a means to address the age-related decline in satellite cell abundance and potentially confer an improved regenerative capacity to aging skeletal muscle. (Adapted from Bellamy, L.M. et al., *PLoS One*, 9, e109739, 2014; McKay, B.R. et al., *FASEB J.*, 26, 2509–2521, 2012; McKay, B.R. et al., *Am. J. Physiol. Cell Physiol.*, 304, C717–C728, 2013; Walker, D.K. et al., *Muscle Nerve*, 46, 51–59, 2012; Hyldahl, R.D. et al., *Front Physiol.*, 5, 485, 2014; Snijders, T. et al., *Age (Dordr)*, 36, 9699, 2014.)

older adults exhibit a delayed response, demonstrating reduced satellite cell proliferation until 72 h postexercise, when satellite cell content is approximately equivalent between younger and older adults. A number of factors may explain the delayed proliferation of satellite cells during postexercise recovery in aged skeletal muscle.

Acute inflammation following resistance exercise promotes satellite cell activation, and chronically low-level inflammation in aged muscle may mask the acute inflammatory response to exercise in older adults, delaying the activation of satellite cells. In addition to promoting muscle protein catabolism and sarcopenia, the chronic, low-level inflammatory state present in aged muscle likely obscures the acute inflammatory response to exercise and diminishes the sensitivity of satellite cells (McKay et al. 2013). Elevated expression of the muscle growth inhibitor myostatin within satellite cells may also interfere with activation and progression through the cell cycle in older adults. Following acute exercise, satellite cells in aged muscle display higher expression of myostatin, contributing to reduced activation and advancement through mitosis (McKay et al. 2012), likely delaying the postexercise increase in satellite cell content. Restricted activation of satellite cells in aged muscle following exercise is likely due to a combination of factors including inflammation, myostatin, or the transition to a senescent state and offer avenues for interventions to restore proper activation following exercise. Even with the delayed response, the proliferative capacity of satellite cells is still fairly robust in aged muscle, demonstrating the restorative potential of resistance exercise.

8.7   CONCLUDING REMARKS

Advancing age elicits profound deleterious effects on the maintenance of skeletal muscle mass and strength, restricting mobility, autonomy, and overall quality of life. Resident stem cells in a variety of tissues exhibit diminished responsiveness to injuries or insults and satellite cells present in skeletal muscle are no different. The past half century has seen incredible advancements in our understanding of satellite cells and their contribution to skeletal muscle regeneration, hypertrophy, and homeostasis. The corresponding decline in satellite cell abundance and activity that occurs during sarcopenia highlights the potential contribution of satellite cells to aging muscle deficits as well as presents the satellite cell as a therapeutic target. The aging process drastically affects the function and phenotype of the satellite cell, and both intrinsic and extrinsic factors account for the age-associated deterioration of satellite cell function. With advancements in health care, increases in longevity necessitate further research efforts in understanding the mechanisms contributing to sarcopenia and the associated decline in satellite cell activity to restore function and reparative capacity, both tightly linked to the quality of life during aging.

REFERENCES

Allbrook, D. B., M. F. Han, and A. E. Hellmuth. 1971. "Population of muscle satellite cells in relation to age and mitotic activity." *Pathology* 3 (3):223–43.

Andres, V., and K. Walsh. 1996. "Myogenin expression, cell cycle withdrawal, and phenotypic differentiation are temporally separable events that precede cell fusion upon myogenesis." *J Cell Biol* 132 (4):657–66.

Arentson-Lantz, E. J., K. L. English, D. Paddon-Jones, and C. S. Fry. 2016. "Fourteen days of bed rest induces a decline in satellite cell content and robust atrophy of skeletal muscle fibers in middle-aged adults." *J Appl Physiol* 120 (8):965–75. doi:10.1152/japplphysiol.00799.2015.

Ballak, S. B., R. T. Jaspers, L. Deldicque, S. Chalil, E. L. Peters, A. de Haan, and H. Degens. 2015. "Blunted hypertrophic response in old mouse muscle is associated with a lower satellite cell density and is not alleviated by resveratrol." *Exp Gerontol* 62:23–31. doi:10.1016/j.exger.2014.12.020.

Bellamy, L. M., S. Joanisse, A. Grubb, C. J. Mitchell, B. R. McKay, S. M. Phillips, S. Baker, and G. Parise. 2014. "The acute satellite cell response and skeletal muscle hypertrophy following resistance training." *PLoS One* 9 (10):e109739. doi:10.1371/journal.pone.0109739.

Bentzinger, C. F., Y. X. Wang, J. von Maltzahn, V. D. Soleimani, H. Yin, and M. A. Rudnicki. 2013. "Fibronectin regulates Wnt7a signaling and satellite cell expansion." *Cell Stem Cell* 12 (1):75–87. doi:10.1016/j.stem.2012.09.015.

Bigot, A., W. J. Duddy, Z. G. Ouandaogo, E. Negroni, V. Mariot, S. Ghimbovschi, B. Harmon et al. 2015. "Age-associated methylation suppresses SPRY1, leading to a failure of re-quiescence and loss of the reserve stem cell pool in elderly muscle." *Cell Rep* 13 (6):1172–82. doi:10.1016/j.celrep.2015.09.067.

Biressi, S., E. H. Miyabara, S. D. Gopinath, P. M. Carlig, and T. A. Rando. 2014. "A Wnt-TGFbeta2 axis induces a fibrogenic program in muscle stem cells from dystrophic mice." *Sci Transl Med* 6 (267):267ra176. doi:10.1126/scitranslmed.3008411.

Blough, E. R., and J. K. Linderman. 2000. "Lack of skeletal muscle hypertrophy in very aged male Fischer 344 x Brown Norway rats." *J Appl Physiol* 88 (4):1265–70.

Brack, A. S., and T. A. Rando. 2007. "Intrinsic changes and extrinsic influences of myogenic stem cell function during aging." *Stem Cell Rev* 3 (3):226–37.

Brack, A. S., M. J. Conboy, S. Roy, M. Lee, C. J. Kuo, C. Keller, and T. A. Rando. 2007. "Increased Wnt signaling during aging alters muscle stem cell fate and increases fibrosis." *Science* 317 (5839):807–10. doi:10.1126/science.1144090.

Budai, Z., L. Balogh, and Z. Sarang. 2018. "Altered gene expression of muscle satellite cells contributes to age-related sarcopenia in mice." *Curr Aging Sci*. doi:10.2174/1874609811666180925104241.

Campisi, J., and L. Robert. 2014. "Cell senescence: Role in aging and age-related diseases." *Interdiscip Top Gerontol* 39:45–61. doi:10.1159/000358899.

Capers, C. R. 1960. "Multinucleation of skeletal muscle *in vitro*." *J Biophys Biochem Cytol* 7:559–66.

Carlson, B. M. 1973. "The regeneration of skeletal muscle: A review." *Am J Anat* 137 (2):119–49. doi:10.1002/aja.1001370202.

Carlson, B. M., and J. A. Faulkner. 1989. "Muscle transplantation between young and old rats: Age of host determines recovery." *Am J Physiol* 256 (6 Pt 1):C1262–6.

Carlson, M. E., M. Hsu, and I. M. Conboy. 2008. "Imbalance between pSmad3 and Notch induces CDK inhibitors in old muscle stem cells." *Nature* 454 (7203):528–32. doi:10.1038/nature07034.

Chakkalakal, J. V., J. Christensen, W. Xiang, M. T. Tierney, F. S. Boscolo, A. Sacco, and A. S. Brack. 2014. "Early forming label-retaining muscle stem cells require p27kip1 for maintenance of the primitive state." *Development* 141 (8):1649–59. doi:10.1242/dev.100842.

Collins, C. A., I. Olsen, P. S. Zammit, L. Heslop, A. Petrie, T. A. Partridge, and J. E. Morgan. 2005. "Stem cell function, self-renewal, and behavioral heterogeneity of cells from the adult muscle satellite cell niche." *Cell* 122 (2):289–301. doi:10.1016/j.cell.2005.05.010.

Conboy, I. M., M. J. Conboy, A. J. Wagers, E. R. Girma, I. L. Weissman, and T. A. Rando. 2005. "Rejuvenation of aged progenitor cells by exposure to a young systemic environment." *Nature* 433 (7027):760–4. doi:10.1038/nature03260.

Cooper, W. G., and I. R. Konigsberg. 1961. "Dynamics of myogenesis *in vitro.*" *Anat Rec* 140:195–205.

Cornelison, D. D., and B. J. Wold. 1997. "Single-cell analysis of regulatory gene expression in quiescent and activated mouse skeletal muscle satellite cells." *Dev Biol* 191 (2):270–83. doi:10.1006/dbio.1997.8721.

Cosgrove, B. D., P. M. Gilbert, E. Porpiglia, F. Mourkioti, S. P. Lee, S. Y. Corbel, M. E. Llewellyn, S. L. Delp, and H. M. Blau. 2014. "Rejuvenation of the muscle stem cell population restores strength to injured aged muscles." *Nat Med* 20 (3):255–64. doi:10.1038/nm.3464.

Dreyer, H. C., C. E. Blanco, F. R. Sattler, E. T. Schroeder, and R. A. Wiswell. 2006. "Satellite cell numbers in young and older men 24 hours after eccentric exercise." *Muscle Nerve* 33 (2):242–53. doi:10.1002/mus.20461.

Duprey, P., and C. Lesens. 1994. "Control of skeletal muscle-specific transcription: Involvement of paired homeodomain and MADS domain transcription factors." *Int J Dev Biol* 38 (4):591–604.

Egerman, M. A., S. M. Cadena, J. A. Gilbert, A. Meyer, H. N. Nelson, S. E. Swalley, C. Mallozzi et al. 2015. "GDF11 Increases with age and inhibits skeletal muscle regeneration." *Cell Metab* 22 (1):164–74. doi:10.1016/j.cmet.2015.05.010.

Enesco, M., and D. Puddy. 1964. "Increase in the number of nuclei and weight in skeletal muscle of rats of various ages." *Am J Anat* 114:235–44. doi:10.1002/aja.1001140204.

Evans, W. 1997. "Functional and metabolic consequences of sarcopenia." *J Nutr* 127 (5 Suppl):998s–1003s.

Fry, C. S., T. J. Kirby, K. Kosmac, J. J. McCarthy, and C. A. Peterson. 2017. "Myogenic progenitor cells control extracellular matrix production by fibroblasts during skeletal muscle hypertrophy." *Cell Stem Cell* 20 (1):56–69. doi:10.1016/j.stem.2016.09.010.

Fry, C. S., J. D. Lee, J. R. Jackson, T. J. Kirby, S. A. Stasko, H. L. Liu, E. E. Dupont-Versteegden, J. J. McCarthy, and C. A. Peterson. 2014. "Regulation of the muscle fiber microenvironment by activated satellite cells during hypertrophy." *FASEB J* 28 (4):1654–65. doi:10.1096/fj.13-239426.

Fry, C. S., J. D. Lee, J. Mula, T. J. Kirby, J. R. Jackson, F. Liu, L. Yang et al. 2015. "Inducible depletion of satellite cells in adult, sedentary mice impairs muscle regenerative capacity without affecting sarcopenia." *Nat Med* 21 (1):76–80. doi:10.1038/nm.3710.

Fry, C. S., B. Noehren, J. Mula, M. F. Ubele, P. M. Westgate, P. A. Kern, and C. A. Peterson. 2014. "Fibre type-specific satellite cell response to aerobic training in sedentary adults." *J Physiol* 592 (Pt 12):2625–35. doi:10.1113/jphysiol.2014.271288.

Fukada, S., A. Uezumi, M. Ikemoto, S. Masuda, M. Segawa, N. Tanimura, H. Yamamoto, Y. Miyagoe-Suzuki, and S. Takeda. 2007. "Molecular signature of quiescent satellite cells in adult skeletal muscle." *Stem Cells* 25 (10):2448–59. doi:10.1634/stemcells.2007-0019.

Garcia-Prat, L., M. Martinez-Vicente, E. Perdiguero, L. Ortet, J. Rodriguez-Ubreva, E. Rebollo, V. Ruiz-Bonilla et al. 2016. "Autophagy maintains stemness by preventing senescence." *Nature* 529 (7584):37–42. doi:10.1038/nature16187.

Gilbert, P. M., K. L. Havenstrite, K. E. Magnusson, A. Sacco, N. A. Leonardi, P. Kraft, N. K. Nguyen, S. Thrun, M. P. Lutolf, and H. M. Blau. 2010. "Substrate elasticity regulates skeletal muscle stem cell self-renewal in culture." *Science* 329 (5995):1078–81. doi:10.1126/science.1191035.

Gutmann, E., and B. M. Carlson. 1976. "Regeneration and transplantation of muscles in old rats and between young and old rats." *Life Sci* 18 (1):109–14.

Halevy, O., Y. Piestun, M. Z. Allouh, B. W. Rosser, Y. Rinkevich, R. Reshef, I. Rozenboim, M. Wleklinski-Lee, and Z. Yablonka-Reuveni. 2004. "Pattern of Pax7 expression during myogenesis in the posthatch chicken establishes a model for satellite cell differentiation and renewal." *Dev Dyn* 231 (3):489–502. doi:10.1002/dvdy.20151.

Hyldahl, R. D., T. Olson, T. Welling, L. Groscost, and A. C. Parcell. 2014. "Satellite cell activity is differentially affected by contraction mode in human muscle following a work-matched bout of exercise." *Front Physiol* 5:485. doi:10.3389/fphys.2014.00485.

Illa, I., M. Leon-Monzon, and M. C. Dalakas. 1992. "Regenerating and denervated human muscle fibers and satellite cells express neural cell adhesion molecule recognized by monoclonal antibodies to natural killer cells." *Ann Neurol* 31 (1):46–52. doi:10.1002/ana.410310109.

Ishido, M., M. Uda, N. Kasuga, and M. Masuhara. 2009. "The expression patterns of Pax7 in satellite cells during overload-induced rat adult skeletal muscle hypertrophy." *Acta Physiol (Oxf)* 195 (4):459–69. doi:10.1111/j.1748-1716.2008.01905.x.

Ishikawa, H. 1966. "Electron microscopic observations of satellite cells with special reference to the development of mammalian skeletal muscles." *Z Anat Entwicklungsgesch* 125 (1):43–63.

Janssen, I., S. B. Heymsfield, Z. M. Wang, and R. Ross. 2000. "Skeletal muscle mass and distribution in 468 men and women aged 18–88 yr." *J Appl Physiol (1985)* 89 (1):81–8.

Kadi, F., N. Charifi, C. Denis, and J. Lexell. 2004. "Satellite cells and myonuclei in young and elderly women and men." *Muscle Nerve* 29 (1):120–27. doi:10.1002/mus.10510.

Keefe, A. C., J. A. Lawson, S. D. Flygare, Z. D. Fox, M. P. Colasanto, S. J. Mathew, M. Yandell, and G. Kardon. 2015. "Muscle stem cells contribute to myofibres in sedentary adult mice." *Nat Commun* 6:7087. doi:10.1038/ncomms8087.

Kirby, T. J., J. J. McCarthy, C. A. Peterson, and C. S. Fry. 2016. "Synergist ablation as a rodent model to study satellite cell dynamics in adult skeletal muscle." *Methods Mol Biol* 1460:43–52. doi:10.1007/978-1-4939-3810-0_4.

Konigsberg, U. R., B. H. Lipton, and I. R. Konigsberg. 1975. "The regenerative response of single mature muscle fibers isolated in vitro." *Dev Biol* 45 (2):260–75.

Kosek, D. J., J. S. Kim, J. K. Petrella, J. M. Cross, and M. M. Bamman. 2006. "Efficacy of 3 days/wk resistance training on myofiber hypertrophy and myogenic mechanisms in young vs. older adults." *J Appl Physiol* 101 (2):531–44. doi:10.1152/japplphysiol.01474.2005.

Kuang, S., S. B. Charge, P. Seale, M. Huh, and M. A. Rudnicki. 2006. "Distinct roles for Pax7 and Pax3 in adult regenerative myogenesis." *J Cell Biol* 172 (1):103–13. doi:10.1083/jcb.200508001.

Kuang, S., K. Kuroda, F. Le Grand, and M. A. Rudnicki. 2007. "Asymmetric self-renewal and commitment of satellite stem cells in muscle." *Cell* 129 (5):999–1010. doi:10.1016/j.cell.2007.03.044.

Lee, J. D., C. S. Fry, J. Mula, T. J. Kirby, J. R. Jackson, F. Liu, L. Yang et al. 2015. "Aged muscle demonstrates fiber-type adaptations in response to mechanical overload, in the absence of myofiber hypertrophy, independent of satellite cell abundance." *J Gerontol A Biol Sci Med Sci.* doi:10.1093/gerona/glv033.

Lepper, C., S. J. Conway, and C. M. Fan. 2009. "Adult satellite cells and embryonic muscle progenitors have distinct genetic requirements." *Nature* 460 (7255):627–31. doi:10.1038/nature08209.

Liu, L., G. W. Charville, T. H. Cheung, B. Yoo, P. J. Santos, M. Schroeder, and T. A. Rando. 2018. "Impaired notch signaling leads to a decrease in p53 activity and mitotic catastrophe in aged muscle stem cells." *Cell Stem Cell.* doi:10.1016/j.stem.2018.08.019.

Mackey, A. L., M. Kjaer, N. Charifi, J. Henriksson, J. Bojsen-Moller, L. Holm, and F. Kadi. 2009. "Assessment of satellite cell number and activity status in human skeletal muscle biopsies." *Muscle Nerve* 40 (3):455–65. doi:10.1002/mus.21369.

Mauro, A. 1961. "Satellite cell of skeletal muscle fibers." *J Biophys Biochem Cytol* 9:493–5.

McCarthy, J. J., J. Mula, M. Miyazaki, R. Erfani, K. Garrison, A. B. Farooqui, R. Srikuea et al. 2011. "Effective fiber hypertrophy in satellite cell-depleted skeletal muscle." *Development* 138 (17):3657–666. doi:10.1242/dev.068858.

McKay, B. R., D. I. Ogborn, J. M. Baker, K. G. Toth, M. A. Tarnopolsky, and G. Parise. 2013. "Elevated SOCS3 and altered IL-6 signaling is associated with age-related human muscle stem cell dysfunction." *Am J Physiol Cell Physiol* 304 (8):C717–28. doi:10.1152/ajpcell.00305.2012.

McKay, B. R., D. I. Ogborn, L. M. Bellamy, M. A. Tarnopolsky, and G. Parise. 2012. "Myostatin is associated with age-related human muscle stem cell dysfunction." *FASEB J* 26 (6):2509–21. doi:10.1096/fj.11-198663.

McLoon, L. K., J. Rowe, J. Wirtschafter, and K. M. McCormick. 2004. "Continuous myofiber remodeling in uninjured extraocular myofibers: Myonuclear turnover and evidence for apoptosis." *Muscle Nerve* 29 (5):707–15. doi:10.1002/mus.20012.

Moss, F. P., and C. P. Leblond. 1971. "Satellite cells as the source of nuclei in muscles of growing rats." *Anat Rec* 170 (4):421–35. doi:10.1002/ar.1091700405.

Murphy, M. M., J. A. Lawson, S. J. Mathew, D. A. Hutcheson, and G. Kardon. 2011. "Satellite cells, connective tissue fibroblasts and their interactions are crucial for muscle regeneration." *Development* 138 (17):3625–637. doi:10.1242/dev.064162.

Oertel, G. 1988. "Morphometric analysis of normal skeletal muscles in infancy, childhood and adolescence: An autopsy study." *J Neurol Sci* 88 (1–3):303–13.

Olguin, H. C., and B. B. Olwin. 2004. "Pax-7 up-regulation inhibits myogenesis and cell cycle progression in satellite cells: A potential mechanism for self-renewal." *Dev Biol* 275 (2):375–88. doi:10.1016/j. ydbio.2004.08.015.

Olguin, H. C., Z. Yang, S. J. Tapscott, and B. B. Olwin. 2007. "Reciprocal inhibition between Pax7 and muscle regulatory factors modulates myogenic cell fate determination." *J Cell Biol* 177 (5):769–79. doi:10.1083/ jcb.200608122.

Pawlikowski, B., C. Pulliam, N. D. Betta, G. Kardon, and B. B. Olwin. 2015. "Pervasive satellite cell contribution to uninjured adult muscle fibers." *Skelet Muscle* 5:42. doi:10.1186/s13395-015-0067-1.

Petrella, J. K., J.-S. Kim, D. L. Mayhew, J. M. Cross, and M. M. Bamman. 2008. "Potent myofiber hypertrophy during resistance training in humans is associated with satellite cell-mediated myonuclear addition: A cluster analysis." *J Appl Physiol* 104 (6):1736–42. doi:10.1152/japplphysiol.01215.2007.

Renault, V., L. E. Thornell, P. O. Eriksson, G. Butler-Browne, and V. Mouly. 2002. "Regenerative potential of human skeletal muscle during aging." *Aging Cell* 1 (2):132–9.

Roth, S. M., G. F. Martel, F. M. Ivey, J. T. Lemmer, E. J. Metter, B. F. Hurley, and M. A. Rogers. 2000. "Skeletal muscle satellite cell populations in healthy young and older men and women." *Anat Rec* 260 (4):351–8.

Schultz, E. 1976. "Fine structure of satellite cells in growing skeletal muscle." *Am J Anat* 147 (1):49–70. doi:10.1002/aja.1001470105.

Schultz, E., M. C. Gibson, and T. Champion. 1978. "Satellite cells are mitotically quiescent in mature mouse muscle: An EM and radioautographic study." *J Exp Zool* 206 (3):451–6. doi:10.1002/jez.1402060314.

Seale, P., J. Ishibashi, A. Scime, and M. A. Rudnicki. 2004. "Pax7 is necessary and sufficient for the myogenic specification of CD45+:Sca1+ stem cells from injured muscle." *PLoS Biol* 2 (5):E130. doi:10.1371/journal.pbio.0020130.

Seale, P., L. A. Sabourin, A. Girgis-Gabardo, A. Mansouri, P. Gruss, and M. A. Rudnicki. 2000. "Pax7 is required for the specification of myogenic satellite cells." *Cell* 102 (6):777–86.

Shefer, G., D. P. Van de Mark, J. B. Richardson, and Z. Yablonka-Reuveni. 2006. "Satellite-cell pool size does matter: Defining the myogenic potency of aging skeletal muscle." *Dev Biol* 294 (1):50–66. doi:10.1016/j. ydbio.2006.02.022.

Shinin, V., B. Gayraud-Morel, D. Gomes, and S. Tajbakhsh. 2006. "Asymmetric division and cosegregation of template DNA strands in adult muscle satellite cells." *Nat Cell Biol* 8 (7):677–87. doi:10.1038/ncb1425.

Sinha, M., Y. C. Jang, J. Oh, D. Khong, E. Y. Wu, R. Manohar, C. Miller et al. 2014. "Restoring systemic GDF11 levels reverses age-related dysfunction in mouse skeletal muscle." *Science* 344 (6184):649–52. doi:10.1126/science.1251152.

Smith, C. K., 2nd, M. J. Janney, and R. E. Allen. 1994. "Temporal expression of myogenic regulatory genes during activation, proliferation, and differentiation of rat skeletal muscle satellite cells." *J Cell Physiol* 159 (2):379–85. doi:10.1002/jcp.1041590222.

Snijders, T., L. B. Verdijk, J. S. Smeets, B. R. McKay, J. M. Senden, F. Hartgens, G. Parise, P. Greenhaff, and L. J. van Loon. 2014. "The skeletal muscle satellite cell response to a single bout of resistance-type exercise is delayed with aging in men." *Age* 36 (4):9699. doi:10.1007/s11357-014-9699-z.

Soleimani, V. D., H. Yin, A. Jahani-Asl, H. Ming, C. E. Kockx, W. F. van Ijcken, F. Grosveld, and M. A. Rudnicki. 2012. "Snail regulates MyoD binding-site occupancy to direct enhancer switching and differentiation-specific transcription in myogenesis." *Mol Cell* 47 (3):457–68. doi:10.1016/j.molcel.2012.05.046.

Sousa-Victor, P., E. Perdiguero, and P. Munoz-Canoves. 2014. "Geroconversion of aged muscle stem cells under regenerative pressure." *Cell Cycle* 13 (20):3183–90. doi:10.4161/15384101.2014.965072.

Sousa-Victor, P., S. Gutarra, L. Garcia-Prat, J. Rodriguez-Ubreva, L. Ortet, V. Ruiz-Bonilla, M. Jardi et al. 2014. "Geriatric muscle stem cells switch reversible quiescence into senescence." *Nature* 506 (7488):316–21. doi:10.1038/nature13013.

Timson, B. F. 1990. "Evaluation of animal models for the study of exercise-induced muscle enlargement." *J Appl Physiol* 69 (6):1935–45.

Urciuolo, A., M. Quarta, V. Morbidoni, F. Gattazzo, S. Molon, P. Grumati, F. Montemurro et al. 2013. "Collagen VI regulates satellite cell self-renewal and muscle regeneration." *Nat Commun* 4:1964. doi:10.1038/ncomms2964.

Venable, J. H. 1966. "Morphology of the cells of normal, testosterone-deprived and testosterone-stimulated levator ani muscles." *Am J Anat* 119 (2):271–301. doi:10.1002/aja.1001190206.

Verdijk, L. B., T. Snijders, M. Drost, T. Delhaas, F. Kadi, and L. J. van Loon. 2014. "Satellite cells in human skeletal muscle; from birth to old age." *Age* 36 (2):545–7. doi:10.1007/s11357-013-9583-2.

Walker, D. K., C. S. Fry, M. J. Drummond, J. M. Dickinson, K. L. Timmerman, D. M. Gundermann, K. Jennings, E. Volpi, and B. B. Rasmussen. 2012. "PAX7+ satellite cells in young and older adults following resistance exercise." *Muscle Nerve* 46 (1):51–9. doi:10.1002/mus.23266.

Yablonka-Reuveni, Z., M. A. Rudnicki, A. J. Rivera, M. Primig, J. E. Anderson, and P. Natanson. 1999. "The transition from proliferation to differentiation is delayed in satellite cells from mice lacking MyoD." *Dev Biol* 210 (2):440–55. doi:10.1006/dbio.1999.9284.

Zammit, P. S., J. P. Golding, Y. Nagata, V. Hudon, T. A. Partridge, and J. R. Beauchamp. 2004. "Muscle satellite cells adopt divergent fates: A mechanism for self-renewal?" *J Cell Biol* 166 (3):347–57. doi:10.1083/jcb.200312007.

Zwetsloot, K. A., T. E. Childs, L. T. Gilpin, and F. W. Booth. 2013. "Non-passaged muscle precursor cells from 32-month old rat skeletal muscle have delayed proliferation and differentiation." *Cell Prolif* 46 (1):45–57. doi:10.1111/cpr.12007.

9 Muscle Stem Cell Microenvironment in Sarcopenia

Neia Naldaiz-Gastesi and Ander Izeta

CONTENTS

9.1 INTRODUCTION

Stem cells represent a fascinating group of cells that have the property of self-renewal and are capable of generating diverse progeny (multipotency). In adults, these cells are of great importance in tissue homeostasis and repair, and for this reason, they are strictly controlled by sophisticated chemical and physical cues that arise from the surrounding environment or "niche" (Das and Zouani 2014). There are overlapping but distinct levels of interaction to this niche: stem cells may respond to systemic signaling, to tissue-specific signals, and to modifications to their immediate microenvironment (Drummond-Barbosa 2008, Morrison and Spradling 2008). Over the lifetime of the individual, spatially and temporally controlled signals are sent to modulate cell behavior affecting stemness and stem cell fate (survival, migration, proliferation, and differentiation processes) (Das and Zouani 2014). Cellular functions may eventually be modified or impaired due to the effects of aging as any small change that happens in the body would at some point reach the effector cells in their protected microenvironments. Thus, it would be useful to understand how the stem cell niche adapts to the milieu and use this knowledge to identify the age-associated changes that affect elderly people and target them using behavioral or pharmacological modulators.

This is particularly important in the case of the satellite cells as the skeletal muscle represents a relevant mass of tissue that carries out important functions, and its failings underlie body function decline with severe consequences for the quality of life and life expectancy of the elderly (Rosenberg 1997). However, satellite cells are not easily modulated because they are not isolated and accessible stem cells. Rather, they are well framed within the complex regulation and hierarchical conformation of the muscle. For these reasons, when studying satellite cells environmental regulation, it is important to bear in mind its complexity due to the coordination of all its modifiers. These include the different connective tissue layers that assemble the muscle structure, the presence or absence of hormones and nitric oxide (NO), and the presence or absence of nonmuscle cell types, for instance, in muscle denervation (Wosczyna and Rando 2018). Furthermore, the extracellular matrix (ECM) components, cell membrane ligands and receptors, and the mixture of soluble cytokines and growth factors that act on satellite cells may be modulated as well by mechanical forces (Lutolf, Gilbert, and Blau 2009, Morrissey et al. 2016, Smith, Cho, and Discher 2017, Yin, Price, and Rudnicki 2013). All these different dimensions of the muscle environment are the stage and also the baton of a very well-orchestrated muscle function. This chapter will summarize the complex interactions that are established between the muscle stem cells and their milieu and thus will complement another chapter in the book that concentrates on satellite stem cell aging (Fry 2019).

9.2 SKELETAL MUSCLE STRUCTURE

There are three types of muscles: cardiac (striated and involuntary), smooth (nonstriated and involuntary), and skeletal (striated and voluntary). The skeletal muscle is a form of striated muscle in which most of the muscles are linked to bones by tendons. Skeletal muscles make up approximately 40% of the total body weight (Frontera and Ochala 2015). Each of these muscles is constituted by muscle fibers wrapped up by a complex network of blood vessels, nerves, and connective tissues.

From whole muscle to single fibers, skeletal muscles have a well-known structural organization (Figure 9.1). All of the skeletal muscles are composed of bundles of cylindrical, elongated, and multinucleated fibers separated in fascicles. These fibers are in turn composed of myofibrils that are repeated sequences of sarcomeres, a single repeat of the banding pattern formed by organized overlapping thick myosin filaments and thin actin filaments. Muscle fibers can be classified under different criteria, but the most common classification differentiates between type I fibers (slow, oxidative, fatigue resistant), type IIA (fast, oxidative, intermediate metabolic properties), and type IIx/d (fastest, glycolytic, fatigable) (Frontera and Ochala 2015).

In between fibers and fascicles, the extracellular space is filled up with proteins and polysaccharides that form the muscle ECM. This space can be completed in myriad different ways, combining diverse macromolecules, and varying their organization, resulting in a high diversity of extracellular matrices adapted to the functional requirements of each muscle group or body region (Thomas, Engler, and Meyer 2015). The local components of the ECM and connective tissue cells constitute the scaffold of the organ and modulate each other in both directions. Muscles are ingrained in the dense connective tissue or fascia of the locomotor system via tendons and ligaments. Moreover, muscles have intramuscular connective tissue that ensheaths its tiered structure as follows: an individual muscle is covered first with a layer called epimysium, which consists mainly of type I collagen and fibroblasts, pericytes, adipocytes, blood vessels, and nerves. Then, the perimysium (which contains type I and type III collagen) wraps each fascicle of fibers and, finally, the endomysium (which contains type III and IV collagen) surrounds muscle fibers.

The endomysium associates with the basement membrane and with the membrane of muscle fibers, the sarcolemma. We describe both membranes in more detail below as they are an essential part of the satellite cell niche. The sarcolemma has a similar biochemical composition and function to the cellular plasma membrane, and it is permeable to water through aquaporin channels. However, it differs from other membranes in its contractile properties.

Excitation and contraction coupling proceeds through interactions between sarcolemma and sarcoplasmic reticulum (SR) proteins. Briefly, brain signals reach neuromuscular junctions, where the neurotransmitter acetylcholine is released and binds to its receptors in the fiber, resulting in a set of electrical changes along the sarcolemma. To maximize depolarization of the membrane and accelerate transportation of the action potential, the sarcolemma forms tunnel-like transverse invaginations (T-tubules) that pass across the muscle cells from side to side. T-tubules are in direct contact with the SR because each tube is surrounded by two SR terminal cisternae, forming triads. When the impulse reaches T-tubules, it activates dihydropyridine receptors that interact with ryanodine receptors from the SR, opening them. Consequently, SR calcium ions are pumped out into the sarcoplasm where they interact with contractile proteins generating power and movement. Once the action potential has ceased and the muscle is about to relax, the sarco/endoplasmic reticulum calcium-ATPase (SERCA) returns calcium ions to the SR, reestablishing basal conditions.

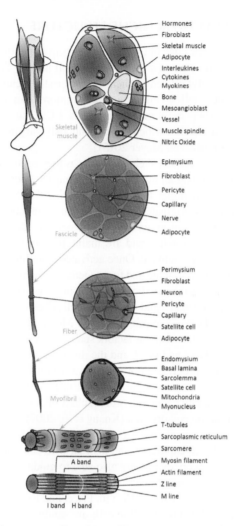

FIGURE 9.1 Skeletal muscle structure. From the macroscopic (top) to the microscopic (bottom) level, the components that constitute skeletal muscle environment are outlined. These include different cell types, their spatial and structural organization through the multiple connective tissue layers, the satellite cell niche, the myonuclei and other specific organelles, and the sarcomeres, the basic contractile unit.

As almost every other tissue, skeletal muscles have an abundant supply of nerves and blood vessels that align with the connective tissue components within a neurovascular bundle. The epimysium includes large caliber blood vessels and nerves and accommodates neuromuscular spindles. Generally, an artery and at least one vein accompany each incoming nerve to the perimysium, and, as they penetrate further into the tissue, they branch into smaller caliber venules and capillaries as well as thinner nerve terminals. As the circumference of each muscle fiber is quite plastic and variable, capillaries have a nonuniform distribution around them, but they are interconnected to homogeneously distribute the oxygen supply. As voluntary muscles, skeletal muscles are controlled by a motor neuron network that not only innervates muscle fibers forming neuromuscular junctions but also contains sensory components. The combination of a single motor neuron attached to the muscle fibers by its branches constitutes a motor unit. A hundred of them control skeletal muscles. In contrast, lymphatic capillaries extend only to the perimysium and epimysium sheaths and then converge in larger veins, thus providing lymphatic drainage to the muscles (Brooks 2003, Frontera and Ochala 2015, Turrina, Martinez-Gonzalez, and Stecco 2013, Alberts et al. 2002, Korthuis 2011, Standring 2016).

9.3   SATELLITE CELLS AND THEIR NICHE STRUCTURE

Muscle stem cells are located on the edge of the muscle fibers in a specific structure delimited by the basal lamina and the sarcolemma, known as "the satellite cell niche." The nature of these cells and their definition itself are essentially related to their characteristic location, as originally described by Mauro (1961). However, the satellite cell niche is much more than a structural compartment; it encompasses the dynamic interplay between satellite cells, ECM, surrounding cells, vascular and neural networks, and diverse molecules, which are continuously varying and adapting to the necessities of the tissue (Thomas, Engler, and Meyer 2015).

Satellite cells are the adult stem cells of skeletal muscle as they promote muscle growth and repair upon injury. It is estimated that among all nuclei present in muscle fibers, 3%–11% correspond to satellite cells (Schmalbruch and Hellhammer 1976, Schmalbruch 2006, Schmalbruch and Hellhammer 1977). Myonuclei in muscle fibers originate from the fusion of myoblasts and become mitotically inactive when the fiber matures during its formation. In contrast, satellite cells are mononucleated cells, do not express myofilament proteins, and are maintained in quiescence under physiological conditions until they are activated by damage. Once activated, satellite cells proliferate to produce a pool of mononucleated cells that differentiate and fuse together to form new multinucleated myofibers. Quiescent satellite cells express several specific membrane proteins, such as integrin $\alpha7$ and $\beta1$, dystroglycan, M-cadherin, CD34, syndecan 3 and 4, vascular cell adhesion molecule 1 (VCAM-1), neural cell adhesion molecule (N-CAM), tyrosine-protein kinase Met, and C-X-C motif chemokine receptor 4 (CXCR4), and also specific transcription factors such as Pax7. During regeneration, a number of transcription factors, such as Pax3 and Pax7, and the myogenic regulatory factors (Myf5, MyoD1, Myogenin, and Mrf4) dictate the progression of cells toward proliferation, differentiation, and self-renewal (Naldaiz-Gastesi et al. 2016, Garcia-Parra et al. 2014). While most newly activated satellite cells differentiate after the proliferative phase, a minor subset of cells renews the original population of muscle stem cells by asymmetric cell division (Kuang et al. 2007). Beyond regeneration of muscle fibers, it is necessary to restore the vessels and nerves that have been damaged upon injury. Generally, the satellite cells are positioned near capillaries and interact with endothelial cells in order to coordinate the angio-myogenic process properly and are also able to guide new axons through Semaphorin 3A secretion, so that the remaining motor neurons can re-innervate nearby fibers after a short-term denervation (Christov et al. 2007, Standring 2016, Yin, Price, and Rudnicki 2013).

It has already been mentioned that the satellite cells are sandwiched between the two-cell membranes, the basal lamina, and the sarcolemma. The endomysium or muscle basement membrane is composed of reticular lamina and basal lamina. Basal lamina is in direct contact with the satellite cells, which express "outside-in" proteins such as vinculin to transmit information to the cell and integrins to link the intracellular space to the ECM. The basal lamina is composed of a network of ECM proteins including collagen type IV and VI, laminin-2, nidogen, perlecan, decorin, biglycan, entactin, fibronectin, and some other proteoglycans and glycoproteins. Proteoglycans store growth factors or myogenic inhibitors such as hepatocyte growth factor (HGF), basic fibroblast growth factor (bFGF), insulin-like growth factor (IGF), Wnt, and transforming growth factor-β (TGF-β) and thus serve as local reservoirs for these. The stored factors may become active by the proteolytic action of matrix metalloproteinases (MMPs) and others, ensuring a rapid response to regulate satellite cell behavior. Moreover, MMPs can degrade other ECM components, and the degraded products themselves become important role-modulating signals for satellite cells. The different protein combinations and the diverse post-translational modifications result in ECM stiffness variations that again modulate satellite cell function. Further, the basal lamina is also linked to the muscle fiber thus stabilizing the niche structure. Laminin from the basal lamina binds to the transmembrane dystroglycan complex, which in turn is connected to dystrophin and the actin cytoskeleton of the fiber. On the apical side, satellite cells have a higher expression of M-cadherin in contact with the sarcolemma, a lipid bilayer plasma membrane covered by an outer thin layer of polysaccharide material, the glycocalyx. This entire protein framework establishes a mechanical and force-driven signaling that will later translate into chemical cues inside the cell (Gopinath and Rando 2008, MacIntosh, Gardiner, and McComas 2006, Thomas, Engler, and Meyer 2015, Yin, Price, and Rudnicki 2013).

Apart from the physical and structural properties of the satellite cell niche, cytokines, growth factors, and other soluble molecules secreted by surrounding cells modulate satellite cell behavior during regeneration and contribute to the ECM maintenance (Figure 9.2):

1. Two distinct Wnt signals are activated through Frizzled receptors and low-density lipo-protein receptor-related proteins LRP5 and LRP6. The Wnt/β-catenin pathway is involved in cell adhesion and morphology and promotes proliferation, differentiation, and cellular plasticity. In contrast, the Wnt/PCP (planar cell polarity) pathway is related to cell polarity and asymmetric divisions.

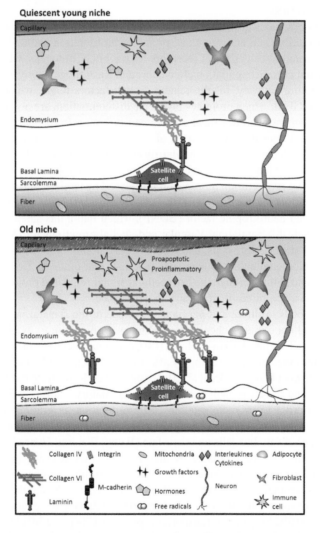

FIGURE 9.2 Satellite cell activation modulators. Satellite cell-specific surface markers are indicated on the cell membrane, as well as the myogenic transcription factors and the presence of miRNAs in the nuclei. Moreover, diverse elements that regulate satellite cell activation such as key signaling pathways, major ECM components, important group of interleukines/cytokines, hormones and growth factors, adjacent cell types, and the mitochondrial metabolism are summarized.

2. In the adult, canonical Notch signaling pathway is critical for sustaining quiescent satellite cells. In these cells, Notch transmembrane receptors (Notch 1, 2, and 3) are cleaved to generate an active form of Notch (NICD) when Delta and Jagged family of ligands bind to them. NICD interacts with downstream effector proteins, such as recombining binding protein-Jκ (RBPJκ), to regulate the transcription of Notch target genes. Through this canonical pathway the expression of Pax7 is induced, and Hes and Hey families of transcription factors suppress the myogenic commitment by inhibiting MyoD1 expression. Moreover, during regeneration, downregulation of Notch is mandatory to allow myogenic cell lineage progression (Almada and Wagers 2016, Dumont, Wang, and Rudnicki 2015, Mourikis and Tajbakhsh 2014).

3. Sphingolipids are a big group of glycolipids with sphingoid backbone, important to control quiescence, proliferation, and proinflammatory and antiapoptotic functions. A whole chapter of this book is dedicated to nutrient sensing and the associated signaling (Graber and Rasmussen 2019).

4. Signaling factors such as growth factors accomplish important regulatory functions: HGF is critical for satellite cell activation and promotes proliferation and cell migration to the site of injury; satellite cells present receptors for the bFGF mitogens, which mediate pro-proliferation and anti-differentiation effects through Ras/mitogen-activated protein kinase (MAPK) signaling pathway. Insulin-related growth factors IGF-I, which has pleiotropic functions, and IGF-II, which promotes myogenic differentiation, are also of key relevance.

5. All of the aforementioned chemical signals can originate from interstitial cells, which secrete FGFs and contribute to deposits of collagen, laminin, fibronectin, among other ECM proteins (Malecova et al. 2018, Madaro et al. 2018). Myogenic cells themselves release ECM proteins and change integrin receptor isoforms and myoblasts secrete fibronectin to promote their own proliferation (fibronectin is replaced by laminin during myotube formation). Motor neurons release nerve growth factor (NGF), involved in muscle regeneration, and brain-derived neurotrophic factor (BDNF), which maintains satellite cells quiescent and prevents their differentiation. Vascular endothelial growth factor (VEGF) is released by endothelial cells, satellite cells, and myoblasts promoting both myogenesis and angiogenesis. Periendothelial cells produce angiopoietin-1 that binds to Tie-2 receptors of the satellite cells, and acting through the extracellular signal-regulated protein kinases 1 and 2 (ERK1/2) pathway promotes their reentry to quiescence.

6. Some immune cells infiltrate and permanently reside in the skeletal muscle, but they are mostly recruited upon injury, chemoattracted by pro-inflammatory cytokines secreted by satellite cells, such as interleukin-1 (IL-1), IL-6, and tumor necrosis factor alpha (TNF-α). The latter contributes to muscle aging by affecting sarcopenia and muscle cell fusion with aging muscle fibers (Wang, Welc, et al. 2018). Immune cells infiltrate the muscle to remove necrotic tissue and secrete growth factors, pro-inflammatory cytokines, ECM components, and MMPs in order to promote satellite cell proliferation, migration, and survival. As the immune cells themselves undergo age-associated changes, old immune cells influence both the number of satellite cells and their fate (Wang, Wehling-Henricks, et al. 2018).

7. Satellite cells can also be regulated by androgens as they express androgen receptors, and androgen administration induces muscle hypertrophy among other implications. Furthermore, thyroid hormones are crucial during myogenesis by regulating the metabolic rate and the genetic expression of several proteins such as sarco/endoplasmic reticulum Ca^{2+}-ATPase (SERCA), glucose transporter type 4 (GLUT4), and MyoD1 (Salvatore et al. 2014). Further, thyroid receptor alpha is involved in the maintenance of the stem cell pool, as well as the progressive loss of the SC pool with aging (Milanesi et al. 2017), and thyroid hormone transporters MCT8 and OATP1C1 (which transport the hormone into the target cell) are upregulated in activated satellite cells, and their depletion impairs both stem cell function and muscle regeneration (Mayerl et al. 2018).

8. NO is an important signaling molecule produced by diverse cell types. It has a protective function in the skeletal muscle for inflammation-induced damage, prevents myofiber lysis, and induces satellite cell activation and migration. Oxidative stress has been covered in another chapter (Brioche et al. 2019).

9. MicroRNAs (miRNAs) also regulate the expression of specific genes to control satellite cell functions; miR-489 is related to quiescence unlike to miR-1/206 and miR-133, which maintain the myogenic program and promote differentiation. Moreover, Myf5 mRNA and its antagonist miR-31 are found sequestered together inside the quiescent satellite cells and delay the myogenesis, controlled in part by miR-31 levels (Crist, Montarras, and Buckingham 2012). We do not expand further on this issue because a related chapter of the book is focused on miRNAs (McGregor 2019).

10. The metabolic status of the satellite cells can affect key pathways during regeneration. It is important to mention that the activated satellite cells present higher metabolic rate and produce nutrients and energy to meet this demand through autophagy (Tang and Rando 2014). Insulin-dependent signaling pathways regulate tightly glucose uptake in skeletal muscle. Autophagy in sarcopenic muscle has been further elaborated in a related chapter (Sakuma, Wakabayashi, and Yamaguchi 2019).

Of course, many of these pathways will interact and inter-regulate among them, which will add another layer of complexity to the issue (Zhang et al. 2018, Siegle et al. 2018). In summary, satellite cells are exposed to a complex and varying microenvironment. The regulatory networks and the role of each cell and molecule involved in the satellite cell fate modulation, as well as how these fit with skeletal muscle physiology, are still incompletely understood (Bentzinger, Wang, Dumont, et al. 2013, Dumont, Wang, and Rudnicki 2015, Yin, Price, and Rudnicki 2013).

9.4 SARCOPENIA AS A MULTIFACTORIAL DISEASE: UNDERLYING CAUSES

From ancient times, loss of vitality and youth have been a major concern for humanity (Narici and Maffulli 2010). Today's extended life expectancy implies a rising number of elders in developed countries (Hayflick 2000a, 2000b). With the increase of elderly population, age-related diseases have become a major issue for our society.

Aging can be defined as the effect of a progressive accumulation of more or less random and deleterious changes as time goes by. These changes cause a functional, anatomical, and physiological decline that leads to the inability to maintain homeostasis under stress (Harman 2003, Mangoni and Jackson 2004). As skeletal muscle is essential for mobility, body stability, and metabolism, any functional decline in muscle entails negative consequences for the individual, which are directly reflected in the greater need of health care (Lang et al. 2010) and may result in frailty. Moreover, perhaps with the only exception of exercise training (Snijders et al. 2017, Snijders and Parise 2017, Verdijk et al. 2016), no effective treatments are available to "rejuvenate" the aged human muscle as attractive target as this is for the biomedical community.

Loss of muscle mass is the most remarkable age-related decline of the skeletal muscle, and it is known as primary sarcopenia (Burks and Cohn 2011, Cruz-Jentoft, Baeyens, et al. 2010, Rosenberg 1997). In contrast, skeletal muscle decay as a result of disuse, disease, or malnutrition is defined as secondary sarcopenia (Cruz-Jentoft, Baeyens, et al. 2010). Primary sarcopenia was originally defined by Rosenberg as an age-related progressive loss of lean body mass. However, the term has evolved to diagnose a geriatric syndrome that presents both loss of skeletal muscle mass and loss of strength, which compromise significantly skeletal muscle function (Cruz-Jentoft, Landi, et al. 2010, Landi et al. 2015, Rosenberg 1997). This definition is not exempted of controversy as actually sarcopenia (age-related loss of the muscle mass) and dynapenia (age-related loss of the muscle strength) are two distinct and nonequivalent processes with different mechanisms and pathophysiology. Some researchers consider that the strength parameter should be removed from the description of sarcopenia (Kwan 2013, Menant et al. 2017, Narici and Maffulli 2010, Visser and Schaap 2011). In any case, primary sarcopenia encompasses all the distinct age-related changes that occur in several skeletal muscle-related processes and the clinical outcomes derived from these situations (Kwan 2013, Lang et al. 2010).

As already described, skeletal muscle represents a complex system where different processes work synergistically and orderly to orchestrate function. Any alteration in any of these processes, although they may not happen uniformly or continuously, will contribute to the clinical manifestations of sarcopenia. Hence, the development of sarcopenia or the contribution to its pathophysiology could be explained by multiple independent and synergistic triggers: genetic factors, mitochondrial dysfunction, elevation of oxidative stress, inflammation, hormonal dysregulation, malnutrition, decline in physical activity, denervation, degeneration of neuromuscular junctions, synapse atrophy, reduced or impaired cell function, altered rate of protein turnover, decreased growth factors, and other diseases such as neurodegenerative or autoimmune diseases. These factors undergo age-related changes that result in a set of clinical outcomes such as decreased muscle strength and mobility, increased fatigue, risk of metabolic disorders and falls, and frailty. This scenario presents many possibilities to address the syndrome, but muscle regeneration and, specifically, modulation of satellite cells seem particularly promising as satellite cells are responsible of muscle repair, and they might be impaired in a similar way as in neuromuscular diseases (Cruz-Jentoft, Landi, et al. 2010, Kwan 2013, Lang et al. 2010).

9.5   NICHE REMODELING IN YOUNG AND OLD MUSCLE

A stem cell niche should be adequate to maintain the quiescent state and the special properties of adult stem cells in the absence of a proliferative signal but then switch to support the activated state of these cells and follow dynamically the local and systemic influences (Gopinath and Rando 2008). Age-related deterioration influences normal cellular responses to stress and injuries, and these chronic degenerative conditions may disturb the stem cell niche and its capacity to respond to external stimuli (Bentzinger, Wang, Dumont, et al. 2013). Muscle regeneration is dramatically disturbed without the presence or correct function of the satellite cells (Naldaiz-Gastesi et al. 2018, Munoz-Canoves et al. 2016). Thus, since any perturbation of the satellite cell niche is predicted to modify their function, we will now focus on the age-related alterations in components of the satellite cell microenvironment (Figure 9.3).

Satellite cells themselves (as an essential part of their own niche) undergo a functional decline in large part attributable not only to extrinsic factors but also due to intrinsic defects (Sousa-Victor, Garcia-Prat, and Munoz-Canoves 2018). There are some contradictory reports about the quantity of satellite cells in old muscles. It is still unclear if the number of satellite cells decreases with aging, but it is certain that they present a delayed response, reduce proliferative capacity, and are more sensitive to apoptosis. Moreover, aged satellite cells show altered expression levels of S100B, Pax7, and other factors, which affect the proliferation and differentiation of the stem cells. Specifically, it seems that the commitment of satellite cells is unaffected, but the ability to maintain or return to quiescence (i.e., self-renewal) is impaired.

On the one hand, FGF signaling pathway is involved in the activation and expansion of satellite cells, and it is disrupted in the aged muscle as the following aspects demonstrate: (i) FGF signaling inhibitor Spry1 is downregulated; (ii) the downstream regulator p38α/βMAPK is overstimulated; and (iii) FGF2 secreted by the myofibers is accumulated in the aged satellite cell niche, and at the same time, the aged satellite cells undergo a desensitization of the FGFR1 receptors. These changes configure an unfavorable environment for the stem cell self-renewal: the FGFR1 receptors do not respond to the presence of FGF2 and are unable to establish cell polarity; consequently the overactivated p38α/βMAPK is distributed symmetrically throughout the dividing cell leading to a higher number of cells committed to differentiation. As the asymmetric self-renewal is impaired, there are less stem cells returning to quiescence, which leads to an exhaustion of this population and a reduction of the regenerative capacity of the aged muscle.

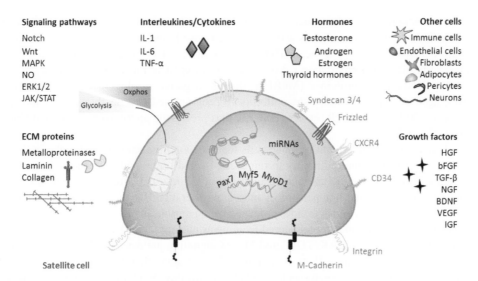

Signaling pathways
Notch
Wnt
MAPK
NO
ERK1/2
JAK/STAT

ECM proteins
Metalloproteinases
Laminin
Collagen

Satellite cell

Interleukines/Cytokines
IL-1
IL-6
TNF-α

Oxphos
Glycolysis

miRNAs

Pax7 Myf5 MyoD1

Hormones
Testosterone
Androgen
Estrogen
Thyroid hormones

Syndecan 3/4
Frizzled
CXCR4
CD34

Integrin
M-Cadherin

Other cells
Immune cells
Endothelial cells
Fibroblasts
Adipocytes
Pericytes
Neurons

Growth factors
HGF
bFGF
TGF-β
NGF
BDNF
VEGF
IGF

FIGURE 9.3 Age-related satellite cell niche remodeling. Differences between a quiescent young niche and an aged one are shown in this diagram. The complexity of the aged-related niche dysfunction is exemplified by the shown changes on several factors that are found deregulated in the old niche, affecting directly to the satellite cell function.

On the other hand, Janus kinase/signal transducers and activators of transcription (JAK/STAT) signaling are augmented in aged satellite cells. It represses myogenic differentiation and promotes myoblast proliferation by regulating cell cycle inhibitors p21 and p27. Thus, JAK/STAT signaling upregulation changes the satellite cell division pattern reducing symmetric divisions toward stem cell production. More symmetric divisions yield more committed cells and a reduced number of satellite cells, that is, reduced capacity to self-renew and to repair the tissue (Barberi et al. 2013, Bentzinger and Rudnicki 2014, Bentzinger, Wang, Dumont, et al. 2013, Dumont, Wang, and Rudnicki 2015, Gopinath and Rando 2008, Kwan 2013).

In spite of these satellite cell defects, the impairment of muscle regeneration is more likely due to a suboptimal environment and signaling not permissive for muscle regeneration. The number and size of the myofibers themselves also decline together with an accumulation of abnormal (smaller) fibers and, accordingly, muscle mass and strength decay. This loss of muscle power correlates with a reduction in excitation-contraction coupling-related receptors causing a dysregulation of the Ca^{2+} levels, which become insufficient for proper contraction and mechanical function. Fewer mitochondria are produced, and the increased oxidative stress and proapoptotic proteins and other protein alterations cause mitochondrial dysfunction, affecting their oxidative function, energy production, and cell viability. Accumulation of fat, fibrosis, and chronic inflammation occur in the deteriorating muscle. Moreover, aged myofibers express less Delta1, affecting Notch signaling pathway and thus satellite cell activation and regeneration potential. Accordingly, Notch signaling is antagonized by Smad3 that is found activated in aged muscles because of an overexpression of TGF-β. Smad3 activation upregulates various cyclin-dependent kinase inhibitors, such as p16 and p21, which are related to the cell cycle regulation and senescence. For its part, TGF-β, a negative regulator of myogenesis detected in serum of elderly humans and mice, stimulates the expansion of fibroblasts as a profibrotic factor inhibiting the myogenic differentiation. The high concentration of Wnt in the aged satellite cell niche, detected also in serum from aged mice, enhances conversion of myogenic cells toward the fibrogenic lineage (Bentzinger, Wang, von Maltzahn, et al. 2013, Dumont, Wang, and Rudnicki 2015, Kwan 2013, Mitchell et al. 2012, Yin, Price, and Rudnicki 2013).

Aging is much related to an increase in pro-apoptotic signals: TNF-α, a cytokine associated to inhibition of differentiation and to myoblast apoptosis, is increased in the aged muscle environment, as well as pro-apoptotic signaling such as caspase 2 and c-Jun N-terminal kinase (JNK). TNF-α interacts with death domain receptors activating procaspase 8 and activates nuclear factor kappa-light-chain-enhancer of activated B cells (NF-κB), also more active in aging per se, which upregulates myostatin/growth/differentiation factor 8, calcium-insensitive nitric oxide synthases (iNOSs), and muscle RING-finger protein-1 (MuRF1), that is, negative trophic factors for the muscle. Moreover, aged cells present more susceptibility to programmed cell death as they express less stress-induced 70-kD heat shock protein (HSP70), thus promoting apoptosome formation (Barberi et al. 2013, Kwan 2013).

There is a decline in functional immune cells that surround the satellite cell niche in the aged muscle. These result in impaired ability to remove damaged tissue and to generate free radicals; they have a reduced phagocytic and chemotactic activity. These perturbations affect satellite cell activation and migration and hamper proper muscle regeneration. Furthermore, the basal level of proinflammatory molecules and cytokines produced by the muscle and by the infiltrating inflammatory cells during regeneration is found altered in aging as nuclear factor kappa light chain enhancer of activated B cells (NF-κB), JAK/STAT, and MAPK signaling pathways is altered. Persistent elevation of myostatin, osteopontin, IL-6, IL-1, and TNF-α, among other cytokines, leads to a higher catabolic rate, more reactive oxygen species, and muscle atrophy (Bentzinger, Wang, Dumont, et al. 2013, Dumont, Wang, and Rudnicki 2015, Gopinath and Rando 2008, Kwan 2013, Yin, Price, and Rudnicki 2013).

Continuing with the age-related structural changes, the vascular endothelium ultrastructure is affected, and the myofibers lose their close contact with microvessels. Consequently, there is a decline in vascular endothelial function with a concomitant decrease in VEGF and endothelial NOS (eNOS) levels. Moreover, there are associated changes in blood circulation, exacerbated by the reduction in capillarization and vasodilatory capacity, which compromise oxygen exchange and metabolite transport, and produce heat and energetic alterations as a result.

Similarly, the aged muscle presents less motor units, specially the α-motor neurons, synaptic vesicles, and more denervated regions. This denervation is observed preferably in fast fibers and causes a motor unit remodeling as these areas are reinnervated by axons from slow motor neurons. If denervation is too severe, the affected area of fibers starts to lose trophic factors and expresses proapoptotic and atrophic factors, which will finally result in the muscle degeneration (Kwan 2013, Yin, Price, and Rudnicki 2013).

With regard to the niche ECM proteins, aging is characterized by a pronounced thickening of the basal lamina and some structural changes that turn the ECM into a denser, more irregular, and amorphous sheath. Different ECM proteins and toxic products derived from the degradation of connective tissue components accumulate within it, changing the basal lamina conformation. Collagen type IV concentration increases in slow muscles, and laminin increases in fast muscles, making them more susceptible to necrosis due to the presence of cleaved fibronectin and elastin. It seems that remodeling of the basal lamina impairs the sequestering, storing, and releasing of growth factors and other important regulator molecules. Age-related ECM remodeling results in accumulation of advanced glycation end products (AGEs), lipoxidation end products (ALEs), and nonspecific cross-linking. Consequently, the aged ECM presents much more elastic stiffness and load-resistant structure that markedly diminish the myogenic potential of the satellite cells and myoblasts. Besides extra deposition of proteins, glycated ECM induces an elevated expression of connective tissue growth factor (CTGF) by fibroblasts, which promotes fibrosis. Also, an increased concentration of osteopontin negatively regulates muscle regeneration. All of the above, in turn, affects the neuromuscular junctions and vascular network and may cause functional alterations in the surrounding cell populations such as Schwann cells and pericytes (Gopinath and Rando 2008, Grzelkowska-Kowalczyk 2016, Kwan 2013, Thomas, Engler, and Meyer 2015, Yin, Price, and Rudnicki 2013).

In terms of metabolism, the skeletal muscle is an important organ for the body, where chemical transformations to energy, proteins, lipids, nucleic acids, and carbohydrates occur, and is also involved in the elimination of nitrogenous wastes. These processes are very well regulated, both at rest and during physical activity. Elderly people are gradually more inactive and have less energy intake, two major inducers of insulin resistance, and reduction of protein synthesis. When the rate between a positive metabolism and a negative metabolism gets unbalanced, cells receive more atrophic signals and undergo atrophy or apoptosis. Moreover, aging involves an altered response of muscles to hormonal stimuli and a systemic decline in testosterone, androgen, estrogen, thyroid hormones, IGF-I, vitamin D, and some growth factors related to the loss of muscle mass and gain of fat mass. Inactivity also leads to redox unbalance with a decrease in antioxidant species. Aging is accompanied by higher oxidative stress with more quantity of H_2O_2, iNOS, nitrotyrosine, and catalase. These metabolites in turn downregulate a set of myogenic proteins. In addition to these biochemical imbalance, many miRNAs are also expressed differently during aging and changing the regulation patterns of proteins (Biolo, Cederholm, and Muscaritoli 2014, Kwan 2013, Yin, Price, and Rudnicki 2013).

9.6 CONCLUSIONS

The mechanisms involved in the pathophysiological outcomes of sarcopenia are complex and interrelated. This entire framework presents more than one therapeutic opportunity for rejuvenation/restoration of muscle function. Thus, how to convert the "aged niche" into a "young niche" and (hopefully) reverse the symptoms of sarcopenia will remain an active area of investigation for the foreseeable future.

ACKNOWLEDGMENTS

We thank Adolfo Lopez de Munain for his overall support. Writing of this book chapter was made possible by grants from Ministerio de Economía y Competitividad (RTC-2015-3750-1) and Instituto de Salud Carlos III (PI16/01430, PI17/01841), co-funded by the European Union (ERDF/ESF, "Investing in your future"), CIBERNED, and the Basque Government (2015/11038, RIS3 2017222021 and BIO16/ER/022). NNG received a studentship from the Department of Education, University, and Research of the Basque Government (PRE2013-1-1168).

REFERENCES

Alberts, B., A. Johnson, J. Lewis, M. Raff, K. Roberts, and P. Walter. 2002. *Molecular Biology of the Cell*. 4th ed. New York: Garland Science.

Almada, A. E., and A. J. Wagers. 2016. "Molecular circuitry of stem cell fate in skeletal muscle regeneration, ageing and disease." *Nat Rev Mol Cell Biol* 17 (5):267–79. doi:10.1038/nrm.2016.7.

Barberi, L., B. M. Scicchitano, M. De Rossi, A. Bigot, S. Duguez, A. Wielgosik, C. Stewart et al. 2013. "Age-dependent alteration in muscle regeneration: The critical role of tissue niche." *Biogerontology* 14 (3):273–92. doi:10.1007/s10522-013-9429-4.

Bentzinger, C. F., and M. A. Rudnicki. 2014. "Rejuvenating aged muscle stem cells." *Nat Med* 20 (3):234–5. doi:10.1038/nm.3499.

Bentzinger, C. F., Y. X. Wang, N. A. Dumont, and M. A. Rudnicki. 2013. "Cellular dynamics in the muscle satellite cell niche." *EMBO Rep* 14 (12):1062–72. doi:10.1038/embor.2013.182.

Bentzinger, C. F., Y. X. Wang, J. von Maltzahn, V. D. Soleimani, H. Yin, and M. A. Rudnicki. 2013. "Fibronectin regulates Wnt7a signaling and satellite cell expansion." *Cell Stem Cell* 12 (1):75–87. doi:10.1016/j.stem.2012.09.015.

Biolo, G., T. Cederholm, and M. Muscaritoli. 2014. "Muscle contractile and metabolic dysfunction is a common feature of sarcopenia of aging and chronic diseases: From sarcopenic obesity to cachexia." *Clin Nutr* 33 (5):737–48. doi:10.1016/j.clnu.2014.03.007.

Brioche, T., B. Allan Pagano, A.-C. Coralie, G. Py, C. Angèle, and G.-C. M. Carmen. 2019. "Sarcopenia and oxidative stress: From the bench to therapeutic strategies." In *Molecular and Cellular Aspects of Sarcopenia*, edited by Dominique Meynial-Denis. Boca Raton, FL: CRC Press/Taylor & Francis Group.

Brooks, S. V. 2003. "Current topics for teaching skeletal muscle physiology." *Adv Physiol Educ* 27 (1–4):171–82.

Burks, T. N., and R. D. Cohn. 2011. "One size may not fit all: Anti-aging therapies and sarcopenia." *Aging* 3 (12):1142–53. doi:10.18632/aging.100409.

Christov, C., F. Chretien, R. Abou-Khalil, G. Bassez, G. Vallet, F. J. Authier, Y. Bassaglia et al. 2007. "Muscle satellite cells and endothelial cells: Close neighbors and privileged partners." *Mol Biol Cell* 18 (4):1397–409. doi:10.1091/mbc. E06-08-0693.

Crist, C. G., D. Montarras, and M. Buckingham. 2012. "Muscle satellite cells are primed for myogenesis but maintain quiescence with sequestration of Myf5 mRNA targeted by microRNA-31 in mRNP granules." *Cell Stem Cell* 11 (1):118–26. doi:10.1016/j.stem.2012.03.011.

Cruz-Jentoft, A. J., J. P. Baeyens, J. M. Bauer, Y. Boirie, T. Cederholm, F. Landi, F. C. Martin et al., and People European Working Group on Sarcopenia in Older. 2010. "Sarcopenia: European consensus on definition and diagnosis: Report of the European orking Group on Sarcopenia in Older People." *Age Ageing* 39 (4):412–23. doi:10.1093/ageing/afq034.

Cruz-Jentoft, A. J., F. Landi, E. Topinkova, and J. P. Michel. 2010. "Understanding sarcopenia as a geriatric syndrome." *Curr Opin Clin Nutr Metab Care* 13 (1):1–7. doi:10.1097/MCO.0b013e328333c1c1.

Das, R. K., and O. F. Zouani. 2014. "A review of the effects of the cell environment physicochemical nano-architecture on stem cell commitment." *Biomaterials* 35 (20):5278–93. doi:10.1016/j.biomaterials.2014.03.044.

Drummond-Barbosa, D. 2008. "Stem cells, their niches and the systemic environment: An aging network." *Genetics* 180 (4):1787–97. doi:10.1534/genetics.108.098244.

Dumont, N. A., Y. X. Wang, and M. A. Rudnicki. 2015. "Intrinsic and extrinsic mechanisms regulating satellite cell function." *Development* 142 (9):1572–81. doi:10.1242/dev.114223.

Frontera, W. R., and J. Ochala. 2015. "Skeletal muscle: A brief review of structure and function." *Calcif Tissue Int* 96 (3):183–95. doi:10.1007/s00223-014-9915-y.

Fry, C. 2019. "Satellite cells and skeletal muscle aging." In *Molecular and Cellular Aspects of Sarcopenia*, edited by Dominique Meynial-Denis. Boca Raton, FL: CRC Press/Taylor & Francis Group.

Garcia-Parra, P., N. Naldaiz-Gastesi, M. Maroto, J. F. Padin, M. Goicoechea, A. Aiastui, J. C. Fernandez-Morales et al. 2014. "Murine muscle engineered from dermal precursors: An *in vitro* model for skeletal muscle generation, degeneration, and fatty infiltration." *Tissue Eng Part C Methods* 20 (1):28–41. doi:10.1089/ten. TEC.2013.0146.

Gopinath, S. D., and T. A. Rando. 2008. "Stem cell review series: Aging of the skeletal muscle stem cell niche." *Aging Cell* 7 (4):590–8. doi:10.1111/j.1474-9726.2008.00399.x.

Graber, T. G., and B. B. Rasmussen. 2019. "Nutrient sensing and signaling in aging muscle." In *Molecular and Cellular Aspects of Sarcopenia*, edited by Dominique Meynial-Denis. Boca Raton, FL: CRC Press/Taylor & Francis Group.

Grzelkowska-Kowalczyk, K. 2016. "The importance of extracellular matrix in skeletal muscle development and function." In *Composition and Function of the Extracellular Matrix in the Human Body*, edited by F Travascio. London, UK: InTech Open.

Harman, D. 2003. "The free radical theory of aging." *Antioxid Redox Signal* 5 (5):557–61. doi:10.1089/152308603770310202.

Hayflick, L. 2000a. "The future of ageing." *Nature* 408 (6809):267–9. doi:10.1038/35041709.

Hayflick, L. 2000b. "New approaches to old age." *Nature* 403 (6768):365. doi:10.1038/35000303.

Korthuis, R.J. 2011. *Skeletal Muscle Circulation*. San Rafael, CA: Morgan & Claypool Life Sciences.

Kuang, S., K. Kuroda, F. Le Grand, and M. A. Rudnicki. 2007. "Asymmetric self-renewal and commitment of satellite stem cells in muscle." *Cell* 129 (5):999–1010. doi:10.1016/j.cell.2007.03.044.

Kwan, P. 2013. "Sarcopenia: The gliogenic perspective." *Mech Ageing Dev* 134 (9):349–55. doi:10.1016/j.mad.2013.06.001.

Landi, F., R. Calvani, M. Cesari, M. Tosato, A. M. Martone, R. Bernabei, G. Onder, and E. Marzetti. 2015. "Sarcopenia as the biological substrate of physical frailty." *Clin Geriatr Med* 31 (3):367–74. doi:10.1016/j.cger.2015.04.005.

Lang, T., T. Streeper, P. Cawthon, K. Baldwin, D. R. Taaffe, and T. B. Harris. 2010. "Sarcopenia: Etiology, clinical consequences, intervention, and assessment." *Osteoporos Int* 21 (4):543–59. doi:10.1007/s00198-009-1059-y.

Lutolf, M. P., P. M. Gilbert, and H. M. Blau. 2009. "Designing materials to direct stem-cell fate." *Nature* 462 (7272):433–41. doi:10.1038/nature08602.

MacIntosh, B. R., P. Gardiner, and A. McComas. 2006. *Skeletal Muscle-Form and Function*. 2nd ed. Champaign, IL: Human Kinetics.

Madaro, L., M. Passafaro, D. Sala, U. Etxaniz, F. Lugarini, D. Proietti, M. V. Alfonsi et al. 2018. "Denervation-activated STAT3-IL-6 signalling in fibro-adipogenic progenitors promotes myofibres atrophy and fibrosis." *Nat Cell Biol* 20 (8):917–927. doi:10.1038/s41556-018-0151-y.

Malecova, B., S. Gatto, U. Etxaniz, M. Passafaro, A. Cortez, C. Nicoletti, L. Giordani et al. 2018. "Dynamics of cellular states of fibro-adipogenic progenitors during myogenesis and muscular dystrophy." *Nat Commun* 9 (1):3670. doi:10.1038/s41467-018-06068-6.

Mangoni, A. A., and S. H. Jackson. 2004. "Age-related changes in pharmacokinetics and pharmacodynamics: Basic principles and practical applications." *Br J Clin Pharmacol* 57 (1):6–14.

Mauro, A. 1961. "Satellite cell of skeletal muscle fibers." *J Biophys Biochem Cytol* 9:493–5.

Mayerl, S., M. Schmidt, D. Doycheva, V. M. Darras, S. S. Huttner, A. Boelen, T. J. Visser, C. Kaether, H. Heuer, and J. von Maltzahn. 2018. "Thyroid hormone transporters MCT8 and OATP1C1 control skeletal muscle regeneration." *Stem Cell Rep* 10 (6):1959–74. doi:10.1016/j.stemcr.2018.03.021.

McGregor, R. 2019. "Role of microRNAs in sarcopenia." In *Molecular and Cellular Aspects of Sarcopenia*, edited by Dominique Meynial-Denis. Boca Raton, FL: CRC Press/Taylor & Francis Group.

Menant, J. C., F. Weber, J. Lo, D. L. Sturnieks, J. C. Close, P. S. Sachdev, H. Brodaty, and S. R. Lord. 2017. "Strength measures are better than muscle mass measures in predicting health-related outcomes in older people: Time to abandon the term sarcopenia?" *Osteoporos Int* 28 (1):59–70. doi:10.1007/s00198-016-3691-7.

Milanesi, A., J. W. Lee, A. Yang, Y. Y. Liu, S. Sedrakyan, S. Y. Cheng, L. Perin, and G. A. Brent. 2017. "Thyroid hormone receptor alpha is essential to maintain the satellite cell niche during skeletal muscle injury and sarcopenia of aging." *Thyroid* 27 (10):1316–22. doi:10.1089/thy.2017.0021.

Mitchell, W. K., J. Williams, P. Atherton, M. Larvin, J. Lund, and M. Narici. 2012. "Sarcopenia, dynapenia, and the impact of advancing age on human skeletal muscle size and strength; a quantitative review." *Front Physiol* 3:260. doi:10.3389/fphys.2012.00260.

Morrison, S. J., and A. C. Spradling. 2008. "Stem cells and niches: Mechanisms that promote stem cell maintenance throughout life." *Cell* 132 (4):598–611. doi:10.1016/j.cell.2008.01.038.

Morrissey, J. B., R. Y. Cheng, S. Davoudi, and P. M. Gilbert. 2016. "Biomechanical origins of muscle stem cell signal transduction." *J Mol Biol* 428 (7):1441–54. doi:10.1016/j.jmb.2015.05.004.

Mourikis, P., and S. Tajbakhsh. 2014. "Distinct contextual roles for Notch signalling in skeletal muscle stem cells." *BMC Dev Biol* 14:2. doi:10.1186/1471-213X-14-2.

Munoz-Canoves, P., J. J. Carvajal, A. Lopez de Munain, and A. Izeta. 2016. "Editorial: Role of stem cells in skeletal muscle development, regeneration, repair, aging, and disease." *Front Aging Neurosci* 8:95. doi:10.3389/fnagi.2016.00095.

Naldaiz-Gastesi, N., O. A. Bahri, A. Lopez de Munain, K. J. A. McCullagh, and A. Izeta. 2018. "The panniculus carnosus muscle: An evolutionary enigma at the intersection of distinct research fields." *J Anat.* doi:10.1111/joa.12840.

Naldaiz-Gastesi, N., M. Goicoechea, S. Alonso-Martin, A. Aiastui, M. Lopez-Mayorga, P. Garcia-Belda, J. Lacalle et al. 2016. "Identification and characterization of the dermal panniculus carnosus muscle stem cells." *Stem Cell Rep* 7 (3):411–24. doi:10.1016/j.stemcr.2016.08.002.

Narici, M. V., and N. Maffulli. 2010. "Sarcopenia: Characteristics, mechanisms and functional significance." *Br Med Bull* 95:139–59. doi:10.1093/bmb/ldq008.

Rosenberg, I. H. 1997. "Sarcopenia: Origins and clinical relevance." *J Nutr* 127 (5 Suppl):990S–91S.

Sakuma, K., H. Wakabayashi, and A. Yamaguchi. 2019. "Differential adaptation of ubiquitin-proteasome and lysosome-autophagy signaling in sarcopenic muscle." In *Molecular and Cellular Aspects of Sarcopenia*, edited by Dominique Meynial-Denis. Boca Raton, FL: CRC Press/Taylor & Francis Group.

Salvatore, D., W. S. Simonides, M. Dentice, A. M. Zavacki, and P. R. Larsen. 2014. "Thyroid hormones and skeletal muscle--new insights and potential implications." *Nat Rev Endocrinol* 10 (4):206–14. doi:10.1038/nrendo.2013.238.

Schmalbruch, H. 2006. "The satellite cell of skeletal muscle fibres." *Braz. J. Morphol. Sci.* 23 (2):159–72.

Schmalbruch, H., and U. Hellhammer. 1976. "The number of satellite cells in normal human muscle." *Anat Rec* 185 (3):279–87. doi:10.1002/ar.1091850303.

Schmalbruch, H., and U. Hellhammer. 1977. "The number of nuclei in adult rat muscles with special reference to satellite cells." *Anat Rec* 189 (2):169–75. doi:10.1002/ar.1091890204.

Siegle, L., J. D. Schwab, S. D. Kuhlwein, L. Lausser, S. Tumpel, A. S. Pfister, M. Kuhl, and H. A. Kestler. 2018. "A Boolean network of the crosstalk between IGF and Wnt signaling in aging satellite cells." *PLoS One* 13 (3):e0195126. doi:10.1371/journal.pone.0195126.

Smith, L., S. Cho, and D. E. Discher. 2017. "Mechanosensing of matrix by stem cells: From matrix heterogeneity, contractility, and the nucleus in pore-migration to cardiogenesis and muscle stem cells *in vivo*." *Semin Cell Dev Biol.* doi:10.1016/j.semcdb.2017.05.025.

Snijders, T., J. P. Nederveen, S. Joanisse, M. Leenders, L. B. Verdijk, L. J. van Loon, and G. Parise. 2017. "Muscle fibre capillarization is a critical factor in muscle fibre hypertrophy during resistance exercise training in older men." *J Cachexia Sarcopenia Muscle* 8 (2):267–276. doi:10.1002/jcsm.12137.

Snijders, T., and G. Parise. 2017. "Role of muscle stem cells in sarcopenia." *Curr Opin Clin Nutr Metab Care* 20 (3):186–90. doi:10.1097/MCO.0000000000000360.

Sousa-Victor, P., L. Garcia-Prat, and P. Munoz-Canoves. 2018. "New mechanisms driving muscle stem cell regenerative decline with aging." *Int J Dev Biol* 62 (6-7-8):583–90. doi:10.1387/ijdb.180041pm.

Standring, S. 2016. *Gray's Anatomy*, 41st ed. Philadelphia, PA: Elsevier.

Tang, A. H., and T. A. Rando. 2014. "Induction of autophagy supports the bioenergetic demands of quiescent muscle stem cell activation." *EMBO J* 33 (23):2782–97. doi:10.15252/embj.201488278.

Thomas, K., A. J. Engler, and G. A. Meyer. 2015. "Extracellular matrix regulation in the muscle satellite cell niche." *Connect Tissue Res* 56 (1):1–8. doi:10.3109/03008207.2014.947369.

Turrina, A., M. A. Martinez-Gonzalez, and C. Stecco. 2013. "The muscular force transmission system: Role of the intramuscular connective tissue." *J Bodyw Mov Ther* 17 (1):95–102. doi:10.1016/j.jbmt.2012.06.001.

Verdijk, L. B., T. Snijders, T. M. Holloway, V. A. N. Kranenburg J, and V. A. N. Loon LJ. 2016. "Resistance training increases skeletal muscle capillarization in healthy older men." *Med Sci Sports Exerc* 48 (11):2157–64. doi:10.1249/MSS.0000000000001019.

Visser, M., and L. A. Schaap. 2011. "Consequences of sarcopenia." *Clin Geriatr Med* 27 (3):387–99. doi:10.1016/j.cger.2011.03.006.

Wang, Y., M. Wehling-Henricks, S. S. Welc, A. L. Fisher, Q. Zuo, and J. G. Tidball. 2018. "Aging of the immune system causes reductions in muscle stem cell populations, promotes their shift to a fibrogenic phenotype, and modulates sarcopenia." *FASEB J.* doi:10.1096/fj.201800973R.

Wang, Y., S. S. Welc, M. Wehling-Henricks, and J. G. Tidball. 2018. "Myeloid cell-derived tumor necrosis factor-alpha promotes sarcopenia and regulates muscle cell fusion with aging muscle fibers." *Aging Cell.* doi:10.1111/acel.12828.

Wosczyna, M. N., and T. A. Rando. 2018. "A muscle stem cell support group: Coordinated cellular responses in muscle regeneration." *Dev Cell* 46 (2):135–43. doi:10.1016/j.devcel.2018.06.018.

Yin, H., F. Price, and M. A. Rudnicki. 2013. "Satellite cells and the muscle stem cell niche." *Physiol Rev* 93 (1):23–67. doi:10.1152/physrev.00043.2011.

Zhang, W., Y. Xu, L. Zhang, S. Wang, B. Yin, S. Zhao, and X. Li. 2018. "Synergistic effects of TGFbeta2, WNT9a, and FGFR4 signals attenuate satellite cell differentiation during skeletal muscle development." *Aging Cell.* doi:10.1111/acel.12788.

10 Sarcopenia and Oxidative Stress
From the Bench to Therapeutic Strategies

Coralie Arc-Chagnaud, Allan F. Pagano, and Thomas Brioche

CONTENTS

10.1 INTRODUCTION

Skeletal muscle is the most abundant tissue in the human body representing ~40% of the body weight and ~30% of the basal energy expenditure. Skeletal muscle plays a central role in locomotion enabling a person to perform activities of daily living, posture maintenance, balance, thermogenesis processes, energy supply (this tissue contains the most important glucose and amino acids stocks), and insulin resistance protection. In order to ensure these essential functions, skeletal muscle must have sufficient mass and quality (Brioche et al. 2016).

Skeletal muscle is a very plastic tissue, which can adapt itself to environmental constraints. Indeed, an increase in stimulation, exercise, or nutrition can lead to muscle hypertrophy or improve endurance (Brioche and Lemoine-Morel 2016). Conversely, a decrease in mechanical constraints will lead to muscle deconditioning characterized by atrophied, weak, and fatigable muscles (Brioche et al. 2016).

Muscle deconditioning occurring with aging is commonly called sarcopenia. The definition of sarcopenia is still under debate. However, pooling different existing definitions, sarcopenia could be defined as a geriatric syndrome initially characterized by a decrease in muscle mass that will get worse causing deterioration in strength and physical performance (Brioche et al. 2016). Due to social, technological, and medical progress, life expectancy has been increasing since the nineteenth century in our modern Western societies, leading to the aging of the world population. Currently, it is projected that the number of elderly will double worldwide from 11% of the population to 22% by 2050 (UN 2007). Inevitably, due to this aging population, prevalence of sarcopenia is growing. Currently, it is estimated that one-quarter to one-half of men and women aged 65 years and older are likely sarcopenic (Janssen et al. 2004). The consequences of the increasing prevalence of sarcopenia are generally considered as catastrophic on the public health costs. Thus, the total cost of sarcopenia to the American Health System has been reported to be approximately $18.4 billion (Janssen et al. 2004). These healthcare costs are linked to a general deterioration of the physical condition resulting in an increased risk of falls and fractures, a progressive inability to perform basic activities of the daily life, loss of independence of the elderly, and ultimately death (Cruz-Jentoft 2012; Delmonico et al. 2007; Goodpaster et al. 2006). Numerous factors contribute to sarcopenia, including diet (see Chapter 5), chronic diseases, physical inactivity, and the aging process itself (Brioche et al. 2016).

At the molecular level, muscle deconditioning during sarcopenia is due to a higher proteolysis of contractile proteins by several proteolytic systems compared with their synthesis. Moreover, there would have a higher degradation of mitochondria compared with their synthesis associated with a deregulation of mitochondrial fusion and fission. Finally, the imbalance between apoptosis and regeneration processes in favor of apoptosis participates in the previous mechanisms. These mechanisms are fully involved in skeletal muscle atrophy, which contributes substantially to the loss of muscle strength during aging. However, other molecular mechanisms are also involved like post-translational modifications of contractile proteins and decoupling of excitation–contraction complex (Brioche and Lemoine-Morel 2016). These mechanisms contribute to the onset of sarcopenia and are affected by numerous upstream factors such as a decreased anabolic hormones release, an increased pro-inflammatory cytokines production, and insulin resistance. Links and interactions between all these factors remain partly unknown. However, chronic oxidative stress occurring during aging appears as a serious candidate (Brioche and Lemoine-Morel 2016; Fougere et al. 2018; Derbre et al. 2014; Pierre et al. 2016; Powers et al. 2016) (Figure 10.1).

In this context, this review aims to highlight the major role of oxidative stress in sarcopenia, discuss about the efficiency of different kind of antioxidant strategies, and focus on the antioxidant action of exercise.

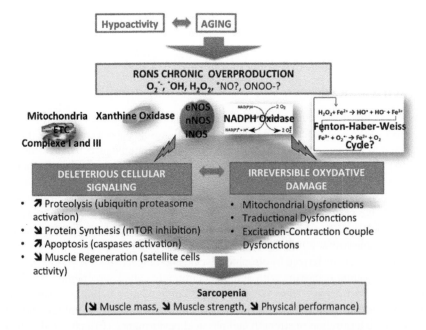

FIGURE 10.1 Main events related to oxidative stress leading to sarcopenia.

10.2   THE CONTRIBUTION OF OXIDATIVE STRESS TO SARCOPENIA

Aging (which is often associated with hypoactivity) leads to a chronic overproduction of reactive oxygen/nitrogen species (RONS) by mitochondria, xanthine oxidase, NO synthase, and NADPH oxidase. Involvement of Fenton–Haber–Weiss Cycle is still under debate, and more data are necessary to conclude (Brioche and Lemoine-Morel 2016; Derbre et al. 2014; Pierre et al. 2016; Powers et al. 2016). RONS overproduction leads to irreversible oxidative damage to cell lipids, proteins, and nucleic acids. This damage can trigger mitochondrial, translational, or excitation–contraction couple dysfunctions. Moreover, RONS chronic overproduction and irreversible oxidative damage appear to disturb cellular and molecular signaling leading to an increase of proteolysis and apoptosis and a decrease of protein synthesis and muscle regenerative capacity (Brioche and Lemoine-Morel 2016; Derbre et al. 2014; Pierre et al. 2016; Powers et al. 2016). Figure 10.1 resumes these mechanisms.

10.2.1   OXIDATIVE STRESS COULD DISTURB PROTEIN TURN-OVER

Maintaining muscle mass is first a balance between protein synthesis and protein degradation systems. It is well established that a negative protein turnover (Combaret et al. 2009) characterized by the reduction of myofibrillar (especially myosin heavy chain noted as MyHC) and mitochondrial protein synthesis (Balagopal et al. 1997; Cuthbertson et al. 2005; Haddad and Adams 2006) and their increased proteolysis via the ubiquitin–proteasome system and the calcium-dependent activation of proteases (i.e., calpains and caspases) are involved in sarcopenia (Gumucio and Mendias 2013; Konopka and Sreekumaran Nair 2013). The decrease in MyHC synthesis is at least due to a decrease in its transcription as the amounts of mRNA of different isoforms decrease during aging in particular MyHC IIa and MyHC IIx isoforms (Balagopal et al. 2001; Short et al. 2005). This may explain in part why MyHC protein content in muscle of old animals is reduced compared with young animals (Haddad and Adams 2006; Thompson et al. 2006) and why MyHC IIa and IIx protein decline by 3% and 1% per decade in humans (Short et al. 2005).

Protein synthesis is mainly controlled by the mechanistic target of rapamycin (mTOR) pathways, and it has been demonstrated in various cell cultures or animal models that overproduction of RONS inhibits key molecule controlling this pathway leading to depress protein synthesis by obstructing mRNA translation at the level of initiation (Shenton et al. 2006; Zhang and Forman 2017; Powers et al. 2016; Zhang et al. 2009; O'Loghlen et al. 2006). This decrease is due to a daily overproduction of RONS especially after lunch and when you give antioxidant to old rats before lunch, it is possible to increases muscle protein synthesis at the same level than young rats (Marzani et al. 2008). Underlying mechanisms need to be studied in sarcopenic elderlies. However, cell culture and animal models showed that RONS such as H_2O_2 (known to increase during sarcopenia) impair mTOR assembly and therefore preventing mTOR-mediated phosphorylation of 4E-BP1 and p70S6K (Zhang et al. 2009). Moreover, oxidative DNA damage is known to activate p53, which is able to inhibit mTOR via AMP-activated protein kinase (AMPK) and Tuberous Sclerosis Complex 2 (TSC2) (Feng et al. 2005).

RONS overproduction impairs muscle protein breakdown by different ways. It is well established that during aging and cachexia, contractile proteins are more degraded by the ubiquitine–proteasome system (implication of this system during sarcopenia is well discussed in Chapter 6). RONS overproduction would lead to activate p38MAPK and IκK-alfa, which would activate the translation of E3 ligase, especially MuRF-1 and MAFbx (Zhang and Forman 2017). These two E3 ligases will ubiquitinate muscle contractile proteins, which will be degraded by the proteasome (Li et al. 2003, 2005).

Aging is also associated with a skeletal muscle cytosolic calcium overload (Fraysse et al. 2006) known to increase calpain activity (Goll et al. 2003). RONS production could play an important role in disturbances in calcium homeostasis (Kandarian and Stevenson 2002). A potential mechanism to link oxidative stress with calcium overload is that RONS-mediated formation of reactive aldehydes (i.e., 4-hydroxy-2,3-trans-nonenal) can inhibit plasma membrane Ca^{+2} ATPase activity, which would lead to intracellular Ca^{+2} accumulation (Siems et al. 2003). In another way, increased lipids peroxidation in old rodents skeletal muscle is associated with an increased caspase-3 activity and muscle atrophy (Wohlgemuth et al. 2010). Recent reports indicate that oxidative stress can activate caspase 3 in muscle fibers *in vitro* and *in vivo*. For example, exposing C2C12 myotubes to H_2O_2 (known to increase during aging) has been shown to activate caspase 3 (Siu, Wang, and Alway 2009). Notably, new evidence reveals that antioxidant-mediated protection against inactivity-induced oxidative stress prevents caspase-3 activation in diaphragm muscle *in vivo* (Whidden et al. 2010).

Concerning oxidative stress, sarcopenia, and autophagy, we are in a paradoxical situation. Autophagy occurs in five steps: induction, expansion, autophagosome completion, autophagosome and lysosome fusion, and protein and organelle degradation. In fact, RONS appear to induce autophagy induction, while RONS would inhibit autophagosome formation. Indeed, chronic RONS overproduction leads to increase ULK-1 activation by inhibiting mTOR, which is known to inhibit ULK1 (Powers et al. 2016). Decrease autophagosome formation appears to be due to the inactivation of ATG4 (by H_2O_2), which is necessary to autophagasome formation (Scherz-Shouval et al. 2007). Finally, aging RONS production would lead to impair autophagy, leading to the accumulation of protein aggregates and damaged mitochondria, which could then produce more RONS.

Finally, RONS can also accelerate proteolysis in muscle fibers by oxidizing muscle proteins, which enhances their susceptibility to proteolytic processing. Indeed, using several purified proteases, Davies first demonstrated that RONS accelerate the protease-mediated breakdown of proteins (Davies, Lin, and Pacifici 1987). Others have expanded these observations, and it is now established that oxidized proteins are readily degraded by many proteases, including the 20S proteasome, calpains, and caspase 3 (Smuder et al. 2010; Grune et al. 2003). In particular, oxidation increases myofibrillar protein breakdown in a dose-dependent manner, and following the oxidative modification, Myosin Heavy Chain (MHC), α-actinin, actin, and troponin I are all rapidly degraded by calpains (I and II) and caspase-3 (Smuder et al. 2010).

All these mechanisms need to be confirmed in Elderlies.

10.2.2 LINK BETWEEN OXIDATIVE STRESS AND IMPAIRED SATELLITE CELL ACTIVITY

Equilibrium between apoptosis and regeneration processes is also involved in maintaining muscle mass (Hikida 2011; Marzetti, Calvani, et al. 2012; Snijders, Verdijk, and van Loon 2009). Decreased muscle regeneration capacity (Hikida 2011; Snijders, Verdijk, and van Loon 2009) and an exacerbation of apoptosis (Marzetti, Calvani, et al. 2012) are usually considered to be involved in sarcopenia muscle deconditioning. This is supported by the observation that apoptotic signaling correlates with slow walking speed and reduced muscle volume in older persons (Marzetti, Lees, et al. 2012). Moreover, numerous studies showed that the extent of apoptotic DNA fragmentation increases in skeletal muscle over the course of aging paralleling the development of sarcopenia (Braga et al. 2008; Marzetti, Lees, et al. 2012; Siu et al. 2006; Wohlgemuth et al. 2010; Kovacheva et al. 2010), which is indeed involved in the transcriptional efficiency decrease observed during sarcopenia (Cuthbertson et al. 2005; Roberts et al. 2010). This phenomenon is probably worsened by alterations in muscle regeneration capacity limiting the incorporation of new nuclei in aging muscle fibers by satellite cells (SCs). Chapter 8 explores in detail the involvement of SCs during sarcopenia.

Numerous studies consider that the cellular environment of the old muscle is responsible for alterations in the activity of SC more than the intrinsic myogenic potential of these latter (Carlson and Faulkner 1989; Carlson et al. 2009). Thereby, recent studies demonstrated in C2C12 cells that reduce the redox environment promotes both proliferation (Renault et al. 2002) and myoblast differentiation (Ardite et al. 2004; Hansen et al. 2007) underlying the importance of RONS in these processes. In contrast, studies have also suggested that the decreased activity of SC in the aged muscle may be related to increased oxidative stress within SC (Fulle et al. 2004; Beccafico et al. 2007).

Indeed, Fulle et al. (2005) showed that antioxidant enzyme activity is decreased in SCs extracted from old men (more than 70 years old) compared with young men (30–40 years old). The lipid peroxidation higher in old myotubes obtained from old SC was associated with a decreased myoblast fusion capacity to generate myotubes (Beccafico et al. 2007). Chapter 9 specifically deals with muscle stem cell microenvironment in healthy aging and sarcopenia.

10.2.3 Link between Oxidative Stress and Mitochondria

Given the vital functions carried out by mitochondria in the context of energy provision, redox homeostasis, and regulation of several catabolic and cell death pathways, it is not surprising that age-related alterations of mitochondrial functions are placed at the center of sarcopenia by numerous authors (Calvani et al. 2013; Picca et al. 2018; Konopka and Sreekumaran Nair 2013; Marzetti et al. 2013). One major consequence of the age-associated mitochondrial dysfunction is a decline in bioenergetics that is witnessed by a decrease in both resting and maximal oxygen consumption ($\dot{V}O_2$max) with advancing age in humans (Short et al. 2004) and mice (Lee et al. 2010) and by a decreased endurance capacity in old rats compared with young rats (Derbre, Gomez-Cabrera, et al. 2012). Moreover, perturbations in skeletal muscle mitochondrial energetics have been correlated with reduced $\dot{V}O_2$max (Short et al. 2005), walking capacity (Coen et al. 2013), and maximal isometric strength (Safdar et al. 2010) in older adults and are associated with an increase in muscle fatigability in old rats (Chabi et al. 2008). The bioenergetic failure of the sarcopenic muscle is associated with a reduction in mitochondrial abundance and functions (Calvani et al. 2013; Vina et al. 2009), which is the result of a vicious cycle involving RONS production and damage/depletion of mitochondrial DNA (mtDNA) and defective mitochondria quality control (Marzetti et al. 2013; Picca et al. 2018).

According to the free radical theory of aging, RONS production during the life leads to an accumulation of oxidative damage to the mitochondrial compounds especially to mtDNA leading to mtDNA mutations. These mutations would lead to the synthesis of defective ETC subunits, which would result in an increase of RONS leakage leading to further oxidative damage. In human and rodent muscles, numerous studies have shown that these increased oxidative damage impair mitochondrial lipids, proteins, and overall mtDNA (Braga et al. 2008; Figueiredo et al. 2009; Lee et al. 2010; Short et al. 2005; Wohlgemuth et al. 2010). Moreover, increase in mitochondrial Ca^{2+} concentration observed during aging would result in disruption of mitochondrial membrane potential, which could be involved in the increased production of RONS by mitochondria. The increase in mitochondrial RONS production with aging occurs mainly in the type I muscles as the soleus of older rats (Capel et al. 2004). However, it appears that the basal mitochondrial RONS production is markedly higher in predominantly glycolytic muscles, regardless of age (Capel et al. 2004; Anderson and Neufer 2006).

10.2.4 Link between Oxidative Stress and Muscle Contractile Qualities Impairment

Although skeletal muscle atrophy contributes substantially to the loss of muscle strength during aging, it is not considered the only factor involved in this phenomenon. Indeed, data from animals showed that the specific strength of isolated muscle fibers (i.e., force normalized to cross-sectional area of the fiber) also decreased with age (Thompson 2009; Gonzalez, Messi, and Delbono 2000; Thompson and Brown 1999; Renganathan, Messi, and Delbono 1998), but contradictory results have been found (Claflin et al. 2011). Several mechanisms are proposed to explain these results such as posttranslational modifications of contractile proteins (Lowe et al. 2001), decoupling of the complex excitation–contraction (Wang, Messi, and Delbono 2000; Tieland, Trouwborst, and Clark 2018), and decreased myosin ATPase activity (Lowe et al. 2004). Studies focused on permeabilized muscle fibers have shown that the decrease in the specific strength is also explained by a reduction of the fraction of myosin heads to bind to the actin filaments (Lowe et al. 2001, 2004). By studying the isolated intact muscle fibers, it revealed the involvement of excitation–contraction coupling in the changes with age of muscle contractile properties. Thus, the maximum release of calcium from the sarcoplasmic reticulum is reduced in the aged rodent muscle tissue (Jimenez-Moreno et al. 2008).

Decoupling of the complex excitation–contraction is due to a lesser release of calcium by the sarcoplasmic reticulum. Indeed, during aging, ryanodine receptor 1 is oxidized and nitrosylated by RONS leading to lesser calcium release and finally leading to the muscle strength decrease (Andersson et al. 2011). Moreover, in cell culture, it has been demonstrated that RONS increase 4-hydroxynonenal levels, which inhibit calcium ATPase pumps (SERCA) (Siems et al. 2003). Inhibition of sarco/endoplasmic reticulum Calcium-ATPase (SERCA) would lead to higher intracellular calcium, which in turn would activate two calcium-dependent proteolytic pathways, calpains, and caspase 3 proteases known to degrade myofilaments. However, more studies are needed to confirm these mechanisms in sarcopenic people.

10.3   STRATEGIES AGAINST SARCOPENIA

RONS are involved in cellular and molecular function and signaling with an inverted U relation. The perfect strategies against sarcopenia would allow staying in the optimal zone. However, today, we do not have sufficient knowledge to always give the perfect dose.

10.3.1   CLASSICAL ANTIOXIDANT STRATEGIES

Table 10.1 presents the main antioxidant strategies against sarcopenia. Globally, there are three classical antioxidant strategies:

- The first aim to directly scavenge the RONS presented in the organisms by supplementation with one antioxidant or a cocktail of various antioxidants such as vitamin C, vitamin E, and carotenoids or supplementation with natural compounds (which can be modified to increase their bioavailability) such as resveratrol.
- The second directly target RONS sources with pharmaceutical products such as allopurinol, which is an inhibitor of xanthine oxidase.
- The last strategies will consist in making a supplementation with precursors of the synthesis of antioxidant molecules such as precursors of Glutathione (GSH) synthesis.

Based on studies measuring physical parameter or muscle mass in humans and animals, RON scavengers appear to be not effective to prevent sarcopenia (Brioche and Lemoine-Morel 2016). Indeed,

TABLE 10.1
Main Antioxidant Strategies Against Sarcopenia

Antioxidant Strategies	Examples	Efficiency
RONS Scavengers	Vitamins C and E, carotenoids, resveratrol	Poor efficiency
RONS source inhibitors	Allopurinol, oxypurinol, inhibitors of mitochondrial RONS production	Lack of data in elderlies but very good results in muscle disuse models, need confirmation in elderlies
Precursor of antioxidants	Glutathione precursors	Very good results in rodents, need confirmation in elderlies
Potential targets	G6PDH ·	Very good results in mice overexpressing G6PDH, need confirmation in elderlies
Physical training	Endurance and/or strength	The best strategy. The only solution that can reverse sarcopenia

although an antioxidant supplementation in old rats with an antioxidant cocktail during 7 weeks (vitamin E, vitamin A, zinc, and selenium) was able to improve the ability of leucine to stimulate protein synthesis in muscles of old rats, no clear effect on muscle mass was observed (Marzani et al. 2008). Antioxidant supplementation was probably not long enough. This study highlighted that an optimal redox status would be an important in protein synthesis. Recently, Nalbant et al. and Bobeuf et al. in older people receiving, respectively, only vitamin E or an antioxidant cocktail (vitamin C and vitamin E) during 6 months failed to show improvement in physical performance and muscle strength (Nalbant et al. 2009; Bobeuf et al. 2011). In the same way, old mice receiving a diet supplemented with resveratrol (0.05% of the total diet) during 10 months presented decreased H_2O_2 muscle content and reduced lipid peroxidation levels associated with an increase in Mn-SOD activities. These parameters were comparable with those observed in younger mice. However, muscle weight (gastrocnemius and plantaris) and functions were not improved (Jackson, Ryan, and Alway 2011). Recently, a study showed in 18-month-old aged mice treated with resveratrol (15 mg/kg/day), a significant longer time to exhaustion with decreased blood lactate and free fatty acid levels associated with improved oxidative stress evidenced by decreased gastrocnemius muscle lipid peroxidation and increased antioxidant enzyme activities, catalase and superoxide dismutase, compared with the aged mice control group (Muhammad and Allam 2018). Resveratrol alone in human has not still showed good results. However, recently Always et al. (2017) showed that resveratrol enhances exercise-induced cellular and functional adaptations of skeletal muscle in older men and women.

For RONS inhibitors, few data are available; however, allopurinol or specific inhibitor of mitochondrial RONS showed very good results to prevent muscle disuse in hypoactivity or diaphragm mechanical ventilation models (Pierre et al. 2016; Powers et al. 2016). For example, although they did not directly treat older people with allopurinol (pharmaceutical inhibitor of xanthine oxidase), Beveridge et al. showed in a retrospective observational study that allopurinol use is associated with greater functional gains in older rehabilitation patients (Beveridge et al. 2013). Moreover, Derbre, Ferrando, et al. (2012) showed in young animal that allopurinol protected against muscle atrophy induced by hind limbs suspension. Recently, Ferrando et al. (2018) showed similar results in humans with lower leg immobilization following ankle sprain in humans. However, allopurinol has side effects with prolonged treatment, and we do not have enough data for specific inhibitor of mitochondrial RONS to definitively conclude.

Supplementations with precursors of antioxidant could be a solution when physical exercise is not possible. Indeed, few years ago, Sinha-Hikim et al. (2013) showed that a cocktail based on glutathione precursors during 6 months prevents muscle mass decrease. In this study, they supplemented old mice from 18 months old to 23 months old with a GSH precursor cocktail containing L-cystine, selenomethionine, and L-glutamine. Old control animals presented gastrocnemius atrophy (attested by weight and Cross Sectional Area (CSA)) associated with increased Oxidative Stress (OS) and decreased antioxidant enzyme activities, exacerbated apoptosis, reduced regenerative potential of skeletal muscle, and maybe impaired protein turnover (supposed by a decreased phosphorylation of Akt). In contrast, old rats treated with this GSH precursor cocktail presented an increased GSH/GSSG ratio associated with an increase in glucose-6-phosphate dehydrogenase (G6PDH) muscle protein content. Moreover, the age-related decline in SOD activity was totally reversed and lipid peroxidation. These beneficial effects on OS were concomitant to an improved regenerative potential of skeletal muscle (attested by an upregulation of the principal compounds of the Notch signaling), a decreased apoptosis index, and an increased Akt phosphorylation. Finally, old treated mice presented a higher muscle mass measured through a higher muscle weight and CSA.

10.3.2 Exercise as an Antioxidant to Reverse Sarcopenia

Physical training is currently the best strategy against sarcopenia because it can increase muscle mass and muscle strength and improve physical performance (Brioche and Lemoine-Morel 2016; Marzetti et al. 2017; Simioni et al. 2018). Moreover, exercise has numerous other beneficial effects on metabolism, cardiovascular, and reproductive systems (Montero-Fernandez and Serra-Rexach 2013; Pillard et al. 2011; Wang and Bai 2012) and does not have any side effects. Furthermore, exercise is known to improve quality of life and psychological health; is associated with better mental health and social integration; and improves anxiety, depression, and self-efficacy in older adults (Mather et al. 2002).

Exercise during aging improves protein turnover (Kim et al. 2013; Agergaard et al. 2017; Konopka et al. 2011; Luo et al. 2013; Short et al. 2004; Wohlgemuth et al. 2010), decreases apoptosis (Luo et al. 2013; Liao et al. 2017; Song, Kwak, and Lawler 2006; Wohlgemuth et al. 2010; Marzetti et al. 2008), stimulates SCs (Leenders et al. 2013; Pugh et al. 2018; Mackey et al. 2007; Shefer et al. 2010; Verdijk et al. 2009; Verney et al. 2008), and improves mitochondrial functions and dynamics (Koltai et al. 2010; Seldeen et al. 2018; Konopka and Sreekumaran Nair 2013; Lanza et al. 2008). As previously presented, all these mechanisms are impaired by oxidative stress, and their improvement could be facilitate by the restoration of a "young redox status" by exercise. Endurance training will restore a young redox status (Brioche and Lemoine-Morel 2016). Endurance exercise activates Peroxisome proliferator-activated receptor gamma coactivator (PGC1-α) leading to higher antioxidant defenses (Mason et al. 2016). Strength training is known to increases insulin-like growth factor-1 (IGF-1) levels, which would activate the KEAP1/NRF2 pathway leading to higher antioxidant defenses. However, this hypothesis needs to more studied. Although physical activity has many well-established health benefits, aging and strenuous exercise (especially exhaustive exercise) are associated with increased free radical generation (Ji 2001). Studies support a notion that heavy training may cause a deficit in muscle antioxidant reserve and protective margin (Gohil et al. 1987; Ji 2001; Starnes et al. 1989). Then, training programs in elderlies have to be progressive, and hence they have to avoid exhaustive exercise. Currently, it would be better to perform multimodal interventions, including exercise programs and nutritional supplementation (Nascimento et al. 2019).

10.4 GLUCOSE-6-PHOSPHATE DEHYDROGENASE AS A POTENTIAL NEW TARGET

G6PDH could be a target to fight sarcopenia. G6PDH was first described in 1931 (Kornberg 1955). Its role in skeletal muscle have been poorly studied, whereas several clinical cases of rhabdomyolysis due to G6PDH deficiency have been reported more than 15 years ago (Kimmick and Owen 1996). Moreover, studies have shown since the 1980s that deregulation of its activity is associated with myopathies (Meijer and Elias 1984; Elias and Meijer 1983). G6PDH controls the entry of glucose-6-phosphate (G6P) into the pentose phosphate pathway (PPP), which is also known as hexose monophosphate shunt. The major products of the PPP are ribose-5-phosphate (R5P) and nicotinamide adenine dinucleotide phosphate (NADPH) generated from NADP by G6PDH and the next enzyme in the pathway, 6-phosphogluconate dehydrogenase (PGD). In the following paragraphs, it will be exposed why through NADPH and R5P, G6PDH may be involved in sarcopenia and why enhancing their production by G6PDH would help to combat sarcopenia.

10.4.1 G6PDH AND ANTIOXIDANT DEFENSES

Several antioxidant systems depend on the production of NADPH for proper function. The first is the glutathione system dependent on the production of reduced glutathione by glutathione reductase (GR) that depends on NADPH (Scott, Wagner, and Chiu 1993). Catalase does not need NADPH to convert hydrogen peroxide to water but has an allosteric binding site for NADPH that maintains catalase in its active conformation (Scott, Wagner, and Chiu 1993). Superoxide dismutase does not use NADPH to convert superoxide to hydrogen peroxide; however, if this is not adequately reduced chemically by catalase or glutathione, the increased hydrogen peroxide levels will quantitatively increase and inhibit the SOD activity (Stanton 2012). It has been shown in various studies that during sarcopenia and aging, decreased G6PDH activity and/or muscle protein content are

associated with a depletion of GSH, an increase in the GSSG/GSH ratio associated with GR, glutathione peroxidase (Gpx), catalase, and SOD decreased activity or protein content (Brioche et al.
2014; Kovacheva et al. 2010; Kumaran et al. 2004; Sinha-Hikim et al. 2013). In contrast, in response
to different antioxidant strategies, testosterone or GH treatment in rats, G6PDH protein content or
activity was increased in skeletal muscle, and a concomitant increase in GSH, GR, Gpx, Cat, and
SOD activities was observed leading to a reduce oxidative damage (Brioche et al. 2014; Kovacheva
et al. 2010; Kumaran et al. 2004; Sinha-Hikim et al. 2013).

10.4.2   G6PDH AND APOPTOSIS

Various studies in cell culture have shown a direct negative relation between G6PDH activity and/or protein content and apoptosis (Fico et al. 2004; Nutt et al. 2005; Tian et al. 1999;
Salvemini et al. 1999). For instance, G6PDH-deleted embryonic stem cells a more sensitive to
H_2O_2-induced apoptosis associated with GSH depletion and increased caspase 3 and 9 protein
content (Fico et al. 2004). In contrast, Nutt et al. (2005) have shown that inhibition G6PDH
by Dehydroepiandrosterone (DHEA) activated caspase 2 and promote oocyte apoptosis. In old
rodents, G6PDH decreased activity and/or protein content in skeletal muscle are associated with
increased apoptosis and atrophy (Braga et al. 2008; Kovacheva et al. 2010; Sinha-Hikim et al.
2013). Moreover, Braga et al. (2008) confirmed in old mice that depletion in G6PDH protein content is associated with enhancement of caspase 2 and caspase 9 protein content in skeletal muscle.
In contrast, in response to different strategies to fight against sarcopenia, increased G6PDH activity or protein content is associated with decreased apoptosis and muscle hypertrophy (Brioche
et al. 2014; Kovacheva et al. 2010; Sinha-Hikim et al. 2013).

10.4.3   G6PDH AND SATELLITE CELL LINKS

G6PDH activity would have an important role in muscle hypertrophy and regeneration by acting
on the potential proliferation of SCs, RNA, and protein synthesis. Indeed, in various old works
studying the muscle degeneration–regeneration cycle, it has been shown that during regeneration
(known to involved SCs), G6PDH activity is dramatically increased (Wagner, Kauffman, and Max
1978; Wagner, Carlson, and Max 1977) while protein synthesis and RNA synthesis were increased
(Wagner et al. 1978). Moreover, inhibition of mRNA and protein synthesis was associated with
G6PDH inhibition (Wagner et al. 1978). Thus, it was argued that G6PDH would play an important
role in mRNA and DNA synthesis since it is the rate-limiting enzyme of the PPP, which is the main
pathway synthesizing R5P, an essential compound of nucleic acid. Through this role, G6PDH would
indirectly impact protein synthesis. These various hypotheses were confirmed by studies *in vitro*
that have shown that overexpression of G6PDH accelerates proliferation of numerous cell lines
associated with increases in DNA and protein synthesis (Tian et al. 1998; Kuo et al. 2000). In contrast, G6PDH-deficient cells presented lower growth rate (Ho et al. 2000). Furthermore, inhibition
of G6PDH caused cells to be more susceptible to the growth inhibitory effects of H_2O_2 due to
NADPH decrease leading to reduce the GSH content (Tian et al. 1998). As inhibition of G6PDH
in cultured cells leads to decrease their proliferation due to a decreased protein and DNA synthesis
associated with an impaired redox status, it could be hypothesized that G6PDH decrease (activity and protein content) observed in skeletal muscle during aging would participate to reduce the
regenerative capacity of skeletal muscle. In contrast, increased G6PDH activity would improve
this mechanism. Data in this way have been published by Kovacheva et al. (2010) and Brioche
et al. (2014), which showed impaired SC proliferation associated with a decrease in skeletal muscle G6PDH protein content and increase in oxidative damage in old sarcopenic mice. Conversely
treated mice or rats with testosterone or GH presented an increased G6PDH muscle protein content
associated with SC proliferation and decreased oxidative damage (Brioche et al. 2014; Kovacheva
et al. 2010). G6PDH decreased during aging would participate to decrease protein synthesis. Until

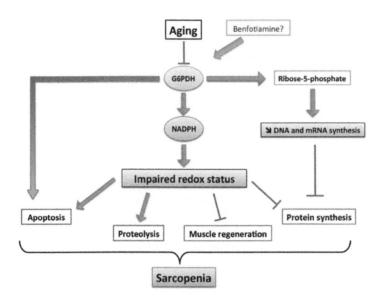

FIGURE 10.2 Mechanisms by which G6PDH could prevent sarcopenia.

now, only a decrease in Akt and p70S6K activation associated with decreased G6PDH activity and atrophy would support this hypothesis in skeletal muscle (Brioche et al. 2014; Kovacheva et al. 2010; Sinha-Hikim et al. 2013).

10.4.4   G6PDH AND PROTEIN BALANCE

Finally, decreased G6PDH activity and/or protein content in skeletal muscle observed during aging would participate in sarcopenia by decreasing the antioxidant capacity, which would impair the phosphoinositide 3-kinase (PI3K)/Akt/mTOR pathway leading to decrease protein synthesis. There is no exiting data about G6PDH and proteolysis. However, by decreasing antioxidant defense, RONS would accumulate their self and promote the activation of several proteolysis pathway (Derbre et al. 2014). In contrast, the parallel decrease in Akt phosphorylation and G6PDH activity leads to apoptosis through caspase activation. Decrease in G6PDH activity would reduce the regenerative potential of skeletal muscle by limiting SC proliferation. Activate G6PDH would restore an optimal redox status and reverse these adverse effects. All these mechanisms are shown in Figure 10.2.

Finally, G6PDH should be more studied since its upregulation in transgenic flies extends their life span (Legan et al. 2008) and since we showed that moderate G6PDH overexpression increased NADPH and GSH associated with lower oxidative damage to DNA and lipids. These effects were associated with a longer median life expectancy in female and with better neuromuscular functions in very old mice (Nobrega-Pereira et al. 2016). Our current work is showing that overexpression of G6DPH in mice delays occurrence of frailty by maintaining muscle strength and muscle mass (Coralie Arc-Chagnaud 2018). Currently, benfotiamine, a synthetic S-acyl derivative of thiamine (vitamin B1), would be the only nonpharmaceutical strategies to activate G6PDH, which could be try to fight sarcopenia.

10.5   CONCLUSION AND PERSPECTIVES

This review clearly outlines the role of oxidative stress in age-related sarcopenia. However, more studies have to be performed in elderlies to confirm all the molecular mechanisms found in cell cultures and animal models. The identification of cost-effectiveness interventions to maintain muscle mass and physical functions in the elderly is one of the most important public health challenges. Since oxidative stress appears to be a major actor in the onset of sarcopenia, antioxidant strategies would be a very potent tool to prevent these later. However, it seems that directly provide RONS scavengers is not a good solution, whereas provide precursors of antioxidants can slow sarcopenia development. Exercise, which act in part by antioxidant mechanisms, appears to be the perfect strategy against sarcopenia because it can lead to an increase in muscle mass, strength, and physical performance. Antioxidant defense systems are largely dependent on the reductive power of NAPDH whose levels are mostly determined by the enzyme G6PDH. G6PDH could be a new target since it is upregulated in various effective strategies against sarcopenia, its overexpression in flies extends lifespan, and its overexpression in mice extends median life expectancy in female with better neuromuscular functions in very old mice. To conclude, any efforts to develop strategies to prevent muscle dysfunction leading to sarcopenia will be a major advance in the public health. Any significant improvement in the loss of muscle function will be a major breakthrough in the health and welfare of the population.

ACKNOWLEDGMENTS

Our studies are supported by the Centre National d'Etudes Spatiales (CNES). The authors are grateful to the Réseau d'Histologie Expérimentale de Montpellier (RHEM) platform. We also thank the animal-facility staff and our METAMUS platform facility, which belongs to the Montpellier Animal Facilities Network (RAM) and the Montpellier RIO Imaging (MRI).

REFERENCES

Agergaard, J., J. Bulow, J. K. Jensen, et al. 2017. Light-load resistance exercise increases muscle protein synthesis and hypertrophy signaling in elderly men. *Am J Physiol Endocrinol Metab* 312 (4):E326–E338.

Alway, S. E., J. L. McCrory, K. Kearcher, et al. 2017. Resveratrol enhances exercise-induced cellular and functional adaptations of skeletal muscle in older men and women. *J Gerontol A Biol Sci Med Sci* 72 (12):1595–1606.

Andersson, D. C., M. J. Betzenhauser, S. Reiken, et al. 2011. Ryanodine receptor oxidation causes intracellular calcium leak and muscle weakness in aging. *Cell Metab* 14 (2):196–207.

Anderson, E. J., and P. D. Neufer. 2006. Type II skeletal myofibers possess unique properties that potentiate mitochondrial H_2O_2 generation. *Am J Physiol Cell Physiol* 290 (3):C844–51.

Ardite, E., J. Albert Barbera, J. Roca, and J. C. Fernández-Checa. 2004. Glutathione depletion impairs myogenic differentiation of murine skeletal muscle C2C12 cells through sustained NF-κB activation. *Am J Pathol* 165 (3):719–28.

Balagopal, P., O. E. Rooyackers, D. B. Adey, P. A. Ades, and K. S. Nair. 1997. Effects of aging on *in vivo* synthesis of skeletal muscle myosin heavy-chain and sarcoplasmic protein in humans. *Am J Physiol* 273 (4 Pt 1):E790–800.

Balagopal, P., J. C. Schimke, P. Ades, D. Adey, and K. S. Nair. 2001. Age effect on transcript levels and synthesis rate of muscle MHC and response to resistance exercise. *Am J Physiol Endocrinol Metab* 280 (2):E203–8.

Beccafico, S., C. Puglielli, T. Pietrangelo, R. Bellomo, G. Fano, and S. Fulle. 2007. Age-dependent effects on functional aspects in human satellite cells. *Ann N Y Acad Sci* 1100:345–52.

Beveridge, L. A., L. Ramage, M. E. McMurdo, J. George, and M. D. Witham. 2013. Allopurinol use is associated with greater functional gains in older rehabilitation patients. *Age Ageing* 42 (3):400–4.

Bobeuf, F., M. Labonte, I. J. Dionne, and A. Khalil. 2011. Combined effect of antioxidant supplementation and resistance training on oxidative stress markers, muscle and body composition in an elderly population. *J Nutr Health Aging* 15 (10):883–9.

Braga, M., A. P. Sinha Hikim, S. Datta, et al. 2008. Involvement of oxidative stress and caspase 2-mediated intrinsic pathway signaling in age-related increase in muscle cell apoptosis in mice. *Apoptosis* 13 (6):822–32.

Brioche, T., R. A. Kireev, S. Cuesta, et al. 2014. Growth hormone replacement therapy prevents sarcopenia by a dual mechanism: Improvement of protein balance and of antioxidant defenses. *J Gerontol A Biol Sci Med Sci* 69 (10):1186–98.

Brioche, T., and S. Lemoine-Morel. 2016. Oxidative stress, sarcopenia, antioxidant strategies and exercise: Molecular aspects. *Curr Pharm Des* 22 (18):2664–78.

Brioche, T., A. F. Pagano, G. Py, and A. Chopard. 2016. Muscle wasting and aging: Experimental models, fatty infiltrations, and prevention. *Mol Aspects Med* 50:56–87.

Calvani, R., A. M. Joseph, P. J. Adhihetty, et al. 2013. Mitochondrial pathways in sarcopenia of aging and disuse muscle atrophy. *Biol Chem* 394 (3):393–414.

Capel, F., C. Buffiere, P. Patureau Mirand, and L. Mosoni. 2004. Differential variation of mitochondrial H_2O_2 release during aging in oxidative and glycolytic muscles in rats. *Mech Ageing Dev* 125 (5):367–73.

Carlson, B. M., and J. A. Faulkner. 1989. Muscle transplantation between young and old rats: Age of host determines recovery. *Am J Physiol* 256 (6 Pt 1):C1262–6.

Carlson, M. E., C. Suetta, M. J. Conboy, et al. 2009. Molecular aging and rejuvenation of human muscle stem cells. *EMBO Mol Med* 1 (8–9):381–91.

Chabi, B., V. Ljubicic, K. J. Menzies, J. H. Huang, A. Saleem, and D. A. Hood. 2008. Mitochondrial function and apoptotic susceptibility in aging skeletal muscle. *Aging Cell* 7 (1):2–12.

Claflin, D. R., L. M. Larkin, P. S. Cederna, et al. 2011. Effects of high- and low-velocity resistance training on the contractile properties of skeletal muscle fibers from young and older humans. *J Appl Physiol* 111 (4):1021–30.

Coen, P. M., S. A. Jubrias, G. Distefano, et al. 2013. Skeletal muscle mitochondrial energetics are associated with maximal aerobic capacity and walking speed in older adults. *J Gerontol A Biol Sci Med Sci* 68 (4):447–55.

Combaret, L., D. Dardevet, D. Bechet, D. Taillandier, L. Mosoni, and D. Attaix. 2009. Skeletal muscle proteolysis in aging. *Curr Opin Clin Nutr Metab Care* 12 (1):37–41.

Coralie A.-C., S.-P. Andrea, C. Aitor, T. Brioche, C. Angèle, P. J. Fernandez-Marcos, M. Serrano, G.-C. M. Carmen , V. José. 2018. Overexpression of G6PD as a model of robustness. *Free Radic Biol Med* 120:S81.

Cruz-Jentoft, A. 2012. *Sarcopenia*. Wiley-Blackwell ed Hoboken, NJ.

Cuthbertson, D., K. Smith, J. Babraj, et al. 2005. Anabolic signaling deficits underlie amino acid resistance of wasting, aging muscle. *FASEB J* 19 (3):422–4.

Davies, K. J., S. W. Lin, and R. E. Pacifici. 1987. Protein damage and degradation by oxygen radicals. IV. Degradation of denatured protein. *J Biol Chem* 262 (20):9914–20.

Delmonico, M. J., T. B. Harris, J. S. Lee, et al. 2007. Alternative definitions of sarcopenia, lower extremity performance, and functional impairment with aging in older men and women. *J Am Geriatr Soc* 55 (5):769–74.

Derbre, F., B. Ferrando, M. C. Gomez-Cabrera, et al. 2012. Inhibition of xanthine oxidase by allopurinol prevents skeletal muscle atrophy: Role of p38 MAPKinase and E3 ubiquitin ligases. *PLoS One* 7 (10):e46668.

Derbre, F., M. C Gomez-Cabrera, A. L. Nascimento, et al. 2012. Age associated low mitochondrial biogenesis may be explained by lack of response of PGC-1alpha to exercise training. *Age (Dordr)* 34 (3):669–79.

Derbre, F., A. Gratas-Delamarche, M. C. Gomez-Cabrera, and J. Vina. 2014. Inactivity-induced oxidative stress: A central role in age-related sarcopenia? *Eur J Sport Sci* 14 (1):S98–108.

Elias, E. A., and A. E. Meijer. 1983. The increase in activity of the NADPH-regenerating enzymes of the pentose phosphate pathway in vitamin E deficiency induced myopathy in rabbits: A histochemical and biochemical study. *Cell Mol Biol* 29 (1):27–37.

Feng, Z., H. Zhang, A. J. Levine, and S. Jin. 2005. The coordinate regulation of the p53 and mTOR pathways in cells. *Proc Natl Acad Sci U S A* 102 (23):8204–9.

Ferrando, B., M. C. Gomez-Cabrera, A. Salvador-Pascual, et al. 2018. Allopurinol partially prevents disuse muscle atrophy in mice and humans. *Sci Rep* 8 (1):3549.

Fico, A., F. Paglialunga, L. Cigliano, et al. 2004. Glucose-6-phosphate dehydrogenase plays a crucial role in protection from redox-stress-induced apoptosis. *Cell Death Differ* 11 (8):823–31.

Figueiredo, P. A., S. K. Powers, R. M. Ferreira, H. J. Appell, and J. A. Duarte. 2009. Aging impairs skeletal muscle mitochondrial bioenergetic function. *J Gerontol A Biol Sci Med Sci* 64 (1):21–33.

Fougere, B., G. A. van Kan, B. Vellas, and M. Cesari. 2018. Redox Systems, Antioxidants and Sarcopenia. *Curr Protein Pept Sci* 19 (7):643–48.

Fraysse, B., J. F. Desaphy, J. F. Rolland, et al. 2006. Fiber type-related changes in rat skeletal muscle calcium homeostasis during aging and restoration by growth hormone. *Neurobiol Dis* 21 (2):372–80.

Fulle, S., S. Di Donna, C. Puglielli, et al. 2005. Age-dependent imbalance of the antioxidative system in human satellite cells. *Exp Gerontol* 40 (3):189–97.

Fulle, S., F. Protasi, G. Di Tano, et al. 2004. The contribution of reactive oxygen species to sarcopenia and muscle ageing. *Exp Gerontol* 39 (1):17–24.

Gohil, K., L. Rothfuss, J. Lang, and L. Packer. 1987. Effect of exercise training on tissue vitamin E and ubiquinone content. *J Appl Physiol* 63 (4):1638–41.

Goll, D. E., V. F. Thompson, H. Li, W. Wei, and J. Cong. 2003. The calpain system. *Physiol Rev* 83 (3):731–801.

Gonzalez, E., M. L. Messi, and O. Delbono. 2000. The specific force of single intact extensor digitorum longus and soleus mouse muscle fibers declines with aging. *J Membr Biol* 178 (3):175–83.

Goodpaster, B. H., S. W. Park, T. B. Harris, et al. 2006. The loss of skeletal muscle strength, mass, and quality in older adults: The health, aging and body composition study. *J Gerontol A Biol Sci Med Sci* 61 (10):1059–64.

Grune, T., K. Merker, G. Sandig, and K. J. Davies. 2003. Selective degradation of oxidatively modified protein substrates by the proteasome. *Biochem Biophys Res Commun* 305 (3):709–18.

Gumucio, J. P., and C. L. Mendias. 2013. Atrogin-1, MuRF-1, and sarcopenia. *Endocrine* 43 (1):12–21.

Haddad, F., and G. R. Adams. 2006. Aging-sensitive cellular and molecular mechanisms associated with skeletal muscle hypertrophy. *J Appl Physiol* 100 (4):1188–203.

Hansen, J. M., M. Klass, C. Harris, and M. Csete. 2007. A reducing redox environment promotes C2C12 myogenesis: implications for regeneration in aged muscle. *Cell Biol Int* 31 (6):546–53.

Hikida, R. S. 2011. Aging changes in satellite cells and their functions. *Curr Aging Sci* 4 (3):279–97.

Ho, H. Y., M. L. Cheng, F. J. Lu, et al. 2000. Enhanced oxidative stress and accelerated cellular senescence in glucose-6-phosphate dehydrogenase (G6PD)-deficient human fibroblasts. *Free Radic Biol Med* 29 (2):156–69.

Jackson, J. R., M. J. Ryan, and S. E. Alway. 2011. Long-term supplementation with resveratrol alleviates oxidative stress but does not attenuate sarcopenia in aged mice. *J Gerontol A Biol Sci Med Sci* 66 (7):751–64.

Janssen, I., R. N. Baumgartner, R. Ross, I. H. Rosenberg, and R. Roubenoff. 2004. Skeletal muscle cutpoints associated with elevated physical disability risk in older men and women. *Am J Epidemiol* 159 (4):413–21.

Ji, L. L. 2001. Exercise at old age: Does it increase or alleviate oxidative stress? *Ann N Y Acad Sci* 928:236–47.

Jimenez-Moreno, R., Z. M. Wang, R. C. Gerring, and O. Delbono. 2008. Sarcoplasmic reticulum Ca^{2+} release declines in muscle fibers from aging mice. *Biophys J* 94 (8):3178–88.

Kandarian, S. C., and E. J. Stevenson. 2002. Molecular events in skeletal muscle during disuse atrophy. *Exerc Sport Sci Rev* 30 (3):111–16.

Kim, Y. A., Y. S. Kim, S. L. Oh, H. J. Kim, and W. Song. 2013. Autophagic response to exercise training in skeletal muscle with age. *J Physiol Biochem* 69 (4):697–705.

Kimmick, G., and J. Owen. 1996. Rhabdomyolysis and hemolysis associated with sickle cell trait and glucose-6-phosphate dehydrogenase deficiency. *South Med J* 89 (11):1097–8.

Koltai, E., Z. Szabo, M. Atalay, et al. 2010. Exercise alters SIRT1, SIRT6, NAD and NAMPT levels in skeletal muscle of aged rats. *Mech Ageing Dev* 131 (1):21–8.

Konopka, A. R., and K. Sreekumaran Nair. 2013. Mitochondrial and skeletal muscle health with advancing age. *Mol Cell Endocrinol* 379 (1–2):19–29.

Konopka, A. R., T. A. Trappe, B. Jemiolo, S. W. Trappe, and M. P. Harber. 2011. Myosin heavy chain plasticity in aging skeletal muscle with aerobic exercise training. *J Gerontol A Biol Sci Med Sci* 66 (8):835–41.

Kornberg, A., Horecker, B.L. & Smyrniotis, P.Z. 1955. Glucose-6-phosphate dehydrogenase 6-phosphogluconic dehydrogenase. *Methods Enzymol* 1:323–327.

Kovacheva, E. L., A. P. Hikim, R. Shen, I. Sinha, and I. Sinha-Hikim. 2010. Testosterone supplementation reverses sarcopenia in aging through regulation of myostatin, c-Jun NH_2-terminal kinase, Notch, and Akt signaling pathways. *Endocrinology* 151 (2):628–38.

Kumaran, S., S. Savitha, M. Anusuya Devi, and C. Panneerselvam. 2004. L-carnitine and DL-alpha-lipoic acid reverse the age-related deficit in glutathione redox state in skeletal muscle and heart tissues. *Mech Ageing Dev* 125 (7):507–12.

Kuo, W., J. Lin, and T. K. Tang. 2000. Human glucose-6-phosphate dehydrogenase (G6PD) gene transforms NIH 3T3 cells and induces tumors in nude mice. *Int J Cancer* 85 (6):857–64.

Lanza, I. R., D. K. Short, K. R. Short, et al. 2008. Endurance exercise as a countermeasure for aging. *Diabetes* 57 (11):2933–42.

Lee, H. Y., C. S. Choi, A. L. Birkenfeld, et al. 2010. Targeted expression of catalase to mitochondria prevents age-associated reductions in mitochondrial function and insulin resistance. *Cell Metab* 12 (6):668–74.

Leenders, M., L. B. Verdijk, L. van der Hoeven, J. van Kranenburg, R. Nilwik, and L. J. van Loon. 2013. Elderly men and women benefit equally from prolonged resistance-type exercise training. *J Gerontol A Biol Sci Med Sci* 68 (7):769–79.

Legan, S. K., I. Rebrin, R. J. Mockett, et al. 2008. Overexpression of glucose-6-phosphate dehydrogenase extends the life span of *Drosophila melanogaster*. *J Biol Chem* 283 (47):32492–9.

Li, Y. P., Y. Chen, J. John, et al. 2005. TNF-alpha acts via p38 MAPK to stimulate expression of the ubiquitin ligase atrogin1/MAFbx in skeletal muscle. *FASEB J* 19 (3):362–70.

Li, Y. P., Y. Chen, A. S. Li, and M. B. Reid. 2003. Hydrogen peroxide stimulates ubiquitin-conjugating activity and expression of genes for specific E2 and E3 proteins in skeletal muscle myotubes. *Am J Physiol Cell Physiol* 285 (4):C806–12.

Liao, Z. Y., J. L. Chen, M. H. Xiao, et al. 2017. The effect of exercise, resveratrol or their combination on Sarcopenia in aged rats via regulation of AMPK/Sirt1 pathway. *Exp Gerontol* 98:177–83.

Lowe, D. A., J. T. Surek, D. D. Thomas, and L. V. Thompson. 2001. Electron paramagnetic resonance reveals age-related myosin structural changes in rat skeletal muscle fibers. *Am J Physiol Cell Physiol* 280 (3):C540–7.

Lowe, D. A., G. L. Warren, L. M. Snow, L. V. Thompson, and D. D. Thomas. 2004. Muscle activity and aging affect myosin structural distribution and force generation in rat fibers. *J Appl Physiol (1985)* 96 (2):498–506.

Luo, L., A. M. Lu, Y. Wang, et al. 2013. Chronic resistance training activates autophagy and reduces apoptosis of muscle cells by modulating IGF-1 and its receptors, Akt/mTOR and Akt/FOXO3a signaling in aged rats. *Exp Gerontol* 48 (4):427–36.

Mackey, A. L., B. Esmarck, F. Kadi, et al. 2007. Enhanced satellite cell proliferation with resistance training in elderly men and women. *Scand J Med Sci Sports* 17 (1):34–42.

Marzani, B., M. Balage, A. Venien, et al. 2008. Antioxidant supplementation restores defective leucine stimulation of protein synthesis in skeletal muscle from old rats. *J Nutr* 138 (11):2205–11.

Marzetti, E., R. Calvani, R. Bernabei, and C. Leeuwenburgh. 2012. Apoptosis in skeletal myocytes: A potential target for interventions against sarcopenia and physical frailty—A mini-review. *Gerontology* 58 (2):99–106.

Marzetti, E., R. Calvani, M. Cesari, et al. 2013. Mitochondrial dysfunction and sarcopenia of aging: From signaling pathways to clinical trials. *Int J Biochem Cell Biol* 45 (10):2288–301.

Marzetti, E., R. Calvani, M. Tosato, et al. 2017. Physical activity and exercise as countermeasures to physical frailty and sarcopenia. *Aging Clin Exp Res* 29 (1):35–42.

Marzetti, E., L. Groban, S. E. Wohlgemuth, et al. 2008. Effects of short-term GH supplementation and treadmill exercise training on physical performance and skeletal muscle apoptosis in old rats. *Am J Physiol Regul Integr Comp Physiol* 294 (2):R558–67.

Marzetti, E., H. A. Lees, T. M. Manini, et al. 2012. Skeletal muscle apoptotic signaling predicts thigh muscle volume and gait speed in community-dwelling older persons: An exploratory study. *PLoS One* 7 (2):e32829.

Mason, S. A., D. Morrison, G. K. McConell, and G. D. Wadley. 2016. Muscle redox signalling pathways in exercise: Role of antioxidants. *Free Radic Biol Med* 98:29–45.

Mather, A. S., C. Rodriguez, M. F. Guthrie, A. M. McHarg, I. C. Reid, and M. E. McMurdo. 2002. Effects of exercise on depressive symptoms in older adults with poorly responsive depressive disorder: Randomised controlled trial. *Br J Psychiatry* 180:411–15.

Meijer, A. E., and E. A. Elias. 1984. The inhibitory effect of actinomycin D and cycloheximide on the increase in activity of glucose-6-phosphate dehydrogenase and 6-phosphogluconate dehydrogenase in experimentally induced diseased skeletal muscles. *Histochem J* 16 (9):971–82.

Montero-Fernandez, N., and J. A. Serra-Rexach. 2013. Role of exercise on sarcopenia in the elderly. *Eur J Phys Rehabil Med* 49 (1):131–43.

Muhammad, M. H., and M. M. Allam. 2018. Resveratrol and/or exercise training counteract aging-associated decline of physical endurance in aged mice; targeting mitochondrial biogenesis and function. *J Physiol Sci* 68 (5):681–88.

Nalbant, O., N. Toktas, N. F. Toraman, et al. 2009. Vitamin E and aerobic exercise: Effects on physical performance in older adults. *Aging Clin Exp Res* 21 (2):111–21.

Nascimento, C. M., M. Ingles, A. Salvador-Pascual, M. R. Cominetti, M. C. Gomez-Cabrera, and J. Vina. 2019. Sarcopenia, frailty and their prevention by exercise. *Free Radic Biol Med* 132:42–9.

Nobrega-Pereira, S., P. J. Fernandez-Marcos, T. Brioche, et al. 2016. G6PD protects from oxidative damage and improves healthspan in mice. *Nat Commun* 7:10894.

Nutt, L. K., S. S. Margolis, M. Jensen, et al. 2005. Metabolic regulation of oocyte cell death through the CaMKII-mediated phosphorylation of caspase-2. *Cell* 123 (1):89–103.

O'Loghlen, A., M. I. Perez-Morgado, M. Salinas, and M. E. Martin. 2006. N-acetyl-cysteine abolishes hydrogen peroxide-induced modification of eukaryotic initiation factor 4F activity via distinct signalling pathways. *Cell Signal* 18 (1):21–31.

Picca, A., R. Calvani, M. Bossola, et al. 2018. Update on mitochondria and muscle aging: All wrong roads lead to sarcopenia. *Biol Chem* 399 (5):421–36.

Pierre, N., Z. Appriou, A. Gratas-Delamarche, and F. Derbre. 2016. From physical inactivity to immobilization: Dissecting the role of oxidative stress in skeletal muscle insulin resistance and atrophy. *Free Radic Biol Med* 98:197–207.

Pillard, F., D. Laoudj-Chenivesse, G. Carnac, et al. 2011. Physical activity and sarcopenia. *Clin Geriatr Med* 27 (3):449–70.

Powers, S. K., A. B. Morton, B. Ahn, and A. J. Smuder. 2016. Redox control of skeletal muscle atrophy. *Free Radic Biol Med* 98:208–17.

Pugh, J. K., S. H. Faulkner, M. C. Turner, and M. A. Nimmo. 2018. Satellite cell response to concurrent resistance exercise and high-intensity interval training in sedentary, overweight/obese, middle-aged individuals. *Eur J Appl Physiol* 118 (2):225–38.

Renault, V., L. E. Thornell, P. O. Eriksson, G. Butler-Browne, and V. Mouly. 2002. Regenerative potential of human skeletal muscle during aging. *Aging Cell* 1 (2):132–9.

Renganathan, M., M. L. Messi, and O. Delbono. 1998. Overexpression of IGF-1 exclusively in skeletal muscle prevents age-related decline in the number of dihydropyridine receptors. *J Biol Chem* 273 (44):28845–51.

Roberts, M. D., C. M. Kerksick, V. J. Dalbo, S. E. Hassell, P. S. Tucker, and R. Brown. 2010. Molecular attributes of human skeletal muscle at rest and after unaccustomed exercise: An age comparison. *J Strength Cond Res* 24 (5):1161–8.

Safdar, A., M. J. Hamadeh, J. J. Kaczor, S. Raha, J. Debeer, and M. A. Tarnopolsky. 2010. Aberrant mitochondrial homeostasis in the skeletal muscle of sedentary older adults. *PLoS One* 5 (5):e10778.

Salvemini, F., A. Franze, A. Iervolino, S. Filosa, S. Salzano, and M. V. Ursini. 1999. Enhanced glutathione levels and oxidoresistance mediated by increased glucose-6-phosphate dehydrogenase expression. *J Biol Chem* 274 (5):2750–7.

Scherz-Shouval, R., E. Shvets, E. Fass, H. Shorer, L. Gil, and Z. Elazar. 2007. Reactive oxygen species are essential for autophagy and specifically regulate the activity of Atg4. *EMBO J* 26 (7):1749–60.

Scott, M. D., T. C. Wagner, and D. T. Chiu. 1993. Decreased catalase activity is the underlying mechanism of oxidant susceptibility in glucose-6-phosphate dehydrogenase-deficient erythrocytes. *Biochim Biophys Acta* 1181 (2):163–8.

Seldeen, K. L., G. Lasky, M. M. Leiker, M. Pang, K. E. Personius, and B. R. Troen. 2018. High intensity interval training improves physical performance and frailty in aged mice. *J Gerontol A Biol Sci Med Sci* 73 (4):429–37.

Shefer, G., G. Rauner, Z. Yablonka-Reuveni, and D. Benayahu. 2010. Reduced satellite cell numbers and myogenic capacity in aging can be alleviated by endurance exercise. *PLoS One* 5 (10):e13307.

Shenton, D., J. B. Smirnova, J. N. Selley, et al. 2006. Global translational responses to oxidative stress impact upon multiple levels of protein synthesis. *J Biol Chem* 281 (39):29011–21.

Short, K. R., M. L. Bigelow, J. Kahl, et al. 2005. Decline in skeletal muscle mitochondrial function with aging in humans. *Proc Natl Acad Sci U S A* 102 (15):5618–23.

Short, K. R., J. L. Vittone, M. L. Bigelow, D. N. Proctor, and K. S. Nair. 2004. Age and aerobic exercise training effects on whole body and muscle protein metabolism. *Am J Physiol Endocrinol Metab* 286 (1):E92–101.

Siems, W., E. Capuozzo, A. Lucano, C. Salerno, and C. Crifo. 2003. High sensitivity of plasma membrane ion transport ATPases from human neutrophils towards 4-hydroxy-2,3-trans-nonenal. *Life Sci* 73 (20):2583–90.

Simioni, C., G. Zauli, A. M. Martelli, et al. 2018. Oxidative stress: Role of physical exercise and antioxidant nutraceuticals in adulthood and aging. *Oncotarget* 9 (24):17181–98.

Sinha-Hikim, I., A. P. Sinha-Hikim, M. Parveen, et al. 2013. Long-term supplementation with a cystine-based antioxidant delays loss of muscle mass in aging. *J Gerontol A Biol Sci Med Sci* 68 (7):749–59.

Siu, P. M., E. E. Pistilli, Z. Murlasits, and S. E. Alway. 2006. Hindlimb unloading increases muscle content of cytosolic but not nuclear Id2 and p53 proteins in young adult and aged rats. *J Appl Physiol* 100 (3):907–16.

Siu, P. M., Y. Wang, and S. E. Alway. 2009. Apoptotic signaling induced by H_2O_2-mediated oxidative stress in differentiated C2C12 myotubes. *Life Sci* 84 (13–14):468–81.

Smuder, A. J., A. N. Kavazis, M. B. Hudson, W. B. Nelson, and S. K. Powers. 2010. Oxidation enhances myofibrillar protein degradation via calpain and caspase-3. *Free Radic Biol Med* 49 (7):1152–60.

Snijders, T., L. B. Verdijk, and L. J. van Loon. 2009. The impact of sarcopenia and exercise training on skeletal muscle satellite cells. *Ageing Res Rev* 8 (4):328–38.

Song, W., H. B. Kwak, and J. M. Lawler. 2006. Exercise training attenuates age-induced changes in apoptotic signaling in rat skeletal muscle. *Antioxid Redox Signal* 8 (3-4):517–28.

Stanton, R. C. 2012. Glucose-6-phosphate dehydrogenase, NADPH, and cell survival. *IUBMB Life* 64 (5):362–9.

Starnes, J. W., G. Cantu, R. P. Farrar, and J. P. Kehrer. 1989. Skeletal muscle lipid peroxidation in exercised and food-restricted rats during aging. *J Appl Physiol* 67 (1):69–75.

Thompson, L. V. 2009. Age-related muscle dysfunction. *Exp Gerontol* 44 (1–2):106–11.

Thompson, L. V., and M. Brown. 1999. Age-related changes in contractile properties of single skeletal fibers from the soleus muscle. *J Appl Physiol* 86 (3):881–6.

Thompson, L. V., D. Durand, N. A. Fugere, and D. A. Ferrington. 2006. Myosin and actin expression and oxidation in aging muscle. *J Appl Physiol* 101 (6):1581–7.

Tian, W. N., L. D. Braunstein, K. Apse, et al. 1999. Importance of glucose-6-phosphate dehydrogenase activity in cell death. *Am J Physiol* 276 (5 Pt 1):C1121–31.

Tian, W. N., L. D. Braunstein, J. Pang, et al. 1998. Importance of glucose-6-phosphate dehydrogenase activity for cell growth. *J Biol Chem* 273 (17):10609–17.

Tieland, M., I. Trouwborst, and B. C. Clark. 2018. Skeletal muscle performance and ageing. *J Cachexia Sarcopenia Muscle* 9 (1):3–19.

UN. 2007. World Population Ageing. In Department of Economic and Social Affairs. New York: United Nation.

Verdijk, L. B., B. G. Gleeson, R. A. Jonkers, et al. 2009. Skeletal muscle hypertrophy following resistance training is accompanied by a fiber type-specific increase in satellite cell content in elderly men. *J Gerontol A Biol Sci Med Sci* 64 (3):332–9.

Verney, J., F. Kadi, N. Charifi, et al. 2008. Effects of combined lower body endurance and upper body resistance training on the satellite cell pool in elderly subjects. *Muscle Nerve* 38 (3):1147–54.

Vina, J., M. C. Gomez-Cabrera, C. Borras, et al. 2009. Mitochondrial biogenesis in exercise and in ageing. *Adv Drug Deliv Rev* 61 (14):1369–74.

Wagner, K. R., B. M. Carlson, and S. R. Max. 1977. Developmental patterns of glycolytic enzymes in regenerating skeletal muscle after autogenous free grafting. *J Neurol Sci* 34 (3):373–90.

Wagner, K. R., F. C. Kauffman, and S. R. Max. 1978. The pentose phosphate pathway in regenerating skeletal muscle. *Biochem J* 170 (1):17–22.

Wang, C., and L. Bai. 2012. Sarcopenia in the elderly: basic and clinical issues. *Geriatr Gerontol Int* 12 (3):388–96.

Wang, Z. M., M. L. Messi, and O. Delbono. 2000. L-Type Ca^{2+} channel charge movement and intracellular Ca^{2+} in skeletal muscle fibers from aging mice. *Biophys J* 78 (4):1947–54.

Whidden, M. A., A. J. Smuder, M. Wu, M. B. Hudson, W. B. Nelson, and S. K. Powers. 2010. Oxidative stress is required for mechanical ventilation-induced protease activation in the diaphragm. *J Appl Physiol* 108 (5):1376–82.

Wohlgemuth, S. E., A. Y. Seo, E. Marzetti, H. A. Lees, and C. Leeuwenburgh. 2010. Skeletal muscle autophagy and apoptosis during aging: effects of calorie restriction and life-long exercise. *Exp Gerontol* 45 (2):138–48.

Zhang, H., and H. J. Forman. 2017. 4-hydroxynonenal-mediated signaling and aging. *Free Radic Biol Med* 111:219–25.

Zhang, L., S. R. Kimball, L. S. Jefferson, and J. S. Shenberger. 2009. Hydrogen peroxide impairs insulin-stimulated assembly of mTORC1. *Free Radic Biol Med* 46 (11):1500–9.

Section IV

Alterations in Muscle Protein Turnover in the Aging Process

11 Muscle Protein Turnover and Sarcopenia in the Elderly
The Effects of Nutrition

Paolo Tessari

CONTENTS

11.1 INTRODUCTION: AGE, SARCOPENIA, AND MUSCLE PROTEIN TURNOVER

The decrease of muscle mass and strength, defined as *sarcopenia*, occurring with ageing, is a "hot" issue today, because of the worldwide increase of life span, population of elderly subjects, and incidence and prevalence of ageing-associated chronic diseases and disabilities.

With ageing, body composition markedly changes. There is a progressive decrease of lean body mass, mostly muscle, and a nearly simultaneous increase of body fat mass, which often exceeds in absolute terms the loss of body cell mass (Roubenoff 1999). Lean muscle mass contributes to about 40% of total body weight (BW) in young adults, but it may decline to about one-fourth in a 75- to 80-year-old subject. Conversely, in a typical human being of ~20 years of age and with a BW of 80 kg, fat mass on average accounts for ~15% of BW, whereas in an 80-year-old man of comparable BW, it can increase up to ~30% of BW. Thus, in the aged, the body muscle apparatus is decreased in mass and weaker, but it has to sustain and move a greater body mass, because of increased total as well as fat mass (see also Chapter 13). These changes end up in the objective demand for augmented physical and psychological efforts. As a consequence, aged people may become more and more reluctant to move, thus reducing their daily activity and thus being involved in a vicious cycle, with progressive impairment of body function and mobility, and increased risk of falls.

Muscle protein is one of the key components of lean body mass. The decrease of muscle mass is the net results of two opposite underlying processes, i.e. protein degradation and synthesis. These processes are continuously operating, and they can be differently modulated by a variety of factors, such as nutrition, hormones, age, exercise, coexisting diseases, and drugs.

Therefore, the understanding of these mechanism(s) is a prerequisite to gain essential information to combat efficiently the causes of changes of muscle mass.

Changes of both protein synthesis and degradation are difficult to measure *in vivo*, because they are subtle and of little magnitude in the short term. As a matter of fact, skeletal muscle is turning over at a relatively slow rate (about 1%–1.5% per day) (Smith et al. 2011), and individual types of muscle proteins turn over a different rates as well. Therefore, the experimental methods employed to measure these processes need to be sensitive enough to be able to pick up small changes occurring over the time span of the measurements, as well as to compare data obtained in subjects at different ages.

In this chapter, the applied methodologies and current data on the changes in protein degradation and synthesis occurring with ageing are summarized and discussed.

11.2   METHODS TO MEASURE MUSCLE PROTEIN TURNOVER

The most largely used methods to measure these processes are briefly listed in Table 11.1. By the arterio-venous (A–V) amino acid balance approach (#1 on Table 11.1), combining the arterio-venous catheterization and amino acid concentration sampling with measurements of blood flow and muscle mass, either the net uptake or the net release of a given amino acid across the sampled district can be determined. This approach has been used by several landmark studies (Wahren et al. 1975). For most amino acids, the A–V concentration difference is small, that is, within few micromoles per liter, but these differences are usually amplified following either oral protein intake or iv amino acid infusions. Nevertheless, precise analytical methods are required. For any chosen amino acid that can be either *de novo* produced or catabolized by muscle, the net A–V balance does not indicate net changes of muscle protein balance. Only in the case that an amino acid is either not irreversibly catabolized by muscle (like phenylalanine) or, conversely, is produced de novo locally from a precursor amino acid (like 3-methyl histidine (3-MH) (Trappe et al. 2004), a post-translational muscle product of histidine) (#2 in Table 11.1), data on net protein synthesis or degradation, respectively, can be derived from those of the tracer amino acid, after correction for the amino acid abundance in muscle.

The most largely used techniques are, however, those listed as #3 to #5 in Table 11.1.

By means of the A–V approach combined with isotope infusion of an essential amino acid (EAA) as a tracer of protein turnover (#3 in Table 11.1), with arterial and deep-venous measurements of labeled and unlabeled amino acid(s), as well as of their catabolic products, implemented with appropriate models (Gelfand and Barrett 1987; Cheng et al. 1987; Tessari et al. 1995), a more complete picture of muscle protein turnover can be derived (Figure 11.1).

By method #4, the direct determination of the incorporation of an amino acid tracer into muscle proteins by biopsy (Phillips et al. 1997) allows to calculate muscle fractional protein synthesis rate (FSR).

Method #5 is based on a combination of methods listed as #3 and #4 (Biolo et al. 1994), and rates of amino acid transport are also estimated by a three-pool model.

Every method has a specific domain of validity, limitations, assumption, and invasivity, which cannot be discussed in detail in this chapter. The common feature of these approaches is that they

---

TABLE 11.1

Methods to Measure Muscle PS and PD In Vivo

1. **Arterial-Venous balance** of non metabolized amino acids (Phe, Tyr):
   - Net extraction: ~ disposal into proteins (**PS**);
2. **Net release** by muscle of amino acid-derived substances synthesized *in situ*:
   - A–V balance of 3-MH: (**PD**);
3. **A–V** differences combined with **isotope** infusions:
   - Disposal into proteins (**PS**) + release from proteins (**PD**);
   - Use of non-compartmental and compartmental models;
4. **Tracer** incorporation **into muscle** proteins (by biopsy):
   - Fractional muscle protein synthesis rate (FSR) (**PS**);
   - In mixed muscle samples or subfractions: FSR: $[\Delta E/\Delta t]/E_{prec}$
5. **A–V** differences, **isotope** infusions, and **muscle biopsy**.

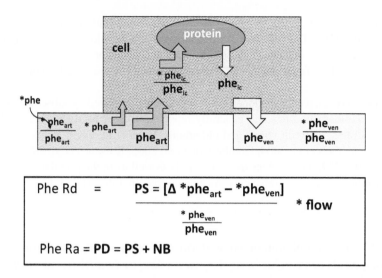

FIGURE 11.1 The figure shows the A–V approach combined with isotope infusion of an essential amino acid (in this instance, phenylalanine) as a tracer to model protein turnover across a sampled district (in this instance, skeletal muscle). The model requires measurements of both the labeled (denoted with an asterisk *) and the unlabeled amino acid, in the artery (dark grey block) and the vein (light grey block), as well as of blood flow. In the lower section of the figure, the equations/calculation used are reported. The venous labeled/unlabeled amino acid ratio is taken as representative of the intracellular pool (dotted block). The use of phenylalanine is particularly advantageous as this essential amino acid is not catabolized by muscle. Abbreviations: Phe: phenylalanine; $_{ic}$: intracellular; $_{art}$: arterial measurements; $_{ven}$: deep venous measurements; Ra: rate of appearance (from protein breakdown); Rd: rate of disappearance; PS: (phenylalanine) use for protein synthesis; NB: net balance; Δ = net difference.

attempt to dissect measurements of muscle protein turnover into protein degradation and synthesis separately, from the kinetic on an EAA.

These techniques have also been repeatedly applied to the study of the mechanism(s) of sarcopenia in ageing. The changes of protein degradation vs. synthesis in ageing, as well as in many experimental conditions or pathological states, are not on/off phenomena, since they are not either entirely suppressed or maximally stimulated but are usually finely modulated. Furthermore, it is increasingly being appreciated that protein degradation is itself an advantageous mechanism in the remodeling of (muscle) proteins, that is, for removal altered/aged proteins, as well as in the recycling of the amino acids derived from proteolysis to favor new protein synthesis (Sandri 2013). Therefore, the knowledge of each of these two processes, and the resulting changes in net protein balance, are key issues to understand the pathogenesis of sarcopenia in ageing.

11.3 PROTEIN SYNTHESIS IN AGEING: "HISTORICAL" DATA

Since 1990, many studies, using any of the techniques outlined above, have been carried out to investigate the pathogenesis of sarcopenia in ageing. For the purpose of clarity, data on skeletal muscle protein synthesis and degradation will be here presented separately.

Several authors had initially reported a decrease of *basal* muscle protein synthesis with ageing (Welle et al. 1993; Yarasheski et al. 1993; Welle et al. 1995; Rooyackers et al. 1996; Balagopal et al. 1997; Hasten et al. 2000; Yarasheski et al. 2002; Short et al. 2004). These results were mainly derived from isotope-determined FSRs (either as total myofibrillar proteins or as the myosin heavy chain [MHC] fraction) in skeletal muscle following biopsy. A decrease of muscle protein synthesis was reported to occur even in middle age (Balagopal et al. 1997).

With regard to the MHC isoforms, the MHC-I mRNA levels did not significantly change with age, but those of the MHC-IIa isoform was decreased by 38% in middle age compared with young subjects, and it was further decreased by 50% in older age (Balagopal et al. 2001). An age-related decrease of mitochondrial function and ATP production was also reported (Short et al. 2005) (see also Chapter 12).

These results, however, were questioned and/or could not confirmed by subsequent studies, which failed to reproduce these findings and generally reported little or no differences in basal muscle protein synthesis rates, between young and old adults (Volpi et al. 1999; Volpi et al. 2000; Volpi et al. 2001; Paddon-Jones et al. 2004; Cuthbertson et al. 2005; Katsanos et al. 2005; Katsanos et al. 2006; Kumar et al. 2009). Differences in the methods as well as in the population studied could be invoked to account for these discrepancies. As an example, in Table 11.2, a detailed comparison of

TABLE 11.2
Defective or Normal Stimulation of Skeletal Muscle Fractional Protein Synthesis Rate (FSR) by Amino Acids or Proteins in Ageing, According to Two Different Studies

	Defective (A)	Normal (B)	Comment
Study group	20 young and 24 elderly healthy men; groups of four	5 men and 5 women in each group	Sex differences
Age (years)	Young 28 ± 6; elderly 70 ± 6 years	Young 41 ± 8; elderly 70 ± 5 years	Smaller age difference in B
Subjects' characteristics	No previous or current disease; absence of muscle wasting >20%	No biochemical or clinical index of current disease	
Previous diet	No dietary control	No dietary control	
Body mass (BMI)	Young 24 ± 3; elderly 26 ± 4 (<28 in all)	Young 29.5 ± 3.6; elderly 24.1 ± 2.0.	A: lower BMI in young than in elderly. The opposite in B
Muscle mass/physical activity	Appendicular muscle mass (kg): young 25 ± 3; elderly 21 ± 3	Physically active but not athletes	
Lean body mass (LBM), kg		Young 60.9 ± 3.7; elderly 48.4 ± 3.3	
Experimental details	Under infusion of octreotide and insulin replacement (~10 µU/mL)		Suppression of the insulin response to amino acids in A
Type of tracer	[1-^{13}C]-Ketoisocaproic acid iv, + oral [1-^{13}C]-Leucine with the AA.	L-[ring-^{13}C$_6$] Phe iv infusion for 10 h	Differences in tracers
Way of amino acid/protein administration	Intake of 500 mL water with 0, 2.5, 5, 10, 20, and (for the elderly only) 40 g of EAA (Arg, His, Ile, Met, Leu, Phe, Threo, Try, Val) +5% oral leucine tracer	113.4-g (~25 g protein) lean ground-beef ingested at $t = 0$ h in 5′	Differences in the administered nutrients: A: oral amino acids; B: oral beef protein
Biopsies	Biopsies (150–300 mg) of vastus lateralis (VL) before and +3 hour after EAA, skin and fascia incisions, different sites, 1% lignocaine	Three biopsies (–3 h, 0 h +5 h); 50–100 mg (Bergstom n.), lateral portion of VL, 2% lidocaine	Longer time span between biopsies in B than in A

Source: (A): Cuthbertson, D. et al., *FASEB J.*, 19, 422–424, 2005; (B): Symons, T.B., et al., *Am. J. Clin. Nutr.*, 86, 451–456, 2007.

the methodological aspects between two studies (Cuthbertson et al. 2005; Symons et al. 2007), ending up into different results as regard to muscle protein synthesis with ageing, is reported. As it can be seen, the reasons for these differences are several, and they also underline the complexity of the methodologies employed in these kinds of studies.

One of the basic reasons for questioning the initial study results that reported (marked) differences of basal muscle protein synthesis between aged and young subjects is that, should basal muscle protein fractional synthesis rates in the aged subjects be diminished by ≈20%–30% as originally reported, these rates, projected over years, would theoretically end up with a complete muscle loss or even to negative numbers (clearly unrealistic results).

11.4 BASAL PROTEIN DEGRADATION IN AGEING: "HISTORICAL" DATA

Data on the effects of ageing in muscle protein degradation are limited, because the measurements of muscle proteolysis are itself more difficult than that of protein synthesis. Using the rate of release into the interstitial space of the nonrecyclable amino acid 3-methylhistidine as a marker of proteolysis, it was initially reported that basal muscle protein breakdown was elevated in old compared with young adults, by up to 50% (Trappe et al. 2004). In contrast with these findings, more recent data, based on a three compartmental model of skeletal muscle involving arterio-venous sampling and muscle biopsy, failed to show an increase of skeletal muscle basal proteolysis with ageing (Volpi et al. 1999).

However, since it is increasingly recognized that the etiology of sarcopenia in ageing is multifaceted, not yet fully defined, and rather complex, a variety of other factors should be involved, which could also explain some of the inconsistencies found in the literature.

11.5 CURRENT VIEWS ON THE EFFECT OF AGE ON MUSCLE PROTEIN TURNOVER: THE RESPONSE TO NUTRITION

As discussed above, a basic criticism of findings reporting marked changes in either protein muscle synthesis or degradation with ageing is that should the reported magnitude of changes be valid, they would determine a much more marked muscle wasting than that is typically observed. that is, in the range of 3%–8% per decade (Koopman 2011).

This also means that the relatively slow rate of muscle loss during ageing, reflecting the unbalance between the average diurnal rate of muscle protein synthesis and breakdown, may be difficult to be measured precisely with the currently available techniques. The possible causes of these variations have been recently reviewed (Smith et al. 2011), and some methodological issues have been highlighted, focused on data derived from the biopsy of the *vastus lateralis*, in healthy, nonobese, untrained adults ≤50 years of age, studied in the postabsorptive state at rest, and using primed, constant tracer amino acid infusion. Potential sources of variation possibly accounting for inconsistency among published data were the following: (a) the choice of the precursor pool to calculate FSR (i.e., the muscle vs. the plasma free AA enrichment): in this aspect, the population variance might be somewhat smaller in studies using plasma amino acid/ketoacid enrichments vs. muscle-free amino acid enrichment; (b) the choice of mixed muscle protein vs. myofibrillar/MHC FSR values; (c) the time interval between biopsies; (d) considerable, perhaps unavoidable, within-subject variability (Table 11.3).

Therefore, current views are that basal fasting protein synthesis as well as breakdown rates in skeletal muscle are not substantially different between young and elderly human subjects (Volpi et al. 1999, 2000, 2001; Hasten et al. 2000; Katsanos et al. 2006; Rasmussen et al. 2006; Kumar et al. 2009). However, the skeletal muscle of aged people may be less responsive to nutrition, that is, to anabolic stimuli such as protein ingestion, both at the whole body level (Koopman et al. 2009) and in

TABLE 11.3

Apparent Sources of Variations Among Published Data, of the Effect of Ageing on Muscle Protein Synthesis

1. Mixed muscle protein FSR values were greater when the muscle vs. the plasma free AA Enrichment was used as the surrogate

2. Average mixed muscle protein FSR rates were greater than the myofibrillar/myosin heavy chain FSR values

3. Within-study variability (i.e., population variance) might be somewhat smaller in studies that used plasma AA/ketoacid enrichments vs. muscle free amino acid enrichment (~24% vs. ~31%)

4. The between-study consistency of measured FSR values (i.e., interquartile range) was positively correlated with average duration between biopsies

5. The variability in reported values is in part due to (1) differences in exp. design (e.g., choice of precursor pool) and (2) considerable within-subject variability

Source: Smith, G.I. et al., *J. Appl. Physiol., (1985)*, 110, 480–491, 2011.

the skeletal muscle (Volpi et al. 2000). Such a relative unresponsiveness to nutrition has been defined as "anabolic resistance" (Rasmussen et al. 2006). This concept is relatively new with regard to ageing and has given a comprehensive frame to scattered observations previously collected. Notably, perhaps the first demonstration of anabolic resistance *in vivo* in humans, to the combined infusion of amino acid and insulin, has been reported by us in subjects with liver cirrhosis (Tessari et al. 1996).

Another possible indirect site of regulation of protein synthesis could be that of transamination/deamination. These steps are the first ones, although reversible, in amino acid metabolism/disposal. Both deamination and reamination rates of leucine are reduced in middle-aged subjects compared with young controls, both in the postabsorptive state and following euglycemic hyperinsulinemia (Tessari 2017). This finding can be interpreted as a potential sparing mechanism to restrain leucine catabolism with ageing.

However, nutrition with proteins and/or amino acids is just one of the factors that can enhance muscle protein anabolism. Others are exercise, anabolic hormones, the enhancement of muscle blood flow (by pharmacological agents), the (direct) stimulation of molecular mediators of protein synthesis (such as protein kinase B (PKB)/Akt$_{ser473}$, 4EBP1, mammalian target of rapamycin (mTOR), S6K1$_{Thr389}$), a decrease of ET-1, and various combinations of the above. In this chapter, just the role of nutrition is discussed.

Historically, it has been shown years ago that the peripheral infusion of amino acids was capable to enhance muscle protein synthesis (Bennet et al. 1990). Using the arterio-venous catheterization of the forearm, we showed that (muscle) protein synthesis is stimulated following the ingestion of a mixed meal in healthy volunteers (Tessari et al. 1996). Some of the mechanism(s) whereby amino acid/protein nutrition stimulates protein synthesis (Smith et al. 1992; Norton and Layman 2006) are summarized in Table 11.4.

The existence of an anabolic resistance to nutrition in the elderly is supported by several experimental data. When elderly subjects (72 year old) ingested a mixed glucose and amino acid meal, no stimulation of muscle FSR was observed, as opposed to that detected in younger controls (30 year old) (Volpi et al. 2000). An impaired response of whole-body protein synthesis was also reported by Koopman following casein ingestion (Koopman et al. 2009). Such a resistance in the elderly is confirmed by a rightward and downward shift of the dose–response relation between myofibrillar protein synthesis and plasma leucine (an index of amino acid availability to sustain protein synthesis), which could not be overcome even by a large amino acid load (Cuthbertson et al. 2005).

The molecular cause(s) for such an anabolic resistance could be numerous. A reduction of the signaling protein kinase B (PKB)/mammalian Target of rapamycin complex 1 (mTORC 1)

TABLE 11.4
Proofs and Mechanisms of the Anabolic Response in Skeletal Muscle to Food Intake

- Ingestion of amino acid (AA) and/or protein stimulates muscle protein synthesis
- Besides serving as a substrate for polypeptide biosynthesis, AA directly activates regulatory proteins in mRNA translation
- While nonessential AAs do not induce an increase in muscle protein synthesis, essential amino acids (EAAs) increase muscle protein synthesis in the absence of an increase in nonessential AA availability
- The branched-chain amino acid, leucine, directly stimulates signaling through mTOR and its downstream targets 4E-binding protein and S6K1 and ribosomal S6
- Therefore, the EAA and leucine, in particular, seem to represent the main anabolic signals responsible for the postprandial increase in muscle protein synthesis

Source: Smith, K. et al., *Am. J. Physiol.*, 262, E372–E376, 1992; Norton, L.E., & Layman, D.K., *J. Nutr.*, 136, 533S–537S, 2006.

pathway in the muscle of aged subjects and an attenuated rise in the activation of other key signaling proteins in this pathway (Cuthbertson et al. 2005) have been evidenced. This defect of muscle protein synthesis and mTORC1 signaling has been confirmed in a large cohort of young and older men and women more recently (Markofski et al. 2015).

Also, a reduction of S6K1 phosphorylation following combined AA and glucose infusions has been shown in the elderly (Guillet et al. 2004), as well as a diminished skeletal muscle microRNA expression, which was associated with the inhibition of Insulin-like Growth Factor-1 (IGF-1) signaling and an attenuated muscle plasticity (Rivas et al. 2014).

It should, however, be acknowledged that not all researchers reported an impaired muscle protein synthetic response to protein intake in the elderly, as similar protein synthetic rates were observed in young and elderly human subjects after ingestion of large amounts of carbohydrate and proteins after exercise (Koopman et al. 2008), or following ingestion of large and small amount of beef (Symons et al. 2007). Furthermore, essential amino acids (EAAs) acutely stimulated muscle protein synthesis in young and elderly people to a similar extent (Paddon-Jones et al. 2004). The cause(s) for the inconsistencies among current data on the existence of anabolic resistance in the elderly cannot be attributed to unaccounted deviations from recommended protein intake among the study populations (Yarasheski et al. 2011).

Interestingly, the anabolic resistance to amino acids at skeletal muscle level in the elderly can be overcome by ingestion of leucine-rich EAA mixture, containing 41% leucine vs. 26% leucine in the younger group (Katsanos et al. 2006). Thus, just increasing leucine in an amino acid/protein meal may be sufficient to normally sustain muscle protein synthesis in old subjects. The specific effect of leucine in stimulating muscle protein synthesis has been known for years (Hong and Layman 1984) and is substantiated by well-defined molecular mechanisms (Dodd and Tee 2012). The effects of leucine-enriched amino acid mixtures are also to prolong myofibrillar protein synthesis and amino acid transporter expression in older men after resistance exercise (Dickinson et al. 2014). This observation is in agreement with a dose-dependent response of myofibrillar protein synthesis with increasing loads of beef ingestion, also in elderly subjects, a response that could reach maximum values and is enhanced by resistance exercise in middle-aged men (Robinson et al. 2013). However, the stimulation of myofibrillar protein synthesis requires a greater relative protein intake in older men than in younger men (Moore et al. 2015), thus substantiating the existence of an anabolic-resistant status in the elderly.

While both the EAA and whey protein supplements stimulated muscle protein synthesis, EAAs may provide a more efficient nutritional supplement to elderly individuals (Paddon-Jones et al. 2006).

A contribution to the anabolic resistance to nutrients in ageing could be due also to an impairment/delay in dietary protein digestion and/or absorption (Boirie et al. 1997; Dangin et al. 2001). Such impairment will reduce and/or blunt the appearance rate of dietary AA in the bloodstream, thereby reducing AA delivery to the muscle and attenuating the muscle protein synthetic response. Therefore, soluble proteins, such as those of the whey fraction of milk, that are absorbed more rapidly and give rise to earlier and greater peaks of amino acid concentrations may enhance muscle protein synthesis particularly in the elderly (Boirie and Guillet 2018). The same advantage for elderly people has been suggested also for a "pulse" protein feeding pattern, that is, that of protein administration as acute loads rather than with a sub-continuous pattern (Arnal et al. 1999). Addition of whey protein of vitamin D and omega 3 fatty acids or combined with exercise (Boirie and Guillet 2018) may enhance protein accretion in skeletal muscle of the elderly.

11.6 CONCLUSIONS

Taken together, the bulk of available experimental data suggests that of the most meaningful alteration of protein turnover in the elderly is a resistance to the anabolic effect of amino acids (and of dietary proteins) on skeletal muscle protein synthesis. This defect can be overcome by an abundant protein intake, possibly enriched with EAAs, particularly leucine, and complemented with other strategies that can amplify such an anabolic effect, above all physical exercise.

REFERENCES

Arnal, M.A., Mosoni, L., Boirie, Y., Houlier, M.L., Morin, L., Verdier, E., Ritz, P. et al. 1999. Protein pulse feeding improves protein retention in elderly women. *Am J Clin Nutr* 69(6):1202–1208.
Balagopal, P., Rooyackers, O.E., Adey, D.B., Ades, P.A., Nair, K.S. 1997. Effects of aging on in vivo synthesis of skeletal muscle myosin heavy-chain and sarcoplasmic protein in humans. *Am J Physiol* 273(4 Pt 1): E790–E800.
Balagopal, P., Schimke, J.C., Ades, P., Adey, D., Nair, K.S. 2001. Age effect on transcript levels and synthesis rate of muscle MHC and response to resistance exercise. *Am J Physiol* 280(2):E203–E208.
Bennet, W.M., Connacher, A.A., Scrimgeour, C.M., Jung, R.T., Rennie, M.J. 1990. Euglycemic hyperinsulinemia augments amino acid uptake by human leg tissues during hyperaminoacidemia. *Am J Physiol* 259(2 Pt 1):E185–E194.
Biolo, G., Gastaldelli, A., Zhang, X.J., Wolfe, R.R. 1994. Protein synthesis and breakdown in skin and muscle: A leg model of amino acid kinetics. *Am J Physiol* 267(3 Pt 1):E467–E474.
Boirie, Y., Dangin, M., Gachon, P., Vasson, M.P., Maubois, J.L., Beaufrère, B. 1997. Slow and fast dietary proteins differently modulate postprandial protein accretion. *Proc Natl Acad Sci USA* 94:14930–14935.
Boirie, Y., Guillet, C. 2018. Fast digestive proteins and sarcopenia of aging. *Curr Opin Clin Nutr Metab Care* 21(1):37–41. doi:10.1097/MCO.0000000000000427.
Cheng, K.N., Pacy, P.J., Dworzak, F., Ford, G.C., Halliday, D. 1987. Influence of fasting on leucine and muscle protein metabolism across the human forearm determined using L-[1–13 C, ¹⁵N]leucine as the tracer. *Clin Sci* (Lond);73(3):241–246.
Cuthbertson, D., Smith, K., Babraj, J., Leese, G., Waddell, T., Atherton, P., Wackerhage, H., Taylor, P.M., Rennie, M.J. 2005. Anabolic signaling deficits underlie amino acid resistance of wasting, aging muscle. *FASEB J* 19(3):422–424.
Dangin, M., Boirie, Y., Garcia-Rodenas, C., Gachon, P., Fauquant, J., Callier, P., Ballèvre, O., Beaufrère, B. 2001. The digestion rate of protein is an independent regulating factor of postprandial protein retention. *Am J Physiol* 280:E340–48.
Dickinson, J.M., Gundermann, D.M., Walker, D.K., Reidy, P.T., Borack, M.S., Drummond, M.J., Arora, M., Volpi, E., Rasmussen, B.B. 2014. Leucine-enriched amino acid ingestion after resistance exercise prolongs myofibrillar protein synthesis and amino acid transporter expression in older men. *J Nutr* 144(11):1694–1702. doi:10.3945/jn.114.198671.
Dodd, K.M., Tee, A.R. 2012. Leucine and mTORC1: A complex relationship. *Am J Physiol* 302(11):E1329–E1342. doi:10.1152/ajpendo.00525.2011.
Gelfand, R.A., Barrett, E.J. 1987. Effect of physiologic hyperinsulinemia on skeletal muscle protein synthesis and breakdown in man. *J Clin Invest* 80(1):1–6.

Guillet, C., Prod'homme, M., Balage, M., Gachon, P., Giraudet, C., Morin, L., Grizard, J., Boirie, Y. 2004. Impaired anabolic response of muscle protein synthesis is associated with S6K1 dysregulation in elderly humans. *FASEB J* 18:1586–1587.

Hasten, D.L., Pak-Loduca, J., Obert, K.A., Yarasheski, K.E. 2000. Resistance exercise acutely increases MHC and mixed muscle protein synthesis rates in 78–84 and 23–32 yr. olds. *Am J Physiol* 278:E620–E626.

Hong, S.O., Layman, D.K. 1984. Effects of leucine on in vitro protein synthesis and degradation in rat skeletal muscles. *J Nutr* 114(7):1204–1212.

Katsanos, C.S., Kobayashi, H., Sheffield-Moore, M., Aarsland, A., Wolfe, R.R. 2005. Aging is associated with diminished accretion of muscle proteins after the ingestion of a small bolus of essential amino acids. *Am J Clin Nutr* 82(5):1065–1073.

Katsanos, C.S., Kobayashi, H., Sheffield-Moore, M., Aarsland, A., Wolfe, R.R. 2006. A high proportion of leucine is required for optimal stimulation of the rate of muscle protein synthesis by essential amino acids in the elderly. *Am J Physiol* 291(2):E381–E387.

Koopman, R. 2011. Dietary protein and exercise training in ageing. *Proc Nutr Soc* 70(1):104–113. doi:10.1017/S0029665110003927.

Koopman, R., Verdijk, L.B., Beelen, M., Gorselink, M., Kruseman, A.N., Wagenmakers, A.J., Kuipers, H., van Loon, L.J. 2008. Co-ingestion of leucine with protein does not further augment post-exercise muscle protein synthesis rates in elderly men. *Br J Nutr* 99(3):571–580.

Koopman, R., Walrand, S., Beelen, M., Gijsen, A.P., Kies, A.K., Boirie, Y., Saris, W.H., van Loon, L.J. 2009. Dietary protein digestion and absorption rates and the subsequent postprandial muscle protein synthetic response do not differ between young and elderly men. *J Nutr* 139:1707–1713.

Kumar, V., Selby, A., Rankin, D., Patel, R., Atherton, P., Hildebrandt, W., Williams, J., Smith, K., Seynnes, O., Hiscock, N., Rennie, M.J. 2009. Age-related differences in the dose-response relationship of muscle protein synthesis to resistance exercise in young and old men. *J Physiol* 587(1):211–217. doi:10.1113/jphysiol.2008.164483.

Markofski, M.M., Dickinson, J.M., Drummond, M.J., Fry, C.S., Fujita, S., Gundermann, D.M., Glynn, E.L. et al. 2015. Effect of age on basal muscle protein synthesis and mTORC1 signaling in a large cohort of young and older men and women. *Exp Gerontol* 65:1–7. doi:10.1016/j.exger.2015.02.015.

Moore, D.R., Churchward-Venne, T.A., Witard, O., Breen, L., Burd, N.A., Tipton, K.D., Phillips, S.M. 2015. Protein ingestion to stimulate myofibrillar protein synthesis requires greater relative protein intakes in healthy older versus younger men. *J Gerontol A Biol Sci Med Sci* 70(1):57–62. doi:10.1093/gerona/glu103.

Norton, L.E., & Layman, D.K. 2006. Leucine regulates translation initiation of protein synthesis in skeletal muscle after exercise. *J Nutr* 136, 533S–537S.

Paddon-Jones, D., Sheffield-Moore, M., Katsanos, C.S., Zhang, X.J., Wolfe, R.R. 2006. Differential stimulation of muscle protein synthesis in elderly humans following isocaloric ingestion of amino acids or whey protein. *Exp Gerontol* 41(2):215–219.

Paddon-Jones, D., Sheffield-Moore, M., Zhang, X.J., Volpi, E., Wolf, S.E., Aarsland, A., Ferrando, A.A., Wolfe, R.R. 2004. Amino acid ingestion improves muscle protein synthesis in the young and elderly. *Am J Physiol* 286(3):E321–E328.

Phillips, S.M., Tipton, K.D., Aarsland, A., Wolf, S.E., Wolfe, R.R. 1997. Mixed muscle protein synthesis and breakdown after resistance exercise in humans. *Am J Physiol* 273 (1 Pt 1):E99–E107.

Rasmussen, B.B., Fujita, S., Wolfe, R.R., Mittendorfer, B., Roy, M., Rowe, V.L., Volpi, E. 2006. Insulin resistance of muscle protein metabolism in aging. *FASEB J* 20:768–769.

Rivas, D.A., Lessard, S.J., Rice, N.P., Lustgarten, M.S., So, K., Goodyear, L.J., Parnell, L.D., Fielding, R.A. 2014. Diminished skeletal muscle microRNA expression with aging is associated with attenuated muscle plasticity and inhibition of IGF-1 signaling. *FASEB J* 28(9):4133–4147. doi:10.1096/fj.14-254490.

Robinson, M.J., Burd, N.A., Breen, L., Rerecich, T., Yang, Y., Hector, A.J., Baker, S.K., Phillips, S.M. 2013. Dose-dependent responses of myofibrillar protein synthesis with beef ingestion are enhanced with resistance exercise in middle-aged men. *Appl Physiol Nutr Metab* 38(2):120–125. doi:10.1139/apnm-2012-0092.

Rooyackers, O.E., Adey, D.B., Ades, P.A., Nair, K.S. 1996. Effect of age on in vivo rates of mitochondrial protein synthesis in human skeletal muscle. *Proc Natl Acad Sci USA* 93:15364–15369.

Roubenoff, R. 1999. The pathophysiology of wasting in the elderly. *J Nutr* 129(1S Suppl):256S–259S.

Sandri, M. 2013. Protein breakdown in muscle wasting: Role of autophagy-lysosome and ubiquitin-proteasome. *Int J Biochem Cell Biol* 45(10):2121–2129. doi:10.1016/j.biocel.2013.04.023.

Short, K.R., Bigelow, M.L., Kahl, J., Singh, R., Coenen-Schimke, J., Raghavakaimal, S, Nair, K.S. 2005. Decline in skeletal muscle mitochondrial function with aging in humans. *Proc Natl Acad Sci USA* 12;102(15):5618–5623.

Short, K.R., Vittone, J.L., Bigelow, M.L., Proctor, D.N., Nair, K.S. 2004. Age and aerobic exercise training effects on whole body and muscle protein metabolism. *Am J Physiol* 286(1):E92–E101.

Smith, GI., Patterson, B.W., Mittendorfer, B. 2011. Human muscle protein turnover—why is it so variable? *J Appl Physiol (1985)* 110(2):480–491. doi:10.1152/japplphysiol.00125.2010.

Smith, K, Barua, J.M., Watt, P.W., Scrimgeour, C.M., Rennie, M.J. 1992. Flooding with L-[1–13°C]leucine stimulates human muscle protein incorporation of continuously infused L-[1–13°C]valine. *Am J Physiol* 262(3 Pt 1), E372–E376.

Symons, T.B., Schutzler, S.E., Cocke, T.L., Chinkes, D.L., Wolfe, R.R., Paddon-Jones, D. 2007. Aging does not impair the anabolic response to a protein-rich meal. *Am J Clin Nutr* 86(2):451–456.

Tessari, P. 2017. Leucine transamination is lower in middle-aged compared with younger adults. *J Nutr* 147(11):2025–2030. doi:10.3945/jn.117.250852.

Tessari, P., Inchiostro, S., Zanetti, M., Barazzoni, R. 1995. A model of skeletal muscle leucine kinetics measured across the human forearm. *Am J Physiol* 269(1 Pt 1):E127–E136.

Tessari, P., Zanetti, M., Barazzoni, R., Biolo, G., Orlando, R., Vettore, M., Inchiostro, S., Perini, P., Tiengo, A. 1996. Response of phenylalanine and leucine kinetics to branched chain-enriched amino acids and insulin in patients with cirrhosis. *Gastroenterology* 111(1):127–137.

Tessari, P., Zanetti, M., Barazzoni, R., Vettore, M., Michielan, F. 1996. Mechanisms of postprandial protein accretion in human skeletal muscle. Insight from leucine and phenylalanine forearm kinetics. *J Clin Invest* 98(6):1361–1372.

Trappe, T., Williams, R., Carrithers, J., Raue, U., Esmarck, B., Kjaer, M., Hickner, R. 2004. Influence of age and resistance exercise on human skeletal muscle proteolysis: A microdialysis approach. *J Physiol* 554(Pt 3):803–813.

Volpi, E., Mittendorfer, B., Rasmussen, B.B., Wolfe, R.R. 2000. The response of muscle protein anabolism to combined hyperaminoacidemia and glucose-induced hyperinsulinemia is impaired in the elderly. *J Clin Endocrinol Metab* 85:4481–4490.

Volpi, E., Mittendorfer, B., Wolf, S.E., Wolfe, R.R. 1999. Oral amino acids stimulate muscle protein anabolism in the elderly despite higher first-pass splanchnic extraction. *Am J Physiol* 277(3 Pt 1):E513–E520.

Volpi, E., Sheffield-Moore, M., Rasmussen, B.B., Wolfe, R.R. 2001. Basal muscle amino acid kinetics and protein synthesis in healthy young and older men. *JAMA* 286(10):1206–1212.

Wahren, J., Hagenfeldt, L., Felig, P. 1975. Splanchnic and leg exchange of glucose, amino acids, and free fatty acids during exercise in diabetes mellitus. *J Clin Invest* 55(6):1303–1314.

Welle, S., Thornton, C., Jozefowicz, R., Statt, M. 1993. Myofibrillar protein synthesis in young and old men. *Am J Physiol* 264(5 Pt 1): E693–E698.

Welle, S., Thornton, C., Statt, M. 1995. Myofibrillar protein synthesis in young and old human subjects after three months of resistance training. *Am J Physiol* 268(3 Pt 1): E422–E427.

Yarasheski, K.E., Castaneda-Sceppa, C., He, J., Kawakubo, M., Bhasin, S., Binder, E.F., Schroeder, E.T., Roubenoff, R., Azen, S.P., Sattler, F.R. 2011. Whole-body and muscle protein metabolism are not affected by acute deviations from habitual protein intake in older men: The Hormonal Regulators of Muscle and Metabolism in Aging (HORMA) Study. *Am J Clin Nutr* 94(1):172–181. doi:10.3945/ajcn.110.010959.

Yarasheski, K.E., Welle, S., Nair, K.S. 2002. Muscle protein synthesis in younger and older men. *JAMA* 287(3): 317–318.

Yarasheski, K.E., Zachwieja, J.J., Bier, D.M. 1993. Acute effects of resistance exercise on muscle protein synthesis rate in young and elderly men and women. *Am J Physiol* 265(2 Pt 1):E210–E214.

12 The Relationship between Muscle Mitochondrial Turnover and Sarcopenia

Heather N. Carter, Nashwa Cheema, and David A. Hood

CONTENTS

12.1 INTRODUCTION: MITOCHONDRIA AND THEIR IMPORTANCE TO SKELETAL MUSCLE

Long known for their vital roles in cellular energy production, mitochondria are now established participants in calcium handling, cellular signaling, and organelle-mediated apoptosis. In addition, mitochondrial dysfunction brought about by either nuclear or mitochondrial DNA (mtDNA) mutations can lead to a wide variety of pathophysiological conditions affecting a number of organ systems, such as skeletal muscle, the heart, or the brain (Wallace 2010). Thus, both basic and clinical scientists have become keenly interested in how mitochondrial function and dysfunction contribute to cellular health and disease.

Exercise is a powerful metabolic stress. When performed repeatedly over the course of several weeks, exercise leads to adaptations in skeletal muscle designed to meet that increased metabolic demand (Hood 2001; Holloszy 1967). One of the most dramatic examples of this phenotypic adaptation is the increase in mitochondrial content within muscle, which is coincident with a reduction in fatigability and an improvement in endurance performance. In contrast to the positive effects that exercise brings about to increase mitochondria, aging and inactivity result in the opposite change. Chronic physical inactivity leads to decrements in organelle content and function within muscle, poor performance, and an increase in apoptotic susceptibility (Chabi et al. 2008; Adhihetty et al. 2007). The metabolic derangements associated with inactivity and the loss of mitochondria include a greater storage, rather than oxidation of lipids (Crane et al. 2010) and a tendency for obesity and insulin resistance (Bharadwaj et al. 2015). Aging also brings about phenotypic changes in mitochondria within muscle, attributed to both inactivity, as well as inherent, aging-induced changes in organelle synthesis and degradation pathways. Thus, opposing changes in the mitochondrial network within muscle cells induced by chronic exercise, aging, or disuse are now recognized to have implications for a broad range of health issues. Notably, regular exercise can counteract many of the aging- and inactivity-induced detrimental effects observed on the metabolism associated with defective mitochondria, and research efforts must continue to seek the molecular underpinnings of how this is brought about.

The steady-state mitochondrial content of muscle at any time, along with the quality of the organelle pool, is a product of complex pathways of synthesis (biogenesis) and degradation (mitophagy) (Hood et al. 2016). The attainment of a new, higher level of functional organelles in response to exercise is the cumulative result of a series of events that begin with the very first bout of exercise in a training program. Each acute exercise bout initiates a new cascade of signaling events involving the activation of protein kinases, acetylases, and phosphatases that modify the activity of downstream proteins such as transcription factors that impact the expression of nuclear genes encoding mitochondrial proteins (NuGEMPs). The result is an increase in the level of mRNAs, which encode precursor proteins destined for mitochondrial localization. Once translated in the cytoplasm, the resulting nuclear-encoded proteins are chaperoned to the mitochondria and imported into the different compartments, such as the matrix space or the inner or outer membrane. A subgroup of these proteins includes transcription factors, such as mitochondrial transcription factor A (Tfam), that act directly on mtDNA to increase the mRNA expression of mitochondrial gene products and mtDNA copy number. mtDNA is critical because it encodes 13 critical proteins involved in the mitochondrial respiratory chain. Thus, mitochondrial biogenesis requires the coordinated cooperation of both the nuclear and the mitochondrial genomes to produce an organelle that is functional in providing cellular ATP. Two subpopulations of mitochondria are found in skeletal muscle, the subsarcolemmal (SS), and intermyofibrillar (IMF), which possess different compositional and biochemical characteristics (Cogswell, Stevens, and Hood 1993), and each of these populations relies on the cooperation of both genomes.

In contrast to the process of biogenesis, regulation of mitochondrial content and quality is also exerted at the level of mitophagy, the selective degradation of mitochondria by the autophagosome-lysosome system. As discussed below, mitophagy is activated by similar signaling reactions as biogenesis, suggesting that a coordinated regulation of these pathways exists. Rates of mitophagy are enhanced during physiological conditions of increased reactive oxygen species (ROS) production or when the mitochondrial membrane potential decreases (Narendra et al. 2008; Li et al. 2015). When this occurs, the affected portion of the organelle reticulum undergoes fission (Twig et al. 2008) to separate it from the remaining "healthy" mitochondrion, and the organelle fragment is labeled and targeted for degradation. Considerably less is known about the regulation of mitophagy, in comparison to biogenesis, particularly in skeletal muscle either during exercise or the aging process.

12.2   ROLES OF MITOCHONDRIA IN SKELETAL MUSCLE

Mitochondria are well renowned as the major source for the provision of intracellular ATP. These organelles are highly adaptable in skeletal muscle to changes in physical activity, with increasing exercise stimulating the synthesis of new organelles (mitochondrial biogenesis) while inactivity (such as with denervation or aging) leads to the degradation of mitochondria presumably through the selective process of mitophagy.

As mentioned, mitochondria also participate in other cellular processes besides ATP production. Mitochondria are reservoirs for the storage of calcium in skeletal muscle (Raffaello et al. 2016), a critical ion for muscle contraction. These organelles also participate in cellular signaling pathways through ROS production (Radak et al. 2013) as well as being mediators for apoptosis (Quadrilatero, Alway, and Dupont-Versteegden 2011), colloquially known as cell death. Normally, during the process of respiration, H^+ pumps into the intermembrane space to facilitate the flow of electrons between cytochrome complexes. At Complex IV, the electrons will be paired with $\frac{1}{2} O_2$ and water will be produced. However, prior to this terminal step, electrons can be prematurely donated to oxygen (Goncalves et al. 2015) causing the formation of a volatile product, known as ROS. At low levels, ROS can be handled adequately by intramitochondrial antioxidants, protecting the organelle. Yet when levels exceed the handling capacity of the antioxidants, damage to neighboring lipids, proteins, and mtDNA may ensue. This damage may perpetuate the cycle of ROS production. For example, if subunits of the electron transport chain (ETC) are harmed, the electron flow may be impeded, generating a cycle where more electrons are capable of leaving the ETC early (i.e., at Complex I or III) to produce ROS. Accumulation of damage to the mitochondria may ultimately trigger apoptosis through the loss of the membrane potential and the release of proteins that can trigger caspase-dependent or -independent cell death pathways, ultimately leading to DNA fragmentation. Thus, ensuring a healthy population of mitochondria will preserve cellular function, and this necessitates intact pathways for mitochondrial biogenesis and degradation.

12.2.1 Mitochondrial Biogenesis

With enhanced periods of muscular contractions, such as endurance-type exercise, mitochondrial biogenesis is stimulated as a means to fulfill the energy requirements demanded by the muscle (Figure 12.1). The process of mitochondrial biogenesis is highly complex and requires refined intracellular communication to ensure the production of a fully functioning organelle. Communication between two sets of genetic material, nuclear and mitochondrial, is required for the production of high-quality organelles (Scarpulla, Vega, and Kelly 2012). The mitochondrial genome is approximately 16.5 kb, with variations in length among species and circular in structure due to its primitive origins. This genome contains the code for 13 genes that generate polypeptides, which are subunits of the ETC complexes. For example, cytochrome oxidase complex IV subunit I (COX I) is a mtDNA-derived product that incorporates into Complex IV and serves as the regulatory subunit of the complex. Moreover, this genome encodes for mitochondrial-specific tRNAs and rRNAs, which allow for the synthesis of mitochondrial proteins. In order to transcribe this genome, mitochondrial transcription factor A (TFAM) is required, and interestingly, there is only one mtDNA transcription factor, as opposed to the plethora of factors for nuclear DNA. TFAM is coded from nDNA and is synthesized in the cytosol before it is imported into the mitochondria. Thus, communication from the mitochondria to the nucleus is imperative for generating competent mitochondria, through the production of ETC subunits, matrix enzymes, and transcriptional regulators for mtDNA. While not completely understood, retrograde signaling from the mitochondria to the nucleus may be achieved through calcium, ROS, energy status, or the unfolded protein response (Quirós, Mottis, and Auwerx 2016).

With exercise, numerous cellular metabolites and molecules undergo fluctuations in concentration, which stimulates the activation of a plethora of signaling kinases. Indeed, changes in the concentration of Ca^{2+}, the ATP:AMP ratio, the NAD^+:NADH ratio, and ROS are the most well-described signaling molecules that lead to the adaptation of mitochondrial biogenesis in muscle (Figure 12.1). Calcium is capable of signaling to calcium/calmodulin kinase (CAMK); increased concentrations of AMP activate AMP-activated kinase (AMPK), while ROS is sensed by p38. Elevations in the concentration of NAD^+ serve as the cofactor for activating the deactylase, SIRT1. The activation of these factors impinges on a crucial target for mitochondrial biogenesis, peroxisome proliferator-activated receptor coactivator 1-alpha (PGC-1α) (Puigserver et al. 2001; Jäger et al. 2007). Interestingly, the cellular activation of these pathways acts on the promoter region of the *Pgc-1α* gene, leading to enhanced mRNA production of the coactivator (Akimoto, Li, and Yan 2008; Irrcher et al. 2008; Irrcher, Ljubicic, and Hood 2009; Amat et al. 2009; Wu et al. 2002). Adding to this, PGC-1α has been shown to behave in a feed-forward mechanism since it is capable of coactivating transcription factors found on its own promoter (Handschin et al. 2003). Thus, PGC-1α is considered by many to be a pivotal factor to promote the expression of numerous NuGEMPs following exercise, including TFAM and subunits of the ETC (Scarpulla 2011; Virbasius and Scarpulla 1994).

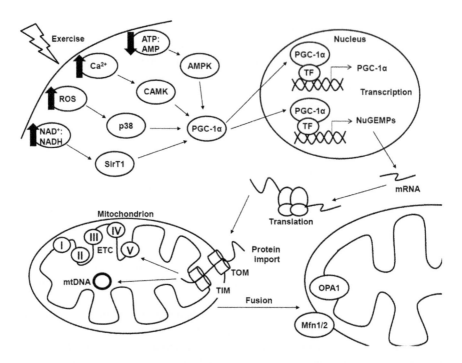

FIGURE 12.1 Mitochondrial biogenesis with exercise. In untrained subjects, an acute bout of exercise elicits signals to activate molecular signaling pathways for the activation gene expression to ultimately elicit adaptation. During exercise, ATP is consumed through muscle contractions leading to a rise in intracellular AMP. NADH is oxidized to NAD^+, and increased amounts of calcium (Ca^{2+}) and reactive oxygen species (ROS) also occur. Increased concentrations of AMP activate the energy-sensing kinase AMPK. NAD^+ serves as a cofactor for the activity of the deacetylase SIRT1. Increased concentrations of ROS are sensed by the stress-responsive kinase p38, while elevations in Ca^{2+} activate calcium/calmodulin kinase (CAMK). The increased activity of AMPK, SIRT1, p38, and CAMK all converge on a crucial factor for mitochondrial biogenesis and peroxisome proliferator-activated receptor coactivator 1-alpha (PGC-1α). PGC-1α acts as a transcriptional coactivator on transcription factors to increase the expression of nuclear genes encoding mitochondrial proteins (NuGEMPs). Additionally, PGC-1α acts on transcription factors found on its own promoter, in an autoregulatory fashion. AMPK will phosphorylate PGC-1α while SIRT1 will deacetylate the protein, priming it for full activation. PGC-1α protein is largely found in the cytosol during resting conditions; however, upon an exercise stimulus, it will translocate to the nucleus. AMPK, p38, and CAMK have indirect roles for PGC-1α, in that they stimulate increased activity of the PGC-1α promoter, leading to greater gene expression of the coactivator. Once mRNAs are produced, they will be moved to the cytosol and translated into protein products. From here, mitochondrial-destined proteins need to be brought into the organelle through the protein import machinery (PIM). Protein complexes found on the outer mitochondrial membranes (TOMs), and inner mitochondrial membrane (TIMs) will facilitate the unfolded proteins to enter the organelle, and then they can be refolded by locally contained chaperones. Once inside the mitochondrion, the proteins can be incorporated into the electron transport chain (ETC) or serve as transcriptional regulators for mtDNA. These events allow for the expansion of the organelle and provide a greater capacity for energy production. Furthermore, adjacent organelles can fuse together with the assistance of Mitofusin 1 and 2 and OPA1, thereby creating a greater reticular organelle environment.

Activation of PGC-1α protein occurs through its phosphorylation by AMPK (Jäger et al. 2007) and deacetylation by SIRT1 (Nemoto, Fergusson, and Finkel 2005; Rodgers et al. 2005). Thereafter, it will translocate from the cytosol to the nucleus where it can dock on target transcription factors, such as nuclear respiratory factor 1 or 2, estrogen-related receptor family members, among many others (Scarpulla 2011). With acute exercise, the translocation of PGC-1α is an early response that occurs prior to the upregulation of gene expression (Wright et al. 2007; Little et al. 2010). Along with other chromatin modifying proteins (Wallberg et al. 2003; Puigserver et al. 1999), PGC-1α will transactivate NuGEMPs leading to a rise in mRNA content of these targets, which can be translated into protein within the cytosol. Once the protein products are made, they need to be delivered to the mitochondria. This occurs through the protein import machinery utilizing translocases of the outer membrane (TOMs) and inner membrane (TIMs). Chronic contractile activity (CCA), a model of exercise in rodents, has been documented to induce increased protein expression of protein import components in conjunction with increased import rates into mitochondria (Joseph and Hood 2012). Interestingly, in aging skeletal muscle, there is no change in protein import rate into the mitochondria, suggesting that this step is not a limiting factor in the aged muscle (Huang et al. 2010).

Once inside the organelle, the imported proteins will be refolded by locally contained chaperones, such as HSP60, and shuttled to their appropriate destination to merge with mtDNA-derived products to cause the ultimate expansion of the organelle. Indeed, mitochondria in skeletal muscle are often found as long reticular structures (Ogata and Yamasaki 1985; Kirkwood, Munn, and Brooks 1986; Kirkwood, Packer, and Brooks 1987), which is biochemically advantageous for the sharing of substrates and for increasing the surface area to house more ETCs along the cristae folds, thereby raising the capacity for ATP synthesis (Liesa and Shirihai 2013). Furthermore, exercise increases the production of fusion factors such as Mitofusin 2 (MFN2) and OPA1 (Iqbal et al. 2013), which are capable of tethering adjacent mitochondria and fusing their membranes together to expand the size of the organelle, creating elongated organelles.

12.2.2 Mitophagy

Mitochondria are capable of becoming dysfunctional when they become damaged or dysfunctional. As such, it is necessitated that these poor-quality mitochondria are removed to liberate space and resources for replenishment/remodeling with new and/or healthy organelles.

Macroautophagy (hereafter autophagy) is the process by which identified bulk cellular components, such as cytosol, protein aggregates, or organelles, are sequestered in a double-membrane vesicle, the autophagosome, which is transported to fuse with the lysosome (Feng, Yao, and Klionsky 2015). Upon fusion with the lysosome, the contents are degraded by the lysosomal pH-dependent enzymes down to basic molecular constituents to be reused in the construction of new organelles or other cellular components. Selectivity in autophagy has been shown to be mediated by receptors unique to different organelles (Rogov et al. 2014). In the case of mitochondria, selective targeting of these organelles is referred to as mitophagy. In skeletal muscle, proper function of autophagy has been reported to be necessary to maintain muscle quality and quantity (Masiero et al. 2009). Indeed, genetic inhibition of upstream autophagy regulators, such as ATG7 or ATG5, in skeletal muscle results in an accumulation of poor-quality mitochondria concomitant with the loss in muscle mass (Masiero et al. 2009; Raben et al. 2008).

Well-functioning mitochondria will have an optimal membrane potential ($\Delta\Psi_m$) and produce low levels of ROS. If an organelle experiences a drop in $\Delta\Psi_m$ and/or produces greater amounts of ROS that the antioxidant system is unable to cope with, then these signals serve to trigger the removal of that mitochondria from the cellular environment (Figure 12.2). One well-described pathway for the removal of mitochondria through mitophagy is through the stabilization of the kinase PTEN-induced putative kinase-1 (PINK1) and recruitment of the E3 ubiquitin ligase, PARKIN. Under regular conditions, PINK1 is rapidly imported into the organelle and cleaved by the protease presenilins-associated rhomboid-like protein (PARL) (Jin et al. 2010; Meissner et al. 2011; Deas et al. 2011). However, with the loss of $\Delta\Psi_m$, PINK1 is no longer imported and becomes stabilized on the outer mitochondrial membrane (Jin et al. 2010). From there, PINK1 will recruit the E3 ubiquitin ligase PARKIN through phosphorylation of both PARKIN and ubiquitin molecules (Koyano et al. 2014; Kane et al. 2014) (Figure 12.2). PINK1 has also been recently demonstrated to recruit other mitophagy receptors to damaged organelles (Lazarou et al. 2015).

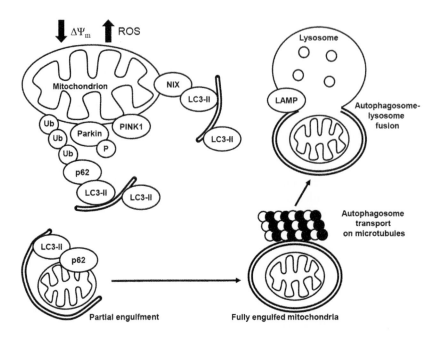

FIGURE 12.2 Removal of mitochondria through mitophagy. When mitochondria becomes defective, they need to be sequestered and removed from the cellular environment. Signals that initiate this process include a loss of membrane potential ($\Delta\Psi_m$) and an increase in ROS production. The loss of $\Delta\Psi_m$ inhibits the import and degradation of kinase PINK1 and instead it becomes stabilized on the outer mitochondrial membrane (OMM). PINK1 phosphorylates Parkin as well as ubiquitin molecules, priming Parkin to add ubiquitin chains to OMM proteins, such as MFN2 or VDAC. The ubiquitin chains are sensed by the adaptor protein p62, which contains an ubiquitin-binding domain as well as an LC3-interacting region (LIR). Thus, p62 will act as a tether between the ubiquitin-tagged mitochondrion and LC3-II. The lipidated form of LC3, LC3-II, is required for forming the autophagosomal double-membrane vesicle around the organelle identified for removal. The PINK1/Parkin pathway is not exclusive for the removal of mitochondria. Other receptors that can embed in the OMM, such as NIX, contain LIR, which can facilitate encapsulation of the organelle. Once the mitochondrion becomes fully encircled by the autophagosomal membrane, it reaches the lysosome by transport on microtubules. Lysosomal-associated membrane proteins (LAMPs) will allow the autophagosome to fuse with the lysosome. Thereafter, the contents of the autophagosome will be released into the lysosome and digested to basic constituents by pH-sensitive enzymes.

These post-translational modifications serve to prime PARKIN for full activation to add ubiquitin chains to numerous outer mitochondrial membrane proteins, such as VDAC1, MFN1, or MFN2 (Chen and Dorn 2013; Geisler et al. 2010; Sun et al. 2012; Gegg et al. 2010). These polyubiquitin chains act as a signaling scaffold to recruit ubiquitin binding proteins that also have domain regions for interacting with microtubule-associated protein 1A/1B-light chain 3-II (LC3-II), which is a critical factor for generating the autophagosomal membrane. Such binding proteins include p62 (Pankiv et al. 2007; Isogai et al. 2011), neighbor of BRCA-1 (NBR1) (Kirkin et al. 2009; Waters et al. 2009), nuclear dot protein 52 (NDP52) (Lazarou et al. 2015), and Optineurin (OPTN) (Wong and Holzbaur 2015), among others that continue to be identified (Chen et al. 2016; Wei et al. 2017). The PINK1/PARKIN pathway is not exclusive for the removal of mitochondria. Other receptors, such as NIX, can be inserted in the outer mitochondrial membrane and interact with LC3-II directly to facilitate the removal of mitochondria (Novak et al. 2010) (Figure 12.2). The targeted organelle will then become fully encapsulated by the double membrane and subsequently be shuttled to the lysosome along microtubules (Xie et al. 2010). Upon reaching the lysosome, the membranes of the lysosome and autophagosome will merge with the assistance of lysosomal-associated membrane proteins (LAMPs) (Fortunato et al. 2009). The sequestered cargo of the autophagosome will then be given to the lysosome for the degradation by pH-sensitive enzymes (Figure 12.2). Upon degradation, the basic materials will be released from the lysosome for reuse in the synthesis of new cellular components.

The processes of autophagy and mitophagy are highly dynamic, making them difficult to measure. Guidelines in the area of autophagy research have stressed the importance of employing autophagy inhibitors (such as colchicine, bafilomycin A1, or choloroquine) to block the degradation and then comparing the accumulation of proteins such as LC3-II and p62 to basal, untreated conditions (Klionsky et al. 2016; J.-S. Ju et al. 2010; Mizushima and Yoshimori 2007). This type of experimental design allows for the quantification of autophagy or mitophagy flux, a gold-standard measurement in determining the amount of cellular degradation through autophagy. Unfortunately, autophagy "flux" is often insinuated through other measures including the LC3-II/LC3-I ratio and the abundance of p62 protein determined by western blotting. These overly simplified measures are not enough to infer autophagic degradation, as interpretations may be skewed by either upstream modifiers (transcription, translation, post-translational modifications) or downstream (autophagy or proteasomal degradation processes). Thus, employing the appropriate methodology will lead to the clearest understanding of how autophagy is altered with different experimental conditions, including aging.

12.3 LOSS OF MITOCHONDRIA WITH AGING AND THE INFLUENCE ON MUSCLE MASS

Mitochondrial content is not static and is related to the energy demands of the cell. In conditions where mitochondrial content or quality is diminished, this may lead to issues with meeting cellular energy demand. Furthermore, examination of pathways regulating mitochondrial biogenesis is often noted to be impaired, suggesting improper maintenance of existing organelles. Dysfunctional mitochondria can have negative consequences for the muscle, and thus mitochondria may serve as trigger points for the loss of muscle mass observed with aging. Thus, maintaining quality organelles may be a useful strategy to mitigate muscle atrophy with aging while also preserving endurance capacity. The following sections will cover the role of mitochondria in sarcopenia and specifically focus on apoptotic signaling, mtDNA mutations, mitochondrial biogenesis, and mitophagy.

12.3.1 ROLE OF MITOCHONDRIA IN SARCOPENIA

The age-related decline in skeletal muscle mass, quality, and function is termed as sarcopenia (Rosenberg 1997). The word is derived from the Greek language where Sarco- means flesh and penia- means poverty. Sarcopenia is a ubiquitous hallmark of aging. It has been observed in nematodes (Herndon et al. 2002), flies (Demontis et al. 2013), rats (Bua et al. 2008), monkeys (McKiernan et al. 2009), and humans (Lexell et al. 1988; Lee et al. 2006). The prevalence of sarcopenia in humans was determined to be greater than 50% after the age of 80 years (Baumgartner et al. 1998). Thus, muscle wasting is an important aspect of aging as it is highly prevalent in old animals and highly conserved in a wide range of organisms.

However, the extent of muscle wasting is not uniform among the different types of muscle groups. Muscles that exhibit atrophy and fiber loss have a higher abundance of mitochondrial abnormalities than muscles that are resistant to sarcopenia (Bua et al. 2002). Similarly these muscles also exhibit a higher degree of apoptosis (Alway et al. 2011; Cheema et al. 2015). The first causative association of mitochondria and sarcopenia was delineated by Trifunovic and colleagues in 2004, wherein they developed transgenic mice expressing a proofreading-deficient mtDNA polymerase. They showed that somatic mtDNA mutations in transgenic mutator mice resulted in premature aging phenotypes such as shortened life span, accelerated osteoporosis, alopecia, weight loss, kyphosis, myocardial hypertrophy, and sarcopenia (Trifunovic et al. 2004). Kujoth and colleagues (Kujoth et al. 2005) used a similar mutator mouse that exhibited extreme muscle wasting and had significant apoptotic signaling. These studies suggest that mitochondria may contribute to decline in muscle function, atrophy, and fiber loss observed in sarcopenic muscles.

Mitochondria exist as a dynamic network within the cell. This morphology is dependent with advancing age. Smaller and age-dependent fragmentation of mitochondria has been observed in nematodes (Regmi et al. 2014), rats (Iqbal et al. 2013), and humans (Joseph et al. 2012). In advancing age, there are low levels of mitochondrial fusion proteins (Joseph et al. 2012) and an increase in fission proteins (Iqbal et al. 2013). Furthermore, inhibition of the mitochondrial machinery protected mice from muscle atrophy. Conversely, upregulating mitochondrial machinery resulted in muscle wasting (Romanello et al. 2010). These studies suggest that mitochondria has an important role in regulating muscle mass.

Abnormalities in mitochondrial structure have been associated with impaired function. Aged mice that have low levels of mitochondrial fusion proteins also have loss of mitochondrial function (Sebastián et al. 2016). Fission often precedes the degradation of dysfunctional mitochondria through mitophagy. Mitochondria with a dysfunctional ETC result in an increase in reactive oxygen species (ROS) generation (Chabi et al. 2008) and diminished capacity for ATP production (Short et al. 2005). This drastically increases the susceptibility of mitochondria to cell death (Chabi et al. 2008).

12.3.2 Mitochondrially Mediated Apoptosis

Mitochondrially mediated apoptosis is a form of programmed cell death. Release of pro-apoptotic proteins from the mitochondria, such as cytochrome c, apoptosis-inducing factor (AIF), or endonuclease G (ENDO G), will trigger caspase-independent or -dependent signaling cascades that result in DNA fragmentation, the hallmark feature of apoptosis. Skeletal muscle is unique as it is a multinucleated tissue, thus removal of one nucleus will not cause loss of the complete fiber but will cause localized muscular atrophy, thus leading to a loss of muscle mass (Alway and Siu 2008). Alterations in skeletal muscle activity patterns can sensitize or desensitize the mitochondria for apoptotic initiation. Indeed, with chronic muscle use, mitochondria are less sensitive to apoptotic stimuli and also produce reduced levels of ROS (Adhihetty, Ljubicic, and Hood 2007). The opposite is also true, as conditions of muscle disuse, such as denervation or aging, lead to a greater susceptibility of the mitochondria to trigger myonuclear decay (Adhihetty et al. 2007; Chabi et al. 2008). In the aging muscle milieu, mitochondria have been shown to produce more reactive oxygen species, which can damage local structures, like mtDNA or proteins within the ETC, while also sensitizing mitochondria to release pro-apoptotic proteins (Chabi et al. 2008; Ljubicic et al. 2009). Furthermore, mitochondria in the aged muscle have been shown to have uncoupled respiration, indicating a reduced efficiency to produce ATP. These mitochondria retain less calcium, which further sensitizes the organelles to pro-apoptotic release (Gouspillou et al. 2014). Indeed, in a rodent model of aging, it has been identified that myonuclei have a threefold greater presence of ENDO G, which would cause DNA fragmentation, myonuclear decay, and ultimately muscle atrophy (Gouspillou et al. 2014). Furthermore, work by Aiken and colleagues have proposed a model where accumulation of mtDNA mutations in focal regions of single fibers drives mitochondrial insufficiency, which activates apoptotic and necrotic pathways, causing local atrophy that expands to myofiber loss (Herbst et al. 2016). When analyzing individual fibers in aged rodent muscle, majority of the fibers with a dysfunctional ETC were apoptotic and necrotic. Fibers with a larger apoptotic and necrotic segment broke in the middle, suggesting that these two cell death pathways play a role in age-related fiber loss (Cheema et al. 2015). A contrasting theory from work by Hepple and coworkers has implicated that in early aging, mitochondria are dysfunctional through altered membrane potential, which may be the triggering factor for apoptotic processes and muscle atrophy. However, in their investigation of aged human muscle as well as experimental animal models of sporadic denervation, they concluded that in advanced age, muscle atrophy is mediated by denervation events at the neuromuscular junction and not by the mitochondria (Spendiff et al. 2016). They suggest that at this later stage, mitochondria may not be the most viable target for recovering muscle mass, although the organelles likely still remain dysfunctional.

Since mitochondria are prone to releasing pro-apoptotic substrates in the aged muscle environment, thereby triggering muscle loss, is there a stimulus that could prevent the initiation of these negative signaling cascades? Increases in chronic contractile activity have been shown to decrease the apoptotic sensitivity of mitochondria in young, healthy muscle, as well as increase anti-apoptotic cytosolic markers (Adhihetty, Ljubicic, and Hood 2007). Application of the CCA model to aging muscle has been shown to reduce Complex II-stimulated ROS production as well as decrease DNA fragmentation, the hallmark measure of apoptosis (Ljubicic et al. 2009). Taken together, it appears that chronic exercise adaptations can improve mitochondrial quality to prevent the release of apoptosis-inducing factors, enhance cellular defense, and diminish DNA degradation, thus preserving muscle mass. This is important for the maintenance of muscle mass because the retention of nuclei (rather than their apoptotic decay) keeps and enhances the capacity of this region of the cell for muscle-specific gene expression and subsequent protein synthesis. In addition, the retention of the nucleus in this area certainly facilitates the synthesis of nuclear-encoded gene products, which could be incorporated into SS mitochondria, allowing for regionally quality control of mitochondria.

12.3.3 mtDNA Mutations

As previously mentioned, mitochondria house their own circular genome, which encodes 13 critical polypeptides of the ETC. Mitochondria carry multiple copies of mtDNA per organelle, and some can be healthy, while others may have deletions or mutations. This is called heteroplasmy. Normally, all individuals have some degree of heteroplasmy; however, when the degree of mutated mtDNA exceeds an acceptable threshold, features of mitochondrial disease or aging may develop. Research has found that mtDNA mutations accumulate with aging and start rapidly expanding in the seventh decade in human subjects (Bua et al. 2006; Kadenbach et al. 1995). The most frequently identified change in mtDNA is the 4977 bp deletion (Cortopassi and Arnheim 1990; Cortopassi et al. 1992; Linnane et al. 1990; Melov et al. 1995). Also, the promoter region of mtDNA, termed the D-loop, manifests age-related reductions in transcription and replication (Wang et al. 2001). Interestingly, mtDNA that is mutated or deleted is often shorter in length and appears to be given preference for replication, thus enhancing the degree of heteroplasmy toward mutated or deleted mtDNA (Kowald and Kirkwood 2014).

mtDNA harboring mutations can lead to the synthesis of electron transport chain components that are defective. This can exacerbate the production of reactive oxygen species and lead to mitochondrial permeability transition pore opening, organelle swelling, and the release of pro-apoptotic proteins (see above). These proteins have the potential to induced DNA fragmentation, nuclear decay, and regional muscle atrophy. Therefore, a long-standing question has been whether or not exercise training, which normally increases mtDNA copy number, can be beneficial or detrimental to patients with mtDNA mutations. A detrimental outcome would be an exercise-induced proliferation of mtDNA copy number with those deleterious mutations. However, the heteroplasmic nature of mtDNA also suggests that "healthy," nonmutated forms of mtDNA could also be replicated as a result of training. Early work by Taivassalo et al. (2001) not only indicated many beneficial adaptations of training on work capacity, VO_2 max, and mitochondrial content in patients with mtDNA mutations but also indicated a preferential increase in mutated mtDNA species within muscle. However, more recent research by Jeppesen et al. (2009) indicated that prolonged training is a safe treatment to alleviate exercise intolerance in patients with mitochondrial DNA (mtDNA) mutations, since training of up to 12 months did not induce adverse effects on clinical symptoms, muscle morphology, and mtDNA mutation load in muscle. Thus, in aged skeletal muscle where increased levels of mutated or deleted mtDNA have been observed, engaging in exercise likely would not be detrimental and may instigate replication of healthy mtDNA to assist in organelle quality control.

Unlike nuclear DNA, mtDNA is also not protected from endogenous damage due to differing structural features, such as the lack of histones. Moreover, mtDNA has only one main proof-reading enzyme, polymerase γ, that if not functional could lead to the expansion of missense mtDNA. Indeed, in an animal model devoid of polymerase γ, mtDNA mutations were significantly enhanced, and the animals presented with numerous features of aging, including sarcopenia (Safdar et al. 2011). After these animals had been trained on the treadmill for 5 months, the premature aging features were attenuated. mtDNA quality was improved in skeletal muscle, and this translated into improvements in organelle function and morphology (Safdar et al. 2011). This study suggests that exercise is an effective strategy to attenuate, if not reverse, the features of aging.

12.3.4   PGC-1α

PGC-1α appears to be a pivotal coactivator for NuGEMPs and for driving mitochondrial biogenesis with exercise. In aging skeletal muscle, reduced levels of this coactivator have been found in both slow and fast-twitch type muscles (Chabi et al. 2008), correlating with the decline in the mitochondrial content.

Recently, more insight has emerged into the diversity of PGC-1α isoforms and how they may regulate different aspects of muscular adaptation with exercise. Indeed, a smaller variant, termed PGC-1α4 by Spiegelman and coworkers (Ruas et al. 2012), was found to induce muscular hypertrophy following resistance exercise training, while the original PGC-1α remained important for endurance exercise adaptations, such as mitochondrial biogenesis. However, how the divergent properties of this coactivator family may apply to aging skeletal muscle has not been well investigated. One study using an acute bout of resistance exercise in young and aged men reported that both the PGC-1α4 variants, as well as PGC-1α1 (the original coactivator), increased 3 h following the exercise bout; however age-related differences were not presented (Ogborn et al. 2015). Resistance exercise is often recommended to build muscle mass, especially in older individuals where preservation of muscle mass will assist in ambulation and stability. However, increasing mitochondrial quality will also promote endurance capacity and metabolic homeostasis. Thus, if the PGC-1α family serves a diverse role in maintaining muscle mass and mitochondrial content with differing exercise stimuli, this would be an ideal therapeutic target through exercise for aged individuals.

Numerous studies have noted decrements in the expression of PGC-1α with age. A recent study from our laboratory has reported that the transcriptional drive for mRNA expression of PGC-1α is reduced in aged rodent muscle (Carter et al. 2018b). In this context, DNA conformation is highly important for transcription factor access to target promoter regions to instigate transcript synthesis. DNA is tightly compressed into chromatin that is wound around protective histone protein complexes. To gain access, DNA must be demethylated and the histones are often acetylated, thereby allowing for an open conformation. Methylation of cytosine nucleotides on promoter regions of genes has been linked to decreases in transcriptional activation. Indeed, an increase in the global methylation of nDNA has been documented in aged skeletal muscle of human subjects (Zykovich et al. 2014), suggesting that the hypermethylated state may impair gene transcription. In support of this, numerous microarray assays have consistently reported decreases in expression of a plethora of mitochondrial metabolic transcripts in aging muscle (Pattison et al. 2003; Ibebunjo et al. 2013; Melov et al. 2007).

12.3.5   Mitochondrial Turnover

Mitochondrial turnover comprises both the processes of biogenesis as well as mitophagy. Mitochondria are not generated *de novo* but are maintained and built upon existing structures. For elimination, portions of the organelle are segmented through fission and can be selectively targeted for degradation. To attain a steady-state level of mitochondria, there must be a relative balance

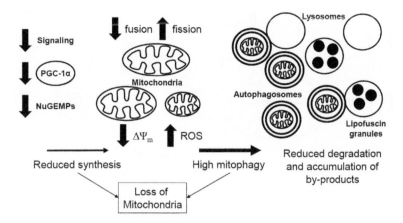

FIGURE 12.3 How are mitochondria lost in aging skeletal muscle? Many studies have reported that mitochondrial content and/or function is compromised in skeletal muscle with aging, even in subjects that are carefully matched for physical activity. How mitochondria may be depleted with aging is not fully understood, but this is likely a consequence of an imbalance between biogenesis and mitophagy. In this schematic, we speculate on how mitochondria may be compromised in aging muscle. Literature has documented a reduced drive for mitochondrial biogenesis. Upstream signaling regulators, such as kinases, often exhibit reduced activity in aged muscle. Indeed, loss of PGC-1α and decreased expression of NuGEMPs are consistently noted in aged muscle, which may partly be a consequence of reduced signaling. Furthermore, mitochondria from aged muscle produce more ROS and have compromised membrane potential. Finally, the morphology of mitochondria in aged muscle often appears more fragmented and may be a consequence of an imbalance between fission and fusion proteins. Given the defective nature of mitochondria in aged muscle, the organelles likely signal for mitophagy protein recruitment in an attempt to clear the cellular milieu. Increased expression of PARKIN has been noted on isolated aged mitochondria, concomitant with increased upstream autophagy protein expression. Therefore, we speculate that the dysfunctional organelles are likely encapsulated in autophagosomes, in a futile attempt to clear them from the muscle. However, when delivered to the lysosome, the contents are incapable of being digested and accumulate as a by-product called lipofuscin. This accretion of lipofuscin incapacitates the lysosome, rendering it unable to digest further material. Since organelles continue to become dysfunctional, there is continued engulfment of mitochondria; however, given the low drive for biogenesis, they are unable to be replaced with new, healthy organelles. Thus, a skewed balance occurs where biogenesis is low and mitophagy is high, which ultimately leads to downsizing of the mitochondrial pool, which we observe in aged muscle.

between the amount of incoming organelles (biogenesis) and the exit of mitochondria (mitophagy). To induce alterations in mitochondrial content, it is therefore necessary to have an imbalance between synthesis and removal (Figures 12.3 and 12.4). Furthermore, the degree to which either of these processes would be affected is likely stimulus dependent. As previously stated, mitochondrial content is, by and large, considered to be decreased in aged skeletal muscle. Thus, with regard to biogenesis and mitophagy, there would likely need to be some combination of reduced synthesis and/or increased mitophagy. This would skew the balance to reduce the content of organelles.

In aging muscle, biogenesis is low as a result of impaired signaling mechanisms, depleted levels of PGC-1α, and reduced NuGEMP concentration (Ljubicic et al. 2009; Ljubicic and Hood 2009). As such, mitochondrial quality is not maintained, as shown by increased ROS production, loss of membrane potential, and increased apoptotic protein release (Chabi et al. 2008). The increase in ROS can signal to the autophagy/mitophagy pathways to clear these poor-quality mitochondria from the muscle (Figure 12.3). Indeed, in aged muscle, it has been documented that upstream signaling proteins for autophagy such as ATG7 are increased in addition to greater localization of autophagy and mitophagy markers to the mitochondria such as LC3-II and PARKIN (O'Leary et al. 2013). Additionally, aged muscle has been identified to harbor more fragmented mitochondria,

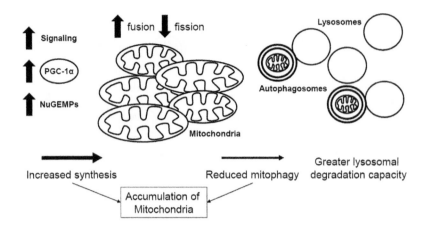

FIGURE 12.4 How do mitochondria accumulate with exercise? As depicted in Figure 12.1, acute exercise causes the activation of cellular signaling pathways to stimulate mitochondrial biogenesis. With regular bouts of exercise (training), the repeated stimulation of these signaling pathways leads to enhanced production of PGC-1α and NuGEMPs. The organelles increase in size with the addition of extra components, in addition to fusing with neighboring mitochondria, expanding the reticulum. These pathways have been understood for many years; however, less is known regarding mitophagy. We speculate that once skeletal muscle is in the trained state, there will be a greater population of organelles that are much healthier and cause limited detriment to the muscle. Therefore, the need to sequester and devour organelles through mitophagy is low. However, we believe that lysosomal biogenesis would increase leaving the muscle with a greater capacity for degradation, potentially for other aspects of autophagy.

likely due to the increased expression of fission proteins (FIS1, DRP1) in comparison with fusion proteins (OPA1, MFN1, or MFN2) (Iqbal et al. 2013). Fission of mitochondria often precedes their degradation by mitophagy; thus the aged muscle mitochondrial environment may be primed for the removal of dysfunctional organelles. Separating dysfunctional portions of mitochondria from the larger reticular network aids in their encapsulation in the autophagosome and degradation via mitophagy. The enhanced fission, higher autophagy pathway protein expression, and greater mitochondrial localization of mitophagy factors could be interpreted to mean that mitophagy is higher in aged muscle.

A report using an established rodent model of aging, the FBN344 rat, employed colchicine treatment as an autophagic inhibitor to compare LC3-II flux in young and aged fast- and slow-twitch muscle with or without reloading after hind limb suspension (Baehr et al. 2016). Interpretation of the differences between young and aged control animals reveals that autophagic flux appears to be markedly elevated in the aged group compared with the young group in both the soleus and tibialis anterior muscles. Recent studies in our laboratory evaluated mitophagic flux in young and aged mice (Chen et al. 2018) and rats (Carter et al. 2018a) by isolating mitochondrial fractions from whole muscle. Mitophagy flux was measured via LC3-II and p62 localization to mitochondria and was found to be enhanced in aged muscle (Carter et al. 2018a). Along with the increased flux, aged mitochondria also had elevated levels of mitophagy markers, which suggested that the mitophagic pathway may be a contributing factor to reduced mitochondrial content observed in aging.

Enhance mitophagy may overwhelm the lysosomal capacity for degradation, resulting in lipofuscin accumulation, indicative of lysosomal dysfunction. Lipofuscin has been shown to be composed of lipids, metals, and reaction by-products that are indigestible. Indeed, it has been shown that CATHEPSIN D, a lysosomal enzyme, is increased in aged muscle; however, the activity of the enzyme is lower (Wiederanders and Oelke 1984), suggesting an impaired ability to digest delivered cargo. Thus, collectively it could be considered that mitochondria are sequestered at a high rate due to their dysfunctional nature (i.e., high ROS, low $\Delta\varPsi_m$). These mitochondrially loaded autophagosomes would be delivered to lysosomes, but lipofuscin may accumulate, rendering a portion of lysosomes nonfunctional. Moreover, due to the reduced activity of cellular signaling pathways and kinases, biogenesis is unable to replace or repair the mitochondria that were sequestered or remain dysfunctional. Altogether this would generate a net loss of organelles, which we often observe in aged muscle (Figure 12.3). If chronic exercise could reduce the rates of mitophagy concomitant with the enhancement of biogenesis pathways (Figure 12.4), there would be the potential for the recuperation of mitochondrial content and quality in aged muscle; however, this possibility remains to be elucidated.

12.4   EXERCISE BENEFITS FOR MITOCHONDRIA AND MUSCLE WITH AGING

Profound skeletal muscle mitochondrial adaptations occur in young, healthy muscle following engagement in a regular exercise regimen and, as described previously, are mediated through PGC-1α and other important transcription factors. However, models that are deficient in PGC-1α, whether *in vivo* (Adhihetty et al. 2009) or *in vitro* (Uguccioni and Hood 2011), and subsequently exposed to an exercise stimulus are able to compensate for the loss of the coactivator. For instance, the deficit in mitochondrial respiration can be repaired in young PGC-1α-deficient mice with exposure to exercise (Adhihetty et al. 2009). How this occurs at a molecular level is still not understood, but may involve redundancy between the PGC-1α family members, additional PGC-1α isoforms or independent pathways. Since PGC-1α is expressed at reduced levels in aged muscle, this raises the question as to whether this affects exercise adaptations in aging muscle. Few data exist on the topic, but one study utilizing PGC-1α-deficient mice exposed to long-term wheel running found that PGC-1α was necessary to confer the benefits of training to mitochondrial adaptation into later age (Leick et al. 2010).

Research from our laboratory has noted that *Pgc-1α* transcript content is reduced in an aged rodent model, but with the instigation of chronic low-frequency muscle stimulation (CLFS) to induce mitochondrial adaptations, the diminished levels of *Pgc-1α* mRNA could be attenuated (Carter et al. 2013). We observed that *Pgc-1α* mRNA in aged animals was comparable to levels observed in a young, untrained muscle after CLFS. Combined, these data suggest that chronic exercise may be a useful strategy to assist in repairing the deficits in mitochondrial content and/or quality found in aged muscle, likely through the restoration of *Pgc-1α* expression.

Whether the mitochondrial adaptations due to exercise training are comparable in aging or senescent muscle as they are in young muscle remains controversial. To be able to interpret the changes in mitochondria with training in aged muscle, initial comparisons of the basal levels of mitochondrial content are necessary to accurately quantify the adaptations that may ensue. Unfortunately, not all studies have employed young vs. aged group comparisons to be able to explore the magnitude of adaptation. Nonetheless, observations using longitudinal or cross-sectional design of exclusively aged subjects have noted mitochondrial retention in muscle from older subjects. For example, in a study by Joseph et al. (2012), older human subjects were stratified into low- and high-functioning groups based on their scores of a standardized test. The subjects that were high functioning exhibited improved markers of mitochondrial content, including greater PGC-1α protein expression. Interestingly, all of the subjects that were recruited for the study were sedentary, thus physical activity levels did not explain the changes in mitochondrial markers, providing further support for the link between mitochondrial health and skeletal muscle function.

In another study, groups of young and older men were analyzed for their transcriptome (Melov et al. 2007). It was observed that numerous transcripts involved in mitochondrial function were downregulated in the older individuals compared with the young group. The subjects were then asked to engage in a 6 months of resistance exercise training. After training, reevaluation of the transcriptome revealed that the gene signature reverted to resemble that of the younger group, including improvements. More recently, Robinson et al. (2017) demonstrated that both high-intensity interval and combined training modes led to improvements muscle strength and mitochondrial respiration. This appeared to be mediated at the level of translation, rather than transcriptionally.

The basal comparison of mitochondrial content and quality in young versus old muscle has generated controversial results. Many studies have reported numerous markers of mitochondrial content and quality to be reduced in aged muscle, such as maximal ATP production rate, citrate synthase activity, COX activity, and histochemistry (Rooyackers et al. 1996; Short et al. 2003; Ljubicic et al. 2009; Chabi et al. 2008; Lee et al. 1998; Short et al. 2005; Trounce, Byrne, and Marzuki 1989; Boffoli et al. 1994; Larsen et al. 2012). However, discussion has led to the adamant conclusion that when comparing young and old muscle, the levels of physical activity must be closely harmonized as changes in physical activity will alter mitochondrial status. However, even with subjects carefully matched for physical activity levels, the results remain controversial, as some report no changes in mitochondrial function (Kent-Braun and Ng 2000; Johnson, Robinson, and Nair 2013; Picard et al. 2011) while others continue to report diminished content and function with age (Ghosh et al. 2011; Johannsen et al. 2012; Rooyackers et al. 1996; Short et al. 2005). Thus, when extending observations to exercise training, it is important to regard the initial baseline level of mitochondria, in making comparisons between young and old. Finally, when developing interpretations on the adaptability of young and aged skeletal muscle mitochondria, it is critical to institute a training paradigm that is equal in relative intensity or stimulus between the groups being studied. By having an equal stimulus between both young and aged groups, this allows for greater insight on the ability of the muscle to adapt and can provide conclusions on whether or not aged muscle displays reduced or equal adaptive capacity to exercise treatments.

In our opinion, studies with equivalent exercise intensity/stimulus between young and aged groups have provided evidence that mitochondrial adaptations in aged muscle are possible. However, the extent or magnitude of this adaptation is often observed to be less than young muscle or may require a greater duration to reach the same adapted level (Carter, Chen, and Hood 2015; Hood et al. 2016). For instance, in a study by Walters et al. (Walters, Sweeney, and Farrar 1991) who used CLFS, aged animals displayed reduced adaptations in the early time course of the experiment compared with the younger group. However, when the protocol was extended, aged muscle was capable of developing similar levels of mitochondrial adaptation. This suggests that the deficits in mitochondrial health are partly age related and not related solely to changes in physical activity. Indeed, we and others have observed impaired upstream signaling mechanisms in older muscle (Ljubicic and Hood 2009; Reznick et al. 2007). For instance, basal levels of AMPK phosphorylation are lower in aged muscle. With the addition of an exercise stimulus, aged muscle continues to display hampered phosphorylation of the kinase, likely diminishing the activation of its downstream targets, including those involved in mitochondrial biogenesis.

To induce mitochondrial biogenesis in aging rodent models, CLFS is a commonly used paradigm in our laboratory, which eliminates the behavioral and motivational aspects associated with wheel running or treadmill training (Ljubicic, Adhihetty, and Hood 2005). Through the use of this model, we have found that adaptations to chronic exercise in aged muscle are attenuated with aging (Ljubicic et al. 2009). Indeed, the increase in COX activity was attenuated in aged muscle versus young. Investigation of the key regulatory events that govern synthesis of new mitochondria, such as protein import pathway regulators and apoptosis pathways following CCA, revealed the magnitude adaptive change for regulatory proteins such as HSP70, TIM23, and TIM17 was decreased with aging. However, increased defense against apoptosis was noted through diminished DNA fragmentation. Interestingly, improvements in organelle quality were noted via decreased SS and IMF ROS production and enhanced oxygen consumption, suggesting that exercise has a beneficial role in remodeling dysfunctional organelles, although the overall content still remains lower than that seen in young muscle. Whether these attenuated adaptations observed in the aged animals would have been enhanced with a greater duration of CCA has yet to be determined.

12.5 CONCLUSIONS AND PERSPECTIVES

The proportion of aging individuals is on the rise; therefore investigation and understanding of strategies to preserve muscle function are necessary. Maintaining quality of life is closely correlated with preserved muscle function to engage in activities of daily living and retain independence. Inherent to the quality of skeletal muscle are the mitochondria that supply energy and participate in metabolic homeostasis. Numerous publications have observed reductions in the overall content and function of these organelles. Dysfunctional mitochondria in aging muscle present with the loss of their membrane potential, limited ATP generating capacity, mtDNA damage, or enhanced ROS production. These events could trigger pro-apoptotic protein release triggering myonuclear decay, although some controversy exists regarding the exact molecular mechanisms. As discussed, maintenance of high-quality organelles requires a delicate balance of synthesis and degradation through biogenesis and mitophagy, respectively. With aging, it appears that defects arise in both biogenesis and mitophagy; however, a comprehensive understanding of autophagy and mitophagy flux in aging muscle has yet to be described. It could be theorized that autophagy/mitophagy is deficient in aging muscle, since accumulation of dysfunctional organelles is often noted, suggesting a reduced capacity to clear mitochondria from the cellular milieu. In contrast, preliminary reports seem to suggest the autophagy may be increased with aging (Baehr et al. 2016). This would lead to the speculation that damaged organelles are, in fact, being degraded. Therefore, perhaps the deficit in mitochondrial maintenance during aging lies in the pathway of suppressed biogenesis, which has been well documented in the literature.

The effects of acute and chronic exercise on autophagy and mitophagy are now becoming delineated in the skeletal muscle. Acute exercise induces organelle remodeling through enhanced mitophagy. Preliminary reports on exercise training suggest autophagy (and potentially mitophagy) remain elevated upon cessation of training; however, further work using autophagy flux inhibitors would help to confirm the initial experiments. Exercise is also well known to effectively lead to mitochondrial biogenesis, which can be achieved in aged muscle, although often to a lesser extent than in young muscle. Thus, if mitochondrial biogenesis and autophagy/mitophagy pathways are both dysregulated in the aged muscle environment, exercise appears to be an ideal strategy to improve mitochondrial quality control.

To gain a fuller appreciation of the mechanisms involved in mitochondrial maintenance in aging muscle, future efforts would benefit from examining (1) autophagy and mitophagy flux in resting aged muscle, (2) autophagy and mitophagy flux with exercise training in young vs. old muscle, and (3) a time-course relation between mitochondrial biogenesis and mitophagy pathways in young and senescent muscle.

REFERENCES

Adhihetty, Peter J., Giulia Uguccioni, Lotte Leick, Juan Hidalgo, Henriette Pilegaard, and David A Hood. 2009. "The Role of PGC-1alpha on Mitochondrial Function and Apoptotic Susceptibility in Muscle." *American Journal of Physiology: Cell Physiology* 297 (1): C217–25.

Adhihetty, Peter J., Michael F N O'Leary, Beatrice Chabi, Karen L. Wicks, and David A. Hood. 2007. "Effect of Denervation on Mitochondrially Mediated Apoptosis in Skeletal Muscle." *Journal of Applied Physiology* 102 (3): 1143–51.

Adhihetty, Peter J., Vladimir Ljubicic, and David A. Hood. 2007. "Effect of Chronic Contractile Activity on SS and IMF Mitochondrial Apoptotic Susceptibility in Skeletal Muscle." *American Journal of Physiology. Endocrinology and Metabolism* 292 (3): E748–55.

Akimoto, Takayuki, Ping Li, and Zhen Yan. 2008. "Functional Interaction of Regulatory Factors with the Pgc-1alpha Promoter in Response to Exercise by in Vivo Imaging." *American Journal of Physiology: Cell Physiology* 295 (1): C288–92.

Alway, Stephen E., Michael R. Morissette, and Parco M. Siu. 2011. "Chapter 4: Aging and apoptosis in muscle. *In Handbook of the Biology of Aging* (7th ed.)." E.J. Masoro and S.N. Austad, editors. Academic Press, San Diego, CA, pp. 63–118.

Alway, Stephen E., and Parco M. Siu. 2008. "Nuclear Apoptosis Contributes to Sarcopenia." *Exercise and Sport Sciences Reviews* 36 (2): 51–7.

Amat, Ramon, Anna Planavila, Shen Liang Chen, Roser Iglesias, Marta Giralt, and Francesc Villarroya. 2009. "SIRT1 Controls the Transcription of the Peroxisome Proliferator-Activated Receptor-Gamma Co-Activator-1alpha (PGC-1alpha) Gene in Skeletal Muscle through the PGC-1alpha Autoregulatory Loop and Interaction with MyoD." *The Journal of Biological Chemistry* 284 (33): 21872–80.

Baehr, Leslie M., Daniel W. D. West, George Marcotte, Andrea G. Marshall, Luis Gustavo De Sousa, Keith Baar, and Sue C. Bodine. 2016. "Age-Related Deficits in Skeletal Muscle Recovery Following Disuse Are Associated with Neuromuscular Junction Instability and ER Stress, Not Impaired Protein Synthesis." *Aging* 8 (1): 127–46.

Baumgartner, Richard N., Kathleen M. Koehler, Dympna Gallagher, Linda Romero, Steven B. Heymsfield, Robert R. Ross, Philip J. Garry, and Robert D. Lindeman. 1998. "Epidemiology of Sarcopenia among the Elderly in New Mexico." *American Journal of Epidemiology* 147:755–63.

Bharadwaj, Manish S., Daniel J. Tyrrell, Iris Leng, Jamehl L. Demons, Mary F. Lyles, J. Jeffrey Carr, Barbara J. Nicklas, and Anthony J. A. Molina. 2015. "Relationships between Mitochondrial Content and Bioenergetics with Obesity, Body Composition and Fat Distribution in Healthy Older Adults." *BMC Obesity* 2 (1): 40.

Boffoli, Domen, S., Salvatore C. Scacco, Rosaria Vergari, Giuseppe Solarino, Luigi Santacroce, Sergio Papa. 1994. "Decline with Age of the Respiratory Chain Activity in Human Skeletal Muscle." *Biochimica et Biophysica Acta* 1226 (1): 73–82.

Bua, Entela, Jody K. Johnson, Debbie McKenzie, and Judd M. Aiken. 2008. "Sarcopenia Accelerates at Advanced Ages in Fisher 344×Brown Norway Rats." *The Journals of GerontologySeries A: Biological Sciences and Medical Sciences.* 63:921–27.

Bua, Entela, Susan H. McKiernan, Jonathan Wanagat, Debbie McKenzie, and Judd.M. Aiken. 2002. "Mitochondrial Abnormalities Are More Frequent in Muscles Undergoing Sarcopenia." *Journal of Applied Physiology* 92:2617–24.

Bua, Entela, Jody K. Johnson, Allen Herbst, Bridget Delong, Debbie McKenzie, Shahriar Salamat, and Judd M Aiken. 2006. "Mitochondrial DNA-Deletion Mutations Accumulate Intracellularly to Detrimental Levels in Aged Human Skeletal Muscle Fibers." *American Journal of Human Genetics* 79 (3): 469–80.

Carter, Heather N., Ayesha Saleem, Michael F. N. O'Leary, Olga Ostojic, and David A. Hood. 2013. "Expression of Sestrins in Skeletal Muscle with Acute Exercise and Aging." *The FASEB Journal* 27 (1 Supplement):939.8–939.8.

Carter, Heather N., Chris C.W. Chen, and David A. Hood. 2015. "Mitochondria, Muscle Health and Exercise with Advancing Age." *Physiology* 30 (3): 208–23.

Carter, Heather N., Yuho Kim, Avi T. Erlich, Dorrin Zarrin-khat, and David A. Hood. 2018a. "Autophagy and Mitophagy Flux in Young and Aged Skeletal Muscle Following Chronic Contractile Activity." *Journal of Applied Physiology* 596:3567–84.

Carter, Heather N., Marion Pauly, Liam D. Tryon, and David A. Hood. 2018b. "Effect of Contractile Activity on PGC-1α Transcription in Young and Aged Skeletal Muscle." *Journal of Applied Physiology* 124:1605–15.

Chabi, Béatrice, Vladimir Ljubicic, Keir J. Menzies, Julianna H. Huang, Ayesha Saleem, and David A. Hood. 2008. "Mitochondrial Function and Apoptotic Susceptibility in Aging Skeletal Muscle." *Aging Cell* 7 (1): 2–12.

Cheema, Nashwa, Allen Herbst, Debbie McKenzie, and Judd M. Aiken. 2015. "Apoptosis and Necrosis Mediate Skeletal Muscle Fiber Loss in Age-induced Mitochondrial Enzymatic Abnormalities." *Aging Cell.* 14:1085–93.

Chen, Chris C.W., Avi T. Erlich, and David A. Hood. 2018. "Role of Parkin and Endurance Training on Mitochondrial Turnover in Skeletal Muscle." *Skeletal Muscle.* 8:10.

Chen, Ming, Ziheng Chen, Yueying Wang, Zheng Tan, Chongzhuo Zhu, Yanjun Li, Zhe Han, et al. 2016. "Mitophagy Receptor FUNDC1 Regulates Mitochondrial Dynamics and Mitophagy." *Autophagy* 12 (4): 689–702.

Chen, Yun, and Gerald W. Dorn. 2013. "PINK1-Phosphorylated Mitofusin 2 Is a Parkin Receptor for Culling Damaged Mitochondria." *Science* 340 (6131): 471–75.

Cogswell, Andria M., Rebecca J. Stevens, and David A. Hood. 1993. "Properties of Skeletal Muscle Mitochondria Isolated from Subsarcolemmal and Intermyofibrillar Regions." *The American Journal of Physiology* 264 (2 Pt 1): C383–9.

Cortopassi, Gino A., and Norman Arnheim. 1990. "Detection of a Specific Mitochondrial DNA Deletion in Tissues of Older Humans." *Nucleic Acids Research* 18 (23): 6927–33.

Cortopassi, Gino A., Daisuke Shibata, New Soong, and Norman Arnheim. 1992. "A Pattern of Accumulation of a Somatic Deletion of Mitochondrial DNA in Aging Human Tissues." *Proceedings of the National Academy of Sciences of the United States of America* 89 (16): 7370–74.

Crane, Justin D., Michaela C. Devries, Adeel Safdar, Mazen J. Hamadeh, and Mark A. Tarnopolsky. 2010. "The Effect of Aging on Human Skeletal Muscle Mitochondrial and Intramyocellular Lipid Ultrastructure." *The Journals of Gerontology. Series A, Biological Sciences and Medical Sciences* 65 (2): 119–28.

Deas, Emma, Helene Plun-Favreau, Sonia Gandhi, Howard Desmond, Svend Kjaer, Samantha H. Y. Loh, Alan E. M. Renton, et al. 2011. "PINK1 Cleavage at Position A103 by the Mitochondrial Protease PARL." *Human Molecular Genetics* 20 (5): 867–79.

Demontis, Fabio, Rosanna Piccirillo, Alfred L. Goldberg, and Norbert Perrimon. 2013. "Mechanisms of Skeletal Muscle Aging: Insights from *Drosophila* and Mammalian Models." *Disease Models & Mechanisms*. 6:1339–52.

Feng, Yuchen, Zhiyuan Yao, and Daniel J. Klionsky. 2015. "How to Control Self-Digestion: Transcriptional, Post-Transcriptional, and Post-Translational Regulation of Autophagy." *Trends in Cell Biology* 25 (6): 354–63.

Fortunato, Franco, Heinrich Bürgers, Frank Bergmann, Peter Rieger, Markus W. Büchler, Guido Kroemer, and Jens Werner. 2009. "Impaired Autolysosome Formation Correlates with Lamp-2 Depletion: Role of Apoptosis, Autophagy, and Necrosis in Pancreatitis." *Gastroenterology* 137 (1): 350–60.e5.

Gegg, Matthew E., J. Mark Cooper, Kai-Yin Chau, Manuel Rojo, Anthony H. V. Schapira, and Jan-Willem Taanman. 2010. "Mitofusin 1 and Mitofusin 2 Are Ubiquitinated in a PINK1/parkin-Dependent Manner upon Induction of Mitophagy." *Human Molecular Genetics* 19 (24): 4861–70.

Geisler, Sven, Kira M. Holmström, Diana Skujat, Fabienne C. Fiesel, Oliver C. Rothfuss, Philipp J. Kahle, and Wolfdieter Springer. 2010. "PINK1/Parkin-Mediated Mitophagy Is Dependent on VDAC1 and p62/SQSTM1." *Nature Cell Biology* 12 (2): 119–31.

Ghosh, Sangeeta, Raweewan Lertwattanarak, Natalie Lefort, Marjorie Molina-Carrion, Joaquin Joya-Galeana, Benjamin P Bowen, Jose de Jesus Garduno-Garcia, et al. 2011. "Reduction in Reactive Oxygen Species Production by Mitochondria from Elderly Subjects with Normal and Impaired Glucose Tolerance." *Diabetes* 60 (8): 2051–60.

Goncalves, Renata L. S., Casey L. Quinlan, Irina V. Perevoshchikova, Martin Hey-Mogensen, and Martin D. Brand. 2015. "Sites of Superoxide and Hydrogen Peroxide Production by Muscle Mitochondria Assessed Ex Vivo under Conditions Mimicking Rest and Exercise." *The Journal of Biological Chemistry* 290 (1): 209–27.

Gouspillou, Gilles, Nicolas Sgarioto, Sophia Kapchinsky, Fennigje Purves-Smith, Brandon Norris, Charlotte H. Pion, Sébastien Barbat-Artigas, et al. 2014. "Increased Sensitivity to Mitochondrial Permeability Transition and Myonuclear Translocation of Endonuclease G in Atrophied Muscle of Physically Active Older Humans." *FASEB Journal: Official Publication of the Federation of American Societies for Experimental Biology* 28 (4): 1621–33.

Handschin, Christoph, James Rhee, Jiandie Lin, Paul T. Tarr, and Bruce M. Spiegelman. 2003. "An Autoregulatory Loop Controls Peroxisome Proliferator-Activated Receptor Gamma Coactivator 1alpha Expression in Muscle." *Proceedings of the National Academy of Sciences of the United States of America* 100 (12): 7111–16.

Herbst, Allen, Jonathan Wanagat, Nashwa Cheema, Kevin Widjaja, Debbie McKenzie, and Judd M. Aiken. 2016. "Latent Mitochondrial DNA Deletion Mutations Drive Muscle Fiber Loss at Old Age." *Aging Cell* 15 (6): 1132–39.

Herndon, Laura A., Peter J. Schmeissner, Justyna M. Dudaronek, Paula A. Brown, Kristin M. Listner, Yuko Sakano, Marie C. Paupard, David H. Hall, and Monica Driscoll. 2002. "Stochastic and Genetic Factors Influence Tissue-Specific Decline in Ageing *C. elegans*." *Nature*. 419:808–14.

Holloszy, John O. 1967. "Biochemical Adaptations in Muscle. Effects of Exercise on Mitochondrial Oxygen Uptake and Respiratory Enzyme Activity in Skeletal Muscle." *The Journal of Biological Chemistry* 242 (9): 2278–82.

Hood, David A. 2001. "Invited Review: Contractile Activity-Induced Mitochondrial Biogenesis in Skeletal Muscle." *Journal of Applied Physiology* 90 (3): 1137–57.

Hood, David A., Liam D. Tryon, Heather N. Carter, Yuho Kim, and Chris C. W. Chen. 2016. "Unravelling the Mechanisms Regulating Muscle Mitochondrial Biogenesis." *Biochemical Journal* 473 (15): 2295–2314.

Huang, Julianna H., Anna-Maria Joseph, Vladimir Ljubicic, Sobia Iqbal, and David A. Hood. 2010. "Effect of Age on the Processing and Import of Matrix-Destined Mitochondrial Proteins in Skeletal Muscle." *The Journals of Gerontology. Series A: Biological Sciences and Medical Sciences* 65 (2): 138–46.

Ibebunjo, Chikwendu, Joel M. Chick, Tracee Kendall, John K. Eash, Christine Li, Yunyu Zhang, Chad Vickers, et al. 2013. "Genomic and Proteomic Profiling Reveals Reduced Mitochondrial Function and Disruption of the Neuromuscular Junction Driving Rat Sarcopenia." *Molecular and Cellular Biology* 33 (2): 194–212.

Iqbal, Sobia, Olga Ostojic, Kaustabh Singh, Anna-Maria Joseph, and David A. Hood. 2013. "Expression of Mitochondrial Fission and Fusion Regulatory Proteins in Skeletal Muscle during Chronic Use and Disuse." *Muscle & Nerve* 48 (6): 963–70.

Irrcher, Isabella, Vladimir Ljubicic, and David A. Hood. 2009. "Interactions between ROS and AMP Kinase Activity in the Regulation of PGC-1alpha Transcription in Skeletal Muscle Cells." *American Journal of Physiology. Cell Physiology* 296 (1): C116–23.

Irrcher, Isabella, Vladimir Ljubicic, Angie F. Kirwan, and David A. Hood. 2008. "AMP-Activated Protein Kinase-Regulated Activation of the PGC-1alpha Promoter in Skeletal Muscle Cells." *PloS One* 3 (10): e3614.

Isogai, Shin, Daichi Morimoto, Kyohei Arita, Satoru Unzai, Takeshi Tenno, Jun Hasegawa, Yu-shin Sou, et al. 2011. "Crystal Structure of the Ubiquitin-Associated (UBA) Domain of p62 and Its Interaction with Ubiquitin." *Journal of Biological Chemistry* 286 (36): 31864–74.

Jäger, Sibylle, Christoph Handschin, Julie St-Pierre, and Bruce M. Spiegelman. 2007. "AMP-Activated Protein Kinase (AMPK) Action in Skeletal Muscle via Direct Phosphorylation of PGC-1alpha." *Proceedings of the National Academy of Sciences of the United States of America* 104 (29): 12017–22.

Jeppesen, Tina D., Morten Dunø, Marianne Schwartz, Thomas Krag, Jabin Rafiq, Flemming Wibrand, John Vissing. 2009. "Short- and Long-Term Effects of Endurance Training in Patients with Mitochondrial Myopathy." *European Journal of Neurology: The Official Journal of the European Federation of Neurological Societies* 16 (12): 1336–39.

Jin, Seok Min, Michael Lazarou, Chunxin Wang, Lesley A. Kane, Derek P. Narendra, and Richard J. Youle. 2010. "Mitochondrial Membrane Potential Regulates PINK1 Import and Proteolytic Destabilization by PARL." *The Journal of Cell Biology* 191 (5): 933–42.

Johannsen, Darcy L., Kevin E. Conley, Sudip Bajpeyi, Mark Punyanitya, Dympna Gallagher, Zhengyu Zhang, Jeffrey Covington, Steven R. Smith, and Eric Ravussin. 2012. "Ectopic Lipid Accumulation and Reduced Glucose Tolerance in Elderly Adults Are Accompanied by Altered Skeletal Muscle Mitochondrial Activity." *The Journal of Clinical Endocrinology and Metabolism* 97 (1): 242–50.

Johnson, Matthew L., Matthew M. Robinson, and K. Sreekumaran Nair. 2013. "Skeletal Muscle Aging and the Mitochondrion." *Trends in Endocrinology and Metabolism: TEM* 24 (5): 247–56.

Joseph, Anna-Maria, Peter J. Adhihetty, Thomas W. Buford, Stephanie E. Wohlgemuth, Hazel A. Lees, Linda M-D. Nguyen, Juan M. Aranda, et al. 2012. "The Impact of Aging on Mitochondrial Function and Biogenesis Pathways in Skeletal Muscle of Sedentary High- and Low-Functioning Elderly Individuals." *Aging Cell* 11 (5): 801–9.

Joseph, Anna-Maria, and David A. Hood. 2012. "Plasticity of TOM Complex Assembly in Skeletal Muscle Mitochondria in Response to Chronic Contractile Activity." *Mitochondrion* 12 (2): 305–12.

Ju, Jeong-Sun, Arun S. Varadhachary, Sara E. Miller, and Conrad C. Weihl. 2010. "Quantitation of 'Autophagic Flux' in Mature Skeletal Muscle." *Autophagy* 6 (7): 929–35.

Kadenbach, Bernhard, Christof Münscher, Viola Frank, Josef Müller-Höcker, and Jorg Napiwotzki. 1995. "Human Aging Is Associated with Stochastic Somatic Mutations of Mitochondrial DNA." *Mutation Research* 338 (1–6): 161–72.

Kane, Lesley A., Michael Lazarou, Adam I. Fogel, Yan Li, Koji Yamano, Shireen A. Sarraf, Soojay Banerjee, and Richard J. Youle. 2014. "PINK1 Phosphorylates Ubiquitin to Activate Parkin E3 Ubiquitin Ligase Activity." *The Journal of Cell Biology* 205 (2): 143–53.

Kent-Braun, Jane A., and Alexander V. Ng. 2000. "Skeletal Muscle Oxidative Capacity in Young and Older Women and Men." *Journal of Applied Physiology* 89 (3): 1072–78.

Kirkin, Vladimir, Trond Lamark, Terje Johansen, and Ivan Dikic. 2009. "NBR1 Cooperates with p62 in Selective Autophagy of Ubiquitinated Targets." *Autophagy* 5 (5): 732–33.

Kirkwood, Susan P., Ed A. Munn, and George A. Brooks. 1986. "Mitochondrial Reticulum in Limb Skeletal Muscle." *The American Journal of Physiology* 251 (3 Pt 1): C395–402.

Kirkwood, S. P., L. Packer, and G. A. Brooks. 1987. "Effects of Endurance Training on a Mitochondrial Reticulum in Limb Skeletal Muscle." *Archives of Biochemistry and Biophysics* 255 (1): 80–88.

Klionsky, Daniel J., Kotb Abdelmohsen, Akihisa Abe, Md Joynal Abedin, Hagai Abeliovich, Abraham Acevedo Arozena, Hiroaki Adachi, et al. 2016. "Guidelines for the Use and Interpretation of Assays for Monitoring Autophagy (3rd ed)." *Autophagy* 12 (1): 1–222.

Kowald, Axel, and Thomas B. Kirkwood. 2014. "Transcription Could Be the Key to the Selection Advantage of Mitochondrial Deletion Mutants in Aging." *Proceedings of the National Academy of Sciences of the United States of America* 111 (8): 2972–77.

Koyano, Fumika, Kei Okatsu, Hidetaka Kosako, Yasushi Tamura, Etsu Go, Mayumi Kimura, Yoko Kimura, et al. 2014. "Ubiquitin Is Phosphorylated by PINK1 to Activate Parkin." *Nature* 510 (7503): 162–66.

Kujoth, Gregory C., Asimina Hiona, Thomas D. Pugh, Shinichi Someya, Ken Panzer, Stephanie E. Wohlgemuth, Tim Hofer, et al. 2005. "Mitochondrial DNA Mutations, Oxidative Stress, and Apoptosis in Mammalian Aging." *Science.* 309:481–84.

Larsen, Ryan G., Damien M. Callahan, Stephen A. Foulis, and Jane A. Kent-Braun. 2012. "Age-Related Changes in Oxidative Capacity Differ between Locomotory Muscles and Are Associated with Physical Activity Behavior." *Applied Physiology, Nutrition, and Metabolism: Physiologie Appliquée, Nutrition et Métabolisme* 37 (1): 88–99.

Lazarou, Michael, Danielle A. Sliter, Lesley A. Kane, Shireen A. Sarraf, Chunxin Wang, Jonathon L. Burman, Dionisia P. Sideris, Adam I. Fogel, and Richard J. Youle. 2015. "The Ubiquitin Kinase PINK1 Recruits Autophagy Receptors to Induce Mitophagy." *Nature* 524 (7565): 309–14.

Lee, Connie M., Marisol E. Lopez, Richard Weindruch, and Judd M. Aiken. 1998. "Association of Age-Related Mitochondrial Abnormalities with Skeletal Muscle Fiber Atrophy." *Free Radical Biology & Medicine* 25 (8): 964–72.

Lee, Wing-Sze, Wing-Hoi Cheung, Ling Qin, Ning Tang, and Kwok Sui Leung. 2006. "Age-Associated Decrease of Type IIA/B Human Skeletal Muscle Fibers." *Clinical Orthopaedics and Related Research* 450:231–37.

Leick, Lotte, Stine Secher Lyngby, Jørgen F. P. Wojtaszewski, and Henriette Pilegaard. 2010. "PGC-1alpha Is Required for Training-Induced Prevention of Age-Associated Decline in Mitochondrial Enzymes in Mouse Skeletal Muscle." *Experimental Gerontology* 45 (5): 336–42.

Lexell, Jan, Chales C. Taylor, and Michael Sjöström. 1988. "What Is the Cause of the Ageing Atrophy?: Total Number, Size and Proportion of Different Fiber Types Studied in Whole Vastus Lateralis Muscle from 15- to 83-Year-Old Men." *Journal of the Neurological Sciences.* 84:275–94.

Li, Lulu, Jin Tan, Yuyang Miao, Ping Lei, and Qiang Zhang. 2015. "ROS and Autophagy: Interactions and Molecular Regulatory Mechanisms." *Cellular and Molecular Neurobiology* 35 (5): 615–21.

Liesa, Marc, and Orian S. Shirihai. 2013. "Mitochondrial Dynamics in the Regulation of Nutrient Utilization and Energy Expenditure." *Cell Metabolism* 17 (4): 491–506.

Linnane, Anthony W., Alessandra Baumer, Ronald J. Maxwell, Henry Preston, Chunfang F. Zhang and Sangkot Marzuki. 1990. "Mitochondrial Gene Mutation: The Ageing Process and Degenerative Diseases." *Biochemistry International* 22 (6): 1067–76.

Little, Jonathan P., Adeel Safdar, Naomi Cermak, Mark A. Tarnopolsky, and Martin J. Gibala. 2010. "Acute Endurance Exercise Increases the Nuclear Abundance of PGC-1alpha in Trained Human Skeletal Muscle." *American Journal of Physiology: Regulatory, Integrative and Comparative Physiology* 298 (4): R912–7.

Ljubicic, Vladimir, and David A. Hood. 2009. "Diminished Contraction-Induced Intracellular Signaling towards Mitochondrial Biogenesis in Aged Skeletal Muscle." *Aging Cell* 8 (4): 394–404.

Ljubicic, Vladimir, Peter J Adhihetty, and David A Hood. 2005. "Application of Animal Models: Chronic Electrical Stimulation-Induced Contractile Activity." *Canadian Journal of Applied Physiology: Revue Canadienne de Physiologie Appliquée* 30 (5): 625–43.

Ljubicic, Vladimir, Anna-Maria Joseph, Peter J. Adhihetty, Julianna H. Huang, Ayesha Saleem, Giulia Uguccioni, and David A. Hood. 2009. "Molecular Basis for an Attenuated Mitochondrial Adaptive Plasticity in Aged Skeletal Muscle." *Aging* 1 (9): 818–30.

Masiero, Eva, Lisa Agatea, Cristina Mammucari, Bert Blaauw, Emanuele Loro, Masaaki Komatsu, Daniel Metzger, Carlo Reggiani, Stefano Schiaffino, and Marco Sandri. 2009. "Autophagy Is Required to Maintain Muscle Mass." *Cell Metabolism* 10 (6): 507–15.

McKiernan, Susan H., Ricki Colman, Marisol E. Lopez, Timothy M. Beasley, Richard Weindruch and Judd M. Aiken. 2009. "Longitudinal Analysis of Early Stage Sarcopenia in Aging Rhesus Monkeys." *Experimental Gerontology.* 44:170–76.

Meissner, Cathrin, Holger Lorenz, Andreas Weihofen, Dennis J. Selkoe, and Marius K. Lemberg. 2011. "The Mitochondrial Intramembrane Protease PARL Cleaves Human Pink1 to Regulate Pink1 Trafficking." *Journal of Neurochemistry* 117 (5): 856–67.

Melov, Simon, John. M. Shoffner, Allan Kaufman and Douglas C. Wallace. 1995. "Marked Increase in the Number and Variety of Mitochondrial DNA Rearrangements in Aging Human Skeletal Muscle." *Nucleic Acids Research* 23 (20): 4122–26.

Melov, Simon, Mark A. Tarnopolsky, Kenneth Beckman, Krysta Felkey, and Alan Hubbard. 2007. "Resistance Exercise Reverses Aging in Human Skeletal Muscle." *PloS One* 2 (5): e465.

Mizushima, Noboru, and Tamotsu Yoshimori. 2007. "How to Interpret LC3 Immunoblotting." *Autophagy* 3 (6) 542–45.

Narendra, Derek, Atsushi Tanaka, Der-Fen Suen, and Richard J. Youle. 2008. "Parkin Is Recruited Selectively to Impaired Mitochondria and Promotes Their Autophagy." *The Journal of Cell Biology* 183 (5): 795–803.

Nemoto, Shino, Maria M. Fergusson, and Toren Finkel. 2005. "SIRT1 Functionally Interacts with the Metabolic Regulator and Transcriptional Coactivator PGC-1alpha." *The Journal of Biological Chemistry* 280 (16): 16456–60.

Novak, Ivana, Vladimir Kirkin, David G. McEwan, Ji Zhang, Philipp Wild, Alexis Rozenknop, Vladimir Rogov, et al. 2010. "Nix Is a Selective Autophagy Receptor for Mitochondrial Clearance." *EMBO Reports* 11 (1): 45–51.

O'Leary, Michael F., Anna Vainshtein, Sobia Iqbal, Olga Ostojic, and David A. Hood. 2013. "Adaptive Plasticity of Autophagic Proteins to Denervation in Aging Skeletal Muscle." *American Journal of Physiology. Cell Physiology* 304 (5): C422–30.

Ogata, Takuro, and Yuichi Yamasaki. 1985. "Scanning Electron-Microscopic Studies on the Three-Dimensional Structure of Mitochondria in the Mammalian Red, White and Intermediate Muscle Fibers." *Cell and Tissue Research* 241 (2): 251–56.

Ogborn, Daniel I., Bryon R McKay, Justin D. Crane, Adeel Safdar, Mahmood Akhtar, Gianni Parise, and Mark A. Tarnopolsky. 2015. "Effects of Age and Unaccustomed Resistance Exercise on Mitochondrial Transcript and Protein Abundance in Skeletal Muscle of Men." *American Journal of Physiology. Regulatory, Integrative and Comparative Physiology* 308 (8): R734–41.

Pankiv, Serhiy, Terje Høyvarde Clausen, Trond Lamark, Andreas Brech, Jack-Ansgar Bruun, Heidi Outzen, Aud Øvervatn, Geir Bjørkøy, and Terje Johansen. 2007. "p62/SQSTM1 Binds Directly to Atg8/LC3 to Facilitate Degradation of Ubiquitinated Protein Aggregates by Autophagy." *Journal of Biological Chemistry* 282 (33): 24131–45.

Pattison, J. Scott, Lillian C. Folk, Richard W. Madsen, Thomas E. Childs, and Frank W. Booth. 2003. "Transcriptional Profiling Identifies Extensive Downregulation of Extracellular Matrix Gene Expression in Sarcopenic Rat Soleus Muscle." *Physiological Genomics* 15 (1): 34–43.

Picard, Martin, Darmyn Ritchie, Melissa M. Thomas, Kathryn J. Wright, and Russell T. Hepple. 2011. "Alterations in Intrinsic Mitochondrial Function with Aging Are Fiber Type-Specific and Do Not Explain Differential Atrophy between Muscles." *Aging Cell* 10 (6): 1047–55.

Puigserver, Pere, Guillaume Adelmant, Zhidan Wu, Melina Fan, Jianming Xu, Bert O'Malley, and Bruce M. Spiegelman. 1999. "Activation of PPARgamma Coactivator-1 through Transcription Factor Docking." *Science* 286 (5443): 1368–71.

Puigserver, Pere, James Rhee, Jiandie Lin, Zhidan Wu, J. Cliff Yoon, Chen-Yu Zhang, Stefan Krauss, Vamsi K. Mootha, Bradford B. Lowell, and Bruce M. Spiegelman. 2001. "Cytokine Stimulation of Energy Expenditure through p38 MAP Kinase Activation of PPARgamma Coactivator-1." *Molecular Cell* 8 (5): 971–82.

Quadrilatero, Joe, Stephen E. Alway, and Esther E. Dupont-Versteegden. 2011. "Skeletal Muscle Apoptotic Response to Physical Activity: Potential Mechanisms for Protection." *Applied Physiology, Nutrition, and Metabolism: Physiologie Appliquée, Nutrition et Métabolisme* 36 (5): 608–17.

Quirós, Pedro M., Adrienne Mottis, and Johan Auwerx. 2016. "Mitonuclear Communication in Homeostasis and Stress." *Nature Reviews Molecular Cell Biology* 17 (4): 213–26.

Raben, Nina., Victoria Hill, Lauren Shea, Shoichi Takikita, Rebecca Baum, Noboru Mizushima, Evelyn Ralston, and Paul Plotz. 2008. "Suppression of Autophagy in Skeletal Muscle Uncovers the Accumulation of Ubiquitinated Proteins and Their Potential Role in Muscle Damage in Pompe Disease." *Human Molecular Genetics* 17 (24): 3897–908.

Radak, Zsolt, Zhongfu Zhao, Erika Koltai, Hideki Ohno, and Mustafa Atalay. 2013. "Oxygen Consumption and Usage During Physical Exercise: The Balance Between Oxidative Stress and ROS-Dependent Adaptive Signaling." *Antioxidants & Redox Signaling* 18 (10): 1208–46.

Raffaello, Anna, Cristina Mammucari, Gaia Gherardi, and Rosario Rizzuto. 2016. "Calcium at the Center of Cell Signaling: Interplay between Endoplasmic Reticulum, Mitochondria, and Lysosomes." *Trends in Biochemical Sciences* 41 (12): 1035–49.

Regmi, Saroj G., Rolland, Stephane G. and Conradt, Barbara. 2014. "Age Dependent Changes in Mitochondrial Morphology and Volume Are Not Predictors of Lifespan." *Aging.* 6:118–30.

Reznick, Richard M, Haihong Zong, Ji Li, Katsutaro Morino, Irene K. Moore, Hannah J. Yu, Zhen-Xiang Liu, et al. 2007. "Aging-Associated Reductions in AMP-Activated Protein Kinase Activity and Mitochondrial Biogenesis." *Cell Metabolism* 5 (2): 151–56.

Robinson, Matthew M., Surendra Dasari, Adam R. Konopka, Matthew L. Johnson, S. Manjunatha, Raul Ruiz Esponda, Rickey E Carter, Ian R Lanza, and K Sreekumaran Nair. 2017. "Enhanced Protein Translation Underlies Improved Metabolic and Physical Adaptations to Different Exercise Training Modes in Young and Old Humans." *Cell Metabolism* 25 (3): 581–92.

Rodgers, Joseph T., Carlos Lerin, Wilhelm Haas, Steven P. Gygi, Bruce M. Spiegelman, and Pere Puigserver. 2005. "Nutrient Control of Glucose Homeostasis through a Complex of PGC-1alpha and SIRT1." *Nature* 434 (7029): 113–18.

Rogov, Vladimir, Volker Dötsch, Terje Johansen, and Vladimir Kirkin. 2014. "Interactions between Autophagy Receptors and Ubiquitin-Like Proteins Form the Molecular Basis for Selective Autophagy." *Molecular Cell* 53 (2): 167–78.

Romanello, Vanina, Eleonora Guadagnin, Ligia Gomes, Ira Roder, Claudia Sandri, Yvonne Petersen, Giulia. Milan, et al. 2010. "Mitochondrial Fission and Remodelling Contributes to Muscle Atrophy." *The EMBO Journal.* 29:1774–85.

Rooyackers, Olav E., Deborah B. Adey, Philip A. Ades, and K. Sreekumaran Nair. 1996. "Effect of Age on in Vivo Rates of Mitochondrial Protein Synthesis in Human Skeletal Muscle." *Proceedings of the National Academy of Sciences of the United States of America* 93 (26): 15364–69.

Rosenberg, I.H. 1997. Sarcopenia: Origins and Clinical Relevance. *The Journal of Nutrition.* 127:990S–91S.

Ruas, Jorge L., James P. White, Rajesh R. Rao, Sandra Kleiner, Kevin T. Brannan, Brooke C. Harrison, Nicholas P. Greene, et al. 2012. "A PGC-1α Isoform Induced by Resistance Training Regulates Skeletal Muscle Hypertrophy." *Cell* 151 (6): 1319–31.

Safdar, Adeel, Jacqueline M. Bourgeois, Daniel I. Ogborn, Jonathan P. Little, Bart P. Hettinga, Mahmood Akhtar, James E. Thompson, et al. 2011. "Endurance Exercise Rescues Progeroid Aging and Induces Systemic Mitochondrial Rejuvenation in mtDNA Mutator Mice." *Proceedings of the National Academy of Sciences of the United States of America* 108 (10): 4135–40.

Scarpulla, Richard C. 2011. "Metabolic Control of Mitochondrial Biogenesis through the PGC-1 Family Regulatory Network." *Biochimica et Biophysica Acta* 1813 (7): 1269–78.

Scarpulla, Richard C., Rick B. Vega, and Daniel P. Kelly. 2012. "Transcriptional Integration of Mitochondrial Biogenesis." *Trends in Endocrinology & Metabolism* 23 (9): 459–66.

Sebastián, David, Eleonora Sorianello, Jessica Segalés, Andrea Irazoki, Vanessa Ruiz-Bonilla, David Sala, Evarist Planet, et al. 2016. "Mfn2 Deficiency Links Age-related Sarcopenia and Impaired Autophagy to Activation of an Adaptive Mitophagy Pathway." *The EMBO Journal* 35 (15): 1677–93.

Short, Kevin R., Janet L. Vittone, Maureen L. Bigelow, David N. Proctor, Robert A. Rizza, Jill M. Coenen-Schimke, and K. Sreekumaran Nair. 2003. "Impact of Aerobic Exercise Training on Age-Related Changes in Insulin Sensitivity and Muscle Oxidative Capacity." *Diabetes* 52 (8): 1888–96.

Short, Kevin R., Maureen L. Bigelow, Jane Kahl, Ravinder Singh, Jill Coenen-Schimke, Sreekumar Raghavakaimal, and K. Sreekumaran Nair. 2005. "Decline in Skeletal Muscle Mitochondrial Function with Aging in Humans." *Proceedings of the National Academy of Sciences of the United States of America* 102 (15): 5618–23.

Spendiff, Sally, Madhusudanarao Vuda, Gilles Gouspillou, Sudhakar Aare, Anna Perez, José A. Morais, Robert T. Jagoe, et al. 2016. "Denervation Drives Mitochondrial Dysfunction in Skeletal Muscle of Octogenarians." *The Journal of Physiology* 594 (24): 7361–79.

Sun, Yu, A. Ajay Vashisht, Jason Tchieu, James A. Wohlschlegel, and Lars Dreier. 2012. "Voltage-Dependent Anion Channels (VDACs) Recruit Parkin to Defective Mitochondria to Promote Mitochondrial Autophagy." *Journal of Biological Chemistry* 287 (48): 40652–60.

Taivassalo, Tanja, Eric A. Shoubridge, Jacqueline Chen, Nancy G. Kennaway, Salvatore DiMauro, Douglas L. Arnold, and Ronald G. Haller. 2001. "Aerobic Conditioning in Patients with Mitochondrial Myopathies: Physiological, Biochemical, and Genetic Effects." *Annals of Neurology* 50 (2): 133–41.

Trifunovic, Aleksandra, Anna Wredenberg, Maria Falkenberg, Johannes N. Spelbrink, Anja T. Rovio, Carl E. Bruder, Mohammed Bohlooly-Y., et al. 2004. "Premature Ageing in Mice Expressing Defective Mitochondrial DNA Polymerase." *Nature.* 429: 417–23.

Trounce, Ian, Edward Byrne, and Sangkot Marzuki. 1989. "Decline in Skeletal Muscle Mitochondrial Respiratory Chain Function: Possible Factor in Ageing." *Lancet* 1 (8639): 637–39.

Twig, Gilad, Alvaro Elorza, Anthony J. A. Molina, Hibo Mohamed, Jakob D. Wikstrom, Gil Walzer, Linsey Stiles, et al. 2008. "Fission and Selective Fusion Govern Mitochondrial Segregation and Elimination by Autophagy." *The EMBO Journal* 27 (2): 433–46.

Uguccioni, Giulia, and David A. Hood. 2011. "The Importance of PGC-1α in Contractile Activity-Induced Mitochondrial Adaptations." *American Journal of Physiology. Endocrinology and Metabolism* 300 (2): E361–71.

Virbasius, Joseph V., and Richard C. Scarpulla. 1994. "Activation of the Human Mitochondrial Transcription Factor A Gene by Nuclear Respiratory Factors: A Potential Regulatory Link between Nuclear and Mitochondrial Gene Expression in Organelle Biogenesis." *Proceedings of the National Academy of Sciences of the United States of America* 91 (4): 1309–13.

Wallace, Douglas C. 2010. "Mitochondrial DNA Mutations in Disease and Aging." *Environmental and Molecular Mutagenesis* 51 (5): 440–50.

Wallberg, Annika E., Soichiro Yamamura, Sohail Malik, Bruce M. Spiegelman, and Robert G. Roeder. 2003. "Coordination of p300-Mediated Chromatin Remodeling and TRAP/Mediator Function through Coactivator PGC-1α." *Molecular Cell* 12 (5): 1137–49.

Walters, Thomas J., H. Lee Sweeney, and Roger P. Farrar. 1991. "Influence of Electrical Stimulation on a Fast-Twitch Muscle in Aging Rats." *Journal of Applied Physiology* 71 (5): 1921–28.

Wang, Yan, Yuichi Michikawa, Con Mallidis, Yan Bai, Linda Woodhouse, Kevin E. Yarasheski, Carol A. Miller, et al. 2001. "Muscle-Specific Mutations Accumulate with Aging in Critical Human mtDNA Control Sites for Replication." *Proceedings of the National Academy of Sciences of the United States of America* 98 (7): 4022–27.

Waters, Sarah, Katie Marchbank, Ellen Solomon, Caroline Whitehouse, and Mathias Gautel. 2009. "Interactions with LC3 and Polyubiquitin Chains Link nbr1 to Autophagic Protein Turnover." *FEBS Letters* 583 (12): 1846–52.

Wei, Yongjie, Wei-Chung Chiang, Rhea Sumpter, Prashant Mishra, and Beth Levine. 2017. "Prohibitin 2 Is an Inner Mitochondrial Membrane Mitophagy Receptor." *Cell* 168 (1–2): 224–238.e10.

Wiederanders, Bernd, and Barbel Oelke. 1984. "Accumulation of Inactive Cathepsin D in Old Rats." *Mechanisms of Ageing and Development* 24 (3): 265–71.

Wong, Yvette C., and Erika L. F. Holzbaur. 2015. "Temporal Dynamics of PARK2/parkin and OPTN/optineurin Recruitment during the Mitophagy of Damaged Mitochondria." *Autophagy* 11 (2): 422–24.

Wright, David C., Dong-Ho Han, Pablo M. Garcia-Roves, Paige C. Geiger, Terry E. Jones, and John O. Holloszy. 2007. "Exercise-Induced Mitochondrial Biogenesis Begins before the Increase in Muscle PGC-1alpha Expression." *The Journal of Biological Chemistry* 282 (1): 194–99.

Wu, Hai, Shane B. Kanatous, Frederick A. Thurmond, Teresa Gallardo, Eiji Isotani, Rhonda Bassel-Duby, and R. Sanders Williams. 2002. "Regulation of Mitochondrial Biogenesis in Skeletal Muscle by CaMK." *Science* 296 (5566): 349–52.

Xie, Rui, Susan Nguyen, Wallace L. McKeehan, and Leyuan Liu. 2010. "Acetylated Microtubules Are Required for Fusion of Autophagosomes with Lysosomes." *BMC Cell Biology* 11 (1): 89.

Zykovich, Artem, Alan Hubbard, James M. Flynn, Mark Tarnopolsky, Mario F. Fraga, Chad Kerksick, Dan Ogborn, Lauren MacNeil, Sean D. Mooney, and Simon Melov. 2014. "Genome-Wide DNA Methylation Changes with Age in Disease-Free Human Skeletal Muscle." *Aging Cell* 13 (2): 360–66.

13 Skeletal Muscle Fat Infiltration with Aging
An Important Factor of Sarcopenia

Allan F. Pagano, Coralie Arc-Chagnaud, Thomas Brioche, Angèle Chopard, and Guillaume Py

CONTENTS

13.1 INTRODUCTION

Muscle deconditioning occurring with aging is considered as a geriatric syndrome called sarcopenia, and multiple factors contribute to it, including diet, chronic diseases, physical inactivity, and the aging process itself (Brioche et al. 2016). Sarcopenia is a well-described phenomenon, mainly characterized by a loss of muscle mass, together with a loss of strength and muscle power (Cruz-Jentoft et al. 2010; Fielding et al. 2011; Brioche et al. 2016). As described in previous chapters, muscle protein turnover (Chapter 11) and mitochondrial turnover (Chapter 12) play a major role in this process. However, there is now a growing body of evidence that the loss of strength and power mostly exceeds the loss of muscle mass observed (Di Prampero and Narici 2003; Delmonico et al. 2009; Jubrias et al. 1997), suggesting that other factors are involved. Among these factors, the phenomenon of fatty infiltration accumulation may play a critical role. Intermuscular adipose tissue (more commonly called IMAT) fat cells are located under the epimysium, that is between muscle fibers and bundles fibers. They represent real adipocyte clusters, that is ectopic fat depot, localized outside muscle cells, and do not be confused with intramyocellular triglyceride accumulation (Addison et al. 2014; Vettor et al. 2009). Although IMAT may be a variable part of healthy human skeletal muscle, its increase and accumulation are linked to muscle dysfunction, deconditioning, and even perturbed regeneration (Addison et al. 2014; Marcus et al. 2010; Sciorati et al. 2015; Uezumi, Ikemoto-Uezumi, and Tsuchida 2014; Pagano et al. 2015). An accumulation of these fatty infiltrations was observed in many conditions closely linked to muscle deconditioning: inactivity (Leskinen et al. 2009; Manini et al. 2007; Tuttle et al. 2011; Pagano et al. 2017), denervation (Dulor et al. 1998; Kim et al. 2012), diabetes (Goodpaster et al. 2003), chronic obstructive pulmonary disease (Robles et al. 2015), cachexia (Gray et al. 2011), tenotomy (Kuzel et al. 2013; Laron et al. 2012), or even sarcopenia (Song et al. 2004; Cree et al. 2004). Recognition of the role of IMAT in muscle function seems to increase year after year, and scientific knowledge will certainly continue

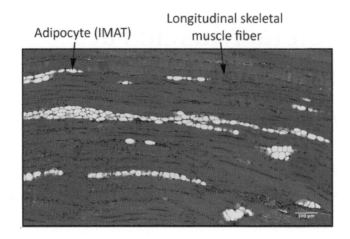

FIGURE 13.1 Example of IMAT development in murine tibialis anterior muscle. Longitudinal paraffin-embedded muscle section, stained with hematoxylin-eosin-saffron, 21 days after intramuscular injection of glycerol into the murine tibialis anterior muscle. (Adapted from Brioche, T. et al., *Mol. Aspects Med.*, 50, 56–87, 2016.)

to expand around this subject. This chapter will focus on IMAT accumulation with aging and its impact on muscle function. We will also develop the major role of increasing physical activity as a countermeasure and finally highlight some cellular players potentially responsible for IMAT development with advancing age (Figure 13.1).

13.2   IMAT ACCUMULATION ALTERS MUSCLE FUNCTION

To our knowledge, Aherne (1965) was the first to describe fat infiltration development in various tissues of a newborn infant, including skeletal muscle. Approximately 20 years after, the studies of Borkan et al. (1983) and Rice et al. (1989) highlighted higher IMAT levels in the elderly population. Since then, numerous works confirmed this result and showed that IMAT content increases strongly with age (Addison et al. 2014; Goodpaster, Thaete, and Kelley 2000; Song et al. 2004; Kirkland et al. 2002; Guglielmi et al. 2014; Cree et al. 2004; Brioche et al. 2016). An emerging body of evidences supports IMAT as a significant predictor of muscle dysfunction in older adults (Miljkovic-Gacic et al. 2008; Visser et al. 2005; Goodpaster, Thaete, and Kelley 2000; Delmonico et al. 2009; Marcus et al. 2012; Yoshida, Marcus, and Lastayo 2012; Goodpaster et al. 2008). Indeed, older men and women with higher IMAT contents in the leg have lower muscle strength and performance (Goodpaster et al. 2001; Visser et al. 2002; Delmonico et al. 2009). Nevertheless, this phenomenon is not only strictly specific to the aging population but also affects subjects with diabetes, obesity, or peripheral neuropathy (Tuttle, Sinacore, and Mueller 2012; Hilton et al. 2008; Granados et al. 2018; Lim et al. 2018) and even healthy young adults after an inactivity period (Manini et al. 2007). In addition, it has been hypothesized that higher levels of IMAT in rotator cuff muscles lead to impairment of fiber pennation angle. An alteration of this angle results in an unfavorable mechanical angle and hence alters force production (Gerber et al. 2007; Meyer et al. 2004). Currently, there is no study available on this aspect in elderlies, but we could hypothesize that it could be found in all populations prone to IMAT accumulation. Another important parameter associated with IMAT accumulation is the metabolism alteration. Indeed, several studies demonstrated that IMAT levels were positively linked to insulin resistance (Simoneau et al. 1995; Goodpaster et al. 1997, 1999; Goodpaster, Thaete, and Kelley 2000; Kelley, Slasky, and Janosky 1991; Goodpaster, Thaete, and Kelley 2000; Goodpaster et al. 2003) and cardiovascular diseases risks (Ryan and Nicklas 1999; Yim et al. 2007).

All these muscle alterations associated with IMAT development and accumulation are known to impair muscle function and are therefore linked to mobility limitations. The relation between IMAT and mobility has been observed in a variety of population including individuals suffering from diabetes (Hilton et al. 2008; Tuttle, Sinacore, and Mueller 2012), chronic obstructive pulmonary disease (Roig et al. 2011), or sarcopenia (Addison et al. 2014; Brioche et al. 2016). Most of the time, mobility impairment is measured by performance tests such as different walk tests, up and go test, gait speed, or chair stands. In the specific context of sarcopenia, numerous studies revealed that IMAT infiltration into the mid-thigh muscle remains an independent risk factor of mobility limitations (Beavers et al. 2013; Tuttle, Sinacore, and Mueller 2012; Visser et al. 2005; Marcus et al. 2012; Murphy et al. 2014; Visser et al. 2002). High amount of IMAT also correlates positively with higher risk of fracture. Indeed, the study of Lang et al. (2010) illustrates that an increase in IMAT content raised the risk of hip fracture by 40% in people aged 70–79 years. This element should be taken into account because, independently of IMAT, aging is accompanied by osteoporosis, which weakens skeletal apparatus and increases risk of fractures (Reginster et al. 2016; Edwards et al. 2015). Once again, the study of Kim et al. (2013) highlighted a negative correlation between IMAT accumulation, reflecting muscle deconditioning severity, and bone mineral density.

13.3 THE FIGHT AGAINST INACTIVITY: THE REAL PURPOSE

Although fatty infiltrations are widely known to be strongly linked to sarcopenia, it seems that IMAT accumulation in muscles is more attributable to inactivity than aging per se (Manini et al. 2007; Tuttle et al. 2011; Leskinen et al. 2009; Farr et al. 2012). Indeed the study of Manini et al. (2007) highlighted, in healthy young adults, an increase in IMAT content of 15% in the thigh and 20% in the calf, after 4 weeks of unilateral lower limb suspension. In a study exclusively conducted on girls (8–14 years), the authors showed that lower physical activity was associated with higher IMAT levels (Farr et al. 2012). The longitudinal study of Leskinen et al. (2009) revealed that IMAT accumulation was greater in inactive co-twins compared with their active counterparts. Furthermore, a study conducted on a stroke victim population demonstrated that fat deposition within muscles was higher in the hemiparetic limb of chronic stroke patients (Ryan et al. 2002). Other studies observed similar results illustrating the correlation between physical activity levels and IMAT accumulation, whether in people with diabetes and peripheral neuropathy (Tuttle et al. 2011), spinal cord injury (Gorgey and Dudley 2007), or even older adults (Murphy et al. 2014). Taken together, these results indicate the highly plausible role of physical activity interventions in aging, in order to prevent IMAT accumulation.

A quarter century after the first study highlighting higher IMAT levels in the elderly population (Borkan et al. 1983), several studies emphasized, in the same year, the potential protective effect of physical activity against IMAT accumulation. Goodpaster et al. (2008) were the first to demonstrate such beneficial effect with a 1-year study of increased overall physical activity (mostly by walking) among older adults. Marcus et al. (2008) observed similar findings on IMAT accumulation in a type 2 diabetes population, with 16-week of aerobic/eccentric resistance, or aerobic training only. The results of Durheim et al. (2008) showed a reduced IMAT content through 8 months of aerobic training in sedentary dyslipidemic individuals in the same year. Later, 12 weeks of resistance training induced a decrease of 11% in IMAT content of the thigh of subjects aged at least 55 years and also allowed an increase of thigh lean tissue (Marcus et al. 2010). One year of increased overall physical activity also reduced the amount of IMAT in the thigh of sedentary men and women with approximately the same age (Murphy et al. 2012). Recently, the study of Bang et al. (2018) showed that a 5-year daily walking program prevented age-related IMAT accumulation in middle-aged and elderly adults. The study of Nicklas et al. (2015) also demonstrated lower IMAT area, as well as strength improvement, after 5 months of resistance training in obese older adults. Moreover, it has been proved that cessation of resistance exercise in trained older persons increases IMAT, while resumption of exercise decreases it (Taaffe et al. 2009). Finally, an interesting study by Wroblewski et al. (2011) showed that old elite athletes, performing high levels of chronic exercise (four to five training sessions per week), do not exhibit any loss of lean muscle mass, strength, and any increase in IMAT amount with aging. The last two studies strengthen the notion that some muscular alterations observed in sarcopenia are a direct consequence of a gradual decrease in physical activity among older adults. Numerous studies also failed to detect any decrease in IMAT content after an exercise program (Christiansen et al. 2009; Jung et al. 2012; Ku et al. 2010; Prior et al. 2007; Santanasto et al. 2011; Walts et al. 2008; Gallagher et al. 2014; Jacobs et al. 2014), even though studies demonstrating no increase in IMAT content after an exercise intervention among the elderly may also be a good result.

To conclude on this issue, increasing physical activity, whether through an overall increase in physical activity or by specific aerobic/resistance exercises, seems to be essential to decrease, or at least stabilize, IMAT accumulation with advancing age. Further studies are obviously needed to precisely define the most effective way to train the elderly, whether in terms of intensity, duration, frequency, and the type of exercise used.

13.4 IMAT DEVELOPMENT: CELLULAR AND MOLECULAR PLAYERS

13.4.1 THE ROLE OF MESENCHYMAL STEM CELLS

Currently, the mechanisms leading to IMAT development during sarcopenia and, more widely, during a sedentary lifestyle, are still poorly understood. To our knowledge, a review by Kirkland et al. (2002) was the first to give some clues to IMAT accumulation during aging. They highlighted a possible leptin deficiency or resistance, which would be associated with fat accumulation in nonadipose tissues (Wang et al. 2001; Koteish and Diehl 2001). This review also rightly pointed out the potential involvement of different muscle-resident stem cells, including mesenchymal cell types, turning them into cells with an adipocyte-like default phenotype, which may lead to IMAT development and accumulation with increasing age.

Among the different muscle-resident stem cells, the literature mainly distinguishes mesenchymal stem cells (MSCs) and satellite cells (SCs). These later, located beneath the basal lamina of muscle fibers (Mauro 1961), are known to have a pool stably maintained until around 70 years of age (Dreyer et al. 2006; Roth et al. 2000), before declining (Verdijk et al. 2014; Kadi et al. 2004; Renault et al. 2002). Verdijk et al. (2014) observed an age-related specific reduction in type II muscle fiber SC content, accompanied by smaller type II fiber size with increasing age. This phenomenon emphasized the preferential development of sarcopenia in the fast-twitch myofibers. Moreover, the proliferation potential of SCs decreases with age, as it was first showed *in vitro* on rats cells by Schultz and Lipton (1982), and this impaired proliferation capacity is concomitant with a decrease in regenerative ability of muscles (Collins-Hooper et al. 2012; Sadeh 1988; Marsh et al. 1997; Gallegly et al. 2004). Recently, numerous reviews focused on SC-specific alterations in sarcopenia and suggested that the decline in muscle SC function with aging is the consequence of both extrinsic and intrinsic factors (Blau, Cosgrove, and Ho 2015; Brack and Munoz-Canoves 2015; Sousa-Victor et al. 2015). This chapter will not elaborate on SC function in sarcopenia nor on transdifferentiation as already discussed in other reviews (Sciorati et al. 2015; Vettor et al. 2009) but will instead discuss the potential role of a specific type of MSCs in IMAT accumulation in conjunction with advancing age.

There are a wide variety of tissue-resident progenitors within the muscle environment (Judson, Zhang, and Rossi 2013), and fibro-adipogenic progenitors (FAPs) represent one of the most studied populations. FAPs are a specific type of MSCs located in the muscle's interstitial space and express the cell surface marker platelet-derived growth factor receptor α (PDGFRα or CD140a). They have the characteristic of having multiple differentiation possibilities: osteogenic, chondrogenic, fibrogenic, or adipogenic lineages (Boppart et al. 2013; Judson, Zhang, and Rossi 2013; Penton et al. 2013; Uezumi, Ikemoto-Uezumi, and Tsuchida 2014). After a quick proliferation following injury,

FAPs participate in regeneration process firstly by promoting phagocytosis of necrotic debris, secondly by enhancing SCs proliferation, and do not differentiate into adipocytes (Heredia et al. 2013; Joe et al. 2010; Lemos et al. 2015). However, in pathological conditions or even muscle disuse, FAPs proliferate and differentiate into adipose and/or fibrous tissue, leading to IMAT accumulation (Uezumi et al. 2011; Uezumi et al. 2014). An experimental mouse model, based on an injection of a glycerol solution (50%) in the tibialis anterior muscle, has been developed in order to investigate IMAT development and accumulation (Pisani et al. 2010). This "glycerol" model of regeneration is commonly used and allows a better characterization of muscle-resident adipocyte precursors (Joe et al. 2010; Lukjanenko et al. 2013; Pagano et al. 2015; Uezumi et al. 2010). Even though no study has been conducted to characterize the role of FAPs in sarcopenia, the literature highlights a possible important function of FAPs in age-related development of IMAT (Blau, Cosgrove, and Ho 2015; Farup et al. 2015). Furthermore, the two recent studies of Arrighi et al. (2015) and Laurens et al. (2016) demonstrated *in vitro* that adipocytes derived from FAPs within skeletal muscle were insulin resistant or induced myofiber insulin resistance, which supports the correlation between IMAT accumulation and insulin sensitivity alteration (Ryan and Nicklas 1999; Goodpaster, Thaete, and Kelley 2000; Goodpaster et al. 1997). These results therefore strengthen the hypothesis that FAPs may be the main stem cell population involved in sarcopenia-related IMAT development.

Otherwise, aging is associated with deregulation of numerous systems including inflammatory and immune systems (Butcher et al. 2001; Lord et al. 2001), both of them are known to play a critical role in the efficient muscle regeneration (Aurora and Olson 2014; Sciorati et al. 2015; Maffioletti et al. 2014). Butcher et al. (2001) have already underscored neutrophil dysfunction in elderly humans, and this result, in light of those of Heredia et al. (2013), may indicate a defective neutrophil-induced interleukin-4/interleukin-13 release and further signaling toward FAPs promoting their adipogenic differentiation. Furthermore, during the regeneration process, M1 macrophages are first recruited and release numerous pro-inflammatory cytokines, including tumor necrosis factor α (TNFα) (Villalta et al. 2009; Saclier et al. 2013; Lemos et al. 2015; Arnold et al. 2007; Kharraz et al. 2013; Perdiguero et al. 2011). Next, M1 macrophages undergo toward a M2 phenotype transition and these M2 macrophages, which have anti-inflammatory proprieties, release, among others, transforming growth factor-β1 (TGF-β1) (Lu et al. 2011; Arnold et al. 2007; Gordon 2003; Perdiguero et al. 2011; Lemos et al. 2015). The study of Lemos et al. (2015) showed the crucial effect of the balance between TNFα and TGF-β1 on FAPs. Indeed, TNFα induces FAP apoptosis, whereas their survival is promoted by TGF-β1. Authors also insist on the essential timing transition from pro-inflammatory (TNFα) to anti-inflammatory (TGF-β1) cytokine production. A deregulation of macrophage function (Przybyla et al. 2006), as well as abnormally high levels of TNFα and TGF-β1, has also been highlighted with advancing age (Carlson et al. 2009; Fagiolo et al. 1993; Merritt et al. 2013). Therefore, the possible overlap between TNFα and TGF-β1 signaling on the one hand and the dysfunction of inflammatory system on the other hand may promote FAP survival and their adipogenic differentiation, leading to IMAT accumulation during sarcopenia (Figure 13.2).

13.4.2 OTHER SIGNALING PATHWAYS

Nevertheless, some other molecular pathways could modulate the fate of FAPs during aging and will be detailed in this section. Numerous studies by the same group evaluated the effect of histone deacetylase inhibitor (HDACi) on dystrophin-deficient *mdx* mice (characterized by permanent muscle lesion-regeneration cycles). These researchers first showed that HDACi treatment was able to promote morphological and functional recovery of *mdx* mice skeletal muscles, certainly mediated by an increase in the expression of follistatin, a major myostatin antagonist (Minetti et al. 2006). Later, they showed that HDACi inhibited FAPs adipogenic potential in young *mdx* mice while promoting SC differentiation. By contrast, FAPs from old *mdx* mice were resistant to HDACi-mediated inhibition of adipogenesis (Mozzetta et al. 2013). This epigenetic intervention prevents the

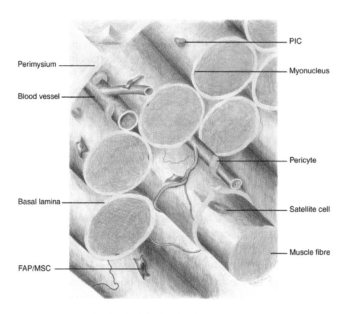

FIGURE 13.2 Illustration of the various populations of tissue-resident progenitors in skeletal muscle. PICs (PW1+/Pax7-interstitial cells, green), FAPs/MSCs (fibro-adipogenic progenitors/mesenchymal stem/progenitor cells, blue), vessel-associated pericytes (yellow), and satellite cells (orange) beneath the basal lamina. (From Judson, R. N., et al., *FEBS J.*, 280, 4100–4108, 2013.)

adipogenic potential of FAPs and also, unexpectedly, promotes FAPs' myogenic lineage in young mice. In contrast, FAPs from old *mdx* mice fail to adopt a myogenic phenotype (Saccone et al. 2014; Faralli and Dilworth 2014). It is important to note, however, that the population isolated in that study was not only purified FAPs: it may contain other populations such as PW1 interstitial cells (PICs), side population (SP) cells, or even type 1 pericytes (Boppart et al. 2013; Judson, Zhang, and Rossi 2013); this methodological point may explain the unexpected FAPs development along the myogenic lineage. Altogether, these studies emphasized the influence of environmental changes, occurring with aging, on FAP fate modulation.

The Notch signaling pathway, known to be involved in the maintenance of SC quiescence and the promotion of their proliferation after injury (Bjornson et al. 2012; Conboy et al. 2003; Mourikis et al. 2012; Wen et al. 2012), is also probably implicated in FAP-driven IMAT development with aging. Conboy et al. (2005) demonstrated that, with aging, impaired SC proliferation as well as muscle regeneration was due to a diminished activation of Notch. This study also revealed that a forced activation of Notch was able to restore regenerative potential of old muscles. It is interesting to note that TGF-β is a Notch inhibitor upregulated during aging (Buas and Kadesch 2010; Carlson, Hsu, and Conboy 2008), which could partly explain muscle regeneration mechanisms decline with aging. Moreover, it has been shown that an altered Notch signaling in the adult heart leads to abnormal development of fibrosis through mesenchymal cardiac progenitors (Croquelois et al. 2008; Nemir et al. 2014). Nevertheless, there are no studies concerning the function of the Notch signaling pathway in FAPs, and further research is clearly needed to identify a highly possible role of this pathway in aged skeletal muscle.

A large number of studies have already highlighted the important role of Wnt signaling pathway in the inhibition of MSC adipogenic differentiation as well as their myogenic potential promotion (Brunt et al. 2012; Bennett et al. 2002; Chung et al. 2012; Moldes et al. 2003; Ross et al. 2000). The study of Vertino et al. (2005) illustrated that alteration in Wnt signaling in aged myoblasts may contribute to increased muscle adiposity. Indeed, they demonstrated that Wnt10b downregulation during aging increases expression of key adipogenic genes, therefore contributing to IMAT development. However, the Wnt signaling pathway is particularly complex, on account of the great diversity of Wnt proteins and their differential variations with aging. Thus, further studies are required to investigate the mechanisms underlying age-related deregulation of Wnt signaling and IMAT development during sarcopenia.

The nitric oxide (NO) pathway is also emerging as a potential mechanism behind FAPs driving adipogenesis with aging. Altered synthesis of NO has already been identified in skeletal muscle of older humans (Nyberg et al. 2012; Samengo et al. 2012). Samengo et al. (2012) showed that muscle aging is accompanied by a loss of neuronal nitric oxide synthase (nNOS), the primary source of NO in skeletal muscle. Moreover, it has been demonstrated *in vitro* that NO regulates FAP fate through inhibition of their differentiation into adipocytes (Cordani et al. 2014), and these findings reveal another potential mechanism behind FAPs' driving adipogenesis with advancing age.

Finally, the fibroblast growth factor 2 protein (FGF-2) also seems to play an important role in age-related impairment of skeletal muscle regeneration. FGF-2 expression increases with age, due to decreased expression of its inhibitor Sprouty 1. This mechanism induces a loss of stem cell quiescence, driving SCs to enter the cell cycle, and then most likely contributes to the loss of SCs observed during aging (Chakkalakal et al. 2012). No studies have been conducted on the effect of FGF-2 on FAPs even though this growth factor seems to also perform significant functions in differentiation of MSCs (Bae et al. 2015; Cai et al. 2013; Lee et al. 2013).

13.5 CONCLUSION

To conclude, IMAT accumulation is a new important factor related to sarcopenia, which appears closely linked to inactivity. Further studies are required to define the most effective training in order to fight against the progressive sedentary lifestyle leading to sarcopenia and IMAT development, both affecting muscle function and overall health of older adults. We tried to emphasize new clues concerning FAPs and aging by linking them to some cellular players. The role of FAPs needs to be more studied, as well as the many signaling pathways potentially involved in aging-induced IMAT development and accumulation.

ACKNOWLEDGMENTS

Our studies are supported by the Centre National d'Etudes Spatiales (CNES).

REFERENCES

Addison, O., R. L. Marcus, P. C. Lastayo, and A. S. Ryan. 2014. Intermuscular fat: A review of the consequences and causes. *Int J Endocrinol* 2014:309570.

Aherne, W. 1965. Fat Infiltration in the Tissues of the Newborn Infant. *Arch Dis Child* 40:406–410.

Arnold, L., A. Henry, F. Poron, et al. 2007. Inflammatory monocytes recruited after skeletal muscle injury switch into antiinflammatory macrophages to support myogenesis. *J Exp Med* 204 (5):1057–1069.

Arrighi, N., C. Moratal, N. Clement, et al. 2015. Characterization of adipocytes derived from fibro/adipogenic progenitors resident in human skeletal muscle. *Cell Death Dis* 6:e1733.

Aurora, A. B., and E. N. Olson. 2014. Immune modulation of stem cells and regeneration. *Cell Stem Cell* 15 (1):14–25.

Bae, S. H., H. Ryu, K. J. Rhee, et al. 2015. L-ascorbic acid 2-phosphate and fibroblast growth factor-2 treatment maintains differentiation potential in bone marrow-derived mesenchymal stem cells through expression of hepatocyte growth factor. *Growth Factors* 33 (2):71–78.

Bang, E., K. Tanabe, N. Yokoyama, S. Chijiki, T. Tsuruzono, and S. Kuno. 2018. Effects of daily walking on intermuscular adipose tissue accumulation with age: A 5-year follow-up of participants in a lifestyle-based daily walking program. *Eur J Appl Physiol* 118 (4):785–793.

Beavers, K. M., D. P. Beavers, D. K. Houston, et al. 2013. Associations between body composition and gait-speed decline: Results from the health, aging, and body composition study. *Am J Clin Nutr* 97 (3):552–560.

Bennett, C. N., S. E. Ross, K. A. Longo, et al. 2002. Regulation of Wnt signaling during adipogenesis. *J Biol Chem* 277 (34):30998–31004.

Bjornson, C. R., T. H. Cheung, L. Liu, P. V. Tripathi, K. M. Steeper, and T. A. Rando. 2012. Notch signaling is necessary to maintain quiescence in adult muscle stem cells. *Stem Cells* 30 (2):232–242.

Blau, H. M., B. D. Cosgrove, and A. T. Ho. 2015. The central role of muscle stem cells in regenerative failure with aging. *Nat Med* 21 (8):854–862.

Boppart, M. D., M. De Lisio, K. Zou, and H. D. Huntsman. 2013. Defining a role for non-satellite stem cells in the regulation of muscle repair following exercise. *Front Physiol* 4:310.

Borkan, G. A., D. E. Hults, S. G. Gerzof, A. H. Robbins, and C. K. Silbert. 1983. Age changes in body composition revealed by computed tomography. *J Gerontol* 38 (6):673–677.

Brack, A. S., and P. Munoz-Canoves. 2015. The ins and outs of muscle stem cell aging. *Skelet Muscle* 6:1.

Brioche, T., A. F. Pagano, G. Py, and A. Chopard. 2016. Muscle wasting and aging: Experimental models, fatty infiltrations, and prevention. *Mol Aspects Med* 50:56–87.

Brunt, K. R., Y. Zhang, A. Mihic, et al. 2012. Role of WNT/beta-catenin signaling in rejuvenating myogenic differentiation of aged mesenchymal stem cells from cardiac patients. *Am J Pathol* 181 (6):2067–2078.

Buas, M. F., and T. Kadesch. 2010. Regulation of skeletal myogenesis by Notch. *Exp Cell Res* 316 (18):3028–3033.

Butcher, S. K., H. Chahal, L. Nayak, et al. 2001. Senescence in innate immune responses: Reduced neutrophil phagocytic capacity and CD16 expression in elderly humans. *J Leukoc Biol* 70 (6):881–886.

Cai, T. Y., W. Zhu, X. S. Chen, S. Y. Zhou, L. S. Jia, and Y. Q. Sun. 2013. Fibroblast growth factor 2 induces mesenchymal stem cells to differentiate into tenocytes through the MAPK pathway. *Mol Med Rep* 8 (5):1323–1328.

Carlson, M. E., C. Suetta, M. J. Conboy, et al. 2009. Molecular aging and rejuvenation of human muscle stem cells. *EMBO Mol Med* 1 (8–9):381–391.

Carlson, M. E., M. Hsu, and I. M. Conboy. 2008. Imbalance between pSmad3 and Notch induces CDK inhibitors in old muscle stem cells. *Nature* 454 (7203):528–532.

Chakkalakal, J. V., K. M. Jones, M. A. Basson, and A. S. Brack. 2012. The aged niche disrupts muscle stem cell quiescence. *Nature* 490 (7420):355–360.

Christiansen, T., S. K. Paulsen, J. M. Bruun, et al. 2009. Comparable reduction of the visceral adipose tissue depot after a diet-induced weight loss with or without aerobic exercise in obese subjects: A 12-week randomized intervention study. *Eur J Endocrinol* 160 (5):759–767.

Chung, S. S., J. S. Lee, M. Kim, et al. 2012. Regulation of Wnt/beta-catenin signaling by CCAAT/enhancer binding protein beta during adipogenesis. *Obesity (Silver Spring)* 20 (3):482–487.

Collins-Hooper, H., T. E. Woolley, L. Dyson, et al. 2012. Age-related changes in speed and mechanism of adult skeletal muscle stem cell migration. *Stem Cells* 30 (6):1182–1195.

Conboy, I. M., M. J. Conboy, A. J. Wagers, E. R. Girma, I. L. Weissman, and T. A. Rando. 2005. Rejuvenation of aged progenitor cells by exposure to a young systemic environment. *Nature* 433 (7027):760–764.

Conboy, I. M., M. J. Conboy, G. M. Smythe, and T. A. Rando. 2003. Notch-mediated restoration of regenerative potential to aged muscle. *Science* 302 (5650):1575–1577.

Cordani, N., V. Pisa, L. Pozzi, C. Sciorati, and E. Clementi. 2014. Nitric oxide controls fat deposition in dystrophic skeletal muscle by regulating fibro-adipogenic precursor differentiation. *Stem Cells* 32 (4):874–885.

Cree, M. G., B. R. Newcomer, C. S. Katsanos, et al. 2004. Intramuscular and liver triglycerides are increased in the elderly. *J Clin Endocrinol Metab* 89 (8):3864–3871.

Croquelois, A., A. A. Domenighetti, M. Nemir, et al. 2008. Control of the adaptive response of the heart to stress via the Notch1 receptor pathway. *J Exp Med* 205 (13):3173–3185.

Cruz-Jentoft, A. J., J. P. Baeyens, J. M. Bauer, et al. 2010. Sarcopenia: European consensus on definition and diagnosis: Report of the European working group on sarcopenia in older people. *Age Ageing* 39 (4):412–423.

Delmonico, M. J., T. B. Harris, M. Visser, et al. 2009. Longitudinal study of muscle strength, quality, and adipose tissue infiltration. *Am J Clin Nutr* 90 (6):1579–1585.

Di Prampero, P. E., and M. V. Narici. 2003. Muscles in microgravity: From fibres to human motion. *J Biomech* 36 (3):403–412.

Dreyer, H. C., C. E. Blanco, F. R. Sattler, E. T. Schroeder, and R. A. Wiswell. 2006. Satellite cell numbers in young and older men 24 hours after eccentric exercise. *Muscle Nerve* 33 (2):242–253.

Dulor, J. P., B. Cambon, P. Vigneron, et al. 1998. Expression of specific white adipose tissue genes in denervation-induced skeletal muscle fatty degeneration. *FEBS Lett* 439 (1–2):89–92.

Durheim, M. T., C. A. Slentz, L. A. Bateman, S. K. Mabe, and W. E. Kraus. 2008. Relationships between exercise-induced reductions in thigh intermuscular adipose tissue, changes in lipoprotein particle size, and visceral adiposity. *Am J Physiol Endocrinol Metab* 295 (2):E407–E412.

Edwards, M. H., E. M. Dennison, A. Aihie Sayer, R. Fielding, and C. Cooper. 2015. Osteoporosis and sarco-
    penia in older age. *Bone* 80:126–130.
Fagiolo, U., A. Cossarizza, E. Scala, et al. 1993. Increased cytokine production in mononuclear cells of healthy
    elderly people. *Eur J Immunol* 23 (9):2375–2378.
Faralli, H., and F. J. Dilworth. 2014. Dystrophic muscle environment induces changes in cell plasticity. *Genes
    Dev* 28 (8):809–811.
Farr, J. N., M. D. Van Loan, T. G. Lohman, and S. B. Going. 2012. Lower physical activity is associated with
    skeletal muscle fat content in girls. *Med Sci Sports Exerc* 44 (7):1375–1381.
Farup, J., L. Madaro, P. L. Puri, and U. R. Mikkelsen. 2015. Interactions between muscle stem cells, mesen-
    chymal-derived cells and immune cells in muscle homeostasis, regeneration and disease. *Cell Death
    Dis* 6:e1830.
Fielding, R. A., B. Vellas, W. J. Evans, et al. 2011. Sarcopenia: An undiagnosed condition in older adults.
    Current consensus definition: Prevalence, etiology, and consequences. International working group on
    sarcopenia. *J Am Med Dir Assoc* 12 (4):249–256.
Gallagher, D., S. Heshka, D. E. Kelley, et al. 2014. Changes in adipose tissue depots and metabolic markers
    following a 1-year diet and exercise intervention in overweight and obese patients with type 2 diabetes.
    Diabetes Care 37 (12):3325–3332.
Gallegly, J. C., N. A. Turesky, B. A. Strotman, C. M. Gurley, C. A. Peterson, and E. E. Dupont-Versteegden.
    2004. Satellite cell regulation of muscle mass is altered at old age. *J Appl Physiol (1985)* 97
    (3):1082–1090.
Gerber, C., A. G. Schneeberger, H. Hoppeler, and D. C. Meyer. 2007. Correlation of atrophy and fatty infiltra-
    tion on strength and integrity of rotator cuff repairs: A study in thirteen patients. *J Shoulder Elbow Surg*
    16 (6):691–696.
Goodpaster, B. H., C. L. Carlson, M. Visser, et al. 2001. Attenuation of skeletal muscle and strength in the
    elderly: The Health ABC Study. *J Appl Physiol (1985)* 90 (6):2157–2165.
Goodpaster, B. H., D. E. Kelley, R. R. Wing, A. Meier, and F. L. Thaete. 1999. Effects of weight loss on
    regional fat distribution and insulin sensitivity in obesity. *Diabetes* 48 (4):839–847.
Goodpaster, B. H., F. L. Thaete, and D. E. Kelley. 2000b. Thigh adipose tissue distribution is associated with
    insulin resistance in obesity and in type 2 diabetes mellitus. *Am J Clin Nutr* 71 (4):885–892.
Goodpaster, B. H., F. L. Thaete, J. A. Simoneau, and D. E. Kelley. 1997. Subcutaneous abdominal fat and thigh
    muscle composition predict insulin sensitivity independently of visceral fat. *Diabetes* 46 (10):1579–1585.
Goodpaster, B. H., P. Chomentowski, B. K. Ward, et al. 2008. Effects of physical activity on strength and
    skeletal muscle fat infiltration in older adults: A randomized controlled trial. *J Appl Physiol (1985)* 105
    (5):1498–1503.
Goodpaster, B. H., S. Krishnaswami, H. Resnick, et al. 2003. Association between regional adipose tissue dis-
    tribution and both type 2 diabetes and impaired glucose tolerance in elderly men and women. *Diabetes
    Care* 26 (2):372–379.
Goodpaster, B. H., F. L. Thaete, and D. E. Kelley. 2000a. Composition of skeletal muscle evaluated with com-
    puted tomography. *Ann N Y Acad Sci* 904:18–24.
Gordon, S. 2003. Alternative activation of macrophages. *Nat Rev Immunol* 3 (1):23–35.
Gorgey, A. S., and G. A. Dudley. 2007. Skeletal muscle atrophy and increased intramuscular fat after incom-
    plete spinal cord injury. *Spinal Cord* 45 (4):304–309.
Granados, A., A. Gebremariam, S. S. Gidding, et al. 2018. Association of abdominal muscle composition with
    prediabetes and diabetes: The CARDIA study. *Diabetes Obes Metab* 21 (2): 267–275.
Gray, C., T. J. MacGillivray, C. Eeley, et al. 2011. Magnetic resonance imaging with k-means clustering
    objectively measures whole muscle volume compartments in sarcopenia/cancer cachexia. *Clin Nutr* 30
    (1):106–11.
Guglielmi, V., L. Maresca, M. D'Adamo, et al. 2014. Age-related different relationships between ectopic adi-
    pose tissues and measures of central obesity in sedentary subjects. *PLoS One* 9 (7):e103381.
Heredia, J. E., L. Mukundan, F. M. Chen, et al. 2013. Type 2 innate signals stimulate fibro/adipogenic progeni-
    tors to facilitate muscle regeneration. *Cell* 153 (2):376–388.
Hilton, T. N., L. J. Tuttle, K. L. Bohnert, M. J. Mueller, and D. R. Sinacore. 2008. Excessive adipose tissue
    infiltration in skeletal muscle in individuals with obesity, diabetes mellitus, and peripheral neuropathy:
    Association with performance and function. *Phys Ther* 88 (11):1336–1344.
Jacobs, J. L., R. L. Marcus, G. Morrell, and P. Lastayo. 2014. Resistance exercise with older fallers: Its impact
    on intermuscular adipose tissue. *Biomed Res Int* 2014:398960.
Joe, A. W., L. Yi, A. Natarajan, et al. 2010. Muscle injury activates resident fibro/adipogenic progenitors that
    facilitate myogenesis. *Nat Cell Biol* 12 (2):153–163.

Jubrias, S. A., I. R. Odderson, P. C. Esselman, and K. E. Conley. 1997. Decline in isokinetic force with age: Muscle cross-sectional area and specific force. *Pflugers Arch* 434 (3):246–253.

Judson, R. N., R. H. Zhang, and F. M. Rossi. 2013. Tissue-resident mesenchymal stem/progenitor cells in skeletal muscle: Collaborators or saboteurs? *FEBS J* 280 (17):4100–4108.

Jung, J. Y., K. A. Han, H. J. Ahn, et al. 2012. Effects of aerobic exercise intensity on abdominal and thigh adipose tissue and skeletal muscle attenuation in overweight women with type 2 diabetes mellitus. *Diabetes Metab J* 36 (3):211–221.

Kadi, F., N. Charifi, C. Denis, and J. Lexell. 2004. Satellite cells and myonuclei in young and elderly women and men. *Muscle Nerve* 29 (1):120–127.

Kelley, D. E., B. S. Slasky, and J. Janosky. 1991. Skeletal muscle density: Effects of obesity and non-insulin-dependent diabetes mellitus. *Am J Clin Nutr* 54 (3):509–515.

Kharraz, Y., J. Guerra, C. J. Mann, A. L. Serrano, and P. Munoz-Canoves. 2013. Macrophage plasticity and the role of inflammation in skeletal muscle repair. *Mediators Inflamm* 2013:491497.

Kim, H. M., L. M. Galatz, C. Lim, N. Havlioglu, and S. Thomopoulos. 2012. The effect of tear size and nerve injury on rotator cuff muscle fatty degeneration in a rodent animal model. *J Shoulder Elbow Surg* 21 (7):847–858.

Kim, J. H., S. H. Choi, S. Lim, et al. 2013. Thigh muscle attenuation measured by computed tomography was associated with the risk of low bone density in community-dwelling elderly population. *Clin Endocrinol (Oxf)* 78 (4):512–517.

Kirkland, J. L., T. Tchkonia, T. Pirtskhalava, J. Han, and I. Karagiannides. 2002. Adipogenesis and aging: Does aging make fat go MAD? *Exp Gerontol* 37 (6):757–767.

Koteish, A., and A. M. Diehl. 2001. Animal models of steatosis. *Semin Liver Dis* 21 (1):89–104.

Ku, Y. H., K. A. Han, H. Ahn, et al. 2010. Resistance exercise did not alter intramuscular adipose tissue but reduced retinol-binding protein-4 concentration in individuals with type 2 diabetes mellitus. *J Int Med Res* 38 (3):782–791.

Kuzel, B. R., S. Grindel, R. Papandrea, and D. Ziegler. 2013. Fatty infiltration and rotator cuff atrophy. *J Am Acad Orthop Surg* 21 (10):613–623.

Lang, T., J. A. Cauley, F. Tylavsky, et al. 2010. Computed tomographic measurements of thigh muscle cross-sectional area and attenuation coefficient predict hip fracture: The health, aging, and body composition study. *J Bone Miner Res* 25 (3):513–519.

Laron, D., S. P. Samagh, X. Liu, H. T. Kim, and B. T. Feeley. 2012. Muscle degeneration in rotator cuff tears. *J Shoulder Elbow Surg* 21 (2):164–174.

Laurens, C., K. Louche, C. Sengenes, et al. 2016. Adipogenic progenitors from obese human skeletal muscle give rise to functional white adipocytes that contribute to insulin resistance. *Int J Obes (Lond)* 40 (3):497–506.

Lee, T. J., J. Jang, S. Kang, et al. 2013. Enhancement of osteogenic and chondrogenic differentiation of human embryonic stem cells by mesodermal lineage induction with BMP-4 and FGF2 treatment. *Biochem Biophys Res Commun* 430 (2):793–797.

Lemos, D. R., F. Babaeijandaghi, M. Low, et al. 2015. Nilotinib reduces muscle fibrosis in chronic muscle injury by promoting TNF-mediated apoptosis of fibro/adipogenic progenitors. *Nat Med* 21 (7):786–794.

Leskinen, T., S. Sipila, M. Alen, et al. 2009. Leisure-time physical activity and high-risk fat: A longitudinal population-based twin study. *Int J Obes (Lond)* 33 (11):1211–1218.

Lim, J. P., M. S. Chong, L. Tay, et al. 2018. Inter-muscular adipose tissue is associated with adipose tissue inflammation and poorer functional performance in central adiposity. *Arch Gerontol Geriatr* 81:1–7.

Lord, J. M., S. Butcher, V. Killampali, D. Lascelles, and M. Salmon. 2001. Neutrophil ageing and immunesenescence. *Mech Ageing Dev* 122 (14):1521–1535.

Lu, D. Huang, N. Saederup, I. F. Charo, R. M. Ransohoff, and L. Zhou. 2011. Macrophages recruited via CCR2 produce insulin-like growth factor-1 to repair acute skeletal muscle injury. *FASEB J* 25 (1):358–369.

Lukjanenko, L., S. Brachat, E. Pierrel, E. Lach-Trifilieff, and J. N. Feige. 2013. Genomic profiling reveals that transient adipogenic activation is a hallmark of mouse models of skeletal muscle regeneration. *PLoS One* 8 (8):e71084.

Maffioletti, S. M., M. Noviello, K. English, and F. S. Tedesco. 2014. Stem cell transplantation for muscular dystrophy: The challenge of immune response. *Biomed Res Int* 2014:964010.

Manini, T. M., B. C. Clark, M. A. Nalls, B. H. Goodpaster, L. L. Ploutz-Snyder, and T. B. Harris. 2007. Reduced physical activity increases intermuscular adipose tissue in healthy young adults. *Am J Clin Nutr* 85 (2):377–384.

Marcus, R. L., O. Addison, J. P. Kidde, L. E. Dibble, and P. C. Lastayo. 2010. Skeletal muscle fat infiltration: Impact of age, inactivity, and exercise. *J Nutr Health Aging* 14 (5):362–366.

Marcus, R. L., O. Addison, L. E. Dibble, K. B. Foreman, G. Morrell, and P. Lastayo. 2012. Intramuscular adipose tissue, sarcopenia, and mobility function in older individuals. *J Aging Res* 2012:629637.

Marcus, R. L., S. Smith, G. Morrell, et al. 2008. Comparison of combined aerobic and high-force eccentric resistance exercise with aerobic exercise only for people with type 2 diabetes mellitus. *Phys Ther* 88 (11):1345–1354.

Marsh, D. R., D. S. Criswell, J. A. Carson, and F. W. Booth. 1997. Myogenic regulatory factors during regeneration of skeletal muscle in young, adult, and old rats. *J Appl Physiol (1985)* 83 (4):1270–1275.

Mauro, A. 1961. Satellite cell of skeletal muscle fibers. *J Biophys Biochem Cytol* 9:493–495.

Merritt, E. K., M. J. Stec, A. Thalacker-Mercer, et al. 2013. Heightened muscle inflammation susceptibility may impair regenerative capacity in aging humans. *J Appl Physiol (1985)* 115 (6):937–948.

Meyer, D. C., H. Hoppeler, B. von Rechenberg, and C. Gerber. 2004. A pathomechanical concept explains muscle loss and fatty muscular changes following surgical tendon release. *J Orthop Res* 22 (5):1004–1007.

Miljkovic-Gacic, I., X. Wang, C. M. Kammerer, et al. 2008. Fat infiltration in muscle: New evidence for familial clustering and associations with diabetes. *Obesity (Silver Spring)* 16 (8):1854–1860.

Minetti, G. C., C. Colussi, R. Adami, et al. 2006. Functional and morphological recovery of dystrophic muscles in mice treated with deacetylase inhibitors. *Nat Med* 12 (10):1147–1150.

Moldes, M., Y. Zuo, R. F. Morrison, et al. 2003. Peroxisome-proliferator-activated receptor gamma suppresses Wnt/beta-catenin signalling during adipogenesis. *Biochem J* 376 (Pt 3):607–613.

Mourikis, P., R. Sambasivan, D. Castel, P. Rocheteau, V. Bizzarro, and S. Tajbakhsh. 2012. A critical requirement for notch signaling in maintenance of the quiescent skeletal muscle stem cell state. *Stem Cells* 30 (2):243–252.

Mozzetta, C., S. Consalvi, V. Saccone, et al. 2013. Fibroadipogenic progenitors mediate the ability of HDAC inhibitors to promote regeneration in dystrophic muscles of young, but not old Mdx mice. *EMBO Mol Med* 5 (4):626–639.

Murphy, J. C., J. L. McDaniel, K. Mora, D. T. Villareal, L. Fontana, and E. P. Weiss. 2012. Preferential reductions in intermuscular and visceral adipose tissue with exercise-induced weight loss compared with calorie restriction. *J Appl Physiol (1985)* 112 (1):79–85.

Murphy, R. A., I. Reinders, T. C. Register, et al. 2014. Associations of BMI and adipose tissue area and density with incident mobility limitation and poor performance in older adults. *Am J Clin Nutr* 99 (5):1059–1065.

Nemir, M., M. Metrich, I. Plaisance, et al. 2014. The Notch pathway controls fibrotic and regenerative repair in the adult heart. *Eur Heart J* 35 (32):2174–2185.

Nicklas, B. J., E. Chmelo, O. Delbono, J. J. Carr, M. F. Lyles, and A. P. Marsh. 2015. Effects of resistance training with and without caloric restriction on physical function and mobility in overweight and obese older adults: A randomized controlled trial. *Am J Clin Nutr* 101 (5):991–999.

Nyberg, M., J. R. Blackwell, R. Damsgaard, A. M. Jones, Y. Hellsten, and S. P. Mortensen. 2012. Lifelong physical activity prevents an age-related reduction in arterial and skeletal muscle nitric oxide bioavailability in humans. *J Physiol* 590 (21):5361–5370.

Pagano, A. F., R. Demangel, T. Brioche, et al. 2015. Muscle regeneration with intermuscular adipose tissue (IMAT) accumulation is modulated by mechanical constraints. *PLoS One* 10 (12):e0144230.

Pagano, A. F., T. Brioche, C. Arc-Chagnaud, R. Demangel, A. Chopard, and G. Py. 2017. Short-term disuse promotes fatty acid infiltration into skeletal muscle. *J Cachexia Sarcopenia Muscle* 9 (2): 335–347.

Penton, C. M., J. M. Thomas-Ahner, E. K. Johnson, C. McAllister, and F. Montanaro. 2013. Muscle side population cells from dystrophic or injured muscle adopt a fibro-adipogenic fate. *PLoS One* 8 (1):e54553.

Perdiguero, E., P. Sousa-Victor, V. Ruiz-Bonilla, et al. 2011. p38/MKP-1-regulated AKT coordinates macrophage transitions and resolution of inflammation during tissue repair. *J Cell Biol* 195 (2):307–322.

Pisani, D. F., C. D. Bottema, C. Butori, C. Dani, and C. A. Dechesne. 2010. Mouse model of skeletal muscle adiposity: A glycerol treatment approach. *Biochem Biophys Res Commun* 396 (3):767–773.

Prior, S. J., L. J. Joseph, J. Brandauer, L. I. Katzel, J. M. Hagberg, and A. S. Ryan. 2007. Reduction in midthigh low-density muscle with aerobic exercise training and weight loss impacts glucose tolerance in older men. *J Clin Endocrinol Metab* 92 (3):880–886.

Przybyla, B., C. Gurley, J. F. Harvey, et al. 2006. Aging alters macrophage properties in human skeletal muscle both at rest and in response to acute resistance exercise. *Exp Gerontol* 41 (3):320–327.

Reginster, J. Y., C. Beaudart, F. Buckinx, and O. Bruyere. 2016. Osteoporosis and sarcopenia: Two diseases or one? *Curr Opin Clin Nutr Metab Care* 19 (1):31–36.

Renault, V., L. E. Thornell, P. O. Eriksson, G. Butler-Browne, and V. Mouly. 2002. Regenerative potential of human skeletal muscle during aging. *Aging Cell* 1 (2):132–139.

Rice, C. L., D. A. Cunningham, D. H. Paterson, and M. S. Lefcoe. 1989. Arm and leg composition determined by computed tomography in young and elderly men. *Clin Physiol* 9 (3):207–220.

Robles, P. G., M. S. Sussman, A. Naraghi, et al. 2015. Intramuscular fat infiltration contributes to impaired muscle function in COPD. *Med Sci Sports Exerc* 47 (7):1334–1341.

Roig, M., J. J. Eng, D. L. MacIntyre, J. D. Road, and W. D. Reid. 2011. Deficits in muscle strength, mass, quality, and mobility in people with chronic obstructive pulmonary disease. *J Cardiopulm Rehabil Prev* 31 (2):120–124.

Ross, S. E., N. Hemati, K. A. Longo, et al. 2000. Inhibition of adipogenesis by Wnt signaling. *Science* 289 (5481):950–953.

Roth, S. M., G. F. Martel, F. M. Ivey, et al. 2000. Skeletal muscle satellite cell populations in healthy young and older men and women. *Anat Rec* 260 (4):351–358.

Ryan, A. S., and B. J. Nicklas. 1999. Age-related changes in fat deposition in mid-thigh muscle in women: Relationships with metabolic cardiovascular disease risk factors. *Int J Obes Relat Metab Disord* 23 (2):126–132.

Ryan, A. S., C. L. Dobrovolny, G. V. Smith, K. H. Silver, and R. F. Macko. 2002. Hemiparetic muscle atrophy and increased intramuscular fat in stroke patients. *Arch Phys Med Rehabil* 83 (12):1703–1707.

Saccone, V., S. Consalvi, L. Giordani, et al. 2014. HDAC-regulated myomiRs control BAF60 variant exchange and direct the functional phenotype of fibro-adipogenic progenitors in dystrophic muscles. *Genes Dev* 28 (8):841–857.

Saclier, M., H. Yacoub-Youssef, A. L. Mackey, et al. 2013. Differentially activated macrophages orchestrate myogenic precursor cell fate during human skeletal muscle regeneration. *Stem Cells* 31 (2):384–396.

Sadeh, M. 1988. Effects of aging on skeletal muscle regeneration. *J Neurol Sci* 87 (1):67–74.

Samengo, G., A. Avik, B. Fedor, et al. 2012. Age-related loss of nitric oxide synthase in skeletal muscle causes reductions in calpain S-nitrosylation that increase myofibril degradation and sarcopenia. *Aging Cell* 11 (6):1036–1045.

Santanasto, A. J., N. W. Glynn, M. A. Newman, et al. 2011. Impact of weight loss on physical function with changes in strength, muscle mass, and muscle fat infiltration in overweight to moderately obese older adults: A randomized clinical trial. *J Obes* 2011.

Schultz, E., and B. H. Lipton. 1982. Skeletal muscle satellite cells: Changes in proliferation potential as a function of age. *Mech Ageing Dev* 20 (4):377–383.

Sciorati, C., E. Clementi, A. A. Manfredi, and P. Rovere-Querini. 2015. Fat deposition and accumulation in the damaged and inflamed skeletal muscle: Cellular and molecular players. *Cell Mol Life Sci* 72 (11):2135–2156.

Simoneau, J. A., S. R. Colberg, F. L. Thaete, and D. E. Kelley. 1995. Skeletal muscle glycolytic and oxidative enzyme capacities are determinants of insulin sensitivity and muscle composition in obese women. *FASEB J* 9 (2):273–278.

Song, M. Y., E. Ruts, J. Kim, I. Janumala, S. Heymsfield, and D. Gallagher. 2004. Sarcopenia and increased adipose tissue infiltration of muscle in elderly African American women. *Am J Clin Nutr* 79 (5):874–880.

Sousa-Victor, P., L. Garcia-Prat, A. L. Serrano, E. Perdiguero, and P. Munoz-Canoves. 2015. Muscle stem cell aging: Regulation and rejuvenation. *Trends Endocrin Met* 26 (6):287–296.

Taaffe, D. R., T. R. Henwood, M. A. Nalls, D. G. Walker, T. F. Lang, and T. B. Harris. 2009. Alterations in muscle attenuation following detraining and retraining in resistance-trained older adults. *Gerontology* 55 (2):217–223.

Tuttle, L. J., D. R. Sinacore, and M. J. Mueller. 2012. Intermuscular adipose tissue is muscle specific and associated with poor functional performance. *J Aging Res* 2012:172957.

Tuttle, L. J., D. R. Sinacore, W. T. Cade, and M. J. Mueller. 2011. Lower physical activity is associated with higher intermuscular adipose tissue in people with type 2 diabetes and peripheral neuropathy. *Phys Ther* 91 (6):923–930.

Uezumi, A., M. Ikemoto-Uezumi, and K. Tsuchida. 2014. Roles of nonmyogenic mesenchymal progenitors in pathogenesis and regeneration of skeletal muscle. *Front Physiol* 5:68.

Uezumi, A., S. Fukada, N. Yamamoto, et al. 2014. Identification and characterization of PDGFRalpha+ mesenchymal progenitors in human skeletal muscle. *Cell Death Dis* 5:e1186.

Uezumi, A., S. Fukada, N. Yamamoto, S. Takeda, and K. Tsuchida. 2010. Mesenchymal progenitors distinct from satellite cells contribute to ectopic fat cell formation in skeletal muscle. *Nat Cell Biol* 12 (2):143–152.

Uezumi, A., T. Ito, D. Morikawa, et al. 2011. Fibrosis and adipogenesis originate from a common mesenchymal progenitor in skeletal muscle. *J Cell Sci* 124 (Pt 21):3654–3664.

Verdijk, L. B., T. Snijders, M. Drost, T. Delhaas, F. Kadi, and L. J. van Loon. 2014. Satellite cells in human skeletal muscle; from birth to old age. *Age (Dordr)* 36 (2):545–547.

Vertino, A. M., J. M. Taylor-Jones, K. A. Longo, et al. 2005. Wnt10b deficiency promotes coexpression of myogenic and adipogenic programs in myoblasts. *Mol Biol Cell* 16 (4):2039–2048.

Vettor, R., G. Milan, C. Franzin, et al. 2009. The origin of intermuscular adipose tissue and its pathophysiological implications. *Am J Physiol Endocrinol Metab* 297 (5):E987–E998.

Villalta, S. A., H. X. Nguyen, B. Deng, T. Gotoh, and J. G. Tidball. 2009. Shifts in macrophage phenotypes and macrophage competition for arginine metabolism affect the severity of muscle pathology in muscular dystrophy. *Hum Mol Genet* 18 (3):482–496.

Visser, M., B. H. Goodpaster, S. B. Kritchevsky, et al. 2005. Muscle mass, muscle strength, and muscle fat infiltration as predictors of incident mobility limitations in well-functioning older persons. *J Gerontol A Biol Sci Med Sci* 60 (3):324–333.

Visser, M., S. B. Kritchevsky, B. H. Goodpaster, et al. 2002. Leg muscle mass and composition in relation to lower extremity performance in men and women aged 70–79: The health, aging and body composition study. *J Am Geriatr Soc* 50 (5):897–904.

Walts, C. T., E. D. Hanson, M. J. Delmonico, L. Yao, M. Q. Wang, and B. F. Hurley. 2008. Do sex or race differences influence strength training effects on muscle or fat? *Med Sci Sports Exerc* 40 (4):669–676.

Wang, Z. W., W. T. Pan, Y. Lee, T. Kakuma, Y. T. Zhou, and R. H. Unger. 2001. The role of leptin resistance in the lipid abnormalities of aging. *FASEB J* 15 (1):108–114.

Wen, Y., P. Bi, W. Liu, A. Asakura, C. Keller, and S. Kuang. 2012. Constitutive Notch activation upregulates Pax7 and promotes the self-renewal of skeletal muscle satellite cells. *Mol Cell Biol* 32 (12):2300–2311.

Wroblewski, A. P., F. Amati, M. A. Smiley, B. Goodpaster, and V. Wright. 2011. Chronic exercise preserves lean muscle mass in masters athletes. *Phys Sportsmed* 39 (3):172–178.

Yim, J. E., S. Heshka, J. Albu, et al. 2007. Intermuscular adipose tissue rivals visceral adipose tissue in independent associations with cardiovascular risk. *Int J Obes (Lond)* 31 (9):1400–1405.

Yoshida, Y., R. L. Marcus, and P. C. Lastayo. 2012. Intramuscular adipose tissue and central activation in older adults. *Muscle Nerve* 46 (5):813–816.

Section V

Recent Advances Limiting Sarcopenia and Supporting Healthy Aging

14 Nutritional Modulation of Mitochondria-Associated Death Signaling in Sarcopenia

Stephen E. Alway

CONTENTS

14.1 INTRODUCTION

Several age-associated changes have been proposed as potential mechanisms that mediate sarcopenia. However, it is difficult to determine whether the changes in signaling trigger sarcopenia or rather if they are a consequence of signaling for muscle atrophy. Examples of changes that might contribute to sarcopenia include an increase in low grade but constant systemic inflammation, elevated production, or reduced buffering of reactive oxygen species (Smith et al. 2018, Brown et al. 2017, Bak et al. 2019, Parousis et al. 2018, Jackson 2016) altered or impaired innervation (Pigna et al. 2018, Hepple and Rice 2016), loss of motor units, and alpha motor neurons

(Tomlinson et al. 1977), reduced regenerative capability (Feige et al. 2018, Sousa-Victor, Garcia-Prat, and Munoz-Canoves 2018), and decreased mitochondrial function (Picca, Calvani, Bossola, et al. 2018, Marzetti et al. 2018, Picca, Lezza, et al. 2018). Low physical activity can exacerbate sarcopenia, whereas exercise training can at least partially attenuate some of the aging-associated alterations in mitochondrial function in aging (Pietrangelo et al. 2015, Muhammad and Allam 2018, Picca, Calvani, Leeuwenburgh, et al. 2018, Picca, Calvani, Bossola, et al. 2018, Carter et al. 2018, Kim, Triolo, and Hood 2017, Hood et al. 2016).

Emerging evidence suggests that dysfunctional mitochondria play an important role in regulating the loss of muscle function that is associated with sarcopenia (Nichols et al. 2015, Alway, Morissette, and Siu 2011, Del Campo et al. 2018, Picca, Lezza, et al. 2018, Brown et al. 2017, Moulin and Ferreiro 2017). This is due in part because optimal mitochondrial function is critical for energy delivery and expenditure, but mitochondrial function is impaired in aging (Hood et al. 2018, Chen et al. 2018, Kim, Zheng, et al. 2018, Kim, Triolo, et al. 2018, Alway, Mohamed, and Myers 2017).

In this chapter, we will provide evidence that the loss of mitochondrial function is central to the dysregulation of apoptosis, autophagy, proteasome, and lysosomal pathways. These same pathways are key to muscle and motor neuronal loss that contributes to sarcopenia. The manner in which nutrition could be expected to alter the signaling from these pathways to attenuate sarcopenia will be discussed.

14.2   AGING ATTENUATES MITOCHONDRIAL FUNCTION

Aging decreases muscle mitochondrial function (Elfawy and Das 2018, Hood et al. 2018, Popa-Wagner et al. 2018), mitochondrial content, and enzyme activity (Parousis et al. 2018, Picca, Lezza, et al. 2018, Picca, Calvani, Leeuwenburgh, et al. 2018, Picca, Calvani, Bossola, et al. 2018, Del Campo et al. 2018, Campbell et al. 2018, Kim, Zheng, et al. 2018, Rygiel, Grady, and Turnbull 2014, Rygiel, Picard, and Turnbull 2016), lowers respiration (Ron-Harel et al. 2018, Porter et al. 2015), and reduces mitochondrial biogenesis (Hood et al. 2018, Dai, Rabinovitch, and Ungvari 2012, Ungvari et al. 2008). Part of the decline in mitochondrial content in aging may be the result of an imbalance between mitochondrial removal and mitochondrial biogenesis. Mitochondrial biogenesis is a complex process consisting of both synthesis, assembly, growth by fusion, and division of pre-existing mitochondria by fission dynamics. However, it should be noted that the aging-associated loss of mitochondrial respiration and function has been challenged in other studies (Distefano et al. 2017, Picard et al. 2013, Picard et al. 2010, Picard et al. 2011), and reductions in mitochondrial function have been attributed more to lack of use than aging per se (Konopka et al. 2014). Furthermore, improvements in mitochondrial function in aging can also be achieved by nutritional intervention without changes in mitochondrial biogenesis (Lalia et al. 2017, Lanza et al. 2012). Moreover, it has been suggested that changes (or lack of changes) in mitochondrial respiratory function with age may be influenced by sex (Khalifa et al. 2017, Drake and Yan 2017, Drake et al. 2013), muscle fiber type, or motor unit recruitment patterns (Capitanio et al. 2016, Milanese et al. 2014).

14.3   POTENTIAL SOURCES OF MITOCHONDRIAL DYSFUNCTION IN AGING

14.3.1   Reactive Oxygen Species-Induced Damage

Aging tends to increase reactive oxygen species (ROS) production, and this is magnified during reduced muscle activity that typically accompanies aging. There is a large data base indicating that mitochondrial ROS production results in widespread oxidative damage to cells (Radak et al. 2018, Ren et al. 2018, Dogru et al. 2018, Brand 2016). Although the origin of ROS can be from multiple sources, generally there is proposed to be an increase in cytokines that occur with aging.

Furthermore, there are data to support a role for myostatin as a pro-oxidant, which generates ROS in muscle cells through tumor necrosis factor-alpha (TNF-alpha) signaling via NF-kappaB and NADPH oxidase (Sriram et al. 2011, Alway 2017). This is supported by observations that myostatin null mice have lower ROS and sarcopenia (Sriram et al. 2011). Reducing ROS production would be expected to attenuate oxidant damage to proteins, lipids, and DNA and reduce mitochondrial insult (Lee, Lim, and Choi 2017).

At least as important as ROS production in determining the accumulation of ROS for potential cell damage, is the level of antioxidants present in the mitochondria and the cytosol of a muscle cell. In fact, most measurements may underestimate ROS production because there is a large antioxidant potential in these compartments (Munro et al. 2016). The importance of antioxidants in sarcopenia is highlighted by observations that both neural and muscle losses of the cytosolic antioxidant CuZn-superoxide dismutase (CuZnSOD) recapitulated sarcopenic muscle loss in a mouse model (Qaisar et al. 2019, Sataranatarajan et al. 2015). It is, however, noteworthy that the loss of CuZnSOD in only neural or muscle cells did not manifest full sarcopenic muscle loss (Sataranatarajan et al. 2015), but this does emphasize the cross talk that probably occurs between these two tissues in aging. Nevertheless, as aging muscles and neurons are associated with an increased ROS accumulation and a reduction of mRNA and protein for several antioxidant enzymes (Haramizu et al. 2017, Alway et al. 2015, Alway, Myers, and Mohamed 2014, Takahashi et al. 2017, Alway, Mohamed, and Myers 2017, Alway 2017), aging likely increases the potential for greater ROS-induced damage to mitochondrial components in motor neurons and muscle cells.

The interactions between aging and inactivity with ROS production, along with lower antioxidant levels (Radak et al. 2018, Radak et al. 2017, Hao et al. 2011, Ryan, Jackson, et al. 2010, Ryan et al. 2011, Alway et al. 2017, Jackson et al. 2010), increase the potential for mitochondrial damage in aged muscles compared with young skeletal muscle cells. It has also been recently suggested that ROS production might be secondary to denervation that occurs in aged muscles (Liang et al. 2018, Qiu et al. 2018, Pollock et al. 2017, Jackson, Ryan, and Alway 2011, Spendiff et al. 2016), again highlighting the potential communication between neural and muscle cells in aging. Whatever the initial source(s) of ROS production, it is clear that accumulation of excessive ROS leads to damaged mitochondria in muscle and neural cells, which in turn can result in mitochondria that are more dysfunctional.

14.3.2 Mitochondrial DNA Damage and Aging

Mitochondrial DNA (mtDNA) deletions or DNA mutations have been proposed as contributors to mitochondrial dysfunction that leads to aging-related muscle fiber loss and atrophy (Picca et al. 2019, Picca, Calvani, Bossola, et al. 2018, Picca, Lezza, et al. 2018, Herbst et al. 2017, Herbst et al. 2016). Indeed, there is evidence of increased mtDNA mutations in areas of muscle oxidative damage (Herbst et al. 2017, Herbst et al. 2016, Tarry-Adkins et al. 2016). Similarly, in neurons, DNA damage precedes neuronal apoptosis (Shi et al. 2018, Khan et al. 2018, Laaper et al. 2017, Cheema et al. 2015), whereas forced repair of DNA damage rescues neurons from elimination by apoptosis (Martin and Wong 2017, Martin and Chang 2018). Although not all increases in ROS production are the result of mtDNA deletions or lead to mtDNA mutations (Wanagat et al. 2015), it is clear that such DNA deletions provide another strong contribution to mitochondrial ROS production in aging muscles and neurons (Kleme et al. 2018, Martin et al. 2016, Cheema et al. 2015). Furthermore, aging-induced mtDNA deletions are linked closely to the loss of mitochondrial function in motor neurons (Rygiel, Grady, and Turnbull 2014, Rygiel, Picard, and Turnbull 2016) and neuronal malfunction in neural diseases such as Parkinson's disease (Robinson, Yousefzadeh, et al. 2018, Robinson, Aguiar, et al. 2018) and in muscle loss with aging (Herbst et al. 2016, Herbst et al. 2017). Together, these observations highlight the important role of mitochondria in maintaining neural and muscle function in aging.

14.3.3 Altered Mitochondrial Dynamics with Aging

Mitochondria are not static structures, as they can form individual organelles or an extensive reticulum. Mitochondrial morphology is regulated by interactions between fusion and fission regulatory proteins. Fusion proteins such as Mitofusins 1 and 2 (Mfn1 and 2) and Optic atrophy 1 (Opa1) can join mitochondrial membranes together to form larger mitochondria or increase the size of the mitochondrial reticular network (Leduc-Gaudet et al. 2015, St-Jean-Pelletier et al. 2017). Fission proteins such as dynamin-related protein 1 (Drp1) and fission protein 1 (Fis1) promote mitochondrial fission, which can result in smaller, individual, or fragmented mitochondria (Leduc-Gaudet et al. 2015, St-Jean-Pelletier et al. 2017, Bak et al. 2019, Manczak et al. 2019, Arribat et al. 2018).

These processes of fission and fusion are important in healthy mitochondria because they allow for exchange of the matrix proteins between individual mitochondria (Bak et al. 2019, Feng et al. 2011, Chen et al. 2010). However, abnormal mitochondrial dynamics may negatively influence mitochondrial health. For example, both the mRNA level and the protein abundance of important fusion and fission proteins have been reported to be lower in old compared with young adult skeletal muscle (Ibebunjo et al. 2013), and this would suggest that the potential for mitochondria to respond to changing environments might be reduced compared with young mitochondria. Indeed, this appears to be the case, because electron microscopy and biochemical analyses have shown distinct mitochondrial profiles in aging muscle. Small, more fragmented mitochondria have been identified in muscles of old animals as compared with mitochondria from muscles in young adult hosts (Iqbal and Hood 2015, 2014), although very large mitochondria have also been noted in muscles of old animals (Leduc-Gaudet et al. 2015; Navratil et al. 2008).

Consistent with the fragmented mitochondrial phenotype in muscles from old hosts, there is evidence to suggest that muscles of aged rodents and humans have a greater overall rate of fission vs. fusion (Iqbal and Hood 2015, Ibebunjo et al. 2013, Iqbal et al. 2013) and lower levels of the fusion protein optic atrophy 1 (Opa1) (Lopez-Lluch 2017, Joseph et al. 2012, Tezze et al. 2017) compared with younger muscles. Fragmented mitochondria tend to have a lower respiratory capacity, and an increased production of ROS, which increases the susceptibility of mitochondria to release its contents and activate the intrinsic caspase apoptotic pathway. Thus, it is not surprising that aging and disuse, which both have excessively fragmented mitochondria, are accompanied by muscle loss (Iqbal and Hood 2015, Iqbal et al. 2013). It is interesting that a knockout of Mfn 1/2 in skeletal muscle, which prevents mitochondrial fusion, increases the accumulation of mtDNA defects and results in muscle atrophy (Kruger et al. 2018, Sebastian et al. 2016, Iqbal et al. 2013). Together, these observations support the idea that muscle mitochondria are important regulators of muscle size. However, to provide a balanced perspective, it is important to point out that other studies have found higher fusion profiles in muscles of humans (Bori et al. 2012), prematurely aged mice (Joseph and Hood 2014, Joseph, Adhihetty, et al. 2013), and larger mitochondria in muscles of old mice (St-Jean-Pelletier et al. 2017, Leduc-Gaudet et al. 2015) or no age-induced change in mitochondrial dynamics (Distefano et al. 2017). Nevertheless, it is interesting that even in studies reporting a higher fusion index in aged muscles as shown by ratios of Mfn1/Mfn2 (Joseph and Hood 2014, Joseph, Adhihetty, et al. 2013) or Mfn2/Drp1 (Leduc-Gaudet et al. 2015), the protein contents of Mfn1, Mfn2, Opa1, or Drp1 did not change. This means that even when higher fusion indexes are recorded there still may be more fragmented mitochondria in aged muscles. Nevertheless, type I muscle fibers with a high mitochondria content tend to have better fission/fusion functions than type II fibers (Crupi et al. 2018), and this supports the idea that mitochondrial fission/fusion might support the resistance of type I fibers to atrophy and apoptosis in aging muscles.

The changes in fission and fusion protein-mediated functions may be the result of damage to mitochondria such as accumulation of mtDNA defects, increased ROS production, and/or inappropriate import or assembly of electron transport proteins. Indeed, high-intensity exercise induces mitochondrial damage and dysregulation of mitochondrial fusion and fission proteins, thereby altering mitochondrial structure (Lee et al. 2015). Such oxidant damage can lead to increased mitochondria permeability (producing leaky mitochondria) and the release of mitochondria-specific proteins, including apoptosis-inducing factor (AIF) and cytochrome c, into the cytosol through the mitochondrial transition pore, which triggers death-signaling pathways. mtDNA deletions or DNA mutations have been proposed contributors to aging-related muscle fiber loss and atrophy (Tezze et al. 2017, Safdar, Annis, et al. 2016, Safdar, Khrapko, et al. 2016). Indeed, there is evidence of increased mtDNA mutations in areas of muscle oxidative damage (Cheema et al. 2015, Herbst et al. 2016). Although not all increases in ROS production are the result of mtDNA deletions (Wanagat et al. 2015) and/or lead to mtDNA mutations (Wanagat et al. 2015), it is clear that such deletions provide another strong contribution to mitochondrial ROS in aging muscles (Cheema et al. 2015, Herbst et al. 2016). Furthermore, mtDNA deletions are linked closely to muscle loss with aging (McKiernan et al. 2009), again highlighting the important role of mitochondria in maintaining muscle mass in aging. Improving mitochondria structure and increasing mitochondrial biogenesis by supplementing old mice with growth differentiation factor 11 (Sinha et al. 2014) or caloric restriction (McKiernan et al. 2012, McKiernan et al. 2009) further support the idea that healthy mitochondria have a pivotal role in maintaining muscle mass and function and that unhealthy mitochondria can eventually contribute to the cell destruction.

14.4   MITOCHONDRIA ARE INITIATORS OF CELL DEATH SIGNALING IN AGING MUSCLES

Two independent pathways are involved in regulating signaling for cell death. These are nuclear apoptosis (Alway et al. 2015; Chabi et al. 2008; Hao et al. 2011; Saleem et al. 2009; Zhang et al. 2013) and autophagy (Carnio et al. 2014; O'Leary et al. 2012; O'Leary et al. 2013; Vainshtein et al. 2016, 2015b), which are pathways that can be activated in response to dysfunctional or damaged mitochondria in aging. While proper balance between these signaling pathways is important for optimizing the health of the muscle fiber, altered signaling and dysregulation of one or both pathways are common in sarcopenia (Alway et al. 2015; Hao et al. 2011; O'Leary et al. 2013).

Low physical activity can exacerbate sarcopenia, whereas exercise training can at least partially attenuate some of the aging-associated alterations including improvements of mitochondrial function in aging (Picca et al. 2019, Carter, Chen, and Hood 2015). While healthy mitochondria provide proper cellular metabolism and production of ATP, dysfunctional or damaged mitochondria can initiate intrinsic (mitochondrial) death pathways that result in the removal of muscle and neural nuclei via nuclear apoptosis (Haramizu et al. 2017, Alway, Mohamed, and Myers 2017, Alway et al. 2015, Alway et al. 2014, Alway, Morissette, and Siu 2011, Hao et al. 2011, Alway et al. 2013) (Figure 14.1). Although autophagy can be upregulated during periods of muscle wasting, the overall pattern to dismantle a dysfunctional mitochondria by autophagy (mitophagy) is to eliminate the source of the death signaling and ultimately to save the muscle cell from complete removal by the apoptotic death pathway (Picca et al. 2019, Picca, Calvani, Bossola, et al. 2018, Picca, Lezza, et al. 2018, Marzetti et al. 2017, Joseph, Adhihetty, et al. 2013). While proper balance between these signaling pathways is important for optimizing the health of the muscle fiber, altered signaling and dysregulation of apoptotic and/or autophagy pathways are common in sarcopenia (Alway, Mohamed, and Myers 2017, Potes et al. 2017).

14.4.1   MITOCHONDRIA-INDUCED NUCLEAR APOPTOSIS IN AGING MUSCLE

The decline of mitochondrial function with aging is associated with sarcopenia. Losses of mitochondrial function may limit the synthesis of sufficient adenosine triphosphate (ATP) for homeostasis, and this could be a contributing factor to apoptosis signaling (Biala, Dhingra, and Kirshenbaum 2015). There are many similarities in mediation of death in single cells that contain only one nucleus

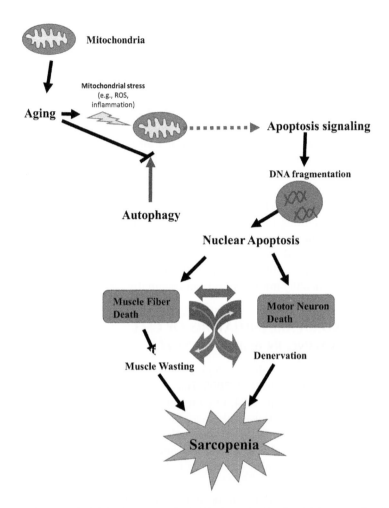

FIGURE 14.1 Mitochondria-initiated sarcopenia. Initiation of mitochondrial stress that includes ROS, inflammation signals, and cytokines (lightning bolt) can result in dysfunctional mitochondria. Damaged mitochondria are removed by mitophagy signaling and proteasome/lysosomal activation in healthy young muscle and motor neurons. However, in aging, the increased ROS and other mitochondrial stresses cause mPTP opening (red dashed arrow). The resulting release of the mitochondrial contents (e.g., cytochrome *c*) to the cell cytosol leads to an apoptotic signaling cascade that ends with DNA fragmentation and removal of nuclei (apoptosis). Sufficient nuclear death in muscle cells will result in the death and removal of the entire muscle cell. Motor neuron death occurs when apoptosis removal of the alpha motor neuron nucleus occurs in the same series of events that occur in skeletal muscle. The interdependence of muscle cells and motor neurons suggests a potential feedback loop (anterograde and retrograde) communication between the muscle and the motor neuronal compartments, which exacerbates death in both compartments. Death in these cell compartments leads to loss of muscle mass and function in aging. Thus, dysfunctional mitochondria provide the signal to initiate the signaling pathways that lead to sarcopenia.

and multinucleated skeletal muscle cells. However, an important difference is that loss of a single nucleus (e.g., via apoptosis) does not result in complete cell death in multinucleated muscle cells, whereas the single nucleated cell will die once its nucleus is eliminated. In skeletal muscle, targeting one nucleus in a muscle region but not another suggests that signaling for cell death is not controlled systemically but rather it is controlled by local targeted signaling networks.

There are three primary pathways for apoptosis, but the intrinsically mediated mitochondrial signaling pathway is attractive for explaining much of the apoptotic signaling associated with sarcopenia. Nuclear apoptosis is characterized by increases in DNA fragmentation and by increases in pro-apoptotic proteins such as Bax, caspase-3, apoptosis protease activating factor-1 (Apaf-1), and AIF (Siu, Wang, and Alway 2009, Alway, Mohamed, and Myers 2017, Haramizu et al. 2017, Nichols et al. 2015, Alway et al. 2015, Alway et al. 2014, Alway, Myers, and Mohamed 2014, Quadrilatero, Alway, and Dupont-Versteegden 2011). There are multiple levels of evidence from our lab and other research groups that apoptosis has an important role in regulating muscle mass in aging (Yoo et al. 2018, Haramizu et al. 2017, Saini et al. 2017, Alway, Mohamed, and Myers 2017, Pardo et al. 2017, Dalle, Rossmeislova, and Koppo 2017, Brioche et al. 2016, Li et al. 2016, Spendiff et al. 2016) and denervation (Spendiff et al. 2016, Siu and Alway 2009, 2005, 2006). Furthermore, there is evidence that downregulating apoptotic signaling can accelerate the loss of muscle mass and function in aging animals or after aging-associated conditions such as denervation (Hepple and Rice 2016, Yoo et al. 2018, Dethlefsen et al. 2018, Siu and Alway 2006). In addition, an upregulation of apoptotic signaling has been identified in premature aging models that exhibit accelerated sarcopenia (Jang et al. 2018, Jang et al. 2010). Increased levels of caspase-3 and DNA fragmentation have also been reported in aged rat muscles.

Figure 14.2 summarizes the main elements of mitochondrial-associated (intrinsic) apoptotic signaling, which contributes to nuclear apoptosis in skeletal muscle.

Increased mitochondria permeability is an important regulator of intrinsic apoptosis signaling in skeletal muscle. This is initiated by Bax:Bax (or Bax:Bak) dimerization, which creates a pore in the outer mitochondrial membrane. Alternatively, mitochondrial permeability can occur via a greater sensitization of the muscle mitochondrial permeability pore opening (mPTP) in response to ROS or calcium. When these channels open, they increase mitochondrial permeability, which allows for the release of mitochondrial housed proteins such as cytochrome c to the cytosol in muscles of old animals (Marzetti et al. 2013, Lopez-Dominguez et al. 2013, Siu, Wang, and Alway 2009, Jackson, Ryan, and Alway 2011). In the cytoplasm, cytochrome c acts as a pro-apoptotic protein by binding dATP and Apaf-1, forming an apoptosome that cleaves caspase-9. An aging-associated increase in the pro-apoptotic Apaf-1 protein and in the level of the pro-apoptotic cleaved caspase-9 protein, along with increased DNA fragmentation, have been reported in gastrocnemius muscles of old rats. This is not surprising because cleaved caspase-9 will activate caspase-3, the final effector caspase in this apoptotic pathway. Mitochondrial-associated caspase signaling is clearly important in muscle loss, but it is not the only source of mitochondrial-associated apoptotic signaling. Caspase-independent signaling associated with mitochondrial dysfunction and permeability has been shown to occur in aging-associated muscle loss. The release of endonuclease G (EndoG), AIF, second mitochondria-derived activator of caspase/direct inhibitor of apoptosis-binding protein with low pI (Smac/DIABLO), and X-linked inhibitor of apoptosis protein (XIAP) from the mitochondria to the cytosol begins the initiation of apoptotic signaling in a noncaspase-dependent signaling manner. Thus, it is clear that mitochondria are intimately involved in the initiation of the signaling cascades leading to apoptosis of a nucleus in aging skeletal muscle. However, it is interesting that nuclear apoptosis in skeletal muscle involves cell signaling that is so precise that specific individual myonuclei can be targeted for elimination in multinucleated skeletal myofiber without targeting other nuclei. While apoptosis signaling can be somewhat general in a single-nucleated nonmuscle cell as there is only one target in that cell, in multinucleated skeletal muscle, the apoptotic signaling requires amazingly precise targeting of some nuclei but not others. It is likely that local signaling from individual dysfunctional mitochondria will only target nuclei within its vicinity.

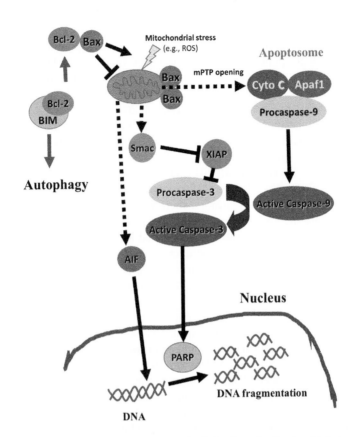

FIGURE 14.2 Mitochondria-induced nuclear apoptosis. Induction of mitochondrial stress (lightning bolt) results in dissociation of the anti-apoptotic B-cell lymphoma (Bcl)-2 with Bcl-2-associated X protein (Bax) and forming a Bax:Bax pore in the outer membrane and a mitochondrial pore (not shown) in the inner mitochondrial membrane. This opens the mPTP pore and allows mitochondria-housed contents to leak into the cytosol, forming an apoptosome, which activates caspase-9 and subsequently activation of the effector caspase-3. Activated caspase-3 enters the nucleus that is in proximity to the dysfunctional mitochondria, activates enzyme poly ADP ribose polymerase (PARP), which cleaves DNA. DNA fragmentation can also be caused by direct activation of mitochondria-housed components such as AIF. Mitochondria-housed second mitochondria-derived activator of caspase/direct inhibitor of apoptosis-binding protein with low pI (Smac/DIABLO) can promote caspase-9 activation via inhibition of the anti-apoptotic XIAP. Although not all DNA damage that occurs in this fashion will result in nuclear removal, sufficient damage will result in the elimination of the nucleus that is targeted by apoptosis. The intersections of extrinsic and endoplasmic reticulum pathways that exist for inducing apoptosis are not shown.

14.4.2 MITOCHONDRIAL PERMEABILITY TRANSITION ACCELERATES APOPTOSIS IN AGING MUSCLES

It has been well established that apoptosis occurs in aging skeletal muscle (Alway, Mohamed, and Myers 2017, Alway, Morissette, and Siu 2011, Alway, Myers, and Mohamed 2014, Alway et al. 2014, Wang, Mohamed, and Alway 2013, Chabi et al. 2008, Herbst et al. 2016, Beyfuss et al. 2018, Hood et al. 2015, Zhang et al. 2013, Ziaaldini et al. 2017, Calvani et al. 2013, Cheema et al. 2015, Alway et al. 2013) including human muscles (Gouspillou et al. 2014, Spendiff et al. 2016). Much of the greater susceptibility of aged skeletal muscle to apoptosis is related to the elevated mitochondrial sensitivity to open the mitochondria permeability transition pore (mPTP) (Xiao et al. 2018, Gouspillou et al. 2014, Spendiff et al. 2016), thereby releasing the mitochondrial contents to the cytosol to initiate apoptotic signaling (Ziaaldini et al. 2017, Picca, Calvani, Leeuwenburgh, et al. 2018, Nicassio et al. 2017, Marzetti et al. 2017).

It could be argued that inactivity that typically accompanies aging might explain some of the increased mPTP susceptibility for opening. However, mPTP opening in aging cannot be solely the function of inactivity, because increased mitochondrial susceptibility to permeability transition opening has also been observed in muscles from active humans (Gouspillou et al. 2014). Nevertheless, exercise, which should improve mitochondria number and function, can at least partially reverse aging-associated apoptosis as exercise training has been reported to decrease catabolic and apoptotic signaling in muscles of aged rodents (Ziaaldini et al. 2017, Ziaaldini, Hosseini, and Fathi 2017). It is also important to note that the susceptibility for mPTP opening is exacerbated by an imbalance of Ca^{2+} homeostasis that likely results from leaky ryanodine receptors that occur in aged skeletal muscle (Andersson et al. 2011, White et al. 2016, Musci, Hamilton, and Miller 2018, Hamilton et al. 2018). Sensitization of the mPTP in aging skeletal muscle may be an important contributor to the initiation of apoptosis and muscle loss leading to sarcopenia. Aging-associated muscle denervation may also contribute to increased mPTP that in turn induces muscle apoptosis and muscle loss (Karam et al. 2017, Lee 2016, Lee, Huttemann, and Malek 2016, Konokhova et al. 2016, Spendiff et al. 2016).

14.5 LOCALIZED APOPTOTIC SIGNALING SPREADS TO REMOVE THE ENTIRE FIBER IN SARCOPENIA

Failure to remove dysfunctional mitochondria by mitophagy results is an accumulation of damaged mitochondria, which together magnify the death-signaling pathway and accelerate sarcopenia (Figure 14.3). Thus, progression of the steps to eliminate muscle fibers leading to sarcopenia hinges upon whether dysfunctional mitochondria can be removed or if they are permitted to continue with their apoptotic death signaling cascade.

A model for the process by which single muscle fibers are eliminated in sarcopenia is shown in Figure 14.4. This model assumes that local mitochondrial damage (potentially via ROS, high calcium loads, mtDNA damage) causes dysfunctional and leaky mitochondria, which are not removed via mitophagy. The accumulation of dysfunctional mitochondria adversely affects other mitochondria within close proximity (via apoptotic signaling, increased Bax, etc.). These adjacent, healthy mitochondria will then have increased mPTP opening to accelerate the wave of apoptotic signaling. Thus, dysfunctional permeabilized mitochondria will initiate a local apoptosis cascade, which when left unchecked will result in widespread cellular destruction. This concept of regional cellular disruption that increases to encompass the entire muscle fiber is strongly supported by the elegant work done by Cheema and colleagues (Herbst et al. 2016). They show that regionalized degeneration of individual muscle fibers is associated with mitochondrial dysfunction (as indicated by dysregulation of succinate dehydrogenase and cytochrome c oxidase enzyme levels) in aging. Furthermore, once the apoptotic signals have been accelerated, and perhaps also by the contribution of more widespread ubiquitin initiated proteolysis and widespread autophagy (rather than selective mitophagy of dysfunctional mitochondria), it is possible that genes of necrosis may be activated to contribute to the overall campaign of cellular destruction (Herbst et al. 2016). This observation provides an important link to our overall hypothesis, by showing that regionalized atrophy occurs concurrent with mitochondrial dysfunction. We speculate that failure of mitophagy to remove these malfunctioning mitochondria initiates cell degradation and cell death pathways including nuclear apoptosis. The initial signaling occurs in a region around the damaged mitochondria, and when this signaling accumulates, the death signaling cascades in a local area of the fiber result in loss of nuclei and other cellular contents within this region. This model explains how specific nuclei might be targeted for apoptotic dismantling, while other nuclei are not.

As mitochondrial and contractile protein synthesis rates are depressed in aged muscles along with some level of disruption in proper mitochondrial protein import and translation, there is a reduced anabolic signal, which would result in localized atrophy even if degradation rates remained constant (but degradation increases with aging). Furthermore, an aging-induced reduction in satellite cell function contributes to the inability to repair this area of local damage, and the satellite

Induction of Mitophagy

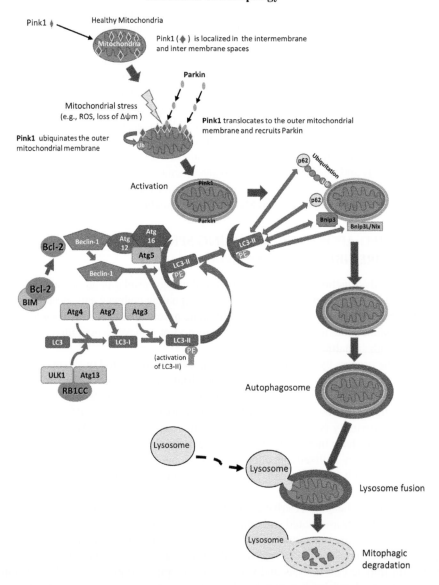

FIGURE 14.3 Induction of mitophagy. Pink1 is imported to mitochondria where it is localized to the intermembrane and intermembrane spaces or degraded under in healthy mitochondria under basal conditions. Excessive ROS or other stresses cause a reduction in the mitochondrial membrane potential. Upon mitochondrial depolarization, Pink1 is recruited to the outer mitochondrial membrane, and transport inside the mitochondria is prevented, resulting in an accumulation of Pink along the outer mitochondrial membrane (depicted by the ring around the outer mitochondrial membrane). The E3 ligase Parkin is recruited to insert into the outer mitochondrial membrane. Parkin ubiquitinates outer mitochondrial membrane proteins (shown by a ring around the mitochondria), resulting in recruitment of ubiquitin-binding autophagy receptors such as p62, Bnip3, and Bnip3L/Nix to the mitochondria. B-cell lymphoma (Bcl)-2 prevents the induction of autophagy by binding Beclin-1. Displacement of Bcl-2 leads to activation of the phagophore nucleation. Elongation of the membrane requires 2 ubiquitin-like conjugation systems: autophagy proteins (ATG) and activation of microtubule-associated protein 1 light chain 3 from its inactive form LC3-I to LC3-II where it is conjugated to phosphatidylethanolamine (PE) to form lipidated (activated) LC3-II (LC3-II-PE)

(Continued)

FIGURE 14.3 (Continued) LC3-II-PE. LC3-II-PE can bind autophagy receptors, which contain transmembrane domains and localize to the outer mitochondrial membrane. The p62 receptor contains a ubiquitin-binding domain, which localizes it to Parkin-ubiquitinated mitochondria. Furthermore, LC3-II-PE participates in autophagosomal membrane elongation (shown in light purple). Bnip3 and its homologue, Bnip3L/Nix, are also mitophagy receptors that, when expressed, localize to the outer mitochondria membrane. It is not clear if it is necessary, but some data suggest that p62 can also bind to ubiquitinated chains (Ub) that are attached to the mitochondria. LC3-II-PE binds to the autophagy receptors (e.g., p62, Bnip3, Bnip3L/Nix, and others; as depicted by the red double arrows), which then tethers the developing membrane of the autophagosome to the outer mitochondria membrane. LC3-II-PE participates in elongation of the double-layered membrane that surrounds the damaged mitochondria. After maturation of the autophagosome, it fuses with a lysosome, which injects its contents into the mitochondria, to degrade it.

Proposed model for local disassembly then more widespread removal of muscle fibers in sarcopenia

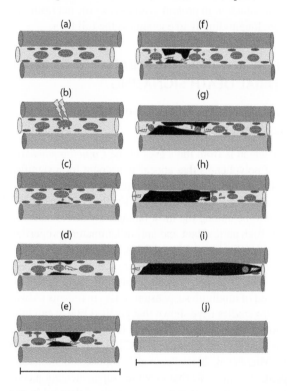

FIGURE 14.4 Hypothetical model for eliminating muscle fibers in sarcopenia via localized mitochondria-associated dysfunction—mitophagy and apoptosis. (a) In healthy muscle, activation of mitophagy eliminates dysfunctional nuclei so that they cannot continue death signaling. (b) Dysfunctional mitochondria that leak their contents to the cytosol will occur in muscle that has received a significant mitochondrial stress (e.g., ROS, inflammatory mediators). (c,d) This initiates the apoptotic signaling cascades. (e,f) If the dysfunctional mitochondria are not eliminated, apoptotic death signaling may be activated to eliminate myonuclei, and this may concurrently or independently result in wide-spread activation of autophagy and the ubiquitin ligase pathway and also trigger the necrosis signaling pathway to remove muscle proteins, mitochondria, and nuclei within the domain of the initial dysfunctional mitochondria (g,h). (i,j) If the death signaling continues to activate other mitochondria, this will increase the extent of dysfunctional mitochondria, which, in turn, will perpetuate signaling for apoptosis, which will remove nuclei, and signaling to eliminate contractile and non-contractile tissue in the fiber segment. This results in eventual elimination of the portion of the fiber and potentially the entire fiber if the signaling spreads unchecked throughout the fiber. We further hypothesize that the initiation of the disassembly and removal of the fiber could be blocked if the dysfunctional mitochondria that initiate the process are removed (e.g., via exercise) and replaced by healthy mitochondria.

cells themselves may be targets of cellular removal. If this cellular destruction is not corrected, the death signaling cascades may then spread to adjacent areas, resulting in more widespread fiber atrophy and cellular removal. With sufficient cellular destruction, the entire fiber may be eliminated. This model assumes that failure to regulate proper mitophagy is upstream of sarcopenic muscle fiber loss.

Ubiquitin ligases and general autophagy or cellular degradation occurs to remove proteins within that local area, which in turn results in localized atrophy. Although loss of a few nuclei will not result in death of the entire multinucleated fiber, widespread nuclear loss could lead to the removal of the entire myocyte if clearing of the dying mitochondria and the associated death signaling are not corrected. The final steps of degradation of proteins within the muscle fiber occur directly via lysosomal cathepsins and cytosolic protease calpains, which progress toward ATP-dependent ubiquitin–proteasome activation via FOXO3a-associated muscle-specific MuRF1 and MAFbx/Atrogin-1 regulation (Cecile et al. 2018, Lu et al. 2017, Rom and Reznick 2015, Cheema et al. 2015). Concomitant with the increased degradation is a reduction in protein synthesis, so the net result is loss of the fiber proteins first locally and then more longitudinally to encompass the full fiber.

14.6 POTENTIAL FOR NUTRITIONAL STRATEGIES TO REVERSE MITOCHONDRIAL DEATH SIGNALING

14.6.1 THE POLYPHENOL GREEN TEA EXTRACT REDUCES APOPTOSIS SIGNALING IN AGED MUSCLES

Green tea extract (GTE), which is from the leaves of the *Camellia sinensis* plant, contains high levels of catechins, a class of polyphenols that appear to account for a significant level of its biological activity. Green tea catechins that are of particular interest include epicatechin, gallocatechin, epigallocatechin, epicatechin-3-gallate, and epigallocatechin-3-gallate (EGCg). EGCg is by far the most abundant and is often believed to be among the most biologically active catechins in green tea, although this is largely based on its high antioxidant and anti-inflammatory properties (Zhu et al. 2017, James, Kennett, and Lambert 2018, Oliveira and Hood 2018, Oliveira et al. 2016). Green tea supplementation reduces oxidative stress and muscle damage after eccentric exercise upon reloading following a period of disuse (da Silva et al. 2018), and green tea catechins were shown to reduce the loss of soleus muscle force during a period of hindlimb suspension (HLS) in rodents (Alway et al. 2014, Alway et al. 2015). Furthermore, *in vitro* studies have shown that purified EGCg can inhibit serum starvation and staurosporine-induced apoptosis in human myoblasts (Zeng and Tan 2015), attenuate muscle protein breakdown, and increase muscle protein synthesis in C2C12 myotubes in response to serum starvation or incubation in TNF-α (Mirza et al. 2014). As oxidative stress, inflammation and muscle wasting are all hallmarks of sarcopenia, it is possible that green tea supplementation could suppress death-signaling pathways in sarcopenic muscle. Furthermore, apoptosis is elevated in response to muscle wasting conditions including muscle disuse and early recovery after disuse (Alway et al. 2014, Alway et al. 2015). Green tea was able to reduce muscle wasting and loss of function during hindlimb unloading in sarcopenic rats, but it did not provide any further improvement during the reloading period (Alway et al. 2014, Alway et al. 2015, Takahashi et al. 2017). As green tea has only ~50% EGCg, it is possible that either low levels of EGCg are needed to slow muscle wasting, or perhaps, along with EGCg, one or more of the other catechins that are present in green tea (Meador et al. 2015) may also have an effect on regulating muscle wasting during disuse in old rodents (Table 14.1).

The rationale behind a role for apoptosis is that a reduced number of nuclei and decreased DNA accretion during reloading (Alway et al. 2015, Alway, Mohamed, and Myers 2017, Alway, Myers, and Mohamed 2014, Mozdziak, Pulvermacher, and Schultz 2001, Alway et al. 2014) could prevent or at least contribute to the lack of muscle recovery during reloading with aging. We have found that the pro-apoptotic proteins AIF and Bax were elevated in response to reloading after HLS in muscles from vehicle-treated rats. This observation is similar to that from other

TABLE 14.1
Polyphenols Compounds Regulating Cellular Signaling in Sarcopenia

Nutritional Compound	Model	Tissue	Muscle Responses	Markers of Oxidative Stress	Regulators of Apoptosis	Regulators of Autophagy	Apoptosis Markers	Autophagy Markers	UPS Signaling	References
EGCg (1.5 mg/kg)	17 mo. BALB/cByJ mice	GAS		SOD1↓ HNE↓	PCG1-α↑ SIRT1↑	PCG1-α↑ SIRT1↑ FOXO3→				Pence et al. (2016)
EGCg (200 mg/kg diet)	20 mo. Sprague-Dawley rats	GAS	Muscle weight ↑		IGF-1↑ IL-15↑				MuRF1↓ MAFbox↓ Myostatin↑	Meador et al. (2015)
EGCg (50 mg/kg body weight)	32 mo. Fisher Brown Norway Rats	Plantaris and recovery after HLS	Muscle weight ↑ Force↑ Fiber CSA↑ Satellite cells↑				Bax↓ Caspase 9↓ Caspase 3↓ AIF↓ FADD↓ TUNEL↓			Alway et al. (2014)
GTE (50 mg/kg body wt.)	32 mo. Fisher Brown Norway Rats	Plantaris & Soleus—recovery after HLS	Fiber CSA↑ Satellite cells↑	SOD↓ 8-iso-PGF2α↓ protein carbonyls↓	p-AKT↑ AKT↑ GSK3-β↑		AIF↓ Bcl-xL↓ Bax↓			Alway et al. (2015)

studies (Takahashi et al. 2017, Hao et al. 2011, Bennett, Mohamed, and Alway 2013, Alway et al. 2014, Alway et al. 2015). GTE suppressed Bax protein abundance in reloaded plantaris and soleus muscles (Alway et al. 2015). Bcl-xL, a mitochondria associated anti-apoptotic protein, decreased with reloading in the soleus, although it was unchanged in plantaris muscles after HLS or reloading, and GTE did not change this response in either muscle after HLS or reloading. Nevertheless, the lower Bax abundance (or the greater Bcl-xL/Bax ratio) in muscles from old animals given GTE would be expected to promote a lower mitochondria-induced apoptotic environment than that in vehicle-treated muscles (Alway et al. 2015). While we have previously found that deletion of Bax-attenuated denervation-induced muscle wasting (Siu and Alway 2006), it was interesting to note that while GTE suppressed pro-apoptotic Bax protein abundance in plantaris and soleus muscles of old rats after reloading following HLS, this was unable to improve muscle recovery in the reloaded muscles compared with vehicle-treated animals (Table 14.1). While increases in individual or multiple signaling proteins in apoptotic pathways likely increase the susceptibility for apoptosis, such changes in signaling cannot be interpreted to mean that removal of nuclei (nuclear apoptosis) will absolutely occur. Moreover, the importance of apoptosis in contributing to a loss of nuclei in post-mitotic tissue is not clear, and apoptosis has not been reported universally in young muscle (Bruusgaard et al. 2012, Bruusgaard et al. 2010, Gundersen and Bruusgaard 2008), although this may differ in aging.

14.6.2 EGCg Reduces Signaling for Apoptosis in Aged Muscles

EGCg is the predominant catechin in green tea. EGCg has strong antioxidant and anti-inflammatory properties, and it is believed to be responsible for most of the health benefits linked to green tea. Both disuse and reloading greatly increase the oxidative stress in the affected muscles of rodents (Andrianjafiniony et al. 2010, Casanova et al. 2014, McArdle et al. 2018, Powers et al. 2011, Liu et al. 2014, Jackson, Ryan, and Alway 2011, Ryan et al. 2011). Therefore, reducing the high basal levels of oxidative stress in aging could potentially reduce muscle loss during disuse conditions and/or improve muscle recovery during reloading after HLS or after exercise in muscles from old animals (Ryan et al. 2011, Alway et al. 2015, Jackson, Ryan, and Alway 2011, Jackson et al. 2010). Oxidative stress is reduced after eccentric exercise upon supplementation with green tea catechins (Haramizu et al. 2011, Ota et al. 2011). Furthermore, green tea catechins reduce the loss of soleus muscle force during a period of HLS in mice (Ota et al. 2011).

EGCg lowers myonuclear apoptosis in the hindlimb muscles of aged rats in response to disuse and improves muscle recovery following reloading (Table 14.2). In these studies, old Fischer 344 x Brown Norway inbred rats (age 34 mo.) that were sarcopenic received either EGCg (50 mg/kg body weight) or water daily by gavage (Alway et al. 2014). One group of animals received HLS for 14 days, and a second group of rats received 14 days of HLS, then the HLS was removed, and they recovered from this forced disuse for 2 weeks. Animals that received EGCg over the HLS period followed by 14 days of recovery had a 14% greater plantaris muscle weight ($p < 0.05$) compared with the animals treated with the vehicle over this same period. Plantaris fiber area was greater after recovery in EGCg (2715.2 ± 113.8 µm^2) vs. vehicle-treated animals (1953.0 ± 41.9 µm^2). Importantly, compared with the vehicle treatment, greater Bcl-2 and lower Bax protein abundance (Table 14.2) explain at least part of the lower apoptotic potential in the plantaris of EGCg-treated animals during reloading. The apoptotic index was in fact lower (0.24% vs. 0.52%), and the abundance of pro-apoptotic proteins Bax (–22%) and Fas-associated protein with death domain (FADD) (–77%) was lower in EGCg-treated plantaris muscles after recovery. Lower AIF and lower FADD abundance likely contributed to preserving the number of new nuclei (activated myogenic precursor cells) that survived during the reloading period. Presumably, maintenance of a greater pool of these myogenic precursor cells in our model (e.g., satellite cells), which can be activated during periods of muscle reloading after atrophy, should improve the ability for muscle to recover from muscle atrophy.

TABLE 14.2

Resveratrol Regulation of Cellular Signaling in Sarcopenia

Nutritional Compound	Model	Tissue	Muscle Responses	Markers of Oxidative Stress	Regulators of Apoptosis	Apoptosis Markers	References
RSV (125 mg/kg/day)	32 mo. Fisher Brown Norway Rats	Plantaris, recovery after HLS	Force → Fatigue ↓ Fiber CSA → Satellite cell proliferation →		p-AMPK ↑ AMPK ↑ PCG1-α ↑ SIRT1→	Bcl-2 ↑ Bax ↑ Bcl-xL → Cleaved caspase 3 ↓ Cleaved caspase 9 ↓	Bennett, Mohamed, and Alway (2013)
RSV (50 mg/kg/day), 6 wk.	27 mo. Fisher Brown Norway Rats	WG	Mitochondrial respiration→		SIRT1↑	Bcl-2 ↑ Bax ↑ Bcl-2/Bax → Cleaved DNA →	Joseph, Malamo, et al. (2013)
RSV (0.05%; ~46–60 g/kg/d), 10 mo.	18 and 28 mo. C57BL/6 mice	GAS & Plantaris	Mass → Force →	MnSOD↑ CuZnSOD↑ H_2O_2 ↓ HNE ↓ Protein carbonyls → Nuclear 8-OHdG →	SIRT1↑ PCG1-α → Citrate synthase → Cytochrome c →		Jackson, Ryan, and Alway (2011)
RSV (500 mg/d)—12 wks. + exercise	Men and women 65–80 yrs.	Vastus lateralis	Torque ↑ Power ↑ Fiber CSA ↑ Total myonuclei ↑ Satellite cells ↑ COX10 ↑			BAK1 ↓	Alway et al. (2017)
RSV (50 mg/kg/day) + 20% CR—6 wk.	27 mo. Fisher Brown Norway Rats	WG	Mitochondrial respiration →	SIRT1 →	SIRT1 →	Bcl-2 ↑ Bax ↑ Bcl-2/Bax → Cleaved DNA →	Joseph, Adhihetty, et al. (2013)

While EGCg did not prevent unloading-induced atrophy, it improved muscle recovery after the atrophic stimulus in fast plantaris muscles from rats with sarcopenia. These data represent potentially important observations with clinical implications for the population of elderly persons who suffer from acute disuse (e.g., hospitalization) and then go through some period of rehabilitation in an attempt to recover function. Subsequent studies should be conducted to test if like our observations in old rats, elderly humans have similar benefits from consuming EGCg during a period of rehabilitation following hospitalization or other disuse. Daily ranges of EGCg between 400 and 800 mg have been reported to be safe and mild enough to be consumed by humans with gastric ulcers (Clifford, van der Hooft, and Crozier 2013), and ranges up to 3000 mg of green tea catechins have been used in human studies apparently without negative side effects (Kim et al. 2017, Kim et al. 2006, Kim, Quon, and Kim 2014). Assuming an 80 kg human, 3000 mg/d would be equivalent to 37.5 mg/kg of EGCg, but this is less than the 50 mg/kg used in our study in rats. It is possible that very high doses could have undesired effects, because mice that consumed a diet that was very high in EGCg (1% w/w) showed evidence of increased inflammation (Pae, Meydani, and Wu 2012, Pae et al. 2012). Thus, while EGCg shows significant promise, the optimal doses of EGCg should be established in elderly humans to obtain the desired biological effects that improve muscle mass after periods of disuse, without incurring undesired side effects.

14.6.3 Resveratrol Activation of SIRT1 Reduces Apoptosis and Oxidative Stress in Aged Muscles

Resveratrol (3,5,4′-trihydroxystilbene; RSV) is found largely in the skins of red grapes. RSV has been shown to inhibit protein degradation and attenuate atrophy of skeletal muscle fibers in several *in vitro* studies (Asami et al. 2018, Wang, Sun, Song, et al. 2018, Dugdale et al. 2018, Sun et al. 2017). A relatively high dose (400 mg/kg/day) of RSV *in vivo* has been shown to attenuate slow muscle fiber atrophy following HLS (Momken et al. 2011) in rodents (Table 14.2). In agreement with these data, we have found that a low dose (12.5 mg/kg/day) of RSV (Ryan, Jackson, et al. 2010, Jackson et al. 2010) had a trend ($p = 0.06$) to blunt fast muscle losses during HLS-induced muscle wasting. Our lab tested the possibility that RSV has the potential to improve recovery of muscle after wasting conditions by reducing apoptotic signaling in aging (Bennett, Mohamed, and Alway 2013, Alway et al. 2017). This seems plausible because RSV has been shown to improve noncancer cell viability, reduce apoptosis (Liang et al. 2018, Haramizu et al. 2017), and promote stem cell proliferation and muscle cell repair (Haramizu et al. 2017, Alway, Mohamed, and Myers 2017, Alway et al. 2017), but this is cell type and dose dependent. Sirtuin 1 (SIRT1), a presumed target of RSV, has been reported to increase proliferation of skeletal muscle stem cells/myoblasts in culture (Wang, et al. 2016, Alway et al. 2017), whereas RSV treatment has been found to increase differentiation of myoblasts rather than induce cell proliferation *in vitro* (Dugdale et al. 2018, Rathbone, Booth, and Lees 2009). It is possible that RSV will function differently in animal and human cells and in skeletal muscle satellite cells *in vivo* compared to *in vitro*.

In sarcopenic rodents, the pro-apoptotic Bax protein abundance was lower in RSV than vehicle-treated plantaris muscles (Bennett, Mohamed, and Alway 2013). In a similar pattern, pro-apoptotic proteins cleaved caspase 3 and cleaved caspase 9 were increased with muscle disuse, but the increases in cleaved caspase 9, a mitochondria-associated pro-apoptotic protein, was suppressed by RSV. RSV suppressed both cleaved caspase 3 and cleaved caspase 9 in the plantaris muscles of the animals that were allowed to recover from disuse compared with the vehicle-treated group (Table 14.2). Bcl-2 was elevated in a similar fashion in vehicle-treated and RSV-treated plantaris muscles during HLS (Bennett, Mohamed, and Alway 2013). During recovery, the Bcl-2 protein abundance returned to control levels in the vehicle-treated plantaris muscle, but it remained elevated in plantaris muscles that were treated with RSV. The protein abundance of the anti-apoptotic Bcl-xL was significantly increased in the RSV group following HLS, while no changes were observed in the vehicle-treated group (0.20 ± 0.16) compared with the cage control (0.28 ± 0.11) animals (Bennett, Mohamed, and Alway 2013). During recovery, the abundance of Bcl-xL was elevated in muscles from both old vehicle (0.83 ± 0.11) and old RSV (0.91 ± 0.09)-treated animals, although there were no differences observed between these two groups (Bennett, Mohamed, and Alway 2013).

During recovery after disuse in old sarcopenic rats, RSV tended to increase in the apoptotic index by 29% in the vehicle-treated animals (1.35 ± 0.49) compared with the RSV-treated animals (0.96 ± 0.50). The increases in Bcl-2 and the reductions in Bax, cleaved caspase 3 and cleaved caspase 9, create a muscle environment that is less favorable for apoptosis, and this may have prevented the loss of some nuclei. More nuclei that are available for transcription may improve muscle recovery, especially in type II fibers, after disuse. These data support the idea that nutritional intervention with RSV that might both reduce oxidative stress and apoptotic signaling and therefore be beneficial for reducing sarcopenic effects in aging (Alway et al. 2017).

Dietary supplementation with RSV has the potential to exert beneficial effects both through its ability to scavenge free radicals (Simioni et al. 2018, Xia et al. 2017) and to upregulate the endogenous antioxidant system (Alway et al. 2017) and by its capacity to modulate the signal transduction and gene expression of pathways regulating cellular proliferation (Rathbone, Booth, and Lees 2009, Ryan, Dudash, et al. 2010), mitochondrial biogenesis (Muhammad and Allam 2018, Wang, Sun, Song, et al. 2018), metabolism (Muhammad and Allam 2018, Truong, Jun, and Jeong 2018), and apoptosis (Haramizu et al. 2017).

Specifically, RSV administration significantly increased MnSOD activity and protein content but not CuZnSOD activity in muscles of old HLS animals (Ryan, Jackson, et al. 2010, Jackson et al. 2010). This is analogous with data showing that RSV protected against oxidative stress through the specific induction of MnSOD (Gan et al. 2016, Xu et al. 2012). Moreover, we have found that RSV administration increased the content and activity of catalase in muscles from old nonsuspended animals, but it did not increase catalase activity or expression further following HLS (Jackson, Ryan, and Alway 2011, Jackson et al. 2010). This may be in part because catalase activity is already increased with aging and HLS, so it may have reached a maximal point of induction. Perhaps most importantly, RSV administration reduced indices of oxidative stress in gastrocnemius muscles of old HLS animals as estimated by H_2O_2 levels and the lipid peroxidation by-products (MDA and HNE). This is congruent with data from our laboratory and others that have shown that RSV protects against H_2O_2-mitigated lipid peroxidation *in vivo* (Ryan, Jackson, et al. 2010, Jackson et al. 2010) and *in vitro* (Brito et al. 2006). The induction of catalase by RSV has previously been demonstrated (Ryan, Jackson, et al. 2010) and, along with increases in MnSOD, may represent important mechanisms by which RSV acts to reduce H_2O_2 and H_2O_2-mediated damage. However, these protective effects of increases in antioxidant enzyme activity and concomitant decreases in markers of oxidant load were not seen in young animals administered RSV, suggesting that there is a differential effect of RSV in young and old animals. The fact that RSV only seems to have an effect in old animals is interesting and is possibly due to different underlying signaling mechanisms that may occur in old animals during disuse. It is also probable that younger animals are able to handle the stress of HLS and the subsequent detrimental effects that are associated with skeletal muscle disuse, and therefore the preconditioning effect of RSV administration helps to augment the oxidative stress response in mitochondria from old but not young animals where it is not needed. This is congruent with our finding that H_2O_2 concentrations were not increased in muscles from young animals following HLS. Although lipid peroxidation markers were increased in muscles from young suspended animals, despite no increases in H_2O_2 concentrations, this might indicate a temporal role of oxidative stress in young animals during muscle disuse (Table 14.2).

In looking at the effects of RSV in humans, older men ($n = 12$) and women ($n = 18$) 65–80 years of age who completed exercise and took either a placebo or an RSV (500 mg/d) were evaluated to test the hypothesis that RSV treatment combined with exercise would increase mitochondrial density, muscle fatigue resistance, and cardiovascular function more than exercise alone

(Alway et al. 2017, Jackson et al. 2010). Twelve weeks of aerobic and resistance exercise coupled with RSV treatment did not further improve cardiovascular risk, indices of mitochondrial density, or muscle fatigue resistance more than placebo and exercise-only treatments. However, RSV mediated an increase in knee extensor muscle peak torque (8%), average peak torque (14%), and power (14%) after training, whereas exercise did not increase these parameters in the placebo-treated older subjects (Alway et al. 2017). Nevertheless, the impact of RSV on sarcopenia was quite interesting. Exercise combined with RSV significantly improved mean fiber area and total myonuclei by 45.3% and 20%, respectively, in muscle fibers from the vastus lateralis of older subjects (Alway et al. 2017). Together, these data indicate a novel anabolic role of RSV in exercise-induced adaptations of older persons. However, it might be that for RSV to impact sarcopenia; it must first have an underlying anabolic stimulus such as exercise. If a "priming" stimulus is necessary, it might be the case that RSV cannot mimic exercise but rather requires exercise to be effective. Thus, RSV combined with exercise might provide a better approach for reversing sarcopenia than exercise alone (Table 14.2).

14.6.4 BETA-HYDROXY-BETA-METHYLBUTYRATE REDUCES MUSCLE MASS LOSS AND MYONUCLEAR APOPTOSIS IN SARCOPENIA

It should be noted that other compounds have also been shown to reduce apoptotic signaling in the HLS model of unloading and reloading after HLS. This includes beta-hydroxy-beta-methylbutyrate (HMB) (Hao et al. 2011), and therefore, mitochondrial regulation of apoptotic signaling might be a good target to reduce muscle wasting and/or improve muscle recovery after disuse in aging. HMB is a metabolite of the essential branched-chain amino acid leucine. HMB has been found to reduce muscle wasting associated with trauma and cancer cachexia (Holecek 2017, Holecek and Vodenicarovova 2018, Cruz-Jentoft 2018, Hasselgren 2014). Furthermore, HMB supplementation has been reported to attenuate fiber atrophy and damage during limb immobilization of adult rodents (Yakabe et al. 2018) and improve recovery after disuse in aged rats (Hao et al. 2011, Alway et al. 2013), reduce glucocorticoid-induced muscle loss (Aversa et al. 2012), and improve muscle mass and function in elderly subjects (Rossi et al. 2017, Woo 2018, Hsieh et al. 2010). HMB has been shown to reduce apoptosis in human myoblasts under conditions of serum starvation or staurosporine (Rossi et al. 2017, Kornasio et al. 2009). HMB supplementation has also been shown to lower myonuclear apoptosis in the hindlimb muscles of aged rats in response to disuse and reloading following HLS-induced muscle disuse (Hao et al. 2011, Alway et al. 2013). This was associated with reduced myonuclear apoptosis and abundance of pro-apoptotic proteins Bax and caspase-3 in skeletal muscle (Table 14.3).

HMB supplementation has been previously shown to blunt muscle loss in critically ill patients (Hsieh et al. 2006, Hsieh et al. 2010) as a result of reduced proteasome activity (Giron et al. 2015). Moreover, there is evidence that HMB has the potential to improve strength and muscle mass in elderly men and women (Malafarina et al. 2017, Holecek 2017). HMB can buffer oxidative stress in muscle cells (Chodkowska et al. 2018) and suppress oxidative stress-induced apoptotic signaling pathways including elevated caspase-3 that are induced by lipopolysaccharide-initiated oxidative stress (Russell and Tisdale 2009) and circulating TNF-α is reduced in human subjects who exercised and also took HMB (Townsend et al. 2013). HMB treatment suppresses apoptosis in both type I and type II fibers, but it may have been more effective at suppressing apoptotic pathways associated with muscle loss in the type II fibered plantaris compared with pathways regulating muscle loss in the type I-fibered soleus muscle (Table 14.3).

Additional studies are needed to determine whether HMB affects apoptosis by stabilizing the function of mitochondria in skeletal muscle under conditions of disuse and reloading. Although apoptosis is an important signaling process that occurs during unloading, clearly apoptosis is not the only contributor to muscle wasting, especially in the soleus muscle where muscle loss is typically more severe than in the plantaris during disuse.

TABLE 14.3

Beta-Hydroxy-beta-methylbutyrate Regulation of Cellular Signaling in Sarcopenia

Nutritional Compound	Model	Tissue	Muscle Responses	Regulators of Apoptosis	Apoptosis Markers	References
HMB (340 mg/kg) 35 days	34 mo. Fisher Brown Norway Rats	Soleus, reloading after HLS	Fiber CSA → Satellite cells →	p-AKT → mTOR → 4EPBP1 → p-4EBP1 →		Alway et al. (2013)
HMB (340 mg/kg), 35 days	34 mo. Fisher Brown Norway Rats	Plantaris Reloading after HLS	Force ↑ Fiber CSA ↑ Satellite cells ↑ MyoD labeling ↑ Id2 ↑	p-AKT → mTOR → 4EPBP1 → p-4EBP1 →		Alway et al. (2013)
HMB (40 mg/kg body wt.)	34 mo. Fisher Brown Norway Rats	Plantaris Reloading after HLS	Force ↑ Fiber CSA ↑		TUNEL ↓ Bcl-2 ↑ Bax Bcl-2 ↓ Bax ↓ Cleaved caspase 3 ↓ Cleaved caspase 9 ↓	Hao et al. (2011)
HMB (40 mg/kg body wt.)	34 mo. Fisher Brown Norway Rats	Soleus Reloading after HLS	Force ↑ Fiber CSA ↑		TUNEL ↓ Bcl-2 ↑ Bax Bcl-2 ↓ Bax ↓ Cleaved caspase 3 ↓ Cleaved caspase 9 ↓	Hao et al. (2011)
HMB (2 g/d), arginine, (2 g/d) lysine (1.5 g/d)	Women 65–90 yrs		Functional tests ↑ Muscle Force ↑ Protein Synthesis ↑			Flakoll et al. (2004)
3 g HMB, 14 g arginine, 14 g glutamine, 6 mo.	Men and women 65–89 years		Total lean mass ↑			Ellis et al. (2018)

14.6.5 Antioxidant Therapies to Reduce Mitochondria-Induced Apoptosis Signaling in Sarcopenia

Mitochondria are targets for ROS damage and are high producers of ROS in aging muscles (Gomez-Cabrera et al. 2012, Yeo et al. 2018, Olaso-Gonzalez et al. 2014, Kang, Yeo, and Ji 2016). Reducing mitochondrial ROS production attenuates mitochondria oxidative damage and improves mitochondrial function (Javadov et al. 2015).

Vitamin E (i.e., α-tocopherol) and vitamin C (i.e., ascorbic acid) are antioxidants that appear to have a protective effect by either reducing or preventing oxidative damage in skeletal muscle of old animals. Lipid-soluble vitamin E prevents lipid peroxidation chain reactions in cellular membranes by interfering with the propagation of lipid radicals. Vitamin C and E antioxidant supplementation improved indices of oxidative stress, mass, and work output in dorsiflexor muscles of aged animals associated with repetitive loading exercise and aging (Ryan, Dudash, et al. 2010) (Table 14.4). Thus, antioxidants may provide a good strategy for reducing ROS damage and improve recovery from sarcopenia especially under conditions of loading or exercise where we might expect an acute elevation in ROS accumulation.

Like other membranes, mitochondrial membrane composition is affected by the content and type of dietary lipids. Dietary intake of long-chain *n*-3 polyunsaturated fatty acids (PUFAs) is widely recommended for cardiovascular health. However, PUFAs are highly prone to oxidation, producing potentially deleterious 4-hydroxy-2-alkenals. Thus, cell membranes with a high PUFA content are more susceptible to oxidative damage, and therefore diets with a low PUFA are more likely to have membranes that will resist ROS damage. Nevertheless, dietary consumption of fish oil that accompanied caloric restriction was shown to improve muscle fiber cross-sectional area and reduce apoptotic signaling or caspase-3, caspase-8/10, caspase 9, and Bax and an increase Bcl-2/Bax ratio in 10-week-old C57BL/6 mice (Lopez-Dominguez et al. 2013). The lower cytosolic cytochrome *c* levels with dietary addition of fish oil is consistent with a reduced mPTP opening. Together these data suggest that dietary fats such as fish oil can reduce mitochondria-associated apoptotic signaling in muscles from young animals (Lopez-Dominguez et al. 2013), and we would anticipate that dietary fats should also improve mitochondrial membranes in muscles from old animals and attenuate apoptotic signaling in sarcopenia.

However, studies in humans are more mixed. For example, a study that provided a diet with a high PUFA level that included 660 mg eicosapentaenoic acid (EPA), 440 mg docosahexaenoic acid (DHA), 200 mg other omega-3 fatty acids, and 10 mg of vitamin E during or immediately after meals (Krzyminska-Siemaszko et al. 2015) found no differences in muscle mass or performance of elderly subjects compared with subjects that did not take the PUFA diet (Table 14.4). In contrast, healthy, but inactive, elderly individuals (mean age: 71.0 years) that consumed a PUFA diet of 1.86 g EPA and 1.5 g DHA daily for 8 weeks improved muscle protein synthesis (Smith et al. 2015). Furthermore, men and women between the ages of 60–85 years who were given a fish oil-derived PUFA diet (1.86 g EPA and 1.50 g DHA had greater thigh muscle mass, hand grip strength, knee extension torque, muscle power than placebo fed subjects (Smith et al. 2015; Table 14.4). This suggested that antioxidants could reduce or reverse sarcopenic muscle loss. The importance of antioxidants could also be due in part to their actions on satellite cells. This is because PUFA supplementation was shown to reduce TNF-α-induced inflammation, caspase-3 activity in the apoptotic pathway, reduce cell death, and improve muscle regeneration and muscle transcription factors that regulate muscle regeneration under a high fat (palmitate)-induced conditions in muscle cell cultures. However, we recognize that *in vitro* data exposures may not be the same as *in vivo* exposures to the antioxidants in animals or humans. Nevertheless, the data are supportive of the idea that antioxidants provide a potential improve to the muscle environment and might reduce the impact of sarcopenia in aging, especially in conditions, which recruit satellite cells such as loading exercises.

The importance of reducing ROS production has been demonstrated in old rats that were given a mitochondria-targeted ROS and electron scavenger, XJB-5-131 (XJB), and ROS damage to

TABLE 14.4
Antioxidant Regulation of Cellular Signaling in Sarcopenia

Nutritional Compound	Model	Tissue	Muscle Responses	Markers of Oxidative Stress	Regulators of Apoptosis	Apoptosis Markers	References
XJB, 4 wks.	29 mo. old F344/ BN rats	GAS	Muscle Vo ↑ Power ↑ Mitochondria complex activity ↑	Muscle & mitochondria Carbonyl levels ↓ MnSOD →			Javadov et al. (2015)
Dietary fish oil (PUFAs: 18% EPA, 12%) + CR	10-week-old C57BL/6[a]	GAS	Fiber CSA	Lipid hydro-peroxides ↓		Caspase-3 ↓ Caspase ↓ 8/10 Caspase-9 Bax ↓ BCl-2 BCl-2/Bax ↑ AIF ↓ Cytosolic cytochrome C ↓	Lopez-Dominguez et al. (2013)
5% sodium butyrate for 10 mo.	C57Bl/6J mice. Aged 26 mo.	GAS	Grip strength →	Catalase ↑ SOD1 ↑ Carbonyls ↓		DNA fragmentation ↓ XIAP ↓	Walsh et al. (2015)
Cystine (3mg/kg/d)	C57BL6 mice aged 17 mo. (middle aged)	GAS	Muscle mass ↑ Fiber CSA ↑ Serum IL6 ↓ FAS ↓	MnSOD ↑ GSH/GSSG ↑ HNE ↓	AMPK ↑ p-AMPK ↑ p-AKT ↑	TUNEL ↓	Sinha-Hikim et al. (2013)

(Continued)

TABLE 14.4 (*Continued*)

Antioxidant Regulation of Cellular Signaling in Sarcopenia

Nutritional Compound	Model	Tissue	Muscle Responses	Markers of Oxidative Stress	Regulators of Apoptosis	Apoptosis Markers	References
Vitamin E (dl-alpha tocopherol acetate; 30,000 mg/kg) and Vitamin C (l-ascorbic acid; 2% by weight) E and 0% Vitamin C.	F344/BN rats, 30 mo.	TA	Muscle mass ↑ Work ↑	GSH ↑ GSH/GSSG ↑ HNE ↓ DNA damage ↓ GPx activity ↑ GPX-1↑ Catalase activity ↑ Catalase protein ↑ SOD 1 ↑			Ryan, Dudash, et al. (2010)
1.3 g of PUFA and 10 mg of vitamin E	Men and women aged 75 yrs.	Thigh	Muscle mass → Power → Strength →				Krzyminska-Siemaszko et al. (2015)
1.86 g EPA and 1.5 g DHA	Men and women aged 71 yrs.	Thigh	Muscle mass ↑ Power ↑ Strength ↑ Protein Synthesis				Smith et al. (2011,2015)

a Not from aging studies

Tables 14.1 through 14.4 abbreviations: Protein kinase B (AKT); apoptosis-inducing factor (AIF); AMP-activated protein kinase (AMPK); 4-hydroxynonenal (HNE, lipid marker for oxidative stress); BCL2 antagonist/killer 1 (BAK1); caloric restriction (CR); CuZn-SOD, manganese superoxide dismutase 1 (SOD1); eukaryotic initiation factor 4E-binding protein 1 (4E-BP1); total superoxide dismutase (SOD) activity (CuZn-SOD, manganese-SOD, an extracellular-SOD); MnSOD-isoprostaglandin F(2α) (8-iso-PGF2α); Fas-associated protein with death domain (FADD); fatty acid synthesis (FAS); forkhead box protein O3 (FOXO3); Bcl-2-associated X protein (Bcl-2); reduced glutathione (GSH); oxidized glutathione (GSSG); gastrocnemius (GAS); mammalian target of rapamycin (mTOR); mammalian target of rapamycin complex 1 (MTORC1); muscle RING-finger protein-1 (MuRF-1); muscle atrophy F-box (MAFbx/Atrogin-1) tibialis anterior (TA); hindlimb suspension (HLS); beta-hydroxy-beta-methylbutyrate (HMB); insulin growth factor-I (IGF1); interleukin-15 (IL15: transcription factor A, mitochondria (TFAM); terminal deoxynucleotidyl transferase dUTP nick end labeling (TUNEL) a marker for DNA fragmentation in nuclear apoptosis; mitochondria-targeted electron scavenger XJB-5-131 (XJB); maximal unloaded shortening velocity (Vo); X-linked inhibitor of apoptosis protein (XIAP).

mitochondria and skeletal muscle protein were reduced, while complexes I, III, and IV activity of mitochondria were improved in aged muscles. Maximal muscle shortening velocity (Vo) and muscle power are reduced with aging in skeletal muscle from old animals or humans (Alway 2002, 1994a,b, Power et al. 2016, Saini et al. 2017), but XJB improved Vo without improving muscle mass in aging (Javadov et al. 2015). These observations suggest that preserving mitochondria function by reducing ROS improves muscle function in sarcopenia.

A cysteine-based antioxidant given to middle-aged mice reduced the oxidative redox environment of the gastrocnemius muscle as shown by greater MnSOD, GSH/GSSG ratios, and HNE levels of lipid peroxidation (Sinha-Hikim et al. 2013). The cysteine antioxidant also reduced serum inflammatory cytokine IL6 levels and improved apoptotic regulators, AMPK, p-AMPK, and p-AKT, and markedly reduced TUNEL-positive apoptotic nuclei in muscles of middle-aged mice (Sinha-Hikim et al. 2013; Table 14.4). Although these studies were done in middle-aged mice, it is likely that antioxidants provide a powerful tool for reducing ROS-mediated apoptosis and thereby reduce sarcopenia in aging.

14.7   CONCLUSIONS

The data summarized in this chapter build on the hypothesis that optimal mitochondrial function, health, and mitochondrial biogenesis are important for maintaining overall muscle mass and function. Dysfunctional mitochondria contribute to muscle wasting in aging. If this is true, optimizing nutrition (e.g., antioxidants, SIRT1 activators, HMB) could initiate selective targeting and removal of the dysfunctional mitochondria via increasing the abundance of mitophagy-specific proteins and thereby activating mitophagy to reduce apoptotic signaling. Some nutritional interventions might need an underlying anabolic signal to be in place ("priming" the system) such as that invoked from exercise, before they become effective for reducing sarcopenia.

A hypothetical model for the potential intervention sites in the process of regulating mitochondria-initiated signaling for sarcopenia is shown in Figure 14.5. Gray arrows identify the nutritional compounds and their corresponding signaling.

The source of most mitochondrial damage is the induction of ROS. Antioxidants such as vitamins C, E, and PUFUs can, in turn, induce or enhance intrinsic antioxidants including MnSOD, catalase, GPx1, and increase the GSH/GSSG ratio (Javadov et al. 2015, Sinha-Hikim et al. 2013). The inclusion of antioxidants also results in less oxidant damage to DNA, proteins, and lipids including the lipid membranes of mitochondria (Sinha-Hikim et al. 2013) that reduce mitochondrial damage and ROS production. In addition, RSV increases antioxidants such as MnSOD to reduce the oxidant burden of mitochondria (Ryan, Jackson, et al. 2010, Jackson, Ryan, and Alway 2011). EGCg increases signaling for autophagy through PGC1-α, SIRT1, and FOXO3a to induce mitophagy removal of mitochondria that are irreparably damaged (Chen et al. 2017, Zhao et al. 2017, Pence et al. 2016, Alway et al. 2015, Jackson et al. 2010).

A potentially important role for nutritional supplements to treat sarcopenia is by removing damaged mitochondria and potentially repairing or reversing the damage in repairable mitochondria. EGCg, RSV, GTE, and HMB have been shown to reduce mPTP opening, and reduce the apoptotic signaling cascade, which results in a reduction of DNA fragmentation and TUNEL-positive nuclei (Bennett, Mohamed, and Alway 2013, Haramizu et al. 2017, Alway et al. 2013). RSV also increases antioxidants and induces mitochondrial biogenesis (Jackson, Ryan, and Alway 2011, Alway et al. 2017) to replace unhealthy mitochondria that were removed via mitophagy, and this slows muscle wasting. In addition, EGCg, GTE, HMB, and RSV all act to reduce DNA fragmentation (Haramizu et al. 2017, Alway, Mohamed, and Myers 2017, Alway et al. 2015, Alway et al. 2014, Joseph, Malamo, et al. 2013), and EGCg reduces atrophic signaling from the ubiquitin–proteasome system (UPS) signaling to reduce the extent of muscle atrophy in aging (Alway et al. 2014, Meador et al. 2015). Thus, in total, sarcopenia may be slowed because muscle atrophy is

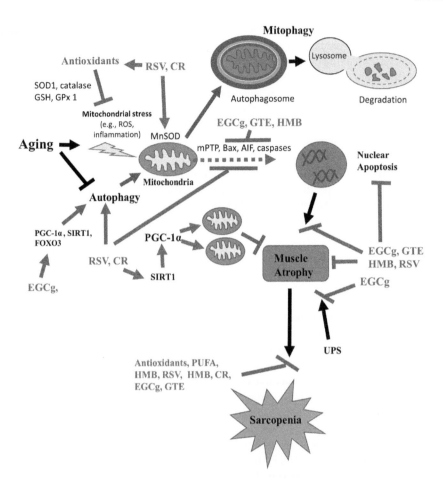

FIGURE 14.5 Nutritional targeting of mitochondria-initiated sarcopenia. Nutritional interventions target different sites along the pathways leading to sarcopenia. Gray arrows identify the nutritional compounds and their corresponding sites where they influence signaling. The site of most mitochondrial damage is the induction of ROS. Antioxidants such as vitamins C, E, and PUFUs can reduce oxidant damage to DNA, proteins, and lipids including the lipid membranes of mitochondria. Resveratrol (RSV) and caloric restriction (CR) increase antioxidants to reduce the oxidant burden of mitochondria. EGCg increases signaling for autophagy. EGCg, GTE, and HMB reduce mPTP opening, and the corresponding apoptotic signaling cascade, which results in a reduction of DNA fragmentation. EGCg reduces UPS signaling to reduce the extent of muscle atrophy in aging.

reduced by nutritional intervention especially if the anabolic signal from loading or exercise first primes the muscle for improved responses to nutritional intervention (Figure 14.6).

Nutritional interventions coupled potentially with exercise could increase selective mitophagy targeting of dysfunctional mitochondria by elevating Bcl-2 and 19 kDa-interacting protein-3 (Bnip3) levels. Bnip3 functions as mitophagy receptor to recruit autophagosomes selectively for clearance. Alternatively, nutrition exercise-induced Pink1 activation of DRP1-associated fission machinery (Pryde et al. 2016) in response to mitochondrial damage (Eid and Kondo 2017) or depolarization (Fallaize, Chin, and Li 2015) would also be expected to result in targeted mitophagy. Indeed, exercise has been shown to increase DRP1 and Bnip3 (Estebanez et al. 2019, Zhao et al. 2018, Yuan and Pan 2018), which suggests a potential mechanism whereby exercise targets dysfunctional mitochondria for elimination. Once mitophagy has been initiated, exercise and nutritional interventions together elevate the abundance of general autophagy proteins, which accelerate the removal of dysfunctional mitochondria and other proteins.

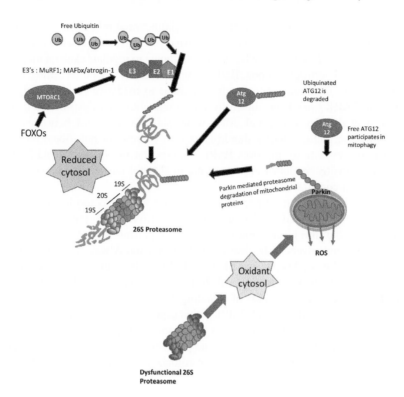

FIGURE 14.6   Ubiquitin–proteasome system. Free ubiquitin conjugates to form a polyubiquitin chain and targets damaged proteins in skeletal muscle. Targeting damaged proteins for proteolysis by the 26S proteasome occurs through Muscle RING Finger 1 (MuRF1) and MAFbx/Atrogin, which are muscle-specific E3 ubiquitin ligases. The 19S subunits recognize and bind the ubiquitinated proteins and begin their adenosine triphosphate (ATP)-dependent disassembly and removal in the 26S proteasome. Proteasome degradation results in small peptides that are further digested into amino acids, while lysosomal degradation directly produces single amino acids. Protein degradation from healthy proteasomes results in a reduced cytosolic environment, but dysfunctional proteasomes induce a highly oxidant cytosolic environment, which in turn damages mitochondrial membranes and causes increased ROS production from the damaged mitochondria. Independent from mitophagy, Parkin ubiquitinates outer mitochondrial membrane components for degradation by the proteasome in depolarized mitochondria. While ATG12 participates in mitophagy, when ubiquitinated, it is degraded in the proteasome to prevent its mitophagy functions, and therefore the proteasome has direct and indirect effects on mitochondria function. Adaptation to exercise can enhance the anabolic signaling pathway and increase the clearing of damaged mitochondria by mitophagy, which in turn, reduces the activation of the ubiquitin–proteasome system. In addition, this "priming" effect of exercise or loading enhances nutritionally based reductions in cytosolic oxidative stress to lower the extent of oxidatively damaged proteins and minimize the need for activation of the ubiquitin–proteasome system, and therefore reduce muscle loss in sarcopenia.

Although speculative, another way that nutrition could target mitochondria for elimination is via the chaperone-mediated autophagy (CMA) pathway. CMA involves the binding of heat shock protein 8 (HSPA8/HSC70) to selected proteins, which are targeted to the lysosomal membrane, where they interact with membrane receptor lysosomal-associated membrane protein 2A (LAMP2A). Aging causes an increased degradation and subsequently decreased availability of LAMP2A in the lysosomal membranes (Catarino, Pereira, and Girao 2017, Issa et al. 2018), which reduces the effectiveness of CMA. Future studies should evaluate the role of LAMP2A and determine whether exercise or nutritional interventions increase CMA and targeted mitochondrial mitophagy in muscles of old animals or humans.

Nutritional interventions could selectively target dysfunctional mitochondria through degradation of PARK7/DJ-1. PARK7 has been studied extensively in Parkinson disease (Wang, Sun, Zhao, et al. 2018, Wang, Cai, et al. 2016), but it has also been shown to regulate skeletal muscle contractile protein synthesis and hypertrophy (Zhou, Barkow, and Freed 2017, Yu et al. 2014). PARK7 is also an antioxidant protein that limits mitochondrial damage to oxidative stress (Zhang et al. 2018). Overexpression of PARK7 has been shown to reduce mitochondrial dysfunction under oxidative stress (Wang, Cai, et al. 2016, Wang et al. 2011), which shows a direct link between autophagy and mitochondrial function. Future studies should identify if exercise or nutritional intervention (e.g., antioxidants, caloric restriction) improves PARK7 expression in muscles of old hosts and if the change in PARK7 directly mediates improvements in skeletal muscle mitochondrial function to attenuate or reverse sarcopenia.

It is important to know whether there becomes a point when it is too late to repair or replace cells that are involved or targeted for destruction. It will also be important to understand whether satellite cells have a role in replacing muscle cells and repairing these fiber sections, as satellite cells will also be affected adversely by death signaling (Siu and Alway 2009, Krajnak et al. 2006, Haramizu et al. 2017, Alway, Myers, and Mohamed 2014, Alway et al. 2013, Alway, Morissette, and Siu 2011, Alway and Siu 2008, Alway et al. 2003, Brooks et al. 2018). Dysfunctional mitochondria are hypothesized to be the initiators of signaling that leads to sarcopenic muscle loss (Alway, Mohamed, and Myers 2017, Alway et al. 2017). If this is proven to be correct, the best approach for reducing the incidence and impact of sarcopenia should include the global goal of understanding the mechanisms by which nutrition targets muscle and particularly muscle mitochondria and optimize this information for clinical outcomes that improve mitochondria health and maximize muscle mass in aging populations.

REFERENCES

Alway, S. E. 1994a. "Characteristics of the elbow flexors in women bodybuilders using anabolic steroids." *J Strength Cond Res* 8:161–169.

Alway, S. E. 1994b. "Force and contractile characteristics after stretch overload in quail anterior latissimus dorsi muscle." *J Appl Physiol (1985)* 77 (1):135–141.

Alway, S. E. 2002. "Attenuation of Ca(2+)-activated ATPase and shortening velocity in hypertrophied fast twitch skeletal muscle from aged Japanese quail." *Exp Gerontol* 37 (5):665–678.

Alway, S. E. 2017. "Inflammation and oxidative stress limit adaptation to stretch-shortening contractions in aging." *Exerc Sport Sci Rev.* doi:10.1249/JES.0000000000000126.

Alway, S. E., B. T. Bennett, J. C. Wilson, N. K. Edens, and S. L. Pereira. 2014. "Epigallocatechin-3-gallate improves plantaris muscle recovery after disuse in aged rats." *Exp Gerontol* 50:82–94.

Alway, S. E., B. T. Bennett, J. C. Wilson, J. Sperringer, J. S. Mohamed, N. K. Edens, and S. L. Pereira. 2015. "Green tea extract attenuates muscle loss and improves muscle function during disuse, but fails to improve muscle recovery following unloading in aged rats." *J Appl Physiol (1985)* 118 (3):319–330. doi:10.1152/japplphysiol.00674.2014.

Alway, S. E., J. K. Martyn, J. Ouyang, A. Chaudhrai, and Z. S. Murlasits. 2003. "Id2 expression during apoptosis and satellite cell activation in unloaded and loaded quail skeletal muscles." *Am J Physiol Regul Integr Comp Physiol* 284 (2):R540–R549.

Alway, S. E., J. L. McCrory, K. Kearcher, A. Vickers, B. Frear, D. L. Gilleland, D. E. Bonner, J. M. Thomas, D. A. Donley, M. W. Lively, and J. S. Mohamed. 2017. "Resveratrol enhances exercise-induced cellular and functional adaptations of skeletal muscle in older men and women." *J Gerontol A Biol Sci Med Sci* 72 (12):1595–1606. doi:10.1093/gerona/glx089.

Alway, S. E., J. S. Mohamed, and M. J. Myers. 2017. "Mitochondria Initiate and Regulate Sarcopenia." *Exerc Sport Sci Rev* 45 (2):58–69. doi:10.1249/JES.0000000000000101.

Alway, S. E., M. R. Morissette, and P. M. Siu. 2011. "Aging and apoptosis in muscle." *Handbook of the Biology of Aging* 7th (4):63–118.

Alway, S. E., M. J. Myers, and J. S. Mohamed. 2014. "Regulation of satellite cell function in sarcopenia." *Front Aging Neurosci* 6 (9):1–15. doi:10.3389/fnagi.2014.00246.

Alway, S. E., S. L. Pereira, N. K. Edens, Y. Hao, and B. T. Bennett. 2013. "B-Hydroxy-B-methylbutyrate (HMB) enhances the proliferation of satellite cells in fast muscles of aged rats during recovery from disuse atrophy." *Exp Gerontol* 48 (9):973–984.

Alway, S. E., and P. M. Siu. 2008. "Nuclear apoptosis contributes to sarcopenia." *Exerc Sport Sci Rev* 36 (2):51–57. doi:10.1097/JES.0b013e318168e9dc.

Andersson, D. C., M. J. Betzenhauser, S. Reiken, A. C. Meli, A. Umanskaya, W. Xie, T. Shiomi, R. Zalk, A. Lacampagne, and A. R. Marks. 2011. "Ryanodine receptor oxidation causes intracellular calcium leak and muscle weakness in aging." *Cell Metab* 14 (2):196–207.

Andrianjafiniony, T., S. Dupre-Aucouturier, D. Letexier, H. Couchoux, and D. Desplanches. 2010. "Oxidative stress, apoptosis, and proteolysis in skeletal muscle repair after unloading." *Am J Physiol Cell Physiol* 299 (2):C307–C315. doi:10.1152/ajpcell.00069.2010.

Arribat, Y., N. T. Broskey, C. Greggio, M. Boutant, S. Conde Alonso, S. S. Kulkarni, S. Lagarrigue, E. A. Carnero, C. Besson, C. Canto, and F. Amati. 2018. "Distinct patterns of skeletal muscle mitochondria fusion, fission and mitophagy upon duration of exercise training." *Acta Physiol (Oxf)*:e13179. doi:10.1111/apha.13179.

Asami, Y., M. Aizawa, M. Kinoshita, J. Ishikawa, and K. Sakuma. 2018. "Resveratrol attenuates denervation-induced muscle atrophy due to the blockade of atrogin-1 and p62 accumulation." *Int J Med Sci* 15 (6):628–637. doi:10.7150/ijms.22723.

Aversa, Z., N. Alamdari, E. Castillero, M. Muscaritoli, F. Rossi Fanelli, and P. O. Hasselgren. 2012. "beta-Hydroxy-beta-methylbutyrate (HMB) prevents dexamethasone-induced myotube atrophy." *Biochem Biophys Res Commun* 423 (4):739–743. doi:10.1016/j.bbrc.2012.06.029.

Bak, D. H., J. Na, S. I. Im, C. T. Oh, J. Y. Kim, S. K. Park, H. J. Han, J. Seok, S. Y. Choi, E. J. Ko, S. K. Mun, S. W. Ahn, and B. J. Kim. 2019. "Antioxidant effect of human placenta hydrolysate against oxidative stress on muscle atrophy." *J Cell Physiol* 234 (2):1643–1658. doi:10.1002/jcp.27034.

Bennett, B. T., J. S. Mohamed, and S. E. Alway. 2013. "Effects of resveratrol on the recovery of muscle mass following disuse in the plantaris muscle of aged rats." *PLoS One* 8 (12):e83518. doi:10.1371/journal.pone.0083518.

Beyfuss, K., A. T. Erlich, M. Triolo, and D. A. Hood. 2018. "The role of p53 in determining mitochondrial adaptations to endurance training in skeletal Muscle." *Sci Rep* 8 (1):14710. doi:10.1038/s41598-018-32887-0.

Biala, A. K., R. Dhingra, and L. A. Kirshenbaum. 2015. "Mitochondrial dynamics: Orchestrating the journey to advanced age." *J Mol Cell Cardiol* 83:37–43. doi:10.1016/j.yjmcc.2015.04.015.

Bori, Z., Z. Zhao, E. Koltai, I. G. Fatouros, A. Z. Jamurtas, I. I. Douroudos, G. Terzis, A. Chatzinikolaou, A. Sovatzidis, D. Draganidis, I. Boldogh, and Z. Radak. 2012. "The effects of aging, physical training, and a single bout of exercise on mitochondrial protein expression in human skeletal muscle." *Exp Gerontol* 47 (6):417–424.

Brand, M. D. 2016. "Mitochondrial generation of superoxide and hydrogen peroxide as the source of mitochondrial redox signaling." *Free Radic Biol Med* 100:14–31. doi:10.1016/j.freeradbiomed.2016.04.001.

Brioche, T., A. F. Pagano, G. Py, and A. Chopard. 2016. "Muscle wasting and aging: Experimental models, fatty infiltrations, and prevention." *Mol Aspects Med* 50:56–87. doi:10.1016/j.mam.2016.04.006.

Brito, P. M., A. Mariano, L. M. Almeida, and T. C. Dinis. 2006. "Resveratrol affords protection against peroxynitrite-mediated endothelial cell death: A role for intracellular glutathione." *Chem Biol Interact* 164 (3):157–166.

Brooks, M. J., A. Hajira, J. S. Mohamed, and S. E. Alway. 2018. "Voluntary wheel running increases satellite cell abundance and improves recovery from disuse in gastrocnemius muscles from mice." *J Appl Physiol (1985)* 124 (6):1616–1628. doi:10.1152/japplphysiol.00451.2017.

Brown, J. L., M. E. Rosa-Caldwell, D. E. Lee, T. A. Blackwell, L. A. Brown, R. A. Perry, W. S. Haynie, J. P. Hardee, J. A. Carson, M. P. Wiggs, T. A. Washington, and N. P. Greene. 2017. "Mitochondrial degeneration precedes the development of muscle atrophy in progression of cancer cachexia in tumour-bearing mice." *J Cachexia Sarcopenia Muscle* 8 (6):926–938. doi:10.1002/jcsm.12232.

Bruusgaard, J. C., I. M. Egner, T. K. Larsen, S. Dupre-Aucouturier, D. Desplanches, and K. Gundersen. 2012. "No change in myonuclear number during muscle unloading and reloading." *J Appl Physiol (1985)* 113 (2):290–296. doi:10.1152/japplphysiol.00436.2012.

Bruusgaard, J. C., I. B. Johansen, I. M. Egner, Z. A. Rana, and K. Gundersen. 2010. "Myonuclei acquired by overload exercise precede hypertrophy and are not lost on detraining." *Proc Natl Acad Sci USA* 107 (34):15111–15116.

Calvani, R., A. M. Joseph, P. J. Adhihetty, A. Miccheli, M. Bossola, C. Leeuwenburgh, R. Bernabei, and E. Marzetti. 2013. "Mitochondrial pathways in sarcopenia of aging and disuse muscle atrophy." *Biol Chem* 394 (3):393–414.

Campbell, M. D., J. Duan, A. T. Samuelson, M. J. Gaffrey, L. Wang, T. K. Bammler, R. J. Moore, C. C. White, T. J. Kavanagh, J. G. Voss, H. H. Szeto, P. S. Rabinovitch, W. J. Qian, and D. J. Marcinek. 2018. "Improving mitochondrial function with SS-31 reverses age-related redox stress and improves exercise tolerance in aged mice." *Free Radic Biol Med.* doi:10.1016/j.freeradbiomed.2018.12.031.

Capitanio, D., M. Vasso, Palma S. De, C. Fania, E. Torretta, F. P. Cammarata, V. Magnaghi, P. Procacci, and C. Gelfi. 2016. "Specific protein changes contribute to the differential muscle mass loss during ageing." *Proteomics* 16 (4):645–656.

Carnio, S., F. LoVerso, M. A. Baraibar, E. Longa, M. M. Khan, M. Maffei, M. Reischl, M. Canepari, S. Loefler, H. Kern, B. Blaauw, B. Friguet, R. Bottinelli, R. Rudolf, and M. Sandri. 2014. "Autophagy impairment in muscle induces neuromuscular junction degeneration and precocious aging." *Cell Rep* 8 (5):1509–1521.

Carter, H. N., C. C. Chen, and D. A. Hood. 2015. "Mitochondria, muscle health, and exercise with advancing age." *Physiology (Bethesda)* 30 (3):208–223.

Carter, H. N., Y. Kim, A. T. Erlich, D. Zarrin-Khat, and D. A. Hood. 2018. "Autophagy and mitophagy flux in young and aged skeletal muscle following chronic contractile activity." *J Physiol* 596 (16):3567–3584. doi:10.1113/JP275998.

Casanova, E., L. Baselga-Escudero, A. Ribas-Latre, A. Arola-Arnal, C. Blade, L. Arola, and M. J. Salvado. 2014. "Epigallocatechin gallate counteracts oxidative stress in docosahexaenoxic acid-treated myocytes." *Biochim Biophys Acta* 1837 (6):783–791. doi:10.1016/j.bbabio.2014.01.014.

Catarino, S., P. Pereira, and H. Girao. 2017. "Molecular control of chaperone-mediated autophagy." *Essays Biochem* 61 (6):663–674. doi:10.1042/EBC20170057.

Cecile, P., A. Julien, A. Andrea, C. Agnes, C. G. Cecile, T. Clara, D. Christiane, C. Lydie, B. Daniel, S. Marco, A. Didier, and T. Daniel. 2018. "UBE2E1 is preferentially expressed in the cytoplasm of slow-twitch fibers and protects skeletal muscles from exacerbated atrophy upon dexamethasone treatment." *Cells* 7 (11). doi:10.3390/cells7110214.

Chabi, B., V. Ljubicic, K. J. Menzies, J. H. Huang, A. Saleem, and D. A. Hood. 2008. "Mitochondrial function and apoptotic susceptibility in aging skeletal muscle." *Aging Cell* 7 (1):2–12.

Cheema, N., A. Herbst, D. McKenzie, and J. M. Aiken. 2015. "Apoptosis and necrosis mediate skeletal muscle fiber loss in age-induced mitochondrial enzymatic abnormalities." *Aging Cell* 14 (6):1085–1093. doi:10.1111/acel.12399.

Chen, C. A., J. M. Chang, E. E. Chang, H. C. Chen, and Y. L. Yang. 2018. "Crosstalk between transforming growth factor-beta1 and endoplasmic reticulum stress regulates alpha-smooth muscle cell actin expression in podocytes." *Life Sci* 209:9–14. doi:10.1016/j.lfs.2018.07.050.

Chen, H., M. Vermulst, Y. E. Wang, A. Chomyn, T. A. Prolla, J. M. McCaffery, and D. C. Chan. 2010. "Mitochondrial fusion is required for mtDNA stability in skeletal muscle and tolerance of mtDNA mutations." *Cell* 141 (2):280–289.

Chen, Y., L. Huang, H. Zhang, X. Diao, S. Zhao, and W. Zhou. 2017. "Reduction in autophagy by (-)-epigallocatechin-3-gallate (EGCG): A potential mechanism of prevention of mitochondrial dysfunction after subarachnoid hemorrhage." *Mol Neurobiol* 54 (1):392–405. doi:10.1007/s12035-015-9629-9.

Chodkowska, K. A., A. Ciecierska, K. Majchrzak, P. Ostaszewski, and T. Sadkowski. 2018. "Effect of beta-hydroxy-beta-methylbutyrate on miRNA expression in differentiating equine satellite cells exposed to hydrogen peroxide." *Genes Nutr* 13:10. doi:10.1186/s12263-018-0598-2.

Clifford, M. N., J. J. van der Hooft, and A. Crozier. 2013. "Human studies on the absorption, distribution, metabolism, and excretion of tea polyphenols." *Am J Clin Nutr* 98 (6 Suppl):1619S–1630S. doi:10.3945/ajcn.113.058958.

Crupi, A. N., J. S. Nunnelee, D. J. Taylor, A. Thomas, J. P. Vit, C. E. Riera, R. A. Gottlieb, and H. S. Goodridge. 2018. "Oxidative muscles have better mitochondrial homeostasis than glycolytic muscles throughout life and maintain mitochondrial function during aging." *Aging (Albany NY)* 10 (11):3327–3352. doi:10.18632/aging.101643.

Cruz-Jentoft, A. J. 2018. "Beta-hydroxy-beta-methyl butyrate (HMB): From experimental data to clinical evidence in sarcopenia." *Curr Protein Pept Sci* 19 (7):668–672. doi:10.2174/1389203718666170529105026.

da Silva, W., A. S. Machado, M. A. Souza, P. B. Mello-Carpes, and F. P. Carpes. 2018. "Effect of green tea extract supplementation on exercise-induced delayed onset muscle soreness and muscular damage." *Physiol Behav* 194:77–82. doi:10.1016/j.physbeh.2018.05.006.

Dai, D. F., P. S. Rabinovitch, and Z. Ungvari. 2012. "Mitochondria and cardiovascular aging." *Circ Res* 110 (8):1109–1124. doi:10.1161/CIRCRESAHA.111.246140.

Dalle, S., L. Rossmeislova, and K. Koppo. 2017. "The role of inflammation in age-related sarcopenia." *Front Physiol* 8:1045. doi:10.3389/fphys.2017.01045.

Del Campo, A., I. Contreras-Hernandez, M. Castro-Sepulveda, C. A. Campos, R. Figueroa, M. F. Tevy, V. Eisner, M. Casas, and E. Jaimovich. 2018. "Muscle function decline and mitochondria changes in middle age precede sarcopenia in mice." *Aging (Albany NY)* 10 (1):34–55. doi:10.18632/aging.101358.

Dethlefsen, M. M., J. F. Halling, H. D. Moller, P. Plomgaard, B. Regenberg, S. Ringholm, and H. Pilegaard. 2018. "Regulation of apoptosis and autophagy in mouse and human skeletal muscle with aging and life-long exercise training." *Exp Gerontol* 111:141–153. doi:10.1016/j.exger.2018.07.011.

Distefano, G., R. A. Standley, J. J. Dube, E. A. Carnero, V. B. Ritov, M. Stefanovic-Racic, F. G. Toledo, S. R. Piva, B. H. Goodpaster, and P. M. Coen. 2017. "Chronological age does not influence ex-vivo mitochondrial respiration and quality control in skeletal muscle." *J Gerontol A Biol Sci Med Sci* 72 (4):535–542. doi:10.1093/gerona/glw102.

Dogru, M., T. Kojima, C. Simsek, and K. Tsubota. 2018. "Potential role of oxidative stress in ocular surface inflammation and dry eye disease." *Invest Ophthalmol Vis Sci* 59 (14):DES163-DES168. doi:10.1167/iovs.17-23402.

Drake, J. C., F. F. Peelor, III, L. M. Biela, M. K. Watkins, R. A. Miller, K. L. Hamilton, and B. F. Miller. 2013. "Assessment of mitochondrial biogenesis and mTORC1 signaling during chronic rapamycin feeding in male and female mice." *J Gerontol. A Biol. Sci Med Sci* 68 (12):1493–1501.

Drake, J. C., and Z. Yan. 2017. "Mitophagy in maintaining skeletal muscle mitochondrial proteostasis and metabolic health with ageing." *J Physiol* 595 (20):6391–6399. doi:10.1113/JP274337.

Dugdale, H. F., D. C. Hughes, R. Allan, C. S. Deane, C. R. Coxon, J. P. Morton, C. E. Stewart, and A. P. Sharples. 2018. "The role of resveratrol on skeletal muscle cell differentiation and myotube hypertrophy during glucose restriction." *Mol Cell Biochem* 444 (1–2):109–123. doi:10.1007/s11010-017-3236-1.

Eid, N., and Y. Kondo. 2017. "Parkin in cancer: Mitophagy-related/unrelated tasks." *World J Hepatol* 9 (7):349–351. doi:10.4254/wjh.v9.i7.349.

Elfawy, H. A., and B. Das. 2018. "Crosstalk between mitochondrial dysfunction, oxidative stress, and age related neurodegenerative disease: Etiologies and therapeutic strategies." *Life Sci.* doi:10.1016/j.lfs.2018.12.029.

Ellis, A. C., G. R. Hunter, A. M. Goss, and B. A. Gower. 2018. "Oral supplementation with beta-hydroxy-beta-methylbutyrate, arginine, and glutamine improves lean body mass in healthy older adults." *J Diet Suppl*:1–13. doi:10.1080/19390211.2018.1454568.

Estebanez, B., O. C. Moreira, M. Almar, J. A. de Paz, J. Gonzalez-Gallego, and M. J. Cuevas. 2019. "Effects of a resistance-training programme on endoplasmic reticulum unfolded protein response and mitochondrial functions in PBMCs from elderly subjects." *Eur J Sport Sci*:1–10. doi:10.1080/17461391.2018.1561950.

Fallaize, D., L. S. Chin, and L. Li. 2015. "Differential submitochondrial localization of PINK1 as a molecular switch for mediating distinct mitochondrial signaling pathways." *Cell Signal.* 27 (12):2543–2554.

Feige, P., C. E. Brun, M. Ritso, and M. A. Rudnicki. 2018. "Orienting muscle stem cells for regeneration in homeostasis, aging, and disease." *Cell Stem Cell* 23 (5):653–664. doi:10.1016/j.stem.2018.10.006.

Feng, Z., L. Bai, J. Yan, Y. Li, W. Shen, Y. Wang, K. Wertz, P. Weber, Y. Zhang, Y. Chen, and J. Liu. 2011. "Mitochondrial dynamic remodeling in strenuous exercise-induced muscle and mitochondrial dysfunction: Regulatory effects of hydroxytyrosol." *Free Radic Biol Med.* 50 (10):1437–1446.

Flakoll, P., R. Sharp, S. Baier, D. Levenhagen, C. Carr, and S. Nissen. 2004. "Effect of beta-hydroxy-beta-methylbutyrate, arginine, and lysine supplementation on strength, functionality, body composition, and protein metabolism in elderly women." *Nutrition* 20 (5):445–451.

Gan, W., Y. Dang, X. Han, S. Ling, J. Duan, J. Liu, and J. W. Xu. 2016. "ERK5/HDAC5-mediated, resveratrol-, and pterostilbene-induced expression of MnSOD in human endothelial cells." *Mol Nutr Food Res* 60 (2):266–277. doi:10.1002/mnfr.201500466.

Giron, M. D., J. D. Vilchez, S. Shreeram, R. Salto, M. Manzano, E. Cabrera, N. Campos, N. K. Edens, R. Rueda, and J. M. Lopez-Pedrosa. 2015. "beta-Hydroxy-beta-methylbutyrate (HMB) normalizes dexamethasone-induced autophagy-lysosomal pathway in skeletal muscle." *PLoS One* 10 (2):e0117520.

Gomez-Cabrera, M. C., F. Sanchis-Gomar, R. Garcia-Valles, H. Pareja-Galeano, J. Gambini, C. Borras, and J. Vina. 2012. "Mitochondria as sources and targets of damage in cellular aging." *Clin Chem Lab Med* 50 (8):1287–1295.

Gouspillou, G., N. Sgarioto, S. Kapchinsky, F. Purves-Smith, B. Norris, C. H. Pion, S. Barbat-Artigas, F. Lemieux, T. Taivassalo, J. A. Morais, M. Aubertin-Leheudre, and R. T. Hepple. 2014. "Increased sensitivity to mitochondrial permeability transition and myonuclear translocation of endonuclease G in atrophied muscle of physically active older humans." *FASEB J* 28 (4):1621–1633.

Gundersen, K., and J. C. Bruusgaard. 2008. "Nuclear domains during muscle atrophy: Nuclei lost or paradigm lost?" *J Physiol* 586 (11):2675–2681. doi:10.1113/jphysiol.2008.154369.

Hamilton, J., T. Brustovetsky, J. E. Rysted, Z. Lin, Y. M. Usachev, and N. Brustovetsky. 2018. "Deletion of mitochondrial calcium uniporter incompletely inhibits calcium uptake and induction of the permeability transition pore in brain mitochondria." *J Biol Chem* 293 (40):15652–15663. doi:10.1074/jbc. RA118.002926.

Hao, Y., J. R. , Y. Wang, N. Edens, S. L. Pereira, and S. E. Alway. 2011. "B-Hydroxy-B-methylbutyrate reduces myonuclear apoptosis during recovery from hind limb suspension-induced muscle fiber atrophy in aged rats." *Am J Physiol Regul Integr Comp Physiol* 301 (3):R701–R715.

Haramizu, S., S. Asano, D. C. Butler, D. A. Stanton, A. Hajira, J. S. Mohamed, and S. E. Alway. 2017. "Dietary resveratrol confers apoptotic resistance to oxidative stress in myoblasts." *J Nutr Biochem* 50:103–115. doi:10.1016/j.jnutbio.2017.08.008.

Haramizu, S., N. Ota, T. Hase, and T. Murase. 2011. "Catechins attenuate eccentric exercise-induced inflammation and loss of force production in muscle in senescence-accelerated mice." *J Appl Physiol (1985)* 111 (6):1654–1663. doi:10.1152/japplphysiol.01434.2010.

Hasselgren, P. O. 2014. "beta-Hydroxy-beta-methylbutyrate (HMB) and prevention of muscle wasting." *Metabolism* 63 (1):5–8. doi:10.1016/j.metabol.2013.09.015.

Hepple, R. T., and C. L. Rice. 2016. "Innervation and neuromuscular control in ageing skeletal muscle." *J Physiol* 594 (8):1965–1978.

Herbst, A., J. Wanagat, N. Cheema, K. Widjaja, D. McKenzie, and J. M. Aiken. 2016. "Latent mitochondrial DNA deletion mutations drive muscle fiber loss at old age." *Aging Cell* 15:1132–1139.

Herbst, A., K. Widjaja, B. Nguy, E. B. Lushaj, T. M. Moore, A. L. Hevener, D. McKenzie, J. M. Aiken, and J. Wanagat. 2017. "Digital PCR quantitation of muscle mitochondrial DNA: Age, fiber type, and mutation-induced changes." *J Gerontol A Biol Sci Med Sci* 72 (10):1327–1333. doi:10.1093/gerona/glx058.

Holecek, M. 2017. "Beta-hydroxy-beta-methylbutyrate supplementation and skeletal muscle in healthy and muscle-wasting conditions." *J Cachexia Sarcopenia Muscle* 8 (4):529–541. doi:10.1002/jcsm.12208.

Holecek, M., and M. Vodenicarovova. 2018. "Effects of beta-hydroxy-beta-methylbutyrate in partially hepatectomized rats." *Physiol Res* 67 (5):741–751.

Hood, D. A., J. M. Memme, A. N. Oliveira, and M. Triolo. 2018. "Maintenance of skeletal muscle mitochondria in health, exercise, and aging." *Annu Rev Physiol.* doi:10.1146/annurev-physiol-020518-114310.

Hood, D. A., L. D. Tryon, H. N. Carter, Y. Kim, and C. C. Chen. 2016. "Unravelling the mechanisms regulating muscle mitochondrial biogenesis." *Biochem J* 473 (15):2295–314. doi:10.1042/BCJ20160009.

Hood, D. A., L. D. Tryon, A. Vainshtein, J. Memme, C. Chen, M. Pauly, M. J. Crilly, and H. Carter. 2015. "Exercise and the regulation of mitochondrial turnover." *Prog Mol Biol Transl Sci* 135:99–127. doi:10.1016/bs.pmbts.2015.07.007.

Hsieh, L. C., S. L. Chien, M. S. Huang, H. F. Tseng, and C. K. Chang. 2006. "Anti-inflammatory and anticatabolic effects of short-term beta-hydroxy-beta-methylbutyrate supplementation on chronic obstructive pulmonary disease patients in intensive care unit." *Asia Pac J Clin Nutr* 15 (4):544–550.

Hsieh, L. C., C. J. Chow, W. C. Chang, T. H. Liu, and C. K. Chang. 2010. "Effect of beta-hydroxy-beta-methylbutyrate on protein metabolism in bed-ridden elderly receiving tube feeding." *Asia Pac J Clin Nutr* 19 (2):200–198.

Ibebunjo, C., J. M. Chick, T. Kendall, J. K. Eash, C. Li, Y. Zhang, C. Vickers, Z. Wu, B. A. Clarke, J. Shi, J. Cruz, B. Fournier, S. Brachat, S. Gutzwiller, Q. Ma, J. Markovits, M. Broome, M. Steinkrauss, E. Skuba, J. R. Galarneau, S. P. Gygi, and D. J. Glass. 2013. "Genomic and proteomic profiling reveals reduced mitochondrial function and disruption of the neuromuscular junction driving rat sarcopenia." *Mol Cell Biol* 33 (2):194–212.

Iqbal, S., and D. A. Hood. 2014. "Oxidative stress-induced mitochondrial fragmentation and movement in skeletal muscle myoblasts." *Am J Physiol Cell Physiol* 306 (12):C1176–C1183.

Iqbal, S., and D. A. Hood. 2015. "The role of mitochondrial fusion and fission in skeletal muscle function and dysfunction." *Front Biosci (Landmark Ed)* 20:157–172.

Iqbal, S., O. Ostojic, K. Singh, A. M. Joseph, and D. A. Hood. 2013. "Expression of mitochondrial fission and fusion regulatory proteins in skeletal muscle during chronic use and disuse." *Muscle Nerve* 48 (6):963–970. doi:10.1002/mus.23838.

Issa, A. R., J. Sun, C. Petitgas, A. Mesquita, A. Dulac, M. Robin, B. Mollereau, A. Jenny, B. Cherif-Zahar, and S. Birman. 2018. "The lysosomal membrane protein LAMP2A promotes autophagic flux and prevents SNCA-induced Parkinson disease-like symptoms in the *Drosophila* brain." *Autophagy* 14 (11):1898–1910. doi:10.1080/15548627.2018.1491489.

Jackson, J. R., M. J. Ryan, and S. E. Alway. 2011. "Long-term supplementation with resveratrol alleviates oxidative stress but does not attenuate sarcopenia in aged mice." *J Gerontol A Biol Sci Med Sci* 66 (7):751–764. doi:10.1093/gerona/glr047.

Jackson, J. R., M. J. Ryan, Y. Hao, and S. E. Alway. 2010. "Mediation of endogenous antioxidant enzymes and apoptotic signaling by resveratrol following muscle disuse in the gastrocnemius muscles of young and old rats." *Am J Physiol Regul Integr Comp Physiol* 299 (6):R1572–R1581.

Jackson, M. J. 2016. "Reactive oxygen species in sarcopenia: Should we focus on excess oxidative damage or defective redox signalling?" *Mol. Aspects Med* 50:33–40. doi:10.1016/j.mam.2016.05.002.

James, K. D., M. J. Kennett, and J. D. Lambert. 2018. "Potential role of the mitochondria as a target for the hepatotoxic effects of (-)-epigallocatechin-3-gallate in mice." *Food Chem Toxicol* 111:302–309. doi:10.1016/j.fct.2017.11.029.

Jang, Y. C., D. J. Hwang, J. H. Koo, H. S. Um, N. H. Lee, D. C. Yeom, Y. Lee, and J. Y. Cho. 2018. "Association of exercise-induced autophagy upregulation and apoptosis suppression with neuroprotection against pharmacologically induced Parkinson's disease." *J Exerc Nutrition Biochem* 22 (1):1–8. doi:10.20463/jenb.2018.0001.

Jang, Y. C., M. S. Lustgarten, Y. Liu, F. L. Muller, A. Bhattacharya, H. Liang, A. B. Salmon, S. V. Brooks, L. Larkin, C. R. Hayworth, A. Richardson, and Remmen H. Van. 2010. "Increased superoxide in vivo accelerates age-associated muscle atrophy through mitochondrial dysfunction and neuromuscular junction degeneration." *FASEB J* 24 (5):1376–1390.

Javadov, S., S. Jang, N. Rodriguez-Reyes, A. E. Rodriguez-Zayas, J. Soto Hernandez, T. Krainz, P. Wipf, and W. Frontera. 2015. "Mitochondria-targeted antioxidant preserves contractile properties and mitochondrial function of skeletal muscle in aged rats." *Oncotarget* 6 (37):39469–39481. doi:10.18632/oncotarget.5783.

Joseph, A. M., P. J. Adhihetty, N. R. Wawrzyniak, S. E. Wohlgemuth, A. Picca, G. C. Kujoth, T. A. Prolla, and C. Leeuwenburgh. 2013. "Dysregulation of mitochondrial quality control processes contribute to sarcopenia in a mouse model of premature aging." *PLoS One* 8 (7):e69327. doi:10.1371/journal.pone.0069327.

Joseph, A. M., and D. A. Hood. 2014. "Relationships between exercise, mitochondrial biogenesis and type 2 diabetes." *Med Sport Sci* 60:48–61. doi:10.1159/000357335.

Joseph, A. M., D. R. Joanisse, R. G. Baillot, and D. A. Hood. 2012. "Mitochondrial dysregulation in the pathogenesis of diabetes: Potential for mitochondrial biogenesis-mediated interventions." *Exp Diabetes Res* 2012:642038. doi:10.1155/2012/642038.

Joseph, A. M., A. G. Malamo, J. Silvestre, N. Wawrzyniak, S. Carey-Love, L. M. Nguyen, D. Dutta, J. Xu, C. Leeuwenburgh, and P. J. Adhihetty. 2013. "Short-term caloric restriction, resveratrol, or combined treatment regimens initiated in late-life alter mitochondrial protein expression profiles in a fiber-type specific manner in aged animals." *Exp Gerontol* 48 (9):858–868. doi:10.1016/j.exger.2013.05.061.

Kang, C., D. Yeo, and L. L. Ji. 2016. "Muscle immobilization activates mitophagy and disrupts mitochondrial dynamics in mice." *Acta Physiol (Oxf)* 218 (3):118–197.

Karam, C., J. Yi, Y. Xiao, K. Dhakal, L. Zhang, X. Li, C. Manno, J. Xu, K. Li, H. Cheng, J. Ma, and J. Zhou. 2017. "Absence of physiological Ca^{2+} transients is an initial trigger for mitochondrial dysfunction in skeletal muscle following denervation." *Skelet Muscle* 7 (1):6. doi:10.1186/s13395-017-0123-0.

Khalifa, A. R., E. A. Abdel-Rahman, A. M. Mahmoud, M. H. Ali, M. Noureldin, S. H. Saber, S. Mohsen, and S. S. Ali. 2017. "Sex-specific differences in mitochondria biogenesis, morphology, respiratory function, and ROS homeostasis in young mouse heart and brain." *Physiol Rep* 5 (6). doi:10.14814/phy2.13125.

Khan, M. M., J. Xiao, D. Patel, and M. S. LeDoux. 2018. "DNA damage and neurodegenerative phenotypes in aged Ciz1 null mice." *Neurobiol Aging* 62:180–190. doi:10.1016/j.neurobiolaging.2017.10.014.

Kim, A. R., K. M. Kim, M. R. Byun, J. H. Hwang, J. I. Park, H. T. Oh, H. K. Kim, M. G. Jeong, E. S. Hwang, and J. H. Hong. 2017. "Catechins activate muscle stem cells by Myf5 induction and stimulate muscle regeneration." *Biochem Biophys Res Commun* 489 (2):142–148. doi:10.1016/j.bbrc.2017.05.114.

Kim, H., J. Deshane, S. Barnes, and S. Meleth. 2006. "Proteomics analysis of the actions of grape seed extract in rat brain: Technological and biological implications for the study of the actions of psychoactive compounds." *Life Sci* 78 (18):2060–2065.

Kim, H. S., M. J. Quon, and J. A. Kim. 2014. "New insights into the mechanisms of polyphenols beyond antioxidant properties; lessons from the green tea polyphenol, epigallocatechin 3-gallate." *Redox Biol* 2:187–195.

Kim, Y., M. Triolo, A. T. Erlich, and D. A. Hood. 2018. "Regulation of autophagic and mitophagic flux during chronic contractile activity-induced muscle adaptations." *Pflugers Arch.* doi:10.1007/s00424-018-2225-x.

Kim, Y., M. Triolo, and D. A. Hood. 2017. "Impact of aging and exercise on mitochondrial quality control in skeletal muscle." *Oxid Med Cell Longev* 2017:3165396. doi:10.1155/2017/3165396.

Kim, Y., X. Zheng, Z. Ansari, M. C. Bunnell, J. R. Herdy, L. Traxler, H. Lee, A. C. M. Paquola, C. Blithikioti, M. Ku, J. C. M. Schlachetzki, J. Winkler, F. Edenhofer, C. K. Glass, A. A. Paucar, B. N. Jaeger, S. Pham, L. Boyer, B. C. Campbell, T. Hunter, J. Mertens, and F. H. Gage. 2018. "Mitochondrial aging defects emerge in directly reprogrammed human neurons due to their metabolic profile." *Cell Rep* 23 (9):2550–2558. doi:10.1016/j.celrep.2018.04.105.

Kleme, M. L., A. Sane, C. Garofalo, E. Seidman, E. Brochiero, Y. Berthiaume, and E. Levy. 2018. "CFTR deletion confers mitochondrial dysfunction and disrupts lipid homeostasis in intestinal epithelial cells." *Nutrients* 10 (7). doi:10.3390/nu10070836.

Konokhova, Y., S. Spendiff, R. T. Jagoe, S. Aare, S. Kapchinsky, N. J. MacMillan, P. Rozakis, M. Picard, M. Aubertin-Leheudre, C. H. Pion, J. Bourbeau, R. T. Hepple, and T. Taivassalo. 2016. "Failed upregulation of TFAM protein and mitochondrial DNA in oxidatively deficient fibers of chronic obstructive pulmonary disease locomotor muscle." *Skelet Muscle* 6:10. doi:10.1186/s13395-016-0083-9.

Konopka, A. R., M. K. Suer, C. A. Wolff, and M. P. Harber. 2014. "Markers of human skeletal muscle mitochondrial biogenesis and quality control: Effects of age and aerobic exercise training." *J Gerontol A Biol Sci Med Sci* 69 (4):371–378.

Kornasio, R., I. Riederer, G. Butler-Browne, V. Mouly, Z. Uni, and O. Halevy. 2009. "Beta-hydroxy-beta-methylbutyrate (HMB) stimulates myogenic cell proliferation, differentiation and survival via the MAPK/ERK and PI3K/Akt pathways." *Biochim Biophys Acta* 1793 (5):755–763. doi:10.1016/j.bbamcr.2008.12.017.

Krajnak, K., S. Waugh, R. Miller, B. Baker, K. Geronilla, S. E. Alway, and R. G. Cutlip. 2006. "Proapoptotic factor Bax is increased in satellite cells in the tibialis anterior muscles of old rats." *Muscle Nerve* 34 (6):720–730.

Kruger, K., K. Geist, F. Stuhldreier, L. Schumacher, L. Blumel, M. Remke, S. Wesselborg, B. Stork, N. Klocker, S. Bormann, W. P. Roos, S. Honnen, and G. Fritz. 2018. "Multiple DNA damage-dependent and DNA damage-independent stress responses define the outcome of ATR/Chk1 targeting in medulloblastoma cells." *Cancer Lett* 430:34–46. doi:10.1016/j.canlet.2018.05.011.

Krzyminska-Siemaszko, R., N. Czepulis, M. Lewandowicz, E. Zasadzka, A. Suwalska, J. Witowski, and K. Wieczorowska-Tobis. 2015. "The effect of a 12-week omega-3 supplementation on body composition, muscle strength and physical performance in elderly individuals with decreased muscle mass." *Int J Environ Res Public Health* 12 (9):10558–10574.

Laaper, M., T. Haque, R. S. Slack, and A. Jahani-Asl. 2017. "Modeling neuronal death and degeneration in mouse primary cerebellar granule neurons." *J Vis Exp* (129). doi:10.3791/55871.

Lalia, A. Z., S. Dasari, M. M. Robinson, H. Abid, D. M. Morse, K. A. Klaus, and I. R. Lanza. 2017. "Influence of omega-3 fatty acids on skeletal muscle protein metabolism and mitochondrial bioenergetics in older adults." *Aging (Albany NY)* 9 (4):1096–1129. doi:10.18632/aging.101210.

Lanza, I. R., P. Zabielski, K. A. Klaus, D. M. Morse, C. J. Heppelmann, H. R. Bergen, 3rd, S. Dasari, S. Walrand, K. R. Short, M. L. Johnson, M. M. Robinson, J. M. Schimke, D. R. Jakaitis, Y. W. Asmann, Z. Sun, and K. S. Nair. 2012. "Chronic caloric restriction preserves mitochondrial function in senescence without increasing mitochondrial biogenesis." *Cell Metab* 16 (6):777–788. doi:10.1016/j.cmet.2012.11.003.

Leduc-Gaudet, J. P., M. Picard, Pelletier F. St-Jean, N. Sgarioto, M. J. Auger, J. Vallee, R. Robitaille, D. H. St-Pierre, and G. Gouspillou. 2015. "Mitochondrial morphology is altered in atrophied skeletal muscle of aged mice." *Oncotarget* 6 (20):17923–17937.

Lee, H., J. Y. Lim, and S. J. Choi. 2017. "Oleate prevents palmitate-induced atrophy via modulation of mitochondrial ROS production in skeletal myotubes." *Oxid Med Cell Longev* 2017:2739721. doi:10.1155/2017/2739721.

Lee, I., M. Huttemann, and M. H. Malek. 2016. "(-)-Epicatechin attenuates degradation of mouse oxidative muscle following hindlimb suspension." *J Strength Cond Res* 30 (1):1–10. doi:10.1519/JSC.0000000000001205.

Lee, M. S. 2016. "Role of mitochondrial function in cell death and body metabolism." *Front Biosci (Landmark. Ed)* 21:1233–1244.

Lee, S., M. Kim, W. Lim, T. Kim, and C. Kang. 2015. "Strenuous exercise induces mitochondrial damage in skeletal muscle of old mice." *Biochem Biophys Res Commun* 461 (2):354–360. doi:10.1016/j.bbrc.2015.04.038.

Li, F. H., H. T. Yu, L. Xiao, and Y. Y. Liu. 2016. "Response of BAX, Bcl-2 Proteins, and SIRT1/PGC-1alpha mRNA expression to 8-week treadmill running in the aging rat skeletal muscle." *Adv Exp Med Biol* 923:283–289.

Liang, Q. X., Y. H. Lin, C. H. Zhang, H. M. Sun, L. Zhou, H. Schatten, Q. Y. Sun, and W. P. Qian. 2018. "Resveratrol increases resistance of mouse oocytes to postovulatory aging in vivo." *Aging (Albany NY)* 10 (7):1586–1596. doi:10.18632/aging.101494.

Liu, J., Y. Peng, Z. Feng, W. Shi, L. Qu, Y. Li, J. Liu, and J. Long. 2014. "Reloading functionally ameliorates disuse-induced muscle atrophy by reversing mitochondrial dysfunction, and similar benefits are gained by administering a combination of mitochondrial nutrients." *Free Radic Biol Med* 69:116–128. doi:10.1016/j.freeradbiomed.2014.01.003.

Lopez-Dominguez, J. A., H. Khraiwesh, J. A. Gonzalez-Reyes, G. Lopez-Lluch, P. Navas, J. J. Ramsey, Cabo R. de, M. I. Buron, and J. M. Villalba. 2013. "Dietary fat modifies mitochondrial and plasma membrane apoptotic signaling in skeletal muscle of calorie-restricted mice." *Age (Dordr.)* 35 (6):2027–2044.

Lopez-Lluch, G. 2017. "Mitochondrial activity and dynamics changes regarding metabolism in ageing and obesity." *Mech Ageing Dev* 162:108–121. doi:10.1016/j.mad.2016.12.005.

Lu, J. J., Q. Wang, L. H. Xie, Q. Zhang, and S. H. Sun. 2017. "Tumor necrosis factor-like weak inducer of apoptosis regulates quadriceps muscle atrophy and fiber-type alteration in a rat model of chronic obstructive pulmonary disease." *Tob Induc Dis* 15:43. doi:10.1186/s12971-017-0148-5.

Malafarina, V., F. Uriz-Otano, C. Malafarina, J. A. Martinez, and M. A. Zulet. 2017. "Effectiveness of nutritional supplementation on sarcopenia and recovery in hip fracture patients. A multi-centre randomized trial." *Maturitas* 101:42–50. doi:10.1016/j.maturitas.2017.04.010.

Manczak, M., R. Kandimalla, X. Yin, and P. H. Reddy. 2019. "Mitochondrial division inhibitor 1 reduces dynamin-related protein 1 and mitochondrial fission activity." *Hum Mol Genet* 28 (2):177–199. doi:10.1093/hmg/ddy335.

Martin, J., M. L. Balmer, S. Rajendran, O. Maurhofer, J. F. Dufour, and M. V. St-Pierre. 2016. "Nutritional stress exacerbates hepatic steatosis induced by deletion of the histidine nucleotide-binding (Hint2) mitochondrial protein." *Am J Physiol Gastrointest Liver Physiol* 310 (7):G497–509. doi:10.1152/ajpgi.00178.2015.

Martin, L. J., and Q. Chang. 2018. "DNA damage response and repair, DNA methylation, and cell death in human neurons and experimental animal neurons are different." *J Neuropathol Exp Neurol* 77 (7):636–655. doi:10.1093/jnen/nly040.

Martin, L. J., and M. Wong. 2017. "Enforced DNA repair enzymes rescue neurons from apoptosis induced by target deprivation and axotomy in mouse models of neurodegeneration." *Mech Ageing Dev* 161 (149–162).

Marzetti, E., R. Calvani, M. Cesari, T. W. Buford, M. Lorenzi, B. J. Behnke, and C. Leeuwenburgh. 2013. "Mitochondrial dysfunction and sarcopenia of aging: From signaling pathways to clinical trials." *Int J Biochem Cell Biol* 45 (10):2288–301. doi:10.1016/j.biocel.2013.06.024.

Marzetti, E., A. C. Hwang, M. Tosato, L. N. Peng, R. Calvani, A. Picca, L. K. Chen, and F. Landi. 2018. "Age-related changes of skeletal muscle mass and strength among Italian and Taiwanese older people: Results from the Milan EXPO 2015 survey and the I-Lan Longitudinal Aging Study." *Exp Gerontol* 102:76–80. doi:10.1016/j.exger.2017.12.008.

Marzetti, E., M. Lorenzi, F. Landi, A. Picca, F. Rosa, F. Tanganelli, M. Galli, G. B. Doglietto, F. Pacelli, M. Cesari, R. Bernabei, R. Calvani, and M. Bossola. 2017. "Altered mitochondrial quality control signaling in muscle of old gastric cancer patients with cachexia." *Exp Gerontol* 87 (Pt A):92–99. doi:10.1016/j.exger.2016.10.003.

McArdle, A., N. Pollock, C. A. Staunton, and M. J. Jackson. 2018. "Aberrant redox signalling and stress response in age-related muscle decline: Role in inter- and intra-cellular signalling." *Free Radic Biol Med.* doi:10.1016/j.freeradbiomed.2018.11.038.

McKiernan, S. H., R. J. Colman, E. Aiken, T. D. Evans, T. M. Beasley, J. M. Aiken, R. Weindruch, and R. M. Anderson. 2012. "Cellular adaptation contributes to calorie restriction-induced preservation of skeletal muscle in aged rhesus monkeys." *Exp Gerontol* 47 (3):229–236.

McKiernan, S. H., R. Colman, M. Lopez, T. M. Beasley, R. Weindruch, and J. M. Aiken. 2009. "Longitudinal analysis of early stage sarcopenia in aging rhesus monkeys." *Exp Gerontol* 44 (3):170–176. doi:10.1016/j.exger.2008.09.014.

Meador, B. M., K. A. Mirza, M. Tian, M. B. Skelding, L. A. Reaves, N. K. Edens, M. J. Tisdale, and S. L. Pereira. 2015. "The green tea polyphenol epigallocatechin-3-gallate (EGCg) attenuates skeletal muscle atrophy in a rat model of sarcopenia." *J Frailty Aging* 4 (4):209–215. doi:10.14283/jfa.2015.58.

Milanese, M., F. Giribaldi, M. Melone, T. Bonifacino, I. Musante, E. Carminati, P. I. Rossi, L. Vergani, A. Voci, F. Conti, A. Puliti, and G. Bonanno. 2014. "Knocking down metabotropic glutamate receptor 1 improves survival and disease progression in the SOD1(G93A) mouse model of amyotrophic lateral sclerosis." *Neurobiol Dis* 64:48–59. doi:10.1016/j.nbd.2013.11.006.

Mirza, K. A., S. L. Pereira, N. K. Edens, and M. J. Tisdale. 2014. "Attenuation of muscle wasting in murine C2C 12 myotubes by epigallocatechin-3-gallate." *J Cachexia Sarcopenia Muscle* 5 (4):339–345. doi:10.1007/s13539-014-0139-9.

Momken, I., L. Stevens, A. Bergouignan, D. Desplanches, F. Rudwill, I. Chery, A. Zahariev, S. Zahn, T. P. Stein, J. L. Sebedio, E. Pujos-Guillot, M. Falempin, C. Simon, V. Coxam, T. Andrianjafiniony, G. Gauquelin-Koch, F. Picquet, and S. Blanc. 2011. "Resveratrol prevents the wasting disorders of mechanical unloading by acting as a physical exercise mimetic in the rat." *FASEB J* 25 (10):3646–3660. doi:10.1096/fj.10-177295.

Moulin, M., and A. Ferreiro. 2017. "Muscle redox disturbances and oxidative stress as pathomechanisms and therapeutic targets in early-onset myopathies." *Semin Cell Dev Biol* 64:213–223. doi:10.1016/j.semcdb.2016.08.003.

Mozdziak, P. E., P. M. Pulvermacher, and E. Schultz. 2001. "Muscle regeneration during hindlimb unloading results in a reduction in muscle size after reloading." *J Appl Physiol* 91 (1):183–190.

Muhammad, M. H., and M. M. Allam. 2018. "Resveratrol and/or exercise training counteract aging-associated decline of physical endurance in aged mice; targeting mitochondrial biogenesis and function." *J Physiol Sci* 68 (5):681–688. doi:10.1007/s12576-017-0582-4.

Munro, D., S. Banh, E. Sotiri, N. Tamanna, and J. R. Treberg. 2016. "The thioredoxin and glutathione-dependent H_2O_2 consumption pathways in muscle mitochondria: Involvement in H_2O_2 metabolism and consequence to H_2O_2 efflux assays." *Free Radic. Biol. Med* 96:334–346.

Musci, R. V., K. L. Hamilton, and B. F. Miller. 2018. "Targeting mitochondrial function and proteostasis to mitigate dynapenia." *Eur J Appl Physiol* 118 (1):1–9. doi:10.1007/s00421-017-3730-x.

Navratil, M., A. Terman, and E. A. Arriaga. 2008. "Giant mitochondria do not fuse and exchange their contents with normal mitochondria." *Exp Cell Res* 314 (1):164–172.

Nicassio, L., F. Fracasso, G. Sirago, C. Musicco, A. Picca, E. Marzetti, R. Calvani, P. Cantatore, M. N. Gadaleta, and V. Pesce. 2017. "Dietary supplementation with acetyl-l-carnitine counteracts age-related alterations of mitochondrial biogenesis, dynamics and antioxidant defenses in brain of old rats." *Exp Gerontol* 98:99–109. doi:10.1016/j.exger.2017.08.017.

Nichols, C. E., D. L. Shepherd, T. L. Knuckles, D. Thapa, J. C. Stricker, P. A. Stapleton, V. C. Minarchick, A. Erdely, P. C. Zeidler-Erdely, S. E. Alway, T. R. Nurkiewicz, and J. M. Hollander. 2015. "Cardiac and mitochondrial dysfunction following acute pulmonary exposure to mountaintop removal mining particulate matter." *Am J Physiol Heart Circ Physiol* 309 (12):H2017–H2030. doi:10.1152/ajpheart.00353.2015.

Olaso-Gonzalez, G., B. Ferrando, A. Derbre, A. Salvador-Pascual, H. Cabo, H. Pareja-Galeano, F. Sabater-Pastor, M. C. Gomez-Cabrera, and J. Vina. 2014. "Redox regulation of E3 ubiquitin ligases and their role in skeletal muscle atrophy." *Free Radic Biol Med* 75 Suppl 1:S43–S44.

O'Leary, M. F., A. Vainshtein, H. N. Carter, Y. Zhang, and D. A. Hood. 2012. "Denervation-induced mitochondrial dysfunction and autophagy in skeletal muscle of apoptosis-deficient animals." *Am J Physiol Cell Physiol* 303 (4):C447–C454.

Oliveira, A. N., and D. A. Hood. 2018. "Effect of Tim23 knockdown in vivo on mitochondrial protein import and retrograde signaling to the UPR(mt) in muscle." *Am J Physiol Cell Physiol* 315 (4):C516–C526. doi:10.1152/ajpcell.00275.2017.

Oliveira, M. R., S. F. Nabavi, M. Daglia, L. Rastrelli, and S. M. Nabavi. 2016. "Epigallocatechin gallate and mitochondria-A story of life and death." *Pharmacol Res* 104:70–85. doi:10.1016/j.phrs.2015.12.027.

Ota, N., S. Soga, S. Haramizu, Y. Yokoi, T. Hase, and T. Murase. 2011. "Tea catechins prevent contractile dysfunction in unloaded murine soleus muscle: A pilot study." *Nutrition* 27 (9):955–959. doi:10.1016/j.nut.2010.10.008.

Pae, M., S. N. Meydani, and D. Wu. 2012. "The role of nutrition in enhancing immunity in aging." *Aging Dis* 3 (1):91–129.

Pae, M., Z. Ren, M. Meydani, F. Shang, D. Smith, S. N. Meydani, and D. Wu. 2012. "Dietary supplementation with high dose of epigallocatechin-3-gallate promotes inflammatory response in mice." *J. Nutr Biochem* 23 (6):526–531. doi: 10.1016/j.jnutbio.2011.02.006.

Pardo, P. S., A. Hajira, A. M. Boriek, and J. S. Mohamed. 2017. "MicroRNA-434-3p regulates age-related apoptosis through eIF5A1 in the skeletal muscle." *Aging (Albany NY)* 9 (3):1012–1029.

Parousis, A., H. N. Carter, C. Tran, A. T. Erlich, Z. S. Mesbah Moosavi, M. Pauly, and D. A. Hood. 2018. "Contractile activity attenuates autophagy suppression and reverses mitochondrial defects in skeletal muscle cells." *Autophagy* 14 (11):1886–1897. doi:10.1080/15548627.2018.1491488.

Pence, B. D., T. E. Gibbons, T. K. Bhattacharya, H. Mach, J. M. Ossyra, G. Petr, S. A. Martin, L. Wang, S. S. Rubakhin, J. V. Sweedler, R. H. McCusker, K. W. Kelley, J. S. Rhodes, R. W. Johnson, and J. A. Woods. 2016. "Effects of exercise and dietary epigallocatechin gallate and beta-alanine on skeletal muscle in aged mice." *Appl. Physiol Nutr Metab* 41 (2):181–190.

Picard, M., D. Ritchie, M. M. Thomas, K. J. Wright, and R. T. Hepple. 2011. "Alterations in intrinsic mitochondrial function with aging are fiber type-specific and do not explain differential atrophy between muscles." *Aging Cell* 10 (6):1047–1055.

Picard, M., D. Ritchie, K. J. Wright, C. Romestaing, M. M. Thomas, S. L. Rowan, T. Taivassalo, and R. T. Hepple. 2010. "Mitochondrial functional impairment with aging is exaggerated in isolated mitochondria compared to permeabilized myofibers." *Aging Cell* 9 (6):1032–1046.

Picard, M., O. S. Shirihai, B. J. Gentil, and Y. Burelle. 2013. "Mitochondrial morphology transitions and functions: Implications for retrograde signaling?" *Am J Physiol Regul Integr Comp Physiol* 304 (6):R393–R406.

Picca, A., R. Calvani, M. Bossola, E. Allocca, A. Menghi, V. Pesce, A. M. S. Lezza, R. Bernabei, F. Landi, and E. Marzetti. 2018. "Update on mitochondria and muscle aging: All wrong roads lead to sarcopenia." *Biol Chem* 399 (5):421–436. doi:10.1515/hsz-2017-0331.

Picca, A., R. Calvani, C. Leeuwenburgh, H. J. Coelho-Junior, R. Bernabei, F. Landi, and E. Marzetti. 2018. "Targeting mitochondrial quality control for treating sarcopenia: Lessons from physical exercise." *Expert Opin Ther Targets* 1–8. doi:10.1080/14728222.2019.1559827.

Picca, A., R. Calvani, C. Leeuwenburgh, H. J. Coelho-Junior, R. Bernabei, F. Landi, and E. Marzetti. 2019. "Targeting mitochondrial quality control for treating sarcopenia: Lessons from physical exercise." *Expert Opin Ther Targets* 23 (2):153–160. doi:10.1080/14728222.2019.1559827.

Picca, A., A. M. S. Lezza, C. Leeuwenburgh, V. Pesce, R. Calvani, M. Bossola, E. Manes-Gravina, F. Landi, R. Bernabei, and E. Marzetti. 2018. "Circulating mitochondrial DNA at the crossroads of mitochondrial dysfunction and inflammation during aging and muscle wasting disorders." *Rejuvenation Res* 21 (4):350–359. doi:10.1089/rej.2017.1989.

Pietrangelo, T., E. S. Di Filippo, R. Mancinelli, C. Doria, A. Rotini, G. Fano-Illic, and S. Fulle. 2015. "Low intensity exercise training improves skeletal muscle regeneration potential." *Front Physiol* 6:399. doi:10.3389/fphys.2015.00399.

Pigna, E., A. Renzini, E. Greco, E. Simonazzi, S. Fulle, R. Mancinelli, V. Moresi, and S. Adamo. 2018. "HDAC4 preserves skeletal muscle structure following long-term denervation by mediating distinct cellular responses." *Skelet Muscle* 8 (1):6. doi:10.1186/s13395-018-0153-2.

Pollock, N., C. A. Staunton, A. Vasilaki, A. McArdle, and M. J. Jackson. 2017. "Denervated muscle fibers induce mitochondrial peroxide generation in neighboring innervated fibers: Role in muscle aging." *Free Radic Biol Med* 112:84–92. doi:10.1016/j.freeradbiomed.2017.07.017.

Popa-Wagner, A., R. E. Sandu, C. Cristin, A. Uzoni, K. A. Welle, J. R. Hryhorenko, and S. Ghaemmaghami. 2018. "Increased degradation rates in the components of the mitochondrial oxidative phosphorylation chain in the cerebellum of old mice." *Front Aging Neurosci* 10:32. doi:10.3389/fnagi.2018.00032.

Porter, C., N. M. Hurren, M. V. Cotter, N. Bhattarai, P. T. Reidy, E. L. Dillon, W. J. Durham, D. Tuvdendorj, M. Sheffield-Moore, E. Volpi, L. S. Sidossis, B. B. Rasmussen, and E. Borsheim. 2015. "Mitochondrial respiratory capacity and coupling control decline with age in human skeletal muscle." *Am J Physiol Endocrinol Metab* 309 (3):E224–E232.

Potes, Y., B. de Luxan-Delgado, S. Rodriguez-Gonzalez, M. R. M. Guimaraes, J. J. Solano, M. Fernandez-Fernandez, M. Bermudez, J. A. Boga, I. Vega-Naredo, and A. Coto-Montes. 2017. "Overweight in elderly people induces impaired autophagy in skeletal muscle." *Free Radic Biol Med* 110:31–41. doi:10.1016/j.freeradbiomed.2017.05.018.

Power, G. A., F. C. Minozzo, S. Spendiff, M. E. Filion, Y. Konokhova, M. F. Purves-Smith, C. Pion, M. Aubertin-Leheudre, J. A. Morais, W. Herzog, R. T. Hepple, T. Taivassalo, and D. E. Rassier. 2016. "Reduction in single muscle fiber rate of force development with aging is not attenuated in world class older masters athletes." *Am J Physiol Cell Physiol* 310 (4):C318–C327.

Powers, S. K., L. L. Ji, A. N. Kavazis, and M. J. Jackson. 2011. "Reactive oxygen species: Impact on skeletal muscle." *Compr. Physiol* 1 (2):941–969.

Pryde, K. R., H. L. Smith, K. Y. Chau, and A. H. Schapira. 2016. "PINK1 disables the anti-fission machinery to segregate damaged mitochondria for mitophagy." *J Cell Biol.* 213 (2):163–171.

Qaisar, R., S. Bhaskaran, R. Ranjit, K. Sataranatarajan, P. Premkumar, K. Huseman, and H. Van Remmen. 2019. "Restoration of SERCA ATPase prevents oxidative stress-related muscle atrophy and weakness." *Redox Biol* 20:68–74. doi:10.1016/j.redox.2018.09.018.

Qiu, J., Q. Fang, T. Xu, C. Wu, L. Xu, L. Wang, X. Yang, S. Yu, Q. Zhang, F. Ding, and H. Sun. 2018. "Mechanistic role of reactive oxygen species and therapeutic potential of antioxidants in denervation- or fasting-induced skeletal muscle atrophy." *Front Physiol* 9:215. doi:10.3389/fphys.2018.00215.

Quadrilatero, J., S. E. Alway, and E. E. Dupont-Versteegden. 2011. "Skeletal muscle apoptotic response to physical activity: Potential mechanisms for protection." *Appl Physiol Nutr Metab* 36 (5):608–617. doi:10.1139/h11-064.

Radak, Z., K. Ishihara, E. Tekus, C. Varga, A. Posa, L. Balogh, I. Boldogh, and E. Koltai. 2017. "Exercise, oxidants, and antioxidants change the shape of the bell-shaped hormesis curve." *Redox Biol* 12:285–290. doi:10.1016/j.redox.2017.02.015.

Radak, Z., F. Torma, I. Berkes, S. Goto, T. Mimura, A. Posa, L. Balogh, I. Boldogh, K. Suzuki, M. Higuchi, and E. Koltai. 2018. "Exercise effects on physiological function during aging." *Free Radic Biol Med.* doi:10.1016/j.freeradbiomed.2018.10.444.

Rathbone, C. R., F. W. Booth, and S. J. Lees. 2009. "Sirt1 increases skeletal muscle precursor cell proliferation." *Eur J Cell Biol* 88 (1):35–44. doi:10.1016/j.ejcb.2008.08.003.

Ren, X., J. T. R. Keeney, S. Miriyala, T. Noel, D. K. Powell, L. Chaiswing, S. Bondada, D. K. St Clair, and D. A. Butterfield. 2018. "The triangle of death of neurons: Oxidative damage, mitochondrial dysfunction, and loss of choline-containing biomolecules in brains of mice treated with doxorubicin. Advanced insights into mechanisms of chemotherapy induced cognitive impairment ('chemobrain') involving TNF-alpha." *Free Radic Biol Med* 134:1–8. doi:10.1016/j.freeradbiomed.2018.12.029.

Robinson, A. R., M. J. Yousefzadeh, T. A. Rozgaja, J. Wang, X. Li, J. S. Tilstra, C. H. Feldman, S. Q. Gregg, C. H. Johnson, E. M. Skoda, M. C. Frantz, et al., 2018. "Spontaneous DNA damage to the nuclear genome promotes senescence, redox imbalance and aging." *Redox Biol* 17:259–273. doi:10.1016/j.redox.2018.04.007.

Robinson, E. J., S. P. Aguiar, W. M. Kouwenhoven, D. S. Starmans, L. von Oerthel, M. P. Smidt, and L. P. van der Heide. 2018. "Survival of midbrain dopamine neurons depends on the Bcl2 factor Mcl1." *Cell Death Discov* 4:107. doi:10.1038/s41420-018-0125-7.

Rom, O., and A. Z. Reznick. 2015. "The role of E3 ubiquitin-ligases MuRF-1 and MAFbx in loss of skeletal muscle mass." *Free Radic Biol Med* I (doi:10.1016/j.freeradbiomed.2015.12.031 [Epub ahead of print]).

Ron-Harel, N., G. Notarangelo, J. M. Ghergurovich, J. A. Paulo, P. T. Sage, D. Santos, F. K. Satterstrom, S. P. Gygi, J. D. Rabinowitz, A. H. Sharpe, and M. C. Haigis. 2018. "Defective respiration and one-carbon metabolism contribute to impaired naive T cell activation in aged mice." *Proc Natl Acad Sci U S A* 115 (52):13347–13352. doi:10.1073/pnas.1804149115.

Rossi, A. P., A. D'Introno, S. Rubele, C. Caliari, S. Gattazzo, E. Zoico, G. Mazzali, F. Fantin, and M. Zamboni. 2017. "The potential of beta-hydroxy-beta-methylbutyrate as a new strategy for the management of sarcopenia and sarcopenic obesity." *Drugs Aging* 34 (11):833–840. doi:10.1007/s40266-017-0496-0.

Russell, S. T., and M. J. Tisdale. 2009. "Mechanism of attenuation by beta-hydroxy-beta-methylbutyrate of muscle protein degradation induced by lipopolysaccharide." *Mol Cell Biochem* 330 (1–2):171–179. doi:10.1007/s11010-009-0130-5.

Ryan, M. J., H. J. Dudash, M. Docherty, K. B. Geronilla, B. A. Baker, G. G. Haff, R. G. Cutlip, and S. E. Alway. 2010. "Vitamin E and C supplementation reduces oxidative stress, improves antioxidant enzymes and positive muscle work in chronically loaded muscles of aged rats." *Exp Gerontol* 45 (11):882–895. doi:10.1016/j.exger.2010.08.002.

Ryan, M. J., J. R. Jackson, Y. Hao, S. S. Leonard, and S. E. Alway. 2011. "Inhibition of xanthine oxidase reduces oxidative stress and improves skeletal muscle function in response to electrically stimulated isometric contractions in aged mice." *Free Radic Biol Med* 51 (1):38–52.

Ryan, M. J., J. R. Jackson, Y. Hao, C. L. Williamson, E. R. Dabkowski, J. M. Hollander, and S. E. Alway. 2010. "Suppression of oxidative stress by resveratrol after isometric contractions in gastrocnemius muscles of aged mice." *J Gerontol A Biol Sci Med Sci* 65 (8):815–831.

Rygiel, K. A., J. P. Grady, and D. M. Turnbull. 2014. "Respiratory chain deficiency in aged spinal motor neurons." *Neurobiol Aging* 35 (10):2230–2238. doi:10.1016/j.neurobiolaging.2014.02.027.

Rygiel, K. A., M. Picard, and D. M. Turnbull. 2016. "The ageing neuromuscular system and sarcopenia—A mitochondrial perspective." *J Physiol* 154 (16):4499–4512.

Safdar, A., S. Annis, Y. Kraytsberg, C. Laverack, A. Saleem, K. Popadin, D. C. Woods, J. L. Tilly, and K. Khrapko. 2016. "Amelioration of premature aging in mtDNA mutator mouse by exercise: The interplay of oxidative stress, PGC-1alpha, p53, and DNA damage. A hypothesis." *Curr Opin Genet Dev* 38:127–132.

Safdar, A., K. Khrapko, J. M. Flynn, A. Saleem, M. De Lisio, A. P. Johnston, Y. Kratysberg, I. A. Samjoo, Y. Kitaoka, D. I. Ogborn, J. P. Little, S. Raha, G. Parise, M. Akhtar, B. P. Hettinga, G. C. Rowe, Z. Arany, T. A. Prolla, and M. A. Tarnopolsky. 2016. "Exercise-induced mitochondrial p53 repairs mtDNA mutations in mutator mice." *Skelet Muscle* 6:7. doi:10.1186/s13395-016-0075-9.

Saini, A., A. P. Sharples, N. Al-Shanti, and C. E. Stewart. 2017. "Omega-3 fatty acid EPA improves regenerative capacity of mouse skeletal muscle cells exposed to saturated fat and inflammation." *Biogerontology* 18 (1):109–129. doi:10.1007/s10522-016-9667-3.

Saleem, A., P. J. Adhihetty, and D. A. Hood. 2009. "Role of p53 in mitochondrial biogenesis and apoptosis in skeletal muscle." *Physiol Genomics* 37 (1):58–66.

Sataranatarajan, K., R. Qaisar, C. Davis, G. K. Sakellariou, A. Vasilaki, Y. Zhang, Y. Liu, S. Bhaskaran, A. McArdle, M. Jackson, S. V. Brooks, A. Richardson, and H. Van Remmen. 2015. "Neuron specific reduction in CuZnSOD is not sufficient to initiate a full sarcopenia phenotype." *Redox Biol* 5:140–148. doi:10.1016/j.redox.2015.04.005.

Sebastian, D., E. Sorianello, J. Segales, A. Irazoki, V. Ruiz-Bonilla, D. Sala, E. Planet, A. Berenguer-Llergo, J. P. Munoz, M. Sanchez-Feutrie, N. Plana, M. I. Hernandez-Alvarez, A. L. Serrano, M. Palacin, and A. Zorzano. 2016. "Mfn2 deficiency links age-related sarcopenia and impaired autophagy to activation of an adaptive mitophagy pathway." *EMBO J* 35 (15):1677–1693. doi:10.15252/embj.201593084.

Shi, Y., X. Guo, J. Zhang, H. Zhou, B. Sun, and J. Feng. 2018. "DNA binding protein HMGB1 secreted by activated microglia promotes the apoptosis of hippocampal neurons in diabetes complicated with OSA." *Brain Behav Immun* 73:482–492. doi:10.1016/j.bbi.2018.06.012.

Simioni, C., G. Zauli, A. M. Martelli, M. Vitale, G. Sacchetti, A. Gonelli, and L. M. Neri. 2018. "Oxidative stress: Role of physical exercise and antioxidant nutraceuticals in adulthood and aging." *Oncotarget* 9 (24):17181–17198. doi:10.18632/oncotarget.24729.

Sinha, M., Y. C. Jang, J. Oh, D. Khong, E. Y. Wu, R. Manohar, C. Miller, S. G. Regalado, F. S. Loffredo, J. R. Pancoast, M. F. Hirshman, J. Lebowitz, J. L. Shadrach, M. Cerletti, M. J. Kim, T. Serwold, L. J. Goodyear, B. Rosner, R. T. Lee, and A. J. Wagers. 2014. "Restoring systemic GDF11 levels reverses age-related dysfunction in mouse skeletal muscle." *Science* 344 (6184):649–652.

Sinha-Hikim, I., A. P. Sinha-Hikim, M. Parveen, R. Shen, R. Goswami, P. Tran, A. Crum, and K. C. Norris. 2013. "Long-term supplementation with a cystine-based antioxidant delays loss of muscle mass in aging." *J Gerontol A Biol Sci Med Sci* 68 (7):749–759.

Siu, P. M., and S. E. Alway. 2005. "Mitochondria-associated apoptotic signalling in denervated rat skeletal muscle." *J Physiol* 565 (Pt 1):309–323.

Siu, P. M., and S. E. Alway. 2006. "Deficiency of the Bax gene attenuates denervation-induced apoptosis." *Apoptosis* 11 (6):967–981. doi:10.1007/s10495-006-6315-4.

Siu, P. M., and S. E. Alway. 2009. "Response and adaptation of skeletal muscle to denervation stress: The role of apoptosis in muscle loss." *Front Biosci* 14:432–452.

Siu, P. M., Y. Wang, and S. E. Alway. 2009. "Apoptotic signaling induced by H_2O_2-mediated oxidative stress in differentiated C_2C_{12} myotubes." *Life Sci* 84 (13–14):468–481.

Smith, G. I., P. Atherton, D. N. Reeds, B. S. Mohammed, D. Rankin, M. J. Rennie, and B. Mittendorfer. 2011. "Dietary omega-3 fatty acid supplementation increases the rate of muscle protein synthesis in older adults: A randomized controlled trial." *Am J Clin Nutr* 93 (2):402–412.

Smith, G. I., S. Julliand, D. N. Reeds, D. R. Sinacore, S. Klein, and B. Mittendorfer. 2015. "Fish oil-derived n-3 PUFA therapy increases muscle mass and function in healthy older adults." *Am J Clin Nutr* 102 (1):115–122.

Smith, N. T., A. Soriano-Arroquia, K. Goljanek-Whysall, M. J. Jackson, and B. McDonagh. 2018. "Redox responses are preserved across muscle fibres with differential susceptibility to aging." *J Proteomics* 177:112–123. doi:10.1016/j.jprot.2018.02.015.

Sousa-Victor, P., L. Garcia-Prat, and P. Munoz-Canoves. 2018. "New mechanisms driving muscle stem cell regenerative decline with aging." *Int J Dev Biol* 62 (6–7-8):583–590. doi:10.1387/ijdb.180041pm.

Spendiff, S., M. Vuda, G. Gouspillou, S. Aare, A. Perez, J. A. Morais, R. T. Jagoe, M. E. Filion, R. Glicksman, S. Kapchinsky, N. J. MacMillan, C. H. Pion, M. Aubertin-Leheudre, S. Hettwer, J. A. Correa, T. Taivassalo, and R. T. Hepple. 2016. "Denervation drives mitochondrial dysfunction in skeletal muscle of octogenarians." *J Physiol* 594 (24):7361–7379. doi:10.1113/JP272487.

Sriram, S., S. Subramanian, D. Sathiakumar, R. Venkatesh, M. S. Salerno, C. D. McFarlane, R. Kambadur, and M. Sharma. 2011. "Modulation of reactive oxygen species in skeletal muscle by myostatin is mediated through NF-kappaB." *Aging Cell* 10 (6):931–948.

St-Jean-Pelletier, F., C. H. Pion, J. P. Leduc-Gaudet, N. Sgarioto, I. Zovile, S. Barbat-Artigas, O. Reynaud, F. Alkaterji, F. C. Lemieux, A. Grenon, P. Gaudreau, R. T. Hepple, S. Chevalier, M. Belanger, J. A. Morais, M. Aubertin-Leheudre, and G. Gouspillou. 2017. "The impact of ageing, physical activity, and pre-frailty on skeletal muscle phenotype, mitochondrial content, and intramyocellular lipids in men." *J Cachexia Sarcopenia Muscle* 8 (2):213–228. doi:10.1002/jcsm.12139.

Sun, L. J., Y. N. Sun, S. J. Chen, S. Liu, and G. R. Jiang. 2017. "Resveratrol attenuates skeletal muscle atrophy induced by chronic kidney disease via MuRF1 signaling pathway." *Biochem Biophys Res Commun* 487 (1):83–89. doi:10.1016/j.bbrc.2017.04.022.

Takahashi, H., Y. Suzuki, J. S. Mohamed, T. Gotoh, S. L. Pereira, and S. E. Alway. 2017. "Epigallocatechin-3-gallate increases autophagy signaling in resting and unloaded plantaris muscles but selectively suppresses autophagy protein abundance in reloaded muscles of aged rats." *Exp Gerontol* 92:56–66. doi:10.1016/j.exger.2017.02.075.

Tarry-Adkins, J. L., D. S. Fernandez-Twinn, J. H. Chen, I. P. Hargreaves, V. Neergheen, C. E. Aiken, and S. E. Ozanne. 2016. "Poor maternal nutrition and accelerated postnatal growth induces an accelerated aging phenotype and oxidative stress in skeletal muscle of male rats." *Dis Model Mech* 9 (10):1221–1229. doi:10.1242/dmm.026591.

Tezze, C., V. Romanello, M. A. Desbats, G. P. Fadini, M. Albiero, G. Favaro, S. Ciciliot, M. E. Soriano, V. Morbidoni, C. Cerqua, S. Loefler, H. Kern, C. Franceschi, S. Salvioli, M. Conte, B. Blaauw, S. Zampieri, L. Salviati, L. Scorrano, and M. Sandri. 2017. "Age-associated loss of OPA1 in muscle impacts muscle mass, metabolic homeostasis, systemic inflammation, and epithelial senescence." *Cell Metab* 25 (6):1374–1389 e6. doi:10.1016/j.cmet.2017.04.021.

Tomlinson, B. E., and D. Irving. 1977. "The numbers of limb motor neurons in the human lumbosacral cord throughout life." *J Neurol Sci* 34 (2):213–219.

Townsend, J. R., M. S. Fragala, A. R. Jajtner, A. M. Gonzalez, A. J. Wells, G. T. Mangine, E. H. th Robinson, W. P. McCormack, K. S. Beyer, G. J. Pruna, C. H. Boone, T. M. Scanlon, J. D. Bohner, J. R. Stout, and J. R. Hoffman. 2013. "Beta-Hydroxy-beta-methylbutyrate (HMB)-free acid attenuates circulating TNF-alpha and TNFR1 expression postresistance exercise." *J Appl Physiol (1985)* 115 (8):1173–1182. doi:10.1152/japplphysiol.00738.2013.

Truong, V. L., M. Jun, and W. S. Jeong. 2018. "Role of resveratrol in regulation of cellular defense systems against oxidative stress." *Biofactors* 44 (1):36–49. doi:10.1002/biof.1399.

Ungvari, Z., N. Labinskyy, S. Gupte, P. N. Chander, J. G. Edwards, and A. Csiszar. 2008. "Dysregulation of mitochondrial biogenesis in vascular endothelial and smooth muscle cells of aged rats." *Am. J Physiol Heart Circ. Physiol* 294 (5):H2121–H2128.

Vainshtein, A., and D. A. Hood. 2016. "The regulation of autophagy during exercise in skeletal muscle." *J Appl Physiol (1985)*. 120 (6):664–673. doi: 10.1152/japplphysiol.00550.2015.

Vainshtein, A., E. M. Desjardins, A. Armani, M. Sandri, and D. A. Hood. 2015. "PGC-1alpha modulates denervation-induced mitophagy in skeletal muscle." *Skelet Muscle* 5:9. doi:10.1186/s13395-015-0033-y.

Vainshtein, A., L. D. Tryon, M. Pauly, and D. A. Hood. 2015. "Role of PGC-1alpha during acute exercise-induced autophagy and mitophagy in skeletal muscle." *Am J Physiol Cell Physiol* 308 (9):C710–C719.

Walsh, M. E., A. Bhattacharya, K. Sataranatarajan, R. Qaisar, L. Sloane, M. M. Rahman, M. Kinter, and H. Van Remmen. 2015. "The histone deacetylase inhibitor butyrate improves metabolism and reduces muscle atrophy during aging." *Aging Cell* 14 (6):957–970. doi:10.1111/acel.12387.

Wanagat, J., N. Ahmadieh, J. H. Bielas, N. G. Ericson, and H. Van Remmen. 2015. "Skeletal muscle mitochondrial DNA deletions are not increased in CuZn-superoxide dismutase deficient mice." *Exp Gerontol* 61:15–19. doi:10.1016/j.exger.2014.11.012.

Wang, B., Z. Cai, K. Tao, W. Zeng, F. Lu, R. Yang, D. Feng, G. Gao, and Q. Yang. 2016. "Essential control of mitochondrial morphology and function by chaperone-mediated autophagy through degradation of PARK7." *Autophagy* 12 (8):1215–1228. doi:10.1080/15548627.2016.1179401.

Wang, D., H. Sun, G. Song, Y. Yang, X. Zou, P. Han, and S. Li. 2018. "Resveratrol improves muscle atrophy by modulating mitochondrial quality control in STZ-induced diabetic mice." *Mol Nutr Food Res* 62 (9):e1700941. doi:10.1002/mnfr.201700941.

Wang, L., T. Zhang, Y. Xi, C. Yang, C. Sun, and D. Li. 2016. "Sirtuin 1 promotes the proliferation of C2C12 myoblast cells via the myostatin signaling pathway." *Mol Med Rep* 14 (2):1309–1315. doi:10.3892/mmr.2016.5346.

Wang, Y., J. S. Mohamed, and S. E. Alway. 2013. "M-cadherin-inhibited phosphorylation of B-catenin augments differentiation of mouse myoblasts." *Cell Tissue Res* 351 (1):183–200.

Wang, Y., Y. Sun, X. Zhao, R. Yuan, H. Jiang, and X. Pu. 2018. "Downregulation of DJ-1 fails to protect mitochondrial complex i subunit NDUFS3 in the testes and contributes to the asthenozoospermia." *Mediators Inflamm* 2018:6136075. doi:10.1155/2018/6136075.

Wang, Z., J. Liu, S. Chen, Y. Wang, L. Cao, Y. Zhang, W. Kang, H. Li, Y. Gui, S. Chen, and J. Ding. 2011. "DJ-1 modulates the expression of Cu/Zn-superoxide dismutase-1 through the Erk1/2-Elk1 pathway in neuroprotection." *Ann Neurol* 70 (4):591–599. doi:10.1002/ana.22514.

White, Z., J. Terrill, R. B. White, C. McMahon, P. Sheard, M. D. Grounds, and T. Shavlakadze. 2016. "Voluntary resistance wheel exercise from mid-life prevents sarcopenia and increases markers of mitochondrial function and autophagy in muscles of old male and female C57BL/6J mice." *Skelet Muscle* 6 (1):45. doi:10.1186/s13395-016-0117-3.

Woo, J. 2018. "Nutritional interventions in sarcopenia: Where do we stand?" *Curr Opin Clin Nutr Metab Care* 21 (1):19–23. doi:10.1097/MCO.0000000000000432.

Xia, N., A. Daiber, U. Forstermann, and H. Li. 2017. "Antioxidant effects of resveratrol in the cardiovascular system." *Br J Pharmacol* 174 (12):1633–1646. doi:10.1111/bph.13492.

Xiao, Y., C. Karam, J. Yi, L. Zhang, X. Li, D. Yoon, H. Wang, K. Dhakal, P. Ramlow, T. Yu, Z. Mo, J. Ma, and J. Zhou. 2018. "ROS-related mitochondrial dysfunction in skeletal muscle of an ALS mouse model during the disease progression." *Pharmacol Res* 138:25–36. doi:10.1016/j.phrs.2018.09.008.

Xu, Y., L. Nie, Y. G. Yin, J. L. Tang, J. Y. Zhou, D. D. Li, and S. W. Zhou. 2012. "Resveratrol protects against hyperglycemia-induced oxidative damage to mitochondria by activating SIRT1 in rat mesangial cells." *Toxicol Appl Pharmacol* 259 (3):395–401. doi:10.1016/j.taap.2011.09.028.

Yakabe, M., S. Ogawa, H. Ota, K. Iijima, M. Eto, Y. Ouchi, and M. Akishita. 2018. "Inhibition of interleukin-6 decreases atrogene expression and ameliorates tail suspension-induced skeletal muscle atrophy." *PLoS One* 13 (1):e0191318. doi:10.1371/journal.pone.0191318.

Yeo, D., C. Kang, M. C. Gomez-Cabrera, J. Vina, and L. L. Ji. 2018. "Intensified mitophagy in skeletal muscle with aging is downregulated by PGC-1alpha overexpression in vivo." *Free Radic Biol Med* 130:361–368. doi:10.1016/j.freeradbiomed.2018.10.456.

Yoo, S. Z., M. H. No, J. W. Heo, D. H. Park, J. H. Kang, S. H. Kim, and H. B. Kwak. 2018. "Role of exercise in age-related sarcopenia." *J Exerc Rehabil* 14 (4):551–558. doi:10.12965/jer.1836268.134.

Yu, H., J. N. Waddell, S. Kuang, and C. A. Bidwell. 2014. "Park7 expression influences myotube size and myosin expression in muscle." *PLoS.One.* 9 (3):e92030.

Yuan, Y., and S. S. Pan. 2018. "Parkin mediates mitophagy to participate in cardioprotection induced by late exercise preconditioning but bnip3 does not." *J Cardiovasc Pharmacol* 71 (5):303–316. doi:10.1097/FJC.0000000000000572.

Zeng, X., and X. Tan. 2015. "Epigallocatechin-3-gallate and zinc provide anti-apoptotic protection against hypoxia/reoxygenation injury in H9c2 rat cardiac myoblast cells." *Mol Med Rep* 12 (2):1850–1856. doi:10.3892/mmr.2015.3603.

Zhang, Y., S. Iqbal, M. F. O'Leary, K. J. Menzies, A. Saleem, S. Ding, and D. A. Hood. 2013. "Altered mitochondrial morphology and defective protein import reveal novel roles for Bax and/or Bak in skeletal muscle." *Am J Physiol Cell Physiol* 305 (5):C502–C511.

Zhang, Y., X. R. Li, L. Zhao, G. L. Duan, L. Xiao, and H. P. Chen. 2018. "DJ-1 preserving mitochondrial complex I activity plays a critical role in resveratrol-mediated cardioprotection against hypoxia/reoxygenation-induced oxidative stress." *Biomed Pharmacother* 98:545–552. doi:10.1016/j.biopha.2017.12.094.

Zhao, L., S. Liu, J. Xu, W. Li, G. Duan, H. Wang, H. Yang, Z. Yang, and R. Zhou. 2017. "A new molecular mechanism underlying the EGCG-mediated autophagic modulation of AFP in HepG2 cells." *Cell Death Dis* 8 (11):e3160. doi:10.1038/cddis.2017.563.

Zhao, Y., Q. Zhu, W. Song, and B. Gao. 2018. "Exercise training and dietary restriction affect PINK1/Parkin and Bnip3/Nix-mediated cardiac mitophagy in mice." *Gen Physiol Biophys* 37 (6):657–666. doi:10.4149/gpb_2018020.

Zhou, W., J. C. Barkow, and C. R. Freed. 2017. "Running wheel exercise reduces alpha-synuclein aggregation and improves motor and cognitive function in a transgenic mouse model of Parkinson's disease." *PLoS One* 12 (12):e0190160. doi:10.1371/journal.pone.0190160.

Zhu, T. T., W. F. Zhang, P. Luo, F. He, X. Y. Ge, Z. Zhang, and C. P. Hu. 2017. "Epigallocatechin-3-gallate ameliorates hypoxia-induced pulmonary vascular remodeling by promoting mitofusin-2-mediated mitochondrial fusion." *Eur J Pharmacol* 809:42–51. doi:10.1016/j.ejphar.2017.05.003.

Ziaaldini, M. M., S. R. Hosseini, and M. Fathi. 2017. "Mitochondrial adaptations in aged skeletal muscle: Effect of exercise training." *Physiol Res* 66 (1):1–14.

Ziaaldini, M. M., E. Marzetti, A. Picca, and Z. Murlasits. 2017. "Biochemical pathways of sarcopenia and their modulation by physical exercise: a narrative review." *Front Med (Lausanne)* 4:167. doi:10.3389/fmed.2017.00167.

15 Beneficial Effects and Limitations of Strategies (Nutritional or Other) to Limit Muscle Wasting due to Normal Aging

Dominique Meynial-Denis

CONTENTS

15.1   INTRODUCTION

The progressive loss of skeletal muscle mass and strength is a hallmark of aging and is referred to as sarcopenia (Rosenberg 1997; Wang and Bai 2012; Palus, von Haehling, and Springer 2014; Cruz-Jentoft et al. 2014). The European Working Group on Sarcopenia in Older People (EWGSOP) recently proposed new cutoff thresholds for skeletal muscle index, handgrip strength (GS), and calf circumference for a clear definition of sarcopenia (Bahat et al. 2016). However, further worldwide studies from different nations and countries are being conducted to obtain better reference values by societies including the Asian Working group for Sarcopenia (AWGS), the European Society for Clinical Nutrition and Metabolism Special Interest group (ESPENSIG), the European Working Group on Sarcopenia in Older People (EWGSOP) the Foundation for the National Institutes of Health (FNIH), the International Working Group on Sarcopenia (IWGS), the Japan Society of Hepatology (JSH), and the Society on Sarcopenia, Cachexia and Wasting Disorders (SSCWD). In 2016, sarcopenia was recognized as a disease state with its own *ICD-10-CM* (*International Classification of Disease, Tenth Revision, Clinical Modification*; see www.prweb.com-prweb13376057).

The impact of sarcopenia on health and well-being is broad and includes impaired function, increased morbidity and incidence of institutionalization, and reduced quality of life and even death. Sarcopenia is the major contributor to frailty in the elderly and increases the risk of age-related diseases and injury. The structural changes induced by aging include a reduction in the muscle mass and muscle fibers and a shift of muscle fibers toward type 1 fibers (Nair 2005). It has been hypothesized that with aging, skeletal muscle tissue becomes less sensitive to the main anabolic stimuli, nutrition, and physical activity. This reduced muscle protein synthetic response (Mosoni et al. 1993) is also referred to as "anabolic resistance" and may be a key factor in the development of sarcopenia. It can also be expressed by an increase in the anabolic threshold toward the stimulatory effect of amino acids (Dardevet et al. 2012) and by decreases in the synthesis rates of many muscle proteins, specifically of myosin heavy chain and mitochondrial proteins. These processes occur with age and are accentuated under stress conditions. Muscle loss is greatly accelerated in the elderly intensive care unit patient (Phillips et al. 2017). In 2018, Deer and Volpi reported that protein requirements are dependent on the stage of critical illness to provide the maximum health and functional benefits in critically ill geriatric patients (Deer and Volpi 2018).

Aging is accompanied by skeletal muscle remodeling that drives a myriad of adverse metabolic changes. Depending on the macronutrient (glucose or lipid or protein or amino acid), these changes are mediated in a sex-specific manner. Despite greater skeletal muscle atrophy in males, both males and females have a blunted anabolic response to dietary protein and amino acid intake. The blunted response diminishes protein accretion, which affects muscle mass. Minimal research supports sex differences in age-related changes in skeletal muscle protein metabolism, but one could speculate that males are affected more by blunted protein synthesis (Gheller et al. 2016). Besides life style-related factors, such as diet and physical activity, sarcopenia also seems to be determined by hormonal dysregulation, chronic inflammatory status, ectopic adipose tissue accumulation, and neurological and vascular changes associated with aging (Budui, Rossi, and Zamboni 2015). Sarcopenia is also defined by a molecular signature that corresponds to 45 genes, of which 27 are upregulated and 18 are downregulated. This molecular signature is useful as a molecular model to judge the effectiveness of exercise and other therapeutic treatments aimed at ameliorating the effects of muscle loss associated with aging (Giresi et al. 2005).

Given that sarcopenia requires a multidimensional approach, the aim of this chapter is to better understand its pathophysiology and to define the molecular targets to combat and/or treat it.

15.1.1   Emerging Stem Cell Strategies to Combat Sarcopenia

This section is an update on recent research investigating the role of muscle satellite cells during muscle fiber regeneration and growth in both animal and human muscles, with an emphasis on the role they play in age-related sarcopenia and how they can intervene to fight sarcopenia.

Because stem cells are among the most long-lived cells, their age-associated decline is a major contributor to organismal aging and associated diseases. Common hallmarks of aging stem cells were recently identified (Garcia-Prat and Munoz-Canoves 2017), such as altered proteostasis (protein homeostasis), DNA damage and reactive oxygen species (ROS) production, epigenetic alterations, mitochondrial dysfunction, and altered metabolism or senescence entry due to the upregulation of regulator gene of senescence p16^{INK4a}. Emerging rejuvenation strategies of stem cells result in muscle regeneration during aging. This could be of interest for the increasing aging human population worldwide. The idea developed in this section on stem cells consists in the use of satellite cell function as a therapeutic target in the treatment of sarcopenia.

15.1.1.1 Hypertrophy

Muscle satellite cells are indispensable for muscle regeneration and for muscle fiber hypertrophy. While it is still debated whether satellite cells are required for hypertrophy, there are a growing number of studies that have consistently demonstrated that muscle fiber hypertrophy is accompanied by a concomitant increase in satellite cell and/or myonuclear content during prolonged resistance-type exercise training (Snijders and Parise 2017).

15.1.1.2 Autophagy

Autophagy refers to the process of self-degradation of cellular or organelles, cytosolic portions and misfolded proteins by autophagosomes, which are delivered to the lysosomes, preventing waste accumulation in the cell. This basal autophagy is essential to maintain the stem-cell quiescent state in mice. Failure of autophagy in physiologically aged satellite cells causes entry into senescence by loss of proteostasis, increased mitochondrial dysfunction (mitophagy), and oxidative stress (increase in ROS), resulting in a decline in the function and number of satellite cells. Re-establishment of autophagy reverses senescence and restores regenerative functions in geriatric satellite cells. Similar processes were observed in human satellite cells from geriatric individuals. Hence, autophagy, which maintains stemness by preventing senescence in sarcopenia, is a decisive stem-cell-fate regulator (Garcia-Prat, Munoz-Canoves, and Martinez-Vicente 2017).

15.1.1.3 Mitochondrial Dysfunction in Aging Muscle

Age-related decrease in muscle mass is characterized by impaired functional properties of mitochondria and an age-associated decrease in lysosomal function (O'Leary et al. 2013). This mitochondrial dysfunction in aging muscle is probably due to alterations in mitophagy (Hepple 2014). The mitochondria accumulate a greater level of autophagy proteins. Mitochondrial dysfunction, protein synthesis/degradation dysregulation, systemic inflammation, and loss of supporting cell types such as satellite cells and motor neurons can be causal or secondary factors contributing to age-related muscle wasting. However, there is evidence that impairments in satellite cell function with aging result in an impaired muscle fiber regenerative response (Snijders and Parise 2017). Moreover, systemic but mild damage to mtDNA, an event previously associated with normal aging, is likely to cause muscle loss not by directly affecting mature muscle but rather by affecting more sensitive myosatellite cell pools (Wang et al. 2013).

Mitochondrial homeostasis is the primary mechanism for modulating cellular and organismal longevity. Hence, induction of mitochondria repair in stem cells what corresponds to a strategy to combat age-related degenerative diseases (Garcia-Prat and Munoz-Canoves 2017). This is in good agreement with previous work by Marzetti and his collaborators, who hypothesized that stem cells are valuable targets for the prevention and treatment of sarcopenia (Marzetti et al. 2013). With advancing age, a significant reason for the observed decline in mitochondrial function is reduced physical activity. Consequently, exercise is an important strategy to stimulate mitochondrial adaptations in older individuals to foster improvements in muscle function and quality of life (Carter, Chen, and Hood 2015).

15.1.1.4 Rejuvenation

Senescence is not an insurmountable barrier to regeneration, as p16^{INK4a} inhibition suffices to reju-venate satellite cells and restore regeneration in physiologically aged mice. p16^{INK4a} silencing in geriatric satellite cells restores quiescence and muscle regenerative functions. As p16^{INK4a} is dys-regulated in human geriatric satellite cells, the gene encoding for this gene provides the basis for stem-cell rejuvenation in sarcopenic muscles (Sousa-Victor et al. 2014). Another theoretical thera-peutic approach is directed toward the rejuvenation of the stem cell niche through its exposure to a "younger environment" (Munoz-Canoves et al. 2016). Marzetti and colleagues showed that exposure of old muscle to a blood supply from a young mouse restored the proliferative and regen-erative capacity of aged satellite cells without recruitment of young cells (Marzetti et al. 2013). Current efforts to rejuvenate stem cells in aged mice include genetic and pharmacological inhi-bition of p16^{INK4a}, STAT3, and p38 mitogen-activated protein kinase (MAPK), augmentation of autophagic flux, NAD$^+$ repletion, and the administration of rejuvenating hormones such as oxyto-cin. Interestingly, stem cell activity has been observed to increase in response to simple lifestyle changes that modify cell metabolism, such as adopting a low-calorie diet. Similarly, exercise has been shown to enhance stem cell number and function and hence to promote better muscle regenera-tion in rodents (Bengal et al. 2017).

15.1.1.5 Intrinsic-Aging Clock

The processes of aging and circadian rhythms are intimately intertwined, but how peripheral clocks involved in metabolic homeostasis contribute to aging remains unknown. Importantly, calorie restriction (CR) extends lifespan in several organisms and rewires circadian metabolism. These age-dependent changes occur in a highly tissue-specific manner, as demonstrated by comparing circadian gene expression in the liver versus epidermal and skeletal muscle stem cells. Cyclic pro-tein acetylation is lost in old mice while CR results in hyperacetylation. Caloric restriction reorga-nizes the circadian metabolic pathway linked to NAD$^+$—SIRT1—AceCS1 in the liver (Sato et al. 2017). Tissue stem cells retain a robust rhythmic circadian machinery during aging. Daily rhythms are reprogrammed in aged stem cells to cope with tissue-specific stress such as DNA damage or inefficient autophagy. Importantly, deletion of circadian clock components does not reproduce the hallmarks of this reprogramming, underscoring that rewiring, rather than arrhythmia, is associated with physiological aging. Rewiring of daily rhythms in aged stem cells is prevented by CR. Indeed, CR has a profound protective effect on many of the rhythmic changes that occur during physiologi-cal aging (Solanas et al. 2017). In short, the intrinsic-aging clock in stem cells can be pharmacologi-cally manipulated and used in antiaging research.

15.1.1.6 Capillarization

Aging has an impact mainly on type II muscle fibers. Moreover, there is a greater distance between capillaries and type II muscle fiber-associated stem cells in old muscle than in young muscle. This dis-tance could be a critical factor in the impaired regulation of the stem cell pool in senescent muscle. However, the idea that satellite cell function is a therapeutic target to treat sarcopenia is debated (Snijders and Parise 2017). The similar capillary distribution indicates that capillary rarefaction dur-ing aging does not occur at random but maintains the distribution of capillaries to preserve the potential for intramuscular oxygenation. In addition, muscle fiber oxidative capacity is the same irre-spective of age (Barnouin et al. 2017). Consequently, to relieve aged muscle of its impaired satellite cell function could be a viable strategy to counteract the ill effect of sarcopenia on skeletal muscle.

15.1.1.7 MicroRNAs

The skeletal muscle satellite cell population is closely regulated by myogenic transcription factors and a subset of miRNAs that control genes governing proliferation and differentiation. At the cel-lular level, miRNAs are involved in the transitioning of muscle stem cells from a quiescent state to either an activated or a senescence state. In future, therapeutic approaches targeting miRNAs in muscle through diet, exercise, or drugs could slow down or prevent age-related changes in muscle and thus promote healthy aging (McGregor, Poppitt, and Cameron-Smith 2014). The activation of inflammatory signal pathways due to aging is suggested to reveal the critical impact on sarcopenia.

Several pro-inflammatory cytokines, especially interleukin-6 (IL-6) and tumor necrosis factor-alpha (TNF-α), play crucial roles in the modulation of inflammatory signaling pathways during the age-related loss of skeletal muscle. MiRNAs have emerged as important regulators for the mass and functional maintenance of skeletal muscle by regulating gene expression of pro-inflammatory cytokines. Clearly, targeting these miRNAs could provide an efficient and noninvasive approach for the diagnosis, prevention, or treatment of sarcopenia via the regulation of pro-inflammatory and anti-inflammatory factors (Fan et al. 2016). The circulating levels can be measured and have been proposed by Siracusa in 2018 as new biomarkers of physiological and pathological muscle processes (Siracusa, Koulmann, and Banzet 2018).

15.1.1.8 Calorie Restriction

CR increases skeletal muscle stem cell frequency in young and aged mice. It has beneficial effects on stem cell function and promotes oxidative over glycolytic metabolism by increasing mitochondria abundance (Cerletti et al. 2012). CR reduces or delays many of the age-related defects that occur in rodent skeletal muscle. However, the effect of CR does not persist over time. CR increases lifespan in both rodents and humans (Boldrin et al. 2017). It induces a more youthful cellular and molecular phenotype in aging stem cells throughout the body, and it enhances stem cell function and promotes healthy aging (Murphy and Thuret 2015).

In summary, the most effective strategy to slow or counter stem cell dysfunction with aging is CR, which promotes longevity.

15.1.2 NUTRITIONAL THERAPIES TO COMBAT SARCOPENIA

15.1.2.1 Protein

Progressive loss of proteostasis is a hallmark of aging that is marked by declines in various components of essential machinery including autophagy, ubiquitin-mediated degradation, and protein synthesis. Other defects in proteostasis have been alleviated in longevity models using overexpression of mitochondrial-targeted catalase, CR-reduced insulin growth factor-1 (IGF-1) signaling, and rapamycin treatment.

Caloric restriction and rapamycin both inhibit the Mammalian target of rapamycin (mTOR) protein, which mediates protein translation and degradation rates in response to nutrient availability and increases lifespan and health span in a variety of animal models (Basisty, Meyer, and Schilling 2018a).

Whey protein isolate and soy-dairy protein blend, particularly at levels above recommended dietary allowance (RDA) (Deer and Volpi 2015), induced similar responses in hyperaminoacidemia, mTORC1 signaling, muscle protein synthesis, and breakdown that counteract sarcopenia (Borack et al. 2016). Frailty, which is a complication of sarcopenia due in part to malnutrition, can be reduced by diet quality throughout life (energy and protein intake) and by physical exercise in older people (Cruz-Jentoft et al. 2017). Yanai advocates a protein dietary intake of 1.0–1.2 g/kg a day with an optimal distribution per meal or 25–30 g of high-quality protein per meal to prevent sarcopenia (Yanai 2015). Wall et al. (2014) suggest that 30–40 g of protein represents an optimal dose to best promote muscle protein synthesis rates in older adults and underline the importance of exercise training in conjunction with these nutritional strategies to maximize gains in muscle mass, strength, and function (Wall, Cermak, and van Loon 2014).

15.1.2.2 Leucine, Other Amino Acids, and Other Metabolites

In older adults, leucine supplementation can improve muscle protein synthesis in response to lower protein meals (Casperson et al. 2012). Leucine supplementation does not reverse the changes in protein status parameters associated with aging, such as the observed decreases in body and muscle protein and total serum protein. Consumption of leucine attenuates body fat gain during aging but does not show signs of prevention of the associated age-related decrease in lean mass (Vianna et al. 2012). Co-ingestion of leucine with carbohydrates and protein following physical activity does not further increase the muscle protein fractional synthetic rate in elderly men when ample amounts of protein and carbohydrate are being ingested (Koopman et al. 2008). Ingestion of a small bolus of essential amino acids (EAAs) results in a reduced responsiveness in the stimulation of muscle protein synthesis in elderly persons. Consequently, a small bolus of EAA does not improve the muscle wasting caused by aging (Katsanos et al. 2005). Ispoglou et al. propose effective nutritional means for addressing protein and total energy deficiencies and for promoting appetite in older women (Ispoglou et al. 2017). This nutritional strategy could become more effective with the use of antioxidant/anti-inflammatory compounds that can decrease the well-known low-grade inflammation associated with aging and which is responsible for blunted response to food intake (Magne et al. 2013). Dysregulated proteostasis occurs with aging, especially beyond the sixth decade of life. Women and men aged 75 y lose muscle mass at a rate of ~0.7% and 1%/y, respectively (sarcopenia), and lose strength two to fivefold faster (dynapenia) as muscle "quality" decreases (Mitchell et al. 2016).

Beta-Hydroxy-Beta-methylbutyrate (HMB) is an active metabolite of the EAA leucine (Nissen and Abumrad 1997). The use of HMB to suppress proteolysis originates from observations that leucine has protein-sparing characteristics (Krebs and Lund 1976; Harper, Benevenga, and Wohlhueter 1970; Nissen and Abumrad 1997; Nissen et al. 1996). The EAA leucine can be transaminated to the α-ketoacid (α-ketoisocaproate, KIC) (Nissen and Abumrad 1997; Krebs and Lund 1976). KIC can then be oxidized to HMB (Sabourin and Bieber 1983, 1981, 1982). About 5% of leucine oxidation proceeds via the second pathway (Van Koevering and Nissen 1992). HMB is the active metabolite of leucine responsible for increased protein synthesis in muscle (Giron et al. 2016). Recently, Wilkinson et al. (2013, 2017) confirmed the acute effects of HMB in stimulating muscle protein synthesis (+70%) and decreasing muscle protein breakdown (–57%) in healthy young men over a 2.5 h period after consumption of HMB and confirmed that the increase in muscle protein synthesis was likely the result of the increased mTOR signaling.

The effect of HMB supplementation in the diet on age-related muscle loss, or sarcopenia, has been extensively reviewed elsewhere (Barillaro et al. 2013; Beaudart et al. 2018; Fitschen et al. 2013; Hickson 2015; Rossi et al. 2017; Woo 2018; Wu et al. 2015). HMB also improved body composition while reducing fat loss in older adults participating in a weight training exercise program (Stout et al. 2015, 2013; Vukovich, Stubbs, and Bohlken 2001; Panton et al. 1998). Even without a training program, HMB helps maintaining healthy muscle mass and muscle function in the general population. Studies have shown the ability of HMB alone or a nutritional cocktail or meal replacement with HMB to reduce the effects of catabolic processes caused by inactivity and sarcopenia and to modestly increase muscle mass and strength in the elderly even when only going about their normal activities (Baier et al. 2009; Flakoll et al. 2004; Wu et al. 2015; Stout et al. 2013; Cramer et al. 2016; Malafarina et al. 2017; Berton et al. 2015; Deutz et al. 2013; Ellis et al. 2018). Sufficient vitamin D status is vital for improved gains in muscular strength in older adults (Fuller et al. 2011).

HMB can help maintain muscle also during inactivity. In a study in which 3 g HMB/d was given to individuals during prolonged forced inactivity (bed rest), HMB was able to significantly decrease the loss of lean and muscle mass (Zachwieia et al. 1999; Rathmacher et al. 1999). Thus, HMB supplementation helps minimize any sarcopenic effect caused by prolonged inactivity. A recent study by Deutz et al. demonstrated that HMB supplementation preserved muscle mass during 10 days of bed rest in healthy older adults (Deutz et al. 2013). The effects of HMB, vitamin D, and protein were studied in elderly women following recovery from hip fracture orthopedic surgery. The combination of HMB and vitamin D improved muscle strength and mobility after surgery, thus lessening the period of immobility or reduced mobility, which can lead to increased muscle loss (Ekinci et al. 2016). Furthermore, a greater number of the patients were mobile on days 15 and 30 after surgery in addition to having significantly increased strength on day 30. The authors concluded that the nutritional combination led to acceleration in wound healing, shortening of the immobilization period, and increased muscle strength (Ekinci et al. 2016). One recent small study of 23 elderly subjects (mean age: 70.5 years) showed that HMB, combined with arginine and glutamine, prevented acute (42 days after surgery) loss of quadriceps strength after total knee arthroscopy (TKA) (Nishizaki et al. 2015). Thus, HMB supplementation helps minimize any sarcopenic effect caused by prolonged inactivity. In other studies of populations with sarcopenic muscle loss (Cramer et al. 2016; Malafarina et al. 2017; Flakoll et al. 2004; Baier et al. 2009; De Luis et al. 2015) and wasting diseases such as cancer (Eubanks May et al. 2002), AIDS (Clark et al. 2000), and trauma (Kuhls et al. 2007), HMB improved protein metabolism and/or lean mass and muscle strength.

15.1.2.3 Calorie Restriction

CR extends the lifespan and health in humans, nonhuman primates, and rodents and also delays dele-terious age-related physiological changes in animals. The Lou/C rat is a valuable model of spontaneous food restriction, with associated improved insulin sensitivity, or a model of successful aging. In addition to its reduced calorie intake, the model shows a preferential channeling of nutrients toward utilization rather than storage (Veyrat-Durebex et al. 2009). Lou/C rats are also a valuable model of spontaneous resistance to diet-induced obesity. IGF-1 signaling could partly mediate the effects of CR on increased longevity in mammals (Marissal-Arvy et al. 2013). Transcription factors SirT1, PGC-1α, AMPK, and TOR may be involved in the lifespan effects of CR. Surprisingly, low body weight in middle-aged and elderly humans is associated with increased mortality. Moreover, enhancement of human longevity may require pharmaceutical interventions (Spindler 2010). Redman et al. confirmed that CR extends maximum lifespan in most species and showed that persistent metabolic slowing is accompanied by reduced oxidative stress. These findings are in good agreement with the rate of living theory for healthy aging (Redman et al. 2018). Burks and Cohn underlined that, because of the complexity of biological mechanisms of sarcopenia, CR is not the only antiaging therapy (Burks and Cohn 2011). Sharples et al. assessed the role of dietary restriction in combination with amino acid supplementation and that of sirtuins in the attenuation of skeletal muscle mass loss with advancing age, while enabling both healthy aging and longevity. Sirtuins, a group of seven highly conserved protein deacetylases involved in the process of chromatic remodeling and gene regulation, also have a pathophysiological role in skeletal muscle via reduced IGF-I signalling and attenuated effect of inflammatory NF-Kb signalling (Sharples et al. 2015). Basisty et al. (2018b) described the role of ubiquitin-mediated protein aggregation with age and suggested that these aggregates, which are attenuated by CR, could be used to identify novel therapeutic targets in aging.

In summary, nutritional strategies to preserve muscle mass, strength, and function should be accompanied by therapeutic concepts to improve bone remodeling, structure, and composition (Compston 2015). In addition, potential high-quality food products should be tailored specifically to enhance the nutritional status and health of older adults (by fortifying foods with selected functional ingredients, vitamins, and minerals) (Baugreet et al. 2017).

15.1.3 PHARMACEUTICAL THERAPIES TO COMBAT SARCOPENIA

Synergies between pharmacological, dietary, and exercise treatments are important for improvements in sarcopenia (Glass and Roubenoff 2010). Drugs that target the mTOR pathway slow the aging process and reduce age-related pathologies in humans (Johnson, Robinson, and Nair 2013).

Alternatively, pharmaceutical therapies with agents such as testosterone, dehydroepiandrosterone, estrogen, growth hormone, ghrelin, vitamin D, angiotensin-converting enzyme inhibitor, and eicosapentaenoic acid can be used to improve muscle mass and strength. However, many of these agents, such as testosterone, growth hormone, and IGF-1, have serious associated side effects that limit their use (Wakabayashi and Sakuma 2014). Moreover, safer and equally effective nonpharmaceutical therapies, such as resistance training, protein, and amino acid supplementation and nonsmoking, can also be used to counteract primary sarcopenia.

Some authors propose that the current understanding of how drugs taken regularly for common conditions in older adults can contribute positively or negatively to sarcopenia and muscle wasting. Indeed, many drugs, including angiotensin converting, enzyme inhibitors, statins, antidiabetic drugs, and vitamin D, can interact with certain mechanisms that alter the balance between protein synthesis and degradation. This can lead to a harmful or beneficial effect on muscle mass and strength. These drugs could play an important role during the time of onset and development of sarcopenia and slow down frailty and the sarcopenic process (Campins et al. 2017).

15.1.3.1 Testosterone

Testosterone increases muscle mass and muscle protein synthesis in rodents and humans (Wakabayashi and Sakuma 2014) without any clear increase in muscle strength in elderly people. However, this observation is vulnerable to confounding factors and experimental design (Chow and Nair 2005). Brioche et al. (2016) confirmed in a review that testosterone stimulates muscle protein synthesis, improves recycling of intracellular amino acids, and decreases the protein breakdown rate. Furthermore, testosterone promotes stem cells via the Notch signaling pathway due to myostatin inhibition and Akt activation. Testosterone also plays an anti-inflammatory role by reducing the plasma concentration of TNF-α and some ILs. However, the testosterone doses currently used are often associated with side effects.

In most obese men with a low to low-normal testosterone level, testosterone treatment increases diet-induced loss of fat mass and prevents loss of muscle mass. Consequently, weight loss during testosterone treatment is almost exclusively due to the loss of body fat (Ng et al. 2016). However, the testosterone doses currently used are often associated with side effects.

15.1.3.2 Vitamin D

Vitamin D deficiency is one cause of proximal myopathy and sarcopenia, which improves with vitamin D supplementation. The association between low vitamin D and low muscle mass has been observed in the elderly, mainly in community-dwelling subjects. A few studies have reported an association between vitamin D deficiency and increased fall risk and nursing home placement (Visser, Deeg, and Lips 2003; Lappe and Binkley 2015). In the Newcastle 85+ Study ($n = 845$), Granic et al. investigated the association between 25(OH)D concentration, grip strength (GS), and physical performance as measured by the timed up-and-go test (TUG) in very old adults (>85 years) over a period of 5 years. They divided the population into four quartiles. They found that low baseline 25(OH)D can contribute to muscle strength decline in the very old and particularly in men (Granic et al. 2017).

A substantial level of vitamin D signaling via vitamin D receptor (VDR) is required for normal muscle growth. VDR expression seems to be affected by aging, suggesting that this could reduce the functional response of the muscle fibers to vitamin D. Consequently, as vitamin D appears to function in primary myoblasts and established myoblast cell lines, understanding how vitamin D signaling contributes to myogenesis will provide a valuable insight into an effective nutritional strategy to moderate sarcopenia (Wagatsuma and Sakuma 2014). The Society on Sarcopenia, Cachexia, and Wasting Diseases recommends checking vitamin D levels and replacing if they are low in all sarcopenic patients (Morley et al. 2010).

15.1.3.3 Vitamin B12

Old age is commonly associated with obesity, and both conditions predispose to a sedentary life style and lack of appropriate nutrition. To maintain muscle mass and functional status, high intakes of folic acid should be added to diet (Johnson et al. 2011). A few studies have reported an association between vitamin B12 deficiency and a decrease in muscle strength and walking speed and with an increased risk of fracture and postural instability (Sato et al. 2005; Mithal et al. 2013). Unfortunately, there are very few studies investigating the link between vitamin B12 status and sarcopenia (Wee 2016; Verlaan et al. 2017), and the results obtained are quite at variance. A recent study by Ates Bulut et al. (2017) of 403 geriatric subjects showed a high association between vitamin B12 deficiency (serum vitamin B12 level <400 pg/mL without methylmalonic acid and homocysteine levels) was associated with sarcopenia and dynapenia (defined as age-associated loss of muscle strength not caused by neurologic or muscular diseases). These authors suggested periodic examination of subjects with vitamin B12 deficiency owing to the potential association with sarcopenia in older adults.

15.1.3.4 Growth Hormone and Myostatin

GH therapy has positive effects by improving body composition and muscle performance but is still not recommended as sarcopenia treatment owing to its side effects (Wakabayashi and Sakuma 2014). However, GH treatment prevents the decrease in muscle mass observed in old rats. These beneficial effects are driven by an increase in protein synthesis concomitant with a decrease in proteolysis, as expected after activation of p70S6K and weaker MuRF1 expression. There is a downregulation of myostatin that increases protein synthesis owing to increased activation of the PI3K/Akt/mTOR pathway (Brioche et al. 2016).

Myostatin, a member of the transforming growth factor-β (TGF-β) superfamily that is highly expressed in skeletal muscle, was first described in 1997. It was known that loss of myostatin function induces an increase in muscle mass in mice, cows, dogs, and humans. Myostatin and its receptor emerged as a therapeutic target for loss of skeletal muscle as in sarcopenia and cachexia and in muscular dystrophies. Myostatin inhibition could therefore have the potential to be a potent therapy for muscle wasting (Saitoh et al. 2017). It is also possible that the use of myostatin inhibitors could be helpful for subjects unable to participate in exercise, and such inhibitors could be used to maintain or rebuild a healthy muscular system (Maurel, Jahn, and Lara-Castillo 2017).

15.1.3.5 Ghrelin

While ghrelin is well known for its stimulation of appetite and lipid accumulation, it has also an anti-inflammatory role and reduces antioxidative stress in inflammatory disorders, notably in sarcopenia. It seems to increase lean body mass but does not necessarily produce changes in muscle strength or physical function (Wakabayashi 2014). Guillory et al. (2017) showed that lifelong deletion of ghrelin prevents the development of age-related obesity along with age-induced decline in muscle strength and endurance and attenuates the decrease in muscle pAMPK seen with aging.

Ghrelin levels are generally lower in the elderly than in younger subjects, and elderly subjects with sarcopenia have lower ghrelin levels than those without sarcopenia. Chronic central ghrelin treatment can also enhance bone mass in rodents. In elderly humans, oral administration of the ghrelin mimetic ibutamoren increased GH and IGF-1 levels to that of younger adults and prevented lean mass loss without severe side effects (Collden, Tschop, and Muller 2017).

15.1.3.6 Apelin

Apelin is a myokine that is expressed and secreted by muscle and adipose tissue (Boucher et al. 2005). Apelin expression and release are stimulated by physical exercise (Vinel et al. 2018; Besse-Patin et al. 2014). Muscle and plasma apelin expression levels decrease with age in both rodents and humans, as does contraction-stimulated muscle release of apelin, and, further, these decreases are statistically associated with sarcopenia (Vinel et al. 2018). Mice deficient in *Apln* or *Aplnr* (the genes coding for apelin and its receptor, respectively) experienced poorer muscle function and accelerated muscle loss with aging, highlighting an important role for apelin in maintaining optimal muscle function (Vinel et al. 2018). Supplementation of aged mice with exogenous apelin enhanced muscle mass, muscle fiber size, muscle fiber force production, and whole muscle function, thus effectively reversing sarcopenia (Vinel et al. 2018). Apelin seems to work via several mechanisms to improve muscle health that include (a) stimulation of muscle anabolic signaling and protein synthesis; (b) inhibition of proteolytic factors; (c) enhanced mitochondrial biogenesis and function; and (d) activation, proliferation, and differentiation of satellite cells. There is also some evidence suggesting that apelin supplementation can improve processes associated with antioxidant enzymes, autophagy, and inflammation (Vinel et al. 2018).

In summary, pharmaceutical therapies can also be used to improve muscle mass and strength; these last strategies are not always recommended for the treatment of sarcopenia owing to their side effects.

15.1.4 EXERCISE

Age-related muscle atrophy is multifactorial in origin and can be related to decreased physical activity and endurance, decreased protein synthetic pathways, and reduced skeletal muscle fiber size and number (Konopka and Harber 2014; Marzetti et al. 2014). These mutiple signaling pathways are associated with enhanced risk factors for various diseases (Piccirillo et al. 2014). Muscle wasting, which is characteristic of age-related muscle atrophy, results from an accelerated breakdown of the normally long-lived myofibrillar proteins (actin and myosin), which make up 60%–70% of muscle protein. Activation of the ubiquitin proteasome pathway seems to be involved in enhanced rates of protein degradation.

Several studies have attributed these changes to a state of chronic inflammation and to age-related decline in muscle function and muscle mitochondrial protein synthesis, all of which lead to a state of altered muscle protein metabolism and loss of muscle function (Rooyackers et al. 1996). In addition, older adults are particularly more prone to the development of insulin resistance (Karakelides et al. 2010). It is estimated that over a third of older adults are obese, and this contributes to the decline in physical function and frailty and exacerbates the existing state of insulin resistance. In a landmark study, Villareal and his collaborators showed that a combination of weight loss and exercise was more effective in improving physical function and amelioration of frailty than either weight loss alone or exercise alone (Villareal et al. 2011).

The physiologic adaptations differ with the type of exercise. Exercise, irrespective of the type used, improves insulin sensitivity and reverses the decreased resting mitochondrial ATP production (Petersen et al. 2003), improves mitochondrial respiratory chain efficiency, and decreases the formation of ROS (Anderson et al. 2009); all of which are implicated in the development of insulin resistance with aging.

15.1.4.1 Aerobic Exercise

Aerobic exercise has been shown to reverse and/or improve muscle function and leads to improved peak oxygen consumption (Lambert and Evans 2005). It results in increased mitochondrial proteins, which are involved in mitochondrial biogenesis, fusion, and fission (Konopka et al. 2011), leading to enhancement of muscle protein synthesis (Short, Nair, and Stump 2004; Harber et al. 2009). There are several studies suggesting that aerobic exercise is also associated with whole muscle hypertrophy in both young and older adults (Harber et al. 2012).

15.1.4.2 Resistance Exercise

Resistance exercise improves neuromuscular adaptations leading to enhanced strength, with minimal effect on peak oxygen consumption. Resistance exercise training reverses sarcopenia and age-related declines in rates of muscle protein synthesis (Balagopal et al. 2001) while strength training significantly improves muscle strength in older adults (Liu et al. 2009; Latham et al. 2004; Fiatarone et al. 1990; Lemmer et al. 2001). This assumes that resistance training establishes an improved balance between anabolic and catabolic factors. However, the effect of resistance exercise on the rate of protein catabolism is not well understood. Fry et al. showed that resistance exercise, in both young and older subjects, increases the ubiquitin–proteasome system, which is suggestive of increased protein breakdown. Aging influences neither muscle protein breakdown nor autophagy (Fry et al. 2013). Earlier studies by Rennie and his group showed that exercise, both resistance and nonresistance types, lead to depression of muscle protein synthesis, while the rates of muscle protein breakdown probably remain unchanged (Kumar et al. 2009). The immediate post-exercise period is characterized by marked elevations in the rates of muscle protein synthesis and muscle protein breakdown in the fasted state, whereas net muscle protein balance remains negative. Rennie and his colleagues found that positive net balance is achieved only when one amino acid is elevated, leading to increased rates of muscle protein synthesis. The rates of muscle protein breakdown are more regulated by the prevailing insulin levels.

15.1.4.3 Resistance Training

Resistance training is a viable and relatively low-cost treatment to counter age-related sarcopenia. Resistance training-induced increases muscle mass, strength, and function; it can be enhanced by certain foods, nutrients, or nutritional supplements such as protein, leucine, βHMB, creatine, and omega-3 PUFAs (Phillips 2015). Dickinson et al. reported how aging affects skeletal muscle protein synthesis and how nutrition and exercise can be strategically employed to overcome age-related protein synthesis impairments and slow the progression of sarcopenia. A practical strategy that has promise for restoring skeletal muscle size and function in older adults was described by Dickinson et al. in 2013 (Dickinson, Volpi, and Rasmussen 2013). Brioche et al. confirmed that performing resistance training cycles and endurance training separately is the best solution to combat sarcopenia (Brioche and Lemoine-Morel 2016). In addition, resistance exercise induces muscle hypertrophy (Snijders and Parise 2017) by specifically targeting the growth of type II muscle fibers in both young and older men. However, adequate muscle fiber perfusion is critical for the maintenance of muscle. Muscle fiber capillarization decreases substantially with advancing age and hence it is a critical factor in muscle fiber hypertrophy during resistance exercise training in older men (Snijders et al. 2017).

More recent studies by Nair et al. showed that high-intensity "aerobic" exercise in young and older adults resulted in improved cardiorespiratory fitness, insulin sensitivity, and mitochondrial respiration, in enhanced fat-free mass (synonymous with enhanced muscle mass) and in muscle strength. These effects were slightly greater than those observed with resistance exercise, which was associated with improved fat-free mass and insulin sensitivity in both groups as well (Robinson et al. 2017). The authors attributed the metabolic and molecular improvements to the regulation of exercise adaptation at the post-translational level in skeletal muscle.

In elderly subjects, protein supplementation increased muscle mass gain only during resistance-type exercise training (Evans, Boccardi, and Paolisso 2013; Tieland et al. 2012) and has been associated with better physical function and muscle strength among elderly women (Isanejad et al. 2016). A recent meta-analysis by Silva et al. showed supplementation with HMB-free acid in combination with resistance training was able to attenuate the markers of muscle damage to enhance training-induced muscle mass and strength and potentially to improve markers of aerobic fitness when combined with high-intensity interval training (Silva et al. 2017). Nair and his colleagues showed that different exercise training modes enhanced the proteins involved in translational machinery irrespective of age (Robinson et al. 2017). Supplementation with HMB (Weihrauch and Handschin 2018) or EAA combined with resistance exercise also increased protein synthesis via mTORC1 signaling (Fry et al. 2011). Indeed, in nonfrail, independent, healthy older adults, aerobic exercise combined with either HMB (Stout et al. 2015) or EAA supplementation increases physical performance and muscle strength: EAA stimulates protein synthesis but without any noticeable increase in muscle mass (Markofski et al. 2018). During a 5-week exercise program, cysteine-rich whey protein increased lean body mass and decreased fat mass in comparison with a control diet, but the authors made no mention of the effect on skeletal muscle mass per se (Droge 2005).

In aged skeletal muscle fibers, there is compromised integrity of the cell membrane, which could contribute to sarcopenia. Thus, the combination of lifelong mild caloric restriction and voluntary wheel running represents a novel method aimed at significantly abrogating age-associated reduction in relative muscle weight and fiber area (Hord, Botchlett, and Lawler 2016). Martone et al. presented evidence showing the beneficial effect of exercise and nutrition on sarcopenia and physical frailty in the SPRINTT cohort (Martone et al. 2017). A more recent study by Villareal et al. showed that weight loss, induced by a weight-management program with combined aerobic and resistance exercise, was the most effective in improving the functional status of obese older adults (Villareal et al. 2017).

In summary, exercise, irrespective of the type used, consists in a way to combat the insulin resistance with aging. Moreover, the combination of exercise with nutritional therapies is more effective in improving muscle loss and physical function.

ACKNOWLEDGMENTS

I am grateful to Naji N. Abumrad for critically reading the manuscript and providing valuable comments to improve the relevance of this chapter. I would also like to thank him for providing personal data on HMB.

ABBREVIATIONS

25(OH)D	25-hydroxy-vitamine D
AceCS1	Acetyl coenzyme A synthetase
AIDS	Acquired immune deficiency syndrome
AMP	Adenosine monophosphate
AMPK	AMP-activated protein kinase
Atg	Autophagy-related genes
ATP	Adenosine triphosphate
Atrogin-1	Atrophy gene-1
CR	Caloric restriction
DNA	Deoxyribonucleic acid
EAA	Essential amino acids
GH	Growth hormone
GS	Grip strength
HMB	Beta-hydroxy-beta-methylbutyrate
IGF-1	Insulin growth factor-1
IL-6	Interleukin-6
KIC	α-Ketoisocaproate
miRNAs	Micro ribonucleic acid
mtDNA	Mitochondrial DNA
mTOR	Mammalian target of rapamycin
mTORC1	mTOR signaling complex1
MuRF-1	Muscle RING finger-1
NAD$^+$	Nicotinamide adenine dinucleotide
NF-κB	Nuclear factor-kappa B
pAMPK	Phosphorylated AMP-activated protein kinase
PGC-1α	Peroxisome proliferator activator receptor γ coactivator-1α

p16INK4a	Senescence gene
p38MAPK	Mitogen-activated protein kinase
p70S6K	Ribosomal protein S6 kinase beta-1
PI3K/Akt/mTOR	Phosphoinositide 3-kinase/protein kinase B/mammalian target of rapamycin
PUFAs	Polyunsaturated fatty acids
RDA	Recommended dietary allowance
ROS	Reactive oxygen species
SIRT1	Sirtuin 1, also known as NAD-dependent deacetylase sirtuin-1
STAT3	Signal transducer and activator of transcription
TGF-β	Transforming growth factor-β
TKA	Total knee arthroscopy
TNF-α	Tumor necrosis factor
UPS	Ubiquitin–proteasome system
VDR	Vitamin D receptor

REFERENCES

Anderson, E. J., M. E. Lustig, K. E. Boyle, et al. 2009. Mitochondrial H_2O_2 emission and cellular redox state link excess fat intake to insulin resistance in both rodents and humans. *J Clin Invest* 119 (3):573–81.

Ates Bulut, E., P. Soysal, A. E. Aydin, O. Dokuzlar, S. E. Kocyigit, and A. T. Isik. 2017. Vitamin B12 deficiency might be related to sarcopenia in older adults. *Exp Gerontol* 95:136–40.

Bahat, G., A. Tufan, F. Tufan, et al. 2016. Cut-off points to identify sarcopenia according to European Working Group on Sarcopenia in Older People (EWGSOP) definition. *Clin Nutr* 35 (6):1557–63.

Baier, S., D. Johannsen, N. N. Abumrad, J. A. Rathmacher, S. L. Nissen, and P. J. Flakoll. 2009. Year-long changes in lean body mass in elderly men and women supplemented with a nutritional cocktail of b-hydroxy-b-methylbutyrate (HMB), arginine, and lysine. *JPEN* 33:71–82.

Balagopal, P., J. C. Schimke, P. Ades, D. Adey, and K. S. Nair. 2001. Age effect on transcript levels and synthesis rate of muscle MHC and response to resistance exercise. *Am J Physiol Endocrinol Metab* 280 (2):E203–8.

Barillaro, C., R. Liperoti, A. M. Martone, G. Onder, and F. Landi. 2013. The new metabolic treatments for sarcopenia. *Aging Clin Exp Res* 25 (2):119–27.

Barnouin, Y., J. S. McPhee, G. Butler-Browne, et al. 2017. Coupling between skeletal muscle fiber size and capillarization is maintained during healthy aging. *J Cachexia Sarcopenia Muscle* 8 (4):647–59.

Basisty, N. B., Y. Liu, J. Reynolds, et al. 2018b. Stable isotope labeling reveals novel insights into ubiquitin-mediated protein aggregation with age, calorie restriction, and rapamycin treatment. *J Gerontol A Biol Sci Med Sci* 73 (5):561–70.

Basisty, N., J. G. Meyer, and B. Schilling. 2018a. Protein turnover in aging and longevity. *Proteomics* 18 (5–6):e1700108.

Baugreet, S., R. M. Hamill, J. P. Kerry, and S. N. McCarthy. 2017. Mitigating nutrition and health deficiencies in older adults: A role for food innovation? *J Food Sci* 82 (4):848–55.

Beaudart, C., V. Rabenda, M. Simmons, et al. 2018. Effects of protein, essential amino acids, B-Hydroxy B-Methylbutyrate, creatine, dehydroepiandrosterone and fatty acid supplementation on muscle mass, muscle strength and physical performance in older people aged 60 years and over: A systematic review on the literature. *J Nutr Health Aging* 22 (1):117–30.

Bengal, E., E. Perdiguero, A. L. Serrano, and P. Munoz-Canoves. 2017. Rejuvenating stem cells to restore muscle regeneration in aging. *F1000Res* 6:76.

Berton, L., G. Bano, S. Carraro, et al. 2015. Effect of oral beta-hydroxy-beta-methylbutyrate (HMB) supplementation on physical performance in healthy old women over 65 years: An open label randomized controlled trial. *PLoS One* 10 (11):e0141757.

Besse-Patin, A., E. Montastier, C. Vinel, et al. 2014. Effect of endurance training on skeletal muscle myokine expression in obese men: Identification of apelin as a novel myokine. *Int J Obes* 38 (5):707–13.

Boldrin, L., J. A. Ross, C. Whitmore, B. Doreste, C. Beaver, and A. Eddaoudi. 2017. The effect of calorie restriction on mouse skeletal muscle is sex, strain and time-dependent. *Sci Rep* 7 (1):5160.

Borack, M. S., P. T. Reidy, S. H. Husaini, et al. 2016. Soy-dairy protein blend or whey protein isolate ingestion induces similar postexercise muscle mechanistic target of rapamycin complex 1 signaling and protein synthesis responses in older men. *J Nutr* 146 (12):2468–75.

Boucher, M., B. P. Wann, S. Kaloustian, et al. 2005. Sustained cardioprotection afforded by A2A adenosine receptor stimulation after 72 hours of myocardial reperfusion. *J Cardiovasc Pharmacol* 45 (5):439–46.

Brioche, T., and S. Lemoine-Morel. 2016. Oxidative stress, sarcopenia, antioxidant strategies and exercise: Molecular aspects. *Curr Pharm Des* 22 (18):2664–78.

Brioche, T., A. F. Pagano, G. Py, and A. Chopard. 2016. Muscle wasting and aging: Experimental models, fatty infiltrations, and prevention. *Mol Aspects Med* 50:56–87.

Budui, S. L., A. P. Rossi, and M. Zamboni. 2015. The pathogenetic bases of sarcopenia. *Clin Cases Miner Bone Metab* 12 (1):22–6.

Burks, T. N., and R. D. Cohn. 2011. One size may not fit all: Anti-aging therapies and sarcopenia. *Aging* 3 (12):1142–53.

Campins, L., M. Camps, A. Riera, E. Pleguezuelos, J. C. Yebenes, and M. Serra-Prat. 2017. Oral drugs related with muscle wasting and sarcopenia: A review. *Pharmacology* 99 (1–2):1–8.

Carter, H. N., C. C. Chen, and D. A. Hood. 2015. Mitochondria, muscle health, and exercise with advancing age. *Physiology* 30 (3):208–23.

Casperson, S. L., M. Sheffield-Moore, S. J. Hewlings, and D. Paddon-Jones. 2012. Leucine supplementation chronically improves muscle protein synthesis in older adults consuming the RDA for protein. *Clin Nutr* 31 (4):512–19.

Cerletti, M., Y. C. Jang, L. W. Finley, M. C. Haigis, and A. J. Wagers. 2012. Short-term calorie restriction enhances skeletal muscle stem cell function. *Cell Stem Cell* 10 (5):515–19.

Chow, L. S., and K. S. Nair. 2005. Sarcopenia of male aging. *Endocrin Metab Clin North America* 34 (4):833–52.

Clark, R. H., G. Feleke, M. Din, et al. 2000. Nutritional treatment for acquired immunodeficiency virus-associated wasting using b-hydroxy-b-methylbutyrate, glutamine and arginine: A randomized, double-blind, placebo-controlled study. *J Parenter Enteral Nutr* 24(3):133–9.

Collden, G., M. H. Tschop, and T. D. Muller. 2017. Therapeutic potential of targeting the ghrelin pathway. *Int J Mol Sci* 18 (4):798.

Compston, J. 2015. Emerging therapeutic concepts for muscle and bone preservation/building. *Bone* 80:150–6.

Cramer, J. T., A. J. Cruz-Jentoft, F. Landi, et al. 2016. Impacts of high-protein oral nutritional supplements among malnourished men and women with sarcopenia: A multicenter, randomized, double-blinded, controlled trial. *J Am Med Dir Assoc* 17 (11):1044–55.

Cruz-Jentoft, A. J., E. Kiesswetter, M. Drey, and C. C. Sieber. 2017. Nutrition, frailty, and sarcopenia. *Aging Clin Exp Res* 29 (1):43–8.

Cruz-Jentoft, A. J., F. Landi, S. M. Schneider, et al. 2014. Prevalence of and interventions for sarcopenia in ageing adults: A systematic review. Report of the International Sarcopenia Initiative (EWGSOP and IWGS). *Age Ageing* 43 (6):748–59.

Dardevet, D., D. Remond, M. A. Peyron, I. Papet, I. Savary-Auzeloux, and L. Mosoni. 2012. Muscle wasting and resistance of muscle anabolism: The "anabolic threshold concept" for adapted nutritional strategies during sarcopenia. *Sci World J* 2012:269531.

De Luis, D. A., O. Izaola, P. Bachiller, and J. Perez Castrillon. 2015. Effect on quality of life and handgrip strength by dynamometry of an enteral specific supplements with beta-hydroxy-beta-methylbutyrate and vitamin d in elderly patients. *Nutr Hosp* 32 (1):202–7.

Deer, R. R., and E. Volpi. 2015. Protein intake and muscle function in older adults. *Curr Opin Clin Nutr Metab Care* 18 (3):248–53.

Deer, R. R., and E. Volpi. 2018. Protein requirements in critically Ill older adults. *Nutrients* 10 (3):378.

Deutz, N. E., S. L. Pereira, N. P. Hays, et al. 2013. Effect of beta-hydroxy-beta-methylbutyrate (HMB) on lean body mass during 10 days of bed rest in older adults. *Clin. Nutr* 32 (5):704–12.

Dickinson, J. M., E. Volpi, and B. B. Rasmussen. 2013. Exercise and nutrition to target protein synthesis impairments in aging skeletal muscle. *Exerc Sport Sci Rev* 41 (4):216–23.

Droge, W. 2005. Oxidative stress and ageing: Is ageing a cysteine deficiency syndrome? *Philos Trans R Soc Lond B Biol Sci* 360 (1464):2355–72.

Ekinci, O., S. Yanik, B. B. Terzioglu, A. E. Yilmaz, A. Dokuyucu, and S. Erdem. 2016. Effect of calcium beta-Hydroxy-beta-Methylbutyrate (CaHMB), vitamin D, and protein supplementation on postoperative immobilization in malnourished older adult patients with hip fracture: A randomized controlled study. *Nutr Clin. Pract* 31:829–835.

Ellis, A. C., G. R. Hunter, A. M. Goss, and B. A. Gower. 2018. Oral supplementation with Beta-Hydroxy-Beta-Methylbutyrate, arginine, and glutamine improves lean body mass in healthy older adults. *J Diet Suppl.* 2019;16(3):281–293.

Eubanks May, P., A. Barber, A. Hourihane, J. T. D'Olimpio, and N. N. Abumrad. 2002. Reversal of cancer-related wasting using oral supplementation with a combination of b-hydroxy-b-methylbutyrate, arginine, and glutamine. *Am J Surgery* 183:471–9.

Evans, W. J., V. Boccardi, and G. Paolisso. 2013. Perspective: Dietary protein needs of elderly people: Protein supplementation as an effective strategy to counteract sarcopenia. *J Am Med Dir Assoc* 14 (1):67–9.

Fan, J., X. Kou, Y. Yang, and N. Chen. 2016. MicroRNA-regulated proinflammatory cytokines in sarcopenia. *Mediators Inflamm* 2016:1438686.

Fiatarone, M. A., E. C. Marks, N. D. Ryan, C. N. Meredith, L. A. Lipsitz, and W. J. Evans. 1990. High-intensity strength training in nonagenarians: Effects on skeletal muscle. *Jama* 263 (22):3029–34.

Fitschen, P. J., G. J. Wilson, J. M. Wilson, and K. R. Wilund. 2013. Efficacy of beta-hydroxy-beta-methylbutyrate supplementation in elderly and clinical populations. *Nutrition* 29 (1):29–36.

Flakoll, P., R. Sharp, S. Baier, D. Levenhagen, C. Carr, and S. Nissen. 2004. Effect of beta-hydroxy-beta-methylbutyrate, arginine, and lysine supplementation on strength, functionality, body composition, and protein metabolism in elderly women. *Nutrition* 20 (5):445–51.

Fry, C. S., M. J. Drummond, E. L. Glynn, et al. 2011. Aging impairs contraction-induced human skeletal muscle mTORC1 signaling and protein synthesis. *Skelet Muscle* 1 (1):11.

Fry, C. S., M. J. Drummond, E. L. Glynn, et al. 2013. Skeletal muscle autophagy and protein breakdown following resistance exercise are similar in younger and older adults. *J Gerontol A Biol Sci Med Sci* 68 (5):599–607.

Fuller, J. C., Jr., S. Baier, P. J. Flakoll, S. L. Nissen, N. N. Abumrad, and J. A. Rathmacher. 2011. Vitamin D status affects strength gains in older adults supplemented with a combination of b-hydroxy-b-methylbutyrate, arginine and lysine: A cohort study. *JPEN* 35:757–62.

Garcia-Prat, L., and P. Munoz-Canoves. 2017. Aging, metabolism and stem cells: Spotlight on muscle stem cells. *Mol Cell Endocrinol* 445:109–17.

Garcia-Prat, L., P. Munoz-Canoves, and M. Martinez-Vicente. 2017. Monitoring autophagy in muscle stem cells. *Methods Mol Biol* 1556:255–80.

Gheller, B. J., E. S. Riddle, M. R. Lem, and A. E. Thalacker-Mercer. 2016. Understanding age-related changes in skeletal muscle metabolism: Differences between females and males. *Annu Rev Nutr* 36:129–56.

Giresi, P. G., E. J. Stevenson, J. Theilhaber, et al. 2005. Identification of a molecular signature of sarcopenia. *Physiol Genomics* 21 (2):253–63.

Giron, M. D., J. D. Vilchez, R. Salto, et al. 2016. Conversion of leucine to beta-hydroxy-beta-methylbutyrate by alpha-keto isocaproate dioxygenase is required for a potent stimulation of protein synthesis in L6 rat myotubes. *J Cachexia Sarcopenia Muscle* 7 (1):68–78.

Glass, D., and R. Roubenoff. 2010. Recent advances in the biology and therapy of muscle wasting. *Ann N Y Acad Sci* 1211:25–36.

Granic, A., T. R. Hill, K. Davies, et al. 2017. Vitamin D status, muscle strength and physical performance decline in very old adults: A prospective study. *Nutrients* 9 (4):379.

Guillory, B., J. A. Chen, S. Patel, et al. 2017. Deletion of ghrelin prevents aging-associated obesity and muscle dysfunction without affecting longevity. *Aging Cell* 16 (4):859–69.

Harber, M. P., J. D. Crane, J. M. Dickinson, et al. 2009. Protein synthesis and the expression of growth-related genes are altered by running in human vastus lateralis and soleus muscles. *Am J Physiol Regul Integr Comp Physiol* 296 (3):R708–14.

Harber, M. P., A. R. Konopka, M. K. Undem, et al. 2012. Aerobic exercise training induces skeletal muscle hypertrophy and age-dependent adaptations in myofiber function in young and older men. *J Appl Physiol* 113 (9):1495–504.

Harper, A. E., N. J. Benevenga, and R. M. Wohlhueter. 1970. Effects of ingestion of disproportionate amounts of amino acids. *Physiol. Rev* 53:428–558.

Hepple, R. T. 2014. Mitochondrial involvement and impact in aging skeletal muscle. *Front Aging Neurosci* 6:13.

Hickson, M. 2015. Nutritional interventions in sarcopenia: A critical review. *Proc Nutr Soc* 74 (4):378–86.

Hord, J. M., R. Botchlett, and J. M. Lawler. 2016. Age-related alterations in the sarcolemmal environment are attenuated by lifelong caloric restriction and voluntary exercise. *Exp Gerontol* 83:148–57.

Isanejad, M., J. Mursu, J. Sirola, et al. 2016. Dietary protein intake is associated with better physical function and muscle strength among elderly women. *Br J Nutr* 115 (7):1281–91.

Ispoglou, T., K. Deighton, R. F. King, H. White, and M. Lees. 2017. Novel essential amino acid supplements enriched with L-leucine facilitate increased protein and energy intakes in older women: A randomised controlled trial. *Nutr J* 16 (1):75.

Johnson, M. A., J. T. Dwyer, G. L. Jensen, et al. 2011. Challenges and new opportunities for clinical nutrition interventions in the aged. *J Nutr* 141 (3):535–41.

Johnson, M. L., M. M. Robinson, and K. S. Nair. 2013. Skeletal muscle aging and the mitochondrion. *Trends Endocrinol Metab* 24 (5):247–56.

Karakelides, H., B. A. Irving, K. R. Short, P. O'Brien, and K. S. Nair. 2010. Age, obesity, and sex effects on insulin sensitivity and skeletal muscle mitochondrial function. *Diabetes* 59 (1):89–97.

Katsanos, C. S., H. Kobayashi, M. Sheffield-Moore, A. Aarsland, and R. R. Wolfe. 2005. Aging is associated with diminished accretion of muscle proteins after the ingestion of a small bolus of essential amino acids. *Am J Clin Nutr* 82 (5):1065–73.

Konopka, A. R., and M. P. Harber. 2014. Skeletal muscle hypertrophy after aerobic exercise training. *Exerc Sport Sci Rev* 42 (2):53–61.

Konopka, A. R., T. A. Trappe, B. Jemiolo, S. W. Trappe, and M. P. Harber. 2011. Myosin heavy chain plasticity in aging skeletal muscle with aerobic exercise training. *J Gerontol A Biol Sci Med Sci* 66 (8):835–41.

Koopman, R., L. B. Verdijk, M. Beelen, et al. 2008. Co-ingestion of leucine with protein does not further augment post-exercise muscle protein synthesis rates in elderly men. *Br J Nutr* 99 (3):571–80.

Krebs, H. A., and P. Lund. 1976. Aspects of the regulation of the metabolism of branched-chain amino acids. *Advan Enzyme Regul* 15:375–94.

Kuhls, D. A., J. A. Rathmacher, M. D. Musngi, et al. 2007. Beta-hydroxy-beta-methylbutyrate supplementation in critically ill trauma patients. *J. Trauma* 62 (1):125–31.

Kumar, A., B. Majumdar, G. Dutta, et al. 2009. The twiddler's plus syndrome: A case report. *Kardiol Pol* 67 (10):1105–6.

Lambert, C. P., and W. J. Evans. 2005. Adaptations to aerobic and resistance exercise in the elderly. *Rev Endocr Metab Disord* 6 (2):137–43.

Lappe, J. M., and N. Binkley. 2015. Vitamin D and sarcopenia/falls. *J Clin Densitom* 18 (4):478–82.

Latham, N. K., D. A. Bennett, C. M. Stretton, and C. S. Anderson. 2004. Systematic review of progressive resistance strength training in older adults. *J Gerontol A Biol Sci Med Sci* 59 (1):48–61.

Lemmer, J. T., F. M. Ivey, A. S. Ryan, et al. 2001. Effect of strength training on resting metabolic rate and physical activity: Age and gender comparisons. *Med Sci Sports Exerc* 33 (4):532–41.

Liu, Y., Z. Sheng, H. Liu, et al. 2009. Juvenile hormone counteracts the bHLH-PAS transcription factors MET and GCE to prevent caspase-dependent programmed cell death in *Drosophila*. *Development* 136 (12):2015–25.

Magne, H., I. Savary-Auzeloux, D. Remond, and D. Dardevet. 2013. Nutritional strategies to counteract muscle atrophy caused by disuse and to improve recovery. *Nutr Res Rev* 26 (2):149–65.

Malafarina, V., F. Uriz-Otano, C. Malafarina, J. A. Martinez, and M. A. Zulet. 2017. Effectiveness of nutritional supplementation on sarcopenia and recovery in hip fracture patients: A multi-centre randomized trial. *Maturitas* 101:42–50.

Marissal-Arvy, N., E. Duron, F. Parmentier, P. Zizzari, P. Mormede, and J. Epelbaum. 2013. QTLs influencing IGF-1 levels in a LOU/CxFischer 344F2 rat population: Tracks towards the metabolic theory of Ageing. *Growth Horm IGF Res* 23 (6):220–8.

Markofski, M. M., K. Jennings, K. L. Timmerman, et al. 2018. Effect of aerobic exercise training and essential amino acid supplementation for 24 weeks on physical function, body composition and muscle metabolism in healthy, independent older adults: A randomized clinical trial. *J Gerontol A Biol Sci Med Sci.* 2019 Sep 15;74(10):1598–1604.

Martone, A. M., E. Marzetti, R. Calvani, et al. 2017. Exercise and protein intake: A synergistic approach against sarcopenia. *Biomed Res Int.* 2017;2017:2672435.

Marzetti, E., R. Calvani, M. Cesari, et al. 2013. Mitochondrial dysfunction and sarcopenia of aging: From signaling pathways to clinical trials. *Int J Biochem Cell Biol* 45 (10):2288–301.

Marzetti, E., F. Landi, F. Marini, et al. 2014. Patterns of circulating inflammatory biomarkers in older persons with varying levels of physical performance: A partial least squares-discriminant analysis approach. *Front Med* 1:27.

Maurel, D. B., K. Jahn, and N. Lara-Castillo. 2017. Muscle-bone crosstalk: Emerging opportunities for novel therapeutic approaches to treat musculoskeletal pathologies. *Biomedicines* 5 (4):62.

McGregor, R. A., S. D. Poppitt, and D. Cameron-Smith. 2014. Role of microRNAs in the age-related changes in skeletal muscle and diet or exercise interventions to promote healthy aging in humans. *Ageing Res Rev* 17:25–33.

Mitchell, W. K., D. J. Wilkinson, B. E. Phillips, J. N. Lund, K. Smith, and P. J. Atherton. 2016. Human skeletal muscle protein metabolism responses to amino acid nutrition. *Adv Nutr* 7 (4):828s–38s.

Mithal, A., J. P. Bonjour, S. Boonen, et al. 2013. Impact of nutrition on muscle mass, strength, and performance in older adults. *Osteoporos Int* 24 (5):1555–66.

Morley, J. E., J. M. Argiles, W. J. Evans, et al. 2010. Nutritional recommendations for the management of sarcopenia. *J Am Med Dir Assoc* 11 (6):391–6.

Mosoni, L., P. Patureau Mirand, M. L. Houlier, and M. Arnal. 1993. Age-related in protein synthesis measured *in vivo* in rat liver and gastrocnemius muscle. *Mech Ageing Dev* 68:209–20.

Munoz-Canoves, P., J. J. Carvajal, A. Lopez de Munain, and A. Izeta. 2016. Editorial: Role of stem cells in skeletal muscle development, regeneration, repair, aging, and disease. *Front Aging Neurosci* 8:95.

Murphy, T., and S. Thuret. 2015. The systemic milieu as a mediator of dietary influence on stem cell function during ageing. *Ageing Res Rev* 19:53–64.

Nair, K. S. 2005. Aging muscle. *Am J Clin Nutr* 81 (5):953–63.

Ng, T. F., M., L. A. Prendergast, P. Dupuis, et al. 2016. Effects of testosterone treatment on body fat and lean mass in obese men on a hypocaloric diet: A randomised controlled trial. *BMC Med* 14 (1):153.

Nishizaki, K., H. Ikegami, Y. Tanaka, R. Imai, and H. Matsumura. 2015. Effects of supplementation with a combination of beta-hydroxy-beta-methyl butyrate, L-arginine, and L-glutamine on postoperative recovery of quadriceps muscle strength after total knee arthroplasty. *Asia Pac J Clin Nutr* 24 (3):412–20.

Nissen, S., R. Sharp, M. Ray, et al. 1996. Effect of leucine metabolite b-hydroxy b-methylbutyrate on muscle metabolism during resistance-exercise training. *J Appl Physiol* 81:2095–104.

Nissen, S. L., and N. N. Abumrad. 1997. Nutritional role of the leucine metabolite b-hydroxy-b-methylbutyrate (HMB). *J Nutr Biochem* 8:300–11.

O'Leary, M. F., A. Vainshtein, S. Iqbal, O. Ostojic, and D. A. Hood. 2013. Adaptive plasticity of autophagic proteins to denervation in aging skeletal muscle. *Am J Physiol Cell Physiol* 304 (5):C422–30.

Palus, S., S. von Haehling, and J. Springer. 2014. Muscle wasting: An overview of recent developments in basic research. *J Cachexia Sarcopenia Muscle* 5 (3):193–8.

Panton, L., J. Rathmacher, J. Fuller, et al. 1998. The effect of b-hydroxy-b-methylbutyrate and resistance training on strength and functional ability in elderly men and women. *Med Sci Sports Exerc* 30:S194.

Petersen, K. F., D. Befroy, S. Dufour, et al. 2003. Mitochondrial dysfunction in the elderly: Possible role in insulin resistance. *Science* 300 (5622):1140–2.

Phillips, S. M. 2015. Nutritional supplements in support of resistance exercise to counter age-related sarcopenia. *Adv Nut* 6 (4):452–60.

Phillips, S. M., R. N. Dickerson, F. A. Moore, D. Paddon-Jones, and P. J. Weijs. 2017. Protein turnover and metabolism in the elderly intensive care unit patient. *Nutr Clin Pract* 32 (1_suppl):112s–20s.

Piccirillo, R., F. Demontis, N. Perrimon, and A. L. Goldberg. 2014. Mechanisms of muscle growth and atrophy in mammals and *Drosophila*. *Dev Dynam* 243 (2):201–15.

Rathmacher, J. A., J. J. Zachwieja, S. R. Smith, G. A. Lovejoy, G. A. Bray. 1999. The effect of the leucine metabolite b-hydroxy-b-methylbutyrate on lean body mass and muscle strength. *FASEB J* 13:A909.

Redman, L. M., S. R. Smith, J. H. Burton, C. K. Martin, D. Il'yasova, and E. Ravussin. 2018. Metabolic slowing and reduced oxidative damage with sustained caloric restriction support the rate of living and oxidative damage theories of aging. *Cell Metab* 27 (4):805–15.e4.

Robinson, M. M., S. Dasari, A. R. Konopka, et al. 2017. Enhanced protein translation underlies improved metabolic and physical adaptations to different exercise training modes in young and old humans. *Cell Metab* 25 (3):581–92.

Rooyackers, O. E., D. B. Adey, P. A. Ades, and K. S. Nair. 1996. Effect of age on in vivo rates of mitochondrial protein synthesis in human skeletal muscle. *Proc Natl Acad Sci USA* 93 (26):15364–9.

Rosenberg, I. H. 1997. Sarcopenia: Origins and clinical relevance. *J Nutr* 127 (5 Suppl):990S–91S.

Rossi, A. P., A. D'Introno, S. Rubele, et al. 2017. The potential of beta-Hydroxy-beta-Methylbutyrate as a new strategy for the management of sarcopenia and sarcopenic obesity. *Drugs Aging* 34 (11):833–40.

Sabourin, P. J., and L. L. Bieber. 1981. Subcellular distribution and partial characterization of an a-ketoisocaproate oxidase of rat liver: Formation of b-hydroxyisovaleric acid. *Arch Biochem Biophys* 206:132–44.

Sabourin, P. J., and L. L. Bieber. 1982. The mechanism of a-ketoisocaproate oxygenase. Formation of b-hydroxyisovalerate from a-ketoisocaproate. *J Biol Chem* 257:7468–71.

Sabourin, P. J., and L. L. Bieber. 1983. Formation of b-hydroxyisovalerate by an a-ketoisocaproate oxygenase in human liver. *Metab* 32:160–4.

Saitoh, M., J. Ishida1, N. Ebner, S. D. Anker, J. Springer, and S. von Haehling1. 2017. Myostatin inhibitors as pharmacological treatment for muscle wasting and muscular dystrophy. *JCSM* 2 (1): 1–x.

Sato, S., G. Solanas, F. O. Peixoto, et al. 2017. Circadian reprogramming in the liver identifies metabolic pathways of aging. *Cell* 170 (4):664–77.e11.

Sato, Y., Y. Honda, J. Iwamoto, T. Kanoko, and K. Satoh. 2005. Effect of folate and mecobalamin on hip fractures in patients with stroke: A randomized controlled trial. *Jama* 293 (9):1082–8.

Sharples, A. P., D. C. Hughes, C. S. Deane, A. Saini, C. Selman, and C. E. Stewart. 2015. Longevity and skel-etal muscle mass: The role of IGF signalling, the sirtuins, dietary restriction and protein intake. *Aging Cell* 14 (4):511–23.

Short, K. R., K. S. Nair, and C. S. Stump. 2004. Impaired mitochondrial activity and insulin-resistant offspring of patients with type 2 diabetes. *New England J Med* 350 (23):2419.

Silva, V. R., F. L. Belozo, T. O. Micheletti, et al. 2017. Beta-hydroxy-beta-methylbutyrate free acid supple-mentation may improve recovery and muscle adaptations after resistance training: A systematic review. *Nutr Res* 45:1–9.

Siracusa, J., N. Koulmann, and S. Banzet. 2018. Circulating myomiRs: A new class of biomarkers to monitor skeletal muscle in physiology and medicine. *J Cachexia Sarcopenia Muscle* 9 (1):20–7.

Snijders, T., J. P. Nederveen, S. Joanisse, et al. 2017. Muscle fibre capillarization is a critical factor in muscle fibre hypertrophy during resistance exercise training in older men. *J Cachexia Sarcopenia Muscle* 8 (2):267–76.

Snijders, T., and G. Parise. 2017. Role of muscle stem cells in sarcopenia. *Curr Opin Clin Nutr Metab Care* 20 (3):186–90.

Solanas, G., F. O. Peixoto, E. Perdiguero, et al. 2017. Aged stem cells reprogram their daily rhythmic functions to adapt to stress. *Cell* 170 (4):678–92.e20.

Sousa-Victor, P., S. Gutarra, L. Garcia-Prat, et al. 2014. Geriatric muscle stem cells switch reversible quies-cence into senescence. *Nature* 506 (7488):316–21.

Spindler, S. R. 2010. Caloric restriction: From soup to nuts. *Ageing Res Rev* 9 (3):324–53.

Stout, J. R., D. H. Fukuda, K. L. Kendall, A. E. Smith-Ryan, J. R. Moon, and J. R. Hoffman. 2015. Beta-Hydroxy-beta-methylbutyrate (HMB) supplementation and resistance exercise significantly reduce abdominal adiposity in healthy elderly men. *Exp Gerontol* 64:33–4.

Stout, J. R., A. E. Smith-Ryan, D. H. Fukuda, et al. 2013. Effect of calcium beta-hydroxy-beta-methylbutyrate (CaHMB) with and without resistance training in men and women 65+yrs: A randomized, double-blind pilot trial. *Exp Gerontol* 48 (11):1303–10.

Tieland, M., M. L. Dirks, N. van der Zwaluw, et al. 2012. Protein supplementation increases muscle mass gain during prolonged resistance-type exercise training in frail elderly people: A randomized, double-blind, placebo-controlled trial. *J Am Med Dir Assoc* 13 (8):713–19.

Van Koevering, M., and S. Nissen. 1992. Oxidation of leucine and a-ketoisocaproate to b-hydroxy-b-methylbutyrate *in vivo*. *Am J Physiol* 262:E27–31.

Verlaan, S., T. J. Aspray, J. M. Bauer, et al. 2017. Nutritional status, body composition, and quality of life in community-dwelling sarcopenic and non-sarcopenic older adults: A case-control study. *Clin Nutr* 36 (1):267–74.

Veyrat-Durebex, C., X. Montet, M. Vinciguerra, et al. 2009. The Lou/C rat: A model of spontaneous food restriction associated with improved insulin sensitivity and decreased lipid storage in adipose tissue. *Am J Physiol Endocrinol Metab* 296 (5):E1120–32.

Vianna, D., G. F. Resende, F. L. Torres-Leal, L. C. Pantaleao, J. Donato, Jr., and J. Tirapegui. 2012. Long-term leucine supplementation reduces fat mass gain without changing body protein status of aging rats. *Nutrition* 28 (2):182–9.

Villareal, D. T., L. Aguirre, A. B. Gurney, et al. 2017. Aerobic or resistance exercise, or both, in dieting obese older adults. *N Engl J Med* 376 (20):1943–55.

Villareal, D. T., S. Chode, N. Parimi, et al. 2011. Weight loss, exercise, or both and physical function in obese older adults. *N Engl J Med* 364 (13):1218–29.

Vinel, C., L. Lukjanenko, A. Batut, et al. 2018. The exerkine apelin reverses age-associated sarcopenia. *Nat Med* 24 (9):1360–71.

Visser, M., D. J. Deeg, and P. Lips. 2003. Low vitamin D and high parathyroid hormone levels as determinants of loss of muscle strength and muscle mass (sarcopenia): The longitudinal aging study Amsterdam. *J Clin Endocrinol Metab* 88 (12):5766–72.

Vukovich, M. D., N. B. Stubbs, and R. M. Bohlken. 2001. Body composition in 70-year old adults responds to dietary b-hydroxy-b-methylbutyrate (HMB) similar to that of young adults. *J Nutr* 131(7):2049–52.

Wagatsuma, A., and K. Sakuma. 2014. Vitamin D signaling in myogenesis: Potential for treatment of sarco-penia. *Biomed Res Int* 2014:121254.

Wakabayashi, H. 2014. Presbyphagia and sarcopenic dysphagia: Association between aging, sarcopenia, and deglutition disorders. *J Frailty Aging* 3 (2):97–103.

Wakabayashi, H., and K. Sakuma. 2014. Comprehensive approach to sarcopenia treatment. *Curr Clin Pharmacol* 9 (2):171–80.

Wakabayashi, H., and K. Sakuma. 2014. Rehabilitation nutrition for sarcopenia with disability: A combination of both rehabilitation and nutrition care management. *J Cachexia Sarcopenia Muscle* 5 (4):269–77.

Wall, B. T., N. M. Cermak, and L. J. van Loon. 2014. Dietary protein considerations to support active aging. *Sports Med* 44 (Suppl 2):S185–94.

Wang, C., and L. Bai. 2012. Sarcopenia in the elderly: Basic and clinical issues. *Geriatr Gerontol Int* 12 (3):388–96.

Wang, X., A. M. Pickrell, S. G. Rossi, et al. 2013. Transient systemic mtDNA damage leads to muscle wasting by reducing the satellite cell pool. *Hum Mol Genet* 22 (19):3976–86.

Wee, A. K. 2016. Serum folate predicts muscle strength: A pilot cross-sectional study of the association between serum vitamin levels and muscle strength and gait measures in patients >65 years old with diabetes mellitus in a primary care setting. *Nutr J* 15 (1):89.

Weihrauch, M., and C. Handschin. 2018. Pharmacological targeting of exercise adaptations in skeletal muscle: Benefits and pitfalls. *Biochem Pharmacol* 147:211–20.

Wilkinson, D. J., T. Hossain, M. C. Limb, et al. 2017. Impact of the calcium form of beta-hydroxy-beta-methylbutyrate upon human skeletal muscle protein metabolism. *Clin Nutr* 37 (6):2068–2075.

Wilkinson, D. J., T. Hossain, D. S. Hill, et al. 2013. Effects of leucine and its metabolite beta-hydroxy-beta-methylbutyrate on human skeletal muscle protein metabolism. *J Physiol* 591 (Pt 11):2911–2923.

Woo, J. 2018. Nutritional interventions in sarcopenia: Where do we stand? *Curr Opin Clin Nutr Metab Care* 21 (1):19–23.

Wu, H., Y. Xia, J. Jiang, et al. 2015. Effect of beta-hydroxy-beta-methylbutyrate supplementation on muscle loss in older adults: A systematic review and meta-analysis. *Arch Gerontol Geriatr* 61 (2):168–75.

Yanai, H. 2015. Nutrition for sarcopenia. *J Clin Med Res* 7 (12):926–31.

Zachwieia, J. J., S. R. Smith, G. A. Bray, et al. 1999. Effect of the leucine metabolite beta-hydroxy-beta-methylbutyrate on muscle protein synthesis during prolonged bedrest. *Faseb J* 13:A1025.

Section VI

Applications

*Part 1: Muscle Impairments or Diseases due
to the Frailty Induced by Sarcopenia*

16 Declines in Whole Muscle Function with Aging

The Role of Age-Related Alterations in Contractile Properties of Single Skeletal Muscle Fibers

Nicole Mazara and Geoffrey A. Power

CONTENTS

16.1 INTRODUCTION: BACKGROUND AND IMPACT OF SARCOPENIA

The biological process of natural adult aging results in a gradual deterioration of the structure and function of the human neuromuscular system that accelerates into very old age (Vandervoort, 2002). The integrity of our cells, regrowth, regeneration, and neural processes are all compromised with aging, and one of the major areas impacted is skeletal muscle. However, it was not until the late 1980s that the marked reductions in skeletal muscle mass in the elderly became an area of concern for Dr. Irwin Rosenberg, who suggested that the age-associated loss of skeletal muscle mass be termed *sarcopenia* (literally translating into a *poverty of flesh*) so as to bring attention to this process (Rosenberg, 1997). Since the classification of sarcopenia, a myriad of causes for the loss of muscle mass over the lifespan have been researched, and as a result, a number of terms have emerged (Muscaratoli et al. 2010). The loss of skeletal muscle mass caused by an underlying disease-driven process—usually cancer—has been termed *cachexia* (Muscaritoli et al. 2010). Moreover, the terms *myopenia*, the loss of skeletal muscle mass or "muscle wasting" (Fearon et al. 2011), and *dynapenia*, the loss of muscle strength (Clark & Manini, 2008), have come into the literature as an effort to better describe and differentiate the alterations in skeletal muscle function (i.e., mass, strength, power, endurance capacity) across various populations. For the purpose of this book chapter, the term *sarcopenia* will refer to the age-related declines in skeletal muscle mass. Whole muscle function will refer to muscle mass, strength, power, and endurance capacity.

At approximately 60 years of age, the rate of skeletal muscle mass loss becomes appreciable (i.e., the onset of the sarcopenia process) and is estimated at ~1%–2% per year, and the associated rate of skeletal muscle strength loss is higher, estimated to be ~3% per year (von Haehling et al. 2010). While the process of sarcopenia affects all older adults, the state of being "sarcopenic" or "frail" has specific diagnostic criteria (Muscaritoli et al. 2010), and the sarcopenic phenotype is often accompanied by muscle weakness, increased fatigability (Power et al. 2013), and/or performance deficits (Muscaritoli et al. 2010; Cruz-Jentoft et al. 2010). On average, it is estimated that ~35% of older adults have diagnosable sarcopenia, and those sarcopenic older adults have an increased risk of falls, fractures, impaired mobility, frailty, metabolic dysfunction (such as type 2 diabetes), and overall lower quality of life (von Haehling et al. 2010).

There is a limited understanding regarding the etiology of sarcopenia, rendering the development of effective treatments limited, and explaining why even active older adults suffer from the sarcopenia process. The progression of sarcopenia has been linked to reorganization of the motor neuron and associated innervation (Deschenes et al. 2010), loss of motor units (Dalton et al. 2010; McNeil et al. 2005; Power et al. 2012), neuromuscular junction degeneration (Jang et al. 2010), mitochondrial dysfunction (Gouspillou et al. 2018; Marzetti et al. 2013), irreversible oxidation damage (Brocca et al. 2017; Jackson & McArdle 2011), and reductions in muscle protein synthesis rates in response to anabolic stimuli (Leenders et al. 2011). Alterations in cellular muscle mechanics, such as the disruption of force-producing characteristics, cross-bridge cycling, and calcium (Ca^{2+}) release and sensitivity, have also been found in aging muscle and could offer another explanation for the seemingly inevitable onset and progression of sarcopenia in healthy adult aging skeletal muscle. The investigation of the role of cellular and molecular mechanisms driving impaired performance of skeletal muscle is vital to the understanding of how sarcopenia impacts the age-related alterations in whole muscle function.

This chapter will discuss:

- The changes in muscle morphology with aging and the association with sarcopenia.
- The testing and quantification of single muscle fiber mechanics.
- The mechanisms underpinning single muscle fiber mechanics.
- How cellular mechanics change with aging and contribute to sarcopenia.

16.2 MUSCLE MORPHOLOGY

Skeletal muscle generates the contractile force necessary to support multiple forms of locomotion, from the explosive power of a jump to the fine motor skill of knitting. In order to enable the execution of these distinctly differe nt physical functions, skeletal muscles exhibit different morphological/architectural features such as pennation angle, fascicle length, muscle length, and muscle fiber innervation (i.e., fast or slow twitch motor units). Similar to whole muscle structure, properties of the muscle cell, such as fiber type and size, fit to match the physical demands of the muscle.

16.2.1 Fiber Type and Myosin Heavy Chain Expression

A motor unit, consisting of the motor neuron and muscle fibers it innervates, can be divided generally into two different types: slow twitch and fast twitch; the fibers of which commonly express myosin heavy chain (MHC) I and II isoforms, respectively. Type I (TI) fibers are generally recruited for low-force producing, repetitive, and accuracy-based tasks, whereas type II (TII) fibers are recruited for high-force producing and more powerful tasks. MHC isoform expression is a commonly used technique to identify fiber type, and it has been shown that MHC expression correlates well with contractile properties of single muscle fibers in young adults (Aagaard & Andersen 1998; Bottinelli et al. 1996, 1999; Harridge et al. 1996). There are three primary MHC isoforms in adult human muscle. MHC I expressing fibers are typically smaller, weaker, slower, and more fatigue resistant (i.e., higher endurance capacity) (Larsson & Moss 1993; Bottinelli et al. 1996) across the three fiber types. Fibers that express MHC IIA are typically larger, faster, stronger, more powerful, and fatigue more quickly than MHC I. There is a small proportion of adult fibers that express MHC IIX, and these fibers are typically very powerful, produce force very quickly, and fatigue the fastest of the three fiber types (Larsson & Moss 1993; Bottinelli et al. 1996). Fibers can also co-express MHC isoforms and are referred to as hybrid fibers; these fibers are common in models of disuse atrophy (Gallagher et al. 2005; Ohira et al. 1999). Hybrid fibers show mixed contractile properties associated with the isoforms being expressed and occur in young, healthy adult muscle in low proportions (Bottinelli et al. 1996).

Age-related changes in fiber type distribution and MHC expression are muscle and species dependent. In a large, cross-sectional study across the ages of 15–83 years, it was determined that the driving force behind whole muscle mass loss with age (i.e., sarcopenia) was a loss of muscle fibers, specifically TII fibers, while TI fibers seemed to remain intact over a lifespan (Lexell et al. 1988). A loss, or lower proportion of, TII/MHC IIA containing fibers with aging has been found in other studies as well (Larsson et al. 1978; Oh et al. 2018; Trappe et al. 1995). The loss of TII fibers with aging can happen as a result of denervation, which either leads to atrophy and cell death or reinnervation by a neighboring motor unit. As reinnervation seems to cause fibers to co-express MHC isoforms (Purves-Smith et al. 2014; Rowan et al. 2012), it is likely that some TII fibers are not "lost" but are changing classification through reinnervation. This is supported in both humans and rodent models where the proportion of hybrid fibers is greater with aging (Andersen et al. 1999; Edstrom & Ulfhake 2005; Klitgaard et al. 1990; Monemi et al. 1999; Rowan et al. 2012). Another characteristic of denervated fibers is the highly angular shape, different from the typical cylindrical shape of muscle fibers, and angular muscle fibers have been found in aging humans (Andersen 2003; Power et al. 2016) and rodent models (Rowan et al. 2011, 2012). The denervation–reinnervation process is a major contributor to the loss of muscle fibers, expressing any MHC isoform, and progression of sarcopenia.

Sarcopenia, being the age-associated loss of muscle mass, is often confounded by disuse. Many older adults become less physically active and are at greater risk of a mobility-limiting injury or illness. Disuse has been shown to cause atrophy and fiber loss (Rowan et al. 2011) and seems to affect TI fibers, conversely to the TII-targeted process of aging (D'Antona et al. 2003; Hortobagyi et al. 2000; Widrick et al. 1999). As aging is often accompanied by disuse, this phenomenon causes fiber type distribution alterations to be unclear in the context of healthy aging compared with disuse atrophy, and how the two interact.

Alterations in MHC isoform expression are likely a symptom of the denervation–reinnervation process that occurs when motor units are lost during aging, and over longitudinal studies, tracking MHC isoform expression of single muscle fibers is a useful tool in quantifying age-related changes in muscle quality.

16.2.2 Fiber Size and Myofibrillar Content

Muscle fiber size is measured as cross-sectional area (CSA) either via diameter and depth measurements of single muscle fibers or area measurements of stained muscle cross-sections. Similar to fiber type distribution, there is evidence that TII fibers are preferentially targeted and atrophy more than TI fibers. It was found that TII fiber size was ~25% smaller in the old as compared with the young group (Lexell et al. 1988). Many studies are in agreement with the above findings from stained muscle cross-sections that TII fibers have a smaller CSA than TI in old compared with young adults (Aniansson et al. 1986; Coggan et al. 1992; Jennekens et al. 1971). A recent study found that older men had smaller TII muscle fiber CSA compared with TI fiber CSA (Snijders et al. 2016), and a study from the same group found a significant reduction in TII muscle fiber CSA between young and older men, but no difference in TI CSA (Nederveen et al. 2016).

However, in a longitudinal study, it was noted that fiber area for all types increased between the ages of 76–80 years suggesting that a "compensatory hypertrophy" was occurring in the fibers causing an increase in fiber size at a critical point in the aging process at about 75 years (Aniansson et al. 1992). This compensatory mechanism is likely associated with the rapid loss of functional motor units, which in turn would cause a loss of muscle fibers, occurring around the same age of ~75 years (Power et al. 2014). At this critical age point, many fibers have been lost to the denervation–reinnervation process and atrophy, and so the fibers left over have greater mechanical tension during any muscular contraction, which likely is the mechanism behind this "compensatory hypertrophy."

Myofilaments (actin and myosin filaments) make up sarcomeres of a skeletal muscle fiber, and if the content of these proteins decrease (i.e., a loss of sarcomeres in parallel), the CSA of the entire fiber also decreases. There is evidence that MHC and myofibrillar protein synthesis rates decrease with age (Balagopal et al. 1997; Welle et al. 1993), as well as the content of other important proteins in muscle function (Gouspillou et al. 2018). In a cross-sectional study, myosin content was measured in young adults, old adults, and old immobilized adults, and it was found that myosin content decreased with age and immobilization, and the greater relative decrease in myosin remained even when normalized to CSA (D'Antona et al. 2003). In rats, it was shown that myosin content and the myosin-to-actin ratio decreased with age in the semimembranosus muscle, which consists primarily of TII fibers (Thompson et al. 2006). However, other studies have not found a lower myosin-to-actin ratio (Callahan et al. 2014) or lower actin content (Miller et al. 2013) in older adults compared with young. While not all studies corroborate a decrease in myofibrillar content, single fiber atrophy is likely progressed by a decrease in myofibrillar content, contributing to the progression of sarcopenia and potentially having an effect on the cellular contraction of single fibers as well.

A disconnect remains between the whole muscle and cellular level, as size decreases at the whole muscle with age, but a consistent result at the cellular level remains to be identified (Table 16.1). This disconnect lends support to the theory that age-related whole muscle atrophy is driven by an overall loss in fibers caused by denervation and not necessarily single fiber atrophy on its own.

16.3 ISOMETRIC AND KINETIC MEASURES OF CONTRACTILE PROPERTIES

Single skeletal muscle fiber analysis offers a model free from neural regulation, which is necessary to avoid confounding activation effects when investigating basic age-related impairments in muscle function. Testing a single muscle fiber biopsied from whole muscle enables direct comparisons between mechanical, structural, and biochemical properties of muscle with the contractile proteins confined to their native environment. Single fiber analysis occurs by permeabilizing the sarcolemma to allow the fiber to be activated via chemical solution (Ca^{2+} and ATP) and employing various length or chemical solution changes to measure multiple aspects of cellular contractile properties.

TABLE 16.1

Summary of Studies Examining Age-related Differences in Cross-Sectional Area (CSA), Maximal Isometric Force (P_o), and/or Specific Force (SF) in Single Muscle Fibers from Humans and Rodent Models

Study	Age/Sex	Muscle	Fiber Type	CSA (μm²)%Δ	P_o (μN)%Δ	SF (P_o/CSA)%Δ	Methods/Notes
D'Antona et al. (2003)	Healthy M n = 7 72.7 ± 2.3 yr		TI	~7000 −22%ᵃ	—	−22%ᵃ	T: 12°C SL: 2.5 μm No CSA swelling correction %Δ from YO M SF units: kN/m²
			TIIA	~7000 −12%	—	−16%ᵃ	
			TIIX	~7300 0%	—	0%	
	Immobilized M n = 2 70 and 72 yr	Vastus lateralis	TI	~4500 −51%ᵃ	—	−55%ᵃ	
			TIIA	~5700 −26%ᵃ	—	−42%ᵃ	
			TIIX	~4900 −24%ᵃ	—	−37%ᵃ	
D'Antona et al. (2007)	Healthy M sedentary n = 3 73 ± 2.5 yr		TI	—	—	32 −26%ᵃ	T: 12°C SL: 2.5 μm No CSA swelling correction %Δ from YO M SF units: kN/m²
			TIIA	—	—	49 −21%ᵃ	
	Healthy M recreationally active, n = 4 73 ± 2.5 yr	Vastus lateralis	TI	—	—	34 −21%ᵃ	
			TIIA	—	—	56 −9%	
	Healthy M endurance trained n = 3 73 ± 2.5 yr		TI	—	—	39 −9%	
			TIIA	—	—	57 −8%	

(Continued)

TABLE 16.1 (Continued)

Summary of Studies Examining Age-related Differences in Cross-Sectional Area (CSA), Maximal Isometric Force (P_o), and/or Specific Force (SF) in Single Muscle Fibers from Humans and Rodent Models

Study	Age/Sex	Muscle	Fiber Type	CSA (μm^2)%Δ	P_o (μN)%Δ	SF (P_o/CSA)%Δ	Methods/Notes
Frontera et al. (2000a)	Healthy M	Vastus lateralis	TI	~4960	505	14.9	T: 15°C
	$n = 12$				-30%[a]	-28%[a]	SL: 2.7–2.85 μm
	74.4 ± 6.1 yr		TIIA	~5110	577	16.4	20% CSA swelling correction
					-27%[a]	-31%[a]	%Δ from YO M
	Healthy F		TI	~4990	472	14.2	Compared O F to YO M
	$n = 12$				-35%[a]	-32%[a]	SF units: N/cm^2
	72.1 ± 4.0 yr		TIIA	~4140	422	15.7	
					-47%[a]	-34%[a]	
Hvid et al. (2011)	Healthy M	Vastus lateralis	TI	6922 ± 609	590 ± 40	77 ± 3.1	T: 22.1°C
	$n = 8$				+15%	-5%	No SL measure: used 120% "slack length"
	66.6 ± 1.5 yr		TIIA	7587 ± 348	790 ± 80	117.4 ± 7.9	No CSA swelling correction
					-1%	-1%	%Δ from YO M
							SF units: kN/m^2
							Mean ± SE
Korhonen et al. (2006)	Healthy M sprint athletes	Vastus lateralis	TI	3200 ± 170	–	26.4 ± 1.8	T: 15°C
	$n = 9$			-26%[a]		+23%	SL: 2.76 ± 0.04 μm
	53–77 yr		TIIA	3000 ± 240	–	36.1 ± 2.8	20% CSA swelling correction
				-36%[a]		+19%	%Δ from YO M
							SF units: N/cm^2
							Mean ± SE
Lamboley et al. (2015)	Healthy M and F	Vastus lateralis	TI	–	–	157 ± 56	T: 23 ± 2°C
	$n = 20$					-8%	SL: 120% resting length
	70 ± 4 yr		TII	–	–	179 ± 43	No CSA swelling correction
						-17%[a]	%ΔYO
							SF units: mN/mm^2
							Mean ± SD

(Continued)

TABLE 16.1 (Continued)

Summary of Studies Examining Age-related Differences in Cross-Sectional Area (CSA), Maximal Isometric Force (P_o), and/or Specific Force (SF) in Single Muscle Fibers from Humans and Rodent Models

Study	Age/Sex	Muscle	Fiber Type	CSA (μm²)%Δ	P_o (μN)%Δ	SF (P_o/CSA)%Δ	Methods/Notes
Larsson et al. (1997a)	Healthy M recreationally active n = 2 73–81 yr	Vastus lateralis	TI	3090 ± 870 +9%	–	0.18 ± 0.06 −5%	T: 12°C SL: 2.79 μm average No CSA swelling correction %Δ from YO M SF units: N/mm² Mean ± SD
			TIIA	2770 ± 740 −28%[a]	–	0.18 ± 0.01 −28%[a]	
	Healthy M very active n = 2 73–81 yr		TI	2870 ± 680 0%	–	0.16 ± 0.05 −15%[a]	
			TIIA	3710 ± 1570 −3%	–	0.21 ± 0.06 −16%[a]	
Larsson et al. (1997b)	Healthy M n = 4 73–81 yr	Vastus lateralis	TI	–	–	Sig decrease from YO	T: 12°C SL: 2.7–2.85 μm Assumed elliptical CSA SF units: unknown No numerical values presented Freeze-dried and chemically skinned
			TIIA	–	–	Sig decrease from YO	
Miller et al. (2013)	Healthy M n = 5 69.5 ± 1.8 yr	Vastus lateralis	TI	~6000 +50%[a]	–	115 0%	T: 15°C Unknown SL No CSA swelling correction %Δ from YO of same sex SF units: mN/mm²
			TIIA	~5800 +5%	–	130 −1%	
	Healthy F n = 7 68.4 ± 1.0 yr		TI	~4500 −10%	–	120 +14%	
			TIIA	~3800 −21%[a]	–	160 +14%[a]	

(Continued)

TABLE 16.1 (Continued)

Summary of Studies Examining Age-related Differences in Cross-Sectional Area (CSA), Maximal Isometric Force (P_o), and/or Specific Force (SF) in Single Muscle Fibers from Humans and Rodent Models

Study	Age/Sex	Muscle	Fiber Type	CSA (μm²)%Δ	P_o (μN)%Δ	SF (P_o/CSA)%Δ	Methods/Notes
Ochala et al. (2007)	Healthy M $n = 6$ 60–74 yr	Vastus lateralis	TI	~3560 ± 573 +12%	–	12 ± 1.5 −25%[a]	T: 15°C SL: 2.75–2.85 μm 20% CSA swelling correction %Δ from YO M SF units: N/cm² Mean ± SD
			TIIA	~3436 ± 486 −6%	–	13.9 ± 1.7 −33%[a]	
Power et al. (2016)	Healthy M nonathlete $n = 5$ 78.2 ± 9.4 yr	Vastus lateralis	"Slow"	~6000 ~5000	130 ± 20 −57%[a]	23.5 ± 5.6 −43%[a] 25.7 ± 3.5 −48%[a]	T: 10°C SL: 2.8 μm Circularity assumed CSA Elliptical assumed CSA %Δ from YO M SF units: mN/mm² Fibers grouped as "slow type" based on V_o Mean ± SE
	Healthy M master's athlete $n = 6$ 78.8 ± 3.6 yr		"Slow"	~8000 ~5000	150 ± 20 −53%[a]	19.06 ± 2.81 −54%[a] 28.39 ± 3.96 −43%[a]	
Raue et al. (2009)	Healthy F 85 ± 1 yr $n = 6$ T1: pre-training	Vastus lateralis	TI TIIA	5383 ± 654 4164 ± 508	570 ± 60 690 ± 70	108.3 ± 3.9 160.8 ± 4.1	T: 15°C SL: 2.5 μm No CSA swelling correction T1 not different from YO F SF units: kN/m² Mean ± SE
	Healthy F 85 ± 1 yr T2: post-training		TI TIIA	5223 ± 538 4694 ± 551	590 ± 60 750 ± 90	110.8 ± 2.9 158.6 ± 1.4	

(Continued)

TABLE 16.1 (Continued)

Summary of Studies Examining Age-related Differences in Cross-Sectional Area (CSA), Maximal Isometric Force (P_o), and/or Specific Force (SF) in Single Muscle Fibers from Humans and Rodent Models

Study	Age/Sex	Muscle	Fiber Type	CSA (μm²)%Δ	P_o (μN)%Δ	SF (P_o/CSA)%Δ	Methods/Notes
Reid et al. (2012)	Healthy M n = 16 73.8 ± 4 yr	Vastus lateralis	TI	4999 ± 931 −6%	512 ± 79 −11%	15.4 ± 3.0 −7%	T: 15°C SL: 2.75–2.85 μm 20% CSA swelling correction %Δ middle-aged group (~45 yr) SF units: N/cm² Fibers per participant averaged – then grouped Mean ± SD
			TIIA	4902 ± 1500 −8%	428 ± 151 −11%	13 ± 3.8 −7%	
	Healthy F n = 7 74.3 ± 4		TI	4407 ± 1174 −10%	453 ± 141 −13%	15.9 ± 4.1 0%	
			TIIA	4619 ± 949 +15%	457 ± 155 +11%	14.9 ± 6.3 +7%	
	Limited mobility M n = 12 78.9 ± 4 yr		TI	4989 ± 1152 −6%	479 ± 142 −17%	14.7 ± 4.5 −11%	
			TIIA	4055 ± 794 −24%	332 ± 87 −31%	12.7 ± 3.8 −9%	
	Limited mobility F n = 13 76.8 ± 5 yr		TI	4747 ± 887 −3%	478 ± 111 −8%	15.4 ± 3.6 −3%	
			TIIA	4110 ± 1646 +2%	391 ± 191 −5%	15.7 ± 8.3 +12%	
Straight et al. (2018)	Healthy M and F Recreationally active n = 10 (3F) 68.8 ± 0.8 yr	Vastus lateralis	TI	~4700 0%	~480 +6%	~100 0%	T: 15°C SL: 2.65 μm No CSA swelling correction %Δ from YO SF units: mn/mm² Mean ± SE
			TIIA	4053 ± 402 −22%[a]	540 ± 50 −11%	138 ± 7 +17%[a]	
Trappe et al. (2003)	Healthy M sedentary n = 6 80 ± 4 yr	Vastus lateralis	TI	–	800 ± 90 +6%	89 ± 7 −4%	T: 15°C SL: 2.5 μm No CSA swelling correction %Δ from YO same sex SF units: kN/m² Mean ± SE
			TIIA		820 ± 90 −13%	129 ± 13 +13%[a]	
	Healthy F Sedentary n = 6 78 ± 2		TI	–	790 ± 1345 +27%[a]	97 ± 7 +2%	
			TIIA		510 ± 70 −33%[a]	140 ± 14 −4%	

(Continued)

TABLE 16.1 (Continued)

Summary of Studies Examining Age-related Differences in Cross-Sectional Area (CSA), Maximal Isometric Force (P_o), and/or Specific Force (SF) in Single Muscle Fibers from Humans and Rodent Models

Study	Age/Sex	Muscle	Fiber Type	CSA (μm^2)%Δ	P_o (μN)%Δ	SF (P_o/CSA)%Δ	Methods/Notes
Venturelli et al. (2015)	Healthy F Mobile (OM) n = 8 87 ± 5 yr	Vastus lateralis	TI	5327 ± 598 +25%	435 ± 41 +9%	93 ± 13 −11%	T: 12°C Unknown SL No CSA swelling correction %Δ from YO F group SF units: mN/mm² Mean ± SE
			TIIA	3854 ± 820 −19%	377 ± 64 −30%	109 ± 12 −20%	
			TIIA/X	4834 ± 545 +16%	561 ± 115 +22%	112 ± 14 −12%	
	Healthy F Immobile (OI) n = 8 88 ± 4 yr		TI	5603 ± 464 +32%	557 ± 42 +40%	111 ± 12 +6%	
			TIIA	2939 ± 521 −38%	378 ± 123 −30%	131 ± 32 −4%	
			TIIA/X	3859 ± 403 −7%	640 ± 75 +39%	177 ± 7 +38%	
	OM	Biceps brachii	TI	4005 ± 333 +17%	375.6 ± 31 +37%	106 ± 13 +20%	
			TIIA	3861 ± 330 −4%	411 ± 52 −26%	112 ± 16 −27%	
			TIIA/X	3827 ± 419 +21%	535 ± 78 +29%	146 ± 23 +1%	
	OI		TI	3069 ± 377 −10%	305 ± 39 +11%	108 ± 8 +23%	
			TIIA	6499 ± 706[a] +61%	701 ± 28 +25%[a]	113 ± 14 −26%	
			TIIA/X	4092 ± 783 +30%	431 ± 65 +4%	119 ± 18 −17%	

(Continued)

TABLE 16.1 (Continued)

Summary of Studies Examining Age-related Differences in Cross-Sectional Area (CSA), Maximal Isometric Force (P_o), and/or Specific Force (SF) in Single Muscle Fibers from Humans and Rodent Models

Study	Age/Sex	Muscle	Fiber Type	CSA (µm²)%Δ	P_o (µN)%Δ	SF (P_o/CSA)%Δ	Methods/Notes
Yu et al. (2007)	Healthy M	Vastus lateralis	TI	3310 ± 1060	860 ± 280	23.4 ± 9.1	T: 12°C
	$n = 17$			−9%	+13%[a]	−23%[a]	SL: 2.61 ± 0.19 µm
	65–85 yr		TIIA	3310 ± 900	710 ± 240	26.0 ± 8.2	20% CSA swelling correction
				−29%[a]	−28%[a]	−14%[a]	Freeze-dried fiber
	Healthy F		TI	3140 ± 1040	550 ± 220	26.56 ± 10.8	preparation
	$n = 5$			+4%	−4%	−20%[a]	%Δ from YO of same sex
	65–85 yr		TIIA	2700 ± 940	520 ± 220	27.7 ± 9.2	SF units: N/cm²
				+5%	−26%[a]	−25%[a]	Mean ± SD
Frontera et al. (2008)	Healthy M and F	Vastus lateralis	TI	5020 ± 1140	493 ± 108	14.8 ± 2.0	T: 15°C
	T1, $n = 24$		(T1)				SL units: 2.7–2.85 µm
	71.1 ± 5.4 yr		TI (T2)	4910 ± 1190	605 ± 191	19 ± 5.7	20% CSA swelling correction
					+23%	+28%	9 yr longitudinal study
	T2, $n = 9$		TIIA (T1)	4230 ± 530	482 ± 113	17 ± 3.5	%Δ from T1
	80 ± 5.3 yr		TIIA (T2)	4840 ± 1260	541 ± 168	17.2 ± 5.0	SF units: N/cm²
				+12.2%	+12.2%	0%	Mean ± SD

(Continued)

TABLE 16.1 (Continued)

Summary of Studies Examining Age-related Differences in Cross-Sectional Area (CSA), Maximal Isometric Force (P_o), and/or Specific Force (SF) in Single Muscle Fibers from Humans and Rodent Models

Study	Age/Sex	Muscle	Fiber Type	CSA (μm²)%Δ	P_o (μN)%Δ	SF (P_o/CSA)%Δ	Methods/Notes
Reid et al. (2014)	Healthy M and F T1: ~74.1 yr T2: ~77 yr n = 14 (5F)	Vastus lateralis	TI (T1)	4787 ± 1063	488 ± 104	15.6 ± 3.4	T: 15°C SL: 2.75–2.85 μm 20% CSA swelling correction %Δ T1 SF units: N/cm² Mean ± SD
			TI (T2)	+4956 ± 1336	+44%[a] 705 ± 245	+40%[a] 21.9 ± 4.8	
			TIIA (T1)	4817 ± 1339	437 ± 149	13.6 ± 4.6	
			TIIA (T2)	+4881 ± 1710	+47%[a] 644 ± 248	+48%[a] 20.1 ± 2.9	
	Limited mobility M and F T1: ~77.2 yr T2: ~80 yr n = 5 (3F)		TI (T1)	4900 ± 930	488 ± 134	15.2 ± 4.1	
			TI (T2)	−4599 ± 646	+41%[a] 689 ± 202	+51%[a] 22.98 ± 4.1	
			TIIA (T1)	4469 ± 1014	386 ± 170	12.7 ± 4.3	
			TIIA (T2)	−3603 ± 869	+31% 504 ± 211	+65%[a] 21.01 ± 3.4	
Brooks and Faulkner (1988)	Mice, healthy M 26–27 mo (OLD)	Soleus	n/a	—	202 ± 9 −5%	18.7 ± 1.05 −9%	T: 25°C Intact electrically stimulated whole muscle preparations %Δ from YO mice Different from ADULT SF units: N/cm² P_o units: mN Mean ± SE
		EDL		—	299 ± 14 −28%[a]	18.6 ± 0.86 −19%[a]	
	Mice, healthy M 9–10 mo (ADULT)	Soleus	n/a	—	259 ± 11 +22%[a]	22.1 ± 0.65 +7%	
		EDL		—	411 ± 13 0%	23.8 ± 0.67 0%	

(Continued)

TABLE 16.1 (Continued)

Summary of Studies Examining Age-related Differences in Cross-Sectional Area (CSA), Maximal Isometric Force (P_o), and/or Specific Force (SF) in Single Muscle Fibers from Humans and Rodent Models

Study	Age/Sex	Muscle	Fiber Type	CSA (μm^2)%Δ	P_o (μN)%Δ	SF (P_o/CSA)%Δ	Methods/Notes
Degens et al. (1998)	Rats, healthy M $n = 5$ 22–24 mo	Soleus	TI	–	–	12.5 ± 2.1 –10%	T: 12°C SL: 2.61 ± 0.19 μm 20% CSA swelling correction Assumed elliptical CSA
		EDL	TIIB and IIX	–	–	14.6 ± 2.0 +13%[a]	
	Rats, healthy F $n = 4$ 22–24 mo	Soleus	TI	–	–	10.6 ± 2.7 0%	Freeze-dried fiber preparation SF units: N/cm^2 Mean ± SD
			TIIB and IIX				
		EDL	TIIA	–	–	10.1 ± 3.0 –2%	
Kim et al. (2013)	Rats, healthy M 32–37 mo	MG	TI	69.0 ± 1.7 –6%	348.8 ± 18.1 –24%[a]	95.8 ± 4.3 –15%[a]	T: 15°C SL: 2.5 μm Fiber diameter (μm) not CSA reported %Δ 5–12 mo M rats SF units: kN/m^2 Mean ± SE
			TII	65.4 ± 1.4 –8%[a]	275.6 ± 14.3 –29%[a]	85.6 ± 3.8 –14%[a]	
Larsson et al. (1997b)	Rats, healthy M and F 20–24 mo	EDL	TII	–	–	n/c from YO rats	T: 12°C SL: 2.45–2.75 μm CSA elliptical assumption No numerical values reported Freeze-dried and chemically skinned
		Soleus	TI	–	–	n/c from YO rats	

(Continued)

TABLE 16.1 (Continued)

Summary of Studies Examining Age-related Differences in Cross-Sectional Area (CSA), Maximal Isometric Force (P_o), and/or Specific Force (SF) in Single Muscle Fibers from Humans and Rodent Models

Study	Age/Sex	Muscle	Fiber Type	CSA (μm^2)%Δ	P_o (μN)%Δ	SF (P_o/CSA)%Δ	Methods/Notes
Li and Larsson (1996)	Rats, healthy M $n = 5$ 20–24 mo	Soleus	TI	2200 ± 440	—	12.4 ± 2.1 -8%[a]	T: 12°C SL: 2.49–2.75 μm 20% CSA swelling correction Assumed elliptical CSA Freeze-dried fiber %Δ from YO M rat SF units: N/cm² Mean ± SD
		EDL	TIIX/B	2800 ± 500[a]	—	14.4 ± 2.1 $+9\%$[a]	
Lowe et al. (2001)	Rats, healthy M $n = 6$ 32–36 mo	SM	TIIB/IIX	5303 ± 692 -10%	659 ± 78 -34%[a]	125 ± 6.6 -27%[a]	T: 25°C SL: 2.5 μm No CSA methods noted %Δ from YO M rat SF units: kN/m² Mean ± SE
Lowe et al. (2002)	Rats, healthy M $n = 7$ 32–37 mo	SM	TIIB TIIX TIIB/IIX	—	—	136 ± 7 -16%[a]	T: 21°C SL: unknown No CSA methods noted %Δ from YO M rat SF units: kN/m² Mean ± SE
Prochniewicz et al. (2005)	Rats, healthy M and F 27–35 mo	SM	TII	—	—	82.7 ± 2.0 -23%[a]	T: 15°C SL: 2.5 μm No CSA methods noted %Δ YO M and F (4–12 mo) SF units: kN/m² Mean ± SE

(Continued)

TABLE 16.1 (*Continued*)

Summary of Studies Examining Age-related Differences in Cross-Sectional Area (CSA), Maximal Isometric Force (P_o), and/or Specific Force (SF) in Single Muscle Fibers from Humans and Rodent Models

Study	Age/Sex	Muscle	Fiber Type	CSA (μm²)%Δ	P_o (μN)%Δ	SF (P_o/CSA)%Δ	Methods/Notes
Thompson and Brown (1999)	Rats, healthy M 30 mo (OLD)	Soleus	TI	6175 ± 284 +11%	440 ± 30 −14%	74 ± 4 −22%[a]	T: 15°C; SL: 2.5 μm; %Δ from 12 mo rats M; Different from 24 mo; Different from 30 mo; SF units: kN/m²; Mean ± SE
	Rats, healthy M 36 mo (OLDER)		TI	3541 ± 249 −36%[a]	250 ± 30 −51%[a]	66 ± 6 −31%[a]	
	Rats, healthy M 37 mo (OLDEST)		TI	4339 ± 280 −22%[a]	230 ± 30 −55%[a]	50 ± 4 −47%[a]	
Thompson et al. (1998)	Rats, healthy M 30 mo	Soleus	TI	87 ± 2 +5%	470 ± 20 −6%	790 ± 50 −14%	T: 15°C; SL: 2.5 μm; Fiber diameter (μm) not CSA reported; %Δ YO M – no stats run between age groups; SF units: in kN/m²; Mean ± SE
Zhong et al. (2006)	Rats, healthy M n = 7 24–25 mo	SM	TII	–	~550 0%	~115 0%	T: 25°C; SL: 2.4 μm; No CSA swelling correction; %Δ YO M rats; Different from 24–25 mo rats; SF units: kN/m²
	Rats, healthy M n = 7 28 mo		TII	–	~340 −35%[a]	~80 −28%[a]	

[a] Significantly different from young in same study.

Notes: Table presents data from the old groups as a percent change (%Δ) from the reference group stated in the "Methods/Notes" section; males (M), females (F), young (YO), old (O); experimental temperature (T), initial sarcomere length (SL), time 1 (T1), time 2 (T2); myosin heavy chain type I (TI), type IIA (TIIA), type IIX (TIIX). Muscle abbreviations: extensor digitorum longus (EDL), medial gastrocnemius (MG), semimembranosus (SM). Methods/Notes section describes pertinent experimental details that could explain some difference in outcomes between studies. Human cross-sectional studies presented first, then human longitudinal, followed by rodent cross-sectional.

To describe cellular contraction briefly, a single muscle cell is made up of thousands of myofibrils and those are made up of myofilaments. The myofibrils are arranged as a string of sarcomeres in series, the smallest fundamental unit of contraction being one sarcomere. There are a variety of proteins that make up the sarcomere, but the two most integral involved in active force production are actin and myosin (Figure 16.1), which interact to form cross-bridges, and produce force. For simplicity, cross-bridge binding can be considered in the context of a three-state model of cross-bridge function: (1) detached, (2) weakly bound, nonforce producing, and (3) strongly bound, force producing (Figure 16.2).

Following a muscle fiber action potential travelling across the sarcolemma, through the transverse tubules (T-tubules) and causing Ca^{2+} release from the sarcoplasmic reticulum (SR), a number of enzymatic interactions and conformational changes cause the cross-bridges to go from one state to the next, which results in the *power stroke*. At the whole muscle level, during voluntary contractions, the neural input together with the cellular and molecular mechanisms produce force and coordinated movement.

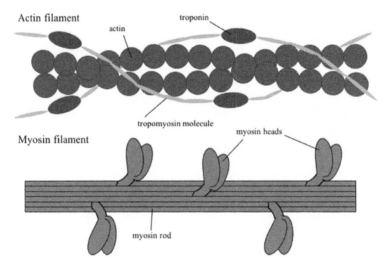

FIGURE 16.1   Actin and myosin filaments and their interaction. The tropomyosin molecule covers the binding sites on actin, until calcium is bound to troponin, changing the conformation of tropomyosin and uncovering the binding sites. This allows the myosin heads to form cross-bridges with actin.

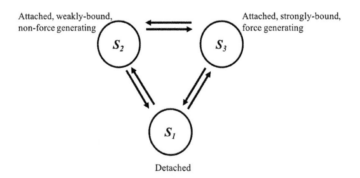

FIGURE 16.2   The figure shows a theoretical three-state model of muscle contraction with the cross-bridges in a detached state (S1), an attached nonforce-generating state (S2), and an attached force-generating state (S3). Arrows denote transitions between states. The way we can think about the generation of muscle force production at the basic cross-bridge level is using this three-state model: (S1) actin and myosin are detached from each other, (S2) actin and myosin are attached in a weakly bound nonforce producing configuration, and (S3) actin and myosin are attached and force producing.

Whether it is form that dictates function or function driving form, natural adult aging is associated with a divergence in muscle contractile properties and the respective structures. This divergence in form and function can be noted in single muscle fiber contractile properties from older adults not "following" what would be expected from a typical fast (MHC II) or slow type (MHC I) fiber (D'Antona et al. 2003, 2007; Power et al. 2016). This is evidenced by a change in contractile properties compared with young adults while expressing the same MHC isoform. This section will discuss basic contractile properties of strength, measured as isometric tension; shortening velocity, a measure of contraction speed; and power, a measure combining force and velocity.

16.3.1 Single Muscle Fiber Force

Single muscle fiber force (i.e., strength) is quantified as maximal Ca^{2+}-activated isometric force (P_o), and is an indirect measurement of cross-bridge interaction and Ca^{2+}-sensitivity. Maximal isometric force is often reported normalized to CSA and is termed specific force (SF) or tension. SF accounts for force per unit of myofibrillar content and often the maximal P_o value is not reported. As whole muscle strength declines with aging, it would seem likely that single muscle fiber P_o and/or SF (based on the proposed decline in CSA) would decrease, but the literature is inconclusive in both rodent models and in humans.

In studies using rodent models, there is consistent evidence for a decrease in P_o between young (~5–12 months) and old (~24–37 months) male rats across muscles and fiber types (Kim & Thompson 2013; Lowe et al. 2001; Thompson & Brown 1999; Thompson et al. 1998; Zhong et al. 2006) and between adult (~9–10 month) and old (~26–27 month) male mice in both the soleus (primarily TI) and extensor digitorum longus (EDL) (primarily TII) muscles (Brooks & Faulkner, 1988). However, when P_o is normalized to CSA, the results are unclear with studies showing an age-related significant decrease in SF across muscles (Kim & Thompson 2013; Lowe et al. 2001, 2002; Prochniewicz et al. 2005; Thompson & Brown 1999; Zhong et al. 2006), increase in EDL SF and decrease in soleus SF (Degens et al. 1998; Li & Larsson 1996), or no difference in the soleus or EDL SF (Larsson et al. 1997b) (Table 16.1).

In studies involving human cross-sectional designs, there is strong support for a decrease in P_o between young (~18–35 years) and old (~60–90 years) adults in both TI and TII fibers of the vastus lateralis (Frontera et al. 2000a; Larsson et al. 1997a, 1997b; Power et al. 2016), while others have reported only a decrease in TII fibers (Yu et al. 2007). Conversely other studies show no significant difference in P_o with age in TI or TII fibers (Hvid et al. 2011; Raue et al. 2009; Trappe et al. 2003; Venturelli et al. 2015) or an increase in single fiber force in older adults (Trappe et al. 2003; Yu et al. 2007).

SF results show similar inconsistencies with studies reporting a decrease with aging in TI and TII fibers (D'Antona et al. 2003; Frontera et al. 2000a; Larsson et al. 1997a, 1997b; Ochala et al. 2006, 2007; Power et al. 2016; Yu et al. 2007), decrease only in TII fibers (Lamboley et al. 2015), and no significant differences (Hvid et al. 2011; Raue et al. 2009; Trappe et al. 2003; Venturelli et al. 2015). Trappe et al. (2003) showed a significant increase in SF of TIIA fibers biopsied from the vastus lateralis of healthy males, but not in TI or in fibers from females (Table 16.1).

When presented with the inconsistencies in age-related changes of single fiber P_o and SF, we need to consider variables outside of aging that could have an effect on cellular contractile function, such as physical activity. Physical activity has a protective effect against the deficits of aging at the whole muscle level and seems to protect single fiber contractile function as well. D'Antona et al. (2007) performed single fiber contractile experiments from sedentary, recreationally active, and endurance trained individuals and determined that lifetime physical activity levels correlated with single fiber SF, whereby the sedentary group showed significant decreases in TI and TIIA fibers of ~26% and ~21%, respectively, from young adults; the recreationally active group showed a significant decrease in TI fibers and no change in TIIA fibers; and the endurance trained group showed no change in both fiber types. A recent study investigating differences in single muscle fiber performance between a group of master's athletes and normal population older adults found that there were no differences between groups; however, this group of older adults was ~80 years, past the "critical" point of motor unit loss and fiber loss. Potentially, life-long physical activity ceases to be protective for single fiber contraction after a certain age (Power et al. 2016), although a robust protection of functional motor units and strength remains (Power et al. 2012).

While cross-sectional studies are important to begin answering basic questions, longitudinal studies are needed in the context of aging, to identify causal relationships of skeletal muscle alterations and the onset and progression of sarcopenia. Longitudinal studies are extremely beneficial but are limited due to the resources and time needed to complete them. Frontera et al. (2008) completed a 9-year longitudinal investigation of older adults with a baseline age range of 62–81 years, and the returning cohort with an age range of 71–90 years. Contrary to previous studies, both P_o and SF of TI fibers were trending toward an increase with average increases of 23% and 28%, respectively. In TIIA fibers, P_o trended toward a significant increase, but there was no change in SF. Reid et al. (2014) completed a three-year longitudinal study with a cohort of healthy, older (mean age 74.1 years) and mobility-limited, older (mean age 77 years) adults. In both cohorts, statistically significant increases in both P_o and SF were found. These longitudinal studies are demonstrating a compensatory effect similar to that of the compensatory hypertrophy (noted in Section 16.2.2), however, with single fiber force. It is likely that a similar critical age is reached when many motor units and muscle fibers have been lost, leaving fewer muscle fibers to perform activities of daily living. The fibers remaining must then produce more force to allow for enough strength to get out of a chair, for example. It should be noted that a longitudinal increase in single fiber P_o or SF within an older adult does not mean maintained contractile function, as there could still be single fiber contractile deficits compared with young adults.

Overall, the results are consistently inconsistent, owing to the large number of variables affecting single fiber P_o and SF. During the aging process, single fibers may become stronger in a compensatory manner, before those fibers are lost in very old age, owing to the death of large remodeled motor units (Dalton et al. 2010; McNeil et al. 2005; Power et al. 2014). This compensatory effect theory for muscle fiber strength (Frontera et al. 2008) likely occurs between the ages of 70 and 80 years. The onset and progression of sarcopenia are certainly tied to both muscle fiber loss and decreases in contractile function, and linking the findings from both cross-sectional and longitudinal studies is important for further understanding.

16.3.2 SINGLE MUSCLE FIBER SHORTENING VELOCITY

Shortening velocity is a measure of contraction speed and can be calculated in two ways. Maximal shortening velocity (V_{max}) is calculated by performing isotonic load steps to a maximally Ca^{2+}-activated fiber and measuring the speed of shortening following release. The data points create a force–velocity curve and V_{max} is calculated. This method allows for power to be calculated, which is the product of the optimal trade off of a submaximal force and velocity. Maximal *unloaded* shortening velocity (V_o) is determined by the slack-test method wherein a fiber is maximally Ca^{2+} activated, and a series of shortenings (slacks) are introduced. The time for force to redevelop after each shortening is determined, and V_o is measured from the slope of the points, measured in muscle fiber lengths per second (FL/s). V_o is the fastest shortening velocity as it occurs at zero load. Based on an older individual's slow muscle phenotype, a slowing of contractile properties at the single muscle fiber level could also be expected, and the literature shows a general trend in that direction; however, inconsistencies still remain.

In rat models, V_o decreases with age in TI fibers from soleus muscle (Degens et al. 1998; Li & Larsson 1996; Thompson et al. 1998, 1999) and less consistently in TII fibers from the EDL and semimembranosus muscles (Degens et al. 1998; Prochniewicz et al. 2005). Other studies have found no significant differences in V_o between age groups (Degens et al. 1998; Li & Larsson 1996; Zhong et al. 2006). V_{max} has been assessed less frequently, and it has been found that in TI fibers from the medial gastrocnemius, there is no significant change between ages, but in TII fibers from the same muscle, there was a significant decrease (Kim & Thompson 2013) (Table 16.2).

TABLE 16.2

Summary of Studies Examining Age-Related Differences in Unloaded Shortening Velocity (V_o) and/or Maximal Shortening Velocity (V_{max}) of Single Muscle Fibers from Humans and Rodent Models

Author/Year	Muscle	Age/Sex	Fiber Type	V_o (FL/s) %Δ	V_{max} (FL/s) %Δ	Methods/Notes
D'Antona et al. (2003)	Vastus lateralis	Healthy M $n = 7$ 72.7 ± 2.3 yr	TI	~0.25 −28%[a]	–	T: 12°C SL: 2.5 μm %Δ from YO M Mean ± SD
			TIIA	~1.10 −12%[a]		
			TIIX	~2.10 −9%		
		Immobilized M $n = 2$ 70 and 72	TI	~0.51 +45%[a]	–	
			TIIA	~1.45 +16%[a]		
			TIIX	~2.70 +17%		
D'Antona et al. (2007)	Vastus lateralis	Healthy M Sedentary 73 ± 2.5 yr	TI	~0.25 −17%[a]	–	T: 12°C SL: 2.5 μm %Δ from YO M Mean ± SE
			TIIA	~1.10 −12%[a]		
		Healthy M Recreationally active 73 ± 2.5 yr	TI	~0.28 0%	–	
			TIIA	~1.05 −16%[a]		
		Healthy M Endurance trained 73 ± 2.5 yr	TI	~0.28 0%	–	
			TIIA	~1.00 −20%[a]		
Korhonen et al. (2006)	Vastus lateralis	Healthy M Sprint athletes $n = 9$ 53–77 yr	TI	−24%[a] 0.48 ± 0.04 −5%	–	T: 15°C SL: 2.76 ± 0.04 μm %Δ from YO M Mean ± SE
			TIIA	1.67 ± 0.13	–	

(Continued)

TABLE 16.2 (Continued)

Summary of Studies Examining Age-Related Differences in Unloaded Shortening Velocity (V_o) and/or Maximal Shortening Velocity (V_{max}) of Single Muscle Fibers from Humans and Rodent Models

Author/Year	Age/Sex	Muscle	Fiber Type	V_o (FL/s) %Δ	V_{max} (FL/s) %Δ	Methods/Notes
Krivickas et al. (2001)	Healthy M $n = 12$ 74.4 ± 5.9	Vastus lateralis	TI	0.77 ± 0.26 0%	–	T: 15°C SL: 2.75–2.85 µm %Δ from YO same sex Mean ± SD
			TIIA	1.78 ± 0.56 −17%[a]	–	
	Healthy F $n = 12$ 72.1 ± 4.3		TI	0.70 ± 0.23 −7%[a]	–	
			TIIA	1.51 ± 0.54 −7%	–	
Larsson et al. (1997a)	Healthy M Recreationally active $n = 2$ 73–81 yr	Vastus lateralis	TI	0.16 ± 0.06 −43%[a]	–	T: 12°C SL: 2.79 µm %Δ from YO M Mean ± SD
			TIIA	0.83 ± 0.22 −28%[a]	–	
	Healthy M Very active $n = 2$ 73–81 yr		TI	0.16 ± 0.05 −43%[a]	–	
			TIIA	0.80 ± 0.33 −30%[a]	–	
Larsson et al. (1997b)	Healthy M $n = 4$ 73–81 yr	Vastus lateralis	TI	−46%[a]	–	T: 12°C SL: 2.7–2.85 µm %Δ from YO M
			TIIA	−30%[a]	–	
Ochala et al. (2007)	Healthy M $n = 6$ 60–74 yr	Vastus lateralis	TI	0.70 ± 0.14 −23%[a]	–	T: 15°C SL: 2.75–2.85 µm %Δ from YO M Mean ± SD
			TIIA	1.30 ± 0.22 −22%[a]	–	

(Continued)

TABLE 16.2 (Continued)

Summary of Studies Examining Age-Related Differences in Unloaded Shortening Velocity (V_o) and/or Maximal Shortening Velocity (V_{max}) of Single Muscle Fibers from Humans and Rodent Models

Author/Year	Age/Sex	Muscle	Fiber Type	V_o (FL/s) %Δ	V_{max} (FL/s) %Δ	Methods/Notes
Power et al. (2016)	Healthy M non-athlete 78.2 ± 9.4 yr n = 5	Vastus lateralis	"Slow"	~0.17 −46%[a]	–	T: 10°C SL: ~2.8 µm %Δ from YO M
	Healthy M master's athlete 78.8 ± 3.6 yr n = 6		"Slow"	~0.13 −62%[a]	–	
Raue et al. (2009)	Healthy F 85 ± 1 yr T1: pre-training	Vastus lateralis	TI	1.35 ± 0.09	0.92 ± 0.05	T: 15°C SL: 2.5 µm T1 not different from YO F
			TIIA	3.35 ± 0.12	3.00 ± 0.10	
	Healthy F 85 ± 1 yr T2: post-training		TI	1.38 ± 0.08	0.86 ± 0.04	Mean ± SE
			TIIA	3.35 ± 0.21	3.18 ± 0.21	
Reid et al. (2012)	Healthy M n = 16 73.8 ± 4 yr	Vastus lateralis	TI	0.62 ± 0.13	–	T: 15°C SL: 2.75–2.85 µm
			TIIA	1.54 ± 0.55	–	%Δ middle-aged group (~45 yr)
	Healthy F n = 7 74.3 ± 4 yr		TI	0.65 ± 0.20	–	Mean ± SD
			TIIA	1.36 ± 0.54	–	
	Limited mobility M n = 12 78.9 ± 4 yr		TI	0.68 ± 0.20	–	
			TIIA	1.59 ± 0.64	–	
	Limited mobility F n = 13 76.8 ± 5 yr		TI	0.62 ± 0.20	–	
			TIIA	1.24 ± 0.64	–	

(Continued)

TABLE 16.2 (Continued)

Summary of Studies Examining Age-Related Differences in Unloaded Shortening Velocity (V_o) and/or Maximal Shortening Velocity (V_{max}) of Single Muscle Fibers from Humans and Rodent Models

Author/Year	Age/Sex	Muscle	Fiber Type	V_o (FL/s) %Δ	V_{max} (FL/s) %Δ	Methods/Notes
Trappe et al. (2003)	Healthy M Sedentary n = 6 80 ± 4 yr	Vastus lateralis	TI	1.4 ± 0.18 −3%	1.0 ± 0.18 +9%	T: 15°C SL: 2.5 μm %Δ from YO same sex Mean ± SE
			TIIA	4.3 ± 0.51 +5%	3.5 ± 0.36 +9%	
	Healthy F Sedentary n = 6 78 ± 2 yr		TI	1.4 ± 0.14 +8	0.96 ± 0.15 +3%	
			TIIA	4.3 ± 0.38 +8%	3.7 ± 0.39 +4%	
Yu et al. (2007)	Healthy M n = 17 65–85 yr	Vastus lateralis	TI	0.50 ± 0.13 −24%[a]	—	T: 12°C SL: 2.70–2.85 μm Unspecified V_o units %Δ from YO same sex Mean ± SD
			TIIA	1.88 ± 0.83 −10%[a]		
	Healthy F n = 5 65–85 yr		TI	0.44 ± 0.14 −20%	—	
			TIIA	1.54 ± 0.54 −30%[a]		
Reid et al. (2014)	Healthy older M and F n = 14 (5F) T1: 74.1 yr T2: ~77	Vastus lateralis	TI (T1)	0.63 ± 0.16		T: 15°C SL: 2.75–2.85 μm %Δ T1 [[a]significantly different from] Mean ± SD
			TI (T2)	0.77 ± 0.21 +22%[a]		
			TIIA (T1)	1.48 ± 0.54		
			TIIA (T2)	2.24 ± 0.71 +51%[a]		
	Limited mobility M and F n = 5 (3F) T1: 77.2 yr T2: ~80		TI (T1)	0.61 ± 0.20		
			TI (T2)	0.97 ± 0.23 +60%[a]		
			TIIA (T1)	1.32 ± 0.52		
			TIIA (T2)	2.0 ± 0.7 +34%[a]		

(Continued)

TABLE 16.2 (Continued)

Summary of Studies Examining Age-Related Differences in Unloaded Shortening Velocity (V_o) and/or Maximal Shortening Velocity (V_{max}) of Single Muscle Fibers from Humans and Rodent Models

Author/Year	Age/Sex	Muscle	Fiber Type	V_o (FL/s) %Δ	V_{max} (FL/s) %Δ	Methods/Notes
Degens et al. (1998)	Rats, healthy M n = 5 22–24 mo	Soleus	TI	0.62 ± 0.27 −44%[a]	–	T: 12°C SL: 2.61 ± 0.19 μm V_o units: muscle length/second %Δ from YO same sex
		EDL	TIIB and IIX	2.12 ± 0.52 −3%	–	
	Rats, healthy F n = 4 22–24 mo	Soleus	TI	0.65 ± 0.53 −36%[a]	–	Mean ± SD
		EDL	TIIB and IIX	2.00 ± 0.66 −19%[a]	–	
Kim et al. (2013)	Rats, healthy M 32–37 mo	MG	TI	–	~1.8	T: 15°C SL: 2.5 μm %Δ from YO M
			TII	–	~2.25 −25%[a]	
Larsson et al. (1997b)	Rats, healthy M and F 20–24 mo	Soleus	TI	0.59 ± 0.28[a] n/c from YO	–	T: 12°C SL: 2.45–2.75 μm
		EDL	TIIB		–	
Li and Larsson (1996)	Rats, healthy M n = 5 20–24 mo	Soleus	TI	0.59 ± 0.28 −47%[a]	–	T: 12°C SL: 2.59 μm %Δ from YO M Mean ± SD
		EDL	TIIX/B	2.17 ± 0.49 0%	–	
Prochniewicz et al. (2005)	Healthy M and F 27–35 mo	SM	TII	7.5 ± 0.5 −22%[a]	–	T: 15°C SL: 2.5 μm %Δ YO M and F; 4–12 mo Mean ± S.D

(Continued)

316 Sarcopenia

TABLE 16.2 (Continued)
Summary of Studies Examining Age-Related Differences in Unloaded Shortening Velocity (V_o) and/or Maximal Shortening Velocity (V_{max}) of Single Muscle Fibers from Humans and Rodent Models

Author/Year	Age/Sex	Muscle	Fiber Type	V_o (FL/s) %Δ	V_{max} (FL/s) %Δ	Methods/Notes
Thompson and Brown (1999)	Rats, healthy M 30 mo (OLD)	Soleus	TI	0.85 ± 0.09 −50%[a]	–	T: 15°C SL: 2.5 μm %Δ from YO M; 12 mo
	Rats, healthy M 36 mo (OLDER)		TI	1.10 ± 0.09 −36%[a]	–	sig diff from 24 mo
	Rats, healthy M 37 mo (OLDEST)		TI	1.04 ± 0.17 −39%[a]	–	sig diff from 30 mo Mean ± SE
Thompson et al. (1998)	Rats, healthy M 30 mo	Soleus	TI	0.90 ± 0.09 −45%[a]		T: 15°C SL: 2.5 μm %Δ YO M; 12 mo Mean ± SE
Zhong et al. (2006)	Rats, healthy M n = 7 24–25 mo	SM	TII	−13% ~7.4	–	T: 25°C SL: 2.4 μm %Δ YO M; 6 mo
	Rats, healthy M n = 7 28 mo		TII	−13% ~7.4	–	Mean ± SD

[a] Significantly different from young in same study.

Notes: Table presents data from the older adult groups as a percent change (%Δ) from the reference group stated in the "Methods/Notes" section; males (M), females (F), young (YO); initial sarcomere length (SL), time 1 (T1), time 2 (T2); myosin heavy chain type I (TI), type IIA (TIIA), type IIX (TIIX). Muscle abbreviations: extensor digitorum longus (EDL), medial gastrocnemius (MG); semimembranosus (SM). Methods/Notes section describes pertinent experimental details that could explain some difference in outcomes between studies. Human cross-sectional studies presented first, then human longitudinal, followed by rodent cross-sectional.

Results from human participants show a similar trend to animal models, with the majority of cross-sectional studies showing a consistent decrease in V_o with aging across fiber types of the vastus lateralis (D'Antona et al. 2003; Larsson 1997a; Ochala et al. 2007; Power et al. 2016; Yu et al. 2007). A decrease in V_o in only TI (Krivickas et al. 2001) or no change in either fiber type (Raue et al. 2009; Reid et al. 2012; Trappe et al. 2003) has also been found. Studies that have assessed and reported V_{max} show no change across age groups or fiber types (Raue et al. 2009; Trappe et al. 2003). One longitudinal study that assessed V_o over ~3 years (Reid et al. 2014) found that in both a healthy and limited mobility, older population V_o significantly increased in both TI and TIIA fibers (Table 16.2).

D'Antona et al. (2007) measured V_o in three separate groups with different levels of physical activity (sedentary, recreationally active, and endurance trained). No differences were found in the endurance trained elderly group or recreationally active elderly group compared with the young controls in TI fibers of the vastus lateralis, whereas a significant decrease was found in the sedentary elderly group. However, in TIIA fibers, a significant decrease was found across all elderly groups compared with the young control group. For comparison, the researchers added data from two elderly immobile participants from a previous study (D'Antona et al. 2003) and found a significant, large increase in V_o in both TI and TIIA fibers compared with young and all other old groups.

The dichotomous interaction between aging and disuse, as was noted with MHC isoform expression (Section 16.2.1), could be a factor in the inconsistencies of age-related alterations of V_o. As demonstrated by D'Antona et al. (2007), the retrospective data from the immobile participants show a drastic increase in V_o, which is opposite to the typical trend, and it has been shown that after 17 days of bed rest, V_o in healthy human TI soleus fibers increased (Widrick et al. 1997). These data suggest that shortening velocity is affected conversely by aging and disuse, which identifies a need for screening and categorization of mobility and physical activity levels of participants to resolve discrepancies across studies. The disuse-aging paradigm likely has a strong influence on research results in this area. While shortening velocity at the whole muscle level is slower in old as compared with young, differential effects of disuse on fiber type shortening velocity may be confounding findings at the cellular level.

16.3.3 SINGLE MUSCLE FIBER POWER

Power is the product of force and velocity. Power of single muscle fibers is calculated using the force–velocity curve fitted with the V_{max} procedure explained above. Both absolute power (μN·FL/s) and normalized power (W/l or kN/m²·FL/s) are often reported, and because power is based on the submaximal trade-off of force and velocity, it is one of the best measurements for complete single fiber function. If there is impairment in the development of P_o and/or V_{max}, decrements will be manifested in power results. Even so, few studies have systematically studied power of single muscle fiber preparations, owing to the greater number of single muscle fiber trials necessary and the greater risk of fiber damage as a result of the increased trial number. Power, like strength and shortening velocity, has varying results with regard to aging.

Absolute power of the medial gastrocnemius of rats was lower in both TI and TII, normalized power was only lower in TII fibers (Kim & Thompson 2013). In studies using human vastus lateralis biopsy samples, Trappe et al. (2003) showed a significant increase in TI fiber and a significant decrease in TIIA absolute power of older compared with young women, while no differences in single fiber power were found between young and old men. And yet, no change in absolute power across young and older women in either fiber type was found (Raue et al. 2009). Normalized power shows similar inconsistencies to absolute power, resulting in a mix of nonsignificant increases and decreases with aging across sex and fiber type (Trappe et al. 2003; Raue et al. 2009). Another group found that absolute power significantly decreased in TI fibers, but not TIIA in older compared with middle-aged adults, and found that normalized power did not change significantly (Table 16.3). In this last study (Reid et al. 2012), it is important to note that

TABLE 16.3

Summary of Studies Examining Age-related Differences in Absolute and/or Normalized Power of Single Muscle Fibers from Humans and Rodent Models

Author/Year	Age/Sex	Muscle	Fiber Type	Absolute Power %Δ	Normalized Power %Δ	Methods/Notes
Krivickas et al. (2006)	Healthy F Sedentary n = 10 65–78 yr	Vastus lateralis	TI	15.3 ± 6.6	4.25 ± 1.83	V_{max} not measured – "specific power" in V_{max} column (kN/m²·FL/s) Peak power µNFL/s No comparison to YO Mean ± SD
			TIIA	47.8 ± 37.7	14.2 ± 13.4	
	Healthy M Sedentary n = 6 73–84 yr		TI	14.2 ± 6.6	3.51 ± 1.83	
			TIIA	25.9 ± 37.7	8.6 ± 13.4	
Trappe et al. (2003)	Healthy M Sedentary 80 ± 4 yr	Vastus lateralis	TI	14.3 ± 2.1 0%	1.58 ± 0.15 –8%	Absolute power = P_o (µN) × V_{max} (FL/s) Normalized power (W/l) = P_o/CSA × V_{max} %Δ from YO same sex Mean ± SE
			TIIA	66.4 ± 14.3 –7%	9.14 ± 1.16 +13%	
	Healthy F Sedentary 78 ± 2		TI	14.2 ± 2.9 +28%[a]	1.74 ± 0.18 0%	
			TIIA	43.9 ± 5.7 –33%[a]	10.64 ± 1.12 +7%	
Raue et al. (2009)	Healthy F 85 ± 1 yr n = 6 T1: pre-training	Vastus lateralis	TI	11.2 ± 1.5 +27%	2.1 ± 0.1 +5%	Absolute power = P_o (µN) × V_{max} (FL/s) Normalized power (W/l) = P_o/CSA × V_{max} %Δ from YO F Mean ± SE
			TIIA	56.2 ± 6.5 +24%	13.1 ± 0.3 +7%	
	Healthy F 85 ± 1 yr T2: post-training		TI	12.4 ± 1.6 +24%	2.4 ± 0.2 +9%	
			TIIA	63.9 ± 10.3 +9%	13.3 ± 0.5 +10%	

(Continued)

TABLE 16.3 (*Continued*)

Summary of Studies Examining Age-related Differences in Absolute and/or Normalized Power of Single Muscle Fibers from Humans and Rodent Models

Author/Year	Age/Sex	Muscle	Fiber Type	Absolute Power %Δ	Normalized Power %Δ	Methods/Notes
Reid et al. (2012)	Healthy M	Vastus lateralis	TI	19.7 ± 5.8	6.05 ± 2.2	V_{max} not measured – "specific power" in V_{max} column
	73.8 ± 4 yr			−20%[a]	−16%	$V_o \times P_o$ /CSA = specific power (kN/m²·FL/s)
	n = 16		TIIA	43.1 ± 21.7	13.2 ± 6.5	%Δ from middle-age same sex
				−21%	−24%	Mean ± SD
	Healthy F		TI	15.7 ± 8.9	5.4 ± 2.6	
	74.3 ± 4			−31%[a]	−17%	
	n = 7		TIIA	47.1 ± 30.5	16.2 ± 12.3	
				+6%	−1%	
	Limited mob M		TI	18.1 ± 8.1	5.65 ± 2.0	
	78.9 ± 4 yr			−26%[a]	−22%	
	n = 12		TIIA	36.4 ± 17	16.2 ± 7.6	
				−33%	−6%	
	Limited mob F		TI	17.2 ± 4.3	5.74 ± 1.8	
	76.8 ± 5 yr			−25%[a]	−17%	
	n = 13		TIIA	36.7 ± 13	18.3 ± 17.5	
				−17%	+12%	

(Continued)

TABLE 16.3 (Continued)
Summary of Studies Examining Age-related Differences in Absolute and/or Normalized Power of Single Muscle Fibers from Humans and Rodent Models

Author/Year	Age/Sex	Muscle	Fiber Type	Absolute Power %Δ	Normalized Power %Δ	Methods/Notes
Reid et al. (2014)	Healthy older M and F	Vastus lateralis	TI (T1)	18.1 ± 7.1	5.8 ± 2.4	V_{max} not measured – "specific power" in V_{max} column $V_o \times P_o$/CSA = specific power (kN/m²·FL/s)
	T1: 74.1 yr		TI (T2)	47.9 ± 29.1 $+164\%^a$	16.3 ± 13.5 $+181\%^a$	Mean $\pm$ SD
	T2: ~77		TIIA (T1)	44.3 ± 23.9	14.1 ± 8.4	
	$n = 14$ (5F)		TIIA (T2)	99.3 ± 58.5 $+124\%^a$	30.4 ± 10.6 $+115\%^a$	
	Limited mob M and F		TI (T1)	18.1 ± 7	5.73 ± 2.2	
	T1: 77.2 yr		TI (T2)	45.2 ± 14.8 $+148\%^a$	14.9 ± 4 $+160\%^a$	
	T2: ~80		TIIA (T1)	35.5 ± 13.3	14.5 ± 7.1	
	$n = 5$ (3F)		TIIA (T2)	152.7 ± 188 $+330\%^a$	35.1 ± 21.8 $+142\%^a$	
Kim et al. (2013)	Rats, healthy M 32–37 mo	MG	TI	~8.5 $-32\%^a$	~2.5 -5%	Absolute power = P_o (μN) $\times V_{max}$ (FL/s)
			TII	~16.8 $-47\%^a$	~4.30 $-36\%^a$	Normalized power (kN/m²·FL/s) = P_o/CSA $\times V_{max}$ %Δ from 5–12 mo

^{a} Significantly different from young in same study.

Notes: Table presents data from the older adult groups as a percent change (%Δ) from the reference group stated in the "Methods/Notes" section; males (M), females (F), young (YO); time 1 (T1), time 2 (T2); myosin heavy chain type I (TI), type IIA (TIIA), type IIX (TIIX). Muscle abbreviations: medial gastrocnemius (MG). Methods/Notes section describes units and methods of calculation for either absolute, normalized, or specific power. Human cross-sectional studies presented first, then human longitudinal, followed by rodent cross-sectional.

the methodology used to measure and calculate single fiber power was not reported in the paper and so these results need to be interpreted and used with caution.

There are an insufficient number of studies in humans to draw conclusions regarding single muscle fiber power alterations with aging. The data seem to be as inconclusive as both strength and shortening velocity, and again factors such as physical activity, age of experimental and control group, and individual variation must be taken into consideration in future research exploring age-related single muscle fiber power alterations.

16.4 MOLECULAR MECHANISMS OF CONTRACTILE PROPERTIES

The force-producing contractile properties discussed above are driven by molecular mechanisms that serve to start and mediate force production independently from the motoneuron. As skinned human single muscle fibers offer a model free from neural control, different facets of cross-bridge mechanics and ion-sensitivity during contraction can be investigated. Alterations in attachment and detachment rates, the bound-state of myosin, and the influence of Ca^{2+} concentration can be studied in a way where most of the variation can be attributed to the factor being manipulated.

16.4.1 CROSS-BRIDGE MECHANICS

The cycling of cross-bridges in skeletal muscle produces force, and the rate at which cross-bridges attach affects force production. The rate of force redevelopment (k_{tr} – k representing the rate constant of *t*ension *r*edevelopment) is one measure used as a proxy for rate of cross-bridge attachment, and the measure ensures that regulatory protein "interference" is minimized, and so the development of force can be attributed primarily to cross-bridge attachment (Colombini et al. 2005). The faster the rate of transition from the weakly bound nonforce producing to force-producing state, the greater the rate of single fiber force production, and potentially whole muscle force production. A recent study has shown that k_{tr} decreases in older adults' single fibers as compared with young, and this decrease was also present in aged-match master's athletes alongside general population older adults (Power et al. 2016). Using a different method of quantifying rate of force redevelopment, it was found that TII fibers from older adults have slower force redevelopment than young TII fibers (Ochala et al. 2007). At a whole muscle level, rate of force development being slowed would seriously impair an individual's ability to produce force quickly making them unable to make a rapid leg adjustment to avoid falling or tripping. Therefore, this measure of single fiber rapid force production can provide useful insight for functional activities of daily living in older adults as compared with single fiber force measurements.

16.4.2 CROSS-BRIDGE BOUND STATE

A decrease in myofibrillar content, namely myosin, contributes to the decrease in overall muscle mass and likely contributes to decreases in force and power production of whole muscle (see Section 16.2.2). Another theory regarding decreases in skeletal muscle function that has been posited is that a higher proportion of cross-bridges formed during contraction are in the weakly bound, nonforce producing state in old adults compared with young adults. Evidence for this has been found in animal models using electron paramagnetic resonance (EPR) where ~30% fewer myosin heads were found in the strongly bound conformation in aged compared with young animals (Lowe et al. 2001). EPR is the only direct way of measuring strongly and weakly bound cross-bridges; however, there are a few indirect methods. A mechanical "stiffness" test—often denoted by k—uses a rapid stretch and shorten to measure the proportion of cross-bridges attached and then compares force production to determine greater or fewer strongly bound cross-bridges. For example, it was found that older adults' muscle fibers maintain single fiber stiffness (k) while P_o is reduced compared with a young adult single fiber, and this was owing to a greater proportion of weakly bound cross-bridges (Power et al. 2016). Another method to indirectly quantify the transition of myosin from weakly to strongly bound uses a mathematical analysis to estimate the rate of change of the cross-bridge bound state (Miller et al. 2013). Using this method, it has been shown that the single fibers of older adults have a decreased rate of change from weakly to strongly bound cross-bridge states, which indicates more cross-bridges remain in the weakly bound nonforce producing state during force production than in young adults (Miller et al. 2013). Although there is no direct evidence (i.e., identifying cross-bridge conformational changes using EPR) for changes in the proportion of weakly bound cross-bridge attachment in humans, the indirect stiffness measurements and changes in force production offer promising and consistent results that the bound state of cross-bridges is likely influencing decreased force production at the whole muscle level in older adults.

16.4.3 CALCIUM REGULATION

To initiate a muscle contraction, an action potential travels from the motoneuron to the sarcolemma and is distributed along the T-tubules, through the dihydropyridine receptors, which then activates the ryanodine receptors (RYRs) on the SR. The RYR opens and releases Ca^{2+}, which is referred to as voltage-gated calcium release. Ca^{2+} can also be released through calcium-induced calcium release, and this phenomenon is employed with skinned fibers, as there is no intact sarcolemma for an action potential to propagate. The fiber is activated by submersing it in a solution, containing free Ca^{2+} ions and ATP, which permeates the fiber binding to troponin C and causing any remaining Ca^{2+} to be released from intact portions of the SR. Calcium bound to troponin C induces a conformational change in tropomyosin to expose the myosin binding sites on actin, allowing cross-bridges to form (see Figure 16.1). As can be gleaned from above, Ca^{2+} is an integral part of cellular and whole muscle contraction, and it has been found that Ca^{2+}-handling—sensitivity, release, re-uptake—is impaired in aged single fibers compared with young (Lamboley et al. 2015, 2016; Straight et al. 2018). A lower Ca^{2+}-sensitivity means that a higher concentration of Ca^{2+} ($[Ca^{2+}]$) needs to be present for a certain amount of force production than in a muscle fiber with higher sensitivity. Another way $[Ca^{2+}]$ is presented is pCa (pCa = $-\log_{10}[Ca^{2+}]$), and this pCa value is simpler to work with than the very small absolute numbers of $[Ca^{2+}]$ (i.e., if $[Ca^{2+}] = 0.00005$, then pCa $\cong 4.3$). Small pCa values (i.e., 4) are very high $[Ca^{2+}]$, whereas low pCa values (i.e., 7) are very low $[Ca^{2+}]$, because of the nature of log transformations.

There are currently only a few studies that have investigated changes in Ca^{2+}-sensitivity in the context of aging in humans. Hvid et al. (2011) investigated differences in force-pCa curves between young and older men and found no difference in Ca^{2+}-sensitivity between age groups. However, other investigations have found that both MHC I and MHC IIA fibers (Straight et al. 2018) and only MHC II fibers (Lamboley et al. 2015) have a lower Ca^{2+}-sensitivity than young fibers. MHC II fibers seem to have a more pronounced reduction in Ca^{2+}-sensitivity with aging than MHC I.

The age effect on storage and release of Ca^{2+} from the SR have also been investigated, and it was found that in aged single fibers, endogenous and releasable Ca^{2+} content of the SR was reduced in both fiber types, and maximal uptake of Ca^{2+} was reduced in TII fibers compared with young (Lamboley et al. 2015). In an examination of SR Ca^{2+} loading, TI fibers of older adults have lower maximal storage relative to fibers from young adults, and both TI and TII fibers from older adults have a lower maximal release of Ca^{2+} from the SR relative to maximal force elicited than young adult fibers (Lamboley et al. 2015). Furthermore, evidence for greater leakage from the RYRs/Ca^{2+} channels was found in TI fibers of older adults (Lamboley et al. 2016). In aged rats, a similar impairment in SR Ca^{2+} release was found in the gastrocnemius muscle (Russ et al. 2011).

These studies provide compelling evidence that aging impairs Ca^{2+}-handling in single muscle fibers, whether it be storage, release, or sensitivity. Ca^{2+}-sensitivity would have the greatest effect on P_o, as it would affect the number of actin binding sites available for cross-bridge formation, and few cross-bridges attached would result in lower force and peak power production. Sensitivity to Ca^{2+} also affects the rate of cross-bridge attachment, as fewer binding sites would decrease how quickly myosin can attach to actin to form a cross-bridge and produce force. Storage and release of

Ca^{2+} are another factor that would impact cross-bridge formation and force production because if less Ca^{2+} is released and sensitivity is lowered than there may always be a lack of Ca^{2+} binding to troponin to make cross-bridge binding possible. Continuing to investigate age-related impairments in Ca^{2+}-handling is imperative, and the next major step in uncovering the mechanisms of muscle fiber atrophy, loss, and function decrements that occur with aging and progress sarcopenia.

16.5 WHOLE MUSCLE VERSUS CELLULAR FACTORS

Age-related alterations to single muscle fiber contractile properties are inconsistent; however, whole muscle size, strength, power, and endurance decrease consistently in older adults. There are few studies that have performed both whole muscle and single fiber performance tests and attempted to relate the two in some ways. A positive relationship was found between single fiber P_o and whole muscle torque with aging (Frontera et al. 2008); single fiber Ca^{2+}-sensitivity has been associated with knee extensor power in older adults (Straight et al. 2018); and reductions in knee extensor torque were accompanied by reductions in SF (Yu et al. 2007). However, for every study demonstrating a relationship between single fiber and whole muscle function, there is one that does not (Frontera et al. 2008; Reid et al. 2012, 2014; Venturelli et al. 2015). These relationships can be muddled by interactions with muscle function from the neural or cardiovascular system causing inconsistencies in associations between the cell and the whole.

Age-related single fiber size changes relative to size changes of whole muscle are another area of inconsistency, as we know whole muscle size decreases with age (i.e., sarcopenia), but single fiber size is maintained or even increased in older adults (Aniansson et al. 1992; D'Antona et al. 2003; Miller et al. 2013; Nederveen et al. 2016). One theory for this discrepancy is compensatory hypertrophy (see Section 16.2.2) occurring in the late seventies (Aniansson et al. 1992). As whole muscle mass/size decreases with age, the total number of muscle fibers also decreases, leaving fewer fibers to perform muscle contractions for activities of daily living.

As discussed in Section 16.3.1, this theory of compensation may also hold true for other aspects of cellular muscle mechanics. Whole muscle gets smaller, weaker, and fatigues more quickly, depending on the task (Power et al. 2013). In contrast, the single muscle fibers of that same whole muscle may be compensating for the loss in fibers and innervation by getting larger and stronger across certain age ranges (Degens et al. 1998; Li et al. 1996; Trappe et al. 2003; Venturelli et al. 2015); however, we only have speculation based on cross-sectional studies as only a longitudinal study would be able to capture these changes. In the few longitudinal studies that have been conducted, both P_o and SF have trended toward significant increases, along with fiber CSA (Frontera et al. 2008; Reid et al. 2014), which may indicate cellular adaptations are compensating for the loss of muscle fibers to maintain whole body function—such as rising from a chair. Across the literature, V_o shows more consistent age-related decreases compared with force and power. At the whole muscle level, shortening velocity is almost always reduced in old age (Dalton et al. 2010; McNeil et al. 2007; Petrella et al. 2005). Therefore, the cellular mechanisms responsible for the attachment and detachment rates of myosin to actin may provide clues to the basis of impaired function in older adults.

There are a few investigations that highlight some of the probable reasons for the ongoing conflict in the changes of cellular muscle mechanics with aging. A study comparing older adults with middle-aged adults showed nonsignificant decreases in whole muscle function (Reid et al. 2012). This study comparing old with middle-aged adults highlights the gradual age-related decline in cellular muscle function. Another study comparing young adults with a mobile and immobile oldest old group (85–92 years) found that for the biceps brachii—a muscle completely mobile in wheelchair use—there were no differences in P_o or SF between the age groups in spite of the approximate 60-year age gap (Venturelli et al. 2015). In this instance, the continual everyday use of biceps brachii likely had a protective effect on single fiber contractile properties, while a muscle like the vastus lateralis, is typically used less with aging.

Importantly, the normal age-related loss in muscle mass does not have to progress to sarcopenia to impair the performance of activities of daily living (e.g., being able to get out of a chair, climb stairs, right oneself from a trip, etc.), as demonstrated by the cross-sectional studies that determined skeletal muscle strength, power, and endurance capacity of nonsarcopenic older adults decline in comparison with young adults (Bassey et al. 1992; Clark et al. 2011; Frontera et al. 2000a, 2000b; Goodpaster et al. 2006; Trappe et al. 2003; Power et al. 2016). Much of the work performed in the cellular muscle mechanics field has focused on nonsarcopenic older adults, and yet we are still finding alterations that may be worsened by the progression of sarcopenia. Separating study groups of older adults by age, physical activity, and/or the presence of sarcopenia could provide avenues to the elusive answer regarding age-related alterations of cellular muscle mechanics. Future work will be needed to identify the critical thresholds or appropriate tests at the cellular level to potentially identify those individuals at risk of future mobility impairment.

16.6   CONCLUSION

At the whole muscle level, deficits in mass and performance are indisputable and consistent in older adults. However, the mechanisms at the cellular level driving whole body impairment remain elusive in older adults, even when attempting to control for confounding variables such as physical activity and myofibrillar protein content. Motor unit reorganization and loss drive many of the morphological changes in muscle cells and contribute to the progression of sarcopenia during the aging process. Contractile properties of the cell are certainly affected by the loss of muscle mass—largely from the loss of muscle fibers—and likely compensate in very old age in order to maintain muscle function needed for activities of daily living. Sedentarism and mobility impairments certainly accelerate the progression of sarcopenia and alterations of single muscle fibers with aging.

REFERENCES

Aagaard, P., & Andersen, J. L. (1998). Correlation between contractile strength and myosin heavy chain isoform composition in human skeletal muscle. *Medicine and Science in Sports and Exercise*, *30*(8), 1217–1222.

Andersen, J. L. (2003). Muscle fibre type adaptation in the elderly human muscle. *Scandinavian Journal of Medicine & Science in Sports*, *13*(1), 40–47.

Andersen, J. L., Terzis, G., & Kryger, A. (1999). Increase in the degree of coexpression of myosin heavy chain isoforms in skeletal muscle fibers of the very old. *Muscle & Nerve*, 22, 449–454.

Aniansson, A., Grimby, G., & Hedberg, M. (1992). Compensatory muscle fiber hypertrophy in elderly men. *Journal of Applied Physiology*, *73*(3), 812–816.

Aniansson, A., Hedberg, M., Henning, G.-B., & Grimby, G. (1986). Muscle morphology, enzymatic activity, and muscle strength in elderly men: A follow-up study. *Muscle & Nerve*, *9*(7), 585–591.

Balagopal, P., Rooyackers, O. E., Adey, D. B., Ades, P. A., & Nair, K. S. (1997). Effects of aging on in vivo synthesis of skeletal muscle myosin heavy-chain and sarcoplasmic protein in humans. *American Journal of Physiology: Endocrinology and Metabolism*, *273*(4), E790–E800.

Bassey, E. J., Fiatarone, M. A., O'neill, E. F., Kelly, M., Evans, W. J., & Lipsitz, L. A. (1992). Leg extensor power and functional performance in very old men and women. *Clinical Science*, *82*(3), 321–327.

Bottinelli, R., Canepari, M., Pellegrino, M. A., & Reggiani, C. (1996). Force-velocity properties of human skeletal muscle fibres: Myosin heavy chain isoform and temperature dependence. *The Journal of Physiology*, *495*(2), 573–586.

Bottinelli, R., Pellegrino, M. A., Canepari, M., Rossi, R., & Reggiani, C. (1999). Specific contributions of various muscle fibre types to human muscle performance: An *in vitro* study. *Journal of Electromyography and Kinesiology*, *9*(2), 87–95.

Brocca, L., McPhee, J. S., Longa, E., Canepari, M., Seynnes, O., De Vito, G., Pellegrino, M. A., Narici, M., & Bottinelli, R. (2017). Structure and function of human muscle fibres and muscle proteome in physically active older men. *The Journal of Physiology*, *595*(14), 4823–4844.

Brooks, S. V., & Faulkner, J. A. (1988). Contractile properties of skeletal muscles from young, adult and aged mice. *The Journal of Physiology, 404*(1), 71–82.

Callahan, D. M., Bedrin, N. G., Subramanian, M., Berking, J., Ades, P. A., Toth, M. J., & Miller, M. S. (2014). Age-related structural alterations in human skeletal muscle fibers and mitochondria are sex specific: Relationship to single-fiber function. *Journal of Applied Physiology, 116*(12), 1582–1592.

Clark, B. C., & Manini, T. M. (2008). Sarcopenia ≠ Dynapenia. *The Journals of Gerontology Series A: Biological Sciences and Medical Sciences, 63*(8), 829–834.

Clark, D. J., Patten, C., Reid, K. F., Carabello, R. J., Phillips, E. M., & Fielding, R. A. (2011). Muscle performance and physical function are associated with voluntary rate of neuromuscular activation in older adults. *The Journals of Gerontology: Series A, 66A*(1), 115–121.

Coggan, A. R., Spina, R. J., King, D. S., Rogers, M. A., Rogers, M. A., Brown, M., Nemeth, P. M., & Holloszy, J. O. (1992). Histochemical and enzymatic comparison of the gastrocnemius muscle of young and elderly men and women. *Journal of Gerontology, 47*(3), B71–B76.

Colombini, B., Bagni, M. A., Palmini, R. B., & Cecchi, G. (2005). Crossbridge formation detected by stiffness measurements in single muscle fibres. In H. Sugi (Ed.), *Sliding Filament Mechanism in Muscle Contraction* (Vol. 565, pp. 127–140). Boston, MA: Springer US.

Cruz-Jentoft, A. J., Baeyens, J. P., Bauer, J. M., Boirie, Y., Cederholm, T., Landi, F., Matrin, F. C.,...Zamboni, M. (2010). Sarcopenia: European consensus on definition and diagnosis Report of the European Working Group on Sarcopenia in Older People. *Age and Ageing, 39*(4), 412–423.

D'Antona, G., Pellegrino, M. A., Adami, R., Rossi, R., Carlizzi, C. N., Canepari, M., Saltin, B., & Bottinelli, R. (2003). The effect of ageing and immobilization on structure and function of human skeletal muscle fibres. *The Journal of Physiology, 552*(2), 499–511.

D'Antona, G., Pellegrino, M. A., Carlizzi, C. N., & Bottinelli, R. (2007). Deterioration of contractile properties of muscle fibres in elderly subjects is modulated by the level of physical activity. *European Journal of Applied Physiology, 100*(5), 603–611.

Dalton, B. H., Jakobi, J. M., Allman, B. L., & Rice, C. L. (2010). Differential age-related changes in motor unit properties between elbow flexors and extensors. *Acta Physiologica, 200*(1), 45–55.

Degens, H., Yu, F., Li, X., & Larsson, L. (1998). Effects of age and gender on shortening velocity and myosin isoforms in single rat muscle fibres. *Acta Physiologica Scandinavica, 163*(1), 33–40.

Deschenes, M. R., Roby, M. A., Eason, M. K., & Harris, M. B. (2010). Remodeling of the neuromuscular junction precedes sarcopenia related alterations in myofibers. *Experimental Gerontology, 45*(5), 389–393.

Edström, E., & Ulfhake, B. (2005). Sarcopenia is not due to lack of regenerative drive in senescent skeletal muscle. *Aging Cell, 4*(2), 65–77.

Fearon, K., Evans, W. J., & Anker, S. D. (2011). Myopenia-a new universal term for muscle wasting. *Journal of Cachexia, Sarcopenia and Muscle, 2*(1), 1–3.

Frontera, W. R., Reid, K. F., Phillips, E. M., Krivickas, L. S., Hughes, V. A., Roubenoff, R., & Fielding, R. A. (2008). Muscle fiber size and function in elderly humans: A longitudinal study. *Journal of Applied Physiology, 105*(2), 637–642.

Frontera, W. R., Suh, D., Krivickas, L. S., Hughes, V. A., Goldstein, R., & Roubenoff, R. (2000a). Skeletal muscle fiber quality in older men and women. *American Journal of Physiology: Cell Physiology, 279*(3), C611–C618.

Frontera, W. R., Hughes, V. A., Fielding, R. A., Fiatarone, M. A., Evans, W. J., & Roubenoff, R. (2000b). Aging of skeletal muscle: A 12-yr longitudinal study. *Journal of Applied Physiology, 88*(4), 1321–1326.

Gallagher, P., Trappe, S., Harber, M., Creer, A., Mazzetti, S., Trappe, T., Alkner, B., & Tesch, P. (2005). Effects of 84-days of bedrest and resistance training on single muscle fibre myosin heavy chain distribution in human vastus lateralis and soleus muscles. *Acta Physiologica Scandinavica, 185*(1), 61–69.

Goodpaster, B. H., Park, S. W., Harris, T. B., Kritchevsky, S. B., Nevitt, M., Schwartz, A. V., Simonsick, E. M.,...Newman, A. B. (2006). The loss of skeletal muscle strength, mass, and quality in older adults: The health, aging and body composition study. *The Journals of Gerontology Series A: Biological Sciences and Medical Sciences, 61*(10), 1059–1064.

Gouspillou, G., Godin, R., Piquereau, J., Picard, M., Mofarrahi, M., Mathew, J., Puves-Smith, F. M.,...Hussain, S. N. A. (2018). Protective role of Parkin in skeletal muscle contractile and mitochondrial function. *The Journal of Physiology, 596*(13), 2565–2579.

Harridge, S. D. R., Bottinelli, R., Canepari, M., Pellegrino, M. A., Reggiani, C., Esbjörnsson, M., & Saltin, B. (1996). Whole-muscle and single-fibre contractile properties and myosin heavy chain isoforms in humans. *Pflügers Archiv, 432*(5), 913–920.

Hortobágyi, T., Dempsey, L., Fraser, D., Zheng, D., Hamilton, G., Lambert, J., & Dohm, L. (2000). Changes in muscle strength, muscle fibre size and myofibrillar gene expression after immobilization and retraining in humans. *The Journal of Physiology, 524*(1), 293–304.

Hvid, L. G., Ørtenblad, N., Aagaard, P., Kjaer, M., & Suetta, C. (2011). Effects of ageing on single muscle fibre contractile function following short-term immobilisation: Immobilisation and ageing impairs single fibre contractile function. *The Journal of Physiology, 589*(19), 4745–4757.

Jackson, M. J., & McArdle, A. (2011). Age-related changes in skeletal muscle reactive oxygen species generation and adaptive responses to reactive oxygen species. *The Journal of Physiology, 589*(9), 2139–2145.

Jang, Y. C., Lustgarten, M. S., Liu, Y., Muller, F. L., Bhattacharya, A., Liang, H., Salmon, A. B.,,…Remmen, H. V. (2010). Increased superoxide in vivo accelerates age-associated muscle atrophy through mitochondrial dysfunction and neuromuscular junction degeneration. *The FASEB Journal, 24*(5), 1376–1390.

Jennekens, F. G. I., Tomlinson, B. E., & Walton, J. N. (1971). The sizes of the two main histochemical fibre types in five limb muscles in man. *Journal of the Neurological Sciences, 13*(3), 281–292.

Kim, J.-H., & Thompson, L. V. (2013). Inactivity, age, and exercise: Single-muscle fiber power generation. *Journal of Applied Physiology, 114*(1), 90–98.

Klitgaard, H., Zhou, M., Schiaffino, S., Betto, R., Salviati, G., & Saltin, B. (1990). Ageing alters the myosin heavy chain composition of single fibres from human skeletal muscle. *Acta Physiologica Scandinavica, 140*(1), 55–62.

Korhonen, M. T., Cristea, A., Alén, M., Häkkinen, K., Sipilä, S., Mero, A., … Suominen, H. (2006). Aging, muscle fiber type, and contractile function in sprint-trained athletes. *Journal of Applied Physiology, 101*(3), 906–917.

Krivickas, L. S., Fielding, R. A., Murray, A., Callahan, D., Johansson, A., Dorer, D. J., & Frontera, W. R. (2006). Sex differences in single muscle fiber power in older adults. *Medicine and Science in Sports and Exercise, 38*(1), 57–63.

Krivickas, L. S., Suh, D., Wilkins, J., Hughes, V. A., Roubenoff, R., & Frontera, W. R. (2001). Age- and gender-related differences in maximum shortening velocity of skeletal muscle fibers. *American Journal of Physiology: Medical Rehabilitation, 80*(6), 447–455.

Lamboley, C. R., Wyckelsma, V. L., Dutka, T. L., McKenna, M. J., Murphy, R. M., & Lamb, G. D. (2015). Contractile properties and sarcoplasmic reticulum calcium content in type I and type II skeletal muscle fibres in active aged humans: Muscle contractile properties and SR Ca^{2+} content in aged subjects. *The Journal of Physiology, 593*(11), 2499–2514.

Lamboley, C. R., Wyckelsma, V. L., McKenna, M. J., Murphy, R. M., & Lamb, G. D. (2016). Ca^{2+} leakage out of the sarcoplasmic reticulum is increased in type I skeletal muscle fibres in aged humans. *The Journal of Physiology, 594*(2), 469–481.

Larsson, L., Li, X., & Frontera, W. R. (1997a). Effects of aging on shortening velocity and myosin isoform composition in single human skeletal muscle cells. *American Journal of Physiology: Cell Physiology, 272*(2), C638–C649.

Larsson, L., Li, X., Yu, F., & Degens, H. (1997b). Age-related changes in contractile properties and expression of myosin isoforms in single skeletal muscle cells. *Muscle & Nerve, 20*(s 5), 74–78.

Larsson, L., & Moss, R. L. (1993). Maximum velocity of shortening in relation to myosin isoform composition in single fibres from human skeletal muscles. *The Journal of Physiology, 472*, 595–614.

Larsson, L., Sjödin, B., & Karlsson, J. (1978). Histochemical and biochemical changes in human skeletal muscle with age in sedentary males, age 22–65 years. *Acta Physiologica Scandinavica, 103*(1), 31–39.

Leenders, M., Verdijk, L. B., van der Hoeven, L., van Kranenburg, J., Hartgens, F., Wodzig, W. K. W. H., Saris, W. H. M., & van Loon, L. J. C. (2011). Prolonged leucine supplementation does not augment muscle mass or affect glycemic control in elderly type 2 diabetic men. *The Journal of Nutrition, 141*(6), 1070–1076.

Lexell, J., Taylor, C. C., & Sjöström, M. (1988). What is the cause of the ageing atrophy?: Total number, size and proportion of different fiber types studied in whole vastus lateralis muscle from 15- to 83-year-old men. *Journal of the Neurological Sciences, 84*(2–3), 275–294.

Li, X., & Larsson, L. (1996). Maximum shortening velocity and myosin isoforms in single muscle fibers from young and old rats. *American Journal of Physiology: Cell Physiology, 270*(1), C352–C360.

Lowe, D. A., Surek, J. T., Thomas, D. D., & Thompson, L. V. (2001). Electron paramagnetic resonance reveals age-related myosin structural changes in rat skeletal muscle fibers. *American Journal of Physiology: Cell Physiology, 280*(3), C540–C547.

Lowe, D. A., Thomas, D. D., & Thompson, L. V. (2002). Force generation, but not myosin ATPase activity, declines with age in rat muscle fibers. *AJP: Cell Physiology, 283*(1), C187–C192.

Marzetti, E., Calvani, R., Cesari, M., Buford, T. W., Lorenzi, M., Behnke, B. J., & Leeuwenburgh, C. (2013). Mitochondrial dysfunction and sarcopenia of aging: From signaling pathways to clinical trials. *The International Journal of Biochemistry & Cell Biology, 45*(10), 2288–2301.

McNeil, C. J., Doherty, T. J., Stashuk, D. W., & Rice, C. L. (2005). Motor unit number estimates in the tibialis anterior muscle of young, old, and very old men. *Muscle & Nerve, 31*(4), 461–467.

McNeil, C. J., Vandervoort, A. A., & Rice, C. L. (2007). Peripheral impairments cause a progressive age-related loss of strength and velocity-dependent power in the dorsiflexors. *Journal of Applied Physiology, 102*(5), 1962–1968.

Miller, M. S., Bedrin, N. G., Callahan, D. M., Previs, M. J., Jennings, M. E., Ades, P. A.,...Toth, M. J. (2013). Age-related slowing of myosin actin cross-bridge kinetics is sex specific and predicts decrements in whole skeletal muscle performance in humans. *Journal of Applied Physiology, 115*(7), 1004–1014.

Monemi, M., Eriksson, P.-O., Kadi, F., Butler-Browne, G. S., & Thornell, L.-E. (1999). Opposite changes in myosin heavy chain composition of human masseter and biceps brachii muscles during aging. *Journal of Muscle Research & Cell Motility, 20*(4), 351–361.

Muscaritoli, M., Anker, S. D., Argilés, J., Aversa, Z., Bauer, J. M., Biolo, G., Boirie, Y.,...Sieber, C. C. (2010). Consensus definition of sarcopenia, cachexia and pre-cachexia: Joint document elaborated by Special Interest Groups (SIG) "cachexia-anorexia in chronic wasting diseases" and "nutrition in geriatrics." *Clinical Nutrition, 29*(2), 154–159.

Nederveen, J. P., Joanisse, S., Snijders, T., Ivankovic, V., Baker, S. K., Phillips, S. M., & Parise, G. (2016). Skeletal muscle satellite cells are located at a closer proximity to capillaries in healthy young compared with older men. *Journal of Cachexia, Sarcopenia and Muscle, 7*(5), 547–554.

Ochala, J., Dorer, D. J., Frontera, W. R., & Krivickas, L. S. (2006). Single skeletal muscle fiber behavior after a quick stretch in young and older men: A possible explanation of the relative preservation of eccentric force in old age. *Pflügers Archiv, 452*(4), 464–470.

Ochala, J., Frontera, W. R., Dorer, D. J., Hoecke, J. V., & Krivickas, L. S. (2007). Single skeletal muscle fiber elastic and contractile characteristics in young and older men. *The Journals of Gerontology Series A: Biological Sciences and Medical Sciences, 62*(4), 375–381.

Oh, S.-L., Yoon, S. H., & Lim, J.-Y. (2018). Age- and sex-related differences in myosin heavy chain isoforms and muscle strength, function, and quality: A cross sectional study. *Journal of Exercise Nutrition & Biochemistry, 22*(2), 43–50.

Ohira, Y., Yoshinaga, T., Ohara, M., Nonaka, I., Yoshioka, T., Yamashita-Goto, K., Shenkman, B. S., Kozlovskaya, I. B.,...Edgerton, V. R. (1999). Myonuclear domain and myosin phenotype in human soleus after bed rest with or without loading. *Journal of Applied Physiology, 87*(5), 1776–1785.

Petrella, J. K., Kim, J., Tuggle, S. C., Hall, S. R., & Bamman, M. M. (2005). Age differences in knee extension power, contractile velocity, and fatigability. *Journal of Applied Physiology, 98*(1), 211–220.

Power, G. A., Allen, M. D., Booth, W. J., Thompson, R. T., Marsh, G. D., & Rice, C. L. (2014). The influence on sarcopenia of muscle quality and quantity derived from magnetic resonance imaging and neuromuscular properties. *AGE, 36*(3), 9642.

Power, G. A., Dalton, B. H., Behm, D. G., Doherty, T. J., Vandervoort, A. A., & Rice, C. L. (2012). Motor unit survival in lifelong runners is muscle dependent. *Medicine & Science in Sports & Exercise, 44*(7), 1235–1242.

Power, G. A., Dalton, B. H., & Rice, C. L. (2013). Human neuromuscular structure and function in old age: A brief review. *Journal of Sport and Health Science, 2*(4), 215–226.

Power, G. A., Minozzo, F. C., Spendiff, S., Filion, M.-E., Konokhova, Y., Purves-Smith, M. F., Pion, C.,...Rassier, D. E. (2016). Reduction in single muscle fiber rate of force development with aging is not attenuated in world class older masters athletes. *American Journal of Physiology-Cell Physiology, 310*(4), C318–C327.

Prochniewicz, E., Thomas, D. D., & Thompson, L. V. (2005). Age-related decline in actomyosin function. *The Journals of Gerontology Series A: Biological Sciences and Medical Sciences, 60*(4), 425–431.

Purves-Smith, F. M., Sgarioto, N., & Hepple, R. T. (2014). Fiber typing in aging muscle. *Exercise and Sport Sciences Reviews, 42*(2), 45–52.

Raue, U., Slivka, D., Minchev, K., & Trappe, S. (2009). Improvements in whole muscle and myocellular function are limited with high-intensity resistance training in octogenarian women. *Journal of Applied Physiology, 106*(5), 1611–1617.

Reid, K. F., Doros, G., Clark, D. J., Patten, C., Carabello, R. J., Cloutier, G. J., Phillips, E. M.,...Fielding, R. A. (2012). Muscle power failure in mobility-limited older adults: Preserved single fiber function despite lower whole muscle size, quality and rate of neuromuscular activation. *European Journal of Applied Physiology, 112*(6), 2289–2301.

Reid, K. F., Pasha, E., Doros, G., Clark, D. J., Patten, C., Phillips, E. M., Frontera, W. R., & Fielding, R. A. (2014). Longitudinal decline of lower extremity muscle power in healthy and mobility-limited older adults: Influence of muscle mass, strength, composition, neuromuscular activation and single fiber contractile properties. *European Journal of Applied Physiology, 114*(1), 29–39.

Rosenberg, I. H. (1997). Sarcopenia: Origins and clinical relevance. *The Journal of Nutrition*, *127*(5), 990S–991S.

Rowan, S. L., Purves-Smith, F. M., Solbak, N. M., & Hepple, R. T. (2011). Accumulation of severely atrophic myofibers marks the acceleration of sarcopenia in slow and fast twitch muscles. *Experimental Gerontology, 46*(8), 660–669.

Rowan, S. L., Rygiel, K., Purves-Smith, F. M., Solbak, N. M., Turnbull, D. M., & Hepple, R. T. (2012). Denervation causes fiber atrophy and myosin heavy chain co-expression in senescent skeletal muscle. *PLoS ONE, 7*(1), e29082.

Russ, D. W., Grandy, J. S., Toma, K., & Ward, C. W. (2011). Ageing, but not yet senescent, rats exhibit reduced muscle quality and sarcoplasmic reticulum function. *Acta Physiologica*, *201*(3), 391–403.

Snijders, T., Nederveen, J. P., Joanisse, S., Leenders, M., Verdijk, L. B., van Loon, L. J. C., & Parise, G. (2016). Muscle fibre capillarization is a critical factor in muscle fibre hypertrophy during resistance exercise training in older men. *Journal of Cachexia, Sarcopenia and Muscle, 8*(2), 267–276.

Straight, C. R., Ades, P. A., Toth, M. J., & Miller, M. S. (2018). Age-related reduction in single muscle fiber calcium sensitivity is associated with decreased muscle power in men and women. *Experimental Gerontology, 102*, 84–92.

Thompson, L. V., & Brown, M. (1999). Age-related changes in contractile properties of single skeletal fibers from the soleus muscle. *Journal of Applied Physiology*, *86*(3), 881–886.

Thompson, L. V., Durand, D., Fugere, N. A., & Ferrington, D. A. (2006). Myosin and actin expression and oxidation in aging muscle. *Journal of Applied Physiology*, *101*(6), 1581–1587.

Thompson, L. V., Johnson, S. A., & Shoeman, J. A. (1998). Single soleus muscle fiber function after hindlimb unweighting in adult and aged rats. *Journal of Applied Physiology*, *84*(6), 1937–1942.

Trappe, S. W., Costill, D. L., Fink, W. J., & Pearson, D. R. (1995). Skeletal muscle characteristics among distance runners: A 20-yr follow-up study. *Journal of Applied Physiology*, *78*(3), 823–829.

Trappe, S., Gallagher, P., Harber, M., Carrithers, J., Fluckey, J., & Trappe, T. (2003). Single muscle fibre contractile properties in young and old men and women. *The Journal of Physiology*, *552*(1), 47–58.

Vandervoort, A. A. (2002). Aging of the human neuromuscular system. *Muscle & Nerve*, *25*(1), 17–25.

Venturelli, M., Saggin, P., Muti, E., Naro, F., Cancellara, L., Toniolo, L., Tarperi, C.,...Schena, F. (2015). In vivo and in vitro evidence that intrinsic upper- and lower-limb skeletal muscle function is unaffected by ageing and disuse in oldest-old humans. *Acta Physiologica*, *215*(1), 58–71.

von Haehling, S., Morley, J. E., & Anker, S. D. (2010). An overview of sarcopenia: Facts and numbers on prevalence and clinical impact. *Journal of Cachexia, Sarcopenia and Muscle*, *1*(2), 129–133.

Welle, S., Thornton, C., Jozefowicz, R., & Statt, M. (1993). Myofibrillar protein synthesis in young and old men. *American Journal of Physiology: Endocrinology and Metabolism*, *264*(5), E693–E698.

Widrick, J. J., Knuth, S. T., Norenberg, K. M., Romatowski, J. G., Bain, J. L. W., Riley, D. A., Karhanek, M.... Fitts, R. H. (1999). Effect of a 17 day spaceflight on contractile properties of human soleus muscle fibres. *The Journal of Physiology*, *516*(3), 915–930.

Widrick, J. J., Romatowski, J. G., Bain, J. L. W., Trappe, S. W., Trappe, T. A., Thompson, J. L., Costill, D. L.... Fitts, R. H. (1997). Effect of 17 days of bed rest on peak isometric force and unloaded shortening velocity of human soleus fibers. *American Journal of Physiology: Cell Physiology*, *273*(5), C1690–C1699.

Yu, F., Hedström, M., Cristea, A., Dalén, N., & Larsson, L. (2007). Effects of ageing and gender on contractile properties in human skeletal muscle and single fibres. *Acta Physiologica*, *190*(3), 229–241.

Zhong, S., Lowe, D. A., & Thompson, L. V. (2006). Effects of hindlimb unweighting and aging on rat semimembranosus muscle and myosin. *Journal of Applied Physiology*, *101*(3), 873–880.

17 Sarcopenic Dysphagia, Presbyphagia, and Rehabilitation Nutrition

Hidetaka Wakabayashi and Kunihiro Sakuma

CONTENTS

17.1 INTRODUCTION

Sarcopenia is characterized by a progressive and generalized loss of skeletal muscle mass and strength and is categorized into primary or age-related sarcopenia and secondary sarcopenia, which can be related to activity, nutrition, or disease. Sarcopenia is an important issue in geriatric rehabilitation medicine because approximately 50% of older people in post-acute care and rehabilitation settings have sarcopenia (Sánchez-Rodríguez et al. 2016). Furthermore, sarcopenia is associated with poorer rehabilitation outcomes and physical function in older patients undergoing rehabilitation (Wakabayashi and Sakuma 2014a). An interest in rehabilitation nutrition combined with nutritional care management to treat sarcopenia and to maximize functionality in disabled people is growing in Japan.

Our special interest is the assessment and treatment of both sarcopenia and dysphagia. Sarcopenic dysphagia is a new concept characterized by a decrease in mass and loss of function of swallowing muscles mass associated with a generalized loss of skeletal muscle mass and function (Wakabayashi and Sakuma 2014a; Wakabayashi 2014; Clave and Shaker 2015; Baijens et al. 2016). Dysphagia management is very important because the condition increases the risk of related complications such as aspiration pneumonia, choking, dehydration, and malnutrition. Pneumonia, mainly aspiration pneumonia in older adults, is the third leading cause of death in Japan following cancer and heart disease. People with severe dysphagia who are nonoral feeding also have a lower quality of life (QOL) as a result of losing the joy of eating. Sarcopenic dysphagia is therefore an important current and future public health issue.

Little attention has been paid to sarcopenic dysphagia in geriatric rehabilitation medicine. Of the 7,132 articles identified in PubMed on 12 February 2019 using the term "sarcopenia," 82 (1.1%) and 28 (0.4%) were retrieved by an additional search using "sarcopenia dysphagia" and "sarcopenic dysphagia," respectively. The reason for this low interest in sarcopenic dysphagia is that sarcopenia has not been considered as a cause of dysphagia. The major causes of dysphagia are stroke, brain injury, neuromuscular diseases, head and neck cancers, and connective tissue diseases. Diseases that directly cause dysphagia are therefore of interest in dysphagia rehabilitation. In contrast, hip fracture is not a direct cause of dysphagia although 34% of older patients were reported to have dysphagia after a hip fracture operation (Love, Cornwell, and Whitehouse 2013). Sarcopenia can be one of the causes of dysphagia in older patients with hip fracture, with a prevalence of 73.6% in men and 67.7% in women (Ho et al. 2016), and is associated with a generalized loss of muscle mass and dysphagia (Kuroda 2014; Kuroda and Kuroda 2012; Murakami, Hirano, Watanabe, Edahiro, et al. 2015; Wakabayashi et al. 2015). Sarcopenic dysphagia may therefore be common in older disabled people with both sarcopenia and dysphagia.

Presbyphagia is characterized by age-related changes in the swallowing mechanism in older people and is different from dysphagia (Wakabayashi 2014). The difference between the two conditions is the necessity for food modification. Thickened drinks and texture-modified foods such as pureed, minced and moist, and soft products (Cichero et al. 2017) are necessary in people with dysphagia to prevent aspiration and choking. In contrast, thickened drinks and texture-modified food are not necessary in older people with presbyphagia. Age-related sarcopenia of the swallowing muscles is one of the causes of presbyphagia. Presbyphagia is associated with a frailty in swallowing and is common in healthy older adults. The number of healthy older people who have an aspiration during flexible endoscopic evaluation of swallowing ranges between 28% and 37% (Butler et al. 2009, 2010, 2011). The prevalence of swallowing problems defined by a questionnaire, and the 30-ml water swallow test in healthy older people was 15.1% (Okamoto et al. 2012). When frail older people with presbyphagia develop aspiration pneumonia, they can simultaneously experience activity-, disease-, and nutrition-related sarcopenia of generalized muscles and swallowing muscles, resulting in the development of sarcopenic dysphagia (Wakabayashi 2014). Early detection of presbyphagia in older people is therefore important.

Therapy for sarcopenic dysphagia includes dysphagia rehabilitation and treatment of sarcopenia (Wakabayashi 2014). Dysphagia rehabilitation should include resistance training of swallowing muscles as it is considered to be the most effective treatment option for sarcopenia (Wakabayashi and Sakuma 2013, 2014). Improvement in nutrition is also very important for treating sarcopenic dysphagia because malnutrition is one of the causes of dysphagia (Veldee and Peth 1992; Hudson, Daubert, and Mills 2000). Furthermore, aggressive nutritional care management that included daily energy accumulation (200–750 kcal/day) to increase body weight and muscle mass has been shown to result in increased body weight and improved swallowing function in patients with sarcopenic dysphagia (Maeda and Akagi 2016; Wakabayashi and Uwano 2016; Hashida et al. 2017). Taken together, these findings indicate rehabilitation nutrition is effective for treating sarcopenic dysphagia. In this review, we discuss sarcopenic dysphagia, rehabilitation nutrition, and presbyphagia.

17.2 SARCOPENIC DYSPHAGIA

17.2.1 Sarcopenia of Swallowing Muscles

Sarcopenic dysphagia is characterized by a loss of swallowing muscle mass and function associated with a generalized loss of skeletal muscle mass and function (Wakabayashi 2014). The function of swallowing is to deliver foods and drinks from the mouth to the stomach. During the swallowing process, numerous muscles are used such as the intrinsic muscle of the tongue and the mimic, masticatory, suprahyoid, infrahyoid, palatal, pharyngeal, and esophageal muscles. Sarcopenia affecting the mass, strength, and function of swallowing muscles has been reported, indicating the presence of sarcopenic dysphagia.

17.2.1.1 Swallowing Muscle Mass

Age-related loss of swallowing muscle mass has been observed in the tongue and geniohyoid and pharyngeal muscles. Age-related loss of intrinsic muscle mass of the tongue assessed by ultrasonography was reported (Tamura et al. 2012), while similar loss of the geniohyoid muscle was observed in healthy older adults by computed tomography (Feng et al. 2013). The mean cross-sectional area of the geniohyoid muscle in the midsagittal plane in female younger adults and older adults is about 490 and 340 mm^2, respectively (31% reduction) and approximately 520 and 440 mm^2 in younger and older adult males, respectively (15% reduction). The geniohyoid muscle has also been reported to be significantly smaller in healthy older men with aspiration compared with older men without aspiration. Pharyngeal wall thickness in young and older women by magnetic resonance imaging showed mean thickness at the level of the anterior inferior second cervical vertebra was 0.25 cm in the 20s, 0.22 cm in the 60s, and 0.19 cm in the 70s (Molfenter et al. 2015). In contrast, there was no association between neck circumference and dysphagia in older people requiring long-term care (Wakabayashi and Matsushima 2016). Age-related loss in skeletal muscle mass occurs in both generalized and swallowing muscles.

17.2.1.2 Swallowing Muscle Strength

Age-related declines of swallowing muscle strength are evaluated using tongue pressure, jaw-opening force, and head lifting strength. Tongue pressure between the front part of the palate and the tongue is the most common method for evaluating swallowing muscle strength. Mean maximum isometric pressure at the anterior location in males aged in their 20s and 80s was reported to be 58.1 and 33.7 kPa, respectively (42% reduction), while the corresponding pressures in females aged in their 20s and 80s were 49.9 and 28.1 kPa, respectively (44% reduction) (Vanderwegen et al. 2013). Decreased tongue pressure associated with dysphagia has been reported in institutionalized older people, stroke patients, and older people without neurological diseases (Yoshida et al. 2006; Konaka et al. 2010; Maeda and Akagi 2015). There is evidence of tongue pressure being associated with grip strength and nutritional status in older inpatients (Sakai et al. 2017). Older healthy adults also have lower lingual isometric pressures and lower swallow pressures than younger healthy adults, assessed by five sensors located on the hard palate (Robbins et al. 2016). It is likely the mechanism of the age-related decline in lingual pressures is due, at least in part, to sarcopenia (Robbins et al. 2016). Jaw-opening force is another method for evaluating swallowing muscle strength. Mean jaw opening force in healthy adult and older males was found to be 9.7 and 7.0 kg, respectively (Iida et al. 2013). In healthy adults and older women, the corresponding values were 5.9 and 4.4 kg, respectively. The effects of sarcopenia on both tongue pressure and jaw-opening force in older adults without dysphagia were described (Machida et al. 2017). Headlifting strength is another tool for assessing the strength of the suprahyoid muscles and is associated with dysphagia and malnutrition in older people requiring long-term care (Wakabayashi, Sashika, and Matsushima 2015). It is therefore clear that there is an age-related decline in the strength of swallowing muscles.

17.2.2 GENERALIZED SARCOPENIA AND SWALLOWING

Loss of generalized skeletal muscle mass and strength affects the physical function of both the entire body and swallowing function. The prevalence of sarcopenia in patients with dysphagia and acute community-acquired pneumonia was reported to be 29.4% (Carrion et al. 2017). Skeletal muscle mass in patients with cancer with dysphagia, assessed by measuring the cross-sectional area of the psoas muscles by abdominal computed tomography, was shown to be associated independently with oral food intake at discharge (Wakabayashi et al. 2015). Swallowing function, activities of daily living (ADL), and nutritional status were also significantly lower in people with sarcopenia than those without the condition (Shiozu, Higashijima, and Koga 2015). Similarly, chewing ability has been shown to be associated independently with sarcopenia in community-dwelling older people (Murakami, Hirano, Watanabe, Sakai, et al. 2015). In patients with Alzheimer's disease, there is evidence of a relationship between decreased skeletal muscle mass and swallowing function (Takagi et al. 2016). A prospective cohort study also demonstrated generalized sarcopenia, low body mass index, and low ADL were independent predictors of dysphagia (Maeda, Takaki, and Akagi 2016). Of 82 older inpatients without dysphagia who had restricted oral intake for more than 2 days, 63 (77%) had sarcopenia, while 21 (26%) developed dysphagia, all of whom had sarcopenia ($p = 0.002$). These results indicated that advanced generalized sarcopenia and low activity of swallowing muscles due to nil per os (no oral intake) are likely to result in sarcopenic dysphagia, because no patients had a stroke or other disease that could directly cause dysphagia other than sarcopenia.

17.3 MECHANISM OF SARCOPENIC DYSPHAGIA

The mechanism of sarcopenic dysphagia in older people is to experience activity-, nutrition-, and disease-related sarcopenia of generalized and swallowing muscles (Wakabayashi 2014; Wakabayashi and Sakuma 2014) and is classified into three patterns. First, generalized sarcopenia occurs before dysphagia. In this pattern, mild sarcopenic dysphagia can occur in older patients without diseases. However, severe sarcopenic dysphagia seems not to occur in older patients without diseases. Sarcopenic dysphagia is likely to occur in frail and disabled older adults with limited functional reserve of generalized and swallowing muscles. Older inpatients with sarcopenia but not dysphagia who had restricted oral intake for more than 2 days may develop sarcopenic dysphagia (Maeda, Takaki, and Akagi 2016). Second, dysphagia occurs before generalized sarcopenia and develops in patients with stroke, brain injury, neuromuscular diseases, head and neck cancer, and connective tissue disorders. For example, there is a published care report of a patient with tongue cancer after a subtotal glossectomy who had dysphagia but not sarcopenia and developed sarcopenic dysphagia due to nutrition- and disease-related sarcopenia (Hashida et al. 2017). Finally, generalized sarcopenia and dysphagia occur simultaneously. For example, a patient with lung cancer who had neither sarcopenia nor dysphagia before surgery developed generalized sarcopenia and sarcopenic dysphagia simultaneously after surgery (Wakabayashi and Uwano 2016).

In a preclinical mouse model, aspiration pneumonia accentuated muscular decomposition and induced muscular atrophy in the tibialis anterior, tongue, and diaphragm (Komatsu et al. 2018). The aspiration challenge activated caspase-3 in all the three muscles examined, whereas calpains were activated in the diaphragm and the tibialis anterior but not in the tongue. Activation of the ubiquitin–proteasome system was detected in all muscles. The aspiration challenge activated autophagy in the tibialis anterior and the tongue, whereas weak or little activation was detected in the diaphragm. The aspiration challenge resulted in a greater proportion of smaller myofibers than in controls in the diaphragm, tibialis anterior, and tongue (Komatsu et al. 2018). We address activity-, nutrition-, and disease-related sarcopenia in the following sections.

17.3.1 ACTIVITY-RELATED SARCOPENIA

Activity-related sarcopenia results from prolonged bed rest, a sedentary lifestyle, and/or deconditioning. The loss of skeletal muscle mass during bed rest occurs at a rate of 0.5% total muscle mass per day (Wall and Van Loon 2013). Prolonged disuse (>10 days) can result in a decline in basal and postprandial rates of muscle protein synthesis, without any apparent change in muscle protein breakdown (Wall and Van Loon 2013).

Activity-related sarcopenia of swallowing muscles occurs as a result of a lower number of chewing and swallowing times due to noneating and less dietary intake. Indeed, 26% of older inpatients without dysphagia who had restricted oral intake for more than 2 days developed dysphagia (Maeda, Takaki, and Akagi 2016). The main cause of dysphagia development is probably sarcopenia. Patients with aspiration pneumonia tend to be prescribed noneating and bed rest during treatment for pneumonia (Wakabayashi 2014). Activity-related sarcopenia develops during noneating and bedridden periods.

Iatrogenic sarcopenia is defined as sarcopenia caused by the activities of medical doctors, nurses, or other health care professionals in health care facilities, especially in acute care hospitals (Nagano, Nishioka, and Wakabayashi 2019). Iatrogenic sarcopenia includes activity-related sarcopenia caused by unnecessary inactivity, nutrition-related sarcopenia caused by inappropriate nutritional care management, and disease-related sarcopenia in cases of iatrogenic diseases. Examples of iatrogenic sarcopenia are noneating and bed rest prescribed in inpatients who can swallow and are generally mobile or activity-related sarcopenia due to unnecessary noneating and bed rest. Iatrogenic sarcopenia due to unnecessary fasting in hospitalized older people should be avoided, because tentative nil per os (no oral intake) in older adults with aspiration pneumonia has been reported to result in prolonged treatment duration and a decline in swallowing ability (Maeda, Koga, and Akagi 2016).

17.3.2 NUTRITION-RELATED SARCOPENIA

Nutrition-related sarcopenia results from inadequate dietary intake of energy and/or protein. Malnutrition is classified as malnutrition in the context of acute illness or injury or invasion, chronic illness or cachexia, and/or social or environmental circumstances that can lead to starvation (White et al. 2012). However, nutrition-related sarcopenia includes only starvation. Older inpatients with hospital-associated deconditioning are at risk of inadequate energy intake that does not meet their basal energy expenditure. A total of 75 (44%) of 169 older inpatients with hospital-associated deconditioning fall into this category (Wakabayashi and Sashika 2014). Indeed, some older inpatients with hospital-associated deconditioning are prescribed peripheral parenteral nutrition that furnishes less than 300 kcal/day in the absence of an oral caloric intake and enteral nutrition (Wakabayashi and Sashika 2014).

Nutrition-related sarcopenia due to inappropriate nutritional management is interpreted as iatrogenic sarcopenia and should be avoided by providing appropriate nutritional support. This strategy is supported by the finding that low energy intake was an independent risk factor for extended aspiration pneumonia treatment duration in older inpatients (Maeda, Koga, and Akagi 2016).

17.3.3 DISEASE-RELATED SARCOPENIA

Disease-related sarcopenia is associated with advanced organ failure, inflammatory disease, malignancy, and/or endocrine disease and results from invasion and cachexia. Invasion or acute illness or injury includes acute infectious diseases such as aspiration pneumonia and septic shock, fractures, burns, and acute trauma. Phases of invasion can be classified as catabolic or anabolic. In the catabolic phase, both muscle and adipose tissue are degraded by proinflammatory cytokines. In patients with critical illnesses, the loss of muscle mass can reach 1 kg/day due to severe inflammatory responses (Wischmeyer and San-Millan 2015). After thoracotomy and laparotomy, the percentage decrease of geniohyoid muscle area measured in patients by ultrasonography was 9.5%–17.5% the postoperative day 7 (Shimizu et al. 2016). In the anabolic phase, both muscle and adipose tissue can be synthesized and increased by appropriate nutritional management and exercise.

Cachexia is included in disease-related sarcopenia. Cachexia is a complex metabolic syndrome associated with an underlying illness characterized by muscle loss with or without fat loss (Evans et al. 2008). The clinical features of cachexia are weight loss, anorexia, inflammation, insulin resistance, and increased muscle protein breakdown (Evans et al. 2008). The causative diseases of cachexia are cancer, chronic heart, renal, respiratory or liver failure, connective tissue diseases and autoimmune diseases, chronic infectious diseases, and inflammatory bowel diseases. Cancer-induced cachexia affects approximately 50%–80% of patients with cancer and may account for up to 20% of cancer deaths (Argiles et al. 2014). The prevalence of cachexia in patients with malignant diseases and chronic heart failure was reported to be 21.8% (Letilovic and Vrhovac 2013). Another study showed the prevalence of protein-energy wasting was 37% in hemodialysis patients, with the loss in muscle mass loss being associated with increased mortality (Carrero et al. 2013).

Neurodegenerative diseases include amyotrophic lateral sclerosis, polymyositis, inclusion body myositis, and muscular dystrophy. In the definition of sarcopenia of the European Working Group on Sarcopenia in Older People (EWGSOP), neurodegenerative diseases are included in the mechanism of sarcopenia (Cruz-Jentoft et al. 2010). However, neurodegenerative diseases should not be included as causes of sarcopenia, because in September 2016, sarcopenia became a medical condition with its own unique *International Classification of Disease, Tenth Revision, Clinical Modification (ICD-10-CM)* code (M62.84) (Anker, Morley, and von Haehling 2016; Cao and Morley 2016), while neurodegenerative diseases had another *ICD-10-CM* codes. Sarcopenia is now considered to be distinctly different from neurodegenerative diseases. Accordingly, disease-related sarcopenia should include only invasion and cachexia.

17.3.4 Overlapping All Causes of Sarcopenia

The mechanism of sarcopenic dysphagia in older people overlaps all causes of sarcopenia. Figure 17.1 shows the mechanism of sarcopenic dysphagia in frail older people with aspiration pneumonia. Complications associated with activity-, nutrition-, or disease-related sarcopenia are common in patients with aspiration pneumonia. For example, all causes of sarcopenia can be complicated in

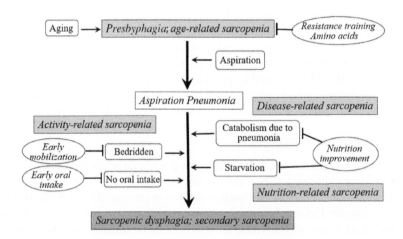

FIGURE 17.1 Mechanism of sarcopenic dysphagia in frail older people with aspiration pneumonia. Presbyphagia and age-related sarcopenia are common in frail older people. If frail older people develop aspiration pneumonia, sarcopenia of generalized skeletal muscles and swallowing muscles rapidly deteriorates, because of activity-, nutrition-, and disease-related sarcopenia, resulting in the development of sarcopenic dysphagia and secondary sarcopenia.

TABLE 17.1

Causes of Sarcopenia in Two Case Reports with Sarcopenic Dysphagia

Causes of Sarcopenia	Maeda and Akagi (2016)	Wakabayashi and Uwano (2016)
Age-related sarcopenia	80-year-old woman	71-year-old man
Activity-related sarcopenia (iatrogenic sarcopenia due to unnecessary noneating and bed rest)	Bedridden after falling Barthel index score 0 points	Bedridden after lung cancer surgery, acute myocardial infarction and pneumonia Barthel Index score 27 points
Nutrition-related sarcopenia (iatrogenic sarcopenia due to inappropriate nutrition management)	Height 153 cm Body weight 22.0 kg Body mass index 9.4 kg/m^2 Weight loss 27.3%/5 months MNA-SF 2 points Enteral nutrition 1200 kcal/day	Height 174.5 cm Body weight 46.6 kg Body mass index 15.3 kg/m^2 Weight loss 18%/80 days MNA-SF 2 points Enteral nutrition 986 kcal/day
Disease-related sarcopenia	Aspiration pneumonia, four times No history of stroke, dementia, or Parkinson's disease	Lung cancer surgery, acute myocardial infarction, pneumonia, chronic obstructive pulmonary disease

Abbreviations: MNA-SF: Mini-Nutritional Assessment Short Form.

frail older patients with a hip fracture and hospital-associated deconditioning (Wakabayashi 2014; Wakabayashi and Sakuma 2014). Indeed, 34% of patients were observed to have dysphagia after hip fracture surgery (Love, Cornwell, and Whitehouse 2013). Table 17.1 shows the causes of sarcopenia in two case reports of sarcopenic dysphagia treated by rehabilitation nutrition (Maeda and Akagi 2016; Wakabayashi and Uwano 2016). Both cases had all causes of sarcopenia and subsequently developed sarcopenic dysphagia. Severe sarcopenic dysphagia does not seem to occur in patients with only one cause of sarcopenia, with the exception of disease-related sarcopenia.

17.4 DIAGNOSIS OF SARCOPENIC DYSPHAGIA

17.4.1 CONSENSUS DIAGNOSTIC CRITERIA FOR SARCOPENIC DYSPHAGIA

A symposium on "sarcopenia and dysphagia rehabilitation" was held during the 19th Annual Meeting of the Japanese Society of Dysphagia Rehabilitation in 2013 (Wakabayashi 2014). Consensus diagnostic criteria for sarcopenic dysphagia were proposed at the symposium. Sarcopenic dysphagia is diagnosed only in patients with dysphagia and loss of mass and strength of both generalized and swallowing muscles. If there are no other causes of dysphagia except sarcopenia in the patient's clinical history, the patient is diagnosed with probable sarcopenic dysphagia. In cases with stroke, brain injury, neuromuscular diseases, head and neck cancer, and connective tissue diseases, the diagnosis is possible sarcopenic dysphagia if the main cause of dysphagia is considered to be sarcopenia. In older stroke patients undergoing enteral nutrition, the risk of severe malnutrition risk independently predicts the achievement of full oral intake (Nishioka et al. 2017). Swallowing muscle mass decreases separately from neurological deficits, and it is known sarcopenic dysphagia may interrupt recovery of swallowing function in stroke patients at risk of developing severe malnutrition (Nishioka et al. 2017). Recently, the position paper of sarcopenia and dysphagia was published by the Japanese Society of Dysphagia Rehabilitation, the Japanese Association of Rehabilitation Nutrition, the Japanese Association on Sarcopenia and Frailty, and the Society of Swallowing and Dysphagia of Japan to consolidate the currently available evidence on the topics of sarcopenia and dysphagia (Fujishima et al. 2019).

17.4.2 Diagnostic Algorithm for Sarcopenic Dysphagia

After the symposium on "sarcopenia and dysphagia rehabilitation," we established a working group on sarcopenic dysphagia (WGSD) comprised multidisciplinary experts and researchers of dysphagia and sarcopenia including doctors specializing in rehabilitation, neurology, and geriatrics, and dentists, speech therapists, registered dietitians, and physical therapists. Several meetings have been held since September 2013 at which the WGSD developed a diagnostic algorithm for sarcopenic dysphagia (Figure 17.2) that referenced the consensus diagnostic criteria for sarcopenic dysphagia (Wakabayashi 2014), the consensus report of the Asian Working Group for Sarcopenia (AWGS) (Chen et al. 2014), and the criteria of the EWGSOP (Cruz-Jentoft et al. 2010). We divided the diagnosis of sarcopenic dysphagia into three categories: probable sarcopenic dysphagia, possible sarcopenic dysphagia, and no sarcopenic dysphagia. We did not include a definite diagnosis of sarcopenic dysphagia because measurement and setting a cutoff value of swallowing muscle mass has not yet been established. We have verified the high reliability of the diagnostic algorithm for sarcopenic dysphagia, with the Kappa coefficient of intraclass and interclass being 0.98 and 0.87, respectively.

People aged 65 years and older who are able to follow commands are eligible for the algorithm because in the EWGSOP criteria assessment of sarcopenia was limited to people aged 65 years and older (Cruz-Jentoft et al. 2010) and requires them to follow commands in order to implement the algorithm. Accordingly, people without generalized sarcopenia and/or normal swallowing function were judged as not having sarcopenic dysphagia. In addition, people with an obvious causative disease of dysphagia were considered not to have sarcopenic dysphagia because this condition should be diagnosed by excluding other obvious causative diseases of dysphagia. Finally, swallowing muscle strength is assessed by tongue pressure, because this measurement is feasible in clinical practice and decreased tongue pressure is known to be associated with sarcopenic dysphagia (Maeda and Akagi 2015). We set the cutoff value for low swallowing muscle strength at <20 kPa tongue pressure based on the finding that mean tongue pressure in older people with and without

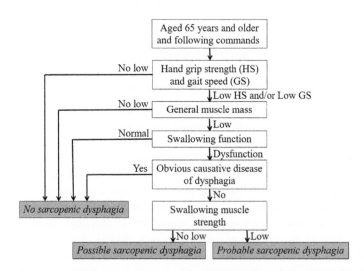

FIGURE 17.2 Diagnostic algorithm for sarcopenic dysphagia. People aged 65 years and older who are able to follow commands are eligible for the algorithm. People without generalized sarcopenia and/or normal swallowing function were judged as not having sarcopenic dysphagia. In addition, people with an obvious causative disease of dysphagia were considered not to have sarcopenic dysphagia. Finally, swallowing muscle strength is assessed by tongue pressure. People with low swallowing muscle strength are judged as probable sarcopenic dysphagia, while people with normal swallowing muscle strength are judged as possible sarcopenic dysphagia.

dysphagia was 14.7 and 25.3 kPa, respectively (Maeda and Akagi 2015). People with low swallowing muscle strength are judged as probable sarcopenic dysphagia, while people with normal swallowing muscle strength are judged as possible sarcopenic dysphagia. In cases lacking tongue pressure measurements, patients with generalized sarcopenia, swallowing dysfunction, and no obvious condition as the main cause of dysphagia are diagnosed with possible sarcopenic dysphagia. This diagnostic algorithm is also introduced in the position paper of sarcopenia and dysphagia (Fujishima et al. 2019).

17.5 TREATMENT FOR SARCOPENIC DYSPHAGIA

Treatment for sarcopenic dysphagia includes dysphagia rehabilitation and treatment of sarcopenia (Wakabayashi 2014). The core components of dysphagia rehabilitation are oral health care, rehabilitative techniques, and food modification (Wakabayashi 2014). Sarcopenia should be treated by rehabilitation nutrition, because people with sarcopenic dysphagia experience activity-, nutrition-, and disease-related sarcopenia of the generalized and swallowing muscles. Rehabilitation alone cannot treat nutrition-related sarcopenia, while nutrition care management alone cannot treat activity-related sarcopenia. Therefore, rehabilitation nutrition is important for treating all causes of sarcopenia.

Rehabilitation nutrition is defined as that (1) evaluates holistically by the International Classification of Functioning, Disability and Health, and the presence and causes of nutritional disorders, sarcopenia and excess or deficiency of nutrient intake, (2) conducts rehabilitation nutrition diagnosis and rehabilitation nutrition goal setting, and (3) elicits the highest body functions, activities, participations, and QOL by improving nutritional status, sarcopenia, and frailty using "nutrition care management in consideration of rehabilitation" and "rehabilitation in consideration of nutrition" in people with a disability and frail older people (Nagano, Nishioka, and Wakabayashi 2019). Rehabilitation nutrition is provided by a rehabilitation nutrition care process (Figure 17.3) that

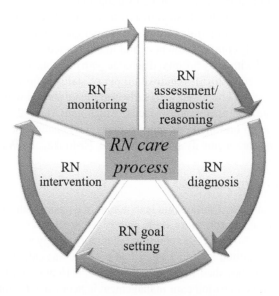

FIGURE 17.3 Rehabilitation nutrition (RN) care process. Rehabilitation nutrition care process includes rehabilitation nutrition assessment and diagnostic reasoning, rehabilitation nutrition diagnosis, rehabilitation nutrition goal setting, rehabilitation nutrition intervention, and rehabilitation nutrition monitoring. Rehabilitation nutrition diagnosis is to identify and describe the presence and causes of nutritional disorders, sarcopenia and excess or deficiency of nutrient intake.

includes (1) rehabilitation nutrition assessment and diagnostic reasoning, (2) rehabilitation nutrition diagnosis, (3) rehabilitation nutrition goal setting, (4) rehabilitation nutrition intervention, and (5) rehabilitation nutrition monitoring (Nagano, Nishioka, and Wakabayashi 2019).

17.5.1 AGE- AND ACTIVITY-RELATED SARCOPENIA

Resistance training of the swallowing muscles such as head lifting exercises and lingual exercises is the most important method to improve age- and activity-related sarcopenia of the swallowing muscles. A systematic review showed head lifting exercises eliminated the symptoms of dysphagia (Antunes and Lunet 2012). The original method of head lifting exercises was that the patients carried out sustained head raisings for 1 min in the supine position, interrupted by a 1-min rest period. These sustained head raisings were performed for 30 consecutive repetitions, three times a day for 6 weeks (Shaker et al. 2002). Lingual exercises have also been shown to improve swallowing pressures and lingual volume in older adults (Robbins et al. 2005). In lingual exercises, people compress an air-filled bulb between the tongue and hard palate 30 times, three times a day, and exercise using a goal of 60% of the baseline maximum pressure for the first week and 80% of the maximum pressure for the remaining 7 weeks (Robbins et al. 2005). Protein and amino acid supplementation immediately after resistance training is also recommended.

Neuromuscular electrical stimulation is another method for improving age- and activity-related sarcopenia of the swallowing muscles. A meta-analysis showed transcutaneous neuromuscular electrical stimulation improved swallowing function in nonstroke patients with dysphagia (Tan et al. 2013). Similarly, a recent systemic review and meta-analysis (Chen et al. 2016) showed that swallowing treatment with neuromuscular electrical stimulation was more effective than no such treatment in stroke patients with dysphagia. Neuromuscular electrical stimulation combined with protein and amino acid supplementation appears to be a good adaptation when people cannot perform resistance training of the swallowing muscles by themselves.

Promoting early oral intake and avoiding prolonged and unnecessary bed rest and no oral intake are good strategies for mitigating activity-related sarcopenia. Early rehabilitation initiated within 3 days of admission administered by physical therapists has also been shown to be associated with a reduction in 30-day in-hospital mortality rates in older patients with severe aspiration pneumonia (Momosaki et al. 2015). Early commencement of oral intake started within 2 days of admission was also found to be associated with early hospital discharge with oral intake in hospitalized older adults with pneumonia (Koyama et al. 2015). Tentative nil per os status at hospital admission resulted in a decline in swallowing ability compared with early oral intake started within 2 days of admission in older adults with aspiration pneumonia (Maeda, Koga, and Akagi 2016). Therefore, avoiding unnecessary bed rest and no oral intake in older patients with pneumonia is useful for preventing iatrogenic sarcopenia and sarcopenic dysphagia (Nagano, Nishioka, and Wakabayashi 2019).

17.5.2 NUTRITION-RELATED SARCOPENIA

The treatment of nutrition-related sarcopenia is to maintain a positive energy and protein balance. To increase body weight and muscle mass, daily energy requirements in nutrition-related sarcopenia should be calculated as daily energy expenditure plus daily energy accumulation (200–750 kcal/day). The excess energy required to gain 1 kg of body weight was calculated to be between 8856 and 22,620 kcal/kg in malnourished nursing home patients (Hebuterne, Bermon, and Schneider 2001). Therefore, aggressive nutrition care management is necessary to improve nutrition-related sarcopenia. Indeed, 47.2–83.2 kcal/kg of present body weight/day was prescribed in two case reports (Table 17.2) (Maeda and Akagi 2016; Wakabayashi and Uwano 2016). This aggressive nutrition care management resulted in an 8.9–11 kg increase in body weight and improved swallowing function.

TABLE 17.2

Treatment of Sarcopenic Dysphagia in Two Case Reports

Causes of sarcopenia	Maeda and Akagi (2016)	Wakabayashi and Uwano (2016)
Age-related sarcopenia	Dysphagia rehabilitation	Dysphagia rehabilitation
	Lingual exercises	Stretch and resistance training of the
	Head lift exercises	swallowing muscles
Activity-related sarcopenia	Sitting up in bed	Resistance training
	Ambulating from the bed during the	Movement exercises
	daytime	Ambulation exercises
Nutrition-related sarcopenia	1830 kcal/day	2200 kcal/day
	83.2 kcal/kg present body weight/day	47.2 kcal/kg present body weight/day
	35.3 kcal/kg ideal body weight/day	32.8 kcal/kg ideal body weight/day
Disease-related sarcopenia	Dysphagia rehabilitation and nutrition	Dysphagia rehabilitation and nutrition
	improvement	improvement
	Oral care to prevent pneumonia	

Maintenance and improvement of nutritional status are important to improve dysphagia. Patients at an acute care hospital with a nutrition intake approximately <22 kcal/kg/day during a follow-up period showed significantly poorer recovery from dysphagia and poor outcome compared with those with an intake approximately >22 kcal/kg/day (Iwamoto et al. 2014). Delayed initiation of total oral intake has been shown to be associated with underweight in older adults with aspiration pneumonia (Momosaki et al. 2016). Furthermore, low body mass index was reported to be an independent predictor of dysphagia in hospitalized older patients who had restricted oral intake without dysphagia (Maeda, Takaki, and Akagi 2016). Therefore, identification of iatrogenic sarcopenia in older hospitalized patients due to nutritional management issues should be addressed as soon as possible to provide earlier appropriate nutritional support.

17.5.3 DISEASE-RELATED SARCOPENIA

Treatment of disease-related sarcopenia differs dependent on the presence of invasion or cachexia. We discuss the treatment of invasion-related sarcopenia resulting from aspiration pneumonia, with appropriate treatment being different in the catabolic phase compared with the anabolic phase. In the catabolic phase, nutrition care management should be modest (15–30 kcal/kg/day), and daily energy expenditure beyond energy intake should be curtailed because of increasing endogenous energy production. Light load activity and exercise should be performed to prevent activity-related sarcopenia. Early oral intake is also important in patients with aspiration pneumonia because early oral intake promotes shorter treatment duration (Maeda, Koga, and Akagi 2016), resulting in less invasion-related sarcopenia. In the anabolic phase, nutrition care management should be tailored to the patient's daily energy expenditure plus daily energy accumulation (200–750 kcal/day) in order to improve muscle mass and strength. Resistance training to improve muscle mass should be performed in patients under aggressive nutrition care management.

A comprehensive approach can be useful for preventing and treating cachexia (Farkas et al. 2013). Treatment of cachexia-related sarcopenia may be effective for improving sarcopenic dysphagia due to the close relationship between generalized sarcopenia and dysphagia in patients with sarcopenic dysphagia. Omega-3 polyunsaturated fatty acids have been shown to have a positive effect on clinical outcomes and survival in patients with pancreatic cancer (Ma et al. 2015). However, in a systematic review of the European Palliative Care Research Collaborative

Cachexia Guidelines project, there was insufficient evidence to support a net benefit of omega-3 polyunsaturated fatty acids in advanced cancer cachexia (Ries et al. 2012). However, exercise may be effective for cachexia because it represents a function-preserving, anti-inflammatory, and metabolism-modulating strategy (Lira et al. 2015). Administration of the selective androgen receptor modulator, enobosarm, and the ghrelin receptor agonist, anamorelin, for cachexia can improve muscle mass (Fearon et al. 2015). Combination of aggressive nutritional care management and rehabilitation may therefore be useful for treating cachexia-related sarcopenia (Someya et al. 2016).

17.6 PRESBYPHAGIA

17.6.1 DEFINITION OF PRESBYPHAGIA

Presbyphagia refers to age-related changes in the swallowing mechanism in older people (Jahnke 1991; Humbert and Robbins 2008) and is characterized by frailty of swallowing rather than dysphagia (Wakabayashi 2014). People who need modification of food texture and liquid thickness to prevent swallowing-related complications are diagnosed as having dysphagia. In contrast, older people who do not require food modification, but sometimes choke and cough on water and food are diagnosed with presbyphagia. Presbyphagia in older people is classified as either primary presbyphagia due to aging only or secondary presbyphagia due to disease (Jahnke 1991). In cases of dysphagia due to diseases, the term presbyphagia should not be used because they are different conditions.

Approximately 40% of healthy older people have presbyphagia (Lever et al. 2015), while 28%–37% have aspiration during swallowing evaluation (Butler et al. 2009, 2010, 2011). The phases of oropharyngeal swallowing and coordination of swallowing with respiration gradually change with aging (Wang et al. 2015). Compared with young mice, old mice had longer pharyngeal and esophageal transit times, swallow larger boluses, with a higher percentage of ineffective primary esophageal swallows (Lever et al. 2015). Early detection of presbyphagia is important because older people with presbyphagia may easily develop sarcopenic dysphagia during hospitalization because of diminished functional reserve.

17.6.2 MECHANISM OF PRESBYPHAGIA

Presbyphagia may present in several ways including a lack of muscle strength complicating bolus propulsion; diminished lingual pressure obstructing bolus driving; halting of the bolus whilst swallowing leading to more difficult cleansing of residues; a decline in taste and smell that makes it more difficult to initiate swallowing; difficulty in controlling the bolus from the anticipatory phase; entry of the bolus into the lower airway; and finally, lack of teeth and wearing or not wearing complete dentures that influence chewing (Nogueira and Reis 2013). Age-related sarcopenia of swallowing muscles is one of the causes of presbyphagia.

Age-related changes in swallowing muscles have been studied mainly in rats and mice. An age-related decrease in expression of myosin heavy chain mRNA and protein was found in rat genioglossus but not masseter and geniohyoid muscles (Kaneko et al. 2014). In the klotho mouse, the deficiency in amino acids caused by the active movement of the masseter muscle and tongue was shown to stimulate the autophagic-lysosomal pathway via downregulation of the mTOR signaling pathway (Iida et al. 2011). The effects of aging on tongue forces in rats produced mixed results (Nagai et al. 2008; Becker, Russell, and Connor 2015). Tongue muscle shifted to more slowly contracting muscle fibers in aged rats (Schaser et al. 2011). Although the number of primary dendrites of hypoglossal motoneurons decreased significantly with age in rats, no age-associated changes were found in the number or size of hypoglossal motoneurons (Schwarz et al. 2009). In older human cadavers, the number of macrophages per striated muscle fiber in the larynx and

TABLE 17.3

10-Item Eating Assessment Tool: EAT-10

To what extent are the following scenarios problematic for you?

Each item is scored from 0 to 4 according to the severity of the problem.

0 = No problem, 4 = Severe problem

1. My swallowing problem has caused me to lose weight.
2. My swallowing problem interferes with my ability to go out for meals.
3. Swallowing liquids takes extra effort.
4. Swallowing solids takes extra effort.
5. Swallowing pills takes extra effort.
6. Swallowing is painful.
7. The pleasure of eating is affected by my swallowing.
8. When I swallow food, it sticks in my throat.
9. I cough when I eat.
10. Swallowing is stressful.

If the EAT-10 score is 3 or higher, you may have problems swallowing efficiently and safely.

pharynx was—five to six times greater than those in the tongue, shoulder, and anus (Rhee et al. 2016). Thinning and death of striated muscle fibers may therefore occur frequently in the larynx and pharynx.

17.6.3 DIAGNOSIS OF PRESBYPHAGIA

To screen for presbyphagia and dysphagia, we recommend using the 10-item Eating Assessment Tool (EAT-10, Table 17.3) (Belafsky et al. 2008). The EAT-10 is a 10-item questionnaire, with each item scored from 0 to 4. An EAT-10 score ≥ 3 is abnormal and indicates the presence of swallowing difficulties. The reliability and validity of the EAT-10 have been confirmed and translated into Japanese (Belafsky et al. 2008; Wakabayashi and Kayashita 2014). We interpret the results of the EAT-10 as follows. People who cannot respond to the EAT-10 are likely to have dysphagia (Wakabayashi and Kayashita 2014). People with an EAT-10 score 0 have normal swallowing function or no awareness of presbyphagia and dysphagia. People with an EAT-10 score 1 or 2 have presbyphagia, while those with an EAT-10 score ≥ 3 are likely to have dysphagia and need further investigation of swallowing.

17.6.4 TREATMENT FOR PRESBYPHAGIA

Treatment for presbyphagia differs depending on its cause. Resistance training of the swallowing muscles combined with protein and amino acid supplementation is effective in cases of age-related sarcopenia of the swallowing muscles. Rehabilitative exercises such as head lifting and lingual exercises for 6 to 8 weeks have beneficial effects on presbyphagia (Ney et al. 2009). Neuromuscular electrical stimulation of the swallowing muscles (Tan et al. 2013; Chen et al. 2016) is another treatment for presbyphagia. Encouraging denture wearing to achieve occlusal support is effective for improving swallowing function in patients with a lack of teeth or those not wearing complete dentures. Tooth loss was reported to affect swallowing and masticatory ability in aged individuals living independently in the community (Okamoto et al. 2012, 2015). There is also evidence that skeletal muscle mass was related with occlusal support of the natural teeth in frail older men (Sagawa et al. 2017). A dental consultation is therefore important for preventing and treating presbyphagia.

17.7 CONCLUSION

The prevalence of sarcopenic dysphagia is likely to increase in developed countries in which older people now represent a larger proportion of the population. The major causes of sarcopenic dysphagia are secondary sarcopenia such as activity-, nutrition-, and disease-related sarcopenia. Sarcopenic dysphagia also increases the risk of aspiration pneumonia, choking, malnutrition, and dehydration. Although sarcopenic dysphagia is a very important public health issue, current pharmacologic therapy for the disorder is very limited. Testosterone can be prescribed in older men with sarcopenia for a few months if no contraindications are encountered during monitoring of side effects (Wakabayashi and Sakuma 2014b). A combination of rehabilitation nutrition and pharmacologic therapy will be the future treatment for sarcopenic dysphagia. Further research is required on sarcopenic dysphagia, especially epidemiological studies, and also basic and intervention studies including investigations on rehabilitation nutrition and pharmacologic therapy.

ACKNOWLEDGMENT

This work was supported by a research grant-in-aid for Scientific Research C (no. 16K01460) from the Ministry of Education, Science, Culture, Sports, Science, and Technology of Japan.

REFERENCES

Anker, S. D., J. E. Morley, and S. von Haehling. 2016. Welcome to the ICD-10 code for sarcopenia. *J Cachexia Sarcopenia Muscle* 7 (5):512–514.

Antunes, E. B., and N. Lunet. 2012. Effects of the head lift exercise on the swallow function: A systematic review. *Gerodontology* 29 (4):247–257.

Argiles, J. M., S. Busquets, B. Stemmler, and F. J. Lopez-Soriano. 2014. Cancer cachexia: Understanding the molecular basis. *Nat Rev Cancer* 14 (11):754–762.

Baijens, L. W., P. Clave, P. Cras, et al. 2016. European society for swallowing disorders—European union geriatric medicine society white paper: Oropharyngeal dysphagia as a geriatric syndrome. *Clin Interv Aging* 11:1403–1428.

Becker, B. J., J. A. Russell, and N. P. Connor. 2015. Effects of aging on evoked retrusive tongue actions. *Arch Oral Biol* 60 (6):966–971.

Belafsky, P. C., D. A. Mouadeb, C. J. Rees, et al. 2008. Validity and reliability of the Eating Assessment Tool (EAT-10). *Ann Otol Rhinol Laryngol* 117 (12):919–924.

Butler, S. G., A. Stuart, X. Leng, C. Rees, J. Williamson, and S. B. Kritchevsky. 2010. Factors influencing aspiration during swallowing in healthy older adults. *Laryngoscope* 120 (11):2147–2152.

Butler, S. G., A. Stuart, X. Leng, et al. 2011. The relationship of aspiration status with tongue and handgrip strength in healthy older adults. *J Gerontol A Biol Sci Med Sci* 66 (4):452–458.

Butler, S. G., A. Stuart, L. Markley, and C. Rees. 2009. Penetration and aspiration in healthy older adults as assessed during endoscopic evaluation of swallowing. *Ann Otol Rhinol Laryngol* 117 (3):190–198.

Cao, L., and J. E. Morley. 2016. Sarcopenia is recognized as an independent condition by an International Classification of Disease, tenth revision, Clinical Modification (ICD-10-CM) Code. *J Am Med Dir Assoc* 17 (8):675–677.

Carrero, J. J., P. Stenvinkel, L. Cuppari, et al. 2013. Etiology of the protein-energy wasting syndrome in chronic kidney disease: A consensus statement from the International Society of Renal Nutrition and Metabolism (ISRNM). *J Ren Nutr* 23 (2):77–90.

Carrion, S., M. Roca, A. Costa, et al. 2017. Nutritional status of older patients with oropharyngeal dysphagia in a chronic versus an acute clinical situation. *Clin Nutr* 36 (4):1110–1116.

Chen, L. K., L. K. Liu, J. Woo, et al. 2014. Sarcopenia in Asia: Consensus report of the Asian working group for sarcopenia. *J Am Med Dir Assoc* 15 (2):95–101.

Chen, Y. W., K. H. Chang, H. C. Chen, W. M. Liang, Y. H. Wang, and Y. N. Lin. 2016. The effects of surface neuromuscular electrical stimulation on post-stroke dysphagia: A systemic review and meta-analysis. *Clin Rehabil* 30 (1):24–35.

Cichero, J. A., P. Lam, C. M. Steele, et al. 2017. Development of International Terminology and Definitions for Texture-Modified Foods and Thickened Fluids Used in Dysphagia Management: The IDDSI Framework. *Dysphagia* 32 (2):293–314.

Clave, P. and R. Shaker. 2015. Dysphagia: Current reality and scope of the problem. *Nat Rev Gastroenterol Hepatol* 12 (5):259–270.

Cruz-Jentoft, A. J., J. P. Baeyens, J. M. Bauer, et al. 2010. Sarcopenia: European consensus on definition and diagnosis: Report of the European Working Group on Sarcopenia in Older People. *Age Ageing* 39 (4):412–423.

Evans, W. J., J. E. Morley, J. Argiles, et al. 2008. Cachexia: A new definition. *Clin Nutr* 27 (6):793–799.

Farkas, J., S. von Haehling, K. Kalantar-Zadeh, J. E. Morley, S. D. Anker, and M. Lainscak. 2013. Cachexia as a major public health problem: frequent, costly, and deadly. *J Cachexia Sarcopenia Muscle* 4 (3):173–178.

Fearon, K., J. M. Argiles, V. E. Baracos, et al. 2015. Request for regulatory guidance for cancer cachexia intervention trials. *J Cachexia Sarcopenia Muscle* 6 (4):272–274.

Feng, X., T. Todd, C. R. Lintzenich, et al. 2013. Aging-related geniohyoid muscle atrophy is related to aspiration status in healthy older adults. *J Gerontol A Biol Sci Med Sci* 68 (7):853–860.

Fujishima, I., M. Fujiu-Kurachi, H. Arai, et al. 2019. Sarcopenia and dysphagia: Position paper by four professional organizations. *Geriatr Gerontol Int* 19 (2):91–97.

Hashida, N., H. Shamoto, K. Maeda, H. Wakabayashi, M. Suzuki, and T. Fujii. 2017. Rehabilitation and nutritional support for sarcopenic dysphagia and tongue atrophy after glossectomy: A case report. *Nutrition* 35:128–131.

Hebuterne, X., S. Bermon, and S. M. Schneider. 2001. Ageing and muscle: The effects of malnutrition, renutrition, and physical exercise. *Curr Opin Clin Nutr Metab Care* 4 (4):295–300.

Ho, A. W., M. M. Lee, E. W. Chan, et al. 2016. Prevalence of pre-sarcopenia and sarcopenia in Hong Kong Chinese geriatric patients with hip fracture and its correlation with different factors. *Hong Kong Med J* 22 (1):23–29.

Hudson, H. M., C. R. Daubert, and R. H. Mills. 2000. The interdependency of protein-energy malnutrition, aging, and dysphagia. *Dysphagia* 15 (1):31–38.

Humbert, I. A., and J. Robbins. 2008. Dysphagia in the elderly. *Phys Med Rehabil Clin N Am* 19 (4):853–866, ix–x.

Iida, R. H., S. Kanko, T. Suga, M. Morito, and A. Yamane. 2011. Autophagic-lysosomal pathway functions in the masseter and tongue muscles in the klotho mouse, a mouse model for aging. *Mol Cell Biochem* 348 (1–2):89–98.

Iida, T., H. Tohara, S. Wada, A. Nakane, R. Sanpei, and K. Ueda. 2013. Aging decreases the strength of suprahyoid muscles involved in swallowing movements. *Tohoku J Exp Med* 231 (3):223–228.

Iwamoto, M., N. Higashibeppu, Y. Arioka, and Y. Nakaya. 2014. Swallowing rehabilitation with nutrition therapy improves clinical outcome in patients with dysphagia at an acute care hospital. *J Med Invest* 61 (3–4):353–360.

Jahnke, V. 1991. Dysphagia in the elderly. *Hno* 39 (11):442–444.

Kaneko, S., R. H. Iida, T. Suga, M. Morito, and A. Yamane. 2014. Age-related changes in rat genioglossus, geniohyoid and masseter muscles. *Gerodontology* 31 (1):56–62.

Komatsu, R., T. Okazaki, S. Ebihara, et al. 2018. Aspiration pneumonia induces muscle atrophy in the respiratory, skeletal, and swallowing systems. *J Cachexia Sarcopenia Muscle* 9 (4):643–653.

Konaka, K., J. Kondo, N. Hirota, et al. 2010. Relationship between tongue pressure and dysphagia in stroke patients. *Eur Neurol* 64 (2):101–107.

Koyama, T., K. Maeda, H. Anzai, Y. Koganei, H. Shamoto, and H. Wakabayashi. 2015. Early commencement of oral intake and physical function are associated with early hospital discharge with oral intake in hospitalized elderly individuals with pneumonia. *J Am Geriatr Soc* 63 (10):2183–2185.

Kuroda, Y., and R. Kuroda. 2012. Relationship between thinness and swallowing function in Japanese older adults: Implications for sarcopenic dysphagia. *J Am Geriatr Soc* 60 (9):1785–1786.

Kuroda, Y. 2014. Relationship between swallowing function, and functional and nutritional status in hospitalized elderly individuals. *Int. J. Speech Lang. Pathol. Audiol.* 2 (1):20–26.

Letilovic, T., and R. Vrhovac. 2013. Influence of additional criteria from a definition of cachexia on its prevalence—Good or bad thing? *Eur J Clin Nutr* 67 (8):797–801.

Lever, T. E., R. T. Brooks, L. A. Thombs, et al. 2015. Videofluoroscopic validation of a translational murine model of presbyphagia. *Dysphagia* 30 (3):328–342.

Lira, F. S., M. Antunes Bde, M. Seelaender, and J. C. Rosa Neto. 2015. The therapeutic potential of exercise to treat cachexia. *Curr Opin Support Palliat Care* 9 (4):317–324.

Love, A. L., P. L. Cornwell, and S. L. Whitehouse. 2013. Oropharyngeal dysphagia in an elderly post-operative hip fracture population: A prospective cohort study. *Age Ageing* 42 (6):782–578.

Ma, Y. J., J. Yu, J. Xiao, and B. W. Cao. 2015. The consumption of omega-3 polyunsaturated fatty acids improves clinical outcomes and prognosis in pancreatic cancer patients: a systematic evaluation. *Nutr Cancer* 67 (1):112–118.

Machida, N., H. Tohara, K. Hara, et al. 2017. Effects of aging and sarcopenia on tongue pressure and jaw-opening force. *Geriatr Gerontol Int* 17 (2):295–301.

Maeda, K., and J. Akagi. 2015. Decreased tongue pressure is associated with sarcopenia and sarcopenic dysphagia in the elderly. *Dysphagia* 30 (1):80–87.

Maeda, K., and J. Akagi. 2016. Treatment of sarcopenic dysphagia with rehabilitation and nutritional support: A comprehensive approach. *J Acad Nutr Diet* 116 (4):573–577.

Maeda, K., T. Koga, and J. Akagi. 2016. Tentative nil per os leads to poor outcomes in older adults with aspiration pneumonia. *Clin Nutr* 35 (5):1147–1152.

Maeda, K., M. Takaki, and J. Akagi. 2016. Decreased skeletal muscle mass and risk factors of sarcopenic dysphagia: A prospective observational cohort study. *J Gerontol A Biol Sci Med Sci* 72 (9):1290–1294.

Molfenter, S. M., M. R. Amin, R. C. Branski, et al. 2015. Age-related changes in pharyngeal lumen size: A retrospective MRI analysis. *Dysphagia* 30 (3):321–327.

Momosaki, R., H. Yasunaga, H. Matsui, H. Horiguchi, K. Fushimi, and M. Abo. 2015. Effect of early rehabilitation by physical therapists on in-hospital mortality after aspiration pneumonia in the elderly. *Arch Phys Med Rehabil* 96 (2):205–209.

Momosaki, R., H. Yasunaga, H. Matsui, H. Horiguchi, K. Fushimi, and M. Abo. 2016. Predictive factors for oral intake after aspiration pneumonia in older adults. *Geriatr Gerontol Int* 16 (5):556–560.

Murakami, K., H. Hirano, Y. Watanabe, et al. 2015. Relationship between swallowing function and the skeletal muscle mass of older adults requiring long-term care. *Geriatr Gerontol Int* 15 (10):1185–1192.

Murakami, M., H. Hirano, Y. Watanabe, K. Sakai, H. Kim, and A. Katakura. 2015. Relationship between chewing ability and sarcopenia in Japanese community-dwelling older adults. *Geriatr Gerontol Int* 15 (8):1007–1012.

Nagai, H., J. A. Russell, M. A. Jackson, and N. P. Connor. 2008. Effect of aging on tongue protrusion forces in rats. *Dysphagia* 23 (2):116–121.

Nagano, A., S. Nishioka S, and H. Wakabayashi. 2019. Rehabilitation nutrition for iatrogenic sarcopenia and sarcopenic dysphagia. *J Nutr Health Aging* 23 (3):256–265.

Ney, D. M., J. M. Weiss, A. J. Kind, and J. Robbins. 2009. Senescent swallowing: Impact, strategies, and interventions. *Nutr Clin Pract* 24 (3):395–413.

Nishioka, S., T. Okamoto, M. Takayama, et al. 2017. Malnutrition risk predicts recovery of full oral intake among older adult stroke patients undergoing enteral nutrition: Secondary analysis of a multicentre survey (the APPLE study). *Clin Nutr* 36 (4):1089–1096.

Nogueira, D. and E. Reis. 2013. Swallowing disorders in nursing home residents: How can the problem be explained? *Clin Interv Aging* 8:221–227.

Okamoto, N., M. Morikawa, M. Yanagi, et al. 2015. Association of tooth loss with development of swallowing problems in community-dwelling independent elderly population: The Fujiwara-kyo study. *J Gerontol A Biol Sci Med Sci* 70 (12):1548–1554.

Okamoto, N., K. Tomioka, K. Saeki, et al. 2012. Relationship between swallowing problems and tooth loss in community-dwelling independent elderly adults: The Fujiwara-kyo study. *J Am Geriatr Soc* 60 (5):849–853.

Rhee, S., M. Yamamoto, K. Kitamura, et al. 2016. Macrophage density in pharyngeal and laryngeal muscles greatly exceeds that in other striated muscles: An immunohistochemical study using elderly human cadavers. *Anat Cell Biol* 49 (3):177–183.

Ries, A., P. Trottenberg, F. Elsner, et al. 2012. A systematic review on the role of fish oil for the treatment of cachexia in advanced cancer: An EPCRC cachexia guidelines project. *Palliat Med* 26 (4):294–304.

Robbins, J., R. E. Gangnon, S. M. Theis, S. A. Kays, A. L. Hewitt, and J. A. Hind. 2005. The effects of lingual exercise on swallowing in older adults. *J Am Geriatr Soc* 53 (9):1483–1489.

Robbins, J., N. S. Humpal, K. Banaszynski, J. Hind, and N. Rogus-Pulia. 2016. Age-related differences in pressures generated during isometric presses and swallows by healthy adults. *Dysphagia* 31 (1):90–96.

Sánchez-Rodríguez, D., A. Calle, A. Contra, et al. 2016. Sarcopenia in post-acute care and rehabilitation of older adults: A review. *Eur Geriatr Med* 7 (3):224–231.

Sagawa, K., T. Kikutani, F. Tamura, and M. Yoshida. 2017. Factors related to skeletal muscle mass in the frail elderly. *Odontology* 105 (1):91–95.

Sakai, K., E. Nakayama, H. Tohara, et al. 2017. Tongue strength is associated with grip strength and nutritional status in older adult inpatients of a rehabilitation hospital. *Dysphagia* 32 (2):241–249.

Schaser, A. J., H. Wang, L. M. Volz, and N. P. Connor. 2011. Biochemistry of the anterior, medial, and posterior genioglossus in the aged rat. *Dysphagia* 26 (3):256–263.

Schwarz, E. C., J. M. Thompson, N. P. Connor, and M. Behan. 2009. The effects of aging on hypoglossal motoneurons in rats. *Dysphagia* 24 (1):40–48.

Shaker, R., C. Easterling, M. Kern, et al. 2002. Rehabilitation of swallowing by exercise in tube-fed patients with pharyngeal dysphagia secondary to abnormal UES opening. *Gastroenterology* 122 (5):1314–1321.

Shimizu, S., K. Hanayama, R. Nakato, T. Sugiyama, and A. Tsubahara. 2016. Ultrasonographic evaluation of geniohyoid muscle mass in perioperative patients. *Kawasaki Med J* 42 (2):47–56.

Shiozu, H., M. Higashijima, and T. Koga. 2015. Association of sarcopenia with swallowing problems, related to nutrition and activities of daily living of elderly individuals. *J Phys Ther Sci* 27 (2):393–396.

Someya, R., H. Wakabayashi, K. Hayashi, E. Akiyama, and K. Kimura. 2016. Rehabilitation nutrition for acute heart failure on inotropes with malnutrition, sarcopenia, and cachexia: A case report. *J Acad Nutr Diet* 116 (5):765–768.

Takagi, D., H. Hirano, Y. Watanabe, et al. 2016. Relationship between skeletal muscle mass and swallowing function in patients with Alzheimer's disease. *Geriatr Gerontol Int.* 17:402–409.

Tamura, F., T. Kikutani, T. Tohara, M. Yoshida, and K. Yaegaki. 2012. Tongue thickness relates to nutritional status in the elderly. *Dysphagia* 27 (4):556–561.

Tan, C., Y. Liu, W. Li, J. Liu, and L. Chen. 2013. Transcutaneous neuromuscular electrical stimulation can improve swallowing function in patients with dysphagia caused by non-stroke diseases: A meta-analysis. *J Oral Rehabil* 40 (6):472–480.

Vanderwegen, J., C. Guns, G. Van Nuffelen, R. Elen, and M. De Bodt. 2013. The influence of age, sex, bulb position, visual feedback, and the order of testing on maximum anterior and posterior tongue strength and endurance in healthy Belgian adults. *Dysphagia* 28 (2):159–166.

Veldee, M. S., and L. D. Peth. 1992. Can protein-calorie malnutrition cause dysphagia? *Dysphagia* 7 (2):86–101.

Wall, B. T., and van Loon, L. J. 2013. Nutritional strategies to attenuate muscle disuse atrophy. *Nutr Rev* 71 (4):195–208.

Wakabayashi, H, and J. Kayashita. 2014. Translation, reliability, and validity of the Japanese version of the 10-item Eating Assessment Tool (EAT-10) for the screening of dysphagia. *JJSPEN* 29:871–876.

Wakabayashi, H. 2014. Presbyphagia and sarcopenic dysphagia: Association between aging, sarcopenia, and deglutition disorders. *J Frailty Aging* 3 (2):97–103.

Wakabayashi, H., and M. Matsushima. 2016. Neck circumference is not associated with dysphagia but with undernutrition in elderly individuals requiring long-term care. *J Nutr Health Aging* 20 (3):355–360.

Wakabayashi, H., M. Matsushima, R. Uwano, N. Watanabe, H. Oritsu, and Y. Shimizu. 2015. Skeletal muscle mass is associated with severe dysphagia in cancer patients. *J Cachexia Sarcopenia Muscle* 6 (4):351–357.

Wakabayashi, H. and K. Sakuma. 2014a. Comprehensive approach to sarcopenia treatment. *Curr Clin Pharmacol* 9 (2):171–180.

Wakabayashi, H. and K. Sakuma. 2014b. Rehabilitation nutrition for sarcopenia with disability: A combination of both rehabilitation and nutrition care management. *J Cachexia Sarcopenia Muscle* 5 (4):269–277.

Wakabayashi, H. and H. Sashika. 2014a. Malnutrition is associated with poor rehabilitation outcome in elderly inpatients with hospital-associated deconditioning a prospective cohort study. *J Rehabil Med* 46 (3):277–282.

Wakabayashi, H. and H. Sashika. 2014b. Response to the "Letter to the editor: Re: Malnutrition is associated with poor rehabilitation outcome in elderly inpatients with hospital-associated deconditioning: A prospective cohort study". *J Rehabil Med* 46 (9):942–943.

Wakabayashi, H., H. Sashika, and M. Matsushima. 2015. Head lifting strength is associated with dysphagia and malnutrition in frail older adults. *Geriatr Gerontol Int* 15 (4):410–416.

Wakabayashi, H. and R. Uwano. 2016. Rehabilitation nutrition for possible sarcopenic dysphagia after lung cancer surgery: A case report. *Am J Phys Med Rehabil* 95 (6):e84–e89.

Wakabayashi, H. and K. Sakuma. 2013. Nutrition, exercise, and pharmaceutical therapies for sarcopenic obesity. *J Nutr Ther* 2 (2):100–111.

Wang, C. M., J. Y. Chen, C. C. Chuang, W. C. Tseng, A. M. Wong, and Y. C. Pei. 2015. Aging-related changes in swallowing, and in the coordination of swallowing and respiration determined by novel non-invasive measurement techniques. *Geriatr Gerontol Int* 15 (6):736–744.

White, J. V., P. Guenter, G. Jensen, A. Malone, and M. Schofield. 2012. Consensus statement of the Academy of Nutrition and Dietetics/American Society for Parenteral and Enteral Nutrition: Characteristics recommended for the identification and documentation of adult malnutrition (undernutrition). *J Acad Nutr Diet* 112 (5):730–738.

Wischmeyer, P. E. and I. San-Millan. 2015. Winning the war against ICU-acquired weakness: New innovations in nutrition and exercise physiology. *Crit Care* 19 (Suppl 3):S6.

Yoshida, M., T. Kikutani, K. Tsuga, Y. Utanohara, R. Hayashi, and Y. Akagawa. 2006. Decreased tongue pressure reflects symptom of dysphagia. *Dysphagia* 21 (1):61–65.

Section VI

Applications

*Part 2: Complications due to Sarcopenia
in Acute or Chronic Diseases*

18 Wasting and Cachexia in Chronic Kidney Disease

Giacomo Garibotto, Daniela Picciotto, and Daniela Verzola

CONTENTS

18.1 INTRODUCTION

Malnutrition, protein-energy wasting (PEW), and cachexia are common in patients with chronic kidney disease (CKD) and are associated with an increased death risk from cardiovascular disease and infection (Kovesdy et al. 2013). Cross-sectional findings show that in patients with CKD, dietary protein and energy intakes and serum and anthropometric measures of protein-energy nutritional status progressively decline as the glomerular filtration rate (GFR) decreases (Kopple et al. 2000). It is interesting that the loss of body weight is not linear during the course of CKD. Body weight is stable until estimated glomerular filtration rate (eGFR) reaches 35 mL/min/1.73 m^2, but it declines at a rate of 1.45 kg for every 10 mL/min/1.73 m^2 further decrease. It is of note that a decrease in body weight in the CKD5 stage is associated with a higher risk for death after the initiation of dialysis therapy (stage CKD5d). In addition, it is also important to recognize that elderly subjects are expected to soon become the majority of those who will need renal replacement therapy. Both sarcopenia and frailty are common in the elderly, mainly in those with end-stage renal disease (ESRD) (Souza et al. 2017) and contribute to the debility and morbidity in this group of patients (Latos 1996; Bao et al. 2012; Nixon et al. 2018; van Loon et al. 2017). Mechanisms causing loss of muscle protein and fat are complex and only in part associated with anorexia. Recent evidence shows that wasting is linked to several abnormalities that stimulate protein degradation and/or decrease protein synthesis. During the course of CKD, the loss of kidney excretory and metabolic functions proceed together with the activation of pathways of endothelial damage, oxidative stress, inflammation, acidosis, altered insulin signaling, and anorexia, which are likely to orchestrate net protein catabolism and wasting. In addition, data from experimental CKD indicate that uremia specifically blunts the regenerative potential in skeletal muscle, by acting on muscle stem cells. In this review, recent findings regarding the mechanisms responsible for wasting in patients with CKD are discussed.

18.2   THE KIDNEY ROLE IN AMINO ACID AND PROTEIN METABOLISM

The human kidney plays a major role in the homeostasis of body amino acid pools. This is achieved through the synthesis, degradation, filtration, reabsorption, and urinary excretion of amino acids and peptides. On quantitative grounds, about 50–70 g of amino acids are filtered and almost completely (97%–98%) reabsorbed by the proximal tubules in a day (Garibotto et al. 2003). Even more importantly, the human kidney is able to finely tunnel the circulating and tissue pools of body amino acids. The kidney is the major organ for the disposal of glutamine and proline from the arterial blood, and for the net release of some amino acids such as serine, tyrosine, and arginine, which are newly generated within the kidney for export to other tissues (Garibotto et al. 2003; Kopple 2007). The magnitude of the net release or uptake of amino acids by the normal kidney can be understood if one considers that in a 70 kg man, the daily net production of serine is ~3–4 g, of which tyrosine ~1 g and arginine ~2 g. Moreover, the human kidney plays also a major role in the removal of Cysteine–Glycine, a peptide deriving from glutathione breakdown (Garibotto et al. 2003), and of S-adenosylhomocysteine (SAH) (Garibotto et al. 2009), a strong inhibitor of methionine-dependent remethylation pathway. Besides that, the kidney also releases smaller amounts (<1 g/day) of threonine, lysine, and leucine into the systemic circulation (Garibotto et al. 2003).

18.2.1   Tyrosine and Phenylalanine Metabolism

While phenylalanine is an essential amino acid (i.e., it cannot be synthesized by humans), tyrosine is considered a conditionally indispensable amino acid, because it can be synthesized by the hydroxylation of phenylalanine by the enzyme phenylalanine hydroxylase. The enzyme is present in the liver, pancreas, and the kidney (Kopple 2007). Recent studies indicate that the human kidney plays a role greater than previously presumed in the metabolism of tyrosine (Tessari and Garibotto 2000; Tessari et al. 1999; Moller et al. 2000). Actually, the phenylalanine hydroxylation activity in the kidney appears to be similar or even greater than that across the splanchnic area (Tessari et al. 1999; Moller et al. 2000). It is noteworthy that the release of tyrosine by the healthy kidney is similar in amounts to those deriving from protein breakdown in muscle (Tessari and Garibotto 2000), and it is markedly diminished in patients with CKD (Tizianello et al. 1980). Furthermore, the whole body phenylalanine hydroxylation rates are also markedly decreased in renal patients (Boirie et al. 2004). These findings may help to explain why plasma and muscle tyrosine are often decreased, plasma and muscle phenylalanine is normal to slightly increased, and the ratio of tyrosine/phenylalanine is almost always reduced in patients with CKD (Garibotto et al. 2003; Kopple 2007). Production of tyrosine may be insufficient in conditions, such as sepsis or liver disease, which are characterized by increased amino acid demand. Tyrosine deficiency could lead to protein depletion and also to impaired synthesis of aromatic amine modulators such as dopamine, norepinephrine, or epinephrine. A number of functions such as blood pressure and the secretion of pituitary hormones could be involved.

In CKD, accumulation of uremic toxins is associated with an increased risk of death. The question whether tyrosine products by microbiome manipulation contribute to toxicity in CKD is still unresolved (Fernandez-Prado et al. 2017). Some uremic toxins are ingested with the diet, such as phosphate. Others result from nutrient processing by gut microbiota. These nutrients include l-carnitine, choline/phosphatidylcholine, tryptophan, and tyrosine. Therefore, in patients with CKD, the use of these supplements may lead to potentially toxic effects.

In patients with CKD, as well as in atherosclerosis, peroxynitrite modifies tyrosine in proteins to create nitrotyrosines, leaving a footprint detectable in vivo. Increased amounts of 3-chlorotyrosine, a specific marker of myeloperoxidase-catalyzed oxidation, and nitrotyrosine have been described in the kidneys of patients with diabetic nephropathy and in plasma proteins and brain of patients with CKD (Kopple 2007). Whether these changes in tyrosine biochemistry represent only a marker of oxidative stress or they also have metabolic/toxic effects is still unknown. Nitration of specific tyrosine residues might affect protein function and be involved in normal regulatory processes and in the pathogenesis of disease.

18.2.2 Arginine and Nitric Oxide Synthesis by the Human Kidney

The kidney is a major site for the synthesis of arginine from citrulline (Tizianello et al. 1980). Arginine, in addition to its role in protein synthesis, plays multiple roles: it is the precursor of nitric oxide (NO), of creatine, agmatine, and other polyamines, and it is an intermediate in the urea cycle. The net production of arginine within the kidney is as much as 10%–12% of the total appearance of arginine within the body (Tessari et al. 1999; Van de Poll et al. 2004). The impact of arginine synthesis by the human kidney on body NO synthesis is not completely understood. On the one hand, plasma arginine appears to contribute to 54% of the substrate used in daily whole body NO synthesis; on the other hand, the fraction of the whole-body arginine flux used for NO formation has been estimated to be 1.2% only of arginine metabolism in healthy subjects (Castillo et al. 1996; Beaumier et al. 1995). Body arginine synthesis is necessary to provide arginine to body pools in special life-threatening conditions, like during endotoxemia (Palm et al. 2009). A diminished availability of NO may be involved in the genesis of hypertension, function, and integrity of endothelial cells, with implications for the atherosclerotic processes (Palm et al. 2009).

NO production is markedly decreased in patients with CKD (Baylis 2012). One obvious cause is that limited arginine provision by diseased kidney hinders tissue availability of this amino acid. In patients with advanced renal disease, anorexia could blunt nutritional intakes, thus further curtailing tissue arginine pools. Furthermore, the accumulation in body fluids of NO synthase inhibitors such as asymmetric dimethylarginine (ADMA) could explain the strong association between circulating levels of ADMA and carotid atherosclerosis (Wang et al. 2018) as well as with overall and cardiovascular mortality (Schepers et al. 2014). Initially, it was assumed that the increased plasma ADMA in ESRD reflected loss of renal clearance; however, very little ADMA is excreted unchanged in the urine and the majority is broken down by the enzymes dimethylarginine dimethylaminohydrolase 1 and 2 (DDAH1 and 2). DDAH is widely distributed and is most abundant in the kidney and also highly expressed in the liver and vasculature (Baylis 2012). Besides low arginine availability in kidney patients, recent observation indicates that citrulline production is severely low in patients with sepsis and is related to diminished de novo arginine and NO production (Luiking et al. 2009). These metabolic alterations, in addition to an increase in circulating ADMA (Winkler et al. 2017), might contribute to a reduced NO production in septic patients.

18.2.3 The Kidney and Sulfur Amino Acid Metabolism

That the human kidney plays a major role in sulfur amino acid metabolism has been known for many years. In 1963, Owen and Robinson showed for the first time that the human kidney releases cyst(e)ine (i.e., cysteine + cystine) into the circulation (Owen and Robinson 1963). About 30 years ago, Wilcken and Wilcken (1976) observed that increased plasma sulfur amino acid levels were associated with atherosclerosis and cardiovascular damage. Among sulfur compounds, homocysteine (Hcy) has been considered to be the culprit for this association.

The rat kidney plays a major role in the maintenance of Hcy plasma homeostasis (Friedman et al. 2001). However, the concept of a major role played by the kidney in Hcy metabolism has been challenged by studies performed in humans, which have not shown the occurrence of any significant arterio-venous gradient of Hcy across this organ (Garibotto et al. 2003; Van Guldener 2005). Besides glomerular filtration, which is restricted in humans because of protein binding, Hcy may be taken up by the peritubular basolateral surface (Friedman et al. 2001). In hypertensive subjects, the fractional extraction of Hcy across the kidney is positively related to renal plasma flow (RPF) but not to GFR (Garibotto et al. 2003). These findings indicate that in humans, the removal of Hcy is less than it occurs in rodents, and that it is limited to conditions characterized by elevated renal blood flow.

Stable-isotope studies in nondiabetic (Van Guldener et al. 1999; Stam et al. 2004) and diabetic patients (Tessari et al. 2005) with CKD have shown impaired metabolic clearance of Hcy by both the trans-sulfuration and the remethylation pathways. Of note, Stam et al. (2004) observed that in patients with end-stage renal disease (ESRD), the methionine remethylation, and transmethylation rates are inversely related to plasma S-adenosylhomocysteine (SAH), suggesting a strong inhibitory action of SAH on methionine-dependent remethylation pathway. We observed that the human kidney plays a unique role in the removal of SAH from the circulation (Garibotto et al. 2009), which suggests that the kidney may have an important role in the control of body transmethylation reactions. SAH plasma level increase is strongly associated with the decline of the filtration rate in CKD, a finding that is not observed for Hcy (Jabs et al. 2006).

The importance of this finding is that intervention trials failed to show a protective effect of lowering plasma Hcy levels (Jamison et al. 2007). Moreover, SAH increase in body fluids, rather than increased Hcy, appears to be the responsible of vascular diseases and tissue damage (Valli et al. 2008; Liu et al. 2008; Zawada et al. 2014). Even more important, there is not a significant reduction of cardiovascular risk by lowering Hcy levels with vitamin B and folate supplementation (Clarke et al. 2010; Albert et al. 2008a). SAH induces early atherosclerosis through oxidative stress and the consequent migration and proliferation of smooth muscle cells (Luo et al. 2012); more recent findings have also identified the presence of epigenetic mechanisms underlying its atherogenic effect (Xiao, Su et al. 2015; Xiao, Huang et al. 2015). Both elevated Hcy and SAH may also be implicated in accelerated cognitive impairment in patients with CKD (Shi et al. 2018).

18.2.4 The Kidney and Leucine Transamination

Transamination of the branched-chain amino acids (BCAAs) such as leucine, valine, and isoleucine plays a key role in nitrogen shunting either to urea formation or to protein synthesis (Albert et al. 2008b).

In particular, leucine represents an important anabolic signal, and its kinetic is often taken as an index of whole body and organ protein turnover (Matthews and Bier 1983). The first steps of leucine utilization are reversible deamination to alfa ketoisocaproic acid (KIC) and irreversible oxidation. With leucine and KIC isotopes, we recently measured leucine deamination and reamination through the whole body and compared it with other already existent data in the skeletal muscle, kidney, and the splanchnic bed (Garibotto et al. 2018b). It was observed that leucine deamination to KIC was greater than KIC reamination to leucine in the whole body, muscles, and the splanchnic area. These rates were not significantly different in the kidneys. Muscle accounted for ≈60%, the splanchnic bed for ≈15%, and the kidney for ≈12%–18%, of whole-body leucine deamination and reamination rates, respectively. In the kidney, percent leucine oxidation over either deamination or reamination was threefold greater than muscle and the splanchnic bed. According to these findings, skeletal muscle contributes by the largest fraction of leucine deamination, reamination, and oxidation. However, in relative terms, the kidney plays a key role in leucine (kic) oxidation. BCAA deamination and transamination are also a potential major topic in patients with CKD. Our finding of preferential leucine/KIC degradation in the human kidney suggests that progressive CKD may be associated with reduced requirement for leucine. This observation may explain, at least in part, the good nutritional status achieved even with very low protein diets in nondialyzed patients with CKD (Bellizzi et al. 2016). Of note, supplemented very low protein diets (0.28–0.40 g·kg⁻¹) containing branched-chain keto acids (BCKAs) are offered to patients with CKD to provide EAA precursors without the nitrogen load from EAAs. These supplemented very low protein diets appear to generate less toxic metabolic products than low protein diets (Gao et al. 2010) and have proven to be effective and safe when postponing dialysis treatment in elderly patients with CKD (Bellizzi et al. 2016; Garibotto et al. 2018a,b).

18.3 SARCOPENIA AND WASTING IN CKD: PREVALENCE AND ASSOCIATIONS

Sarcopenia and wasting are common in patients with CKD, with a reported prevalence of ≥20%–25% in early to moderate CKD and an even higher prevalence (up to 75%) in the dialysis population (Kovesdy et al. 2013; Mehrotra and Kopple 2001; Carrero et al. 2018). Loss of muscle mass is associated with worse clinical outcomes and high mortality owing to cardiovascular and infectious complications (Morley et al. 2013; Kang et al. 2017; Isoyama et al. 2014; Carrero et al. 2008).

The association between wasting and mortality in dialysis patients is underlined by the "reverse epidemiology" phenomenon (Kalantar-Zadeh et al. 2003); while in the general population, a high body mass index (BMI; kg/m²) is associated with an increased cardiovascular risk and with all-cause mortality, in dialysis patients, the effect of overweight or obesity is paradoxically in the opposite direction, with a higher BMI leading to an improved survival (Kalantar-Zadeh et al. 2003). This benefit is more pronounced in incident hemodialysis (HD) patients and those younger than 65 years (Vashistha et al. 2014). Also, during the course of progressive decline in renal function, the profile of risk for death may change because the occurrence of other risk factors, such as progressive wasting (Kovesdy and Kalantar-Zadeh 2009) and inflammation (Stenvinkel 2002), which may play an increasing role and may outweigh the effects of traditional risk factors. Chmielewski et al. (2009) recently showed that apoB/apoA-I ratio is associated with greater survival in HD patients only in the short term (1-year mortality). Similarly, (Chmielewski et al. 2011) the reverse association between hypercholesterolaemia and all-cause mortality declined gradually after the first year of follow up. This is probably due to the temporal impact of competing risks. Liu et al. (2004) recently postulated that the paradoxical relationship between cholesterol level and mortality could be explained by the effect of the presence of the complex malnutrition-inflammation (defined as BMI <23 kg/m² or C-reactive protein >10 mg/L) in the dialysis population. Recently, Contreras et al. (2010) assessed the prevalence of malnutrition-inflammation and its modifying effects on the risk-relationship of cholesterol levels with subsequent CVD events in African American with hypertensive CKD. They showed that the hazard ratio for the primary CVD outcome augmented as total cholesterol increased in subjects without malnutrition-inflammation, whereas it tended to decrease in subjects with malnutrition-inflammation. Inflammation appears to actually play an important role as an effect modifier in the obesity paradox. Stenvinkel et al. showed that a higher BMI was associated with a survival advantage in dialysis patients with inflammation (Stenvinkel et al. 2015). This could lead to the opportunity of identifying subgroups of patients that could actually obtain an advantage from weight loss and could therefore have importance in a clinical setting (Drechsler and Wanner 2015).

According to the International Society of Renal Nutrition and Metabolism (ISRNM) definitions, the clinical syndrome called "protein-energy wasting" (PEW) as the state of loss of protein body mass and energy reserves, which can ultimately evolves to cachexia. Diagnosis is made if three of these four elements are fulfilled: low levels of at least one biochemical indicator (serum albumin, serum prealbumin, or serum cholesterol), reduced body mass, reduced muscle mass or an adequate dietary intake (Fouque et al. 2008). PEW has to be distinguished from malnutrition considering its multiple and complex causes that go far beyond an inadequate diet and are linked to inflammation and progressive kidney dysfunction.

18.3.1 AGING

People over the age of 65 years are expected to soon become the majority of those who will need renal replacement therapy in many developed countries. Nutritional problems are common in elderly patients with ESRD and contribute to the debility and morbidity in this group of dialysis patients (Latos 1996). Low dietary intake and diminished muscle masses are both common in the old individuals and may cause low values of BUN and serum creatinine, even in the presence of advanced renal failure (Lindeman 1990). Selected deficits are specific to certain muscle protein components, such as myosin heavy chain and mitochondrial proteins (Nair 2000). Moreover, in elderly subjects, a decreased sensitivity of insulin action is linked to frailty and cognitive impairment (Hooshmand et al. 2018). These alterations in protein metabolism due to aging likely potentiate those caused by uremia. Weight loss in older adults is highly predictive of increased morbidity and mortality (Oreopoulos et al. 2009; Miller et al. 2008), and it is believed to follow microinflammation and excess cytokine elaboration (inflammageing). Aging is associated with increased concentrations of TNF-alfa, IL-6, IL1 receptor antagonist, and soluble TNF-α receptor. Acute phase proteins such as C-reactive protein and serum amyloid-A are also elevated, which suggests the activation of the entire inflammatory cascade. Therefore, microinflammation occurring with aging could interact with the upregulations of specific catabolic pathways in uremia to promote wasting. In cross-sectional studies, the prevalence of sarcopenia is estimated between 5% and 28.7% in patients with CKD (Souza et al. 2017). Sarcopenia is more prevalent in the more advanced stages of CKD and associated with age, male gender, body mass index, diabetes mellitus, and loop diuretic use (Ishikawa et al. 2018). It is important to recognize that sarcopenia defined as reduced handgrip strength and low skeletal muscle mass index estimated by bio-impedance (BIA) is an independent predictor of mortality in these patients (Pereira et al. 2015).

18.3.2 FRAILTY

In the general elderly population, frailty is a condition characterized by loss of biological reserves, failure of homoeostatic mechanisms, and susceptibility to adverse outcomes, which increases vulnerability for loss of independence and death (Morley et al. 2013). Several aspects of frailty, including its underlying mechanisms, are still obscure. It is remarkable that some of its components are reversible, particularly in the pre-frail state and after acute illnesses (Allen et al. 2017). The prevalence of frailty is considerably higher among dialysis patients than among community-resident elders, and frail CKD patients are also at higher risk of morbidity and mortality. In patients with nondialyzed CKD, lower kidney function is a strong correlate of low urinary creatinine excretion and frailty, independently from comorbidities (Polinder-Bos et al. 2017). As shown by cross-sectional studies, the prevalence of frailty increases as CKD progresses (Kim et al. 2013; Stenvinkel et al. 2016; Johansen et al. 2017), supporting the concept of a series of events in which catabolic factors occurring in CKD (such as anorexia, acidosis, insulin resistance and inflammation) together with the accumulation of neurotoxins (such as homocysteine and ADMA) progressively lead to wasting, cognitive and functional impairment, and inactivity, with deconditioning leading to further inactivity and disability (Chiang et al. 2019; Shi et al. 2018).

However, although CKD and aging contribute to frailty, there may also be potentially reversible contributors (such as acute illnesses and reduced nutritional intakes) and frailty may not always be a permanent or progressive condition. It is interesting that key convergent pathological effects of frailty and wasting include loss of muscle mass and strength, with consequent impact on mobility and activities of daily living.

A reduced testosterone level may contribute to loss of energy, strength, skeletal, and muscle mass; favor muscle wasting and cachexia; and in the end, contribute to frailty and mortality in CKD. Recently, Chiang et al. were able to show that lower levels of serum-free testosterone are associated with higher odds of being frail and of becoming frail over 12 months in a cohort of HD-treated patients (Chiang et al. 2019). Of note, serum-free testosterone was also associated with two major components of the physical function of frailty, that is, grip strength and gait speed. These associations are especially noteworthy as slow gait speed and low grip strength may be important for maintaining independence and are also predictive of mortality in patients on dialysis. This study lends further support to the hypothesis that the measure of free testosterone may be useful for individualized patient care and that androgen replacement may be a feasible therapeutic target toward prevention of frailty in CKD, although studies would be needed to test these possibilities.

18.3.3 COMORBIDITIES

Comorbidities contribute to PEW and frailty by sharing with CKD and ESRD important mechanisms. Diabetes is probably the most important comorbidity leading to wasting by representing a state of insulin resistance and increased protein breakdown (Pupim et al. 2005a, 2005c). Moreover, diabetes is associated with other mechanisms that may induce PEW and frailty, such as gastroparesis and a state of chronic inflammation. Cardiovascular diseases are common in renal patients, and cardiac cachexia due to heart failure is characterized by neurohumoral responses that contribute to PEW (von Haehling et al. 2009). Moreover, the interaction between the increase in parathyroid hormone (PTH) and the downregulation of vitamin D and circulating Klotho in the CKD metabolic bone disease (CKD-MBD) might potentiate catabolism, since both parathyroid hormone and vitamin D are regulators of protein metabolism (Garcia et al. 2011). Of note, correction of hypovitaminosis D might improve muscle mass and muscle strength (Gordon et al. 2007).

18.3.4 ANOREXIA

Finally, CKD5 and CKD5d are characterized by an important change in the perceived quality of life, which ultimately may lead to depression, a condition that contributes to anorexia.

Besides depression, there are other possible causes of anorexia in patients with CKD such as diabetes' gastroparesis and the accumulation of toxic molecules, pro-inflammatory cytokines, and middle-size molecules that can inhibit appetite. Kalantar-Zadeh et al. showed an association between anorexia and higher concentration of proinflammatory cytokines, hypo-responsiveness to erythropoietin, hospitalization rates, and all-cause mortality in HD patients (Kalantar-Zadeh et al. 2004). Carrero et al. recently found a correlation between appetite, malnutrition, inflammation, and worse outcome in HD patients (Carrero et al. 2007). Moreover, they observed a greater susceptibility to inflammation-induced anorexia in men, compared with women, in agreement with different gender-specific orexigenic and anorexigenic response in inflammation obtained in rodents (Gayle et al. 2006).

18.3.5 LEPTIN/MELANOCORTIN SIGNALING

Another mechanism responsible of anorexia in uremia is linked to perturbation of appetite-regulating hormones, such as leptin and ghrelin. Leptin and ghrelin are cleared from the circulation by the kidney (Garibotto et al. 1998; Masuda et al. 2000; Van der Lely et al. 2004), and their plasma levels progressively increase along with the decrease in GFR. In addition, both CKD and inflammation influence leptin and central melanocortin systems in hypothalamus, by increasing the release of alfa melanocyte-stimulating hormone that binds melanocortin receptors 3 and 4 (MC3-R, MC4-R) and inhibits food intake (Mak et al. 2018). In fact, in mice, pharmacological blockade of MC4-R reduces the development of cachexia associated with CKD (Cheung et al. 2005, 2007). Cheung et al. found that the injection of agouti-related peptide (AgRP) in the lateral ventricle of mice with subtotal nephrectomy induced an increase in food intake and body growth, thus suggesting that uremia could cause a defect in the capacity of AgRP to block the melanocortin-4 receptor, MC4-R (Cheung et al. 2005; Mak et al. 2018).

18.3.6 REDUCED PHYSICAL ACTIVITY

There are many factors that can limit physical activity in the CKD population including cardiovascular diseases, anemia, and muscle loss itself. Physical exercise provides a multitude of benefits, including improved physical functioning and exercise tolerance, increased muscle mass, reduced cardiovascular risk and systemic inflammation, as well as improved quality of life (Howden et al. 2015). However, the mechanism by which exercise creates an increase muscle mass, or the molecular responses mounted to exercise by patients with CKD are not yet fully explored. Exercise decreases insulin resistance, and it may therefore be of great importance in ameliorating PEW (LeBrasseur et al. 2011). Wang et al. (2005), in right vastus lateralis muscle of noninflamed HD patients, studied the effects of various types of exercise training on mRNA levels for various growth factors that stimulate or suppress protein synthesis or inhibit protein degradation, including myostatin. Of note, myostatin mRNA fell significantly after exercise, an effect that was associated with a pattern of changes in gene transcription that would be likely to promote protein synthesis and reduce protein degradation. Watson et al. observed that in patients with CKD, unaccustomed exercise caused large inflammatory muscle response, which was blunted after training (Watson et al. 2017). Myostatin mRNA expression was also significantly suppressed following training. Therefore, the few studies available indicate that the decline of myostatin expression in muscle is a component of the anabolic response to physical exercise in CKD.

18.3.7 CATABOLIC EFFECTS OF DIALYSIS

There is evidence that an adequate HD dose is not sufficient to reduce the catabolic pathways activated by uremia (Rocco et al. 2004; Chertow et al. 2010). The reason is that HD itself is responsible of catabolic processes such as the loss of 6–8 g amino acid per session and the release of pro-inflammatory cytokines (Ikizler et al. 1994; Caglar et al. 2002, Garibotto et al. 2012). These effects are caused by decreased protein synthesis and increased proteolysis. Decreased protein synthesis is likely mediated by the significant decrease in plasma amino acid concentrations during HD. In contrast, increased protein degradation is mediated by the HD-associated inflammatory response (Garibotto et al. 2012).

18.4 CKD-RELATED MECHANISMS AND MEDIATORS OF WASTING

The past two decades have witnessed tremendous progress in our understanding of the mechanisms underlying wasting and cachexia in CKD and in other chronic illnesses, such as cancer and heart failure. In all these conditions, wasting is an effect of the activation of protein degradation in muscle, a response that increases the risk of morbidity and mortality. Major recent advances in our knowledge include the mechanisms by which CKD per se, acidosis, and inflammation affect anabolic intracellular cellular signals that decrease protein synthesis and promote protein degradation.

18.4.1 MUSCLE PROTEIN TURNOVER IN CKD: ROLE OF METABOLIC ACIDOSIS

Early studies examining net amino acid exchange across peripheral tissues in nonmalnourished patients with moderate-advanced (stage 4–5) CKD (Tizianello et al. 1988) failed to show an increase in net muscle breakdown, at variance with data obtained in experimental models of uremia in animals. Also Alvestrand et al. (1988) observed that, in the post-absorptive state, in maintenance of HD patients, the release of amino acids from the lower limb was not increased. More recently, mixed muscle protein synthesis and degradation have been evaluated by the measurement of the ^{3}H-phenylalanine kinetics in patients with moderately advanced CKD (Garibotto et al. 1994) and metabolic acidosis (mean $HCO_3 = 20$ mmol/L). Net protein balance across the forearm was negative and was similar to controls, because protein degradation was greater than protein synthesis. However, rates of muscle protein turnover were greater than in control subjects, analogously to what observed in whole body protein turnover studies in patients with CKD and metabolic acidosis (Graham et al. 1996, 1997). Inverse correlations were observed between net protein balance and blood bicarbonate, suggesting that acidosis is responsible for the more negative protein balance. It is of note that, when muscle protein turnover was evaluated in patients with similar degrees of CKD, but without acidemia, rates of muscle protein turnover were similar to controls (Garibotto et al. 1996). These results are in accordance with observations in the whole body, showing that protein degradation is increased in patients with CKD with metabolic acidosis. Moreover, basal rates of synthesis and degradation of mixed muscle proteins appear to be in the normal range, if acidosis is corrected in nonmalnourished patients with CKD and unrestricted nutrient intakes.

The measurements of amino acid kinetics associated with the limb perfusion made in these studies represent the average of synthetic rates of all muscle proteins. Adey et al. measured the synthesis rate of myosin heavy chain (MHC) in muscle biopsy samples in patients with CKD and moderate renal impairment (Adey et al. 2000). Despite the finding of rates of whole body ^{13}C-leucine turnover similar to controls, the authors observed significantly lower synthetic rates of myosin heavy chain as well as mitochondrial protein, muscle cytochrome c-oxidase activity, and citrate synthase, suggesting that in CKD decreased energy availability is responsible for reduced protein synthesis rates.

Another possibility is that muscle protein turnover in uremia is inhibited by some circulating factor(s), which accumulates in blood. Reduced bioavailability of insulin-like growth factor (IGF)-I, due to increased serum binding proteins, may play a role in decreasing muscle protein synthesis. Protein turnover of mixed muscle proteins has been evaluated in a group of nonacidotic HD patients with stabilized, chronic protein-calorie malnutrition (BMI), before and after the administration of rhGH for 6 weeks (Garibotto et al. 1997; Bradley et al. 1990). In these patients, rates of forearm muscle protein synthesis and degradation were in the lower range of the values recorded in normal healthy subjects (Garibotto et al. 1997). However, when patients were given rhGH, both muscle protein synthesis and whole body oxygen consumption increased (Garibotto et al. 2000). It is interesting that changes in the rates of muscle protein synthesis were related to variations in the changes in insulin-like growth factor binding protein (IGFBP)-I (which decreased in blood) and in the IGF-I/IGFBP-3 ratio (which increased in blood), which suggests the dependency of muscle protein synthesis on the availability of "free" IGF-1 levels. Moreover, changes in resting expenditure were for a great part accounted for the increased rates of muscle protein synthesis. These results suggest the occurrence of a causal relationship between protein synthesis rates and energy metabolism.

An emerging issue in studies on muscle perfusion is the role of vasculature, which in conjunction with hormonal effects plays an important role in the control of muscle metabolism and function (Clark et al. 1995). Besides reduced energy metabolism or reduced bioavailability of growth factors, a major cause of wasting could be related to the changes in muscle structure and reduction of nutritive blood flow observed in patients with CKD (Bradley et al. 1990).

In summary, available data on the in vivo measure of protein metabolism suggest that metabolic acidosis and some unknown factors (including the reduced availability of IGF-I), changes in muscle blood flow, some unidentified uremic toxin or reduced oxidative metabolism, have different, additive detrimental effects on muscle protein metabolism in patients with moderate-stage CKD, acting on different pathways of protein turnover and different pools of skeletal muscle proteins.

18.4.2 GH/IGF-1 Resistance

Resistance to growth factor hormone (GH) is responsible for the defective growth of uremic children. Besides its stimulatory effects on growth and anabolism, circulating GH and IGF-1 levels may play a role in the metabolic response to fasting, by contributing to protein conservation (Tessari et al. 1999). GH resistance is common in uremia and together with resistance to insulin-like growth factor-1 (IGF-1) contributes to muscle wasting. Development of GH resistance in uremia stems not only from reduced IGF-1 synthesis (Kaskel 2003), sensitivity (Qing et al. 1999), or bioavailability (Blum et al. 1991), but also from impaired GH signal transduction (Sun et al. 2004). Defective intracellular GH signaling may stem result from impaired GH stimulated Jak2-STAT5 phosphorylation, which may be the result of inflammation-induced SOCS2 expression (Sun et al. 2004). In addition, the hepatic resistance to GH-induced IGF-1 expression in uremia arises due to defects in STAT5b phosphorylation and its impaired binding to DNA, processes further aggravated by inflammation (Chen et al. 2010). In accordance with these findings, previous elegant clinical studies have shown that the response to IGF-1 is impaired in dialysis patients (Fouque et al. 1995). However, when IGF-I has been used in malnourished dialysis patients, an anabolic effect has been observed (Fouque et al. 2000). In addition, several clinical studies have reported that GH has a salutary effect on body composition (Kaskel 2003; Chen et al. 2001) and muscle protein synthesis (Ikizler et al. 1996; Garibotto et al. 1997; Pupim, Flakoll, Yu et al. 2005b) in patients with CKD. Nonetheless, the GH response appears to be variable (Kotzmann et al. 2001). A 3-month trial of GH revealed a significant increase in IGF-1 levels and markers of bone turnover but not lean body and fat mass, in malnourished HD patients (Kotzmann et al. 2001). Conversely, Kopple et al. (2005) observed that GH induced a strong and sustained anabolic effect in HD patients with protein–energy malnutrition. The reasons for the diverging clinical effects of GH are not understandable. Identification of variables predicting the response to GH would offer substantial clinical benefits. Besides uremia per se, the effectiveness of GH might also be attenuated by other factors often found in patients with CKD, such as metabolic acidosis (Cooney et al. 2006) and low nutrient intake (Brungger et al. 1997). We observed (Garibotto et al. 2008) that the GH acute anabolic response regarding two different GH receptor-downstream pathways (potassium and amino acid metabolism) is overall preserved in patients with advanced CKD, while it is blunted in patients displaying evidence of microinflammation, suggesting a role for inflammatory changes in the regulation of skeletal muscle protein balance. Unfortunately, a large-scale randomized clinical trial of growth hormone in HD patients was terminated due to slow recruitment (Kopple et al. 2011).

18.4.3 Insulin Resistance

Insulin resistance is a well-known complication of CKD (Rigalleau and Gin 2005; Fliser et al. 1998; Kobayashi et al. 2005; Alvestrand et al. 1989; Xu and Carrero, 2017). Although not clinically remarkable, insulin resistance appears to be strongly implicated in the pathogenesis of hypertension, accelerated atherosclerosis, and cardiovascular events (DeFronzo et al. 1983; Chen et al. 2003; Shinohara et al. 2002; Teta 2015). A post-receptor defect in muscle responsiveness to insulin is the cause of insulin resistance with regard to glucose metabolism in patients with CKD (Tessari et al. 1999). However, insulin resistance could also extend to the antiproteolytic action of this hormone. Besides its effects on glucose and lipid metabolism, insulin also influences a number of metabolic pathways related to protein metabolism, which lead to anabolic or anticatabolic effects (Tessari et al. 2011). Particularly, even small increases in blood insulin levels, well within the physiological range, are associated with pronounced inhibition of protein breakdown (Louard et al. 1992; Tessari et al. 2011). An emergent hypothesis is that a resistance to the anabolic drive by insulin may contribute to the loss of strength and muscle mass, which are observed with the progression of CKD (Wang and Mitch 2014). This hypothesis is supported by studies obtained in animal models of CKD. Bailey et al. have identified a series of abnormal postreceptor signaling changes in the insulin/IGF-1 pathway in muscle of rats with CKD (Bailey et al. 2006). These include the occurrence of functional abnormalities in the IRS/PI3-K cascade that decrease the phosphorylation of the downstream effector protein kinase (Akt). The low phosphorylated Akt activity has been shown to stimulate the expression of specific E3 ubiquitin conjugating enzymes, atrogin-1/MAFbx and muscle-specific RING finger-1 (MuRF1), to accelerate muscle protein degradation. Further, a decrease in muscle PI3K activity per se activates Bax leading to the stimulation of caspase-3 activity and increased protein degradation (Wang and Mitch 2014; Bailey et al. 2006; Thomas et al. 2013). It is interesting that metabolic acidosis, a common complication observed in patients with CKD, also interferes with insulin-induced intracellular signaling by suppressing PI3-K activity in muscle and thus increases protein degradation through an upregulation of the ubiquitin-requiring pathway (Rajan et al. 2008; Reaich et al. 1993). Administration of high dose of insulin as well as treating CKD-rats with bicarbonate was, respectively, able to restore pAkt content and reduce protein degradation (Bailey et al. 2006). Of note, the defective pAkt signal can be partially recovered by muscle overloading (Wang et al. 2009), suggesting that physical exercise could partially restore the abnormal insulin signaling in uremia.

The evidence that insulin resistance also extends to protein metabolism in CKD is suggested by cross-sectional studies in which protein turnover was compared with basal insulin levels or with "static" indexes of insulin resistance (Siew et al. 2007; Garibotto et al 1994). However, in the few studies (Alvestrand et al. 1988; Castellino et al. 1999) in which amino acid or protein metabolism was evaluated during the euglycemic hyperinsulinemic clamp, a normal antiproteolytic response to insulin was observed, even in the presence of metabolic acidosis (Reaich et al. 1995). It is interesting that in these studies, the effects of insulin were tested in the high (~60–100 μU/mL) insulin physiological range, which is needed to test insulin sensitivity regarding glucose. However, protein metabolism appears to be maximally sensitive at only modestly increased (up to 30 μU/mL) insulin concentrations (Louard et al. 1992). Therefore, the occurrence of defects in muscle insulin's response in uremia could be offset by the high insulin levels attained in previous studies. Recently, we explored the effects of two different insulin levels on two selected endpoints of its action, such as glucose and protein metabolism in muscle of patients with nondiabetic CKD (Garibotto et al. 2015). Forearm glucose balance and protein turnover (^{2}H-phenylalanine kinetics) were measured basally and in response to insulin infused at different rates for 2 h to increase local forearm plasma insulin concentration by approximately 20 and 50 μU/mL. In response to insulin, forearm glucose uptake was significantly increased to a lesser extent (−40%) in patients with CKD than controls. In addition, whereas in the controls, net muscle protein balance and protein degradation were decreased by both insulin infusion rates, in patients with CKD, net protein balance and protein degradation were sensitive to the high (0.035 mU/kg per min) but not the low (0.01 mU/kg per min) insulin infusion. Besides blunting muscle glucose uptake, CKD and acidosis interfere with the normal suppression of protein degradation in response to a moderate rise in plasma insulin. In humans, insulin plays a pivotal role in maintaining muscle protein mass during fasting, during which proteins are lost, and in the fed state, during which protein gains take place. We showed that patients with acidemic CKD need higher than normal levels of insulin to inhibit proteolysis. The degree of insulinemia achieved when a low dose of insulin was administered study (~28–30 μU/mL) was comparable with that expected after a low glycemic index meal, such as breakfast. Thus, it turns from these results that the normal suppression of proteolysis by insulin at low insulin levels, such as during fasting, is impaired in CKD. In contrast, the sensitivity to insulin at higher insulin concentrations, as may occur with meals containing refined carbohydrates, is preserved. Overall, the defects in the regulation by insulin of protein metabolism in uremia on the one hand may lead to changes in body tissue composition, metabolic rates, and individual amino acid metabolic steps that may become clinically evident in conditions characterized by low insulinemia, such as during fasting and/or low energy intakes.

Early studies have suggested that insulin resistance in uremia is not due to impaired insulin binding to its receptor but rather to defects in intracellular signaling processes (Smith and DeFronzo 1982; DeFronzo et al. 1978). The molecular mechanisms by which insulin reduces muscle protein degradation are only in part known. In myotubes, insulin decreases the activity of the ubiquitin–proteasome pathway through changes in the phosphorylation status of members of the Akt-dependent signaling pathway (Sandri et al. 2004). In rats, there is evidence that insulin-stimulated Akt activity induces phosphorylation of the FOXO family of transcription factors (Kim et al. 2008; Tang et al. 1999). In humans, when muscle protein kinetics in association to intracellular signaling was studied in response to insulin, decreases in the amounts of MAFbx protein and the C2 proteasome unit, but not of MuRF-1, were observed. These findings provide a partial mechanistic explanation for the fall in insulin-induced muscle proteolysis (Greenhaff et al. 2008). It is interesting that metabolic acidosis stimulates the activity of different proteases including caspase-3 and the ubiquitin–proteasome system (UPS), which increase muscle protein degradation (Wang and Mitch 2014). Therefore, insulin and acidosis appear to target in an opposite way the same signaling pathway in muscle, as also shown in a cultured muscle cell model (Franch et al. 2004).

Besides acidosis, different complications of CKD, such as inflammation (Thomas et al. 2013) and elevated angiotensin II (Thomas et al. 2013), blunt insulin- and IGF-1-induced intracellular signaling to promote muscle catabolism. Other triggers of insulin resistance include uremic toxins accumulation. P-cresyl sulfate in patients with CKD or in animal models of CKD resulted in impaired insulin-stimulated intracellular signaling (Koppe et al. 2013). An excess of urea has also been implicated as a mechanism causing insulin resistance at least in cultured 3T3-L1 adipocytes (D'Apolito et al. 2010).

18.4.4 Inflammation and Innate Immunity in Uremic Muscle

Although uremic cachexia has for a long time been regarded as a nonimmune disease, an emerging hypothesis is that innate immunity plays a role in its development and progression (Mitch 2006; Vaziri et al. 2012; Anders et al. 2013), in analogy with cancer (Onesti and Guttridge 2014) and cardiac (Egerman and Glass 2014) cachexia. Initial observations have shown that circulating levels of C-reactive protein and proinflammatory cytokines, including interleukin-6 (IL-6) and tumor necrosis factor-α (TNF-α), are increased in blood of patients with CKD and are strongly associated with wasting (Bergström and Lindholm 1998; Kaysen 2014).

Recently, we observed that toll-like receptor 4 (TLR4) is a link between uremia and innate immunity in skeletal muscle. In predialysis patients, an upregulation of TLR4 is predicted by eGFR and subjective global assessment, suggesting that the progressive decline in renal function and wasting mediate TLR4 activation. With this regard, muscle cell myotubes (C2C12) exposed to uremic serum show an increase in TLR4 and TNF-α and a downregulation of pAkt. These effects are prevented by blockade of TLR4 (Verzola et al. 2017). These findings suggest therefore that muscle inflammation may be at least in part due to an upregulation of TLR4 with activation of its downstream inflammatory pathway (Verzola et al. 2017; Esposito et al. 2018).

Although precise mechanisms that contribute to the high prevalence of inflammation in CKD are unknown, ROS have been proposed as a potential contributor. On the one hand, oxidative stress is able to activate transcriptor factors, such as NF-kB, which regulate inflammatory mediator gene expression. On the other hand, chronic inflammation may cause increased oxidative stress, thus creating a vicious circle in the determination of cardiovascular risk in patients with CKD. Notably, CRP and IL-6 levels increase progressively along with the decline of glomerular filtration rate in patients with CKD (Panichi et al. 2002; Eustace et al. 2004). These effects could be due to reduced kidney excretory function, since at least some pro-inflammatory cytokines are excreted through the kidney.

More recently, an upregulation of several genes associated with inflammation in muscle has been demonstrated to occur both in rodent models (Zhang et al. 2011, 2013) and humans with CKD (Garibotto et al. 2006; Shah et al. 2006; Verzola et al. 2011). Pro-inflammatory cytokines, mainly TNF-α and IL-6, have per se catabolic effects on skeletal muscle by accelerating protein degradation or reducing protein synthesis (Argiles et al. 2005; Goodman 1991; Haddad et al. 2005). When we measured net protein catabolism across the forearm of HD patients, an association was found between IL6 and phenylalanine release, suggesting increased catabolism (Garibotto et al. 2006). Recently, Deger et al. used stable isotope kinetic studies to assess whole body and skeletal muscle protein turnover in a large cohort of patients on HD. They were able to show an association between CRP and forearm skeletal muscle negative net protein balance (Deger et al. 2017). Moreover, Zhang et al. could recently demonstrate that high IL-6, by the activation of the JAK-Stat3 pathway, upregulates the gene expression of myostatin, a major negative regulator of muscle protein content and regeneration. Such a mechanism has been observed both in the rodent model (Bailey et al. 1996) and in muscle of patients with CKD (Zhang et al. 2013). Interestingly, Miyamoto et al. (2011) observed that increased circulating levels of follistatin, which neutralizes the biologic activities of myostatin occurred together with increased inflammation and reduced muscle strength, suggesting an involvement of a mechanism including follistatin in the uremic wasting process. In addition, in CKD, proinflammatory cytokines cause a resistance to the action of insulin (Thomas et al. 2013) and GH/IGF-1 (Chen et al. 2010; Garibotto et al. 2008). To understand how inflammation is related to insulin resistance, some studies have examined biochemical responses of macrophages, adipocytes, or skeletal muscle to determine how inflammation can stimulate the production of inflammatory cytokines and how the cytokines might interfere with the functions of insulin (Pereira et al. 1996; Hotamisligil et al. 1993; Feinstein et al. 1993; Kimmel et al. 1998).

Circulating cytokines can interfere with biochemical pathways resulting in insulin resistance and its metabolic consequences. Thomas et al. demonstrated a marked muscle increase in the expression of SIRP-α, a transmembrane glycoprotein, that was shown to interact directly with the insulin receptor IRS-1 phosphorylation and proteolysis (Thomas et al. 2013). IRS-1 phosphorylation was also reduced suggesting that a low value of IRS-1 causes insulin resistance by interrupting insulin-stimulated intracellular signaling. The trigger of increased SIRP-α expression in muscle is the activation of the NF-κB pathway, which indicates a novel mechanisms linking inflammation, insulin resistance, and protein degradation in uremia (Thomas et al. 2013).

18.4.5 PAKT

Insulin binding to its receptor is followed by phosphorylation of specific tyrosines in the insulin receptor and insulin receptor substrate-1 (IRS-1). This is followed by phosphorylation of phosphatidylinositol-3 kinase (PI3K) and the Akt kinase (p-Akt). pAkt plays a crucial role in determining muscle mass in CKD as well as in many other chronic diseases. There are two major branches of Akt signaling pathways that are specifically involved in protein metabolism, namely the pAkt-mTor pathway, which represents the anabolic branch, and the Class 1 PI3K/Akt/FoxO pathway, which instead has catabolic effects (Figure 18.1).

These pathways are influenced in an opposite way by the insulin signal and myostatin. During atrophy, MuRF-1 and MAFbx direct the polyubiquitination of proteins to target them for proteolysis by the 26S proteasome, mediating muscle breakdown (Cohen et al. 2009). In muscle physiology, the activation of the Akt signaling pathway sets the balance between anabolic and catabolic reactions. Inhibition of this pathway upregulates, while its activation downregulates, the expression of the transcription factors MAFbx and MuRF-1 via inhibition of the Foxo family of transcription factors (Bodine et al. 2001) (Figure 18.1). We observed that phosphorylated Akt was markedly downregulated and closely associated with the loss of myonuclei by apoptosis in skeletal muscle of patients at an advanced CKD Stage 5, which suggests that impaired Akt signaling favors the loss of myofibers (Verzola et al. 2011). This observation gives further strength to the hypothesis that altered intracellular signaling causes a shift in the balance between pro-survival and pro-apoptotic pathways in the muscle.

18.4.6 MTOR

The mechanistic target of rapamycin (mTOR) signaling pathway is a major regulator of protein synthesis and thus of muscle growth in adulthood. It is also involved in many other cellular

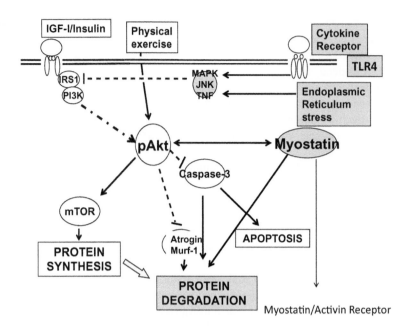

FIGURE 18.1 Intracellular signals activating net catabolism in experimental uremia and in patients with end-stage renal disease.

processes and in an increasing number of chronic diseases such as cancer, type two diabetes, and CKD (Laplante and Sabatini 2012). mTOR forms two large complexes (mTORCs) together with RAPTOR (mTORC1) or RICTOR (mTORC2) (Wullschleger et al. 2006). The effect of mTORC1 on the translation machinery is mediated on the one hand through inhibition of 4E-binding protein1 (4E-BP1), which permits the initiation of translation, and on the other hand through activation of the ribosomal protein S6 kinase (S6K), which has direct influence on ribosome biogenesis (Hay and Sonenberg 2004; Ma and Blenis 2009; Tee and Blenis 2005). There are several physiological conditions, such as feeding (Fujita et al. 2007) and resistance exercise (Coffey et al. 2006; Dreyer et al. 2008), which can activate translation (Kimball and Jefferson 2006) of this pathway.

In the muscle of CKD mice, the level of p-Akt is depressed, as also the p-FoxO1 phosphorylation and p-p70S6K, p-mTOR levels, and protein synthesis. However, these lower levels are reversed either by muscle overloading or treadmill running (Wang et al. 2009). Therefore, physical exercise could counteract, at least in part, the blunted intracellular anabolic signal. It is interesting that uremia-induced changes in muscle metabolism of specific amino acids could compromise muscle regeneration. Intracellular oxidation of leucine due to acidosis could impair the activation of satellite cells and thus myoblasts proliferation through the mTOR signaling (Bonanni et al. 2011). In keeping with this hypothesis, in CKD rats, a suppressed phosphorylation of 4E-BP1 and rpS6 can be partially restored by leucine administration (Chen et al. 2011). It is therefore reasonable to state that a combination of simultaneous exercise and nutritional supplementation could augment the anabolic effects of exercise, at least in the acute setting (Ikizler 2011).

18.4.7 FoxO

Phospho-AKT was identified as a main regulator of the Forkhead box O (FoxO) family of transcription factors in diverse organisms (Tzivion et al. 2011). FOXOs are involved in several cellular processes that influence stress resistance, metabolism, cell cycle, and apoptosis (Martins et al. 2016). Moreover, they are key mediators of the catabolic response during atrophy as they are able to regulate the two major cellular proteolytic mechanisms: the autophagy-lysosome and the UPSs

(Rodriguez et al. 2014). pAkt-mediated dephosphorylation of FoxO leads to an increased expression of atrophy related genes: MuRF1 and MAFbx (also known as atrogin-1). These E3 Ub-ligases stimulate the UPS-mediated degradation of muscle proteins, including contractile proteins (Wang and Mitch 2014) (Figure 18.1).

Recently Liu et al. demonstrated that SCP4, a member of the small C-terminal domain phosphatases, is able to influence FoxO's phosphorylation status. In particular, overexpression of SCP4 in cultured mouse myoblast cells led to accumulation of FoxO1/3a and therefore to an increase in catabolic factors and proteolysis. SCP4 is overexpressed in skeletal muscle of experimental models and in humans with CKD and siRNA-based gene knockdown of SCP4 in mice was able to prevent muscle wasting, thus launching a new therapeutic horizon in PEW (Liu et al. 2017).

18.4.8 miRNA

MicroRNAs are relatively short (21–24 nucleotides), noncoding RNAs that function as negative regulators of gene expression and are involved in a variety of biologic processes and diverse diseases. As a general mechanism, miRNAs bind to target sequences in the 3'-untranslated region (3'-UTR) of a complementary mRNA. This binding leads to an inhibition of translation or degradation of the targeted mRNA. Both muscle-specific and nonmuscle-specific mRNAs are involved in myogenesis and muscle metabolism (Wang 2013). MicroRNAs have been studied in several diseases and conditions associated with muscle atrophy, such as sarcopenia (Drummond et al. 2011), COPD (Lewis et al. 2012), or atrophy denervation (Williams et al. 2009). MicroRNAs in mice with CKD or sham-operated, pair-fed controls rats were studied using a microarray analysis by Wada et al. (2011). Twelve significant changes in microRNA levels were observed in CKD, including a decrease in miR-29a and miR-29b. A decrease in miR-29 level resulted in an upregulation of Yin and Yang 1 (YY1) protein, which inhibited myogenesis. This finding is interesting, since suggests that inhibition of myogenesis contributes to CKD-induced muscle atrophy.

Another possibility is that microRNA could be used for treatment of wasting. As an example, in mice with CKD, the miR-486 mimic blocked dexamethasone-stimulated muscle-protein degradation without influencing protein synthesis. In addition, transfection of miR-486 in muscle of CKD mice leads to suppression of MAFbx/atrogin-1 and MuRF1 as well as to an increase in muscle mass (Xu et al. 2012). Thus miR-486 participates in the checks and balance to control muscle mass in uremia although the underlying mechanisms have not been fully described yet.

More recently, miRNAs were studied with the aim of finding exercise mimetic for treating muscle wasting. Wang et al. have identified miR-23a and miR27a as possible candidates. They were able to demonstrate an increased muscle expression of miR-23a and miR27a with resistance overload exercise in CKD mice. Moreover, in vivo transfection of the miR-23a and miR27a, gene cluster reduced muscle wasting by decreasing atrophy-related signaling and improving muscle regeneration pathways (Wang et al. 2017).

18.4.9 Activation of the UPS and Caspase-3

The principal catabolic mechanism causing muscle atrophy in rodent models of uremia involves the activation of the UPS (Rajan and Mitch 2008). UPS is strongly activated by metabolic acidosis, as shown in in vitro studies of cultured muscle cells, isolated muscle, and in rodents with experimental uremia (Mitch et al. 1994; Isozaki et al. 1996). It is interesting that metabolic acidosis is not associated with a decrease in intracellular pH (Bailey et al. 1995). The effect of acidosis on protein breakdown is likely due to its effects on insulin sensitivity and insulin signaling (Folli et al. 1996; DeFronzo and Beckles 1979).

In experimental uremia, the activities of caspase-3 and the UPS work together to stimulate muscle wasting. Caspases are proteases that participate in cell apoptosis and play a critical role in UPS muscle protein degradation (Chen et al. 2013). The activation of caspase-3 is regulated by

PI3K pathway (Yamaguchi and Wang 2001). In skeletal muscle of uremic rodents, caspase-3 acts to cleave the structure of actomyosin, then yielding substrates for the UPS that can rapidly degrade monomeric myosin or actin. Furthermore, caspase-3 activity can activate directly stimulates the proteolytic activity of the proteasome, increasing the amount of unfolded protein being inserted into channel of the 20S proteasome (Wang et al. 2010).

Caspase-3 activation has been demonstrated in muscle biopsy samples from animal models and patients with catabolic conditions (Workeneh et al. 2006). As a fingerprint of the proteolytic activity of caspase-3, a 14-kDa actin fragment has been proposed as a biomarker of wasting (Workeneh et al. 2006; Wang and Mitch 2014; Du et al. 2004).

18.4.10 Myostatin

In the last few years, myostatin, a structurally related member of the TGFβ superfamily, has emerged as a major regulator of muscle mass. Myostatin is mainly expressed in skeletal muscle, but it can be detected in several other tissues, including heart muscle and adipocytes. Myostatin binds to one of the two activin type II receptors (ActRIIB, and to a lesser extent, ActRIIA), thereby activating the activin type I receptors (ALK4 and ALK5), which in turn leads to the phosphorylation of Smad2 and Smad3. Phosphorylated Smad2 and Smad3 form a heterodimeric complex with the common mediator Smad4. These activated Smad proteins translocate into the nucleus and activate the transcription of the target genes through interaction with DNA and other nuclear factors. Myostatin signaling is controlled by another member of the Smad family, the inhibitory Smad7 protein.

Both myostatin deletion and inhibition cause muscle overgrowth, whereas its overexpression promotes muscle atrophy. Myostatin suppresses growth of skeletal muscles and in animals lacking myostatin, very large increases in muscle mass and strength have been observed (Sartori et al. 2009).

Myostatin has several roles. First of all, myostatin signaling is able to modulate the expression of target genes involved in the control of myoblast proliferation (Rodriguez et al. 2014). Incubation with recombinant myostatin can arrest cultured muscle cells (C2C12 myoblasts, primary bovine myoblasts and mouse satellite cells) in the G1 phase of the cell cycle. The antiproliferative action of myostatin is associated with upregulation of p21 (cyclin-dependent kinase inhibitor) and the subsequent decrease of cyclin-dependent kinases (CDK2, CDK4) and phosphorylated retinoblastoma protein (RB) levels.

A second major effect of myostatin on protein metabolism is the relationship that this protein has with two major branches of the protein kinase B (Akt) signaling pathways (Rodriguez et al. 2014). There is now of evidence of a consistent cross-talk between the pAkt-mTor pathway and myostatin (Rodriguez et al. 2014). The myostatin gene transfer in adult muscle causes downregulation of the Akt-mTOR signaling pathway (Amirouche et al. 2009). Downregulation of the pAkt pathway leads to augmented levels of Foxo1 and thus to an upregulation of atrophy-related genes (McFarlane et al. 2006).

Myostatin's suppression leads in the end to inhibition of muscle differentiation and therefore reduction of myotube size (Trendelenburg et al. 2009). Of note, myostatin inhibition results in skeletal muscle hypertrophy in adult muscle also independently of satellite cells.

It is noteworthy that a change in the balance between IGF-I and myostatin has been observed in muscle of rodents with experimental CKD (Sun et al. 2006). Recently, we analyzed muscle apoptosis and myostatin mRNA and their related intracellular signals in rectus abdominis samples obtained from patients with CKD scheduled for peritoneal dialysis. Apoptotic loss of myonuclei was threefold to fivefold increased. In addition, myostatin and IL-6 gene expressions were enhanced by ~50%, while levels of IGF-I mRNA were lower than in controls. Phosphorylated JNK and its downstream effector, p-c-Jun, were upregulated, while p-Akt was markedly downregulated. Multivariate analysis models revealed p-Akt and IL-6 to contribute individually and significantly to the prediction of apoptosis and myostatin gene expression, respectively. These findings demonstrate the occurrence of the activation of multiple pathways to promote muscle atrophy in skeletal muscle of patients with CKD. Both apoptosis of myonuclei and myostatin are upregulated. However, they appear to be associated with different signals, suggesting that their intracellular pathways are differently regulated in patients with CKD (Verzola et al. 2011). While apoptosis appears to be linked with defective insulin signaling, myostatin appears to be correlated with activation of innate immunity in muscle. Interestingly, Zhang et al. could recently demonstrate that high IL-6, by the activation of the JAK-Stat3 pathway, upregulates the myostatin gene (Zhang et al. 2013). IL-6 stimulates Stat3 to increase the expression of CCAAT/enhancer-binding protein δ (C/EBPδ). Moreover, the myostatin promoter contains recognition sites for members of the C/EBP family of transcription factors. The IL6-p-Stat3-C/EBPδ-myostatin pathway has been detected in muscle of both rodents and patients with CKD (Zhang et al. 2013). The understanding of the role of myostatin in muscle atrophy of patients with CKD opens the doors to translational therapies in these patients. The pharmacological inhibition by subcutaneous injections of an anti-myostatin peptibody into CKD mice reversed the loss of body weight and muscle mass and suppressed circulating inflammatory cytokines. Furthermore, myostatin inhibition also decreased the rate of protein degradation by 16% and increased protein synthesis by 13%. As a further observation, myostatin inhibition enhanced satellite cell function and improved IGF-1 intracellular signaling (Zhang et al. 2011). In addition, preventing extracellular proteolytic activation of myostatin in mice lead to an enhanced muscle function (Pirruccello-Straub et al. 2018).

18.4.11 AUTOPHAGY

Autophagy is an intracellular catabolic process that is used for the degradation and recycling, through the lysosomal system, of long-lived proteins, and organelles (Mizushima and Komatsu 2011). While balanced autophagy may actually prevent muscle wasting by recycling essential nutrients and preventing protein accumulation, excessive autophagy is activated in muscle cells during catabolic conditions (Schiaffino et al. 2013) might cause leads to wasting (Hussain and Sandri 2013). About 35 genes, generically known as autophagy-related genes (Atg), regulate each step of autophagy (Klionsky et al. 2003). Wang et al. demonstrated that several autophagy-related genes (Atg-3, Atg-12, LC3-II, and Beclin-1) were increased in muscles of CKD rats, as well as in C2C12 myotubes treated with TNF-α, which suggests that the activation of ALS mediates the degradation of myofibrillar and nonmyofibrillar proteins in uremia and inflammation. In addition, they observed that myostatin enhances autophagy and proteasomal degradation, which together contribute to the loss of muscle mass (Wang et al. 2015).

Very recent studies suggest that autophagy is linked to hyperphosphatemia. Hyperphosphatemia is very common in patients with CKD and is closely associated with the renal senescent phenotype, including cardiovascular disease and bone resorption and, possibly, cachexia.

Chen et al. (2018) explored the relationship of phosphate levels with muscle strength, dynapenia, and sarcopenia in 7421 participants of National Health and Nutrition Examination Survey. Notably, a significant association between phosphate and muscle strength was existed in elderly but not in young or middle-aged adults. Zhang et al. recently observed that high concentration of phosphate induce cell atrophy in myotubes in a dose- and time-dependent manner, through the activation of autophagy (Zhang et al. 2018). Therefore targeting phosphate levels and/or autophagy could be a therapeutic strategy for preventing muscle wasting in elderly subjects with CKD.

18.5 CONCLUSION

CKD accelerates sarcopenia and frailty in the elderly. At this time, besides nutritional supplementation and regular physical activity, no treatment has been shown to be effective in promoting anabolism in patients with CKD. New mechanisms that lead to wasting in patients with CKD have been recently identified, and new hope for therapeutic strategies to suppress or block protein loss is arising. Available data indicate that the checks and balances regulating muscle protein turnover in normal physiology are altered along with the decline in kidney function. On the one hand, GH/IGF1 resistance and defective insulin signaling favor a decline in muscle protein synthesis. On the other hand, complications associated with CKD, such as metabolic acidosis, defective insulin signaling, inflammation and abnormal appetite regulation, can activate protein breakdown. Inflammation stimulates intracellular signaling pathways that activate myostatin, which accelerates UPS-mediated catabolism and decrease protein synthesis. The use of inhibitors of myostatin or the myostatin-driven pathway is expected to yield new therapeutic options for muscle protein wasting in CKD (Verzola et al. 2019).

REFERENCES

Adey, D., R. Kumar, J. McCarthy, and K.S. Nair. 2000. Reduced synthesis of muscle proteins in chronic renal failure. *Am J Physiol Endocrinol Metab* 278:219–225.

Albert, C.M., N.R. Cook, J.M. Gaziano, et al. 2008a. Effect of folic acid and B vitamins on risk of cardiovascular events and total mortality among women at high risk for cardiovascular disease: A randomized trial. *JAMA* 299(17):2027–2036.

Albert, L.L., D.L. Nelson, and M.M. Cox. 2008b. *Lehninger Principles of Biochemistry* (5th ed.). New York: W.H. Freeman.

Allen, S.C. 2017. Systemic inflammation in the genesis of frailty and sarcopenia: An overview of the preventative and therapeutic role of exercise and the potential for drug treatments. *Geriatrics* 2:3–9.

Alvestrand, A., R.A. DeFronzo, D. Smith, and J. Wahren. 1988. Influence of hyperinsulinaemia on intracellular amino acids levels and amino acid exchange across splanchnic and leg tissues in uremia. *Clin Sci* 74:155–162.

Alvestrand, A., M. Mujagic, A. Wajngot, and S. Efendic. 1989. Glucose intolerance in uremic patients: The relative contributions of impaired beta-cell function and insulin resistance. *Clin Nephrol* 31:175–183.

Amirouche, A., A.C. Durieux, S. Banzet, et al. 2009. Down-regulation of Akt/mammalian target of rapamycin signaling pathway in response to myostatin overexpression in skeletal muscle. *Endocrinology* 150:286–294.

Anders, H.J., K. Andersen, and B. Stecher. 2013. The intestinal microbiota, a leaky gut, and abnormal immunity in kidney disease. *Kidney Int* 83:1010–1016.

Argiles, J.M., M. Busquet, and F.J. Lopez-Soriano. 2005. The pivotal role of cytokines in muscle wasting during cancer. *Int J Biochem Cell Biol* 37:2036–2046.

Bailey, J.L., B.K. England, R.C. Long Jr et al. 1995. Experimental acidemia and muscle cell pH in chronic acidosis and renal failure. *Am J Physiol* 269:706–712.

Bailey, J.L., X. Wang, B.K. England, et al. 1996. The acidosis of chronic renal failure activates muscle proteolysis in rats by augmenting transcription of genes encoding proteins of the ATP – Dependent ubiquitin-proteasome pathway. *J Clin Invest* 97:1447–1453.

Bailey, J.L., B. Zheng, Z. Hu, et al. 2006. Chronic kidney disease causes defects in signalling through the insulin receptor substrate/phosphatidylinositol 3-kinase/Akt pathway: Implications for muscle atrophy. *J Am Soc Nephrol* 17:1388–1394.

Bao, Y., L. Dalrymple, G.M. Chertow, et al. 2012. Frailty, dialysis initiation, and mortality in end-stage renal disease. *Arch Intern Med* 172(14):1071–1077.

Baylis, C. 2012. Nitric oxide synthase derangements and hypertension in kidney disease. *Curr Opin Nephrol Hypertens* 21:1–6.

Beaumier, L., L. Castillo, A.M. Ajami, and V.R. Young. 1995. Urea cycle intermediate kinetics and nitrate excretion at normal and "therapeutic" intakes of arginine in humans. *Am J Physiol* 269:E884–E896.

Bellizzi, V., A. Cupisti, F. Locatelli, et al. 2016. Low-protein diets for chronic kidney disease patients: The Italian experience. *BMC Nephrol* 17(1):77.

Bergström, J., and B. Lindholm. 1998. Malnutrition, cardiac disease, and mortality: An integrated point of view. *Am J Kidney Dis* 32:834–841.

Blum, W.F., M.B. Ranke, K. Kietzmann, et al. 1991. Growth hormone resistance and inhibition of somatomedin activity by excess of insulin-like growth factor binding protein in uremia. *Pediatr Nephrol* 5:539–544.

Bodine, S.C., E. Latres, S. Baumhueter, et al. 2001. Identification of ubiquitin ligases required for skeletal muscle atrophy. *Science* 294:1704–1708.

Boirie, Y., R. Albright, M. Bigelow, and K.S. Nair. 2004. Impairment of phenylalanine conversion to tyrosine in end-stage renal disease causing tyrosine deficiency. *Kidney Int* 66:591–596.

Bonanni, A., I. Mannucci, D. Verzola, et al. 2011. Protein energy wasting and mortality in chronic kidney disease. *Int J Environ Res Public Health* 8:1631–1654.

Bradley, J.R., J.R. Anderson, D.B. Evans, and A.J. Cowley. 1990. Impaired nutritive skeletal muscle blood flow in patients with chronic renal failure. *Clin Sci* 79:239–245.

Brungger, M., H.N. Hulter, and R. Krapf. 1997. Effect of chronic metabolic acidosis on the growth hormone/IGF-1 endocrine axis: New cause of growth hormone insensitivity in humans. *Kidney Int* 51:216–221.

Caglar, K., Y. Peng, L.B. Pupim, et al. 2002. Inflammatory signals associated with hemodialysis. *Kidney Int* 62(4):1408–1416.

Carrero, J.J., M. Chmielewski, J. Axelsson, et al. 2008. Muscle atrophy, inflammation and clinical outcome in incident and prevalent dialysis patients. *Clin Nutr* 27:557–564.

Carrero, J.J., A.R. Qureshi, J. Axelsson, et al. 2007. Comparison of nutritional and inflammatory markers in dialysis patients with reduced appetite. *Am J Clin Nutr* 85:695–701.

Carrero, J.J., F. Thomas, K. Nagy, et al. 2018. Global prevalence of protein-energy wasting in kidney disease: A meta-analysis of contemporary observational studies from the International Society of Renal Nutrition and Metabolism. *J Ren Nutr* 28:380–392.

Castellino, P., L. Luzi, M. Giordano, and R.A. Defronzo. 1999. Effects of insulin and amino acids on glucose and leucine metabolism in CAPD patients. *J Am Soc Nephrol* 10:1050–1058.

Castillo, L., L. Beaumier, A.M. Ajami, and V.R. Young. 1996. Whole body nitric oxide synthesis in healthy men determined from [15N] arginine-to-[15N] citrulline labeling. *Proc Natl Acad Sci USA* 93:11460–11465.

Chen, C.T., S.H. Lin, J.S. Chen, and Y.J. Hsu. 2013. Muscle wasting in hemodialysis patients: New therapeutic strategies for resolving an old problem. *Sci World J* 2013:643954.

Chen, J., P. Muntner, L. Lee Ham, et al. 2003. Insulin resistance and the risk of chronic kidney disease in nondiabetic US adults. *J Am Soc Nephrol* 14:469–477.

Chen, Y., J. Biada, S. Sood, and R. Rabkin. 2010. Uremia attenuates growth hormone-stimulated insulin-like growth factor-1 expression, a process worsened by inflammation. *Kidney Int* 78:89–95.

Chen, Y., F.C. Fervenza, and R. Rabkin. 2001. Growth factors in the treatment of wasting in kidney failure. *J Ren Nutr* 11:62–66.

Chen, Y., S. Sood, K. McIntire, et al. 2011. Leucine-stimulated mTOR signaling is partly attenuated in skeletal muscle of chronically uremic rats. *Am J Physiol Endocrinol Metab* 301(5):E873–E881.

Chen, Y.Y., K. Tung-Wei, W.C. Cheng-Wai et al. 2018. Exploring the link between serum phosphate levels and low muscle strength, dynapenia and sarcopenia. *Sci Rep* 2:23–32.

Chertow, G.M., N.W. Levin, G.J. Beck, et al. 2010. In-center hemodialysis six times per week versus three times per week. *N Engl J Med* 363(24):2287–2300.

Cheung, W., P.X. Yu, B.M. Little, et al. 2005. Role of leptin and melanocortin signaling in uremia-associated cachexia. *J Clin Invest* 115(6):1659–1665.

Cheung, W.W., H.J. Kuo, S. Markison, et al. 2007. Peripheral administration of the melanocortin-4 receptor antagonist NBI-12i ameliorates uremia-associated cachexia in mice. *J Am Soc Nephrol* 18(9):2517–2524.

Chiang, J., G. Kaysen, and M. Segal, et al. 2019. Low testosterone is associated with frailty, muscle wasting, and physical dysfunction among men receiving hemodialysis: A longitudinal analysis. *Nephrol Dial Transpl* 34:802–810.

Chmielewski, M., J.J. Carrero, A.R. Qureshi, et al. 2009. Temporal discrepancies in the association between the apoB/apoA-I ratio and mortality in incident dialysis patients. *J Int Med* 265:708–716.

Chmielewski, M., M. Verduijn, C. Drechsler, et al. 2011. Low cholesterol in dialysis patients-causal factor for mortality or an effect of confounding. *Nephrol Dial Transpl* 10:3325–3331.

Clark, M.G., E.Q. Colquhon, S. Rattigan, et al. 1995. Vascular and endocrine control of muscle metabolism. *Am J Physiol* 268:797–803.

Clarke, R., J. Halsey, S. Lewington, et al. 2010. Effects of lowering homocysteine levels with B vitamins on cardiovascular disease, cancer, and cause-specific mortality: Meta-analysis of 8 randomized trials involving 37 485 individuals. *Arch Intern Med* 170(18):1622–1631.

Coffey, V.G., Z. Zhong, A. Shield, et al. 2006. Early signaling responses to divergent exercise stimuli in skeletal muscle from well-trained humans. *FASEB J* 20:190–192.

Cohen, S., J.J. Brault, S.P. Gygi, et al. 2009. During muscle atrophy, thick, but not thin, filament components are degraded by MuRF1-dependent ubiquitylation. *J Cell Biol* 185:1083–1095.

Contreras, G., B. Hu, B.C. Astor, et al. 2010. African-American study of kidney disease, and hypertension study group. Malnutrition-inflammation modifies the relationship of cholesterol with cardiovascular disease. *J Am Soc Nephrol* 21:2131–2142.

Cooney, R.N., and M. Shumate. 2006. The inhibitory effects of interleukin-1 on growth hormone action during catabolic illness. *Vitam Horm* 74:317–340.

D'Apolito, M., X. Du, H. Zong, et al. 2010. Urea-induced ROS generation causes insulin resistance in mice with chronic renal failure. *J Clin Invest* 120:203–213.

DeFronzo, R.A., and A.D. Beckles. 1979. Glucose intolerance following chronic metabolic acidosis in man. *Am J Physiol* 236:328–334.

DeFronzo, R.A., D. Smith, and A. Alvestrand. 1983. Insulin action in uremia. *Kidney Int* 24:S102–S114.

DeFronzo, R.A., J.D. Tobin, J.W. Rowe, and R. Andres. 1978. Glucose intolerance in uremia: Quantification of pancreatic beta cell sensitivity to glucose and tissue sensitivity to insulin. *J Clin Invest* 62:425–435.

Deger, S.M., A.M. Hung, J.L. Gamboa, et al. 2017. Systemic inflammation is associated with exaggerated skeletal muscle protein catabolism in maintenance hemodialysis patients. *JCI Insight* 2(22):e95185.

Drechsler, C., and C. Wanner. 2015. The obesity paradox and the role of inflammation. *J Am Soc Nephrol* 27(5):1270–1272.

Dreyer, H.C., M.J. Drummond, B. Pennings, et al. 2008. Leucine-enriched essential amino acid and carbohydrate ingestion following resistance exercise enhances mTOR signaling and protein synthesis in human muscle. *Am J Physiol Endocrinol Metab* 294:E392–E400.

Drummond, M.J., J.J. McCarthy, M. Sinha, et al. 2011. Aging and microRNA expression in human skeletal muscle: A microarray and bioinformatics analysis. *Physiol Genomics* 43:595–603.

Du, J., X. Wang, C. Miereles, et al. 2004. Activation of caspase-3 is an initial step triggering accelerated muscle proteolysis in catabolic conditions. *J Clin Invest* 113:115–123.

Egerman, M.A., and D.J. Glass. 2014. Signaling pathways controlling skeletal muscle mass. *Crit Rev Biochem Mol Biol* 49:59–68.

Esposito, P., E. La Porta, M.A. Grignano, et al. 2018. Soluble toll-like receptor 4: A new player in subclinical inflammation and malnutrition in hemodialysis patients. *J Ren Nutr* 28:259–264.

Eustace, J.A., B. Astor, P.M. Muntner, et al. 2004. Prevalence of acidosis and inflammation and their association with low serum albumin in chronic kidney disease. *Kidney Int* 65:1031–1040.

Feinstein, R., H. Kanety, M.Z. Papa, et al. 1993. Tumor necrosis factor-a suppresses insulin-induced tyrosine phosphorylation of insulin receptor and its substrates. *J Biol Chem* 268:26055–26058.

Fernandez-Prado, R., R. Esteras, M.V. Perez-Gomez, et al. 2017. Nutrients turned into toxins: Microbiota modulation of nutrient properties in chronic kidney disease. *Nutrients* 9(5):489.

Fliser, D., G. Pacini, R. Engelleiter, et al. 1998. Insulin resistance and hyperinsulinemia are already present in patients with incipient renal disease. *Kidney Int* 53:1343–1347.

Folli, F., M.J. Saad, and C.R. Kahn. 1996. Insulin receptor/IRS-1/PI 3-kinase signaling system in corticosteroid-induced insulin resistance. *Acta Diabetol* 33:185–192.

Fouque, D., K. Kalantar-Zadeh, J. Kopple, et al. 2008. A proposed nomenclature and diagnostic criteria for protein-energy wasting in acute and chronic kidney disease. *Kidney Int* 73:391–398.

Fouque, D., S.C. Peng, and J.D. Kopple. 1995. Impaired metabolic response to recombinant insulin-like growth factor-1 in dialysis patients. *Kidney Int* 47:876–883.

Fouque, D., S.C. Peng, E. Shamir, and J.D. Kopple. 2000. Recombinant human insulin-like growth factor-1 induces an anabolic response in malnourished CAPD patients. *Kidney Int* 57:646–654.

Franch, H.A., S. Raissi, X. Wang, et al. 2004. Acidosis impairs insulin receptor substrate-1-associated phosphoinositide 3-kinase signaling in muscle cells: Consequences on proteolysis. *Am J Physiol Renal Physiol* 287:F700–F706.

Friedman, A.N., A.G. Bostom, J. Selhub, A.S. Levey, and I.H. Rosenberg. 2001. The kidney and homocysteine metabolism. *J Am Soc Nephrol* 12:2181–2189.

Fujita, S., H.C. Dreyer, M.J. Drummond, et al. 2007. Nutrient signalling in the regulation of human muscle protein synthesis. *J Physiol* 582:813–823.

Gao, X., J. Wu, Z. Dong, et al. 2010. A low-protein diet supplemented with ketoacids plays a more protective role against oxidative stress of rat kidney tissue with 5/6 nephrectomy than a low-protein diet alone. *Br J Nutr* 103:608–616.

Garcia, L.A., K.K. King, M.G. Ferrini, et al. 2011. 1,25(OH)2vitamin D3 stimulates myogenic differentiation by inhibiting cell proliferation and modulating the expression of promyogenic growth factors and myostatin in C2C12 skeletal muscle cells. *Endocrinology* 152:2976–2986.

Garibotto, G., A. Barreca, R. Russo, et al. 1997. Effects of recombinant human growth hormone on muscle protein turnover in malnourished hemodialysis patients. *J Clin Invest* 99:97–105.

Garibotto, G., A. Barreca, A. Sofia, et al. 2000. Effects of growth hormone on leptin metabolism and energy expenditure in hemodialysis patients with protein-calorie malnutrition. *J Am Soc Nephrol* 11:2106–2113.

Garibotto, G., A. Bonanni, and D. Verzola. 2012. Effect of kidney failure and hemodialysis on protein and amino acid metabolism. *Curr Opin Clin Nutr Metab Care* 15:78–84.

Garibotto, G., R. Russo, A. Sofia, et al. 1994. Skeletal muscle protein synthesis and degradation in patients with chronic renal failure. *Kidney Int* 45:1432–1439.

Garibotto, G., R. Russo, A. Sofia, et al. 1996. Muscle protein turnover in CRF patients with metabolic acidosis or normal acid-base balance. *Miner Electrolyte Metab* 22:58–61.

Garibotto, G., R. Russo, A. Sofia, et al. 2008. Effects of uremia and inflammation on growth hormone resistance in patients with chronic kidney disease. *Kidney Int* 74:937–945.

Garibotto, G., R. Russo, R. Franceschini, et al. 1998. Inter-organ leptin exchange in humans. *Biochem Biophys Res Commun* 247(2):504–509.

Garibotto, G., A. Sofia, E.L. Parodi, et al. 2018. Effects of low-protein, and supplemented very low-protein diets, on muscle protein turnover in patients with CKD. *Kidney Int Rep* 3:701–710.

Garibotto, G., A. Sofia, V. Procopio, et al. 2006. Peripheral tissue release of interleukin-6 in patients with chronic kidney diseases: Effects of end-stage renal disease and microinflammatory state. *Kidney Int* 70:384–390.

Garibotto, G., A. Sofia, R. Russo, et al. 2015. Insulin sensitivity of muscle protein metabolism is altered in patients with chronic kidney disease and metabolic acidosis. *Kidney Int* 88:1419–1426.

Garibotto, G., A. Sofia, S. Saffioti, et al. 2003. Interorgan exchange of aminothiols in humans. *Am J Physiol Endocrinol Metab* 284:E757–E763.

Garibotto, G., A. Valli, B. Anderstam, et al. 2009. The kidney is a major site of S-adenosylhomocysteine disposal in humans. *Kidney Int* 76:293–296.

Garibotto, G., D. Verzola, M. Vettore, and P. Tessari. 2018. The contribution of muscle, kidney and splanchnic tissues to leucine transamination in humans. *Can J Physiol Pharmacol* 96(4):382–387.

Gayle, D.A., M. Desai, E. Casillas, R. Belooseski, and M.G. Ross. 2006. Gender specific orexigenic and anorexigenic mechanisms in rats. *Life Sci* 79:1531–1536.

Goodman, M.N. 1991. Tumor necrosis factor induces skeletal muscle protein breakdown in rats. *Am J Physiol Endocrinol Metab* 260:E727–E730.

Gordon, P.L., G.K. Sakkas, J.W. Doyle, T. Shubert, and K.L. Johansen. 2007. Relationship between vitamin d and muscle size and strength in patients on hemodialysis. *J Ren Nutr* 17:397–407.

Graham, K.A., D. Reaich, S.M. Channon, et al. 1996. Correction of acidosis in CAPD decreases whole body protein degradation. *Kidney Int* 49:1396–1400.

Graham, K.A., D. Reaich, S.M. Channon, et al. 1997. Correction of acidosis in hemodialysis decreases wholebody protein degradation. *J Am Soc Nephrol* 8:632–637.

Greenhaff, P.L., L.G. Karagounis, N. Peirce, et al. 2008. Disassociation between the effects of amino acids and insulin on signaling, ubiquitin ligases, and protein turnover in human muscle. *Am J Physiol Endocrinol Metab* 295:E595–E604.

Haddad, F., F. Zaldivar, D.M. Cooper, and G.R. Adams. 2005. IL-6-induced skeletal muscle atrophy. *J Appl Physiol* 98:911–917.

Hay, N., and N. Sonenberg. 2004. Upstream and downstream of mTOR. *Genes Dev* 18(16):1926–1945.

Hooshmand, B., M. Rusanen, T. Nuganda, et al. 2018. Serum insulin and cognitive performance in older adults: A longitudinal study. *Am J Med* 132:367–373.

Hotamisligil, G.S., N.S. Shargill, and B.M. Spiegelman. 1993. Adipose expression of tumor necrosis factor-alpha: Direct role in obesity-linked insulin resistance. *Science* 259:87–91.

Howden, E.J., J.S. Coombes, and N.M. Isbe. 2015. The role of exercise training in the management of chronic kidney disease. *Curr Opin Nephrol Hypertens* 24:480–487.

Hussain, S.N., and M. Sandri. 2013. Role of autophagy in COPD skeletal muscle dysfunction. *J Appl Physiol* 114:1273–1281.

Ikizler, T.A. 2011. Exercise as an anabolic intervention in patients with end-stage renal disease. *J Ren Nutr* 21(1):52–56.

Ikizler, T.A., P.J. Flakoll, R.A. Parker, and R.M. Hakim. 1994. Amino acid and albumin losses during hemodialysis. *Kidney Int* 46(3):830–837.

Ikizler, T.A., R.L. Wingard, P.J. Flakoll, et al. 1996. Effects of recombinant human growth hormone on plasma and dialysate amino acid profiles in CAPD patients. *Kidney Int* 50:229–234.

Ishikawa, S., S. Naito, S. Iimori, et al. 2018. Loop diuretics are associated with greater risk of sarcopenia in patients with non-dialysis-dependent chronic kidney disease. *PLoS One* 13:e0192990.

Isoyama, N., A.R. Qureshi, C.M. Avesani, et al. 2014. Comparative associations of muscle mass and muscle strength with mortality in dialysis patients. *Clin J Am Soc Nephrol* 9(10):1720–1728.

Isozaki, U., W.E. Mitch, B.K. England, and S.R. Price. 1996. Protein degradation and increased mRNAs encoding proteins of the ubiquitin-proteasome proteolytic pathway in BC3H1 myocytes require an interaction between glucocorticoids and acidification. *Proc Natl Acad Sci U S A* 5:1967–1971.

Jabs, K., M.J. Koury, W.D. Dupont, and C. Wagner. 2006. Relationship between plasma S-adenosylhomocysteine concentration and glomerular filtration rate in children. *Metabolism* 55(2):252–257.

Jamison, R.L., P. Hartigan, J.S. Kaufman, et al. 2007. Effect of homocysteine lowering on mortality and vascular disease in advanced chronic kidney disease and end-stage renal disease: A randomized controlled trial. *JAMA* 12(298):1163–1170.

Johansen, K.L., L.S. Dalrymple, and C. Delgado. 2017. Factors associated with frailty and its trajectory among patients on hemodialysis. *Clin J Am Soc Nephrol* 12:1100–1108.

Kalantar-Zadeh, K., G. Block, M.H. Humphreys, and J.D. Kopple. 2003. Reverse epidemiology of cardiovascular risk factors in maintenance dialysis patients. *Kidney Int* 63:793–808.

Kalantar-Zadeh, K., G. Block, C.J. McAllister, M.H. Humphreys, and J.D. Kopple. 2004. Appetite and inflammation, nutrition, anemia, and clinical outcome in hemodialysis patients. *Am J Clin Nutr* 80:299–307.

Kang, S.H., K.H. Cho, J.W. Park, and J.Y. Do. 2017. Low appendicular muscle mass is associated with mortality in peritoneal dialysis patients: A single-center cohort study. *Eur J Clin Nutr* 71(12):1405–1410.

Kaskel, F. 2003. Chronic renal disease: A growing problem. *Kidney Int* 64:1141–1151.

Kaysen, G.A. 2014. Progressive inflammation and wasting in patients with ESRD. *Clin J Am Soc Nephrol* 9:225–226.

Kim, D.H., J.Y. Kim, B.P. Yu, and H.Y. Chung. 2008. The activation of NF-kappaB through Akt-induced FOXO1 phosphorylation during aging and its modulation by calorie restriction. *Biogerontology* 9:33–47.

Kim, J.C., K. Kalantar-Zadeh, and J.D. Kopple. 2013. Frailty and protein-energy wasting in elderly patients with end stage kidney disease. *J Am Soc Nephrol* 24:337–351.

Kimball, S.R., and L.S. Jefferson. 2006. New functions for amino acids: Effects on gene transcription and translation. *Am J Clin Nutr* 83:500–507.

Kimmel, P.L., T.M. Phillips, S.J. Simmens, et al. 1998. Immunologic function and survival in hemodialysis patients. *Kidney Int* 54:236–244.

Klionsky, D.J., J.M. Cregg, W.A. Dunn, et al. 2003. A unified nomenclature for yeast autophagy-related genes. *Dev Cell* 5:539–545.

Kobayashi, S., K. Maesato, H. Moriya, et al. 2005. Insulin resistance in patients with chronic kidney disease. *Am J Kidney Dis* 45:275–280.

Koppe, L., N.J. Pillon, R.E. Vella, et al. 2013. p-Cresyl sulfate promotes insulin resistance associated with CKD. *J Am Soc Nephrol* 24:88–99.

Kopple, J.D. 2007. Phenylalanine and tyrosine metabolism in chronic kidney failure. *J Nutr* 137:1586S–1590S.

Kopple, J.D., G. Brunori, M. Leiserowitz, and D. Fouque. 2005. Growth hormone induces anabolism in malnourished maintenance haemodialysis patients. *Nephrol Dial Transplant* 20:952–958.

Kopple, J.D., A.K. Cheung, J.S. Christiansen, et al. 2011. OPPORTUNITY: A large-scale randomized clinical trial of growth hormone in hemodialysis patients. *Nephrol Dial Transplant* 26(12):4095–4103.

Kopple, J.D., T. Greene, W.C. Chumlea, et al. 2000. Relationship between nutritional status and the glomerular filtration rate: Results from the MDRD study. *Kidney Int* 57(4):1688–1703.

Kotzmann, H., N. Yilmaz, P. Lercher, et al. 2001. Differential effects of growth hormone therapy in malnourished hemodialysis patients. *Kidney Int* 60:1578–1585.

Kovesdy, C.P., and K. Kalantar-Zadeh. 2009. Why is protein-energy wasting associated with mortality in chronic kidney disease? *Semin Nephrol* 29:3–14.

Kovesdy, C.P., J.D. Kopple, and K. Kalantar-Zadeh. 2013. Management of protein-energy wasting in non-dialysis-dependent chronic kidney disease: Reconciling low protein intake with nutritional therapy. *Am J Clin Nutr* 97(6):1163–1177.

Laplante, M., and D.M. Sabatini. 2012. mTOR signaling in growth control and disease. *Cell* 149(2):274–293.

Latos, D.L. 1996. Chronic dialysis in patients over age 65. *J Am Soc Nephrol* 7:637–646.

LeBrasseur, N.K., K. Walsh, and Z. Arany. 2011. Metabolic benefits of resistance training and fast glycolytic skeletal muscle. *Am J Physiol* 300:E3–E10.

Lewis, A., J. Riddoch-Contreras, S.A. Natanek, et al. 2012. Downregulation of the serum response factor/miR-1 axis in the quadriceps of patients with COPD. *Thorax* 67:26–34.

Lindeman, R.D. 1990. Overview: Renal physiology and pathophysiology of aging. *Am J Kidney Dis* 16:275–282.

Liu, C., Q. Wang, H. Guo, et al. 2008. Plasma S adenosylhomocysteine is a better biomarker of atherosclerosis than homocysteine in apolipoprotein E-deficient mice fed high dietary methionine. *J Nutr* 138:311–315.

Liu, X., R. Yu, L. Sun, et al. 2017. The nuclear phosphatase SCP4 regulates FoxO transcription factors during muscle wasting in chronic kidney disease. *Kidney Int* 92(2):336–348.

Liu, Y., J. Coresh, J.A. Eustace, et al. 2004. Association between cholesterol level and mortality in dialysis patients: Role of inflammation and malnutrition. *J Am Med Assoc* 291:451–459.

Louard, R.J., D.A. Fryburg, and R.A. Gelfand. 1992. Insulin sensitivity of protein and glucose metabolism in human forearm skeletal muscle. *J Clin Invest* 90:2348–2354.

Luiking, Y.C., M. Poeze, G. Ramsay, and N.E. Deutz. 2009. Reduced citrulline production in sepsis is related to diminished de novo arginine and nitric oxide production. *Am J Clin Nutr* 89:142–152.

Luo, X., Y. Xiao, F. Song, et al. 2012. Increased plasma S-adenosyl-homocysteine levels induce the proliferation and migration of VSMCs through an oxidative stress-ERK1/2 pathway in apoE(-/-) mice. *Cardiovasc Res* 95(2):241–250.

Ma, X.M., and J. Blenis. 2009. Molecular mechanisms of mTOR-mediated translational control. *Nat Rev Mol Cell Biol* 10(5):307–318.

Mak, R.H., W.W. Cheung, G. Solomon, and A. Gertler. 2018. Preparation of potent leptin receptor antagonists and their therapeutic use in mouse models of uremic cachexia and kidney fibrosis. *Curr Pharm Des* 24:1012–1018.

Martins, R., G.J. Lithgow, and W. Link. 2016. Long live FOXO: Unraveling the role of FOXO proteins in aging and longevity. *Aging Cell* 15(2):196–207.

Masuda, Y., T. Tanaka, N. Inomata, et al. 2000. Ghrelin stimulates gastric acid secretion and motility in rats. *Biochem Biophys Res Commun* 276:905–908.

Matthews, D.E., and D.M. Bier. 1983. Stable isotope methods for nutritional investigation. *Annu Rev Nutr* 3:309–339.

McFarlane, C., E. Plummer, M. Thomas, et al. 2006. Myostatin induces cachexia by activating the ubiquitin proteolytic system through an NF-kappaB-independent, FoxO1-dependent mechanism. *J Cell Physiol* 209:501–514.

Mehrotra, R., and J.D. Kopple. 2001. Nutritional management of maintenance dialysis patients: Why aren't we doing better? *Annu Rev Nutr* 21:343–379.

Miller, S.L., and R.R. Wolfe. 2008. The danger of weight loss in the elderly. *J Nutr Health Aging* 12:487–489.

Mitch, W.E. 2006. Proteolytic mechanisms, not malnutrition, cause loss of muscle mass in kidney failure. *J Ren Nutr* 16:208–211.

Mitch, W.E., R. Medina, S. Grieber, et al. 1994. Metabolic acidosis stimulates muscle protein degradation by activating the adenosine triphosphate-dependent pathway involving ubiquitin and proteasomes. *J Clin Invest* 93:2127–2133.

Miyamoto, T., J.J. Carrero, A.R. Qureshi, et al. 2011. Circulating follistatin in patients with chronic kidney disease: Implications for muscle strength, bone mineral density, inflammation, and survival. *Clin J Am Soc Nephrol* 6:1001–1008.

Mizushima, N., and M. Komatsu. 2011. Autophagy: Renovation of cells and tissues. *Cell* 147:728–741.

Moller, N., S. Meek, M. Bigelow, J. Andrews, and K.S. Nair. 2000. The kidney is an important site for in vivo phenylalanine-to-tyrosine conversion in adult humans: A metabolic role of the kidney. *Proc Natl Acad Sci USA* 97:1242–1246.

Morley, J.E., B. Vellas, G.A. van Kan, et al. 2013. Frailty consensus: A call to action. *J Am Med Dir Assoc* 4:392–397.

Nair, K.S. 2000. Age-related changes in muscle. *Mayo Clin Proc* 75:14–S18.

Nixon, A.C., T.M. Bampouras, N. Pendleton, et al. 2018. Frailty and chronic kidney disease: Current evidence and continuing uncertainties. *Clin Kidney J* 2:236–245.

Onesti, J.K., and D.C. Guttridge. 2014. Inflammation based regulation of cancer cachexia. *Biomed Res Int* 2014:16840.

Oreopoulos, A., K. Kalantar-Zadeh, A.M. Sharma, and G.C. Fonarow, 2009. The obesity paradox in the elderly: Potential mechanisms and clinical implications. *Clin Geriatr Med* 25:643–659.

Owen, E.E., and R.R. Robinson. 1963. Amino acid extraction and ammonia metabolism by the human kidney during the prolonged administration of ammonium chloride. *J Clin Invest* 42:263–276.

Palm, F., T. Teerlink, and P. Hansell. 2009. Nitric oxide and kidney oxygenation. *Curr Opin Nephrol Hypertens* 18:68–73.

Panichi, V., M. Migliori, S. De Pietro, et al. 2002. C-reactive protein and interleukin-6 levels are related to renal function in predialytic chronic renal failure. *Nephron* 91:594–600.

Pereira, B.J., S. Sundaram, B. Snodgrass, et al. 1996. Plasma lipopolysaccharide binding protein and bactericidal/permeability increasing factor in CRF and HD patients. *J Am Soc Nephrol* 7:479–487.

Pereira, R.A., A.C. Cordeiro, C.M. Avesani, et al. 2015. Sarcopenia in chronic kidney disease on conservative therapy: Prevalence and association with mortality. *Nephrol Dial Transplant* 10:1718–1725.

Pirruccello-Straub, M., J. Jackson, S. Wawersik, et al. 2018. Blocking extracellular activation of myostatin as a strategy for treating muscle wasting. *Sci Rep* 8(1):2292.

Polinder-Bos, H.A., H. Nacak, F.W. Dekker, et al. 2017. Low urinary creatinine excretion is associated with self-reported frailty in patients with advanced chronic kidney disease. *Kidney Int Rep* 2017(2):676–685.

Pupim, L.B., P.J. Flakoll, K.M. Majchrzak, et al. 2005a. Increased muscle protein breakdown in chronic hemodialysis patients with type 2 diabetes mellitus. *Kidney Int* 68:1857–1865.

Pupim, L.B., P.J. Flakoll, C. Yu, and T.A. Ikizler. 2005b. Recombinant human growth hormone improves muscle amino acid uptake and whole-body protein metabolism in chronic hemodialysis patients. *Am J Clin Nutr* 82:1235–1243.

Pupim, L.B., O. Heimburger, A.R. Qureshi, et al. 2005c. Accelerated lean body mass loss in incident chronic dialysis patients with diabetes mellitus. *Kidney Int* 68:2368–2374.

Qing, D.P., H.U. Ding, and J. Vagdama. 1999. Elevated myocardial cytosolic calcium impairs insulin-like growth factor-1 stimlated protein synthesis in chronic renal failure. *J Am Soc Nephrol* 10:84–91.

Rajan, V., and W.E. Mitch. 2008. Ubiquitin, proteasomes and proteolytic mechanisms activated by kidney disease. *Biochim Biophys Acta* 1782:795–799.

Reaich, D., S.M. Channon, C.M. Scrimgeour, et al. 1993. Correction of acidosis in humans with CRF decreases protein degradation and amino acid oxidation. *Am J Physiol* 265:230–235.

Reaich, K.A., K.A. Graham, S.M. Channon, et al. 1995. Insulin-mediated changes in PD and glucose uptake after correction of acidosis in humans with CRF. *Am J Physiol Endocrinol Metab* 268:E121–E126.

Rigalleau, V., and H. Gin. 2005. Carbohydrate metabolism in uraemia. *Curr Opin Clin Nutr Metab Care* 8:463–469.

Rocco, M.V., J.T. Dwyer, B. Larive, et al. 2004. The effect of dialysis dose and membrane flux on nutritional parameters in hemodialysis patients: Results of the HEMO Study. *Kidney Int* 65(6):2321–2334.

Rodriguez, J., B. Vernus, I. Chelh, et al. 2014. Myostatin and the skeletal muscle atrophy and hypertrophy signaling pathways. *Cell Mol Life Sci* 71:4361–4371.

Sandri, M., C. Sandri, A. Gilbert, et al. 2004. Foxo transcription factors induce the atrophy-related ubiquitin ligase atrogin-1 and cause skeletal muscle atrophy. *Cell* 117:399–412.

Sartori, R., G. Milan, M. Patron, et al. 2009. Smad2 and 3 transcription factors control muscle mass in adulthood. *Am J Physiol Cell Physiol* 296:1248–1257.

Schepers, E., T. Speer, S.M. Bode-Böger, et al. 2014. Dimethylarginines ADMA and SDMA: The real water-soluble small toxins? *Semin Nephrol* 34:97–105.

Schiaffino, S., K.A. Dyar, S. Ciciliot, B. Blaauw, and M. Sandri. 2013. Mechanisms regulating skeletal muscle growth and atrophy. *FEBS J* 280:4294–4314.

Shah, V.O., E.A. Dominic, P. Moseley, et al. 2006. Hemodialysis modulates gene expression profile in skeletal muscle. *Am J Kidney Dis* 48:616–628.

Shi, Y., Z. Liu, Y. Shen, et al. 2018. A novel perspective linkage between kidney function and Alzheimer's disease. *Front Cell Neurosci* 12:384–397.

Shinohara, K., T. Shoji, M. Emoto, et al. 2002. Insulin resistance as an independent predictor of cardiovascular mortality in patients with end-stage renal disease. *J Am Soc Nephrol* 13:1894–1900.

Siew, E.D., L.B. Pupim, K.M. Majchrzak, et al. 2007. Insulin resistance is associated with skeletal muscle protein breakdown in non-diabetic chronic hemodialysis patients. *Kidney Int* 71:146–152.

Smith, D., and R.A. DeFronzo. 1982. Insulin resistance in uremia mediated by postbinding defects. *Kidney Int* 22:54–62.

Souza, V.A., D. Oliveira, S.R. Barbosa, et al. 2017. Sarcopenia in patients with chronic kidney disease not yet on dialysis: Analysis of the prevalence and associated factors. *PLoS One* 12(4):e0176230.

Stam, F., C. Van Guldener, P.M. ter Wee, et al. 2004. Homocysteine clearance and methylation flux rates in health and end-stage renal disease: Association with S-adenosylhomocysteine. *Am J Physiol Renal Physiol* 287:F215–F223.

Stenvinkel, P. 2002. Inflammation in end-stage renal failure: Could it be treated. *Nephrol Dial Transpl* 17:S33–S38.

Stenvinkel, P., J.J. Carrero, F. von Walden, et al. 2016. Muscle wasting in end-stage renal disease promulgates premature death: Established, emerging and potential novel treatment strategies. *Nephrol Dial Transplant* 31:1070–1077.

Stenvinkel, P., I.A. Gillespie, J. Tunks, et al. 2015. Inflammation modifies the paradoxical association between body mass index and mortality in hemodialysis patients. *J Am Soc Nephrol* 27(5):1479–1486.

Sun, D.F., Y. Chen, and R. Rabkin. 2006. Work-induced changes in skeletal muscle IGF-I and myostatin gene expression in uraemia. *Kidney Int* 70:453–459.

Sun, D.F., Z. Zheng, P. Tummala, et al. 2004. Chronic uremia attenuates growth hormone-induced signal transduction in skeletal muscle. *J Am Soc Nephrol* 15:2630–2646.

Tang, E.D., G. Nunez, F.G. Barr, et al. 1999. Negative regulation of the forkhead transcription factor FKHR. *Akt J Biol Chem* 274:16741–16746.

Tee, A.R., and J. Blenis. 2005. mTOR, translational control and human disease. *Semin Cell Dev Biol* 16(1):29–37.

Tessari, P., D. Cecchet, A. Cosma, et al. 2011. Insulin resistance of amino acid and protein metabolism in type 2 diabetes. *Clin Nutr* 30:267–272.

Tessari, P., A. Coracina, E. Kiwanuka, et al. 2005. Effects of insulin on methionine and homocysteine kinetics in type 2 diabetes with nephropathy. *Diabetes* 54:2968–2976.

Tessari, P., G. Deferrari, C. Robaudo, et al. 1999. Phenylalanine hydroxylation across the kidney in humans. *Kidney Int* 56:2168–2172.

Tessari, P., and G. Garibotto. 2000. Inter-organ amino acid exchange. *Curr Opin Clin Nut Metab Care* 3:51–57.

Teta, D. 2015. Insulin resistance as a therapeutic target for chronic kidney disease. *J Ren Nutr* 25:226–229.

Thomas, S.S., Y. Dong, L. Zhang, and W.E. Mitch. 2013. Signal regulatory protein-α interacts with the insulin receptor contributing to muscle wasting in chronic kidney disease. *Kidney Int* 84:308–316.

Tizianello, A., G. Deferrari, G. Garibotto, et al. 1988. Abnormalities of amino acid and keto acid metabolism in renal failure. In Davison, D., ed., *Nephrology*. London, UK: Baillière, pp. 1011–1019.

Tizianello, A., G. Deferrari, G. Gurreri, and C. Robaudo. 1980. Renal metabolism of amino acids and ammonia in subjects with normal renal function and in patients with chronic renal insufficiency. *J Clin Invest* 65:1162–1173.

Trendelenburg, A.U., A. Meyer, D. Rohner, et al. 2009. Myostatin reduces Akt/TORC1/p70S6K signaling, inhibiting myoblast differentiation and myotube size. *Am J Physiol Cell Physiol* 296:1258–1270.

Tzivion, G., M. Dobson, and G. Ramakrishnan. 2011. FoxO transcription factors; Regulation by AKT and 14-3-3 proteins. *Biochim Biophys Acta* 1813:1938–1945.

Valli, A., J.J. Carrero, A.R. Qureshi, et al. 2008. Elevated serum levels of S-adenosylhomocysteine, but not homocysteine, are associated with cardiovascular disease in stage 5 chronic kidney disease patients. *Clin Chim Acta* 395(1–2):106–110.

Van de Poll, M.C., P.B. Soeters, N.E. Deutz, K.C. Fearon, and C.H. Dejong. 2004. Renal metabolism of amino acids: Its role in interorgan amino acid exchange. *Am J Clin Nutr* 79:185–197.

Van der Lely, A.J., M. Tschöp, M.L. Heiman, and E. Ghigo. 2004. Biological, physiological, pathophysiological, and pharmacological aspects of ghrelin. *Endocr Rev* 25:426–457.

Van Guldener, C. 2005. Homocysteine and the kidney. *Curr Drug Metab* 6:23–26.

Van Guldener, C., C. Kulik, R. Berger, et al. 1999. Homocysteine and methionine metabolism in ESRD: A stable isotope study. *Kidney Int* 56:1064–1071.

van Loon, I.N., N.A. Goto, F.T.J. Boereboom, et al. 2017. Frailty screening tools for elderly patients incident to dialysis. *Clin J Am Soc Nephrol* 12:1480–1488.

Vashistha, T., R. Mehrotra, J. Park, et al. 2014. Effect of age and dialysis vintage on obesity paradox in long-term hemodialysis patients. *Am J Kidney Dis* 63:612–622.

Vaziri, N.D., M.V. Pahl, A. Crum, and K. Norris. 2012. Effect of uremia on structure and function of immune system. *J Ren Nutr* 22:149–156.

Verzola, D., C. Barisione, D. Picciotto, and G. Garibotto. 2019. Emerging role of Myostatin and its inhibition in the setting of chronic kidney disease. *Kidney Int* 95:506–517.

Verzola, D., A. Bonanni, A. Sofia, et al. 2017. Toll-like receptor 4 signalling mediates inflammation in skeletal muscle of patients with chronic kidney disease. *J Cachexia Sarcopenia Muscle* 8(1):131–144.

Verzola, D., V. Procopio, A. Sofia, et al. 2011. Apoptosis and myostatin mRNA are upregulated in the skeletal muscle of patients with chronic kidney disease. *Kidney Int* 79:773–782.

Von Haehling, S., M. Lainscak, J. Springer, and S.D. Anker. 2009. Cardiac cachexia: A systematic overview. *Pharmacol Ther* 121:227–252.

Wada, S., Y. Kato, M. Okutsu, et al. 2011. Translational suppression of atrophic regulators by microRNA-23a integrates resistance to skeletal muscle atrophy. *J Biol Chem* 286:38456–38465.

Wang, B., C. Zhang, A. Zhang, et al. 2017. MicroRNA-23a and MicroRNA-27a mimic exercise by ameliorating CKD-induced muscle atrophy. *J Am Soc Nephrol* 28(9):2631–2640.

Wang, D.T., Y.J. Yang, R.H. Huang, et al. 2015. Myostatin activates the ubiquitin-proteasome and autophagy-lysosome systems contributing to muscle wasting in chronic kidney disease. *Oxid Med Cell Longev* 2015:684965.

Wang, F., R. Xiong, S. Feng, et al. 2018. Association of circulating levels of ADMA with carotid intima-media thickness in patients with CKD: A systematic review and meta-analysis. *Kidney Blood Press Res* 43:25–33.

Wang, H., R. Casaburi, W.E. Taylor, et al. 2005. Skeletal muscle mRNA for IGF-IEa, IGF-II, and IGF-I receptor is decreased in sedentary chronic hemodialysis patients. *Kidney Int* 68:352–361.

Wang, X.H. 2013. MicroRNA in myogenesis and muscle atrophy. *Curr Opin Clin Nutr Metab Care* 16:258–266.

Wang, X.H., J. Du, J.D. Klein, et al. 2009. Exercise ameliorates chronic kidney disease–induced defects in muscle protein metabolism and progenitor cell function. *Kidney Int* 76:751–759.

Wang, X.H., and W.E. Mitch. 2014. Mechanisms of muscle wasting in chronic kidney disease. *Nat Rev Nephrol* 10:504–516.

Wang, X.H., L. Zhang, W.E. Mitch, et al. 2010. Caspase-3 cleaves specific 19 S proteasome subunits in skeletal muscle stimulating proteasome activity. *J Biol Chem* 285:21249–21257.

Watson, E.L., J.L. Viana, D. Wimbury, et al. 2017. The effect of resistance exercise on inflammatory and myogenic markers in patients with chronic kidney disease. *Front Physiol* 8:541–555.

Wilcken, D.E., and B. Wilcken. 1976. The pathogenesis of coronary artery disease. A possible role for methionine metabolism. *J Clin Invest* 57:1079–1082.

Williams, A.H., G. Valdez, V. Moresi, et al. 2009. MicroRNA-206 delays ALS progression and promotes regeneration of neuromuscular synapses in mice. *Science* 326:1549–1554.

Winkler, M.S., S. Kluge, M. Holzmann, et al. 2017. Markers of nitric oxide are associated with sepsis severity: An observational study. *Crit Care* 21:189–199.

Workeneh, B.T., H. Rondon-Berrios, L. Zhang, et al. 2006. Development of a diagnostic method for detecting increased muscle protein degradation in patients with catabolic conditions. *J Am Soc Nephrol* 17:3233–3239.

Wullschleger, S., R. Loewith, and M.N. Hall. 2006. TOR signaling in growth and metabolism. *Cell* 124:471–484.

Xiao, Y., W. Huang, J. Zhang, et al. 2015. Increased plasma S-adenosylhomocysteine-accelerated atherosclerosis is associated with epigenetic regulation of endoplasmic reticulum stress in apoE-/- mice. *Arterioscler Thromb Vasc Biol* 35(1):60–70.

Xiao, Y., X. Su, W. Huang, et al. 2015. Role of S-adenosylhomocysteine in cardiovascular disease and its potential epigenetic mechanism. *Int J Biochem Cell Biol* 67:158–166.

Xu, J., R. Li, B. Workeneh, et al. 2012. Transcription factor FoxO1, the dominant mediator of muscle wasting in chronic kidney disease, is inhibited by microRNA-486. *Kidney Int* 82:401–411.

Xu, H., and J.J. Carrero. 2017. Insulin resistance in chronic kidney disease. *Nephrology (Carlton)* S 4:31–34.

Yamaguchi, H., and H.G. Wang. 2001. The protein kinase PKB/Akt regulates cell survival and apoptosis by inhibiting Bax conformational change. *Oncogene* 20:7779–7786.

Zawada, A.M., K.S. Rogacev, B. Hummel, J.T. Berg, A. Friedrich, H.J. Roth, R. Obeid, J. Geisel, D. Fliser, and G.H. Heine. 2014. S-adenosylhomocysteine is associated with subclinical atherosclerosis and renal function in a cardiovascular low-risk population. *Atherosclerosis* 234(1):17–22.

Zhang, L., J. Pan, Y. Dong, et al. 2013. Stat3 activation links a C/EBPδ to myostatin pathway to stimulate loss of muscle mass. *Cell Metab* 18:368–379.

Zhang, L., V. Rajan, E. Lin, et al. 2011. Pharmacological inhibition of myostatin suppresses systemic inflammation and muscle atrophy in mice with chronic kidney disease. *FASEB J* 25:1653–1663.

Zhang, Y.Y., M. Yang, J.F. Bao, et al. 2018. Phosphate stimulates myotube atrophy through autophagy activation: Evidence of hyperphosphatemia contributing to skeletal muscle wasting in chronic kidney disease. *BMC Nephrol* 9:45–58.

19 Sarcopenia and Parkinson's Disease

Molecular Mechanisms and Clinical Management

Manlio Vinciguerra

CONTENTS

19.1 INTRODUCTION: PARKINSON'S DISEASE AND SARCOPENIA

Parkinson's disease (PD) is a degenerative disorder of the central nervous system that affects the motor system. In the early stages of the disease, major symptoms include shaking, rigidity, slowness of movement, and difficulty with walking and gait (Kalia and Lang 2015). Late appearing symptoms include thinking and behavioral problems, with dementia occurring very often in the advanced stages of PD, and depression as the most common psychiatric co-morbidity. Additional symptoms include also sensory and sleep issues. The main motor symptoms are collectively called parkinsonism (Kalia and Lang 2015). Worldwide, PD affects ~50 million people with ~100,000 deaths each year. The life expectancy of people with PD is significantly reduced, with mortality ratios around twice or more those of people without PD. Mortality risk factors include cognitive decline and dementia, old age at onset, and presence of swallowing impairments. Death from aspiration pneumonia occurs twice as often in individuals with PD as in the healthy population. Sarcopenia is the degenerative loss of skeletal muscle tissue mass, integrity, and strength associated with increasing

---

Dedicated to my beloved father Prof. Domenico M. Vinciguerra (1937–2015), affected by Parkinson's Disease.

age. Sarcopenia is part of the frailty syndrome and often part of cachexia, but it can also exist independently of cachexia. Sarcopenia is a risk factor for falls and consequent fractures in old people with PD. Most adults affected by PD display significantly compromised bone and muscle health. The aim of this review is to summarize the state-of-the-art knowledge about sarcopenia in PD in terms of causes, mechanisms of onset, and therapeutic management.

19.2 PARKINSON DISEASE: CAUSES AND MECHANISMS

Parkinson disease is a disease that has been known about since ancient times. In the ancient Indian Ayurveda medical system, it was termed with the name Kampavata. In Western world, it was described as "shaking palsy" by the Galen in AD 175. The history of PD then found expansion from 1817, when British apothecary James Parkinson published the book titled "An Essay on the Shaking Palsy," to nowadays. Prior to Parkinson's descriptions, others had already described features of the disease, but only the twentieth century greatly improved knowledge of the disease and its treatments. Before twentieth century, PD was known as *paralysis agitans*. The term "Parkinson's disease" was proposed several decades later by French neurologist Jean-Martin Charcot. PD is a degenerative disorder of the central nervous system, which mainly affects the motor system. Charcot managed to distinguish PD from multiple sclerosis and other tremor-associated disorders, and he recognized cases that later would likely be incorporated among the Parkinsonism (Goetz 2011). Early treatments of PD were based on the empirical observation and practice: in the nineteenth century, anticholinergic drugs were the front-line treatment. The subsequent discovery of dopaminergic deficits in PD and the synthetic pathway of dopamine led to the first human trials of levodopa. Early in the course of PD, evident symptoms are movement related (shaking, rigidity, slowness of movement, and difficulty with walking and gait). In advanced stages, thinking and behavioral problems may arise, with dementia and depression occurring very often as psychiatric complications of the disease (Sveinbjornsdottir 2016). Other symptoms include sensory, sleep, and emotional problems. The main motor symptoms are collectively called parkinsonism or parkinsonian syndrome. Signs and symptoms of PD are in fact varied and are summarized in Table 19.1.

Among these, the most important can be classified as motor and neuropsychiatric symptoms (Jankovic 2008, Samii, Nutt, and Ransom 2004). Four motor symptoms are considered crucial in PD development: slowness of movement (bradykinesia), tremor, rigidity, and instability of posture. Very typical for PD is an initial asymmetric distribution of these symptoms, where in the course of

---

TABLE 19.1

PD Signs and Symptoms

- **Tremor.** A tremor, or shaking, usually begins in a limb, often your hand or fingers. Patients may notice a back-and-forth rubbing of thumbs and forefingers. One characteristic of PD is tremor of hands when at rest.
- **Bradykinesia.** Over time, Parkinson's disease may reduce the ability to move and slow movements, making simple tasks difficult and time-consuming. Steps may become shorter when walking, or patients may find it difficult to get out of a chair. Also, patients may drag their feet as they try to walk, making it difficult to move.
- **Rigid muscles.** Muscle stiffness may occur in any part of the body. The stiff muscles can limit the range of motion and cause pain.
- **Impaired posture and balance.** Posture may become stooped, or patients may have balance problems as a result of PD.
- **Loss of automatic movements.** In PD, patients may have a decreased ability to perform unconscious movements, including blinking, smiling, or swinging their arms when they walk.
- **Speech changes.** PD patients may have speech problems. They may speak softly, quickly, slur or hesitate before talking. Their speech may be more of a monotone-boring rather than with the usual inflections.
- **Writing changes.** It may become hard to write for PD patients, and writing may appear small.

the disease, a gradual progression to bilateral symptoms develops although some asymmetry usually persists. Other motor symptoms include posture disturbances such as decreased arm swing, a forward-flexed posture, and the use of tiny steps when walking; speech and swallowing disturbances are frequent; and other symptoms such as a stone-faced expression or small and incoherent handwriting are examples of the range of common motor problems that are likely to appear (Samii, Nutt, and Ransom 2004, Jankovic 2008). Neuropsychiatric symptoms, in contrast, include mainly cognition, mood, and behavior problems and actually can be as disabling as motor symptoms, if not worse (Caballol, Marti, and Tolosa 2007, Parker et al. 2013). Cognitive disturbances can occur even in the initial stages of the disease in some cases. A very high proportion of sufferers will have mild cognitive impairment as the disease advances. It must be noted that most common deficits in nondemented patients are executive dysfunction, which translates into impaired set shifting, poor problem solving, and fluctuations in attention among other difficulties; slowed cognitive speed; memory problems; specifically in recalling learned information, with an important improvement with cues; visual-spatial skill difficulties, which can be seen when the subject with PD is, for instance, asked to perform tests of facial recognition and perception of line orientation. Deficits tend to aggravate with time, developing in many cases into dementia. A person with PD has a sixfold increased risk of suffering it, and the overall rate in PD individuals is around 30%. Cognitive problems and dementia are usually accompanied by behavior and mood alterations, which most frequently include depression, apathy, anxiety, obsessive–compulsive behaviors. Psychotic symptoms are also common in PD, generally associated with dopamine therapy. Symptoms of psychosis, or impaired reality testing, are either visual hallucinations, less commonly auditory, and rarely in other domains including tactile, gustatory or olfactory, or delusions and irrational beliefs. Hallucinations are generally stereotyped and without emotional content. Initially patients usually have insights so that the hallucinations are benign in terms of their immediate impact (they can even be considered "funny") but indicate unfortunately poor prognosis, with increased risk of dementia, worsened psychotic symptoms, and mortality (Shergill, Walker, and Le Katona 1998, Friedman 2010). Hallucinations can occur in Parkinsonian syndromes for different reasons. There is an overlap between PD and Lewy body dementia, so that where Lewy bodies are present in the visual cortex, hallucinations may result. More often instead hallucinations can also be brought about by excessive dopaminergic stimulation: therefore paradoxically, the treatment of motor symptoms induces other cognitive symptoms (Shergill, Walker, and Le Katona 1998, Friedman 2010). Most hallucinations are visual in nature, often formed as familiar people or animals (horses, etc.) and are generally nonthreatening in their nature. Treatment options consist of modifying the dosage of dopaminergic drugs taken each day, adding an antipsychotic drug like quetiapine, or offering carers a psychosocial intervention to help them cope with the hallucinations, also if this type of intervention is rarely used or available in the national health systems. As already mentioned, when other diseases mimic PD, they are categorized as Parkinsonism. Death in patients with PD can be caused by several fatal events: patients with PD are more likely to die from pneumonia and less likely as malignancy or ischemic heart disease, than the control population. Pneumonia is a terminal event in 45% of patients with PD. One dangerous type of pneumonia affecting patients with PD is aspiration pneumonia (*ab ingestis*): this is bronchopneumonia that develops due to the entrance of foreign materials into the bronchial tree, usually oral including food, saliva, or nasal secretions (Martinez-Ramirez, Almeida, et al. 2015). Depending on the nature of the aspirate, a chemical pneumonitis can develop, and particularly anaerobic pathogenic bacteria may trigger inflammation. Aspiration pneumonia is often caused by an incompetent swallowing mechanism in patients with PD. Other terminal events include malignant neoplasms (11.6%), heart diseases (4.1%), cerebral infarction (3.7%), and septicemia (3.3%). Cerebral hemorrhage was the 11_{th} most frequent cause of death, accounting for only 0.8% of deaths among patients with PD (Iwasaki et al. 1990).

PD can be either primary or secondary. Primary PD has no known cause although some atypical cases have a genetic origin (Shadrina, Slominsky, and Limborska 2010). The putative role of genetic factors in the pathogenesis of PD has long been discussed, and the data collected during analyses of family highlight the significance of the genetic component in the progress of PD. For instance, studying twins has demonstrated that even in PD discordant twins, concordance was observed with regard to nigrostriatal function disturbance (45% in monozygotic twins) (Burn et al. 1992). Studies using positron emission tomography have stated a high concordance (75% in monozygotic twins) for the subclinical dopaminergic dysfunction level (Piccini et al. 1999). Analysis of families with PD has shown that the risk of the disease in the relatives of sporadic parkinsonism patients is 3- to 12-fold, given that the genetic backgrounds and the mode of life are similar (Pankratz and Foroud 2004). Families have also been revealed showing Mendelian inheritance of PD, autosomal recessively, and autosomal dominantly. The first direct proof in favor of a significant role of genetic factors in PD pathogenesis was obtained at the end of the twentieth century when the SNCA gene, whose mutations lead to the development of PD, was first identified in studies of the autosomal-dominant form of the disease. This gene encodes the protein α-synuclein, the main protein component of Lewy bodies. To date, several different loci have been discovered, which are, in one way or another, involved in the pathogenesis of PD. However, the incidence of familial PD forms is only 10%–15%, and in most cases, the disease is sporadic and idiopathic. Several issues, determined by the complicated pattern of the disease, exist in the identification of the role of genetic factors in the development of the sporadic PD form. The factors that considerably complicate the investigation of the genetic causes of PD include the delayed start of the disease; in a majority of cases, incomplete penetrance of the genetic factors involved; the genetic heterogeneity of the disease; and the ponderous contribution of environmental factors to PD pathogenesis. In both forms of PD (genetic and sporadic), however, the common pattern of neuropathology is always observed, linked with the degeneration of dopaminergic neurons within the substantia nigra. The familial and the sporadic forms of PD are indistinguishable at the clinical level, and this possibly indicates the existence of shared pathogenic mechanisms. Analysis of all hitherto-revealed loci allows of course several mechanisms to be suggested, which can account for the causes of the selective and the progressive death of dopaminergic neurons (Lill 2016). Those are processes connected with the ubiquitin-dependent proteasomal protein degradation, mitochondrial dysfunction, the differentiation of dopaminergic neurons, the functioning of synapses and lysosomes, the exchange of dopamine, and other processes (Lill 2016).

Secondary parkinsonism is due instead to known causes like toxins. Many risks and protective factors have been identified and investigated so far: the strongest evidence is for an increased risk in people exposed to certain pesticides and a reduced risk in tobacco smokers. Besides toxins, drug-induced parkinsonism (DIP) is the second most common etiology of parkinsonism in the elderly after PD. Many patients with DIP may be misdiagnosed with PD because the clinical features of these two conditions are quite indistinguishable (Shin and Chung 2012). Moreover, neurological deficits in patients with DIP may be severe enough to affect daily activities and may persist for long periods after the cessation of drug taking. In addition to typical antipsychotics, DIP may be caused by calcium channel blockers, atypical antipsychotics, and antiepileptic drugs. The clinical manifestations of DIP are classically described as bilateral and symmetric parkinsonism without tremor at rest. However, about half of patients with DIP show asymmetrical parkinsonism and tremor at rest, making it difficult to differentiate DIP from PD. The pathophysiology of DIP is related to drug-induced changes in the basal ganglia motor circuit secondary to dopaminergic receptor blockade. Since these effects are limited to postsynaptic dopaminergic receptors, it is expected that presynaptic dopaminergic neurons in the striatum will be intact. For this reason, dopamine transporter (DAT) imaging is useful for diagnosing presynaptic parkinsonism. DAT uptake in the striatum is significantly decreased even in the early stage of PD, and this characteristic may help in differentiating PD from DIP. Typical antipsychotics are the most common causes of DIP. However, atypical antipsychotics can also induce parkinsonism. DIP may have a significant and longstanding effect on daily lives of patients with PD, and it is very important that physicians are cautious when giving dopaminergic receptor blockers while they should monitor patients' neurological signs, especially for parkinsonism and other movement disorders.

Either way, primary or secondary PD, the motor symptoms of the disease result from the death of cells in the substantia nigra, in the midbrain (Davie 2008). This results in insufficient dopamine in these areas. The reasons for this cell death are poorly understood but involve the build-up of proteins into Lewy bodies in the neurons (Dickson 2007). Where the Lewy bodies are located is partly related to the expression and degree of the symptoms. At the molecular level, PD is speculated to be caused by some exogenous or endogenous substances that are neurotoxic toward catecholamine neurons, which toxicity leads to mitochondrial dysfunction and subsequent oxidative stress resulting in the programmed cell death (apoptosis or autophagy) of dopaminergic neurons (Shadrina, Slominsky, and Limborska 2010). Recent studies on the causative genes of rare familial PD cases, such as α-synuclein and parkin, suggest that dysfunction of the ubiquitin–proteasome system (UPS) and the resultant accumulation of misfolded proteins and endoplasmic reticulum stress may cause the death of dopaminergic neurons in the substantia nigra (Shadrina, Slominsky, and Limborska 2010). Activated microglia, which accompany an inflammatory process, are present in the substantia nigra of the PD brain, and they produce protective or toxic substances, such as cytokines, neurotrophins, and reactive oxygen or nitrogen species (Shadrina, Slominsky, and Limborska 2010). Activated microglia may be protective at first and later may become neurotoxic owing to toxic change to promote the progression toward the neuronal death (Shadrina, Slominsky, and Limborska 2010). All of these accumulating evidences on PD points to a hypothesis that multiple primary causes of PD may be ultimately linked to a final common signal-transduction pathway leading to programmed cell death, that is, apoptosis or autophagy, of the neurons.

Diagnosis of typical cases is mainly based on symptoms, with tests such as neuroimaging being used for confirmation (Brooks 2010). As it will be discussed more in detail in Section 19.6.1, treatments, typically the anti-PD medications L-DOPA and dopamine agonists, improve the early symptoms of the disease. As the disease progresses and neurons continue to be lost, these medications become ineffective while producing a complication marked by involuntary writhing movements and hallucinations. Diet has shown some effectiveness at improving symptoms (Kones 2010). Surgery to place deep brain stimulation (DBS) has been used to reduce motor symptoms in severe cases where drugs are ineffective (Papageorgiou, Deschner, and Papageorgiou 2016).

19.3 SARCOPENIA

19.3.1 CLINICAL DEFINITION

The term "sarcopenia" (Greek "sarx" or flesh + "penia" or loss) was coined by Rosenber in 1989 to describe the age-related decrease of muscle mass. Sarcopenia is defined as the degenerative loss of skeletal muscle mass (0.5%–1% loss per year after the age of 50 years), quality, and strength associated with aging. Sarcopenia increases from 14% in those aged above 65 years but below 70 years to 53% in those above 80 years of age. The prevalence in 60- to 70-year-olds is reported as 5%–13%, while the prevalence ranges from 11% to 50% in people >80 years. The number of people around the world aged ≥60 years was estimated at 600 million in 2000, a figure that is expected to rise to 1.2 billion by 2025 and 2 billion by 2050 (Dodds et al. 2015, Kim and Choi 2013). Even with a conservative estimate of prevalence, sarcopenia affects >50 million people today and will affect >200 million in the next 40 years. The impact of sarcopenia on older people is enormous; its substantial tools are measured in terms of morbidity, disability, high costs of health care, and mortality (Santilli et al. 2014). Sarcopenia is common and associated with serious health consequences in terms of frailty, disability, morbidity, and mortality. It is often a component of cachexia but can also exist independently of cachexia; whereas cachexia includes malaise and is secondary to an underlying pathology (such as cancer), sarcopenia may occur in healthy people. It is not a disease or a syndrome, and it is not always even a medical sign, because the degree to which it is physiologic versus pathologic is not determined. The three consensus papers, which have published a definition of sarcopenia, were written under the auspices of, respectively, the European Working Group on

Sarcopenia in Older People (EWGSOP) (Cruz-Jentoft et al. 2010), the European Society for Clinical Nutrition and Metabolism Special Interest Groups (ESPEN-SIG) (Muscaritoli et al. 2010), and the International Working Group on Sarcopenia (IWGS) (Fielding et al. 2011). The definitions out of these consensus were as follows: (A) the presence of low skeletal muscle mass and either low muscle strength (handgrip) or low muscle performance (walking speed); when all three conditions are present, severe sarcopenia may be diagnosed (EWGSOP); (B) the presence of low skeletal muscle mass and low muscle strength (which they advised could be assessed by walking speed) (ESPEN-SIG); and (C) the presence of low skeletal muscle mass and low muscle function (which they advised could be assessed by walking speed) and that sarcopenia is associated with muscle mass loss alone or in conjunction with increased fat mass (IWGS).

The diagnosis of sarcopenia can be carried out by assessing walking speed in patients >65 years old. If walking speed is below 0.8 m/s at the 4-m walking test, it is important to measure the muscle mass. A low muscle mass, which is intended as the percentage of muscle mass divided by height squared is below two standard deviations of the normal young mean (<7.23 kg/m^2 and in women at <5.67 kg/m^2) as defined using dual-energy X-ray absorptiometry. If the walking speed at the 4-m walking test is higher than 0.8 m/s, the hand-grip strength should be tested; if this value is lower than 20 kg in women and 30 kg in man, the muscle mass must be analyzed as described previously (Santilli et al. 2014). Sarcopenia is characterized initially by a muscle atrophy—a decrease in the size of the muscle, along with a reduction in muscle tissue quality, accompanied by the replacement of muscle fibers with fat, an increase in fibrosis, changes in muscle metabolism, oxidative stress, and degeneration of the neuromuscular junction and leading to progressive loss of muscle function and frailty (Ryall, Schertzer, and Lynch 2008). Sarcopenia intuitively is determined by two factors: initial amount of muscle mass and rate at which aging decreases muscle mass. In general, because of the loss of independence associated with loss of muscle strength, the threshold at which muscle wasting can be considered a disease is very individual and varies from subject to subject (Marcell 2003). Sarcopenia is also characterized by a decrease in the circumference of distinct types of muscle fibers, identified by expression of distinct myosin variants. During sarcopenia, there is a decrease in type 2 fiber circumference (Type II), with no decrease in type 1 fiber circumference (Type I), and enervated type 2 fibers are often transformed to type 1 fibers by innervation by slow type 1 fiber motor nerves (Doherty 2003) (Figure 19.1).

However, simple circumference measurement does not provide enough information to decide whether or not an individual is suffering from severe sarcopenia. In some individuals, an unequivocal cause of sarcopenia can be identified. In other cases, no evident cause can be identified. Introducing the categories of primary sarcopenia and secondary sarcopenia has been useful in clinical practice. Sarcopenia can be considered "primary"—age-related when no other cause is evident but aging itself, while sarcopenia can be considered "secondary" when one or more other causes are evident (Cruz-Jentoft et al. 2010). In most old people, the etiology of sarcopenia is multifactorial so that it may not be possible to characterize each individual as having a primary or secondary condition. This scenario is quite consistent with recognizing sarcopenia as a multifaceted geriatric syndrome. Among external factors a deficient intake of energy and protein will contribute to loss of muscle and function. Reduced intake of vitamin D has been associated with low functionality in the elderly. Acute and chronic co-morbidities will also contribute to the development of sarcopenia in older persons. Sarcopenia is associated with major co-morbidity such as obesity, osteoporosis and type 2 diabetes, and insulin resistance (Janssen et al. 2004, Gale et al. 2007). Co-morbidities may, on the one hand, lead to reduced physical activity and, on the other hand, to increased generation of proinflammatory cytokines that play important roles in triggering protein turnover. Individuals who have had an active lifestyle throughout their life have more lean body mass and muscle mass when they age (Zamboni et al. 2008, Jensen 2008). Sarcopenia is also a risk factor for other adverse events, and it increases the risk of physical limitation and subsequent disability; recent studies also show that this condition increases the risk of comorbidities. The age-related muscle mass loss is also

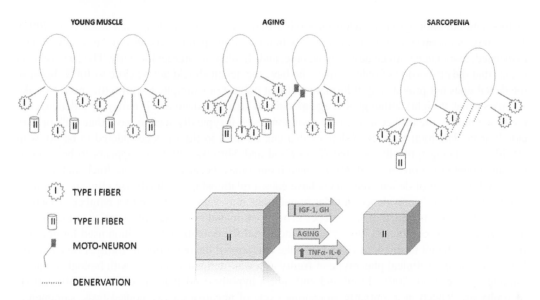

FIGURE 19.1 Effect of age on the motor unit, depicting, young, aged, and aged sarcopenic fibers. This drawing depicts the pronounced denervation of type II fibers and the recruitment of type I fibers into surviving motor units in older subjects, with impairment of recruitment in sarcopenic subjects.

associated with an increased risk of incident disability, all-cause mortality, and higher health-care costs in the older people (Rockwood 2005, Landi et al. 2010). The muscle mass is a predictor of overall mortality after a follow-up of 4 years. However, some recent studies suggested that muscle function may be a more powerful predictor of disability and mortality than the muscle mass (Santilli et al. 2014). Sarcopenia, independent of its causes, may predict negative outcomes, such as falls and/or subsequent difficulty in instrumental and basic activities of daily living (ADL). Furthermore, it has been associated with an increased risk of death, hospitalization, need for long-term care, and higher health care expenditures. The evidence that sarcopenia has a greater effect on survival than other clinical characteristics is significant for clinical practice among old and frail older persons. The traditional medical model should move from a disease-centered perspective to a functioning-centered view. In this respect, the prevention of sarcopenia is one of the major goals of public health professionals and clinicians. There is an established link between inactivity and losses of muscle mass and strength; this suggests that physical activity should be a protective factor for the prevention but also the management of sarcopenia. Great attention is needed to prevent or postpone as much as possible the onset of sarcopenia among older people, to enhance survival and to reduce the demand for long-term care (Landi, Liperoti, Fusco, Mastropaolo, Quattrociocchi, Proia, Tosato, et al. 2012, Chastin et al. 2012, Liu and Latham 2009).

19.3.2 Frailty Syndrome and Aging

Frailty is a common geriatric syndrome that carries an elevated risk of dramatic declines in health and function among older people. Frailty is a condition associated with aging, it has been recognized for many centuries, and it is often a result of loss of muscle with aging. The frailty phenotype defines frailty as a distinct clinical syndrome meeting three or more of five phenotypic criteria: weakness, slowness, low level of physical activity, self-reported exhaustion, and unsought weight loss. The frailty index (FI) defines frailty as cumulative deficits identified in a full geriatric assessment. The FI was developed originally by Rockwood and co-workers by counting the number of deficits accumulated, including diseases, physical and cognitive impairments, psychosocial risk factors, and common geriatric syndromes other than frailty (Jones, Song, and Rockwood 2004). The criteria for a variable to be defined as a deficit is that the parameters need to be acquired, age-associated, associated with an adverse outcome, and should not saturate too early. The last criterion means that the proportion of older adults who have the deficit should not be close to 100%, because the deficit does not provide any information at that point. For example, nocturia (the need to wake and pass urine at night), although it is age-associated, disrupts sleep and is definitely a deficit, cannot be counted in the FI, because it is so common that it is typically seen in more than 90% of very elderly men. The total number of deficits that can be used in the FI is considered to be 80, with 30–70 items being typically counted (Rockwood and Mitnitski 2011). FI appears to be a sensitive and quantitative predictor of adverse health outcomes, because of its more finely defined risk scale and inclusion of deficits that likely have causal relationships with adverse clinical outcomes. Estimates of frailty's prevalence in older populations may vary according to a number of factors, including the place in which the prevalence is being estimated—for example, nursing home (higher prevalence) versus community (lower prevalence), and the operational definition used for defining frailty. In an attempt to standardize and to make operational the definition of frailty, Fried and colleagues proposed a clinical phenotype of frailty as a well-defined syndrome with biological underpinnings (Fried et al. 2001). Fried and colleagues hypothesized that the clinical manifestations of frailty are related in a mutually worsening cycle of negative energy homeostasis, sarcopenia, and diminished strength. Building on this conceptual framework, preliminary evidence has been obtained on the natural history of the clinical phenotype of frailty (Gill et al. 2006, Xue et al. 2008). Using this frailty phenotype framework proposed by Fried, prevalence estimates of 7%–16% have been reported in noninstitutionalized, community-living older adults. The occurrence of frailty increases steadily with advancing age and is more common among those of lower socioeconomic status. Frail older adults are at high risk for major adverse health outcomes, including disability, falls, institutionalization, hospitalization, and mortality. Chronic inflammation is likely a key pathophysiologic process that contributes to the frailty syndrome directly and indirectly through other intermediate physiologic systems, such as the musculoskeletal, endocrine, and cardiovascular systems (Chen, Mao, and Leng 2014, de Lau and Breteler 2006, Pringsheim et al. 2014, Allam, Del Castillo, and Navajas 2005). The complex multifactorial etiologies of frailty also include obesity and specific diseases. Epidemiologic research to date has led to the identification of a number of risk factors for frailty, including (a) chronic diseases, such as cardiovascular disease, diabetes, chronic kidney disease, depression, and cognitive impairment and (b) physiologic impairments, such as activation of inflammation and coagulation systems, anemia, atherosclerosis, autonomic dysfunction, hormonal abnormalities, obesity, hypovitaminosis D in men, and environment-related factors such as life space and neighborhood features (Chen, Mao, and Leng 2014, de Lau and Breteler 2006, Pringsheim et al. 2014, Allam, Del Castillo, and Navajas 2005). As it has been already mentioned, the biological underpinnings of frailty are multifactorial, involving dysregulation across many physiological systems. A proinflammatory state, sarcopenia, anemia, relative hormonal deficiencies or excessive exposure, insulin resistance, compromised altered immune function, micronutrient deficiencies, and oxidative stress are each individually associated with a higher likelihood of frailty (Figure 19.2) (Walston 2012).

Additional findings demonstrate that the risk of frailty augments with the number of dysregulated physiological systems nonlinearly and independently of chronic diseases and chronologic age, suggesting synergistic effects that on their own may be relatively mild (Fried et al. 2009). The clinical implication of these synergies is that interventions that affect multiple systems may yield greater benefits in prevention and treatment of frailty than interventions that affect only one of the systems. Associations between specific disease states with frailty have also been observed, including cardiovascular disease, diabetes mellitus, renal insufficiency, and other inflammatory processes (Walston 2012, Marcell 2003). As well as dysregulation across several physiologic systems underlies the pathogenesis of the frailty, specific disease states are likely concomitant manifestations of the underlying impaired physiologic function. Clinically measurable disease states can manifest

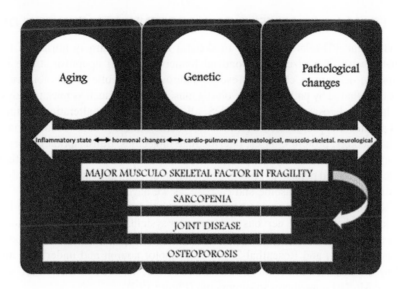

FIGURE 19.2   Factors leading to frailty.

themselves or be captured actually before the onset of frailty. No single disease state is necessary and sufficient for the pathogenesis of frailty, since many individuals with chronic diseases are not frail. In summary, rather than being dependent on the presence of measurable diseases, frailty is an expression of a critical mass of an ensemble of physiologic impairments. Components of frailty syndrome, in addition of sarcopenia, include osteoporosis and muscle weaknesses (Figure 19.2).

19.4   SARCOPENIA IN PARKINSON DISEASE

There are limited data on the prevalence of sarcopenia and frailty in PD. A study including 104 patients with PD and 330 non-PD controls from a population-based cohort aged >65 years showed that the prevalence of sarcopenia was 55.8% in patients with PD and 8.2% in non-PD controls (Peball et al. 2018). Moreover, the severity of sarcopenia parallels the increase in disease stage among patients with PD, compared with that among the control group of the same age (Yazar et al. 2018, Vetrano et al. 2018). Worse PD motor severity is a reliable epidemiological predictor of sarcopenia and frailty in PD (Tan et al. 2018). Geroscience is a new interdisciplinary field that aims to understand the relationship between aging and chronic age-related diseases such as PD. It is based on epidemiological evidence and experimental data that aging is the major risk factor for such pathologies, and assumes that aging and age-associated diseases share a common set of basic biological mechanisms. According to geroscience, the decline in muscle performance in PD, sarcopenia could be a manifestation of accelerated aging (Franceschi et al. 2018). The fact that the co-occurrence of sarcopenia and PD within one individual is higher than expected does not clarify whether one of the syndromes induces the other, or whether there may be common underlying causes. It is not known at present if the higher energy demand due to muscular rigidity or to abnormal movement, or rather the decreased activity typical of the disease, might underlie the aggravation of sarcopenia in patients with PD. A pilot study, conducted on 255 individuals of the TREND cohort, investigated the association of the features of increased risk for PD with early-stage sarcopenia (ESS): the authors found a significant association of the Unified PD Rating Scale (UPDRS-III, a 6 section rating scale used to follow the longitudinal course of PD) score with ESS in this cohort (Drey et al. 2017). No association was found between the other PD-related features and ESS (Drey et al. 2017). This might indicate a common degenerative pathway in both diseases and ESS as a potential prodromal marker of PD.

19.5   MOLECULAR MECHANISMS OF SARCOPENIA IN PARKINSON DISEASE

Since age-related and PD-related changes in skeletal muscle are largely attributed to molecular mediators that govern fiber size, mitochondrial homeostasis, and apoptosis, these mechanisms responsible for the deleterious changes present numerous therapeutic targets for drug discovery. These changes are shared by pathologies involving muscle wasting, such as muscular dystrophies or amyotrophic lateral sclerosis, cancer, and AIDS, all characterized by alterations in metabolic and physiological parameters, progressive weakness in specific muscle groups. Modulation of extracellular agonists, receptors, protein kinases, intermediate molecules, transcription factors, and tissue-specific gene expression collectively compromises the functionality of skeletal muscle tissue, leading to muscle degeneration and persistent protein degradation through activation of proteolytic systems, such as calpain, ubiquitin–proteasome and caspase, as we have reviewed elsewhere (Vinciguerra, Musaro, and Rosenthal 2010). Importantly, sarcopenia may be neurogenic in origin based on the intimate relationship between the nervous and muscular system, for example, the motor neuron and its underlying muscle fibers, although this is a matter of debate (Carter, Chen, and Hood 2015, Kwan 2013, Vinciguerra, Hede, and Rosenthal 2010). Both motor neuron and underlying muscle fibers are affected by the cellular environment. Disruption of Akt-mTOR and RhoA-SRF signaling but not Atrogin-1 or MuRF1 contributes to sarcopenia (Figure 19.3).

Ubiquitin-dependent proteolysis via the 26S proteasome system plays an important role in the regulation of DNA replication, DNA transcription, and cell differentiation in response to either exogenous or endogenous stimuli (Ciechanover and Brundin 2003). Perturbation of proteins normal proteasomal degradation by way of ubiquitin-dependent proteolysis in 26S proteasomes in the pathogenesis of PD is well established, because the three main monogenic forms of the disease are conditioned by gene mutations directly involved in proteasome degradation processes (SNCA, PARK2, and UCHLI). Decrease in the activity of the 26S proteasome complex is observed in the

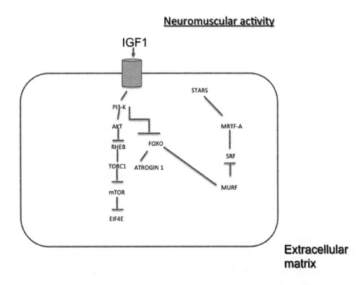

FIGURE 19.3   Molecular mechanisms involved in sarcopenia associated with PD. Mechanical loading (neuromuscular activity) upregulates the amount of IGF-I and then stimulates protein synthesis by activating Akt/mTOR pathway. Akt blocks the nuclear translocation of FOXO to inhibit the expression of atrogin-1 and MuRF and the consequent protein degradation. Myosin–actin interaction by mechanical loading can activate STARS/MRTF-A/SRF signaling. eIF4E, eukaryotic initiation factor 4E; FOXO, Forkhead box O; MRTF-A, myocardin-related transcription factor-A; mTOR, mammalian target of rapamycin; MuRF, muscle ring-finger protein; PI3-K, phosphatidylinositol 3-kinase; Rheb, Ras homolog enriched in brain; SRF, serum response factor; STARS, striated muscle activators of Rho signaling; TORC1, a component of TOR signaling complex 1.

substantia nigra in patients with PD, in turn causing the accumulation of oxidized proteins in cells (Jenner and Olanow 1998). As already mentioned, data are available indicating that genes encoding proteins for ubiquitin-dependent proteasomal degradation (SNCA, PARK2, UCHLI, and SNCAIP) may be involved in the pathogenesis of the sporadic form of PD. For instance, several intron SNPs and dinucleotide polymorphism in the promoter region of SNCA have been uncovered, and their association with PD in different populations is well established. It is hypothesized that these polymorphisms can influence the stability of mRNA and the transcriptional activity of SNCA (Brighina et al. 2008, Cronin et al. 2009, Mellick, Maraganore, and Silburn 2005). Homozygosity or compound heterozygosity at mutations in PARK2 lead to the development of the disease in 10%–20% of sporadic from patients with PD with an early (before 50 years of age) onset of the condition (Hedrich et al. 2004, Shadrina, Slominsky, and Limborska 2010, Sironi et al. 2008). Mutations can cause development of the disease or elevate the organism's sensitivity to environmental factors. Moreover, several point polymorphisms influencing the risk of PD development have been discovered (Lucking et al. 2003, Tan and Skipper 2007). To date, certain mutations in the LRRK2, SNCA, and TMEM230 genes appear to be causative for clinically typical and Lewy body-confirmed PD. Variants that slightly increase PD's risk are very common worldwide, with many of them present in 20%–40% of the healthy population, and none of whom will ever develop the disease. The following genes, among many others, are reported to affect the risk of PD.

TMEM230: mutations in the gene encoding this synaptic vesicle transmembrane protein appear highly penetrant. **EIF4G1**: rs112176450, also known as Arg1205His, is one of several rare mutations in this gene, which are associated with autosomally dominant forms of familial PD. **GBA**: rs35095275, also known as L444P, increases the odds of developing PD by at least fivefold, lowering the age of onset by about 4 years, across all ethnicities analyzed (Sidransky et al. 2009). **GRIN2A**, a glutamate receptor gene: rs4998386 the T allele in combination with coffee drinking is protective against Parkinson's (Hamza et al. 2011). **GSTP1**, a detoxification enzyme: variations have been linked to earlier onset of PD, especially in patients exposed to herbicides (Wilk et al. 2006). **TNF-alpha** promoter (Wu et al. 2007). **COX2** (Hakansson et al. 2007). **LRRK2**: rs34778348; rs34637584, also known as G2019S. **SLC6A3**, encoding a dopamine transporter. With pesticide exposure, the risk associated with these alleles is six times higher (Kelada et al. 2006). **rs287235** and **rs838552** have been linked to PD (Li et al. 2006). **ADH1C**: rs283413 (Buervenich et al. 2005). Various SNPs in the **NOS1** and **NOS2A** genes were associated with sporadic PD and/or age of onset, with a subset also interacting with either smoking or pesticides (Hancock et al. 2008). The **SREBF1** and **SREBF2** genes, which are involved in lipid metabolism, have been implicated in type-2 diabetes and PD (Grarup et al. 2008, Do et al. 2011), and schizophrenia (Le Hellard et al. 2010). **COMT**: SNPs in this gene may influence levodopa dosage used to treat patients with PD (Bialecka et al. 2008). A significant association was seen between caffeine intake and the onset of PD ($P = 2.01\tilde{A}—10(-5)$), with the odds ratio for moderate and high drinkers at 0.71 (95% confidence interval [CI]: 0.50–1.00) and 0.47 (95% CI: 0.34–0.65), respectively, against the low drinkers (Tan et al. 2007) ($P = 0.08$). **CAST**: rs1559085 and other SNPs (Allen and Satten 2010).

Ubiquitin carboxyl hydrolase (UCH-L1), which releases ubiquitin from its complex with protein, plays a key role during the final stage of protein degradation processes. Association between the risk of development of PD and the frequent S18Y polymorphism in the gene's coding region has been revealed for UCH-L1, and it has been demonstrated that the 18Y variant reduces the risk of development of the disease (Belin and Westerlund 2008, Maraganore et al. 1999).

Besides ubiquitin, proteins NEDDS and SUMO-1, the functional analogs of ubiquitin, can take part in the processes of ubiquitin-dependent proteolysis, the former being revealed in Lewy bodies, and SUMO-1 playing a role in DJ-1 protein modification. Genes of these proteins are considered as possible candidate genes for PD. Moreover, parkin target proteins, including sinfilin-1, may be involved in PD pathogenesis. Like a-synuclein, sinfilin-1 forms part of Lewy bodies. Polymorphism R621C, with its established either presence or absence of association with PD in different populations, has been revealed in the sinfilin-1 gene (SNCAIP) (Myhre et al. 2008). Data concerning a possible role of mitochondrial dysfunction in PD pathogenesis were acquired since the 1980s. It was demonstrated that young people who abused crudely manufactured 1-methyl-4-phenyl-4-pro-pionoxypiperidine (MPPP) showed symptoms of parkinsonism because of the neurotoxic action of the 1-methyl-4-phenyl-1,2,3,6-tetrahydropyridine (MPTP) by-product from MPPP "heroin" synthesis, which inhibits the mitochondrial complex I (Betarbet, Porter, and Greenamyre 2000, Mandemakers, Morais, and De Strooper 2007). Insufficient activity of the mitochondrial complex I and a decrease in the amount of this complex subunits in the neurons in the central nervous system in patients with PD have been documented (Swerdlow et al. 1996). Genes connected with the development of familial PD forms SNCA, PARK2, DJ-1, PINK1, and LRRK2 encode pro-teins involved in mitochondria homeostasis and functions. In fact, modulation of mitochondrial function by α-synuclein is responsible for resistance to mitochondrial toxins (Biskup et al. 2005, Klivenyi et al. 2006), possibly because α-synuclein can be localized on the mitochondria's inner membrane at the expense of the cryptic mitochondrial targeting signal discovered at the N terminal end. Aggregated α-synuclein can impede the normal functioning of the mitochondrial complex I (Devi et al. 2008). Several data have been gathered, concerning the direct localization and involve-ment of genes PARK2, PINK1, DJ-1, and LRRK2 in mitochondria functioning. As in the case of PARK2, it was established that individual pathologically significant mutations in heterozygotic states of genes PINK1, DJ-1, and LRRK2 can influence the risk of the development of a sporadic form of PD (Abousleiman and Sikavitsas 2006, Lesage and Brice 2009). HTRA2 and POLG1 are also described as genes whose mutations may lead to mitochondrial dysfunction. HTRA2 codes for the mitochondrial serine protease HTRA2, which shows proteolytic activity, is localized in the intermembrane area of mitochondria and is released into the cytosol during apoptosis. A polymor-phous variant (A141S) of this gene that is associated with an increased risk of PD development has been highlighted. POLG1 is a mitochondrial DNA polymerase localized on the inner membrane of mitochondria. A cluster of rare CAG repeat variants in POLGI1, whose occurrence could be regarded as a PD-predisposing factor, has been revealed (Luoma et al. 2007). Moreover, association between mitochondrial monotypes and the risk of development of sporadic parkinsonism has been discovered, and it has been shown that the risk of the disease is considerably lower in Europeans with monotypes J and K compared with the carriers of monotype H, the most frequently found monotype in Europeans (van der Walt et al. 2003).

19.6 MANAGEMENT OF SARCOPENIA IN PARKINSON DISEASE

Current strategies to counteract this loss in PD include resistance training, myostatin inhibition, and treatment with amino acids or testosterone and calorie restriction. Resistance training in com-bination with amino acid-containing nutrition would be the best candidate to attenuate, prevent, or ultimately reverse age-related muscle wasting and weakness (Sakuma and Yamaguchi 2012).

19.6.1 PHARMACOTHERAPY

One has to distinguish between pharmacotherapy against PD and pharmacotherapy against sar-copenia (within PD). There are no approved pharmaceuticals for the treatment of sarcopenia, but several strategies are available.

19.6.1.1 Pharmacotherapy against PD

The available pharmacotherapies for PD address symptomatology because no agent has been dem-onstrated to provide definite neuroprotection against the disease. Choice of pharmacotherapy must include the consideration of short-term benefits as well as long-term consequences. Patients with mild PD often function adequately without symptomatic treatment. However, recent data suggest that initiation of treatment with a well-tolerated agent, like the monoamine oxidase [MAO]-B inhib-itor rasagiline (Rascol 2005), in the absence of functional impairment is associated with improved long-term outcomes (Chen, Swope, and Dashtipour 2007). Consideration should also be given to many patient-specific factors, including patient expectations, level of disability, employment status, functional as well as chronologic age, expected efficacy and tolerability of drugs, and response to previous PD therapies. Increasingly, initial monotherapy begins with a nondopaminergic agent or, if the patient is considered functionally young, a dopamine agonist. Since PD is a progressive disorder, adjustments to pharmacotherapy must be expected over time. When greater symptomatic relief is desired, or in the more frail elderly patient, levodopa therapy is usually adopted (Jankovic 2002). However, because levodopa therapy is frequently associated with motor complications, such as fluctuations and dyskinesia, there is ongoing debate as to when in the course of PD, it is most appropriate to initiate levodopa therapy. For management of levodopa-induced dyskinesia, addition of amantadine is an option. Amantadine, originally used as an antiviral drug, has been shown to improve the symptoms of PD (Danielczyk 1995). If motor fluctuations develop, addition of a catechol-O-methyltransferase (COMT) inhibitor (for example: entacapone) inhibitor or MAO-B inhibitor is helpful. While this pharmacologic action of the COMT inhibitors may prolong the "on" time without markedly increasing dyskinesia, most studies do report increased levodopa-induced dyskinesia in patients taking COMT inhibitors, requiring a substantial reduction in daily levodopa dosage (Jankovic and Aguilar 2008). Thus, patients benefit from the addition of entacapone to their levodopa treatment. Except for nausea and increased dyskinesia, entacapone is usually well tolerated (Jankovic and Aguilar 2008). Surgery may be considered when patients need additional symptomatic control or are experiencing severe motor complications despite pharmacologically optimized therapy. As surgical approach, DBS is considered a safe and well-tolerated surgical procedure to alleviate PD and other movement disorder symptoms along with some psychiatric conditions. Over the last few decades, DBS has been shown to provide remarkable therapeutic effect on carefully selected patients (Martinez-Ramirez, Hu, et al. 2015). Although its precise mechanism of action is still unknown, DBS improves motor functions and therefore quality of life.

Quite apart from the treatment of PD, DIP (described in Section 19.2) is generally treated by cessation of the detrimental drugs. Patients who cannot stop taking antipsychotic drugs because of their psychiatric diseases, such as those with schizophrenia or major depressive disorders, may be switched to atypical antipsychotics (Shin and Chung 2012). People who are prescribed dopamine antagonists due to simple GI disturbance, headache, dizziness, or insomnia should stop taking the offending drugs as soon as possible. Anticholinergics such as trihexyphenidyl, benztropine, amantadine, and levodopa have been empirically tested for their ability to relieve symptoms of DIP, but this has produced no clear evidence of their effects in patients with DIP.

19.6.1.2 Pharmacotherapy against PD-Related Sarcopenia

Testosterone, secreted by the Leydig cells in men and ovarian thecal cells in women, appears to increase muscle mass, protein synthesis, and the number of satellite cells in both animals and humans (Griggs et al. 1989, Sinha-Hikim et al. 2004). Hypogonadism is frequent in old men. Approximately 20% of men >60 years and 50% men >80 years are categorized as hypogonadal (Harman et al. 2001). Moreover, the total amount of bioavailable testosterone decreases. Testosterone decreases gradually at a rate of 1% per year and bioavailable testosterone by 2% per year in males from the age of 30 years. Global reduction of testosterone is associated with loss of muscle strength, muscle mass, a reduction in bone mineral density, and increased risk of fracture risk following falls (Mellstrom et al. 2006, Srinivas-Shankar et al. 2010). Evidence to support testosterone supplementation is controversial. Some trials found an increase in lean body mass and hand grip strength (Gruenewald and Matsumoto 2003) and up to 25% increase in leg strength in as little as 4 weeks of therapy (Urban et al. 1995). Some other studies have found no increase in muscle strength or function (Burton and Sumukadas 2010). On the negative side, testosterone supplementation has been shown to increase the size of the prostate gland in men, and this could be detrimental to men older than 60 years in which the prevalence of early-stage prostate cancer is already high (Snyder et al. 2000). Importantly, the likelihood of acquiring a high-risk prostate cancer in men >65 years doubles for every 0.1 unit increase in free testosterone. This, together with other potential side effects of testosterone therapy such as fluid retention, gynecomastia, polycythemia, and sleep apnea limit its usefulness as a treatment for sarcopenia. Beta-Hydroxy-beta-methylbutyrate (HMB), a metabolite of leucine, which is sold as a dietary supplement, has demonstrated efficacy in preventing the loss of muscle mass in individuals with sarcopenia. Based on meta-analyses, HMB supplementation helps preserve lean muscle mass in older adults, and it may be useful for preventing muscle atrophy from bed rest (Burton and Sumukadas 2010, Yanai 2015). Growth hormone (GH) is required for maintenance of muscle and bone. GH exerts most of its anabolic actions through IGF-1, which is synthesized in the liver for systemic release. IGF-1 helps improve muscle function by increasing production of muscle satellite cells as well as stimulating production of muscle contractile proteins. Not only do GH and IGF-1 levels decline with age, the amplitude and frequency of pulsatile GH release are also significantly reduced. GH has been shown to have little to no effect in the setting of sarcopenia. GH increases muscle protein synthesis and increases muscle mass but does not lead to gains in strength and function in most studies. This, and the similar lack of efficacy of its IGF-1, may be due to local resistance of IGF-1 in aging muscle, resulting from inflammation. Other approved medications under investigation as possible treatments for sarcopenia include ghrelin, vitamin D, angiotensin-converting enzyme inhibitors, and eicosapentaenoic acid (Burton and Sumukadas 2010, Yanai 2015).

New therapies in clinical development include myostatin and the selective androgen receptor modulators (SARMs). Myostatin is a highly conserved member of the transforming growth factor-β superfamily. It is abundant in skeletal muscle but also expressed to a lesser extent in adipose tissue and cardiac muscle. Myostatin strongly affects skeletal muscle development and postnatal growth. Indeed, deletion and loss of function mutations in myostatin cause skeletal muscle hyperplasia (an increase in the number of skeletal muscle fibers) and hypertrophy (an increase in the size of skeletal muscle fibers) (Parise and Snijders 2015). Therefore, myostatin is a robust regulator of muscle development and postnatal growth whose activity is regulated in a complex fashion (Table 19.2). It is not clear whether or not the abundance or activity of myostatin is affected by aging or if myostatin plays a causal role in sarcopenia. Nonetheless, the inhibition of myostatin in adult and older animals significantly increases muscle mass and confers benefits upon measures of performance and metabolism. Since its discovery, multiple strategies to disrupt myostatin activity have been developed, including neutralizing antibodies, propeptides, soluble ActRIIB receptors, and interacting proteins (GASP-1, follistatin, and FLRG), making myostatin is a unique and desirable therapeutic target for sarcopenia (Parise and Snijders 2015). Nonsteroidal SARMs are of particular interest, given that they exhibit significant selectivity between the anabolic effects of testosterone on muscle, but with little to no evidence of androgenic effects (such as prostate stimulation in men). Research using experimental models has demonstrated that administration of SARM S-4 exerts potent anabolic effects on skeletal muscle and bone and only minimal effects on the prostate (Basualto-Alarcon et al. 2014). SARM can be used for the treatment of late-onset male hypogonadism (Coss et al. 2014).

TABLE 19.2

Myostatin and Its Benefits against Muscle Loss

- Myostatin regulates muscle development and postnatal growth.
- Inhibition of myostatin in adult and older animals significantly increases muscle mass and improves muscle performance and metabolism.
- These effects, along with the relative exclusivity of myostatin to muscle and the effects of its targeted inhibition on muscle, make it a promising drug target for sarcopenia.
- More research is needed to determine the best means by which to disrupt the activity of myostatin and related factors to safely increase muscle mass.

SARM MT-102, the first-in-class anabolic catabolic transforming agent (ACTA), has recently been tested in a Phase-II clinical study for treating cachexia in patients with late-stage cancer. The study shows significant increases in body weight in patients treated with 10 mg of MT-102 twice daily over the study period of 16 weeks compared with significant decrease in body weight in patients receiving placebo treatment. In preclinical models, MT-102 has also shown benefits in reversing sarcopenia in elderly animals (Potsch et al. 2014). MT-102 possesses three potential pharmacological targets in sarcopenia: (1) reduced catabolism through nonselective β-blockade, (2) reduced fatigue and thermogenesis through central antagonism with 5-HT1a, and (3) increased anabolism through partial agonism with β-2 receptor (Stewart Coats et al. 2011).

19.6.2 EXERCISE

Lack of exercise is thought to be a significant risk factor for sarcopenia. Exercise is of interest in treatment of sarcopenia; evidence indicates increased ability and capacity of skeletal muscle to synthesize proteins in response to short-term resistance exercise, even in old individuals. Exercise in its four forms (aerobic exercise, progressive resistance exercise, balance training, and flexibility exercise) is believed to be the most effective of all interventions proposed to improve quality of life and functionality in sarcopenic (old) people. There is a wealth of scientific evidence about the benefits of exercise recommendation for health-related and quality of life benefits (Fiatarone et al. 1989). Nevertheless, there is still skepticism between clinicians and health authorities on the role of exercise for disease prevention and treatment in frail adults, mainly because it cannot really avoid the aging process. Even individuals who were previously sedentary but initiate exercise as late as 85 demonstrate a significant survival benefit in 3 years in comparison with individuals who were sedentary (Wen et al. 2011). The physiological, cardiovascular, and neuromuscular adaptive response to regular exercise in the elderly has been repeatedly demonstrated to be similar to those of the young. Exercise training attenuates the age-related loss of arterial compliance and preserves the endothelial function in a similar way to what happens in younger people. All older adults, including the oldest ones and those with multiple morbidities including PD or are in chronic care facility, benefit from exercise. This occurs in despite the increasing likelihood of comorbidities, frailty, dependence, and even-shortening life expectancy, remaining or starting to exercise increases the chance to live longer and to be functionally independent (Morey et al. 2002). Physically active older adults compared with sedentary have less total and abdominal body fat, greater muscle mass, and higher bone mineral density (Hughes et al. 2001). Exercise also reduces likelihood of falls and related injuries, and it is associated with better mental health (Mather et al. 2002, Tinetti and Kumar 2010).

There is also compelling evidence that exercise can improve fitness among individuals with PD. A comparative, prospective, randomized, single-blinded clinical trial of three types of exercise among patients with PD, and gait impairment was performed (Shulman et al. 2013). In this study, patients were randomized to either lower-intensity treadmill exercise, higher-intensity treadmill exercise, or a combination of stretching and resistance training. The authors concluded that all three types of exercise have benefits, and patients may benefit most from a combination of lower-intensity training, stretching, and resistance. More recently, physical exercise strategies have demonstrated potential benefit for both motor and nonmotor symptoms of PD. A study of Dashtipour et al. (2015) compared the effects of Lee Silverman Voice Therapy BIG (LSVT BIG therapy, designed to overcome amplitude deficits associated with PD; it improves proprioception through increasing amplitude together with sustained attention and cognitive involvement by mentally focusing on individual movements) versus a general exercise program (combined treadmill plus seated trunk and limb exercises) on motor and nonmotor symptoms of PD in 11 patients with early-mid stage PD. In this prospective, double-blinded, randomized clinical trial patients with PD received sixteen 1-h supervised training sessions over 4 weeks and displayed positive effects upon general exercise and LSVT BIG therapy on motor and nonmotor symptoms: however, the number of patients enrolled was small and all the patients were at the early stage of their disease, without certain motor and nonmotor complications such as motor fluctuation, dyskinesia, cognitive decline, gait impairment, and poor balance (Dashtipour et al. 2015). Several other reports demonstrated the value of exercise in PD; Cochrane's update review in 2012 including 39 trials concluded that while differences between physiotherapy and placebo groups in motor performance, and other parameters were small, they were clinically meaningful to patients (Tomlinson et al. 2012). It appears also that forced exercise (FE) improves overall motor function in patients with PD: in a study, 10 enrolled patients with mild-to-moderate PD were randomly assigned to complete 8 weeks of FE or voluntary exercise (VE). With the assistance of a trainer, patients in the FE group pedaled at a rate 30% greater than their preferred voluntary rate, whereas patients in the VE group pedaled at their preferred rate. Aerobic intensity for both groups was identical, 60%–80% of their individualized training heart rate. Strikingly, the control and coordination of grasping forces during the performance of a functional bimanual dexterity task improved significantly for patients in the FE group, whereas no changes in motor performance were observed following VE (Ridgel, Vitek, and Alberts 2009). Improvements in clinical measures of rigidity and bradykinesia and biomechanical measures of bimanual dexterity were maintained 4 weeks after FE cessation (Ridgel, Vitek, and Alberts 2009). Each of these studies demonstrates the importance of exercise in improving the health and well-being of individuals with PD. On top of its benefits on physical health, exercise has the advantage to give patients an active role in the management of their PD. Patients want very much such a role, which is consistent with a patient-centered care model in which health care is closely congruent with and responsive to the needs of patients, who prefer more nonpharmacological interventions. Exercise is by far the only intervention to put the patient with PD—not a pill—at the center of care.

19.6.3 Nutrition

Recent epidemiological studies have revealed the promise of some nutrients in reducing the risk of PD. Moreover, PD-associated sarcopenia is strongly associated with nutritional status (Landi, Liperoti, Fusco, Mastropaolo, Quattrociocchi, Proia, Russo, et al. 2012). Low protein intake is associated with overall mortality in the population older than 65 years (Levine et al. 2014). An optimal dietary protein intake, 1.0–1.2 g/kg (body weight)/day with an optimal repartition over each daily meal or 25–30 g of high-quality protein per meal, has been recommended to prevent age-associated sarcopenia (Robinson et al. 2018). In PD, both the symptoms of the disease and side effects of treatment can compromise nutritional intake. This can cause a person with PD to be at considerable risk of developing weight loss and undernutrition. There are special nutritional concerns in PD, which should be considered in order to ensure adequate intake of nutrients. For example, swallowing and chewing difficulties and increased energy needs associated with tremor or rigidity are taken into account in the nutritional care. It is important to have a well-balanced diet of carbohydrates, fats, proteins, vitamins, minerals, fibers, and water to maintain optimal nutritional status. Having a nutritionally balanced diet can help to prevent weight and muscle loss, minimize practical difficulties associated with eating or swallowing, reduce the risks of hospital admission due to falls, and reduce drug-related side effects such as dehydration, constipation, and being overweight. Moreover, dietary reduction or manipulation of protein intake has been suggested to aid in decreasing the long-term effectiveness of drug therapy and thus help in relieving symptoms of PD. There are three dietary components that deserve special attention in the dietary management of PD, fiber, calcium/vitamin D, and proteins. Constipation is a common feature of PD. This occurs as a result of one of the side effects of some anti-parkinsonian drugs, reduced mobility; reduce bowel movement, and a lack of fiber and fluid intake. Constipation can interfere with the uptake of drugs such as L-dopa. It is therefore important to have sufficient fluids of approximately 2 l per day and to eat fiber-rich foods that are easily managed. Whole-meal bread, soft fruits, vegetables, and legumes should be included as part of a meal. Adequate intake of calcium (through low-fat dairy products such as milk, cheese, and yogurt or soy milk) and vitamin D (through exposure to sunlight and by consuming fortified products such as cereals and bread) helps to prevent the loss of bone density and the risk for fractures. Proteins: as L-dopa is an amino acid, it competes with the amino acids of dietary protein for absorption. This competition may prevent full absorption and reduce the efficacy of L-dopa therapy, especially late in the disease. Dietary recommendations of protein intake is either evenly distributed or redistributed mostly to the evening meal. Patients with PD should not indulge in excessive consumption of protein-rich foods and keep to recommended dietary allowance of protein for most adults of 0.8 g/kg body weight/day.

Nutrients that may be associated with a decreased risk or progression of PD include phytochemicals, omega-3 (docosahexaenoic acid [DHA]), soy (genistein), caffeine, alcohol, and tea. (1) Various subgroups of polyphenols (flavonoids, phenolic acids, stilbenes, and lignanes) and terpenes are the most abundant groups of phytochemicals with well-established antiparkinsonian effects. The health benefits associated with the intake of phytochemicals present in fruits and vegetables lead to decreased functional decline associated with aging and may slow the progression of PD (Liu 2003). Epidemiological studies found that high intake of fruits, vegetables, and fish was inversely associated with the PD risk (Gao et al. 2007). Dietary patterns, characteristic of a Mediterranean diet, are emerging as a potential neuroprotective alternative for PD (Alcalay et al. 2012). Phytochemicals perform their antiparkinsonian effect through several mechanisms of action, including suppressing apoptosis (via the reduction of Bax/Bcl-2, caspase-3, -8, and -9, and α-synuclein accumulation), decreasing dopaminergic neuronal loss and dopamine depletion, reducing the expression of proinflammatory cytokines (such as prostaglandin E2, interleukin-6, interleukin-1β, and nuclear factor-κB), and modulating nuclear and cellular inflammatory signaling, elevation of neurotrophic factors, and improvement of antioxidant status (Shahpiri et al. 2016). (2) Omega-3 polyunsaturated fatty acids (PUFAs) appear to be neuroprotective for several neurodegenerative diseases (Bousquet, Calon, and Cicchetti 2011). Current research focuses specifically on the omega-3 fatty acid DHA. DHA is an essential factor in brain growth and development and has anti-inflammatory potential due to its ability to inhibit cyclooxygenase-2 (Massaro et al. 2006). DHA protects neurons against cytotoxicity, inhibition of nitrogen oxide (NO) production, and calcium (Ca^{2+}) influx. DHA also increases the activities of antioxidant enzymes such as glutathione peroxidase and glutathione reductase. Furthermore, DHA supplementation reduced apoptosis in dopaminergic cells (Ozsoy et al. 2011). Short-term administration of DHA reduced levodopa-induced dyskinesia in parkinsonian primates by up to 40% (Samadi et al. 2006). (3) The primary soybean isoflavone genistein is a source of protein that appears to be neuroprotective and antidiabetes (Cederroth et al. 2008, Cederroth et al. 2007). Genistein pretreatment improved spatial learning and memory in parkinsonian rats (Sarkaki et al. 2009) and restored tyrosine hydroxylase, DAT, and Bcl-2 mRNA expression. Moreover, genistein attenuated rotational behavior, protected neurons, and preserved motor function. Genistein's neuroprotective actions may regulate mitochondria-dependent apoptosis pathways and suppress ROS-induced pathways (Qian et al. 2012). The neuroprotective benefits of genistein have not been tested in clinical trials to date. (4) An inverse association between PD and coffee and caffeine from noncoffee sources has been reported (Wang et al. 2015). In general, animal studies also indicate that caffeine is neuroprotective. The administration of caffeine to paraquat-treated rodents reduced the number of degenerating dopaminergic neurons, microglial cells, and nitrite content while normalizing expression of IL-1β, p38 MAPK, NF-kB, and TK (Kachroo, Irizarry, and Schwarzschild 2010, Yadav et al. 2012). Caffeine treatment provides neuroprotection in rodent models of PD (Xu et al. 2010), thus extending its beneficial effects. It is not fully established that caffeine's neuroprotective role is the reason for reduced risk of PD, nor is it known whether the association is causal. (5) Alcohol may exert neuroprotective effects in PD. One case-controlled study found an inverse association between total alcohol consumption and PD (Ragonese et al. 2003). A recent study suggests that low-to-moderate beer consumption may be associated with a lower PD risk, whereas greater liquor consumption may increase the risk of PD (Liu et al. 2013). However, currently, the association between alcohol consumption and the risk of PD remains poorly understood. Despite conflicting studies, specific components found in red wine including resveratrol and quercetin may elicit neuroprotection against PD. Resveratrol has elicited neuroprotective effects by preventing behavioral, biochemical, and histopathological changes that occur in PD animal models (Zhang et al. 2012). Diets containing resveratrol protects dopaminergic neurons and attenuates motor coordination in rodent (Zhang et al. 2012). (6) Several epidemiological studies have addressed the influence of drinking tea (*Camellia sinensis*) on the risk of PD (Renaud et al. 2015); unfortunately, no distinction between green and black tea was made. Several reports have revealed that both black and green tea exert neuroprotective effects in PD animal models (Bastianetto et al. 2006, Chaturvedi et al. 2006). Polyphenols in green and black tea extracts provide highly potent antioxidant-radical scavenging activities in brain mitochondrial membrane fractions. Moreover, polyphenols in tea reduce occurrence of disease and provide neuroprotection in cell culture and animal models. In black tea, the polyphenol theaflavin possess a wide variety of pharmacological properties including antioxidative, antiapoptotic, and anti-inflammatory effects (Caruana and Vassallo 2015). Tea consumption seems to be a promising lifestyle choice that may slow age-related deficits and neurodegenerative diseases.

19.7 CONCLUSION AND PERSPECTIVES

In this chapter, we made an attempt to outline the clinical definition of sarcopenia within PD. PD is currently incurable, and sarcopenia, progressive general loss of skeletal muscle, accompanies the late stages of the diseases, being a negative prognostic factors. Postponing or retarding muscle loss is of the utmost important for patients with PD. Keeping a good level of physical activity and educating patients of its importance is fundamental. Moderate resistance training is the most effective and safe intervention to attenuate or recover some of the loss of muscle mass and strength that accompanies PD. Myostatin inhibition through gene or antibody therapy could compensate for the loss of muscle tissue in sarcopenia, and it is an area of molecular and clinical investigation holding much promise in the future years.

ACKNOWLEDGMENTS

This work is supported by the European Social Fund and European Regional Development Fund—Project MAGNET (No. CZ.02.1.01/0.0/0.0/15_003/0000492).

REFERENCES

Abousleiman, R. I., and V. I. Sikavitsas. 2006. "Bioreactors for tissues of the musculoskeletal system." *Adv Exp Med Biol* no. 585:243–59.

Alcalay, R. N., Y. Gu, H. Mejia-Santana, L. Cote, K. S. Marder, and N. Scarmeas. 2012. "The association between Mediterranean diet adherence and Parkinson's disease." *Mov Disord* no. 27 (6):771–4. doi:10.1002/mds.24918.

Allam, M. F., A. S. Del Castillo, and R. F. Navajas. 2005. "Parkinson's disease risk factors: Genetic, environmental, or both?" *Neurol Res* no. 27 (2):206–8. doi:10.1179/016164105 × 22057.

Allen, A. S., and G. A. Satten. 2010. "SNPs in CAST are associated with Parkinson disease: A confirmation study." *Am J Med Genet B Neuropsychiatr Genet* no. 153B (4):973–9. doi:10.1002/ajmg.b.31061.

Bastianetto, S., Z. X. Yao, V. Papadopoulos, and R. Quirion. 2006. "Neuroprotective effects of green and black teas and their catechin gallate esters against beta-amyloid-induced toxicity." *Eur J Neurosci* no. 23 (1):55–64. doi:10.1111/j.1460-9568.2005.04532.x.

Basualto-Alarcon, C., D. Varela, J. Duran, R. Maass, and M. Estrada. 2014. "Sarcopenia and androgens: A link between pathology and treatment." *Front Endocrinol (Lausanne)* no. 5:217. doi:10.3389/fendo.2014.00217.

Belin, A. C., and M. Westerlund. 2008. "Parkinson's disease: A genetic perspective." *FEBS J* no. 275 (7):1377–83. doi:10.1111/j.1742-4658.2008.06301.x.

Betarbet, R., R. H. Porter, and J. T. Greenamyre. 2000. "GluR1 glutamate receptor subunit is regulated differentially in the primate basal ganglia following nigrostriatal dopamine denervation." *J Neurochem* no. 74 (3):1166–74.

Bialecka, M., M. Kurzawski, G. Klodowska-Duda, G. Opala, E. K. Tan, and M. Drozdzik. 2008. "The association of functional catechol-O-methyltransferase haplotypes with risk of Parkinson's disease, levodopa treatment response, and complications." *Pharmacogenet Genomics* no. 18 (9):815–21. doi:10.1097/FPC.0b013e328306c2f2.

Biskup, S., J. C. Mueller, M. Sharma, P. Lichtner, A. Zimprich, D. Berg, U. Wullner, T. Illig, T. Meitinger, and T. Gasser. 2005. "Common variants of LRRK2 are not associated with sporadic Parkinson's disease." *Ann Neurol* no. 58 (6):905–8. doi:10.1002/ana.20664.

Bousquet, M., F. Calon, and F. Cicchetti. 2011. "Impact of omega-3 fatty acids in Parkinson's disease." *Ageing Res Rev* no. 10 (4):453–63. doi:10.1016/j.arr.2011.03.001.

Brighina, L., R. Frigerio, N. K. Schneider, T. G. Lesnick, M. de Andrade, J. M. Cunningham, M. J. Farrer et al. 2008. "Alpha-synuclein, pesticides, and Parkinson disease: A case-control study." *Neurology* no. 70 (16 Pt 2):1461–9. doi:10.1212/01.wnl.0000304049.31377.f2.

Brooks, D. J. 2010. "Imaging approaches to Parkinson disease." *J Nucl Med* no. 51 (4):596–609. doi:10.2967/jnumed.108.059998.

Buervenich, S., A. Carmine, D. Galter, H. N. Shahabi, B. Johnels, B. Holmberg, J. Ahlberg et al. 2005. "A rare truncating mutation in ADH1C (G78Stop) shows significant association with Parkinson disease in a large international sample." *Arch Neurol* no. 62 (1):74–8. doi:10.1001/archneur.62.1.74.

Burn, D. J., M. H. Mark, E. D. Playford, D. M. Maraganore, T. R. Zimmerman, Jr., R. C. Duvoisin, A. E. Harding, C. D. Marsden, and D. J. Brooks. 1992. "Parkinson's disease in twins studied with 18F-dopa and positron emission tomography." *Neurology* no. 42 (10):1894–900.

Burton, L. A., and D. Sumukadas. 2010. "Optimal management of sarcopenia." *Clin Interv Aging* no. 5:217–28.

Caballol, N., M. J. Marti, and E. Tolosa. 2007. "Cognitive dysfunction and dementia in Parkinson disease." *Mov Disord* no. 22 (17):S358–66. doi:10.1002/mds.21677.

Carter, H. N., C. C. Chen, and D. A. Hood. 2015. "Mitochondria, muscle health, and exercise with advancing age." *Physiology* no. 30 (3):208–23. doi:10.1152/physiol.00039.2014.

Caruana, M., and N. Vassallo. 2015. "Tea polyphenols in Parkinson's disease." *Adv Exp Med Biol* no. 863:117–37. doi:10.1007/978-3-319-18365-7_6.

Cederroth, C. R., M. Vinciguerra, A. Gjinovci, F. Kuhne, M. Klein, M. Cederroth, D. Caille et al. 2008. "Dietary phytoestrogens activate AMP-activated protein kinase with improvement in lipid and glucose metabolism." *Diabetes* no. 57 (5):1176–85. doi:10.2337/db07-0630.

Cederroth, C. R., M. Vinciguerra, F. Kuhne, R. Madani, D. R. Doerge, T. J. Visser, M. Foti, F. Rohner-Jeanrenaud, J. D. Vassalli, and S. Nef. 2007. "A phytoestrogen-rich diet increases energy expenditure and decreases adiposity in mice." *Environ Health Perspect* no. 115 (10):1467–73. doi:10.1289/ehp.10413.

Chastin, S. F., E. Ferriolli, N. A. Stephens, K. C. Fearon, and C. Greig. 2012. "Relationship between sedentary behaviour, physical activity, muscle quality and body composition in healthy older adults." *Age Ageing* no. 41 (1):111–4. doi:10.1093/ageing/afr075.

Chaturvedi, R. K., S. Shukla, K. Seth, S. Chauhan, C. Sinha, Y. Shukla, and A. K. Agrawal. 2006. "Neuroprotective and neurorescue effect of black tea extract in 6-hydroxydopamine-lesioned rat model of Parkinson's disease." *Neurobiol Dis* no. 22 (2):421–34. doi:10.1016/j.nbd.2005.12.008.

Chen, J. J., D. M. Swope, and K. Dashtipour. 2007. "Comprehensive review of rasagiline, a second-generation monoamine oxidase inhibitor, for the treatment of Parkinson's disease." *Clin Ther* no. 29 (9):1825–49. doi:10.1016/j.clinthera.2007.09.021.

Chen, X., G. Mao, and S. X. Leng. 2014. "Frailty syndrome: An overview." *Clin Interv Aging* no. 9:433–41. doi:10.2147/CIA.S45300.

Ciechanover, A., and P. Brundin. 2003. "The ubiquitin proteasome system in neurodegenerative diseases: Sometimes the chicken, sometimes the egg." *Neuron* no. 40 (2):427–46.

Coss, C. C., A. Jones, M. L. Hancock, M. S. Steiner, and J. T. Dalton. 2014. "Selective androgen receptor modulators for the treatment of late onset male hypogonadism." *Asian J Androl* no. 16 (2):256–61. doi:10.4103/1008-682X.122339.

Cronin, K. D., D. Ge, P. Manninger, C. Linnertz, A. Rossoshek, B. M. Orrison, D. J. Bernard, O. M. El-Agnaf, M. G. Schlossmacher, R. L. Nussbaum, and O. Chiba-Falek. 2009. "Expansion of the Parkinson disease-associated SNCA-Rep1 allele upregulates human alpha-synuclein in transgenic mouse brain." *Hum Mol Genet* no. 18 (17):3274–85. doi:10.1093/hmg/ddp265.

Cruz-Jentoft, A. J., J. P. Baeyens, J. M. Bauer, Y. Boirie, T. Cederholm, F. Landi, F. C. Martin et al., and People European Working Group on Sarcopenia in Older. 2010. "Sarcopenia: European consensus on definition and diagnosis: Report of the European Working Group on Sarcopenia in Older People." *Age Ageing* no. 39 (4):412–23. doi:10.1093/ageing/afq034.

Danielczyk, W. 1995. "Twenty-five years of amantadine therapy in Parkinson's disease." *J Neural Transm Suppl* no. 46:399–405.

Dashtipour, K., E. Johnson, C. Kani, K. Kani, E. Hadi, M. Ghamsary, S. Pezeshkian, and J. J. Chen. 2015. "Effect of exercise on motor and nonmotor symptoms of Parkinson's disease." *Parkinsons Dis* no. 2015:586378. doi:10.1155/2015/586378.

Davie, C. A. 2008. "A review of Parkinson's disease." *Br Med Bull* no. 86:109–27. doi:10.1093/bmb/ldn013.

de Lau, L. M., and M. M. Breteler. 2006. "Epidemiology of Parkinson's disease." *Lancet Neurol* no. 5 (6):525–35. doi:10.1016/S1474-4422(06)70471-9.

Devi, L., V. Raghavendran, B. M. Prabhu, N. G. Avadhani, and H. K. Anandatheerthavarada. 2008. "Mitochondrial import and accumulation of alpha-synuclein impair complex I in human dopaminergic neuronal cultures and Parkinson disease brain." *J Biol Chem* no. 283 (14):9089–100. doi:10.1074/jbc. M710012200.

Dickson, D.V. 2007. Neuropathology of movement disorders. Edited by E Tolosa, Jankovic, JJ, *Parkinson's Disease and Movement Disorders*. Hagerstown, MD: Lippincott Williams & Wilkins.

Do, C. B., J. Y. Tung, E. Dorfman, A. K. Kiefer, E. M. Drabant, U. Francke, J. L. Mountain et al. 2011. "Web-based genome-wide association study identifies two novel loci and a substantial genetic component for Parkinson's disease." *PLoS Genet* no. 7 (6):e1002141. doi:10.1371/journal.pgen.1002141.

Dodds, R. M., H. C. Roberts, C. Cooper, and A. A. Sayer. 2015. "The epidemiology of sarcopenia." *J Clin Densitom* no. 18 (4):461–6. doi:10.1016/j.jocd.2015.04.012.

Doherty, T. J. 2003. "Invited review: Aging and sarcopenia." *J Appl Physiol* no. 95 (4):1717–27. doi:10.1152/japplphysiol.00347.2003.

Drey, M., S. E. Hasmann, J. P. Krenovsky, M. A. Hobert, S. Straub, M. Elshehabi, A. K. von Thaler et al. 2017. "Associations between Early Markers of Parkinson's disease and Sarcopenia." *Front Aging Neurosci* no. 9:53. doi:10.3389/fnagi.2017.00053.

Fiatarone, M. A., J. E. Morley, E. T. Bloom, D. Benton, G. F. Solomon, and T. Makinodan. 1989. "The effect of exercise on natural killer cell activity in young and old subjects." *J Gerontol* no. 44 (2):M37–45.

Fielding, R. A., B. Vellas, W. J. Evans, S. Bhasin, J. E. Morley, A. B. Newman, G. Abellan van Kan et al. 2011. "Sarcopenia: An undiagnosed condition in older adults. Current consensus definition: Prevalence, etiology, and consequences. International working group on sarcopenia." *J Am Med Dir Assoc* no. 12 (4):249–56. doi:10.1016/j.jamda.2011.01.003.

Franceschi, C., P. Garagnani, C. Morsiani, M. Conte, A. Santoro, A. Grignolio, D. Monti, M. Capri, and S. Salvioli. 2018. "The continuum of aging and age-related diseases: Common mechanisms but different rates." *Front Med* no. 5:61. doi:10.3389/fmed.2018.00061.

Fried, L. P., C. M. Tangen, J. Walston, A. B. Newman, C. Hirsch, J. Gottdiener, T. Seeman et al., and Group Cardiovascular Health Study Collaborative Research. 2001. "Frailty in older adults: Evidence for a phenotype." *J Gerontol A Biol Sci Med Sci* no. 56 (3):M146–56.

Fried, L. P., Q. L. Xue, A. R. Cappola, L. Ferrucci, P. Chaves, R. Varadhan, J. M. Guralnik et al. 2009. "Nonlinear multisystem physiological dysregulation associated with frailty in older women: Implications for etiology and treatment." *J Gerontol A Biol Sci Med Sci* no. 64 (10):1049–57. doi:10.1093/gerona/glp076.

Friedman, J. H. 2010. "Parkinson's disease psychosis 2010: A review article." *Parkinsonism Relat Disord* no. 16 (9):553–60. doi:10.1016/j.parkreldis.2010.05.004.

Gale, C. R., C. N. Martyn, C. Cooper, and A. A. Sayer. 2007. "Grip strength, body composition, and mortality." *Int J Epidemiol* no. 36 (1):228–35. doi:10.1093/ije/dyl224.

Gao, X., H. Chen, T. T. Fung, G. Logroscino, M. A. Schwarzschild, F. B. Hu, and A. Ascherio. 2007. "Prospective study of dietary pattern and risk of Parkinson disease." *Am J Clin Nutr* no. 86 (5):1486–94.

Gill, T. M., E. A. Gahbauer, H. G. Allore, and L. Han. 2006. "Transitions between frailty states among community-living older persons." *Arch Intern Med* no. 166 (4):418–23. doi:10.1001/archinte.166.4.418.

Goetz, C. G. 2011. "The history of Parkinson's disease: Early clinical descriptions and neurological therapies." *Cold Spring Harb Perspect Med* no. 1 (1):a008862. doi:10.1101/cshperspect.a008862.

Grarup, N., K. L. Stender-Petersen, E. A. Andersson, T. Jorgensen, K. Borch-Johnsen, A. Sandbaek, T. Lauritzen, O. Schmitz, T. Hansen, and O. Pedersen. 2008. "Association of variants in the sterol regulatory element-binding factor 1 (SREBF1) gene with type 2 diabetes, glycemia, and insulin resistance: A study of 15,734 Danish subjects." *Diabetes* no. 57 (4):1136–42. doi:10.2337/db07-1534.

Griggs, R. C., W. Kingston, R. F. Jozefowicz, B. E. Herr, G. Forbes, and D. Halliday. 1989. "Effect of testosterone on muscle mass and muscle protein synthesis." *J Appl Physiol (1985)* no. 66 (1):498–503.

Gruenewald, D. A., and A. M. Matsumoto. 2003. "Testosterone supplementation therapy for older men: Potential benefits and risks." *J Am Geriatr Soc* no. 51 (1):101–15; discussion 115.

Hakansson, A., O. Bergman, C. Chrapkowska, L. Westberg, A. C. Belin, O. Sydow, B. Johnels, L. Olson, B. Holmberg, and H. Nissbrandt. 2007. "Cyclooxygenase-2 polymorphisms in Parkinson's disease." *Am J Med Genet B Neuropsychiatr Genet* no. 144B (3):367–9. doi:10.1002/ajmg.b.30449.

Hamza, T. H., H. Chen, E. M. Hill-Burns, S. L. Rhodes, J. Montimurro, D. M. Kay, A. Tenesa et al. 2011. "Genome-wide gene-environment study identifies glutamate receptor gene GRIN2A as a Parkinson's disease modifier gene via interaction with coffee." *PLoS Genet* no. 7 (8):e1002237. doi:10.1371/journal. pgen.1002237.

Hancock, D. B., E. R. Martin, J. M. Vance, and W. K. Scott. 2008. "Nitric oxide synthase genes and their interactions with environmental factors in Parkinson's disease." *Neurogenetics* no. 9 (4):249–62. doi:10.1007/s10048-008-0137-1.

Harman, S. M., E. J. Metter, J. D. Tobin, J. Pearson, M. R. Blackman, and Aging Baltimore Longitudinal Study of. 2001. "Longitudinal effects of aging on serum total and free testosterone levels in healthy men: Baltimore Longitudinal Study of Aging." *J Clin Endocrinol Metab* no. 86 (2):724–31. doi:10.1210/jcem.86.2.7219.

Hedrich, K., A. Djarmati, N. Schafer, R. Hering, C. Wellenbrock, P. H. Weiss, R. Hilker et al. 2004. "DJ-1 (PARK7) mutations are less frequent than Parkin (PARK2) mutations in early-onset Parkinson disease." *Neurology* no. 62 (3):389–94.

Hughes, V. A., W. R. Frontera, M. Wood, W. J. Evans, G. E. Dallal, R. Roubenoff, and M. A. Fiatarone Singh. 2001. "Longitudinal muscle strength changes in older adults: influence of muscle mass, physical activity, and health." *J Gerontol A Biol Sci Med Sci* no. 56 (5):B209–17.

Iwasaki, S., Y. Narabayashi, K. Hamaguchi, A. Iwasaki, and M. Takakusagi. 1990. "Cause of death among patients with Parkinson's disease: A rare mortality due to cerebral haemorrhage." *J Neurol* no. 237 (2):77–9.

Jankovic, J. 2002. "Levodopa strengths and weaknesses." *Neurology* no. 58 (4 Suppl 1):S19–32.

Jankovic, J. 2008. "Parkinson's disease: Clinical features and diagnosis." *J Neurol Neurosurg Psychiatry* no. 79 (4):368–76. doi:10.1136/jnnp.2007.131045.

Jankovic, J., and L. G. Aguilar. 2008. "Current approaches to the treatment of Parkinson's disease." *Neuropsychiatr Dis Treat* no. 4 (4):743–57.

Janssen, I., D. S. Shepard, P. T. Katzmarzyk, and R. Roubenoff. 2004. "The healthcare costs of sarcopenia in the United States." *J Am Geriatr Soc* no. 52 (1):80–5.

Jenner, P., and C. W. Olanow. 1998. "Understanding cell death in Parkinson's disease." *Ann Neurol* no. 44 (3 Suppl 1):S72–84.

Jensen, G. L. 2008. "Inflammation: Roles in aging and sarcopenia." *JPEN J Parenter Enteral Nutr* no. 32 (6):656–9. doi:10.1177/0148607108324585.

Jones, D. M., X. Song, and K. Rockwood. 2004. "Operationalizing a frailty index from a standardized comprehensive geriatric assessment." *J Am Geriatr Soc* no. 52 (11):1929–33. doi:10.1111/j.1532-5415.2004.52521.x.

Kachroo, A., M. C. Irizarry, and M. A. Schwarzschild. 2010. "Caffeine protects against combined paraquat and maneb-induced dopaminergic neuron degeneration." *Exp Neurol* no. 223 (2):657–61. doi:10.1016/j. xpneurol.2010.02.007.

Kalia, L. V., and A. E. Lang. 2015. "Parkinson's disease." *Lancet* no. 386 (9996):896–912. doi:10.1016/S0140-6736(14)61393-3.

Kelada, S. N., H. Checkoway, S. L. Kardia, C. S. Carlson, P. Costa-Mallen, D. L. Eaton, J. Firestone et al. 2006. "5' and 3' region variability in the dopamine transporter gene (SLC6A3), pesticide exposure and Parkinson's disease risk: A hypothesis-generating study." *Hum Mol Genet* no. 15 (20):3055–62. doi:10.1093/hmg/ddl247.

Kim, T. N., and K. M. Choi. 2013. "Sarcopenia: Definition, epidemiology, and pathophysiology." *J Bone Metab* no. 20 (1):1–10. doi:10.11005/jbm.2013.20.1.1.

Klivenyi, P., D. Siwek, G. Gardian, L. Yang, A. Starkov, C. Cleren, R. J. Ferrante, N. W. Kowall, A. Abeliovich, and M. F. Beal. 2006. "Mice lacking alpha-synuclein are resistant to mitochondrial toxins." *Neurobiol Dis* no. 21 (3):541–8. doi:10.1016/j.nbd.2005.08.018.

Kones, R. 2010. "Parkinson's disease: Mitochondrial molecular pathology, inflammation, statins, and therapeutic neuroprotective nutrition." *Nutr Clin Pract* no. 25 (4):371–89. doi:10.1177/0884533610373932.

Kwan, P. 2013. "Sarcopenia, a neurogenic syndrome?" *J Aging Res* no. 2013:791679. doi:10.1155/2013/791679.

Landi, F., A. Russo, R. Liperoti, M. Pahor, M. Tosato, E. Capoluongo, R. Bernabei, and G. Onder. 2010. "Midarm muscle circumference, physical performance and mortality: Results from the aging and longevity study in the Sirente geographic area (ilSIRENTE study)." *Clin Nutr* no. 29 (4):441–7. doi:10.1016/j. clnu.2009.12.006.

Landi, F., R. Liperoti, D. Fusco, S. Mastropaolo, D. Quattrociocchi, A. Proia, A. Russo, R. Bernabei, and G. Onder. 2012. "Prevalence and risk factors of sarcopenia among nursing home older residents." *J Gerontol A Biol Sci Med Sci* no. 67 (1):48–55. doi:10.1093/gerona/glr035.

Landi, F., R. Liperoti, D. Fusco, S. Mastropaolo, D. Quattrociocchi, A. Proia, M. Tosato, R. Bernabei, and G. Onder. 2012. "Sarcopenia and mortality among older nursing home residents." *J Am Med Dir Assoc* no. 13 (2):121–6. doi:10.1016/j.jamda.2011.07.004.

Le Hellard, S., T. W. Muhleisen, S. Djurovic, J. Ferno, Z. Ouriaghi, M. Mattheisen, C. Vasilescuc et al. 2010. "Polymorphisms in SREBF1 and SREBF2, two antipsychotic-activated transcription factors controlling cellular lipogenesis, are associated with schizophrenia in German and Scandinavian samples." *Mol Psychiatry* no. 15 (5):463–72. doi:10.1038/mp.2008.110.

Lesage, S., and A. Brice. 2009. "Parkinson's disease: From monogenic forms to genetic susceptibility factors." *Hum Mol Genet* no. 18 (R1):R48–59. doi:10.1093/hmg/ddp012.

Levine, M. E., J. A. Suarez, S. Brandhorst, P. Balasubramanian, C. W. Cheng, F. Madia, L. Fontana et al. 2014. "Low protein intake is associated with a major reduction in IGF-1, cancer, and overall mortality in the 65 and younger but not older population." *Cell Metab* no. 19 (3):407–17. doi:10.1016/j.cmet.2014.02.006.

Li, Y., S. Schrodi, C. Rowland, K. Tacey, J. Catanese, and A. Grupe. 2006. "Genetic evidence for ubiquitin-specific proteases USP24 and USP40 as candidate genes for late-onset Parkinson disease." *Hum Mutat* no. 27 (10):1017–23. doi:10.1002/humu.20382.

Lill, C. M. 2016. "Genetics of Parkinson's disease." *Mol Cell Probes* no. 30 (6):386–96. doi:10.1016/j.mcp.2016.11.001.

Liu, C. J., and N. K. Latham. 2009. "Progressive resistance strength training for improving physical function in older adults." *Cochrane Database Syst Rev* (3):CD002759. doi:10.1002/14651858. CD002759.pub2.

Liu, R. H. 2003. "Health benefits of fruit and vegetables are from additive and synergistic combinations of phytochemicals." *Am J Clin Nutr* no. 78 (3 Suppl):517S–20S.

Liu, R., X. Guo, Y. Park, J. Wang, X. Huang, A. Hollenbeck, A. Blair, and H. Chen. 2013. "Alcohol consumption, types of alcohol, and Parkinson's disease." *PLoS One* no. 8 (6):e66452. doi:10.1371/journal. pone.0066452.

Lucking, C. B., V. Chesneau, E. Lohmann, P. Verpillat, C. Dulac, A. M. Bonnet, F. Gasparini, Y. Agid, A. Durr, and A. Brice. 2003. "Coding polymorphisms in the parkin gene and susceptibility to Parkinson disease." *Arch Neurol* no. 60 (9):1253–6. doi:10.1001/archneur.60.9.1253.

Luoma, P. T., J. Eerola, S. Ahola, A. H. Hakonen, O. Hellstrom, K. T. Kivisto, P. J. Tienari, and A. Suomalainen. 2007. "Mitochondrial DNA polymerase gamma variants in idiopathic sporadic Parkinson disease." *Neurology* no. 69 (11):1152–9. doi:10.1212/01.wnl.0000276955.23735.eb.

Mandemakers, W., V. A. Morais, and B. De Strooper. 2007. "A cell biological perspective on mitochondrial dysfunction in Parkinson disease and other neurodegenerative diseases." *J Cell Sci* no. 120 (Pt 10):1707–16. doi:10.1242/jcs.03443.

Maraganore, D. M., M. K. O'Connor, J. H. Bower, K. M. Kuntz, S. K. McDonnell, D. J. Schaid, and W. A. Rocca. 1999. "Detection of preclinical Parkinson disease in at-risk family members with use of [123I] beta-CIT and SPECT: an exploratory study." *Mayo Clin Proc* no. 74 (7):681–5.

Marcell, T. J. 2003. "Sarcopenia: Causes, consequences, and preventions." *J Gerontol A Biol Sci Med Sci* no. 58 (10):M911–6.

Martinez-Ramirez, D., L. Almeida, J. C. Giugni, B. Ahmed, M. A. Higuchi, C. S. Little, J. P. Chapman et al. 2015. "Rate of aspiration pneumonia in hospitalized Parkinson's disease patients: A cross-sectional study." *BMC Neurol* no. 15:104. doi:10.1186/s12883-015-0362-9.

Martinez-Ramirez, D., W. Hu, A. R. Bona, M. S. Okun, and A. Wagle Shukla. 2015. "Update on deep brain stimulation in Parkinson's disease." *Transl Neurodegener* no. 4:12. doi:10.1186/s40035-015-0034-0.

Massaro, M., A. Habib, L. Lubrano, S. Del Turco, G. Lazzerini, T. Bourcier, B. B. Weksler, and R. De Caterina. 2006. "The omega-3 fatty acid docosahexaenoate attenuates endothelial cyclooxygenase-2 induction through both NADP(H) oxidase and PKC epsilon inhibition." *Proc Natl Acad Sci U S A* no. 103 (41):15184–9. doi:10.1073/pnas.0510086103.

Mather, A. S., C. Rodriguez, M. F. Guthrie, A. M. McHarg, I. C. Reid, and M. E. McMurdo. 2002. "Effects of exercise on depressive symptoms in older adults with poorly responsive depressive disorder: Randomised controlled trial." *Br J Psychiatry* no. 180:411–5.

Mellick, G. D., D. M. Maraganore, and P. A. Silburn. 2005. "Australian data and meta-analysis lend support for alpha-synuclein (NACP-Rep1) as a risk factor for Parkinson's disease." *Neurosci Lett* no. 375 (2):112–6. doi:10.1016/j.neulet.2004.10.078.

Mellstrom, D., O. Johnell, O. Ljunggren, A. L. Eriksson, M. Lorentzon, H. Mallmin, A. Holmberg, I. Redlund-Johnell, E. Orwoll, and C. Ohlsson. 2006. "Free testosterone is an independent predictor of BMD and prevalent fractures in elderly men: MrOS Sweden." *J Bone Miner Res* no. 21 (4):529–35. doi:10.1359/jbmr.060110.

Morey, M. C., C. F. Pieper, G. M. Crowley, R. J. Sullivan, and C. M. Puglisi. 2002. "Exercise adherence and 10-year mortality in chronically ill older adults." *J Am Geriatr Soc* no. 50 (12):1929–33.

Muscaritoli, M., S. D. Anker, J. Argiles, Z. Aversa, J. M. Bauer, G. Biolo, Y. Boirie et al. 2010. "Consensus definition of sarcopenia, cachexia and pre-cachexia: Joint document elaborated by Special Interest Groups (SIG) "cachexia-anorexia in chronic wasting diseases" and "nutrition in geriatrics"." *Clin Nutr* no. 29 (2):154–9. doi:10.1016/j.clnu.2009.12.004.

Myhre, R., H. Klungland, M. J. Farrer, and J. O. Aasly. 2008. "Genetic association study of synphilin-1 in idiopathic Parkinson's disease." *BMC Med Genet* no. 9:19. doi:10.1186/1471-2350-9-19.

Ozsoy, O., Y. Seval-Celik, G. Hacioglu, P. Yargicoglu, R. Demir, A. Agar, and M. Aslan. 2011. "The influence and the mechanism of docosahexaenoic acid on a mouse model of Parkinson's disease." *Neurochem Int* no. 59 (5):664–70. doi:10.1016/j.neuint.2011.06.012.

Pankratz, N., and T. Foroud. 2004. "Genetics of Parkinson disease." *NeuroRx* no. 1 (2):235–42. doi:10.1602/neurorx.1.2.235.

Papageorgiou, P. N., J. Deschner, and S. N. Papageorgiou. 2016. "Effectiveness and adverse effects of deep brain stimulation: Umbrella review of meta-analyses." *J Neurol Surg A Cent Eur Neurosurg.* doi:10.1055/s-0036-1592158.

Parise, G., and T. Snijders. 2015. "Myostatin inhibition for treatment of sarcopenia." *Lancet Diabetes Endocrinol* no. 3 (12):917–8. doi:10.1016/S2213-8587(15)00324-1.

Parker, K. L., D. Lamichhane, M. S. Caetano, and N. S. Narayanan. 2013. "Executive dysfunction in Parkinson's disease and timing deficits." *Front Integr Neurosci* no. 7:75. doi:10.3389/fnint.2013.00075.

Peball, M., P. Mahlknecht, M. Werkmann, K. Marini, F. Murr, H. Herzmann, H. Stockner et al. 2018. "Prevalence and associated factors of sarcopenia and frailty in Parkinson's disease: A cross-sectional study." *Gerontology*:1–13. doi:10.1159/000492572.

Piccini, P., D. J. Burn, R. Ceravolo, D. Maraganore, and D. J. Brooks. 1999. "The role of inheritance in sporadic Parkinson's disease: Evidence from a longitudinal study of dopaminergic function in twins." *Ann Neurol* no. 45 (5):577–82.

Potsch, M. S., A. Tschirner, S. Palus, S. von Haehling, W. Doehner, J. Beadle, A. J. Coats, S. D. Anker, and J. Springer. 2014. "The anabolic catabolic transforming agent (ACTA) espindolol increases muscle mass and decreases fat mass in old rats." *J Cachexia Sarcopenia Muscle* no. 5 (2):149–58. doi:10.1007/s13539-013-0125-7.

Pringsheim, T., N. Jette, A. Frolkis, and T. D. Steeves. 2014. "The prevalence of Parkinson's disease: A systematic review and meta-analysis." *Mov Disord* no. 29 (13):1583–90. doi:10.1002/mds.25945.

Qian, Y., T. Guan, M. Huang, L. Cao, Y. Li, H. Cheng, H. Jin, and D. Yu. 2012. "Neuroprotection by the soy isoflavone, genistein, via inhibition of mitochondria-dependent apoptosis pathways and reactive oxygen induced-NF-kappaB activation in a cerebral ischemia mouse model." *Neurochem Int* no. 60 (8):759–67. doi:10.1016/j.neuint.2012.03.011.

Ragonese, P., G. Salemi, L. Morgante, P. Aridon, A. Epifanio, D. Buffa, F. Scoppa, and G. Savettieri. 2003. "A case-control study on cigarette, alcohol, and coffee consumption preceding Parkinson's disease." *Neuroepidemiology* no. 22 (5):297–304. doi:71193.

Rascol, O. 2005. "Rasagiline in the pharmacotherapy of Parkinson's disease: A review." *Expert Opin Pharmacother* no. 6 (12):2061–75. doi:10.1517/14656566.6.12.2061.

Renaud, J., S. F. Nabavi, M. Daglia, S. M. Nabavi, and M. G. Martinoli. 2015. "Epigallocatechin-3-Gallate, a promising molecule for Parkinson's disease?" *Rejuvenation Res* no. 18 (3):257–69. doi:10.1089/rej.2014.1639.

Ridgel, A. L., J. L. Vitek, and J. L. Alberts. 2009. "Forced, not voluntary, exercise improves motor function in Parkinson's disease patients." *Neurorehabil Neural Repair* no. 23 (6):600–8. doi:10.1177/1545968308328726.

Robinson, S. M., J. Y. Reginster, R. Rizzoli, S. C. Shaw, J. A. Kanis, I. Bautmans, H. Bischoff-Ferrari et al., and ESCEO working group. 2018. "Does nutrition play a role in the prevention and management of sarcopenia?" *Clin Nutr* no. 37 (4):1121–32. doi:10.1016/j.clnu.2017.08.016.

Rockwood, K. 2005. "What would make a definition of frailty successful?" *Age Ageing* no. 34 (5):432–4. doi:10.1093/ageing/afi146.

Rockwood, K., and A. Mitnitski. 2011. "Frailty defined by deficit accumulation and geriatric medicine defined by frailty." *Clin Geriatr Med* no. 27 (1):17–26. doi:10.1016/j.cger.2010.08.008.

Ryall, J. G., J. D. Schertzer, and G. S. Lynch. 2008. "Cellular and molecular mechanisms underlying age-related skeletal muscle wasting and weakness." *Biogerontology* no. 9 (4):213–28. doi:10.1007/s10522-008-9131-0.

Sakuma, K., and A. Yamaguchi. 2012. "Sarcopenia and cachexia: The adaptations of negative regulators of skeletal muscle mass." *J Cachexia Sarcopenia Muscle* no. 3 (2):77–94. doi:10.1007/s13539-011-0052-4.

Samadi, P., L. Gregoire, C. Rouillard, P. J. Bedard, T. Di Paolo, and D. Levesque. 2006. "Docosahexaenoic acid reduces levodopa-induced dyskinesias in 1-methyl-4-phenyl-1,2,3,6-tetrahydropyridine monkeys." *Ann Neurol* no. 59 (2):282–8. doi:10.1002/ana.20738.

Samii, A., J. G. Nutt, and B. R. Ransom. 2004. "Parkinson's disease." *Lancet* no. 363 (9423):1783–93. doi:10.1016/S0140-6736(04)16305-8.

Santilli, V., A. Bernetti, M. Mangone, and M. Paoloni. 2014. "Clinical definition of sarcopenia." *Clin Cases Miner Bone Metab* no. 11 (3):177–80.

Sarkaki, A., M. Badavi, H. Aligholi, and A. Z. Moghaddam. 2009. "Preventive effects of soy meal (± isoflavone) on spatial cognitive deficiency and body weight in an ovariectomized animal model of Parkinson's disease." *Pak J Biol Sci* no. 12 (20):1338–45.

Shadrina, M. I., P. A. Slominsky, and S. A. Limborska. 2010. "Molecular mechanisms of pathogenesis of Parkinson's disease." *Int Rev Cell Mol Biol* no. 281:229–66. doi:10.1016/S1937-6448(10)81006-8.

Shahpiri, Z., R. Bahramsoltani, M. Hosein Farzaei, F. Farzaei, and R. Rahimi. 2016. "Phytochemicals as future drugs for Parkinson's disease: A comprehensive review." *Rev Neurosci* no. 27 (6):651–68. doi:10.1515/revneuro-2016-0004.

Shergill, S. S., Z. Walker, and C. Le Katona. 1998. "A preliminary investigation of laterality in Parkinson's disease and susceptibility to psychosis." *J Neurol Neurosurg Psychiatry* no. 65 (4):610–1.

Shin, H. W., and S. J. Chung. 2012. "Drug-induced parkinsonism." *J Clin Neurol* no. 8 (1):15–21. doi:10.3988/jcn.2012.8.1.15.

Shulman, L. M., L. I. Katzel, F. M. Ivey, J. D. Sorkin, K. Favors, K. E. Anderson, B. A. Smith et al. 2013. "Randomized clinical trial of 3 types of physical exercise for patients with Parkinson disease." *JAMA Neurol* no. 70 (2):183–90. doi:10.1001/jamaneurol.2013.646.

Sidransky, E., M. A. Nalls, J. O. Aasly, J. Aharon-Peretz, G. Annesi, E. R. Barbosa, A. Bar-Shira et al. 2009. "Multicenter analysis of glucocerebrosidase mutations in Parkinson's disease." *N Engl J Med* no. 361 (17):1651–61. doi:10.1056/NEJMoa0901281.

Sinha-Hikim, I., W. E. Taylor, N. F. Gonzalez-Cadavid, W. Zheng, and S. Bhasin. 2004. "Androgen receptor in human skeletal muscle and cultured muscle satellite cells: Up-regulation by androgen treatment." *J Clin Endocrinol Metab* no. 89 (10):5245–55. doi:10.1210/jc.2004-0084.

Sironi, F., P. Primignani, M. Zini, S. Tunesi, C. Ruffmann, S. Ricca, T. Brambilla et al. 2008. "Parkin analysis in early onset Parkinson's disease." *Parkinsonism Relat Disord* no. 14 (4):326–33. doi:10.1016/j.parkreldis.2007.10.003.

Snyder, P. J., H. Peachey, J. A. Berlin, P. Hannoush, G. Haddad, A. Dlewati, J. Santanna et al. 2000. "Effects of testosterone replacement in hypogonadal men." *J Clin Endocrinol Metab* no. 85 (8):2670–7. doi:10.1210/jcem.85.8.6731.

Srinivas-Shankar, U., S. A. Roberts, M. J. Connolly, M. D. O'Connell, J. E. Adams, J. A. Oldham, and F. C. Wu. 2010. "Effects of testosterone on muscle strength, physical function, body composition, and quality of life in intermediate-frail and frail elderly men: A randomized, double-blind, placebo-controlled study." *J Clin Endocrinol Metab* no. 95 (2):639–50. doi:10.1210/jc.2009-1251.

Stewart Coats, A. J., V. Srinivasan, J. Surendran, H. Chiramana, S. R. Vangipuram, N. N. Bhatt, M. Jain et al., and A. C. T. O. N. E. Trial Investigators on behalf of the. 2011. "The ACT-ONE trial, a multicentre, randomised, double-blind, placebo-controlled, dose-finding study of the anabolic/catabolic transforming agent, MT-102 in subjects with cachexia related to stage III and IV non-small cell lung cancer and colorectal cancer: Study design." *J Cachexia Sarcopenia Muscle* no. 2 (4):201–07. doi:10.1007/s13539-011-0046-2.

Sveinbjornsdottir, S. 2016. "The clinical symptoms of Parkinson's disease." *J Neurochem* no. 139 (1):318–24. doi:10.1111/jnc.13691.

Swerdlow, R. H., J. K. Parks, S. W. Miller, J. B. Tuttle, P. A. Trimmer, J. P. Sheehan, J. P. Bennett, Jr., R. E. Davis, and W. D. Parker, Jr. 1996. "Origin and functional consequences of the complex I defect in Parkinson's disease." *Ann Neurol* no. 40 (4):663–71. doi:10.1002/ana.410400417.

Tan, A. H., Y. C. Hew, S. Y. Lim, N. M. Ramli, S. B. Kamaruzzaman, M. P. Tan, M. Grossmann et al. 2018. "Altered body composition, sarcopenia, frailty, and their clinico-biological correlates, in Parkinson's disease." *Parkinsonism Relat Disord* no. 56:58–64. doi:10.1016/j.parkreldis.2018.06.020.

Tan, E. K., and L. M. Skipper. 2007. "Pathogenic mutations in Parkinson disease." *Hum Mutat* no. 28 (7):641–53. doi:10.1002/humu.20507.

Tan, E. K., E. Chua, S. M. Fook-Chong, Y. Y. Teo, Y. Yuen, L. Tan, and Y. Zhao. 2007. "Association between caffeine intake and risk of Parkinson's disease among fast and slow metabolizers." *Pharmacogenet Genomics* no. 17 (11):1001–5. doi:10.1097/FPC.0b013e3282f09265.

Tinetti, M. E., and C. Kumar. 2010. "The patient who falls: "It's always a trade-off"." *JAMA* no. 303 (3):258–66. doi:10.1001/jama.2009.2024.

Tomlinson, C. L., S. Patel, C. Meek, C. E. Clarke, R. Stowe, L. Shah, C. M. Sackley et al. 2012. "Physiotherapy versus placebo or no intervention in Parkinson's disease." *Cochrane Database Syst Rev* (8):CD002817. doi:10.1002/14651858.CD002817.pub3.

Urban, R. J., Y. H. Bodenburg, C. Gilkison, J. Foxworth, A. R. Coggan, R. R. Wolfe, and A. Ferrando. 1995. "Testosterone administration to elderly men increases skeletal muscle strength and protein synthesis." *Am J Physiol* no. 269 (5 Pt 1):E820–6.

van der Walt, J. M., K. K. Nicodemus, E. R. Martin, W. K. Scott, M. A. Nance, R. L. Watts, J. P. Hubble et al. 2003. "Mitochondrial polymorphisms significantly reduce the risk of Parkinson disease." *Am J Hum Genet* no. 72 (4):804–11.

Vetrano, D. L., M. S. Pisciotta, A. Laudisio, M. R. Lo Monaco, G. Onder, V. Brandi, D. Fusco et al. 2018. "Sarcopenia in Parkinson disease: Comparison of different criteria and association with disease severity." *J Am Med Dir Assoc* no. 19 (6):523–27. doi:10.1016/j.jamda.2017.12.005.

Vinciguerra, M., A. Musaro, and N. Rosenthal. 2010. "Regulation of muscle atrophy in aging and disease." *Adv Exp Med Biol* no. 694:211–33.

Vinciguerra, M., M. Hede, and N. Rosenthal. 2010. "Comments on Point:Counterpoint: IGF is/is not the major physiological regulator of muscle mass. IGF-1 is a major regulator of muscle mass during growth but not for adult myofiber hypertrophy." *J Appl Physiol* no. 108 (6):1829–30.

Walston, J. D. 2012. "Sarcopenia in older adults." *Curr Opin Rheumatol* no. 24 (6):623–7. doi:10.1097/BOR.0b013e328358d59b.

Wang, A., Y. Lin, Y. Wu, and D. Zhang. 2015. "Macronutrients intake and risk of Parkinson's disease: A meta-analysis." *Geriatr Gerontol Int* no. 15 (5):606–16. doi:10.1111/ggi.12321.

Wen, C. P., J. P. Wai, M. K. Tsai, Y. C. Yang, T. Y. Cheng, M. C. Lee, H. T. Chan, C. K. Tsao, S. P. Tsai, and X. Wu. 2011. "Minimum amount of physical activity for reduced mortality and extended life expectancy: A prospective cohort study." *Lancet* no. 378 (9798):1244–53. doi:10.1016/S0140-6736(11)60749-6.

Wilk, J. B., J. E. Tobin, O. Suchowersky, H. A. Shill, C. Klein, G. F. Wooten, M. F. Lew et al. 2006. "Herbicide exposure modifies GSTP1 haplotype association to Parkinson onset age: The GenePD Study." *Neurology* no. 67 (12):2206–10. doi:10.1212/01.wnl.0000249149.22407.d1.

Wu, Y. R., I. H. Feng, R. K. Lyu, K. H. Chang, Y. Y. Lin, H. Chan, F. J. Hu, G. J. Lee-Chen, and C. M. Chen. 2007. "Tumor necrosis factor-alpha promoter polymorphism is associated with the risk of Parkinson's disease." *Am J Med Genet B Neuropsychiatr Genet* no. 144B (3):300–4. doi:10.1002/ajmg.b.30435.

Xu, K., Y. H. Xu, J. F. Chen, and M. A. Schwarzschild. 2010. "Neuroprotection by caffeine: Time course and role of its metabolites in the MPTP model of Parkinson's disease." *Neuroscience* no. 167 (2):475–81. doi:10.1016/j.neuroscience.2010.02.020.

Xue, Q. L., K. Bandeen-Roche, R. Varadhan, J. Zhou, and L. P. Fried. 2008. "Initial manifestations of frailty criteria and the development of frailty phenotype in the Women's Health and Aging Study II." *J Gerontol A Biol Sci Med Sci* no. 63 (9):984–90.

Yadav, S., S. P. Gupta, G. Srivastava, P. K. Srivastava, and M. P. Singh. 2012. "Role of secondary mediators in caffeine-mediated neuroprotection in maneb- and paraquat-induced Parkinson's disease phenotype in the mouse." *Neurochem Res* no. 37 (4):875–84. doi:10.1007/s11064-011-0682-0.

Yanai, H. 2015. "Nutrition for sarcopenia." *J Clin Med Res* no. 7 (12):926–31. doi:10.14740/jocmr2361w.

Yazar, T., H. O. Yazar, E. Zayimoglu, and S. Cankaya. 2018. "Incidence of sarcopenia and dynapenia according to stage in patients with idiopathic Parkinson's disease." *Neurol Sci*. doi:10.1007/s10072-018-3439-6.

Zamboni, M., G. Mazzali, F. Fantin, A. Rossi, and V. Di Francesco. 2008. "Sarcopenic obesity: A new category of obesity in the elderly." *Nutr Metab Cardiovasc Dis* no. 18 (5):388–95. doi:10.1016/j.numecd.2007.10.002.

Zhang, F., Y. Y. Wang, H. Liu, Y. F. Lu, Q. Wu, J. Liu, and J. S. Shi. 2012. "Resveratrol produces neurotrophic effects on cultured dopaminergic neurons through prompting astroglial BDNF and GDNF release." *Evid Based Complement Alternat Med* no. 2012:937605. doi:10.1155/2012/937605.

20 Sarcopenic Obesity in the Elderly

Michael Tieland, Inez Trouwborst, Amely Verreijen,
Robert Memelink, and Peter J.M. Weijs

CONTENTS

20.1 INTRODUCTION

The world population is aging rapidly. Since 1980, the number of people aged 60 years and over has doubled to approximately 810 million (United Nations 2012). The elderly population will continue to grow to approximately 2 billion in 2050 (United Nations 2012). It has been predicted that 22% of the total population will be older than 60 years and around 5% will be older than 80 years in 2050 (United Nations 2012). As society ages, the general health problems will increase as well. In western society, as much as 42% of the older adults have difficulties in performing activities of daily living, and 85%–90% need medication for chronic diseases, such as type II diabetes, high blood pressure, or cardiovascular disease (Kaufman et al. 2002; Rozenfeld, Fonseca, and Acurcio 2008). These physical limitations and chronic diseases increase the risk of falls, institutionalization, dependence, and premature death. In addition, these burdens increase the demand on our health care system and have tremendous health care cost.

While there are a number of contributors to physical limitations with advancing age, one of the more prominent contributors is undoubtedly a reduction in skeletal muscle mass, which is commonly referred to as sarcopenia (Janssen, Heymsfield, and Ross 2002; Baumgartner et al. 1998). Sarcopenia also increases the risk for chronic diseases such as type II diabetes and obesity. Obesity, or the increase in fat mass, is another major contributor to physical limitations and also has a major impact on the development of chronic diseases. Coexistence of sarcopenia and obesity, called sarcopenic obesity, may act synergistically and maximizes their health threatening effects (Zamboni et al. 2008; Lee et al. 2016; Stenholm et al. 2008). Importantly, sarcopenic obesity has been reported to increase the risk of metabolic impairment and physical disability more than either sarcopenia or obesity alone (Rolland et al. 2009; Batsis et al. 2015). Therefore, prevention and/or treatment of sarcopenic obesity is of major relevance for public health and healthy aging. This book chapter focuses on the definition, etiology, causes, and consequences of sarcopenic obesity in older adults. Furthermore, we aim to address the impact of sarcopenic obesity in a clinical situation, and finally, we aim to provide some means to counteract sarcopenic obesity and support healthy aging.

20.2   DEFINITION OF SARCOPENIC OBESITY

As proposed by many in the literature, sarcopenic obesity should be defined by the combination of the definitions of sarcopenia and obesity (Stenholm et al. 2008; Zamboni et al. 2008; Batsis et al. 2013). A clear consensus of the definition, however, is lacking (Table 20.1). Many different definitions have been used to define sarcopenic obesity, leading to up to a 26-fold variation in measured prevalence for sarcopenic obesity (Batsis et al. 2013).

The use of different methods to establish both sarcopenia and obesity has led to the inconsistent findings in the prevalence of sarcopenic obesity and makes comparisons between studies difficult. To illustrate, obesity has been defined using various outcomes measures such as body mass index (BMI), skinfold measurements, or fat mass (Baumgartner 2000; Levine and Crimmins 2012; Bouchard, Dionne, and Brochu 2009; Kim et al. 2009; Zoico et al. 2004). Also for sarcopenia, various outcomes and criteria exit. Whereas some solely use (appendicular) lean mass as outcome to characterize sarcopenia among older adults (Baumgartner 2000; Janssen, Heymsfield, and Ross 2002; Davison et al. 2002), others combine appendicular lean mass with handgrip strength and/or gait speed (Cruz-Jentoft et al. 2010; Fielding et al. 2011). Also different lean mass assessment tools, that is, dual-energy X-ray absorptiometry (DXA) and bio-electrical impedance analysis (BIA), lean mass cutoffs, and correction factors (height^2, bodyweight, or BMI), are described. These differences among sarcopenia criteria largely contribute to the inconsistency in findings of the prevalence of sarcopenia and sarcopenic obesity (Lee et al. 2016).

Lee et al. (2016) recommended that several primary sarcopenic obesity parameters should be considered when defining sarcopenic obesity. These are muscle mass, muscle strength, physical performance, fat mass, and waist circumference, using cutoff values as developed by the European Working Group on Sarcopenia in Older People (EWGSOP) (Cruz-Jentoft et al. 2010). Using these outcomes, the average prevalence of sarcopenic obesity is 5%–10% in older adults, with significant higher numbers in people aged over 80 years (Lee et al. 2016).

TABLE 20.1
Overview of the Current Definitions and Their Prevalence

References	Study population	Sarcopenia	Obesity	Prevalence
Baumgartner (2000)	$N = 831$, age ≥ 60	ASM/height^2 = <2.0 SD below young reference population	Fat mass = Men >27% Women >38%	Men = 4% Women = 3%
Davison et al. (2002)	$N = 2917$, age ≥ 70	Muscle mass lowest 40%	Fat mass = top 40%	7.3%
Newman et al. (2003)	$N = 831$, age ≥ 60	1. ASM/height^2 2. ASM adjusted for fat and height	BMI > 30 kg/m^2	1. 0% 2. Men = 11.5%, Women = 14.5%

(Continued)

TABLE 20.1 (*Continued*) Overview of the Current Definitions and Their Prevalence

References	Study population	Sarcopenia	Obesity	Prevalence
Baumgartner et al. (2004)	$N = 451$, mean age $= 72.8$	ASM/height^2 = <2.0 SD below young reference population	Fat mass = Men >27% Women >38%	5.8%
Schrager et al. (2007)	$N = 871$, age ≥ 70	Grip strength = lowest tertile	Waist circumference = upper tertile, BMI $\geq$30 kg/m^2	Waist circumference: Men = 11%, Women = 12% BMI: Men = 5%, Women = 11%
Stenholm et al. (2008)	1. $N = 1826$, mean age = 75.8 2. $N = 1189$, mean age = 75.8 3. $N = 1413$, mean age = 76.4 4. $N = 856$, mean age = 74.3	Grip strength = lowest sex-specific tertile	BMI $\geq$30 kg/m^2	1. Men = 3.5% Women = 6.6% 2. Men = 5.1% Women = 8.9% 3. Men = 6.1% Women = 11% 4. Men = 6.3% Women = 8.7%
Bouchard, Dionne, and Brochu (2009)	$N = 894$, age 62–82	ASM/height^2 = <2.0 SD	Fat mass = Men >28% Women >35%	Men = 19% Women = 11%
Rolland et al. (2009)	$N = 1308$, age ≥ 75	ASM/height^2 = <2.0 SD	Fat mass >40%	Women = 2.7%
Kim et al. (2012)	$N = 3196$, age ≥ 50	1. ASM/height^2 2. ASM/weight Both <2.0 SD	Waist circumference = Men >90 cm, Women >85 cm	1. Men = 0.2%, Women = 0% 2. Men = 7.6%, Women = 9.1%
Lim et al. (2010)	$N = 565$, age ≥ 65	1. ASM/height^2 2. ASM/weight Both <2.0 SD	Visceral fat area $\geq$100 cm	1. 11% 2. 41%
Levine and Crimmins (2012)	$N = 2287$, age ≥ 60	ASM/BMI <2.0 SD	Waist circumference = Men >102 cm, Women >88 cm	10.4%
Meng et al. (2014)	$N = 101$, age ≥ 80	ASM/height^2	BMI >27.5	4.9%
Dufour et al. (2013)	$N = 767$, mean age = 79	ASM/height^2	Fat mass = Men >30% Women >40%	Men = 8% Women = 4%

Source: Batsis, J.A. et al., *J. Am. Geriatr. Soc.* 61, 974–980, 2013.
Abbreviations: ASM = appendicular skeletal muscle, BMI = body mass index, SD = standard deviation.

20.3 ETIOLOGY OF SARCOPENIC OBESITY

20.3.1 AGE-RELATED CHANGES IN BODY COMPOSITION

20.3.1.1 Muscle Tissue

Aging is strongly related to changes in body composition. Both the loss of skeletal muscle mass and the increase in adipose tissue are common features with aging (Baumgartner 2000; St-Onge 2005). The loss of skeletal muscle mass, also described as sarcopenia, starts at the age of 30 years and becomes more significant after the age of 65 and 80 years (Janssen et al. 2000).

A recent quantitative review showed that the median decline in muscle mass throughout the lifespan is 0.37% per year in women and 0.47% per year in men (Mitchell et al. 2012). According to longitudinal studies in people aged 75 years or over, muscle mass is lost at a rate of 0.64%–0.70% per year in women and 0.80%–0.98% per year in men (Mitchell et al. 2012). During periods of physical inactivity, however, skeletal muscle atrophy is substantially accelerated. For instance, data from immobilization and bed rest studies show a substantial 1 kg loss of muscle mass in 10 days, suggesting that multiple episodes of physical inactivity accelerate muscle mass loss during aging (Ferrando et al. 2010; Kortebein et al. 2007; English and Paddon-Jones 2010; Paddon-Jones 2006; Dirks et al. 2014; Wall et al. 2014; Wall and van Loon 2013; Snijders et al. 2014; Wall, Dirks, and van Loon 2013). At the myocellular level, many studies have reported a substantial decrease in the muscle fiber size in the older adults (Dreyer et al. 2006; Larsson 1978; Verdijk et al. 2007). This reduction in muscle fiber size has been shown to be fiber type specific, with 10%–40% smaller type II fibers observed in the older adults compared with young adults. In contrast, type I muscle fiber size seems to be largely sustained with aging (Larsson 1978; Martel et al. 2006; Verdijk et al. 2007; Snijders, Verdijk, and van Loon 2009). Less data are available on the decline in muscle fiber number with aging. Nilwik et al. (2013), however, clearly showed that the number of muscle fibers in the quadriceps muscle did not differ between young and old subjects, but that predominantly type II muscle fiber size is declined with aging. The decline in type II muscle fibers is accompanied with a strong decline in physical performance (den Ouden et al. 2011; Guralnik et al. 1994; Berger and Doherty 2010; Cruz-Jentoft et al. 2010; Fielding et al. 2011). Emerging evidence, however, shows that the decline in skeletal muscle mass is not the sole contributor to the decline in physical performance. For instance, motor coordination, excitation-contraction coupling, energetics, skeletal integrity, and fat infiltration are critically important for physical performance in the older adults. Fat infiltration and its role in skeletal muscle performance are one of the striking features of the interaction between sarcopenia and obesity in aging (De Stefano et al. 2015).

20.3.1.2 Adipose Tissue

A peak in adipose tissue mass is generally observed around the age of 65 years and is mainly characterized by an increase in visceral fat mass (Prentice and Jebb 2001; Cetin, Lessig, and Nasr 2016). The increase in visceral fat may lead to the development of obesity. Current estimates of the prevalence of obesity in older adults range from 20% to 30% (Mathus-Vliegen 2012). Obesity is caused by an imbalance between energy intake and energy expenditure. High caloric intake, low levels of physical activity, and a low basal metabolic rate could therefore largely contribute to the development of obesity (Elia, Ritz, and Stubbs 2000). In addition, hormonal changes can be responsible for the increased storage of adipose tissue with age. A decrease in the growth hormone and testosterone secretion, reduced thyroid hormone responsiveness and leptin resistance, which are commonly seen with age and could, via different mechanisms, contribute to the development of obesity (Villareal et al. 2005).

The ratio between visceral fat and subcutaneous fat is increased with age, illustrating that not only adipose tissue mass but also adipose tissue distribution changes with age (Enzi et al. 1986; Zamboni et al. 1997; Cetin, Lessig, and Nasr 2016). Increased visceral fat is an important risk factor for many health conditions, such as type II diabetes, ischemic heart disease, hypertension, and certain cancers, all contributing to decrease the quality of life and premature mortality (Pucci, Batterham, and Manning 2014).

Since older adults are at risk for both the development of sarcopenia and the development of obesity, a double burden may exist, a condition defined as "sarcopenic obesity" (Roubenoff 2004). Older adults with low muscle mass and strength are more likely to be obese, compared with older adults with normal muscle mass (Stenholm et al. 2008). This could be explained by the risk factors for the development of sarcopenia and obesity, as they are often similar, that is, hormonal changes, physical inactivity, and suboptimal dietary intake (Zamboni et al. 2008; Stenholm et al. 2008).

20.3.2 Causes of Sarcopenic Obesity

The underlying mechanisms that cause the development of sarcopenia and obesity are partly overlapping. The most significant cause of sarcopenia and obesity is the lower energy expenditure with age, which is a result of lower physical activity as well as a lower basal metabolic rate as often seen with older age (Cooper et al. 2013; Elia, Ritz, and Stubbs 2000). Lower physical activity is a great risk factor for the development of sarcopenia and also contributes to a more positive energy balance, leading to the development of obesity (Montero-Fernandez and Serra-Rexach 2013). Maintaining the same dietary pattern in combination with a lower energy expenditure leads to a more positive energy balance, resulting in excess storage of energy in adipose tissue (Elia, Ritz, and Stubbs 2000). Furthermore, physiological factors such as hormone levels, vascular changes, and immunological factors could contribute to the development of both sarcopenia and obesity (Molino et al. 2016), making it not very surprising that the two conditions often coexist.

Not only do obesity and sarcopenia have similar pathological causes the physiological consequences of obesity are also risk factors for the development of sarcopenia (Rolland et al. 2008). For this reason, Kalinkovich and Livshits (2016) state that the condition should be defined as "obese sarcopenia" to better illustrate the direction of the pathological pathway. To explain, obesity may cause resistance to anabolic stimuli such as growth factors, hormones, amino acids, and exercise, a phenomenon called "anabolic resistance" (Nilsson et al. 2013). Especially muscular fat infiltration leads to anabolic resistance by affecting signaling pathways that are involved in the muscle protein synthesis (Hilton et al. 2008; Tardif et al. 2014; Cleasby, Jamieson, and Atherton 2016). The changes in signaling cause a decline in the skeletal muscle protein synthetic response and an increase in protein breakdown response, causing a negative muscle protein balance and a loss of skeletal muscle tissue (Koopman 2007). The decline of muscle protein synthesis is a result of the lower activation of the protein active kinase B (Akt)/mammalian target of rapamycin (mTOR) signaling cascade. Anabolic stimuli normally promote the activation of Akt, which activates the mTOR signaling. Both mTOR(C1) and mTOR(C2) are activated, and both are involved in the maintenance of cell metabolism, growth, and proliferation, but only mTOR(C1) is related to the muscle protein synthesis (Laplante and Sabatini 2009; Schiaffino and Mammucari 2011). Activated mTOR(C1) phosphorylates and activates S6 kinase 1 (S6K1), which is involved in protein synthesis by activation of different transcription and translation processes. Also, mTOR(C1) phosphorylates the eukaryotic translation initiation factor (eIF4E) 4E-binding protein 1 (4EBP1), which promotes the translation of proteins by binding to eIF4E (Schiaffino and Mammucari 2011). Anabolic resistance not only leads to the lower rate of muscle protein synthesis but also is responsible for the increased rate of muscle protein breakdown. Akt is normally also involved in the inhibition of the transcription factor forkhead box class O (FoxO). FoxO enters the nucleus of a muscle cell to activate muscle atrophy F-box (MAFbx) and muscle RING finger-1 (MuRF-1), both stimulators of muscle protein breakdown (Sanchez, Candau, and Bernardi 2014). MAFbx and MuRF-1 bind to ubiquitin, which together bind to a protein to tag them for degradation. When multiple ubiquitin proteins (polyubiquitin) are bound to a protein, 26S proteasome recognizes these tagged proteins and degrades the protein into smaller peptides (Bodine and Baehr 2014). Overall, obesity-induced anabolic resistance leads to changes in protein signaling, leading to a negative muscle protein balance and ultimately to loss of skeletal muscle mass.

Moreover, obesity-induced muscular fat infiltration not only does accelerate anabolic resistance in obese older adults but also affects muscle quality. The storage of adipose tissue in the muscle leads to increased stiffness of the muscle, affecting the shortening and expansion capacity of the muscle fiber (Rahemi, Nigam, and Wakeling 2015). This leads to a decrease in the force per unit skeletal muscle tissue and thus a decrease in muscle quality (Tieland, Trouwborst, and Clark 2017). Finally, obesity is responsible for causing a systemic low-grade inflammation, particularly by visceral fat that excretes several different pro-inflammatory cytokines (Tam et al. 2010; Invitti 2002). To illustrate, interleukin-6 (IL-6) and C-reactive protein (CRP) are positively related to fat mass and negatively to muscle mass (Cesari et al. 2005; Schrager et al. 2007). High levels of these cytokines are predictive for low muscle mass and strength in a 3-year follow-up study (Schaap et al. 2006). Tumor necrosis factor α (TNF-α) is also a cytokine, which is excreted in high amount by adipose tissue (Tzanavari, Giannogonas, and Karalis 2010). TNF-α is mediated by nuclear factor-κB (NF-κB), which upregulates the ubiquitin–proteasome pathway, leading to increased muscle protein breakdown (Reid and Li 2001). Obesity-induced low-grade inflammation consequently may contribute to a decrease in muscle mass and strength (Kalinkovich and Livshits 2016; Batsis et al. 2016; Cesari et al. 2005; Schaap et al. 2006). Obesity-induced low-grade inflammation also has the potential to contribute to anabolic resistance, by leading to several cardiovascular and metabolic complications (van Greevenbroek, Schalkwijk, and Stehouwer 2013; Mraz and Haluzik 2014). Insulin is an anabolic hormone that activates the Akt/mTOR pathway and inhibits the Akt/FOXO pathway, leading to increased muscle protein synthesis and decreased muscle protein breakdown, leading to a positive muscle protein balance. The resistance to insulin consequently results in a negative muscle protein balance, eventually leading to the loss of muscle mass over time (Park et al. 2007, 2009; Abbatecola et al. 2005; Guillet and Boirie 2005; Cleasby, Jamieson, and Atherton 2016). Also, insulin resistance leads to a lower ability of a muscle cell to take up glucose from the blood, leading to a lower storage of glycogen in the muscle and lower mitochondrial function, contributing to a lower endurance capacity over time (Cleasby, Jamieson, and Atherton 2016). Next to insulin, low levels of other anabolic hormones, for example, testosterone, insulin-like growth factor-1 (IGF-1), estrogen, are related to lower functional mobility, lower muscle strength, and lower muscle mass (Michalakis et al. 2013; Schaap et al. 2005). Overall, there are many causes of sarcopenic obesity that are complex, interrelated, and partly not completely understood.

20.3.3 Consequences of Sarcopenic Obesity

Both sarcopenia and obesity are risk factors for many different conditions, ranging from decreased physical performance and lower physical mobility (Auyeung et al. 2013; Stenholm et al. 2008; Goodpaster et al. 2006) to cardiovascular disease (Gusmao-Sena et al. 2016; Atkins et al. 2014), metabolic disorders (Wannamethee and Atkins 2015), muscle-skeletal disorders such as atherosclerosis (Kohara 2014), and increased mortality risk (Batsis et al. 2014).

Although sarcopenia and obesity are both risk factors for many disorders, the combination of the two seems to amplify each other consequences, resulting in even larger health consequences. For example, an 8-year prospective cohort study among 3,366 older adults, showed that cardiovascular risk factors were not significantly increased for sarcopenia and obesity separately, whereas sarcopenic obese older adults did show an increase in these risk factors (Stephen and Janssen 2009). In addition, longitudinal data showed that sarcopenic obese older adults where two to three times more likely to develop functional limitations compared with obese or sarcopenic older adults after 8 years (Baumgartner et al. 2004). Furthermore, several studies suggest that sarcopenic obese older adults are also more likely to develop other conditions such as metabolic and muscle-skeletal disorders (Chung et al. 2013; Kim, Cho, and Park 2015; Wannamethee and Atkins 2015; Molino et al. 2016). These negative outcomes demonstrate the importance of maintaining and increasing muscle mass and preventing fat mass gain. Interventions that target both adipose tissue mass and skeletal muscle mass are increasingly important to support healthy aging.

20.4 SARCOPENIC OBESITY IN A CLINICAL SETTING

20.4.1 Cancer

Before 2008, there was hardly any mention of the combined phenomenon of sarcopenia and obesity. From the research group of Vicky Baracos came the landmark study of Prado et al. (2008) that associated the prevalence of sarcopenic obesity in (respiratory and gastrointestinal) patients with cancer with their survival rate. Also sarcopenic obesity was associated with poorer functional status of these patients with cancer. The muscle mass was assessed in an innovative way by CT scan analysis, as described by Mourtzakis et al. (2008). Moreover, it was found that in this population, the chemotherapy calculated per unit body surface area was extremely different from using fat-free mass (FFM) for volume of distribution for chemotherapy. Therefore, the assessment of body composition became extremely relevant for the issue of cytotoxicity of chemotherapy drugs as well (Prado et al. 2008).

The actual use of body composition analysis in clinical practice has not been advanced as much as possible. A possible cause of this delay may be that body composition analysis in general has not reached its potential in clinical practice, which may be related to factor expertise, cost, appropriate facilities, and time (Tan et al. 2009, 2015; Gonzalez et al. 2014).

Also changes in muscle mass during the course of diagnosis and treatment of the patient with cancer are highly relevant for postoperative outcome (Reisinger et al. 2015). When we intensified the dietician-delivered nutritional support, a decrease in severe postoperative complications was observed in esophageal cancer (Ligthart-Melis et al. 2013). Sarcopenic obesity is a strong predictor of major postoperative complications (Sandini et al. 2016); therefore, preoperative nutrition (and exercise) therapy appears to be indicated. However, sarcopenic obesity was not always shown to be associated with worse outcome (Grotenhuis et al. 2016).

Sarcopenic obesity as well as aging is associated not just by lower skeletal muscle mass but with lower muscle quality as well. Muscle quality declines with fat infiltration, also termed myosteatosis. Myosteatosis results in lower attenuation (or density), expressed as Hounsfield Units, when CT scans are analyzed. In patients with cirrhosis, both sarcopenic obesity and myosteatosis are found to be prognostic for survival (Montano-Loza et al. 2016). However, for patients with metastatic breast cancer, it was found that mainly low muscle attenuation is a prognostic factor for survival and not sarcopenic obesity (Rier et al. 2017). Also in metastatic gastric cancer, low muscle density was found to be more prognostic for poor survival than low muscle mass index (Hayashi et al. 2016).

A recent study showed for colorectal cancer that sarcopenic obesity was associated with the worst survival rate (Caan et al. 2017). As the authors state, this might well be an important aspect of explaining the obesity paradox, in this case for patients with cancer. Most important is the message that the depletion of muscle and the quality of the muscle are important prognostic markers at any given body weight of patients with cancer, as shown by Martin et al. (2013).

20.4.2 INTENSIVE CARE

Muscle wasting is a severe complication of critical illness (Puthucheary et al. 2013). Puthucheary et al. (2013) reported a steady decrease in skeletal muscle mass of almost 20% during the first 10 days of intensive care unit (ICU) admission. Loss of muscle has been associated with longer duration of mechanical ventilation and higher ICU and hospital mortality (De Jonghe et al. 2002, 2007). If patients survive, they exhibit long-term functional disability with great impact on quality of life for up to 5 years after admission (Wieske et al. 2015; Herridge et al. 2011). We were among the first to associate muscle mass during admission to the ICU to mortality in ICU patients, not just for the older adults but for a broad age group (Weijs et al. 2014; Moisey et al. 2013). Many patients already have a low muscle mass upon admission to the ICU (60%–70%), which is associated with a higher mortality rate (Weijs et al. 2014; Moisey et al. 2013). We, therefore, assessed ICU-specific cutoff values for CT-scan derived muscle area: 110 cm^2 for women and 170 cm^2 for men (Weijs et al. 2014).

Along with a decline in muscle mass, fat infiltration of muscles or myosteatosis has been identified as a possible cause of loss of muscle quality (Goodpaster et al. 2001). Myosteatosis can be apparent within muscle fibers and evaluated on CT scans by measuring skeletal muscle density or between muscle fibers and evaluated on CT scans by measuring the amount of adipose tissue between muscles (also termed intermuscular adipose tissue [IMAT]). A lower skeletal muscle density was associated with increased lipid infiltration in muscle biopsies and poor clinical outcomes in non-ICU populations (Martin et al. 2013; Antoun et al. 2013; Goodpaster et al. 2000).

An increased amount of IMAT as assessed on CT scans has been associated with decreased muscle function and increased (systemic) inflammation in non-ICU populations (Addison et al. 2014; Coen and Goodpaster 2012). We showed that both skeletal muscle density and IMAT appear to be relevant for survival in ICU patients, adjusted for disease severity as well as the amount of muscle mass (Looijaard et al. 2016). Both skeletal muscle density and IMAT might be viewed as indicators of sarcopenic obesity.

20.4.3 CLINICAL SETTING IN SUMMARY

While a low level of muscle mass has been identified as an important factor in the worse clinical outcome of any form of malnourished patients, it is now also clear that the sarcopenic obesity phenotype is related to an even worse clinical outcome. Therefore the disproportionate changes in both muscle mass and fat mass during disease contribute to worse outcome and should be treated whenever possible.

20.5 COUNTERMEASURES FOR SARCOPENIC OBESITY

To counteract sarcopenic obesity, several interventions are suggested. Promising interventions include exercise intervention as well as nutritional interventions. These interventions generally aim to preserve or increase skeletal muscle mass and decline fat mass in older adults. First, promising exercise interventions will be described, and thereafter the impact of nutrition with or without exercise will be addressed.

20.5.1 EXERCISE INTERVENTIONS

The key step in prevention or treatment of sarcopenic obesity is to increase physical activity in older adults. Current practice in the treatment of obesity involves lifestyle advice, advocating an increase of physical activity. This can be in the form of everyday activities or deliberate exercise. A recent systematic review on the effect of combined exercise and nutrition interventions in sarcopenia indicated that physical exercise has a positive impact on muscle mass and muscle function in subjects aged 65 years and older (Beaudart et al. 2017). Cross-sectional studies have reported strong relations between physical activity level and prevalence of sarcopenic obesity as well (Ryu et al. 2013; Aggio et al. 2016; Tyrovolas et al. 2016), suggesting that a low amount of physical activity energy expenditure contributes to the development of sarcopenic obesity. Because both obesity and sarcopenia may lead to a further reduction in physical activity, it is hard to overcome this self-strengthening process of worsening physical and metabolic condition using lifestyle advice only. Structured, supervised exercise interventions may be able to break the vicious circle. While resistance and endurance exercise training are well established in different populations, the combination of resistance and endurance exercise seems most promising to counteract sarcopenic obesity. This section aims to outline the reported benefits of resistance exercise, endurance exercise, and the combination of resistance and endurance exercise in older adults with sarcopenic obesity.

20.5.1.1 Resistance Exercise

The majority of studies investigated the impact of resistance-type exercise on sarcopenic obesity outcomes. Resistance exercise training increases skeletal muscle mass (Peterson, Sen, and Gordon 2011), muscle strength (Peterson et al. 2010), improves physical performance (Gine-Garriga et al. 2014), and reduces adipose tissue (Leenders et al. 2013) in older adults. In addition, resistance exercise may diminish the pathological progression of sarcopenic obesity by offering a range of physiological benefits to skeletal muscle in older adults. These benefits include increased testosterone and growth hormone production (Smilios et al. 2007), reduced inflammatory cytokines, and increased mitochondrial function (Molina et al. 2016).

The first intervention study in older adults with sarcopenic obesity was performed by Balachandran and coworkers (2014), comparing 15-week high-speed circuit training with traditional resistance exercise training. The authors found that lower body power was significantly higher after the high-speed circuit training compared with the resistance exercise group. Also, change in function (modified Short Physical Performance Battery (SPPB) score) tended to be 1.1 point higher ($p = 0.08$) in the high-speed circuit training group. Skeletal muscle index, body fat %, and strength parameters showed no statistically significant differences between groups (Balachandran et al. 2014). More recently, Vasconcelos et al. (2016) showed that a 10-week resistance exercise program with a high-speed component was not effective for improving physical function, strength, or power in older women with sarcopenic obesity, compared with the control group. The relatively short duration of the intervention and the small sample size may have contributed to this result. Stoever et al. (2016) showed that older obese persons with sarcopenia were able to increase their physical performance in handgrip strength, gait speed, SPPB, and modified physical performance test (PPT) score after 16 weeks of (traditional) resistance training, without observed changes in muscle mass. Gauche et al. (2016), however, demonstrated significant improvements in both strength (knee extensor muscle strength) and body composition (increased appendicular and total FFM and decreased fat mass) compared with the control group after 24 weeks of traditional resistance training. A recent 12-week pilot study using elastic band resistance exercise significantly reduced fat mass in older women with sarcopenic obesity, compared with the nonexercising control group, while muscle mass was not better preserved in the exercising group. These studies clearly show that resistance exercise is a strong intervention to prevent and/or treat sarcopenic obesity in older adults.

20.5.1.2 Endurance Exercise

Endurance, or aerobic, exercise has been reported to increase mitochondrial enzyme activity (Ormsbee et al. 2007), mitochondrial function (Ipavec-Levasseur et al. 2015), intramyocellular lipid oxidation (Pruchnic et al. 2004), and oxidative capacity (Dube et al. 2008). A review of Bouaziz et al. (2015) summarized the health benefits of endurance training in obese older adults. The authors reported a positive effect of endurance training on body composition, insulin sensitivity, blood pressure, lipid profile, and primary prevention of cardiovascular disease. The effect of endurance exercise in sarcopenic obesity has been evaluated in a study of Chen et al. (2017). In this study, 8 weeks of aerobic exercise significantly increased muscle mass compared with the control group (continuation of current lifestyle), whereas fat mass and strength measures did not increase significantly compared with the control group. In frail obese older adults, Villareal et al. (2017) showed that 6 months of aerobic exercise in combination with a hypocaloric diet significantly decreased body weight (−9%) and increased PPT score (+14%) and peak oxygen consumption (+18%) compared with the control group (no weight management or exercise program). These data show that endurance exercise has a positive impact on health status in (sarcopenic or frail) obese older adults.

20.5.1.3 Concurrent Exercise

Additional benefits of combined aerobic and resistance training (concurrent exercise) on both insulin sensitivity and physical function have been reported in previously sedentary, abdominally obese older adults (Davidson et al. 2009). In frail obese older adults, Villareal et al. (2011) evaluated the effect of a multicomponent exercise program with and without hypocaloric diet. The 1-year multicomponent exercise program, consisting of aerobic, resistance, balance, and flexibility exercises, led to a higher improvement in physical function assessed by the PPT score and to increased lean mass or improved preservation of lean mass, in combination with a hypocaloric diet. In a more recent study from the same group, Villareal and coworkers (2017) evaluated the separate and combined effects of endurance and resistance training, in combination with a hypocaloric diet, in 160 frail obese older adults. After 6 months, body weight decreased by 9% in all exercise groups and did not change significantly in the control group. Lean body mass (LBM) decreased less and strength increased more in the combination and resistance groups (LBM: −3% and −2%, respectively, strength: +18% and +19%, respectively) than in the aerobic group (LBM: −5%, strength: +4%), whereas peak oxygen consumption increased more in the combination and aerobic groups (+17% and +18%, respectively) than in the resistance group (+8%). PPT score increased more in the combination group (+21%) than in the aerobic and resistance groups (+14% both). The authors conclude that combined aerobic and resistance exercise was the most effective in improving functional status of obese older adults, in combination with a hypocaloric diet. It may well be hypothesized that also older adults with sarcopenic obesity will benefit more from combined endurance and resistance exercise than from either exercise modality alone, but this has not been fully shown yet. Chen et al. (2017) studied the effect of 8-week aerobic exercise, resistance exercise, and combined aerobic and resistance exercise, compared with maintenance of current lifestyle without exercise (control group). In all exercise groups, muscle mass increased significantly compared with the control group but without significant differences between exercise groups. Muscle strength parameters in the resistance and the combined exercise groups were superior to the control group, with more pronounced effect in the resistance group. In another study, a 3-month resistance exercise program combined with a 12-min bout of aerobic cycling each session did not show improvements in muscle mass, muscle strength, or walking speed compared with the nonexercising groups (Kim, Kim, et al. 2016).

In summary, the combination of resistance exercise and endurance exercise seems most effective to preserve lean mass, reduce fat mass, and simultaneously increase strength and physical function in sarcopenic obesity. Evidence for more specific exercise modalities (e.g., high speed resistance exercise, high-intensity interval training, functional training) is still lacking.

20.5.2 Nutritional Interventions

In addition to exercise, nutritional interventions may be very effective to prevent muscle loss and increase the loss of fat mass and/or weight loss. Voluntary weight loss in obese older adults has shown to give clinically important benefits with regard to physical function, osteoarthritis, type 2 diabetes mellitus, and coronary heart disease (Li and Heber 2012). Weight loss treatment, however, may also affect the loss of skeletal muscle and bone mass. A staggering 25%–30% of lost weight is lean mass and, as such, accelerating the development of sarcopenia (Mathus-Vliegen 2012). Therefore, nutritional intervention should not only focus on weight loss or the decline in fat mass but should also focus on the preservation of skeletal muscle mass in older adults.

Although there is ample data available on nutritional interventions on voluntary weight loss in obese older adults (Batsis et al. 2017; Waters, Ward, and Villareal 2013), there is very limited data available on nutritional interventions on weight loss and muscle mass preservation in sarcopenic obese older adults. To our knowledge, only one weight loss trial has been performed in a sarcopenic obese study population (Muscariello et al. 2016), and three weight loss trials were performed in frail obese older adults (Porter Starr et al. 2016; Villareal et al. 2006, 2011). Energy restriction in the hypocaloric diets range from −500 to −750 kcal/day and resulted in 13%–33% of weight loss as lean mass. Surprisingly, physical performance improved with weight loss, despite the loss of muscle mass (Porter Starr et al. 2016; Villareal et al. 2006, 2011). Mathus-Vliegen (2012) suggest that for the treatment of sarcopenic obesity a less severe energy restriction (−200 to −500 kcal) is recommended. This is in line with the suggestion of Weinheimer, Sands, and Campbell (2010) that an energy restriction of −250 to −750 kcal/day would have a similar effect on FFM.

Waters et al. (2013 #96) and Batsis et al. (2017 #95) systematically summarized weight loss trials that have been performed in obese older adults. Energy restriction in the hypocaloric diets included in these reviews ranged from −500 to −1000 kcal/day. They concluded that caloric restriction is a valid means to achieve weight loss in older adults. To preserve LBM during caloric restriction, however, an increase in dietary protein intake has been suggested. In a meta-analysis, Kim, O'Connor, et al. (2016) analyzed the effects of protein intake (<25% vs. ≥25% of energy intake or ≥1.0 g/kg/d) during a hypocaloric diet on changes in body mass, lean mass, and fat mass in older adults. They demonstrated that older adults preserved more lean mass (+0.83 kg) and lost more fat mass (−0.53 kg) during weight loss when consuming higher protein diets (intake ≥1.0 g/kg/d). These findings support the expert opinion on protein recommendations for older adults, which ranges from 1.0 to 1.5 g/kg/d (Deutz et al. 2014; Nowson and O'Connell 2015; Bauer et al. 2013). Unfortunately, recommendation specifically for sarcopenic obese older adults do not exist but may be in line with the general protein recommendation of 1.0–1.5 g/kg/d (Mathus-Vliegen 2012; Benton, Whyte, and Dyal 2011; Goisser et al. 2015).

The type of protein, amino acid composition, and timing of protein intake are also suggested to be relevant for muscle mass preservation or gain during weight loss. Whey protein has been shown to be very effective in stimulating postprandial muscle protein accretion in older men (Pennings et al. 2011; Hector et al. 2015), which has been ascribed to its fast digestion and to the high leucine content (Breen and Phillips 2011). However, whether type, content, and timing of protein intake (25–30 g/meal) (Paddon-Jones et al. 2015) really affect the preservation of muscle mass during a weight loss treatment has yet to be elucidated, since the number of weight loss treatments evaluating the effect of a higher protein intake and meal-based threshold strategies in older obese adults are very limited.

Porter Starr et al. (2016) evaluated the effect of a high-protein hypocaloric diet using meal-based protein foods in obese functionally limited older adults over a 6-month period. The intervention group received individual hypocaloric dietary advice (500 kcal below needs) incorporating ≥30 g protein in each meal (mainly animal protein), which result in 1.2 g/kg/d. They found a positive effect on physical performance, but no significant effect on FFM. Backx et al. (2016) randomized overweight and obese older adults into either a high-protein diet group (1.7 g/kg/d) or into a normal-protein diet group (0.9 g/kg/d) during a 12-week hypocaloric diet. No preservation of muscle mass, muscle strength, or physical performance was observed in those consuming a high-protein diet during caloric restriction. Verreijen et al. (2015) studied the effect of a high whey protein-, leucine-, and vitamin D-enriched supplement on muscle mass preservation during a 13-week weight loss program including three times/week resistance exercise in obese older adults. Subjects in the intervention group received a supplement containing 21 g whey protein (10 servings/wk), whereas the control group received an isocaloric control supplement. In this study, the high whey protein-, leucine-, and vitamin D-enriched supplement resulted in a muscle preserving effect of 0.95 kg compared with control. Mojtahedi et al. (2011) conducted a weight loss trial in overweight and obese older women. They compared a high whey protein diet (1.2 g/kg/d) with a control diet (0.9 g/kg/d maltodextrin) during caloric restriction and flexibility/endurance exercise training. In line with the results of Verreijen et al. (2015), this study showed that increasing protein intake during weight loss can counteract the deleterious effects on muscle mass by maintaining more muscle relative to the weight lost. Both Mojtahedi (2011) and Verreijen et al. (2015) included a training component in their studies, which might explain the muscle preserving effect of protein in these studies in comparison with the studies of Porter Starr et al. (2016) and Backx et al. (2016).

In summary, weight loss trials in obese older adults are effective in reducing body weight and fat mass but might also reduce lean mass. Higher protein intakes may help to prevent muscle mass loss especially when combined with an exercise intervention.

20.6 CONCLUDING REMARKS

Sarcopenia and obesity are two conditions that—if combined—cause serious health problems in older adults. Sarcopenic obesity is complex and factors that may play a role are still poorly understood. Yet, more research groups focus on sarcopenic obesity as certainly more studies are needed to understand its etiology. Fortunately, there are interventions available to counteract the negative effects of sarcopenic obesity. Especially, concurrent exercise and nutritional interventions focusing on ample dietary protein intake are considered promising means to prevent and/or treat sarcopenic obesity and, as such, support healthy aging.

REFERENCES

Abbatecola, A. M., L. Ferrucci, G. Ceda, et al. 2005. Insulin resistance and muscle strength in older persons. *J Gerontol A Biol Sci Med Sci* 60 (10):1278–1282.

Addison, O., R. L. Marcus, P. C. Lastayo, and A. S. Ryan. 2014. Intermuscular fat: A review of the consequences and causes. *Int J Endocrinol* 2014:309570.

Aggio, D. A., C. Sartini, O. Papacosta, et al. 2016. Cross-sectional associations of objectively measured physical activity and sedentary time with sarcopenia and sarcopenic obesity in older men. *Prev Med* 91:264–272.

Antoun, S., E. Lanoy, R. Iacovelli, et al. 2013. Skeletal muscle density predicts prognosis in patients with metastatic renal cell carcinoma treated with targeted therapies. *Cancer* 119 (18):3377–3384.

Atkins, J. L., P. H. Whincup, R. W. Morris, L. T. Lennon, O. Papacosta, and S. G. Wannamethee. 2014. Sarcopenic obesity and risk of cardiovascular disease and mortality: A population-based cohort study of older men. *J Am Geriatr Soc* 62 (2):253–260.

Auyeung, T. W., J. S. Lee, J. Leung, T. Kwok, and J. Woo. 2013. Adiposity to muscle ratio predicts incident physical limitation in a cohort of 3,153 older adults—An alternative measurement of sarcopenia and sarcopenic obesity. *Age* 35 (4):1377–1385.

Backx, E. M., M. Tieland, K. J. Borgonjen-van den Berg, P. R. Claessen, L. J. van Loon, and L. C. de Groot. 2016. Protein intake and lean body mass preservation during energy intake restriction in overweight older adults. *Int J Obes* 40 (2):299–304.

Balachandran, A., S. N. Krawczyk, M. Potiaumpai, and J. F. Signorile. 2014. High-speed circuit training vs hypertrophy training to improve physical function in sarcopenic obese adults: A randomized controlled trial. *Exp Gerontol* 60:64–71.

Batsis, J. A., L. K. Barre, T. A. Mackenzie, S. I. Pratt, F. Lopez-Jimenez, and S. J. Bartels. 2013. Variation in the prevalence of sarcopenia and sarcopenic obesity in older adults associated with different research definitions: dual-energy X-ray absorptiometry data from the National Health and Nutrition Examination Survey 1999–2004. *J Am Geriatr Soc* 61 (6):974–980.

Batsis, J. A., L. E. Gill, R. K. Masutani, et al. 2017. Weight loss interventions in older adults with obesity: A systematic review of randomized controlled trials since 2005. *J Am Geriatr Soc* 65 (2):257–268.

Batsis, J. A., T. A. Mackenzie, L. K. Barre, F. Lopez-Jimenez, and S. J. Bartels. 2014. Sarcopenia, sarcopenic obesity and mortality in older adults: Results from the National Health and Nutrition Examination Survey III. *Eur J Clin Nutr* 68 (9):1001–1007.

Batsis, J. A., T. A. Mackenzie, J. D. Jones, F. Lopez-Jimenez, and S. J. Bartels. 2016. Sarcopenia, sarcopenic obesity and inflammation: Results from the 1999–2004 National Health and Nutrition Examination Survey. *Clin Nutr* 35:1472–1483.

Batsis, J. A., A. J. Zbehlik, D. Pidgeon, and S. J. Bartels. 2015. Dynapenic obesity and the effect on long-term physical function and quality of life: Data from the osteoarthritis initiative. *BMC Geriatr* 15:118.

Bauer, J., G. Biolo, T. Cederholm, et al. 2013. Evidence-based recommendations for optimal dietary protein intake in older people: A position paper from the PROT-AGE Study Group. *J Am Med Dir Assoc* 14 (8):542–559.

Baumgartner, R. N. 2000. Body composition in healthy aging. *Ann N Y Acad Sci* 904:437–448.

Baumgartner, R. N., K. M. Koehler, D. Gallagher, et al. 1998. Epidemiology of sarcopenia among the elderly in New Mexico. *Am J Epidemiol* 147 (8):755–763.

Baumgartner, R. N., S. J. Wayne, D. L. Waters, I. Janssen, D. Gallagher, and J. E. Morley. 2004. Sarcopenic obesity predicts instrumental activities of daily living disability in the elderly. *Obes Res* 12 (12):1995–2004.

Beaudart, C., A. Dawson, S. C. Shaw, et al. 2017. Nutrition and physical activity in the prevention and treatment of sarcopenia: Systematic review. *Osteoporos Int* 28:1817–1833.

Benton, M. J., M. D. Whyte, and B. W. Dyal. 2011. Sarcopenic obesity: Strategies for management. *Am J Nurs* 111 (12):38–44; quiz 45–46.

Berger, M. J., and T. J. Doherty. 2010. Sarcopenia: Prevalence, mechanisms, and functional consequences. *Interdiscip Top Gerontol* 37:94–114.

Bodine, S. C., and L. M. Baehr. 2014. Skeletal muscle atrophy and the E3 ubiquitin ligases MuRF1 and MAFbx/atrogin-1. *Am J Physiol Endocrinol Metab* 307 (6):E469–E484.

Bouaziz, W., E. Schmitt, G. Kaltenbach, B. Geny, and T. Vogel. 2015. Health benefits of endurance training alone or combined with diet for obese patients over 60: A review. *Int J Clin Pract* 69 (10):1032–1049.

Bouchard, D. R., I. J. Dionne, and M. Brochu. 2009. Sarcopenic/obesity and physical capacity in older men and women: Data from the Nutrition as a Determinant of Successful Aging (NuAge)-the Quebec longitudinal Study. *Obesity (Silver Spring)* 17 (11):2082–2088.

Breen, L., and S. M. Phillips. 2011. Skeletal muscle protein metabolism in the elderly: Interventions to counteract the "anabolic resistance" of ageing. *Nutr Metab (Lond)* 8:68.

Caan, B. J., J. A. Meyerhardt, C. H. Kroenke, et al. 2017. Explaining the obesity paradox: The association between body composition and colorectal cancer survival (C-SCANS Study). *Cancer Epidemiol Biomarkers Prev* 26:1008–1015.

Cesari, M., S. B. Kritchevsky, R. N. Baumgartner, et al. 2005. Sarcopenia, obesity, and inflammation—Results from the Trial of Angiotensin Converting Enzyme Inhibition and Novel Cardiovascular Risk Factors study. *Am J Clin Nutr* 82 (2):428–434.

Cetin, D., B. A. Lessig, and E. Nasr. 2016. Comprehensive evaluation for obesity: Beyond body mass index. *J Am Osteopath Assoc* 116 (6):376–382.

Chen, T. C., W. C. Tseng, G. L. Huang, H. L. Chen, K. W. Tseng, and K. Nosaka. 2017. Superior effects of eccentric to concentric knee extensor resistance training on physical fitness, insulin sensitivity and lipid profiles of elderly men. *Front Physiol* 8:209.

Chung, J. Y., H. T. Kang, D. C. Lee, H. R. Lee, and Y. J. Lee. 2013. Body composition and its association with cardiometabolic risk factors in the elderly: A focus on sarcopenic obesity. *Arch Gerontol Geriatr* 56 (1):270–278.

Cleasby, M. E., P. M. Jamieson, and P. J. Atherton. 2016. Insulin resistance and sarcopenia: Mechanistic links between common co-morbidities. *J Endocrinol* 229 (2):R67–R81.

Coen, P. M., and B. H. Goodpaster. 2012. Role of intramyocelluar lipids in human health. *Trends Endocrinol Metab* 23 (8):391–398.

Cooper, J. A., T. M. Manini, C. M. Paton, et al. 2013. Longitudinal change in energy expenditure and effects on energy requirements of the elderly. *Nutr J* 12:73.

Cruz-Jentoft, A. J., J. P. Baeyens, J. M. Bauer, et al. 2010. Sarcopenia: European consensus on definition and diagnosis: Report of the European Working Group on Sarcopenia in Older People. *Age Ageing* 39 (4):412–423.

Davidson, L. E., R. Hudson, K. Kilpatrick, et al. 2009. Effects of exercise modality on insulin resistance and functional limitation in older adults: A randomized controlled trial. *Arch Intern Med* 169 (2):122–131.

Davison, K. K., E. S. Ford, M. E. Cogswell, and W. H. Dietz. 2002. Percentage of body fat and body mass index are associated with mobility limitations in people aged 70 and older from NHANES III. *J Am Geriatr Soc* 50 (11):1802–1809.

De Jonghe, B., S. Bastuji-Garin, M. C. Durand, et al. 2007. Respiratory weakness is associated with limb weakness and delayed weaning in critical illness. *Crit Care Med* 35 (9):2007–2015.

De Jonghe, B., T. Sharshar, J. P. Lefaucheur, et al. 2002. Paresis acquired in the intensive care unit: A prospective multicenter study. *JAMA* 288 (22):2859–2867.

De Stefano, F., S. Zambon, L. Giacometti, et al. 2015. Obesity, muscular strength, muscle composition and physical performance in an elderly population. *J Nutr Health Aging* 19 (7):785–791.

den Ouden, M. E., M. J. Schuurmans, I. E. Arts, and Y. T. van der Schouw. 2011. Physical performance characteristics related to disability in older persons: A systematic review. *Maturitas* 69 (3):208–219.

Deutz, N. E., J. M. Bauer, R. Barazzoni, et al. 2014. Protein intake and exercise for optimal muscle function with aging: Recommendations from the ESPEN Expert Group. *Clin Nutr* 33 (6):929–936.

Dirks, M. L., B. T. Wall, R. Nilwik, D. H. Weerts, L. B. Verdijk, and L. J. van Loon. 2014. Skeletal muscle disuse atrophy is not attenuated by dietary protein supplementation in healthy older men. *J Nutr* 144 (8):1196–1203.

Dreyer, H. C., S. Fujita, J. G. Cadenas, D. L. Chinkes, E. Volpi, and B. B. Rasmussen. 2006. Resistance exercise increases AMPK activity and reduces 4E-BP1 phosphorylation and protein synthesis in human skeletal muscle. *J Physiol* 576 (Pt 2):613–624.

Dube, J. J., F. Amati, M. Stefanovic-Racic, F. G. Toledo, S. E. Sauers, and B. H. Goodpaster. 2008. Exercise-induced alterations in intramyocellular lipids and insulin resistance: The athlete's paradox revisited. *Am J Physiol Endocrinol Metab* 294 (5):E882–E888.

Dufour, A. B., M. T. Hannan, J. M. Murabito, D. P. Kiel, and R. R. McLean. 2013. Sarcopenia definitions considering body size and fat mass are associated with mobility limitations: The Framingham Study. *J Gerontol A Biol Sci Med Sci* 68 (2):168–174.

Elia, M., P. Ritz, and R. J. Stubbs. 2000. Total energy expenditure in the elderly. *Eur J Clin Nutr* 54 (Suppl 3):S92–S103.

English, K. L., and D. Paddon-Jones. 2010. Protecting muscle mass and function in older adults during bed rest. *Curr Opin Clin Nutr Metab Care* 13 (1):34–39.

Enzi, G., M. Gasparo, P. R. Biondetti, D. Fiore, M. Semisa, and F. Zurlo. 1986. Subcutaneous and visceral fat distribution according to sex, age, and overweight, evaluated by computed tomography. *Am J Clin Nutr* 44 (6):739–746.

Ferrando, A. A., D. Paddon-Jones, N. P. Hays, et al. 2010. EAA supplementation to increase nitrogen intake improves muscle function during bed rest in the elderly. *Clin Nutr* 29 (1):18–23.

Fielding, R. A., B. Vellas, W. J. Evans, et al. 2011. Sarcopenia: An undiagnosed condition in older adults. Current consensus definition: Prevalence, etiology, and consequences. International working group on sarcopenia. *J Am Med Dir Assoc* 12 (4):249–256.

Gauche, R., J. B. Ferreira-Junior, A. B. Gadelha, et al. 2016. Session perceived exertion following traditional and circuit resistance exercise methods in older hypertensive women. *Percept Mot Skills* 124:166–181.

Gine-Garriga, M., M. Roque-Figuls, L. Coll-Planas, M. Sitja-Rabert, and A. Salva. 2014. Physical exercise interventions for improving performance-based measures of physical function in community-dwelling, frail older adults: A systematic review and meta-analysis. *Arch Phys Med Rehabil* 95 (4):753–769e3.

Goisser, S., W. Kemmler, S. Porzel, et al. 2015. Sarcopenic obesity and complex interventions with nutrition and exercise in community-dwelling older persons—A narrative review. *Clin Interv Aging* 10:1267–1282.

Gonzalez, M. C., C. A. Pastore, S. P. Orlandi, and S. B. Heymsfield. 2014. Obesity paradox in cancer: New insights provided by body composition. *Am J Clin Nutr* 99 (5):999–1005.

Goodpaster, B. H., C. L. Carlson, M. Visser, et al. 2001. Attenuation of skeletal muscle and strength in the elderly: The Health ABC Study. *J Appl Physiol (1985)* 90 (6):2157–2165.

Goodpaster, B. H., D. E. Kelley, F. L. Thaete, J. He, and R. Ross. 2000. Skeletal muscle attenuation determined by computed tomography is associated with skeletal muscle lipid content. *J Appl Physiol (1985)* 89 (1):104–110.

Goodpaster, B. H., S. W. Park, T. B. Harris, et al. 2006. The loss of skeletal muscle strength, mass, and quality in older adults: the health, aging and body composition study. *J Gerontol A Biol Sci Med Sci* 61 (10):1059–1064.

Grotenhuis, B. A., J. Shapiro, S. van Adrichem, et al. 2016. Sarcopenia/muscle mass is not a prognostic factor for short- and long-term outcome after esophagectomy for cancer. *World J Surg* 40 (11):2698–2704.

Guillet, C., and Y. Boirie. 2005. Insulin resistance: A contributing factor to age-related muscle mass loss? *Diabetes Metab* 31 Spec No 2:5S20–5S26.

Guralnik, J. M., E. M. Simonsick, L. Ferrucci, et al. 1994. A short physical performance battery assessing lower extremity function: Association with self-reported disability and prediction of mortality and nursing home admission. *J Gerontol* 49 (2):M85–M94.

Gusmao-Sena, M. H., K. Curvello-Silva, J. M. Barreto-Medeiros, and C. H. Da-Cunha-Daltro. 2016. Association between sarcopenic obesity and cardiovascular risk: Where are we? *Nutr Hosp* 33 (5):592.

Hayashi, N., Y. Ando, B. Gyawali, et al. 2016. Low skeletal muscle density is associated with poor survival in patients who receive chemotherapy for metastatic gastric cancer. *Oncol Rep* 35 (3):1727–1731.

Hector, A. J., G. R. Marcotte, T. A. Churchward-Venne, et al. 2015. Whey protein supplementation preserves postprandial myofibrillar protein synthesis during short-term energy restriction in overweight and obese adults. *J Nutr* 145 (2):246–252.

Herridge, M. S., C. M. Tansey, A. Matte, et al. 2011. Functional disability 5 years after acute respiratory distress syndrome. *N Engl J Med* 364 (14):1293–1304.

Hilton, T. N., L. J. Tuttle, K. L. Bohnert, M. J. Mueller, and D. R. Sinacore. 2008. Excessive adipose tissue infiltration in skeletal muscle in individuals with obesity, diabetes mellitus, and peripheral neuropathy: association with performance and function. *Phys Ther* 88 (11):1336–1344.

Invitti, C. 2002. Obesity and low-grade systemic inflammation. *Minerva Endocrinol* 27 (3):209–214.

Ipavec-Levasseur, S., I. Croci, S. Choquette, et al. 2015. Effect of 1-h moderate-intensity aerobic exercise on intramyocellular lipids in obese men before and after a lifestyle intervention. *Appl Physiol Nutr Metab* 40 (12):1262–1268.

Janssen, I., S. B. Heymsfield, and R. Ross. 2002. Low relative skeletal muscle mass (sarcopenia) in older persons is associated with functional impairment and physical disability. *J Am Geriatr Soc* 50 (5):889–896.

Janssen, I., S. B. Heymsfield, Z. M. Wang, and R. Ross. 2000. Skeletal muscle mass and distribution in 468 men and women aged 18–88 yr. *J Appl Physiol (1985)* 89 (1):81–88.

Kalinkovich, A., and G. Livshits. 2016. Sarcopenic obesity or obese sarcopenia: A cross talk between age-associated adipose tissue and skeletal muscle inflammation as a main mechanism of the pathogenesis. *Ageing Res Rev* 35:200–221.

Kaufman, D. W., J. P. Kelly, L. Rosenberg, T. E. Anderson, and A. A. Mitchell. 2002. Recent patterns of medication use in the ambulatory adult population of the United States: The Slone survey. *JAMA* 287 (3):337–344.

Kim, H., M. Kim, N. Kojima, et al. 2016. Exercise and nutritional supplementation on community-dwelling elderly Japanese women with sarcopenic obesity: A randomized controlled trial. *J Am Med Dir Assoc* 17 (11):1011–1019.

Kim, J. E., L. E. O'Connor, L. P. Sands, M. B. Slebodnik, and W. W. Campbell. 2016. Effects of dietary protein intake on body composition changes after weight loss in older adults: A systematic review and meta-analysis. *Nutr Rev* 74 (3):210–224.

Kim, J. H., J. J. Cho, and Y. S. Park. 2015. Relationship between sarcopenic obesity and cardiovascular disease risk as estimated by the Framingham risk score. *J Korean Med Sci* 30 (3):264–271.

Kim, Y. S., Y. Lee, Y. S. Chung, et al. 2012. Prevalence of sarcopenia and sarcopenic obesity in the Korean population based on the Fourth Korean National Health and Nutritional Examination Surveys. *J Gerontol A Biol Sci Med Sci* 67 (10):1107–1113.

Kim., S. J. Yang, H. J. Yoo, et al. 2009. Prevalence of sarcopenia and sarcopenic obesity in Korean adults: The Korean sarcopenic obesity study. *Int J Obes (Lond)* 33 (8):885–892.

Kohara, K. 2014. Sarcopenic obesity in aging population: Current status and future directions for research. *Endocrine* 45 (1):15–25.

Koopman, R. 2007. Role of amino acids and peptides in the molecular signaling in skeletal muscle after resistance exercise. *Int J Sport Nutr Exerc Metab* 17 (Suppl):S47–S57.

Kortebein, P., A. Ferrando, J. Lombeida, R. Wolfe, and W. J. Evans. 2007. Effect of 10 days of bed rest on skeletal muscle in healthy older adults. *JAMA* 297 (16):1772–1774.

Laplante, M., and D. M. Sabatini. 2009. mTOR signaling at a glance. *J Cell Sci* 122 (Pt 20):3589–3594.

Larsson, L. 1978. Morphological and functional characteristics of the ageing skeletal muscle in man. A cross-sectional study. *Acta Physiol Scand Suppl* 457:1–36.

Lee, D.-C., R. P. Shook, C. Drenowatz, and S. N. Blair. 2016. Physical activity and sarcopenic obesity: Definition, assessment, prevalence and mechanism. *Future Science OA* 2 (3):FSO127.

Leenders, M., L. B. Verdijk, L. Van der Hoeven, et al. 2013. Protein supplementation during resistance-type exercise training in the elderly. *Med Sci Sports Exerc* 45 (3):542–552.

Levine, M. E., and E. M. Crimmins. 2012. The impact of insulin resistance and inflammation on the association between sarcopenic obesity and physical functioning. *Obesity (Silver Spring)* 20 (10):2101–2106.

Li, Z., and D. Heber. 2012. Sarcopenic obesity in the elderly and strategies for weight management. *Nutr Rev* 70 (1):57–64.

Ligthart-Melis, G. C., P. J. Weijs, N. D. te Boveldt, et al. 2013. Dietician-delivered intensive nutritional support is associated with a decrease in severe postoperative complications after surgery in patients with esophageal cancer. *Dis Esophagus* 26 (6):587–593.

Lim, S., J. H. Kim, J. W. Yoon, et al. 2010. Sarcopenic obesity: Prevalence and association with metabolic syndrome in the Korean Longitudinal Study on Health and Aging (KLoSHA). *Diabetes Care* 33 (7):1652–1654.

Looijaard, W. G., I. M. Dekker, S. N. Stapel, et al. 2016. Skeletal muscle quality as assessed by CT-derived skeletal muscle density is associated with 6-month mortality in mechanically ventilated critically ill patients. *Crit Care* 20 (1):386.

Martel, G. F., S. M. Roth, F. M. Ivey, et al. 2006. Age and sex affect human muscle fibre adaptations to heavy-resistance strength training. *Exp Physiol* 91 (2):457–464.

Martin, L., L. Birdsell, N. Macdonald, et al. 2013. Cancer cachexia in the age of obesity: Skeletal muscle depletion is a powerful prognostic factor, independent of body mass index. *J Clin Oncol* 31 (12):1539–1547.

Mathus-Vliegen, E. M. 2012. Obesity and the elderly. *J Clin Gastroenterol* 46 (7):533–544.

Meng, P., Y. X. Hu, L. Fan, et al. 2014. Sarcopenia and sarcopenic obesity among men aged 80 years and older in Beijing: Prevalence and its association with functional performance. *Geriatr Gerontol Int* 14 (Suppl 1):29–35.

Michalakis, K., D. G. Goulis, A. Vazaiou, G. Mintziori, A. Polymeris, and A. Abrahamian-Michalakis. 2013. Obesity in the ageing man. *Metabolism* 62 (10):1341–1349.

Mitchell, W. K., J. Williams, P. Atherton, M. Larvin, J. Lund, and M. Narici. 2012. Sarcopenia, dynapenia, and the impact of advancing age on human skeletal muscle size and strength; a quantitative review. *Front Physiol* 3:260.

Moisey, L. L., M. Mourtzakis, B. A. Cotton, et al. 2013. Skeletal muscle predicts ventilator-free days, ICU-free days, and mortality in elderly ICU patients. *Crit Care* 17 (5):R206.

Mojtahedi, M. C., M. P. Thorpe, D. C. Karampinos, et al. 2011. The effects of a higher protein intake during energy restriction on changes in body composition and physical function in older women. *J Gerontol A Biol Sci Med Sci* 66 (11):1218–1225.

Molina, A. J., M. S. Bharadwaj, C. Van Horn, et al. 2016. Skeletal muscle mitochondrial content, oxidative capacity, and Mfn2 expression are reduced in older patients with heart failure and preserved ejection fraction and are related to exercise intolerance. *JACC Heart Fail* 4 (8):636–645.

Molino, S., M. Dossena, D. Buonocore, and M. Verri. 2016. Sarcopenic obesity: An appraisal of the current status of knowledge and management in elderly people. *J Nutr Health Aging* 20 (7):780–788.

Montano-Loza, A. J., P. Angulo, J. Meza-Junco, et al. 2016. Sarcopenic obesity and myosteatosis are associated with higher mortality in patients with cirrhosis. *J Cachexia Sarcopenia Muscle* 7 (2):126–135.

Montero-Fernandez, N., and J. A. Serra-Rexach. 2013. Role of exercise on sarcopenia in the elderly. *Eur J Phys Rehabil Med* 49 (1):131–143.

Mourtzakis, M., C. M. Prado, J. R. Lieffers, T. Reiman, L. J. McCargar, and V. E. Baracos. 2008. A practical and precise approach to quantification of body composition in cancer patients using computed tomography images acquired during routine care. *Appl Physiol Nutr Metab* 33 (5):997–1006.

Mraz, M., and M. Haluzik. 2014. The role of adipose tissue immune cells in obesity and low-grade inflammation. *J Endocrinol* 222 (3):R113–R127.

Muscariello, E., G. Nasti, M. Siervo, et al. 2016. Dietary protein intake in sarcopenic obese older women. *Clin Interv Aging* 11:133–140.

Newman, A. B., V. Kupelian, M. Visser, et al. 2003. Sarcopenia: Alternative definitions and associations with lower extremity function. *J Am Geriatr Soc* 51 (11):1602–1609.

Nilsson, M. I., J. P. Dobson, N. P. Greene, et al. 2013. Abnormal protein turnover and anabolic resistance to exercise in sarcopenic obesity. *FASEB J* 27 (10):3905–3916.

Nilwik, R., T. Snijders, M. Leenders, et al. 2013. The decline in skeletal muscle mass with aging is mainly attributed to a reduction in type II muscle fiber size. *Exp Gerontol* 48 (5):492–498.

Nowson, C., and S. O'Connell. 2015. Protein Requirements and Recommendations for Older People: A Review. *Nutrients* 7 (8):6874–6899.

Ormsbee, M. J., J. P. Thyfault, E. A. Johnson, R. M. Kraus, M. D. Choi, and R. C. Hickner. 2007. Fat metabolism and acute resistance exercise in trained men. *J Appl Physiol (1985)* 102 (5):1767–1772.

Paddon-Jones, D. 2006. Interplay of stress and physical inactivity on muscle loss: Nutritional countermeasures. *J Nutr* 136 (8):2123–2126.

Paddon-Jones, D., W. W. Campbell, P. F. Jacques, et al. 2015. Protein and healthy aging. *Am J Clin Nutr.* 101:1339S–1345S.

Park, S. W., B. H. Goodpaster, J. S. Lee, et al. 2009. Excessive loss of skeletal muscle mass in older adults with type 2 diabetes. *Diabetes Care* 32 (11):1993–1997.

Park, S. W., B. H. Goodpaster, E. S. Strotmeyer, et al. 2007. Accelerated loss of skeletal muscle strength in older adults with type 2 diabetes: the health, aging, and body composition study. *Diabetes Care* 30 (6):1507–1512.

Pennings, B., Y. Boirie, J. M. Senden, A. P. Gijsen, H. Kuipers, and L. J. van Loon. 2011. Whey protein stimulates postprandial muscle protein accretion more effectively than do casein and casein hydrolysate in older men. *Am J Clin Nutr* 93 (5):997–1005.

Peterson, M. D., M. R. Rhea, A. Sen, and P. M. Gordon. 2010. Resistance exercise for muscular strength in older adults: A meta-analysis. *Ageing Res Rev* 9 (3):226–237.

Peterson, M. D., A. Sen, and P. M. Gordon. 2011. Influence of resistance exercise on lean body mass in aging adults: A meta-analysis. *Med Sci Sports Exerc* 43 (2):249–258.

Porter Starr, K. N., C. F. Pieper, M. C. Orenduff, et al. 2016. Improved function with enhanced protein intake per meal: A pilot study of weight reduction in frail, obese older adults. *J Gerontol A Biol Sci Med Sci* 71 (10):1369–1375.

Prado, C. M., J. R. Lieffers, L. J. McCargar, et al. 2008. Prevalence and clinical implications of sarcopenic obesity in patients with solid tumours of the respiratory and gastrointestinal tracts: A population-based study. *Lancet Oncol* 9 (7):629–635.

Prentice, A. M., and S. A. Jebb. 2001. Beyond body mass index. *Obes Rev* 2 (3):141–147.

Pruchnic, R., A. Katsiaras, J. He, D. E. Kelley, C. Winters, and B. H. Goodpaster. 2004. Exercise training increases intramyocellular lipid and oxidative capacity in older adults. *Am J Physiol Endocrinol Metab* 287 (5):E857–E862.

Pucci, A., R. L. Batterham, and S. Manning. 2014. Obesity: Causes, consequences and patient-centred thera-peutic approaches. *Health Management* 14 (3):21–24.

Puthucheary, Z. A., J. Rawal, M. McPhail, et al. 2013. Acute skeletal muscle wasting in critical illness. *JAMA* 310 (15):1591–1600.

Rahemi, H., N. Nigam, and J. M. Wakeling. 2015. The effect of intramuscular fat on skeletal muscle mechanics: Implications for the elderly and obese. *J R Soc Interface* 12 (109):20150365.

Reid, M. B., and Y. P. Li. 2001. Tumor necrosis factor-alpha and muscle wasting: A cellular perspective. *Respir Res* 2 (5):269–272.

Reisinger, K. W., J. W. Bosmans, M. Uittenbogaart, et al. 2015. Loss of skeletal muscle mass during neoadjuvant chemoradiotherapy predicts postoperative mortality in esophageal cancer surgery. *Ann Surg Oncol* 22 (13):4445–4452.

Rier, H. N., A. Jager, S. Sleijfer, J. van Rosmalen, M. C. Kock, and M. D. Levin. 2017. Low muscle attenuation is a prognostic factor for survival in metastatic breast cancer patients treated with first line palliative chemotherapy. *Breast* 31:9–15.

Rolland, Y., S. Czerwinski, G. Abellan Van Kan, et al. 2008. Sarcopenia: Its assessment, etiology, pathogenesis, consequences and future perspectives. *J Nutr Health Aging* 12 (7):433–450.

Rolland, Y., V. Lauwers-Cances, C. Cristini, et al. 2009. Difficulties with physical function associated with obesity, sarcopenia, and sarcopenic-obesity in community-dwelling elderly women: the EPIDOS (EPIDemiologie de l'OSteoporose) Study. *Am J Clin Nutr* 89 (6):1895–1900.

Roubenoff, R. 2004. Sarcopenic obesity: The confluence of two epidemics. *Obes Res* 12 (6):887–888.

Rozenfeld, S., M. J. Fonseca, and F. A. Acurcio. 2008. Drug utilization and polypharmacy among the elderly: A survey in Rio de Janeiro City, Brazil. *Rev Panam Salud Publica* 23 (1):34–43.

Ryu, M., J. Jo, Y. Lee, Y. S. Chung, K. M. Kim, and W. C. Baek. 2013. Association of physical activity with sarcopenia and sarcopenic obesity in community-dwelling older adults: The Fourth Korea National Health and Nutrition Examination Survey. *Age Ageing* 42 (6):734–740.

Sanchez, A. M., R. B. Candau, and H. Bernardi. 2014. FoxO transcription factors: Their roles in the mainte-nance of skeletal muscle homeostasis. *Cell Mol Life Sci* 71 (9):1657–1671.

Sandini, M., D. P. Bernasconi, D. Fior, et al. 2016. A high visceral adipose tissue-to-skeletal muscle ratio as a determinant of major complications after pancreatoduodenectomy for cancer. *Nutrition* 32 (11–12):1231–1237.

Schaap, L. A., S. M. Pluijm, D. J. Deeg, and M. Visser. 2006. Inflammatory markers and loss of muscle mass (sarcopenia) and strength. *Am J Med* 119 (6):526e9–526e17.

Schaap, L. A., S. M. Pluijm, J. H. Smit, et al. 2005. The association of sex hormone levels with poor mobility, low muscle strength and incidence of falls among older men and women. *Clin Endocrinol (Oxf)* 63 (2):152–160.

Schiaffino, S., and C. Mammucari. 2011. Regulation of skeletal muscle growth by the IGF1-Akt/PKB path-way: Insights from genetic models. *Skelet Muscle* 1 (1):4.

Schrager, M. A., E. J. Metter, E. Simonsick, et al. 2007. Sarcopenic obesity and inflammation in the InCHIANTI study. *J Appl Physiol (1985)* 102 (3):919–925.

Smilios, I., T. Pilianidis, M. Karamouzis, A. Parlavantzas, and S. P. Tokmakidis. 2007. Hormonal responses after a strength endurance resistance exercise protocol in young and elderly males. *Int J Sports Med* 28 (5):401–406.

Snijders, T., L. B. Verdijk, and L. J. van Loon. 2009. The impact of sarcopenia and exercise training on skel-etal muscle satellite cells. *Ageing Res Rev* 8 (4):328–338.

Snijders, T., B. T. Wall, M. L. Dirks, et al. 2014. Muscle disuse atrophy is not accompanied by changes in skeletal muscle satellite cell content. *Clin Sci (Lond)* 126 (8):557–566.

St-Onge, M. P. 2005. Relationship between body composition changes and changes in physical function and metabolic risk factors in aging. *Curr Opin Clin Nutr Metab Care* 8 (5):523–528.

Stenholm, S., T. B. Harris, T. Rantanen, M. Visser, S. B. Kritchevsky, and L. Ferrucci. 2008. Sarcopenic obe-sity: Definition, cause and consequences. *Curr Opin Clin Nutr Metab Care* 11 (6):693–700.

Stephen, W. C., and I. Janssen. 2009. Sarcopenic-obesity and cardiovascular disease risk in the elderly. *J Nutr Health Aging* 13 (5):460–466.

Stoever, K., A. Heber, S. Eichberg, and K. Brixius. 2016. Influences of resistance training on physical function in older, obese men and women with sarcopenia. *J Geriatr Phys Ther.* 41:20–27.

Tam, C. S., K. Clement, L. A. Baur, and J. Tordjman. 2010. Obesity and low-grade inflammation: A paediatric perspective. *Obes Rev* 11 (2):118–126.

Tan, B. H., L. A. Birdsell, L. Martin, V. E. Baracos, and K. C. Fearon. 2009. Sarcopenia in an overweight or obese patient is an adverse prognostic factor in pancreatic cancer. *Clin Cancer Res* 15 (22):6973–6979.

Tan, B. H., K. Brammer, N. Randhawa, et al. 2015. Sarcopenia is associated with toxicity in patients undergoing neo-adjuvant chemotherapy for oesophago-gastric cancer. *Eur J Surg Oncol* 41 (3):333–338.

Tardif, N., J. Salles, C. Guillet, et al. 2014. Muscle ectopic fat deposition contributes to anabolic resistance in obese sarcopenic old rats through eIF2alpha activation. *Aging Cell* 13 (6):1001–1011.

Tieland, M., I. Trouwborst, and B. C. Clark. 2017. Skeletal muscle performance and aging. *J Cachexia, Sarcopenia Muscle* 9 (1):3–19, 2017.

Tyrovolas, S., A. Koyanagi, B. Olaya, et al. 2016. Factors associated with skeletal muscle mass, sarcopenia, and sarcopenic obesity in older adults: A multi-continent study. *J Cachexia Sarcopenia Muscle* 7 (3):312–321.

Tzanavari, T., P. Giannogonas, and K. P. Karalis. 2010. TNF-alpha and obesity. *Curr Dir Autoimmun* 11:145–156.

United Nations. 2012. *Population Ageing and Development.* edited by D. o. E. a. S. Affairs. New York: United Nations.

van Greevenbroek, M. M., C. G. Schalkwijk, and C. D. Stehouwer. 2013. Obesity-associated low-grade inflammation in type 2 diabetes mellitus: Causes and consequences. *Neth J Med* 71 (4):174–187.

Vasconcelos, K. S., J. M. Dias, M. C. Araujo, A. C. Pinheiro, B. S. Moreira, and R. C. Dias. 2016. Effects of a progressive resistance exercise program with high-speed component on the physical function of older women with sarcopenic obesity: A randomized controlled trial. *Braz J Phys Ther* 20 (5):432–440.

Verdijk, L. B., R. Koopman, G. Schaart, K. Meijer, H. H. Savelberg, and L. J. van Loon. 2007. Satellite cell content is specifically reduced in type II skeletal muscle fibers in the elderly. *Am J Physiol Endocrinol Metab* 292 (1):E151–E157.

Verreijen, A. M., S. Verlaan, M. F. Engberink, S. Swinkels, J. de Vogel-van den Bosch, and P. J. Weijs. 2015. A high whey protein-, leucine-, and vitamin D-enriched supplement preserves muscle mass during intentional weight loss in obese older adults: A double-blind randomized controlled trial. *Am J Clin Nutr* 101 (2):279–286.

Villareal, D. T., L. Aguirre, A. B. Gurney, et al. 2017. Aerobic or resistance exercise, or both, in dieting obese older adults. *N Engl J Med* 376 (20):1943–1955.

Villareal, D. T., C. M. Apovian, R. F. Kushner, S. Klein. 2005. American Society for Nutrition; NAASO, The Obesity Society. Obesity in older adults: Technical review and position statement of the American Society for Nutrition and NAASO, The Obesity Society. *Obes Res* 13 (11):1849–1863.

Villareal, D. T., M. Banks, D. R. Sinacore, C. Siener, and S. Klein. 2006. Effect of weight loss and exercise on frailty in obese older adults. *Arch Intern Med* 166 (8):860–866.

Villareal, D. T., S. Chode, N. Parimi, et al. 2011. Weight loss, exercise, or both and physical function in obese older adults. *N Engl J Med* 364 (13):1218–1229.

Wall, B. T., M. L. Dirks, T. Snijders, J. M. Senden, J. Dolmans, and L. J. van Loon. 2014. Substantial skeletal muscle loss occurs during only 5 days of disuse. *Acta Physiol (Oxf)* 210 (3):600–611.

Wall, B. T., M. L. Dirks, and L. J. van Loon. 2013. Skeletal muscle atrophy during short-term disuse: Implications for age-related sarcopenia. *Ageing Res Rev* 12 (4):898–906.

Wall, B. T., and L. J. van Loon. 2013. Nutritional strategies to attenuate muscle disuse atrophy. *Nutr Rev* 71 (4):195–208.

Wannamethee, S. G., and J. L. Atkins. 2015. Muscle loss and obesity: The health implications of sarcopenia and sarcopenic obesity. *Proc Nutr Soc* 74 (4):405–412.

Waters, D. L., A. L. Ward, and D. T. Villareal. 2013. Weight loss in obese adults 65 years and older: A review of the controversy. *Exp Gerontol* 48 (10):1054–1061.

Weijs, P. J., W. G. Looijaard, I. M. Dekker, et al. 2014. Low skeletal muscle area is a risk factor for mortality in mechanically ventilated critically ill patients. *Crit Care* 18 (2):R12.

Weinheimer, E. M., L. P. Sands, and W. W. Campbell. 2010. A systematic review of the separate and combined effects of energy restriction and exercise on fat-free mass in middle-aged and older adults: implications for sarcopenic obesity. *Nutr Rev* 68 (7):375–388.

Wieske, L., D. S. Dettling-Ihnenfeldt, C. Verhamme, et al. 2015. Impact of ICU-acquired weakness on post-ICU physical functioning: A follow-up study. *Crit Care* 19:196.

Zamboni, M., F. Armellini, T. Harris, et al. 1997. Effects of age on body fat distribution and cardiovascular risk factors in women. *Am J Clin Nutr* 66 (1):111–115.

Zamboni, M., G. Mazzali, F. Fantin, A. Rossi, and V. Di Francesco. 2008. Sarcopenic obesity: A new category of obesity in the elderly. *Nutr Metab Cardiovasc Dis* 18 (5):388–395.

Zoico, E., V. Di Francesco, J. M. Guralnik, et al. 2004. Physical disability and muscular strength in relation to obesity and different body composition indexes in a sample of healthy elderly women. *Int J Obes Relat Metab Disord* 28 (2):234–241.

21 Sum Up and Future Research

Dominique Meynial-Denis

CONTENTS

Progressive loss of skeletal muscle mass and strength was first defined by Rosenberg in 1997 as a hallmark of aging and has since been referred to as sarcopenia. As life expectancy continues to increase worldwide, sarcopenia has become a major public health issue. The problem assumes greater importance and worsens in the presence of chronic diseases, which accelerate the progression of sarcopenia. Cachexia is considered as a primary accelerating model of sarcopenia; it occurs in humans irrespective of age and commonly occurs in patients with cancer, especially during advanced stages.

21.1 BASICS OF SARCOPENIA: DEFINITION AND CHALLENGES OF SARCOPENIA RESEARCH

In Chapter 1, Bischoff-Ferrari and Dawson-Hughes deal with the definitions of sarcopenia. The authors acknowledge that, in the future, it would be better to propose a spectrum of operational definitions for sarcopenia. For example, to prevent the progression of sarcopenia and its consequences, it would be better to define a health state for being at risk of sarcopenia (i.e., "presarcopenia") or a health state for disease progression (i.e., "severe sarcopenia") for those at high risk of developing physical frailty with aging. They give a good explanation of what low lean mass and decreased functional performance represent in old people.

21.2 NEW DATA ON SARCOPENIA

Buford et al. report models of accelerated sarcopenia (Chapter 2). The authors explain sarcopenia both mechanistically and morphologically and how chronic diseases can contribute to accelerating the progression of sarcopenia. They establish the definition of cachexia as a primary accelerating model of sarcopenia. Cachexia, unlike sarcopenia, can occur in individuals of all ages and be induced, for example, by cancer. They observe that chronic inflammatory and immune diseases, hypoxia-related diseases, chronic kidney disease, liver failure, diabetes mellitus type 2, and neurodegenerative conditions exacerbate the progression of sarcopenia. Skeletal muscle atrophy in "healthy" older adults should be treated differently from atrophy in other patient populations.

Tabue-Tegueo et al. underline the need to consider sarcopenia in the context of physical frailty. Indeed, sarcopenia and frailty are inseparable. Impairment of physical function in the absence of disability is the conclusive link between the two conditions (Chapter 3).

21.3 MOLECULAR AND CELLULAR ASPECTS OF SARCOPENIA

Prado and Gonzales-Barbosa Silva present the role of imaging techniques in the diagnosis of sarcopenia in Chapter 4. The authors give a good explanation of the difficulties in choosing a body composition assessment method from among ultrasound, dual energy X-ray absorptiometry, computerized tomography, and magnetic resonance imaging in sarcopenia. The best method has yet to be determined.

In Chapter 5, Graber and Rasmussen discuss mTORC1 regulation and sarcopenia. They give a good explanation of the role of aging in the loss of skeletal muscle with age. They provide evidence that mTORC1 signaling is a vital process required for maintaining muscle mass, strength, and function, and that dysregulation of mTORC1 is manifest during aging and can contribute to sarcopenia or the ability to maintain muscle mass. They propose that *periodic and transient* increases in mTORC1 activity in response to anabolic stimuli such as feeding, growth factors, and exercise are a necessary process for maintaining muscle mass, quality, and function. They acknowledge that the use of these interventions, which are designed to enhance the protein anabolic response in older adults, will be essential in counteracting sarcopenia.

In Chapter 6, Sakuma et al. deal with the pathways of protein degradation such as ubiquitin–proteasome and lysosome-autophagy. New interesting concepts developed by the authors (*Pflugers Arch* 2017 and 2018; *J Cachexia Sarcopenia Muscle* 2016) should also be mentioned. For example, sarcopenia induces a marked defect of autophagy signaling but in contrast modifies few mediators of the ubiquitin–proteasome system.

Puppa et al. focus on myokines in aging muscle in Chapter 7. They give a good explanation of the role of skeletal muscle as an endocrine organ in the release of numerous signaling molecules called myokines. They describe the role of myokines in autocrine, paracrine, and endocrine signaling of the skeletal muscle during aging. They propose that the dysregulation of several myokines is involved in sarcopenia. For example, the serum level of myostatin, which is secreted mainly by the skeletal muscle, increases with age and is inversely associated with skeletal muscle mass. The level is also particularly high in frail women. Exercise could be one method to offset the age-related dysregulation of myokines. The authors state that myokines are secreted not only by the skeletal muscle but also by the adipose tissue and the immune system. In conclusion, myokines participate in the regulation of whole-body homeostasis during aging.

Fry presents the contribution of satellite cells to skeletal muscle aging (Chapter 8). It is important to focus on the advances in the understanding of satellite cells and their contribution to skeletal muscle regeneration, hypertrophy, and homeostasis. The corresponding decline in satellite cell abundance and activity that occurs during sarcopenia emphasizes the potential contribution of satellite cells to aging muscle deficits and also presents the satellite cell as a therapeutic target. The aging process drastically affects the function and phenotype of the satellite cell, and both intrinsic and extrinsic factors account for the age-associated deterioration of satellite cell function. Now that life expectancy is continuing to increase worldwide; it is also important to better understand the role of satellite cells in the quality of life during aging.

In Chapter 9, Izeta deals with satellite cells and sarcopenia. The article gives a good description of the satellite cells in skeletal muscle.

Brioche et al. describe in Chapter 10 the role of oxidative stress in sarcopenia. They explain that some studies in cell cultures and animal models need to be confirmed in humans. They show that exercise is always a very good strategy to combat sarcopenia. It exerts its effect via antioxidant mechanisms that are dependent on G6PDH, which is becoming a new target to combat sarcopenia.

21.4 ALTERATIONS IN MUSCLE PROTEIN TURNOVER IN THE AGING PROCESS

In Chapter 11, Tessari deals with the effect of nutrition on muscle protein turnover during sarcopenia. The author gives a good description of anabolic resistance to nutrition in the elderly, explaining the mechanisms involved: reduction of the signaling protein kinase B/mTORC1 pathway, reduction of S6K1 phosphorylation, diminution of skeletal muscle μRNA expression, and inhibition of IGF-1 signaling. The author explains (1) that ingestion of leucine-rich essential amino acid mixture can overcome anabolic resistance and sustain normal muscle protein synthesis in the elderly and (2) that abundant protein intake, possibly enriched with essential amino acids, particularly leucine, is required to increase such an anabolic effect and can be complemented by physical exercise.

In Chapter 12, Hood et al. explain muscle mitochondrial turnover and sarcopenia. The chapter describes the effect of exercise on muscle mitochondria and its impact on the quality of muscle. The authors note that preserved muscle function is necessary to maintain the quality of muscle and so has an important role in the activities of daily living and independence with aging. They conclude that (1) mitochondrial biogenesis and autophagy/mitophagy pathways are both dysregulated in the aged muscle environment and that (2) exercise is an ideal strategy to improve mitochondrial quality control.

In Chapter 13, Pagano et al. focus on skeletal muscle fat infiltration and aging. The authors argue that accumulation of intermuscular adipose tissue (IMAT) is a new important factor involved in sarcopenia that seems to be closely linked to inactivity. It underlines the role of fibro/adipogenic progenitors, which are mesenchymal stem cells, in aging-induced IMAT development and accumulation.

21.5 RECENT ADVANCES LIMITING SARCOPENIA AND SUPPORTING HEALTHY AGING

In Chapter 14, Alway treats dysfunctional mitochondria during sarcopenia. The author explains that dysfunctional mitochondria initiates a signaling cascade that leads to motor neuron and muscle fiber death and culminates in sarcopenia.

In Chapter 15, Meynial-Denis deals with the beneficial effect of nutritional strategies to limit muscle wasting due to normal aging.

21.6 APPLICATIONS

21.6.1 PART 1: MUSCLE IMPAIRMENTS OR DISEASES DUE TO THE FRAILTY INDUCED BY SARCOPENIA

In Chapter 16, Power et al. give a good explanation of the decline in whole muscle function with aging, in particular of the alterations in the contractile properties of single skeletal muscle fibers.

In Chapter 17, Wakabayashi and Sakuma report on sarcopenic dysphagia and sarcopenia. The authors explain that the major causes of sarcopenic dysphagia are secondary sarcopenia such as activity-, nutrition-, and disease-related sarcopenia. They state that although the disorder is a major public health issue, wide-ranging investigations (in the field of both rehabilitation nutrition and pharmacologic therapy) are needed before a treatment for the disease can be proposed. The condition increases the risk of aspiration pneumonia, choking, malnutrition, and dehydration. The authors also discuss presbyphagia, which refers to age-related changes in the swallowing mechanism in older people and is characterized by frailty in swallowing rather than dysphagia. This chapter opens up new research opportunities for investigating the consequences of sarcopenia.

21.6.2 PART 2: COMPLICATIONS DUE TO SARCOPENIA IN ACUTE OR CHRONIC DISEASE

In Chapter 18, Garibotto et al. deal with wasting and cachexia in chronic kidney disease (CKD). They give a comprehensive account of CKD. However, they do not look specifically at sarcopenia.

The real question is whether there are many aged people with CKD and the consequences of CKD on muscle mass. It is likely that CKD increases the frailty of the elderly.

In Chapter 19, Vinciguerra deals with Parkinson's disease (PD) and sarcopenia. The author explains that PD is currently incurable, and sarcopenia, the progressive general loss of skeletal muscle, accompanies the late stages of the disease and is a negative prognostic factor. He points out that postponing or retarding muscle loss is of the utmost important for patients with PD. However, it is clear that maintaining a good level of physical activity and educating patients about its importance is fundamental. The authors suggest that myostatin inhibition could compensate for the loss of muscle in sarcopenia and warrants further study.

In Chapter 20, Weijs et al. discuss sarcopenic obesity in the elderly. The authors explain that sarcopenia and obesity are two conditions that, if combined, cause serious health problems in older adults. They point out that sarcopenic obesity is complex, and factors that may play a role are still poorly understood. However, it is clear that hormonal changes, physical inactivity, and suboptimal dietary intake can contribute to this malfunction. The authors suggest ways to counteract the negative effects of sarcopenic obesity. Exercise and nutritional interventions focusing on ample dietary protein intake are considered promising means to prevent and/or treat sarcopenic obesity and, as such, to support healthy aging.

21.7   FUTURE PERSPECTIVES

The topics of this book encompass the current understanding of sarcopenia in the elderly and the link with an extreme condition resulting in both dramatic loss of muscle mass and muscle strength as cachexia. Since sarcopenia, the loss of muscle mass and physical function with advancing age is an ineluctable process its prevention and management are the only ways to promote healthy aging, which is defined by the World Health Organization as "the process of developing and maintaining the functional ability that enables well-being in older age." It is evident that our understanding of the conditions that ensure healthy aging is incomplete. Perhaps action should be taken in the course of youth to ensure successful ageing. This approach has not yet been investigated but could be but a promising new area of study. Screening biodiversity for new approaches could be another promising means to combat human muscle atrophy. Hibernating bears remain inactive in winter for up to 7 months without eating, drinking, urinating, or defecating. They have a strong and unique ability to preserve their muscle mass and strength under conditions of muscle disuse and food deprivation in which muscle atrophy is observed in humans. Deciphering the underlying mechanisms associated with sparing effects in bears could enable us to identify new therapeutic targets, with the ultimate goal of dramatically mitigating human muscle wasting diseases.

Hence, the future directions of research in this area of sarcopenia should focus on a large number of multidisciplinary and related areas, of which a few examples are given below:

1. To better understand the interaction between sarcopenia, an aging process, and cachexia, an acute disease process induced, for example, by cancer that can appear at any age.

2. To better understand the development of physical frailty and predict its occurrence in the older population. Frailty, which is a complication of sarcopenia, is due to in part to malnutrition and can be reduced with healthy lifelong dietary habits (energy and protein intake) and by physical activity in older people.

3. To develop strategies in order to prevent or limit muscle wasting due to normal aging or to combat sarcopenia to promote healthy aging. These strategies would essentially concern older adults and are described below:

Emerging stem cell strategies to combat sarcopenia

- *Stem cells* should be considered as valuable targets for the prevention and treatment of sarcopenia. Because stem cells are among the most long-lived cells, their age-associated decline is a major contributor to organismal aging and associated diseases. Indeed, the decline in satellite cell abundance and activity highlights the potential contribution of satellite cells to aging muscle deficits.
- Failure of *autophagy* in physiologically aged satellite cells causes entry into senescence by loss of proteostasis, increased mitochondrial dysfunction (mitophagy), and oxidative stress (increase in ROS), resulting in a decline in the function and number of satellite cells. Re-establishment of autophagy reverses senescence and restores regenerative functions in geriatric satellite cells. Autophagy, therefore, which maintains stemness by preventing senescence in sarcopenia, is a decisive stem-cell-fate regulator.
- *Mitochondrial dysfunction* in aging muscle is probably due to alteration in mitophagy.
- The *intrinsic-aging clock* in stem cells can be pharmacologically manipulated and implicated in anti-aging research.
- At the cellular level, *miRNAs* are involved in the transitioning of muscle stem cells from a quiescent state to either an activated or a senescence state. miRNAs have emerged as important regulators for the mass and functional maintenance of skeletal muscle by their regulation of the gene expression of proinflammatory cytokines.
- The most effective strategy to slow or counter stem cell dysfunction with aging is *calorie restriction*, which promotes longevity.
- In brief, common hallmarks of aging stem cells were recently identified such as proteostasis (protein homeostasis) loss, DNA damage and reactive-oxygen species (ROS) production, epigenetic alterations, mitochondrial dysfunction, and altered metabolism or senescence entry due to the upregulation of regulator of senescence p16^{INK4a}. The p16^{INK4a} silencing gene in geriatric satellite cells restores quiescence and muscle regenerative functions.

➤ In conclusion, these interesting developments warrant further investigation.

Nutritional therapies to combat sarcopenia

- *Dysregulation of mTORC1* is manifest during aging and can contribute to sarcopenia or the ability to maintain muscle mass. mTORC1 signaling is a vital process required for maintaining muscle mass, strength, and function. The role of mTORC1 in sarcopenia should be studied in depth.
- Sarcopenia induces *a marked defect of autophagy signaling* but, in contrast, modifies few mediators of the ubiquitin–proteasome system. The autophagy process therefore needs to be better understood.
- *Progressive loss of proteostasis* is a hallmark of aging that is marked by declines in various components of essential machinery, including autophagy, ubiquitin-mediated degradation, and protein synthesis.
- Consumption of *leucine* attenuates body fat gain during aging but does not show signs of prevention of the associated age-related decrease in lean mass. Co-ingestion of leucine with carbohydrates and protein following physical activity does not further elevate muscle protein fractional synthetic rate in elderly men when ample protein is ingested. Ingestion of a small bolus of essential amino acids (EAAs) (histidine, isoleucine, leucine, lysine, methionine, phenylalanine, threonine, and valine) results

in a reduced responsiveness in the stimulation of muscle protein synthesis in elderly persons. However, EAA does not improve muscle wasting due to aging.

- *HMB* is the active metabolite of leucine responsible for the increase in protein synthesis in muscle. After consumption of HMB in humans, it was concluded that the increase in muscle protein synthesis was likely the result of increased mTOR signaling. It has also been shown to result in decreased muscle protein breakdown. Hence, HMB supplementation helps minimize any sarcopenic effect caused by prolonged inactivity. In conclusion, HMB can improve protein metabolism and/or lean mass and muscle strength.
- *Calorie restriction* promotes health and extends lifespan in humans, nonhuman primates, and rodents. It also delays deleterious age-related physiological changes in animals.
- *Myokines* can be used to combat sarcopenia. It was recently suggested that skeletal muscle should be considered as an *endocrine organ*. During contraction of muscle fibers, it produces and releases cytokines, and other peptides, called myokines, into the blood stream. Only a few myokines have been characterized and their potential signaling pathways involved in the muscle cell proliferation, differentiation, and growth to maintain muscle mass, muscle strength, and function, identified. *Certain myokines can contribute to muscle loss with aging and could therefore be an attractive way of combating skeletal muscle atrophy.* Maintenance of *adequate nutrition* in combination with *physical exercises* during the various stages of aging can play a fundamental role maintaining muscle health or even used as preventive measures against sarcopenia. By contrast, *Apelin*, which is a myokine that is expressed in and secreted by muscle and adipose tissue, is stimulated by physical exercise. Apelin supplementation can improve processes associated with antioxidant enzymes, autophagy, and inflammation and certainly with muscle loss with aging.

> ➤ In conclusion, nutritional strategies to preserve muscle mass, strength, and function must be accompanied by therapeutic concepts to improve bone remodeling, structure, and composition.

Pharmaceutical therapies to combat sarcopenia

- Testosterone is known to increase muscle mass and muscle protein synthesis in rodents and humans but without any clear increase in muscle strength in elderly people. Weight loss during testosterone treatment is almost exclusively due to the loss of body fat.
- The association between low vitamin D and low muscle mass has been reported in the elderly, mainly in community-dwelling subjects. *Sufficient vitamin D* status is vital for improved gains in muscular strength in older adults.
- There is an association between *vitamin B12 deficiency* and a decrease in muscle strength and walking speed and with an increased risk of fracture and postural instability.
- *Growth hormone* therapy has positive effects by improving body composition and muscle performance but is still not recommended as sarcopenia treatment owing to its side effects.
- *Myostatin inhibition* has the potential to be a potent therapy for muscle wasting by inducing an increase in muscle mass. It could compensate for the loss of muscle in sarcopenia and warrants further study.
- *Ghrelin* is well known for its stimulation of appetite and lipid accumulation but also has an anti-inflammatory effect and improves antioxidative stress in sarcopenia. Ghrelin

levels are generally lower in the elderly than in younger subjects, and *elderly subjects with sarcopenia have lower ghrelin levels* than those without sarcopenia.

> ➤ In conclusion, pharmaceutical therapies to combat sarcopenia can increase muscle mass and muscle protein synthesis in rodents and humans but are not always recommended for the treatment of sarcopenia treatment owing to their side effects.

Exercise

Exercises that are designed to enhance the protein anabolic response in older adults could be the best method of counteracting sarcopenia. The physiological adaptations differ with the type of exercise. Exercise, irrespective of the type used, improves insulin sensitivity, reverses decreased resting mitochondrial ATP production, improves mitochondrial respiratory chain efficiency, and decreases the formation of reactive oxygen species, all of which are implicated in the development of insulin resistance with aging.

- *Aerobic exercise* has been shown to reverse and/or improve muscle function and leads to improved peak oxygen consumption. Aerobic exercise results in increased mitochondrial proteins, which are involved in mitochondrial biogenesis, fusion and fission, resulting in the enhancement of muscle protein synthesis.
- *Resistance exercise* improves neuromuscular adaptations, which leads to improved muscle strength. Resistance training reverses sarcopenia and age-related declines in rates of muscle protein synthesis and results in significant improvement in the muscle strength in older adults.
- *Resistance training* (RT) is a viable and relatively low-cost treatment to counter age-related sarcopenia. RT-induced increases in muscle mass, strength, and function can be enhanced by certain foods, nutrients, or nutritional supplements such as protein, leucine, βHMB, creatine, and Omega 3 PUFAs.
- *High-intensity aerobic* exercise results in improved cardiorespiratory fitness, insulin sensitivity, and mitochondrial respiration, in enhanced fat-free mass (synonymous with enhanced muscle mass) and muscle strength in young and older adults.

> ➤ In conclusion, exercise is the best method to maintain muscle mass and strength in old humans and deserves to be more fully investigated.

In general conclusion of this chapter *stem cells* can be considered as valuable targets for the prevention and treatment of sarcopenia and warrant further study. The use of *nutritional therapies*, which are important in counteracting sarcopenia, must continue to be explored. Attention should be paid to the adverse effects of *pharmaceutical therapies. Exercise*, which is essential to enhance the protein anabolic response and muscle strength in older adults, seems to be the best strategy to combat sarcopenia. However, combining these methods would perhaps be more effective in preventing or limiting muscle wasting due to normal aging and in counteracting sarcopenia worldwide. It is an option that requires further investigations. However, the best way to fight against sarcopenia might be to intervene at an early age during youth before the occurrence of muscle wasting with aging.

Index

Note: Page numbers in italic and bold refer to figures and tables, respectively.

Printed and bound by CPI Group (UK) Ltd, Croydon, CR0 4YY

17/10/2024

01775698-0010